PLANT PHENOLICS AND HUMAN HEALTH

THE WILEY–IUBMB SERIES ON BIOCHEMISTRY AND MOLECULAR BIOLOGY

Plant Phenolics and Human Health: Biochemistry, Nutrition, and Pharmacology
Editor: Cesar G. Fraga

PLANT PHENOLICS AND HUMAN HEALTH

Biochemistry, Nutrition, and Pharmacology

Edited by

Cesar G. Fraga

WILEY

A JOHN WILEY & SONS, INC., PUBLICATION

Published by John Wiley & Sons, Inc., Hoboken, New Jersey.
Published simultaneously in Canada.

***Library of Congress Cataloging-in-Publication Data*:**

Plant phenolics and human health : biochemistry, nutrition, and pharmacology /
edited, Cesar G. Fraga.
p. ; cm.
Includes bibliographical references and index.
ISBN 978-0-470-28721-7 (cloth)
1. Phenols–Physiological effect. 2. Flavonoids–Physiological effect.
3. Phytochemicals–Physiological effect. 4. Nutrition. I. Fraga, Cesar G. [DNLM:
1. Phenols–pharmacology. 2. Phenols–therapeutic use. 3. Flavonoids–pharmacology.
4. Flavonoids–therapeutic use. 5. Nutritive Value. 6. Plant Extracts–pharmacology.
7. Plant Extracts–therapeutic use. 8. Review Literature as Topic. QV 223 P541 2009]
QP801.P4P46 2009
547'.632–dc22

2009009328

10 9 8 7 6 5 4 3 2 1

CONTENTS

PREFACE

In association with well-known health benefits related to the consumption of fruit- and vegetable-rich diets, research on the protective effects of plant-derived phenolic compounds (polyphenols) has developed notably in recent years. In particular, their antioxidant properties have been the objective of extensive research. However these phenolics are the target of an array of chemical reactions that, if confirmed to occur in vivo, would contribute to their health promoting effects. It is now emerging that both parent compounds and their metabolites produced after ingestion can regulate cell and tissue functions by both antioxidant and nonantioxidant mechanisms. This volume provides the latest evidence supporting these concepts.

The strategy behind the selection of the themes was to provide a comprehensive overview of the basic and applied research on phenolic compounds and their potential protective effects on health. The first chapters are on the identification, metabolism, and basic mechanisms affecting phenolic actions in biological systems. The book then develops in a series of pivotal chapters addressing the effects of flavonoids, stilbenes, and curcuminoids on cardiovascular disease, cancer, and neurodegeneration. The final chapter is on the complex functions that phenolics perform in plants, as a model that can help to better understand their effects on animal physiology. Explanations are essentially centered in applying basic biochemical mechanisms to improve nutrition and/or developing pharmacological strategies.

As being part of a series launched under the umbrella of the IUBMB, the volume was planned to tackle not only the cutting edge research, but also to provide a source for basic, educational information. The target audience includes not only scientists and health professionals but also educators and students, policymakers, food and pharmaceutical developers, and many others interested in understanding how plant-derived phenolic compounds can affect human health and so, in part, explains how fruit and vegetables play a key role in enhancing human health.

The color Figures in this title are posted on the following ftp site: ftp://ftp.wiley.com/public/sci_tech_med/phenolic_compounds.

I want to especially thank the group of outstanding scientist that provided chapters of the highest quality and readability. Particular appreciation is due to prof. Angelo Azzi who was central in the planning and concretion of the book.

I dedicate this volume to my four children,
MAGGIE, MARTÍN, JOAQUÍN, AND IGNACIO

CONTRIBUTORS

BHARAT B. AGGARWAL, Cytokine Research Laboratory, Department of Experimental Therapeutics, The University of Texas M.D. Anderson Cancer Center, Houston, TX 77030, USA

PREETHA ANAND, Cytokine Research Laboratory, Department of Experimental Therapeutics, The University of Texas M.D. Anderson Cancer Center, Houston, TX 77030, USA

CRISTINA ANDRES-LACUEVA, Department of Nutrition and Food Science, XaRTA. INSA, Pharmacy Faculty, University of Barcelona, 08028 Barcelona, Spain

CONSUELO BORRÁS, Department of Physiology, Faculty of Medicine, University of Valencia, 46010 Valencia, Spain

CRISTINA BOSETTI, Istituto di Ricerche Farmacologiche "Mario Negri," 20156 Milan, Italy

VITTORIO CALABRESE, Department of Chemistry, Faculty of Medicine, University of Catania, Catania, Italy

KA LUNG CHEUNG, Graduate Program in Pharmaceutical Science, Ernets-Mario School of Pharmacy, Rutgers, The State University of New Jersey, Piscataway, NJ 08854, USA

CHIARA CINI, Department of Biochemical Sciences, University of Rome "Sapienza," Rome, Italy

MICHAEL N. CLIFFORD, Faculty of Health and Medical Sciences, The University of Surrey, Guildford, Surrey GU2 7XH, United Kingdom

RAFFAELLA COCCIA, Department of Biochemical Sciences, University of Rome "Sapienza," Rome, Italy

DOUGLAS R. COOK, Department of Plant Pathology and, College of Agricultural and Environmental Sciences Genomics Facility, University of California, Davis, CA 95616, USA

CAROLIN CORNELIUS, Department of Chemistry, Biochemistry and Molecular Biology Section, Faculty of Medicine, University of Catania, Catania, Italy

KEVIN D. CROFT, School of Medicine and Pharmacology, University of Western Australia, Perth, Australia

ALAN CROZIER, Plant Products and Human Nutrition Group, Division of Environmental and Evolutionary Biology, Faculty of Biomedical and Life Sciences, University of Glasgow, Glasgow G12 8QQ, United Kingdom

Dipak K. Das, Cardiovascular Research Center, University of Connecticut School of Medicine, Farmington, CT 06030, USA

Samarjit Das, Cardiovascular Research Center, University of Connecticut School of Medicine, Farmington, CT 06030, USA

Fabio Di Domenico, Department of Biochemical Sciences, University of Rome "Sapienza," Rome, Italy

Albena T. Dinkova-Kostova, Biomedical Research Centre, Ninewells Hospital and Medical School, University of Dundee, Scotland, United Kingdom; and Department of Pharmacology and Molecular Sciences and Department of Medicine, Johns Hopkins University, Baltimore, MD 21287, USA

Juan Duarte, Department of Pharmacology, School of Pharmacy, University of Granada, Granada, Spain

Cesar G. Fraga, Physical-Chemistry-PRALIB, School of Pharmacy and Biochemistry, University of Buenos Aires, 1113 Buenos Aires, Argentina; and Department of Nutrition, University of California, Davis, Davis, CA 95616, USA

Adrian A. Franke, Cancer Research Center of Hawaii, Natural Products and Cancer Biology Program, Honolulu, HI 96813, USA

Monica Gallcano, Physical-Chemistry-PRALIB, School of Pharmacy and-Biochemistry, University of Buenos Aires, 1113 Buenos Aires, Argentina

Brunhild M. Halm, Cancer Research Center of Hawaii, Cancer Prevention and Control Program, and Kapi'olani Medical Center for Women and Children, Honolulu, HI 96813, USA

Jonathan M. Hodgson, School of Medicine and Pharmacology, University of Western Australia, Perth, Australia

Alberto B. Landolino, Monsanto, Davis, CA 95616, USA

Indu B. Jaganath, Malaysian Agricultural, Research and Development Institute, 43400 Serdang, Selangor, Malaysia

Kerry Kakazu, Cancer Research Center of Hawaii, Natural Products and Cancer Biology Program, Honolulu, HI 96813, USA

Naghma Khan, Department of Dermatology, University of Wisconsin, Madison, WI 53706, USA

Ah-Ng Kong, Department of Pharmaceutics, Ernest-Mario School of Pharmacy, Rutgers, The State University of New Jersey, Piscataway, NJ 08854, USA

Elisabeth Kravets, Institute for Biochemistry and Molecular Biology I, Heinrich Heine University Düsseldorf, Dusseldorf D-40001, Germany

Joydeb K. Kundu, National Research Laboratory of Molecular Carcinogenesis and Chemoprevention, College of Pharmacy, Seoul National University, Seoul 151-742, South Korea

AJAIKUMAR B. KUNNUMAKKARA, Cytokine Research Laboratory, Department of Experimental Therapeutics, The University of Texas M. D. Anderson Cancer Center, Houston, TX 77030, USA

ROSA M. LAMUELA-RAVENTOS, Department of Nutrition and Food Science, XaRTA, INSA, Pharmacy Faculty, University of Barcelona, Barcelona 08028, Spain

CARLO LA VECCHIA, Istituto di Ricerche Farmacologiche "Mario Negri," Milan 20156, Italy and, Istituto di Statistica Medica e Biometria "G.A. Maccacaro," Universitá degli Studi di Milanovia, Milan 20133, Italy

WENGE LI, Department of Pharmaceutics, Ernest-Mario School of Pharmacy, Rutgers, The State University of New Jersey, Piscataway, NJ 08854, USA

XINGNAN LI, Cancer Research Center of Hawaii, Natural Products and Cancer Biology Program, Honolulu, HI 96813, USA

WAI MUN LOKE, School of Medicine and Pharmacology and School of Biomedical, Biomolecular, and Chemical Sciences, University of Western Australia, Perth, Australia

GIOVANNI E. MANN, Cardiovascular Division, School of Medicine, King's College London, London SEI 9NH, United Kingdom

HASAN MUKHTAR, Department of Dermatology, University of Wisconsin, Madison, WI 53706, USA

EVA NEGRI, Instituto di Ricerche Farmacologiche "Mario Negri," Milan 20156, Italy

PATRICIA I. OTEIZA, Department of Nutrition and Department of Environmental Toxicology, University of California, Davis, Davis, CA 95616

FRANCISCO PÉREZ-VIZCAÍNO, Department of Pharmacology, School of Medicine, Universidad Complutense, Madrid, Spain

CLAUDIO PELUCCHI, Istituto di Ricerche Farmacologiche "Mario Negri," Milan 20156, Italy

GIOVANNI PENNISI, Department of Neurosciences, Faculty of Medicine, University of Catania, Catania, Italy

MARZIA PERLUIGI, Department of Biochemical Sciences, University of Rome "Sapienza," Rome, Italy

LAURE POQUET, Nestlé Research Centre, Vers-chez-les-Blanc, 1000 Lausanne 26, Switzerland

MARTA ROSSI, Istituto di Ricerche Farmacologiche "Mario Negri," Milan 20156, Italy

GULCIN SAGDICOGLU CELEP, Department of Nutrition, University of California, Davis, CA, 95616, USA; and Food and Nutrition Technology, Gazi University, Ankara, Turkey

TANKRED SCHEWE, Institute of Biochemistry and Molecular Biology I, Heinrich Heine University Düsseldorf, Düsseldorf D-40001, Germany

RICHARD C. M. SIOW, Cardiovascular Division, School of Medicine, King's College London, London SEI 9NH, United Kingdom

HELMUT SIES, Institute for Biochemistry and Molecular Biology I, Heinrich Heine University Düsseldorf, D-40001 Düsseldorf, Germany

YVONNE STEFFEN, Institute for Biochemistry and Molecular Biology I, Heinrich Heine University Düsseldorf, D-40001 Düsseldorf, Germany

YOUNG-JOON SURH, National Research Laboratory of Molecular Carcinogenesis and Chemoprevention, College of Pharmacy, Seoul National University, Seoul 151-742 and Cancer Research Institute, Seoul National University, Seoul 110-799, South Korea

JUNJI TERAO, Department of Food Science, Graduate School of Nutrition and Bioscience, The University of Tokushima, Tokushima 770-8503, Japan

MIREIA URPI-SARDA, Department of Nutrition and Food Science, XaRTA, INSA, Pharmacy Faculty, University of Barcelona, 08028 Barcelona, Spain

HANNAH R. VASANTHI, Department of Biochemistry, Sri Ramachandra Medical College and Research Institute, Sri Ramachandra University, Chennai, India

SANDRA V. VERSTRAETEN, Department of Biological Chemistry, IIMHNO (UBA) and IQUIFIB, University of Buenos Aires–CONICET, Buenos Aires, Argentina

JOSE VINA, Department of Physiology, Faculty of Medicine, University of Valencia, 46010 Valencia, Spain

GARY WILLIAMSON, Procter Department of Food Science, University of Leeds, Leeds, LS2 9JT, United Kingdom

SIWANG YU, Department of Pharmaceutics, Ernest-Mario School of Pharmacy, Rutgers, The State University of New Jersey, Piscataway, NJ 08854, USA

RAUL ZAMORA-ROS, Department of Nutrition and Food Science, XaRTA, INSA, Pharmacy Faculty, University of Barcelona, 08028 Barcelona, Spain

BAOLU ZHAO, State Key Laboratory of Brain and Cognitive Sciences, Institute of Biophysics, Academia Sinica, Beijing, People's Republic of China

1 Dietary Flavonoids and Phenolic Compounds

INDU B. JAGANATH[1] AND ALAN CROZIER[2]
[1]Malaysian Agricultural, Research and Development Institute, Kuala Lumpur, Malaysia
[2]Plant Products and Human Nutrition Group, Division of Environmental and Evolutionary Biology, Faculty of Biomedical and Life Sciences, University of Glasgow, Glasgow, United Kingdom

CONTENTS

Plant Phenolics and Human Health: Biochemistry, Nutrition, and Pharmacology,
Edited by Cesar G. Fraga.

INTRODUCTION

Plants synthesize a vast range of secondary metabolites with a significant portion consisting of phenolic compounds and flavonoid compounds [Crozier et al., 2006a]. These phytochemicals are structurally diverse, and many are distributed among a very limited number of species within the plant kingdom. This character allows them to act as biodiagnostic markers in chemotaxonomic studies. Phenolic compounds and flavonoids accumulate in relatively high amounts in plants and appear to have a myriad of supplemental functions in a plant's life cycle. These include structural roles in different supporting or protective tissues, involvement in defense strategies, as attractants for pollinators and seed-dispersing animals, and as allelopathic agents, ultra violet (UV) protectants and signal molecules in the interactions between plants and their environment. One of the most versatile classes of flavonoids, the anthocyanins, protect chloroplasts from photodegradation by absorbing high-energy quanta, while scavenging free radicals and reactive oxygen species (ROS) [Gould, 2004]. Flavonols, as well as providing protection against the damaging effects of UV-B light, are also involved in promoting the growth of pollen tubes down the style to facilitate fertilization. In addition, isoflavonoids play important defense roles against pathogen and insect attack and are key signal molecules in the formation of nitrogen-fixing root nodules in legumes. After the death of plants, phenolic compounds may persist for weeks or months and affect decomposer organisms and decomposition processes in soils. Therefore, their effects are not restricted to only the plant itself but may extend to the functioning of whole ecosystems [Horner et al., 1988].

Secondary metabolites, other than providing plants with unique survival or adaptive strategies, are of commercial significance to humankind. They have been used as dyes, fibers, glues, oils, waxes, flavoring agents, drugs, and perfumes and are viewed as potential sources of new natural drugs, antibiotics, insecticides, and herbicides [Croteau et al., 2000; Dewick, 2002]. In recent years the role of phenolic compounds and flavonoids as protective dietary constituents has become an increasingly important area of human nutrition research. Unlike the traditional vitamins, they are not essential for short-term well-being, but there is increasing evidence that modest long-term intakes may exhibit a potential for modulating human metabolism in a manner favorable for the prevention or reduction in the risk of degenerative diseases such as cardiovascular diseases, diabetes, obesity, and cancer [Anderson et al., 1999].

HEALTH BENEFITS AND MODE OF ACTION OF FLAVONOIDS AND PHENOLIC COMPOUNDS

The rapid rise of degenerative diseases worldwide is threatening economic and social development as well as the lives and health of millions of people. It represents a major health challenge to global development in the coming century.

It is estimated that up to 80% of cardiovascular disease, 90% of Type II diabetes, and one third of cancers can be avoided by changing lifestyle, including diet [WHO, 2003]. Diet-related high cholesterol, high blood pressure, obesity, and insufficient consumption of fruits and vegetables have been cited as significant interlinking risk factors that cause the majority of these diseases. There is, therefore, increasing interest in the role of nutrition and specific dietary constituents in the prevention of such diseases. Flavonoids and phenolic compounds are prominent among dietary constituents that are the focus of such interest.

Since the 1990s a number of epidemiological studies have been carried out attempting to correlate high dietary phenolic compounds and flavonoid intake, through the consumption of fruits and vegetables, with reduced risk of degenerative diseases. Many, but not all, of these studies have indicated some degree of inverse associations between high dietary phenolic/flavonoid intake and reduction of degenerative diseases [Steinmetz and Potter, 1996; Law and Morris, 1998; Riboli and Norat, 2003]. Since oxidative stress imposed by ROS is known to play a crucial role in the pathophysiology associated with neoplasia, atherosclerosis, and neurodegenerative diseases, the potential mechanism of the protective effects of phenolic compounds and flavonoids were thought to be due to direct scavenging of free radicals [see Heim et al., 2002].

Accumulating evidence now indicates the importance of interactions between various phytochemicals in reducing the risk of various degenerative diseases [Chan et al., 2000; Mouria et al., 2002; Mertens-Talcott et al., 2003]. The combination of antioxidative agents with different modes of action is thought to increase efficacy and minimize toxicity. In a recent review by Lee and Lee [2006], the abilities of phenolic-based antioxidant therapies to decrease ROS levels was shown to produce the best health benefits through a diet rich in multiple antioxidants rather than a high dosage of a single supplement. Evidence of the potential benefits of food synergy was provided by Liu et al. [2000] when they demonstrated that a combination of fruits, such as orange, apple, grape, and blueberry, displayed a synergistic effect on antioxidant activity in vitro. The median effective dose (EC_{50}) of each fruit in combination was five times lower than the EC_{50} of each fruit alone, suggesting synergistic effects due to the combination of the four fruits. In another study, Sakamoto [2000] emphasized the importance of consuming black tea together with soybean products as commonly occurs in a typical Japanese diet. In this study, thearubigen in black tea did not alter the in vitro growth of human prostate cancer cells. However, a small amount of thearubigen (0.5 $\mu g\ mL^{-1}$) administered with genistein (20 $\mu g\ mL^{-1}$), the major isoflavone in soybean, synergistically inhibited cell growth and increased the DNA distribution at the G2 M phase of the cell division cycle by 34% compared with genistein alone [Sakamoto, 2000]. Similar conclusions were reached by Temple and Gladwin [2003] when they reviewed 200 cohort and case–control studies that provided risk ratios concerning intake of fruits and vegetables and risk of cancer. Their studies showed that the cancer-preventing action of fruits and vegetables is most probably due to the many bioactive compounds that act in concert to prevent cancer rather than being due to one or two potent anticarcinogens.

Nutrients generally have very specific functions such as being an enzyme cofactor. In contrast, in addition to their additive and synergistic effects, phenolic compounds and flavonoids, often exhibit pleiotropic effects that in combination may reduce the risk of chronic disease. For instance, curcumin, the active constituent of turmeric (*Curcuma longa*), a root vegetable, has been shown to be beneficial in all three stages of carcinogenesis [Thangapazham et al., 2006]. Isoflavones, the bioactive ingredient in leguminous vegetables, not only cause a small reduction in blood cholesterol but also reduce blood pressure, arterial dimensions, and oxidative stress [Anderson et al., 1999]. This combined effect may cause a reduction in the risk of coronary heart disease [Kris-Etherton et al., 2004].

In addition to the complexity mentioned above, the health implications of dietary phenolic compounds and flavonoids are also dependent on the composition of the components of the diet and the bioavailability of the individual compounds under study. Accumulating evidence on the absorption and bioavailability of phenolic compounds and flavonoids in humans reveals that most of these phytochemicals are modified during absorption from the small intestine, through conjugation and metabolism, and by the large intestine, mainly through the actions of the colonic microflora, and by subsequent hepatic metabolism [Graefe et al., 2001; Manach et al., 2004; Mullen et al., 2004, 2006, 2008; Jaganath et al., 2006]. Thus, metabolites that reach the cells and tissues are chemically, and, in many instances, functionally distinct from the dietary form, and such features underlie their bioactivity [Kroon et al., 2004]. This, in addition to the fact that in most instances very low levels of dietary phenolic compounds and flavonoids are actually absorbed and appear in the bloodstream ($< 10\ \mu M$), implies that the concept of these compounds functioning as hydrogen-donating antioxidants in vivo appear to be an oversimplified view of their mode of action [Williams et al., 2004; Williamson and Manach, 2005; Fraga, 2007].

It has been hypothesized that cells respond to phytochemicals through direct interactions with receptors or enzymes involved in signal transduction, or through modifying gene expressions that may result in alteration of the redox status of the cell that may trigger a series of redox-dependent reactions [Williams et al., 2004]. There is now emerging evidence that flavonoids may play an important role in molecular processes especially as modulators of intracellular signaling cascades, which are vital to cellular function [Williams et al., 2004]. For example, in a recent study carried out by Mackenzie and associates (2008), a naturally occurring phenolic compound, curcumin [1,7-bis(4-hydroxy-3-methoxyphenyl)-1,6-heptadiene-3,5-dione,1] was found to deregulate signaling cascades, such as NF-κB, leading to a decreased expression of proteins involved in cell proliferation and apoptosis. In another study on Caco-2 cells, hexameric procyanidins was found to inhibit TNFα-induced NF-κB activation, which is belived to play a central role in inflammation including human intestinal bowel disease [Erlejman et al., 2008].

There is growing evidence from human feeding studies that the absorption and bioavailability and thus bioactivity of phenolic compounds and flavonoids are very much dependent on the nature of their chemical structure. Their chemical classification and dietary occurrence is briefly discussed in the following section.

FLAVONOIDS—STRUCTURE AND THEIR DIETARY OCCURRENCE

To date, more than 6000 different flavonoids have been described and the number continues to increase [Harborne and Williams, 2000]. Flavonoids are polyphenolic compounds comprising of 15 carbons, with 2 aromatic rings connected by a 3-carbon bridge. According to the modifications of the central C-ring, they can be divided into different structural classes including flavonols, flavones, flavan-3-ols, flavanones, isoflavones, and anthocyanidins (Fig. 1.1). In a few cases, the 6-membered heterocyclic ring C occurs in an isomeric open form or is replaced by a 5-membered ring as in the case of chalcone. Other flavonoid groups, which quantitatively are relatively minor dietary components, are dihydroflavones, flavan-3,4-diols, coumarins, and aurones.

Flavonols

The flavonols are the most widespread of the flavonoids in plant food. They vary in color from white to yellow and are closely related in structure to the flavones. They are represented mainly by quercetin, kaempferol, and myricetin while the methylated derivative isorhamnetin is also quite common (Fig. 1.2). Of the various flavonols found in the diet, quercetin is the most ubiquitous. It is present in various fruits and vegetables, with especially high concentrations, 200–1000 $\mu g\ g^{-1}$ fresh weight, occurring in onions (*Allium cepa*) [Hertog et al., 1992; Crozier et al. 1997]. In a recent study by Sultana and Anwar [2008], flavonol levels were determined in 22 plant materials (9 vegetables, 5 fruits, and 8 medicinal plants). The highest concentrations were detected in the medicinal plant, moringa (*Moringa oleifera*) (68 $\mu g\ g^{-1}$ fresh weight), followed by strawberry (*Fragaria* spp.) (40 $\mu g\ g^{-1}$), peepal (*Ficus religious*) (12 $\mu g\ g^{-1}$), spinach *(Spinaceae oleraceae)* (19 $\mu g\ g^{-1}$), and cauliflower *(Brassica oleraceae)* (18 $\mu g\ g^{-1}$).

Flavonols that accumulate in plant tissues are almost always in the form of glycosylated conjugates. The main flavonols in onions are quercetin- 4′-*O*-glucoside and quercetin-3,4′-*O*-,diglucoside with smaller amounts of isorhamnetin-4′-*O*-glucoside (Fig. 1.3) [Mullen et al., 2004].

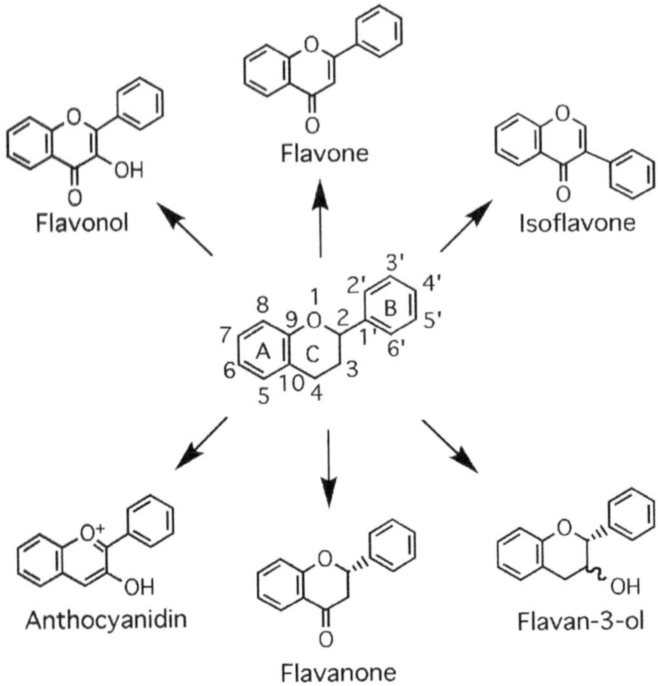

Figure 1.1 Structures of the main flavonoid subgroups.

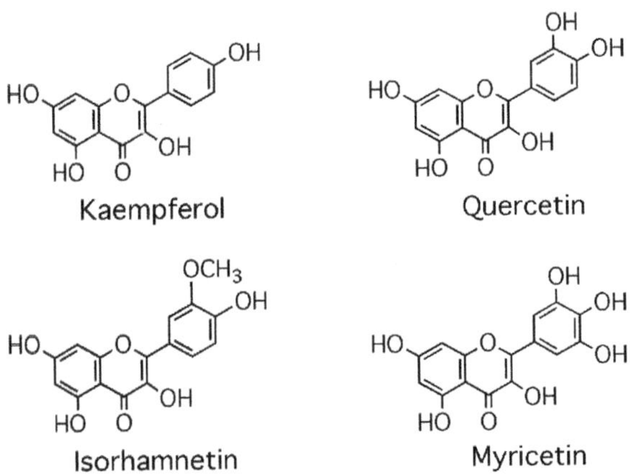

Figure 1.2 Structures of common flavonol aglycones.

A whole range of other quercetin conjugates such as quercetin-3-*O*-galactoside, quercetin-3-*O*-rhamnoside, quercetin-3-*O*-xyloside, quercetin-3-*O*-rutinoside, quercetin-3-*O*-arabinopyranoside, and quercetin-3-*O*-arabinofuranoside are found in apples (*Malus* x *domestica*) (Fig. 1.4) [Marks et al., 2008].

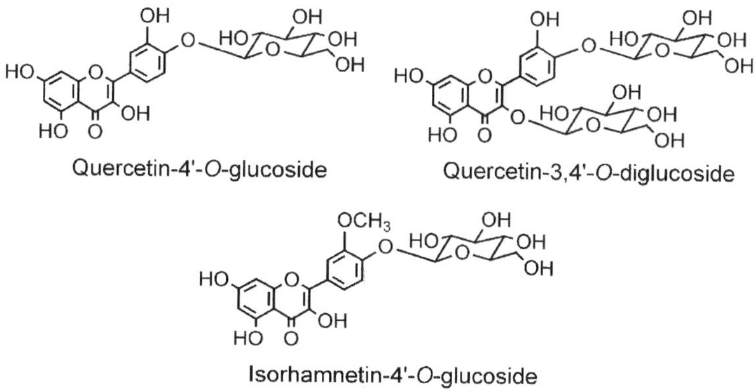

Figure 1.3 Main flavonol glucosides in onion.

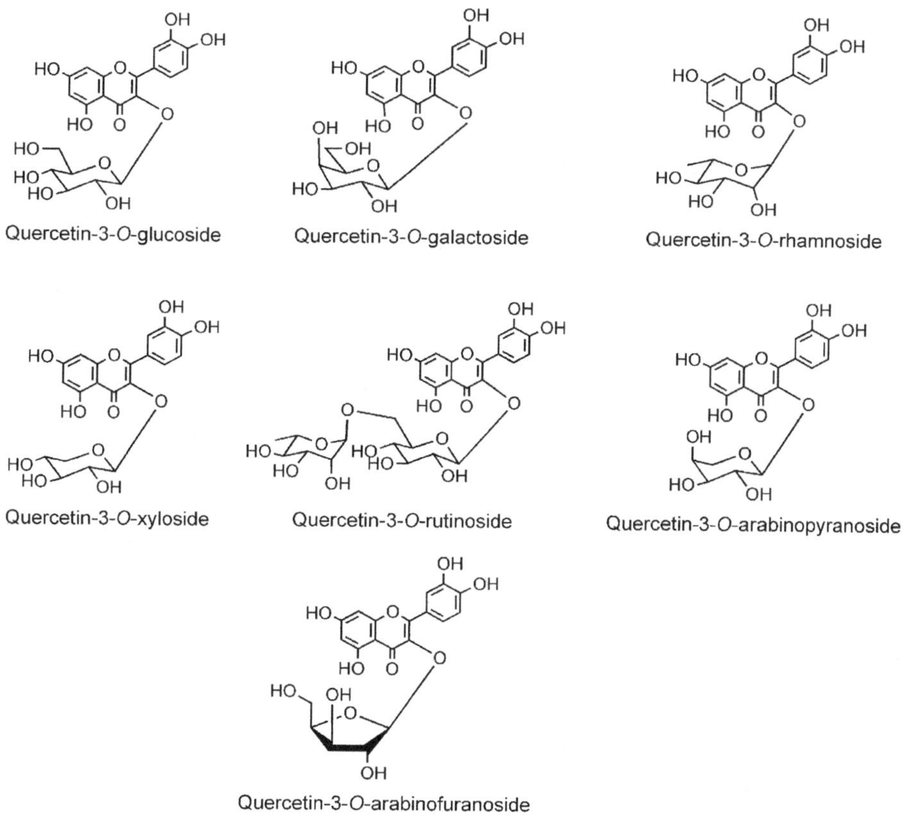

Figure 1.4 Principal flavonol glucosides in apples.

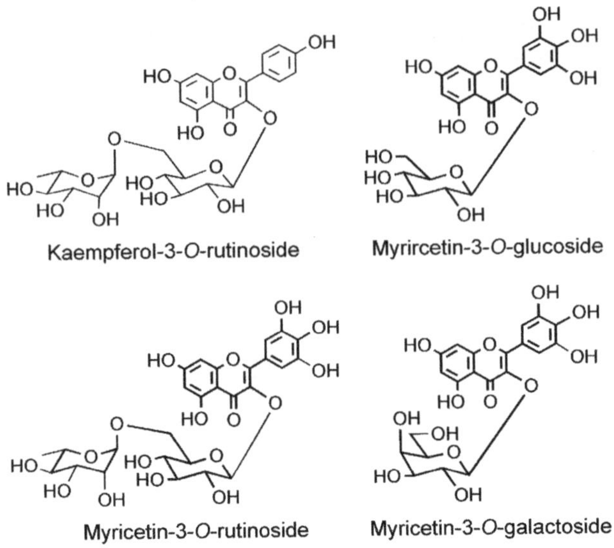

Figure 1.5 Flavonol conjugates found in berries.

Quercetin-3-*O*-rutinoside, on the other hand, is the main flavonol in tomatoes *(Lycopersicon esculentum)*, asparagus *(Asparagus officinalis)*, peaches *(Prunus persica)*, and nectarines *(Prunus persica* var. nectarina*)* [Makris and Rossiter, 2001; Crozier et al., 2006c]. Quercetin-3-*O*-glycoside, quercetin-3-galactoside, and aquercetin arabinoside has also been detected in mangos *(Mangifera indica)* [Schieber et al., 2000]. Other flavonols in the diet include kaempferol-3-*O*-rutinoside in kiwi fruit *(Actinidia deliciosa)* and conjugates of myricetin in berries (Fig. 1.5) (Peterson and Dwyer, 1998].

Grapes of *Vitis vinifera*, grape products, and wines contain a wide range of flavonols such as quercetin, myricetin, kaempferol, isorhamnetin, quercetin-3-*O*-glucoside, quercetin-3-*O*-glucuronide, quercetin-3-*O*-glucoside, quercetin-3-*O*-galactoside, kaempferol-3-*O*-glucoside, and kaempferol-3-*O*-galactoside [Makris et al., 2006]. Tea *(Camellia sinensis)* infusions also contain a diverse spectrum of flavonols linked to mono-, di- and tri-saccharides [Del Rio et al., 2004].

Flavones

Flavones are structurally very similar to flavonols and differ only in the absence of hydroxylation at the 3-position on the C-ring. Flavones are mainly represented in the diet by apigenin and luteolin. Unlike flavonols, they are not widely distributed with significant concentrations being reported in only celery (*Apium graveolens*), parsley (*Petroselinum crispum*), and artichoke (*Cynara scolymus*) [Crozier et al., 2006a]. As a consequence their dietary intake is very low. Flavone conjugates such as the 7-*O*-(2″-*O*-apiosyl)glucosides of apigenin, luteolin, and chrysoeriol (Fig. 1.6) are found in celery [Herrmann, 1976], while artichoke contains luteolin-7-*O*-glucoside, luteolin-7-*O*-rutinoside, and apigenin-7-*O*-rutinoside (Fig. 1.7) [Wang et al., 2003].

Substantial quantities of luteolin-7-*O*-glucuronide, luteolin-7-*O*-glucoside, and luteolin-7-*O*-rutinoside occur in Red Oak Leaf and Lollo Rosso, two red-leaved varieties of lettuce (*Lactuca sativa*) [Llorach et al., 2008]. Polymethoxylated flavones such as nobiletin, scutellarein, sinensetin, and tangeretin (Fig. 1.8) are found exclusively in citrus species [Crozier et al., 2006c], while diosmetin-7-*O*-glucuronide has been isolated from the fruits of a Chinese herb, *Luffa cylindrical*.

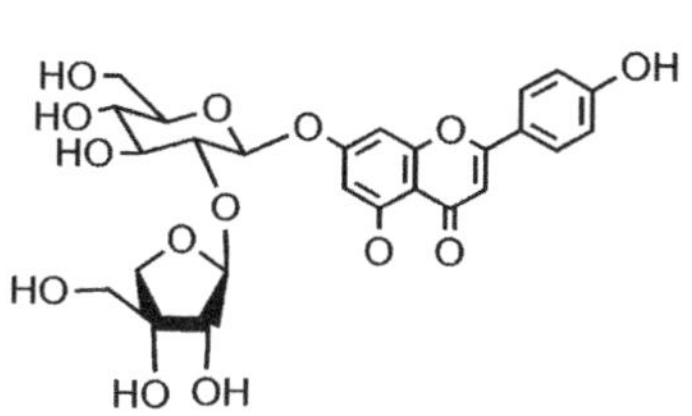

Apigenin-7-*O*-(2"-*O*-apiosyl)glucoside

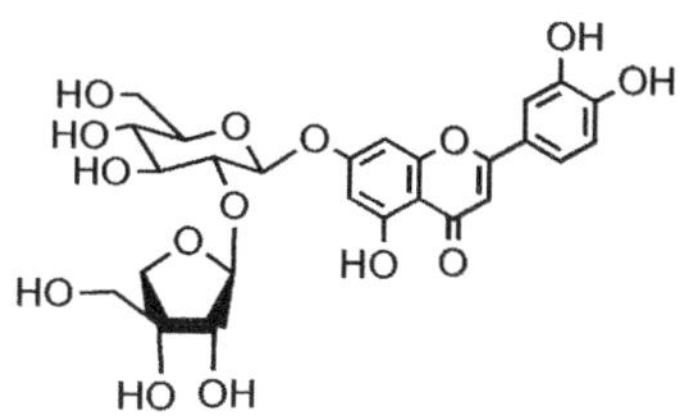

Luteolin-7-*O*-(2"-*O*-apiosyl)glucoside

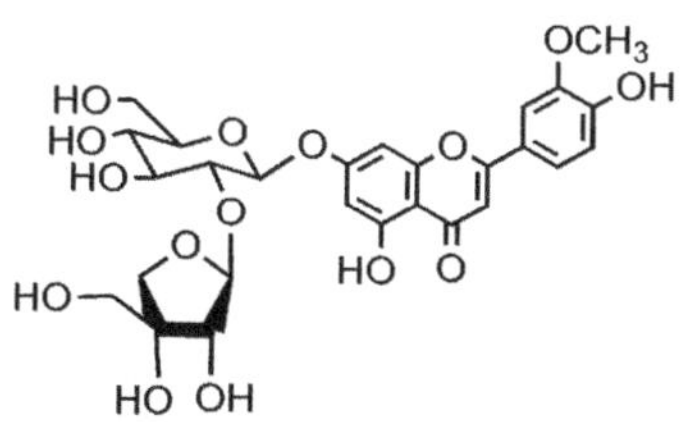

Chrysoeriol-7-*O*-(2"-*O*-apiosyl)glucoside

Figure 1.6 Flavone conjugates occurring in celery.

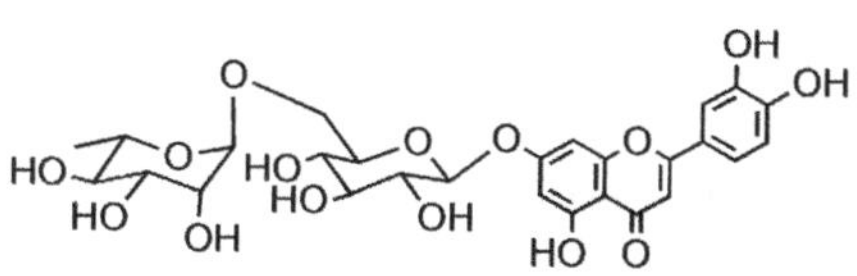

Luteolin-7-*O*-rutinoside

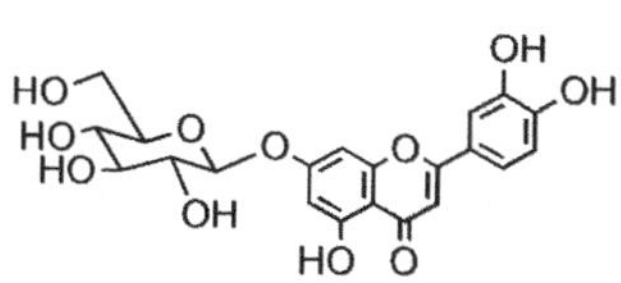

Luteolin-7-*O*-glucoside

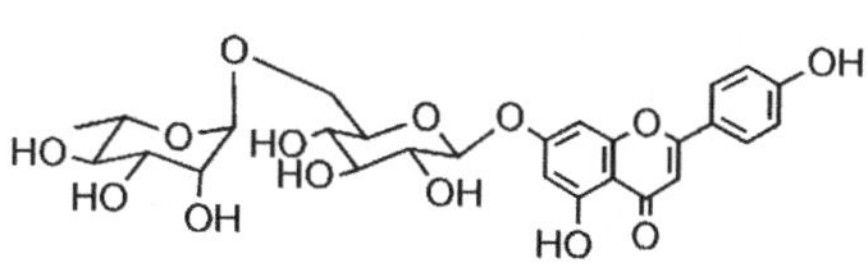

Apigenin-7-*O*-rutinoside

Figure 1.7 Flavone conjugates found in artichoke.

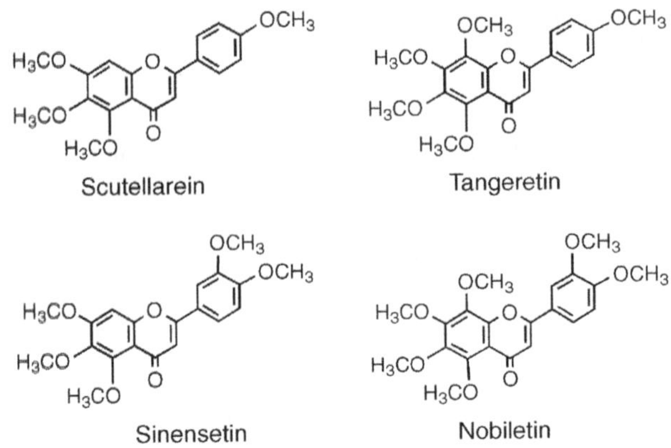

Figure 1.8 Polymethoxylated flavones found in citrus species.

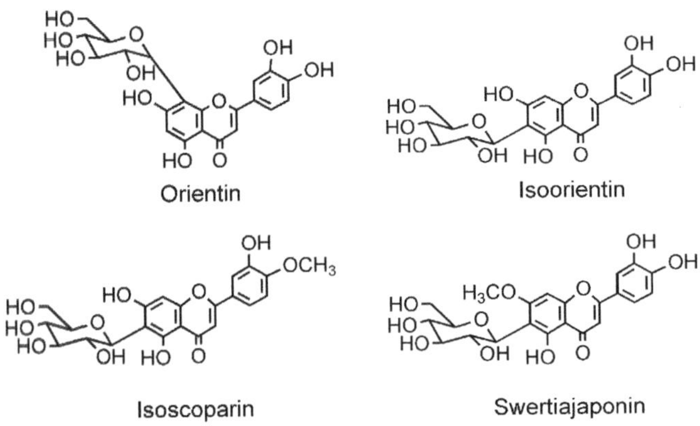

Figure 1.9 Flavones found in rooibos tea.

Red bush or rooibos tea, made from infusions of young leaves and shoots of the South African shrub *Aspalathus linearis,* and popularized by the *The No. 1 Ladies' Detective Agency* novels of the Edinburgh University Emeritus Professor of Medical Law, Alexander McColl Smith [1999, 2000], contains a number of compounds including *C*-flavone glycosides in the form of isoorientin (luteolin-6-*C*-glucoside) and orientin (luteolin-8-*C*-glucoside) [Bramati et al., 2003]. Orientin and isoorientin also occur in lemongrass (*Cymbopogon citrata*) along with two other flavone *C*-glucosides, chrysoeriol-6-*C*-glucoside (isoscoparin) and 7-*O*-methyl-luteolin-6-*C*-glucoside (swertiajaponin) (Fig. 1.9) [Cheel et al., 2005].

Recent observations reveal that when flavones are methoxylated, metabolic stability and membrane transport in the intestine/liver dramatically increases, thus improving oral bioavailability. In addition, methoxyflavones also show increased cancer chemopreventive properties when compared to the more common unmethylated flavones [Walle, 2007].

Flavan-3-ols

Flavan-3-ols represent the most common flavonoid consumed in the American and, most probably, the Western diet and are regarded as functional ingredients in various beverages, whole and processed foods, herbal remedies, and supplements. Their presence in food affects quality parameters such as astringency, bitterness, sourness, sweetness, salivary viscosity, aroma, and color formation [Aron and Kennedy, 2007]. Flavan-3-ols are structurally the most complex subclass of flavonoids ranging from the simple monomers (+)-catechin and its isomer (−)-epicatechin to the oligomeric and polymeric proanthocyanidins (Fig. 1.10), which are also known as condensed tannins [Crozier et al., 2006b].

The most abundant type of proanthocyanidins in plants are the procyanidins, which consist exclusively of (epi)catechin units. The less common proanthocyanidins containing (epi)afzelechin (Fig. 1.11) and (epi)gallocatechin (Fig. 1.10) subunits are called propelargonidins and prodelphinidins, respectively [Balentine et al., 1997].

Flavan-3-ols are found abundantly in fruits such as apricots (*Prunus armeniaca*), sour cherries (*Prunus cerasus*), grapes and blackberries (*Rubus* spp.) [Porter, 1988]. The seeds of grapes contain substantial quantities of (+)-catechin, (−)-epicatechin, procyanidin oligomers, and polymers [Gu et al., 2004]. Apples, on the other hand, are a good source of (−)-epicatechin and procyanidin dimers B_1 and B_2 (Fig. 1.12), while peaches and nectarines contain (+)-catechin, (−)-epicatechin, and proanthocyanidins including procyanidin B_1 [Hong et al., 2004]. Barley, seemingly, is the only common cereal with a significant proanthocyanidin content (0.6–1.3 g kg^{-1}) [Santos-Buelga and Scalbert, 2000].

(+)-Catechin and the proanthocyanidin prodelphinidin B_3 are, respectively, the major monomeric and dimeric flavan-3-ols found in barley and malt where prodelphinidin B_3 is the main contributor for the radical scavenging activity [Dvoráková et al., 2007]. Proanthocyanidins have also been detected in nuts. Hazelnuts (*Corylus avellana*) and pecans (*Carya illinoensis*) are particularly rich in proanthocyanidins containing ca. 5 g kg^{-1}, whereas almonds (*Prunus dulcis*) and pistachios (*Pistachio vera*) contain 1.8–2.4 mg kg^{-1}, walnuts (*Juglans* spp.) ca. 0.67 g kg^{-1}, roasted peanuts (*Arachis hypgaea*) 0.16 g kg^{-1}, and cashews (*Anarcardium occidentale*) 0.09 g kg^{-1} [Crozier et al., 2006c]. Dark chocolate derived from the roasted seeds of cocoa (*Theobroma cacao*) is also a rich source of procyanidins [Gu et al., 2004]. Monomeric flavan-3-ols and the proanthocyanidin B_2, B_5 dimers, and C_1 trimer are found in fresh cocoa beans (Fig. 1.13). Flavan-3-ols have also been detected in mint

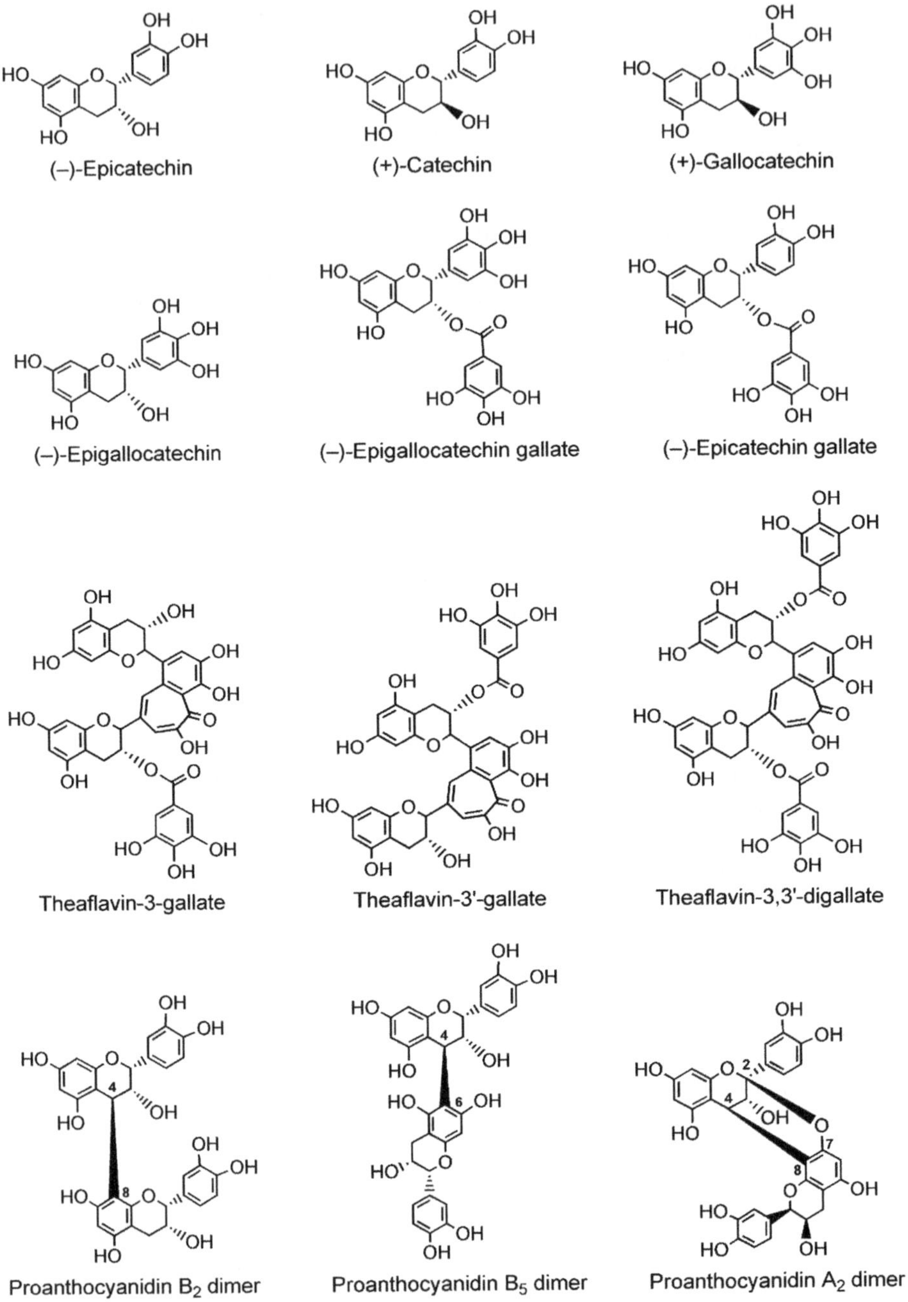

Figure 1.10 Flavan-3-ol structures.

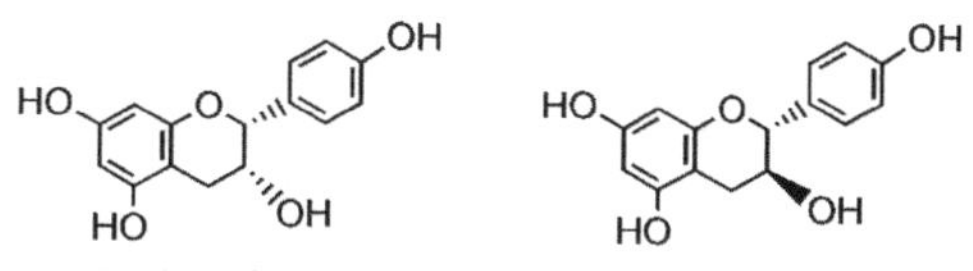

Figure 1.11 Less common flavan-3-ol monomers: (−)-epiafzelechin and (+)-afzelechin.

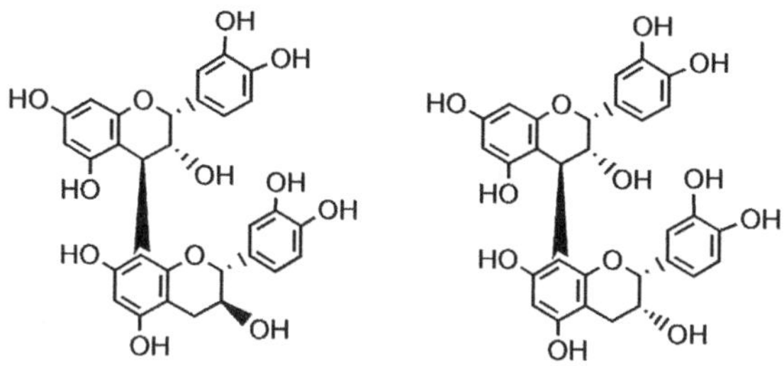

Figure 1.12 Procyanidin dimers occurring in apples.

(Mentha rotundifolia), basil *(Ocimum basilicum)*, rosemary *(Rosemarinus officinalis)*, sage *(Salvia officinalis)*, and dill (*Anethum graveolens*) [Shan et al., 2005].

Flavan-3-ols can undergo esterification with gallic acid to form catechin gallates, and hydroxylation reactions to form gallocatechins (Fig. 1.10). Gallocatechins such as (−)-epigallocatechin, (−)-epigallocatechin gallate, and (−)-epicatechin gallate are abundant in green tea infusions [Stewart et al., 2005]. During fermentation to produce black tea, these compounds polymerize, giving rise to theaflavins and high-molecular-weight thearubigins (Fig. 1.14) [Crozier et al., 2006c]. Other beverages such as red wine and beer are also rich in flavan-3-ols. Red wines contain oligomeric procyanidins and prodelphinidins, originating mainly from the seeds of red grapes [Auger et al., 2004]. Flavan-3-ols such as (+)-catechin and (−)-epicatechin, and the dimers prodelphinidin B_3 and procyanidin B_3 have been detected in beer [Crozier et al., 2006c].

Flavanones and Chalcones

Flavanones are mainly represented by naringenin, hesperetin, and eriodictyol, while a number of minor compounds, including sakuranetin and isosakuranetin, also occur (Fig. 1.15). Two structural features—the absence of a $\Delta^{2,3}$ double

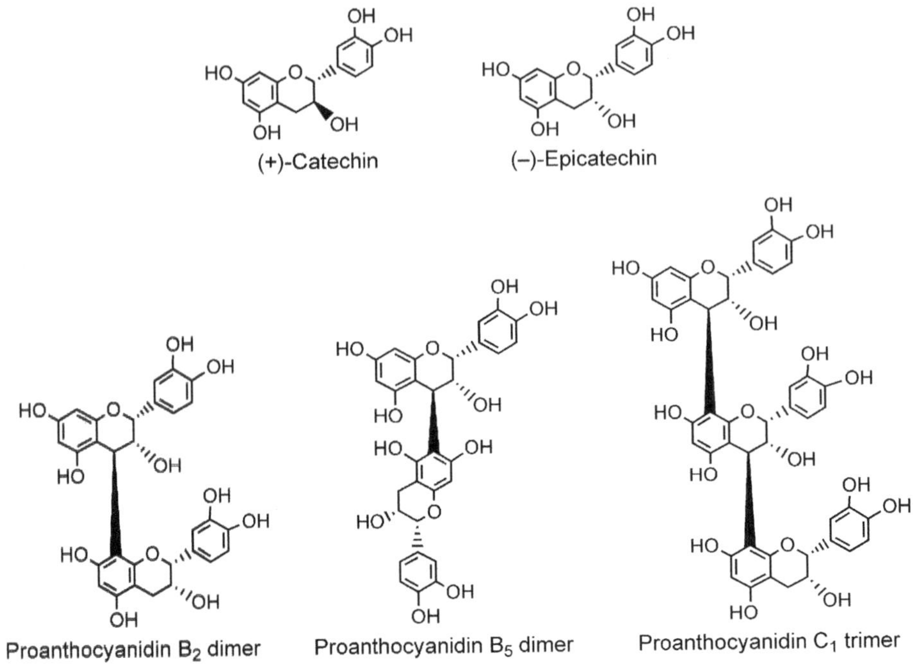

Figure 1.13 Monomeric flavan-3-ols and proanthocyanidin B_2, B_5 dimers, and C_1 trimer found in fresh cocoa beans.

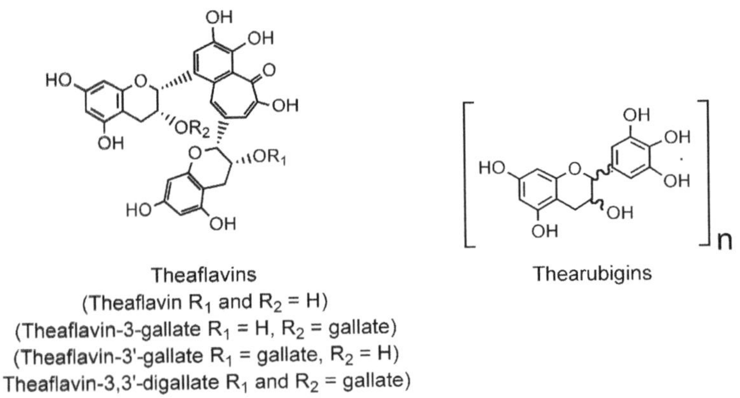

Figure 1.14 Theaflavins and thearubigins present in black tea.

bond and the presence of a chiral center at the carbon-2—characterize flavanones [Iwashina, 2000]. In the majority of naturally occurring flavanones, the C-ring is attached to the B-ring at C_2 in the α configuration.

The flavanone structure is highly reactive and has been reported to undergo hydroxylation, glycosylation, and *O*-methylation reactions. Flavanones are

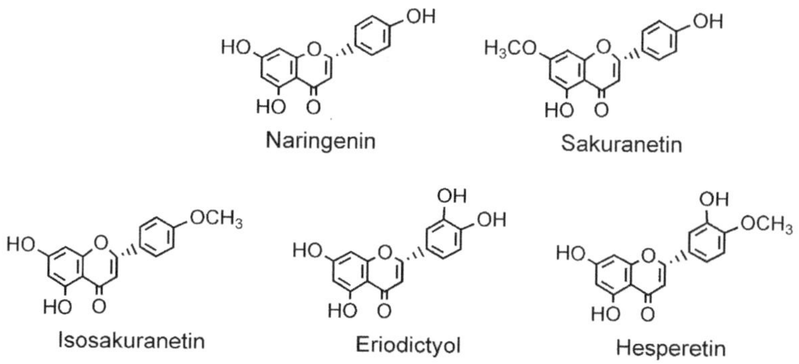

Figure 1.15 Structures of common flavanone aglycones.

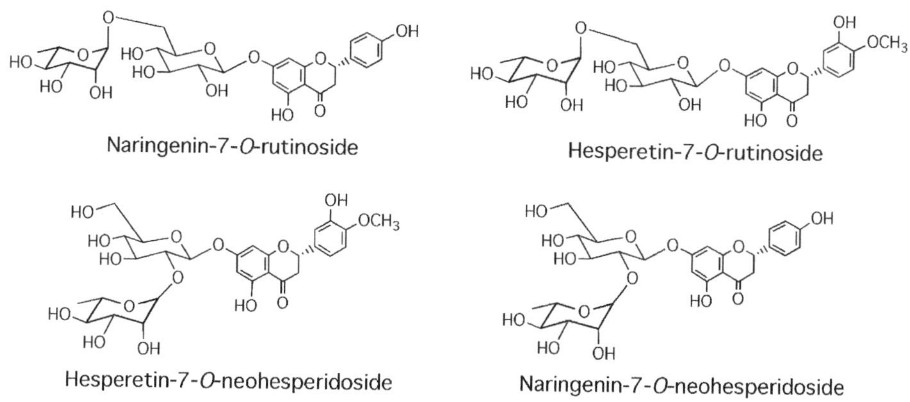

Figure 1.16 Flavanone conjugates in citrus fruit.

exclusively found in citrus fruits in their glycosidic forms. Grapefruit (*Citrus paradisi*) juice contains up to 377 mg L^{-1} of naringin (naringenin-7-*O*-neohesperidoside) and orange juice, 16–84 mg L^{-1} of narirutin (naringenin-7-*O*-rutinoside) [Manach et al., 2004; Tomás-Barberán and Clifford, 2000]. The peel is by far the richest part of citrus fruit in terms of its flavanone content. Substantial quantities of eriodictyol-7-*O*-rutinoside have been reported in lemon (*Citrus limon*) and lime (*Citrus aurantifolia*) [Peterson et al., 2006]. Flavanone rutinosides are tasteless, while neohesperidoside conjugates such as hesperetin-7-*O*-neohesperidoside (neohesperidin) from bitter orange (*Citrus aurantium*) and naringenin-7-*O*-neohesperidoside (naringin) from grapefruit peel (*Citrus paradisi*) have an intensely bitter taste (Fig. 1.16). Naringenin is also found in tomatoes and tomato-based products. Fresh tomatoes, especially the skin, also contain naringenin chalcone, which is converted to naringenin during the manufacture of tomato ketchup [Krause and Galensa, 1992]. Hesperetin-7-*O*-rutinoside has also been detected in kiwi fruit, while

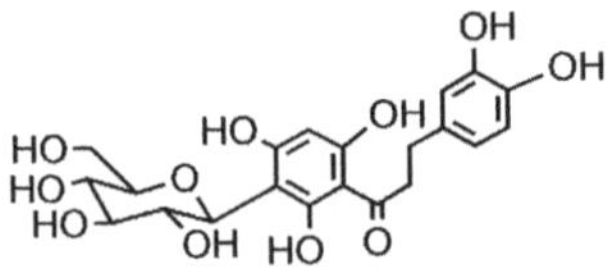

Aspalathin

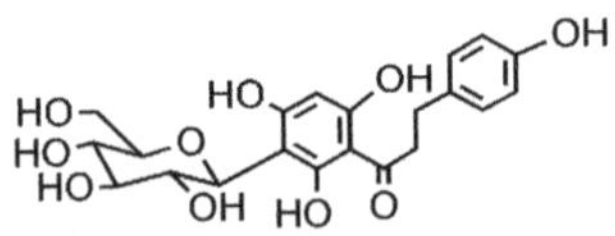

Nothofagin

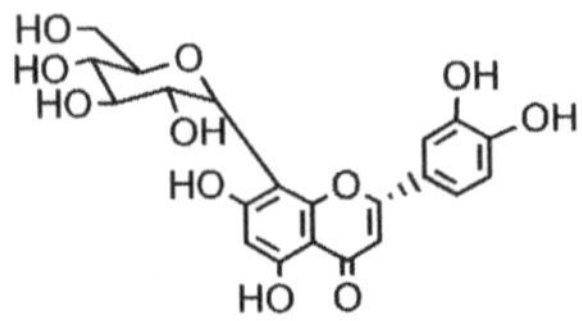

Eriodictyol-8-*C*-glucoside

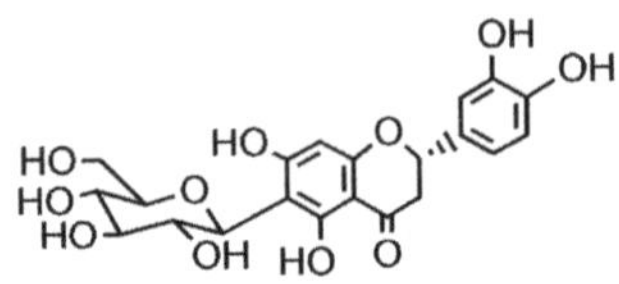

Eriodictyol-6-*C*-glucoside

Figure 1.17 Hydroxychalcones occurring in unfermented rooibos tea and flavanone *C*-glycosides that accumulate during fermentation.

hesperetin-7-*O*-neohesperidoside was reported in bananas (*Musa cavendishii*) [Dégenéve 2004; Kanazawa and Sakakibara, 2000].

As mention earlier, rooibos tea, which is claimed to have a number of medicinal properties [Joubert and Ferreira, 1996; McKay and Blumberg, 2007], contains the flavone *C*-glucosides orientin and isoorientin. It also contains a number of rare dihydrochalcone *C*-glycosides, the main components being 2′,3′,4,4′,6′-pentahydroxy-dihydrochalcone-3-*C*-glucoside (aspalathin) and 2′,4,4′,6′-tetrahydroxy-dihydrochalcone-3-*C*-glucoside (nothofagin). During fermentation aspalathin is oxidized to the flavanone *C*-glycosides eriodictyol-6-*C*-glucoside eriodictyol-8-*C*-glucoside (Fig. 1.17) [Krafczyk and Glomb 2008].

Anthocyanidins/Anthocyanins

Anthocyanins are water-soluble plant pigments and are particularly evident in fruit and flower tissue where they are responsible for a diverse range of red, blue, and purple colors. They occur primarily as glycosides of their respective aglycone anthocyanidin-chromophores (Fig. 1.18), with the sugar moiety typically attached at the 3-position on the C-ring or the 5-position on the A-ring [Prior and Wu, 2006]. They are involved in the protection of plants against excessive light by shading leaf mesophyll cells and also have an important role to play in attracting pollinating insects.

There are about 17 anthocyanidins found in nature, but only 6 — cyanidin, delphinidin, petunidin, peonidin, pelargonidin, and malvidin—are ubiquitously distributed and of dietary importance. The variation of anthocyanins are due to: (i) the number and position of hydroxyl and methoxy groups on the basic anthocyanidin skeleton; (ii) the identity, number, and positions at which sugars are attached; and (iii) the extent of sugar acylation and the identity of the

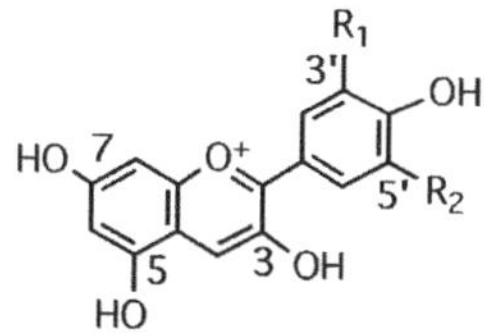

Anthocyanidin	R_1	R_2
Pelargonidin	H	H
Cyanidin	OH	H
Delphinidin	OH	OH
Peonidin	OCH_3	H
Petunidin	OCH_3	OH
Malvidin	OCH_3	OCH_3

Figure 1.18 Structures of major anthocyanins.

acylating agent [Prior and Wu, 2006]. Unlike other subgroups of flavonoids with the same C_6–C_3–C_6 skeleton, anthocyanins have a positive charge in their structure at acidic pH.

The most widespread anthocyanin in fruits is cyanidin-3-glucoside [Kong et al., 2003]. However, malvidin glycosides are the characteristic anthocyanins in red grapes and their derived products [Mazza and Miniati, 1993]. Other anthocyanins that occur in grapes include petunidin-3-*O*-glucoside, malvidin-3-*O*-(6″-*O*-*p*-coumaroyl)glucoside, malvidin-3-*O*-(6″-*O*-acetyl)glucoside, delphinidin-3-*O*-glucoside, and malvidin-3,5-*O*-diglucoside (Fig. 1.19) [Burns et al., 2001, 2002a].

Purple grape juice, from Concord grapes, a native American cultivar *Vitis labrusca,* which have a thicker skin and larger seeds than grapes of *Vitis vinifera*, is a rich source of more than 20 anthocyanins. The main components being 3-*O*-glucosides and 3,5-*O*-diglucosides of cyanidin, peonidin, delphinidin, and malvidin, delphinidin-3-*O*-(6″-*O*-acetyl)glucoside, delphinidin-3-*O*-(6″-*O*-*p*-coumaroyl)-5-*O*-diglucoside, and delphinidin-3-*O*-(6″-*O*-*p*-coumaroyl)glucoside (Fig. 1.20) [Wang et al., 2003; Mullen et al., 2007]

Anthocyanins occur in abundance in berries where they provide the fruits with their distinctive and vibrant palate of colors. Cranberry (*Vaccinium macrocarpon*), blackberry, and elderberry (*Sambucus nigra*) contain derivatives of only one type of anthocyanin (i.e., cyanidin), while a wide array of anthocyanins is found in blueberry (*Vaccinium corymbosum*) and blackcurrant (*Ribes nigrum*) (Fig. 1.21).

Anthocyanins such as cyanidin-3-*O*-rutinoside, cyanidin-3-*O*-glucoside, and peonidin-3-*O*-rutinoside (Fig. 1.22) have also been reported in sweet cherries (*Prunus avium*) and sour cherries *(Prunus cerasus)* [Wu et al., 2004].

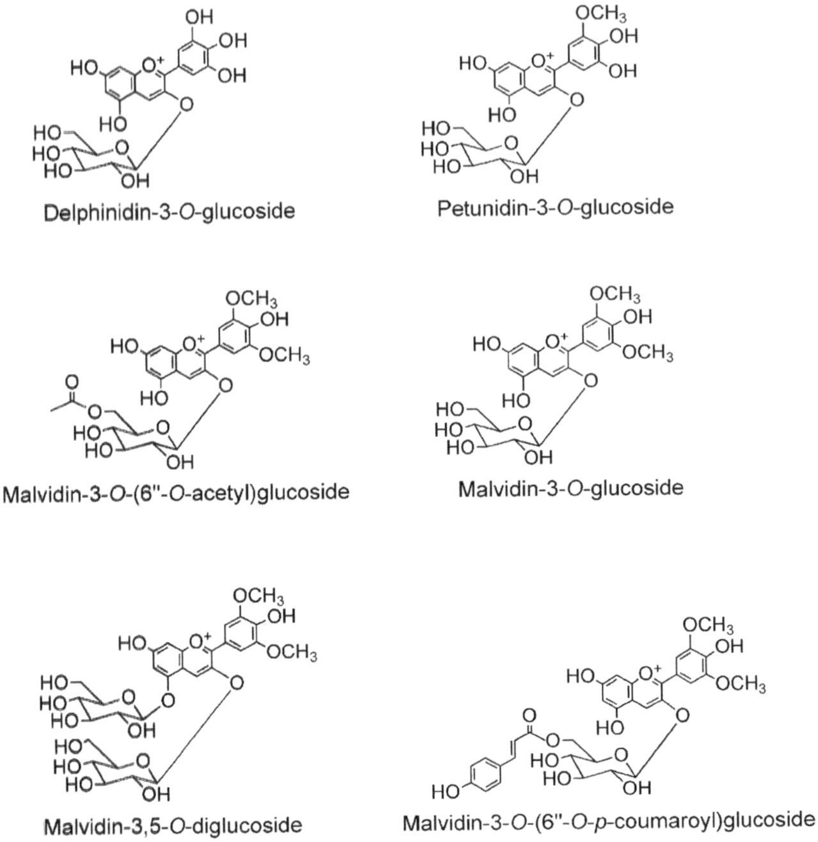

Figure 1.19 Anthocyanins found in red grapes.

Plums *(Prunus domestica)* and peaches are also a rich source of cyanidin-3-*O*-glucoside and cyanidin-3-*O*-rutinoside [Crozier et al., 2006c].

Red onions contain up to 250 mg kg^{-1} anthocyanins [Clifford, 2000]; among the major components are cyanidin-3-*O*-(6″-malonyl)glucoside and cyanidin-3-*O*-(6″-malonyl)laminaribioside (Fig. 1.23) [Donner et al., 1997]. Cyanidin-3-*O*-(6″-malonyl)glucoside is also a component of the red-leaved Lollo Rosso lettuce [Ferreres et al., 1997], while 3-*O*-glucosides and 3,5-*O*-diglucosides of cyanidin and delphinidin have also been detected in pomegranate (*Punica granatum*) juice [Gil et al., 2000].

Isoflavones

In contrast to most other flavonoids, isoflavones are characterized by having the B-ring attached at C3 rather than the C2 position. They have a

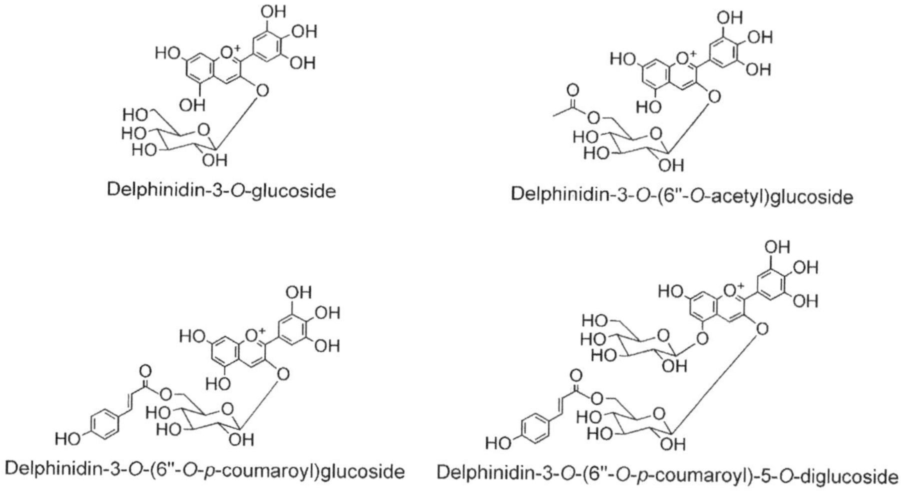

Figure 1.20 Purple grape juice anthocyanins.

very limited distribution in the plant kingdom with substantial quantities being found only in leguminous species [Graham, 1991; Dixon and Steele, 1999]. Isoflavones are known for their estrogenic activity due to their ability to bind to estrogen receptor and have received much attention due to their putative role in the prevention of breast cancer and osteoporosis [Barnes, 2003].

Worldwide, soybeans (*Glycine max*) are almost the sole dietary source of isoflavones. Common isoflavones such as genistein, daidzein, and glycitein (Fig. 1.24), also occur, albeit in low levels, in black beans (*Phaseolus vulgaris*) and green peas (*Pisum sativum*). In plants isoflavones occur predominantly as β-glucosides (genistin, daidzin, glycitin), or as acetyl-β-glucosides and malonyl-β-glucosides, and are therefore polar, water-soluble compounds [Coward et al., 1998]. Isoflavones also undergo various modifications, such as methylation, hydroxylation, or polymerization, and these modifications lead to simple isoflavonoids, such as isoflavanones, isoflavans, and isoflavanols, as well as more complex structures including rotenoids, pterocarpans, and coumestans [Dewick, 1993].

Isoflavones such as diadzein-7-*O*-(6″-*O*-malonyl)glucoside and diadzein-7-*O*-(6″-*O*-acetyl) glucoside (Fig. 1.25) occur in high concentrations in soybean [Barnes et al., 1994]. Formononetin and biochanin A (Fig. 1.24), present as 6″-*O*-malonyl-7-*O*-glucosides, 7-*O*-glucosides, and aglycones, are the most abundant isoflavones in red clover (*Trifolium pretense*), which is one of the ingredients used to extract isoflavones for dietary supplements [Delmonte

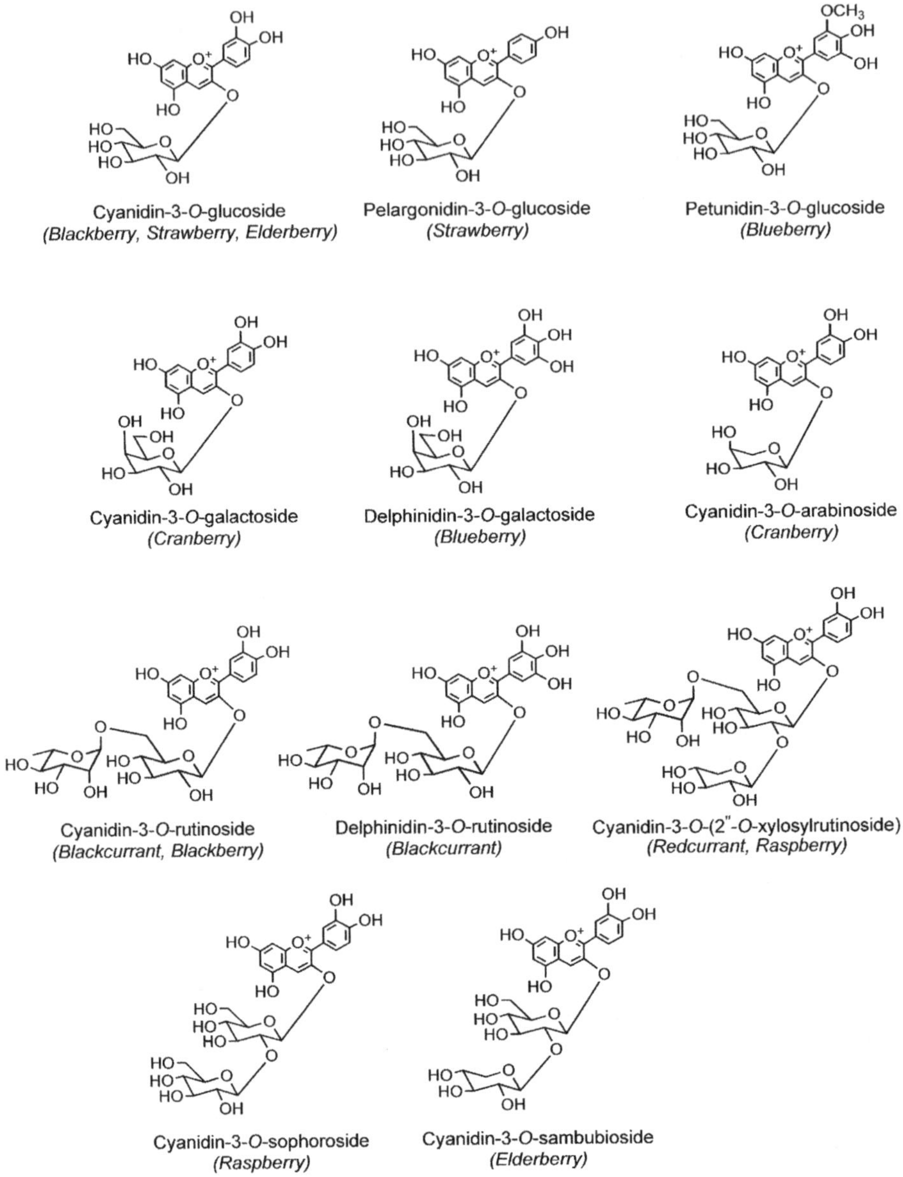

Figure 1.21 Major anthocyanins in berries.

et al., 2006]. Other than soya, *Pueraria lobata* (common name kudzu), a perennial vine native to Japan and China that also grows in the southeastern United States, is another commercial source of isoflavones for dietary supplements. Puerarin (daidzein-7-*C*-glucoside), daidzin (daidzein-7-*O*-glucoside), and daidzein are the main isoflavones in kudzu [Delmonte et al., 2006].

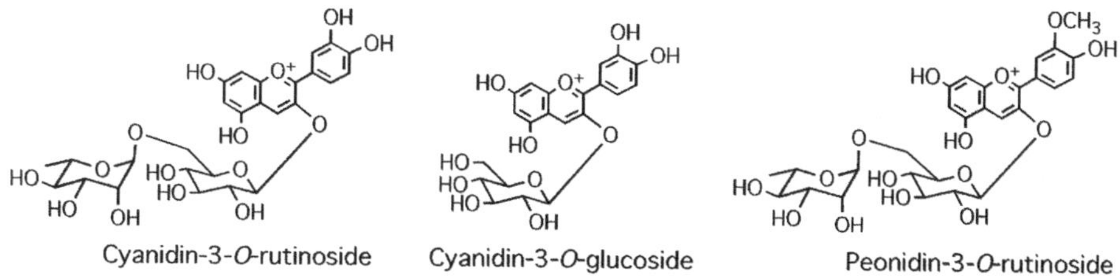

Figure 1.22 Anthocyanins present in cherries.

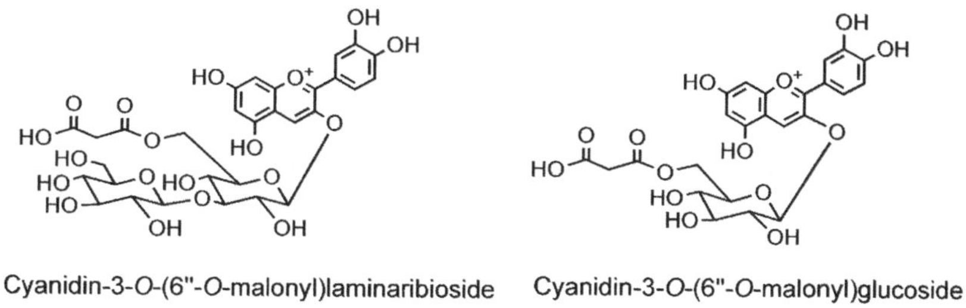

Figure 1.23 Main anthocyanins in red onions.

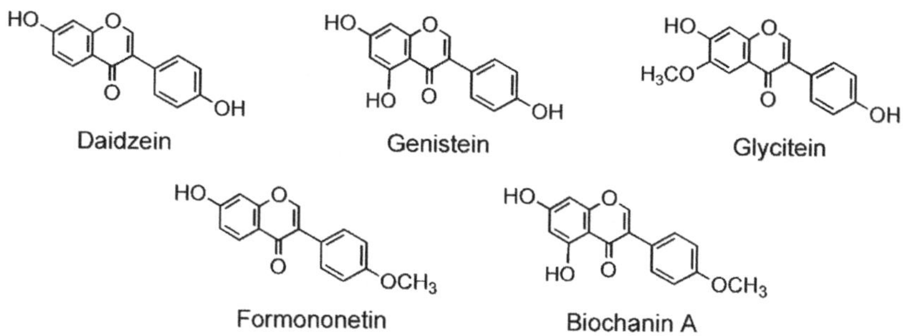

Figure 1.24 Structure of common isoflavone aglycones.

NONFLAVONOID PHENOLIC COMPOUNDS—STRUCTURE AND THEIR DIETARY OCCURRENCE

Phenolics are defined as compounds possessing one or more aromatic rings to which is attached at least one hydroxyl group. Phenolic compounds can be categorized as flavonoids and nonflavonoid phenolic compounds. The main nonflavonoid phenolic compounds of dietary significance are the C_6–C_1

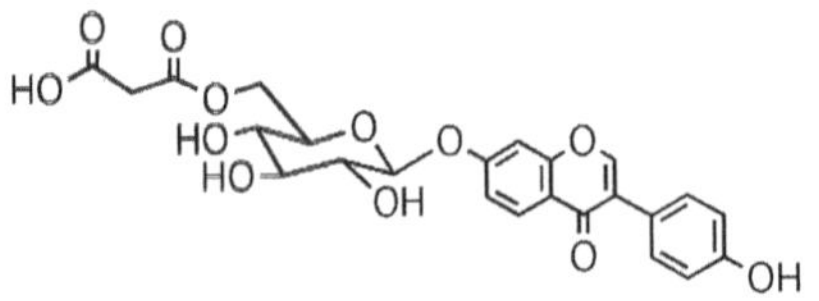
Daidzein-7-*O*-(6''-*O*-malonyl)glucoside

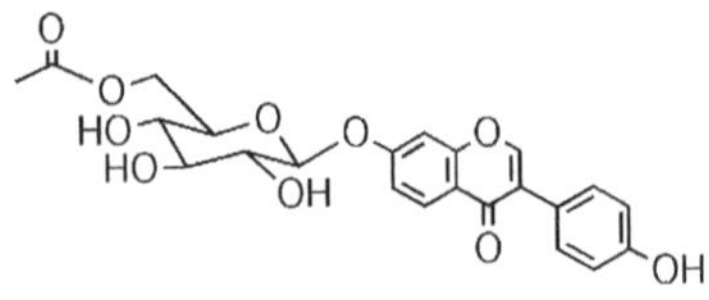
Daidzein-7-*O*-(6''-*O*-acetyl)glucoside

Figure 1.25 Legume isoflavone conjugates.

phenolic acids, the C_6–C_3 hydroxycinammates and their conjugated derivatives, and the polyphenolic C_6–C_2–C_6 stilbenes.

Phenolic Acids

Phenolic acids are also known as hydroxybenzoates, and they are commonly represented by gallic, *p*-hydroxybenzoic, protocatechuic, vanillic, and syringic acids. Phenolic acids are usually present in the bound form and are typically components of complex structures such as lignins and hydrolyzable tannins. They can also be found as derivatives of sugars and organic acids in plant foods. Gallic acid is the base unit of gallotannins, whereas gallic acid and hexahydroxydiphenoyl moieties are both subunits of the ellagitannins, which are classified as hydrolsable tannins.

Ellagic acid has been reported to be present in berries, particularly raspberries (*Rubus idaeus*), strawberries, and blackberries [Amakura et al., 2000]. However, free ellagic acid is normally present in low levels in berries that more commonly contain ellagitannins, such as sanguiin H-6 and lambertianin C, which release ellagic and gallic acid when treated with acid (Fig. 1.26) [Mullen et al., 2002].

Pomegranate juice is increasing in popularity and some, but far from all, commercial juices/drinks have a high ellagitannin and antioxidant content [Mullen et al., 2008]. Pomegranate juice contains gallagic acid, an analog of ellagic acid containing four gallic acid residues, and punicalagin, the principal monomeric, hydrolysable tannin, in which gallagic acid is bound to glucose (Fig. 1.27) [Gil et al., 2000]. Dates (*Phoenix dactylifera*), one of the oldest cultivated fruit, contain protocatechuic acid, vanillic acid, and syringic acid (Fig. 1.28) [Al-Farsi et al., 2005].

Free and bounded phenolic acids are also found in cereals. Different grains such as sorghum *(Sorghum bicolor)*, millet *(Pennisetum americanum)*, barley *(Hordeum vulgare)*, wheat *(Triticum vulgare)*, rice *(Oryza sativa)*, oat *(Avena sativa)*, and rye *(Secale cereale)* contain diverse phenolic acids such as gallic, protocatechuic, *p*-hydroxybenzoic, gentisic, salicylic, vanillic, and syringic acids [see Dykes and Rooney, 2007]. Hydroxybenzoic acid glycosides are also characteristic of some herbs and spices [Tomás-Barberán and Clifford, 2000]. After hydrolysis, protocatechuic acid is the dominant hydroxybenzoate in

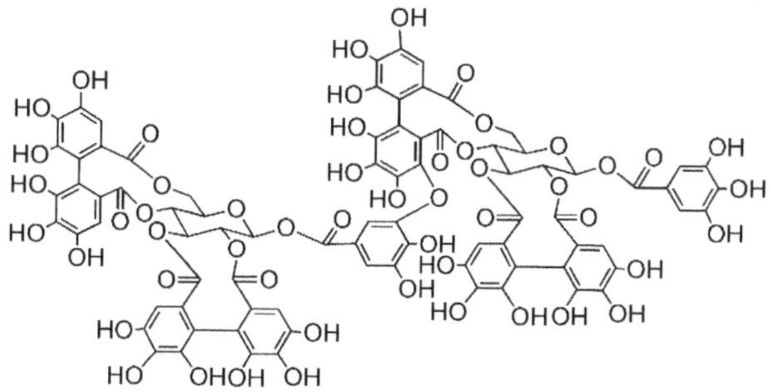

Sanguiin H-6

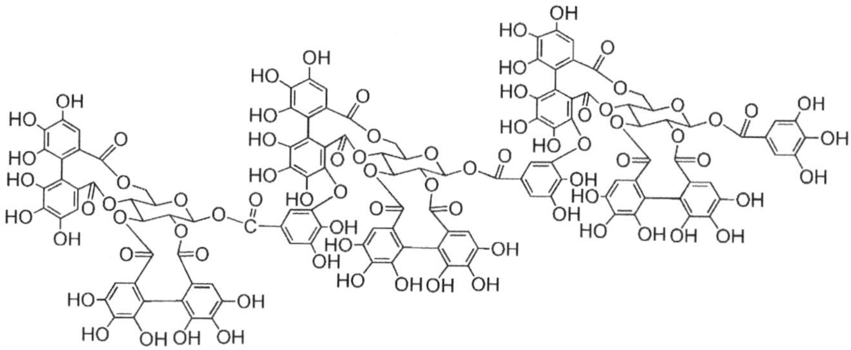

Lambertianin C

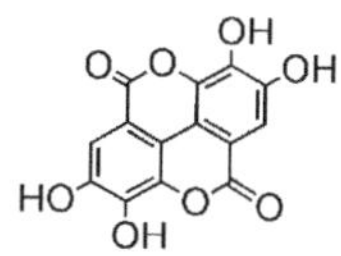

Ellagic acid

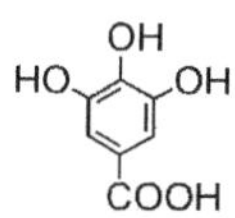

Gallic acid

Figure 1.26 Ellagitannins and trace amounts of gallic acid and ellagic acid occur in raspberries.

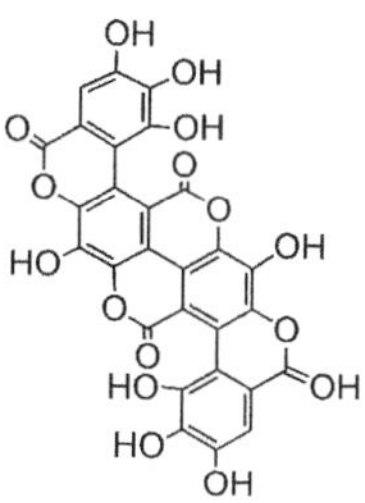

Gallagic acid

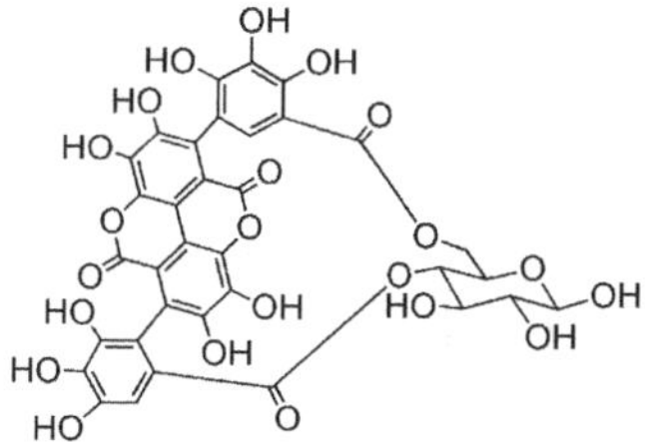

Punicalin

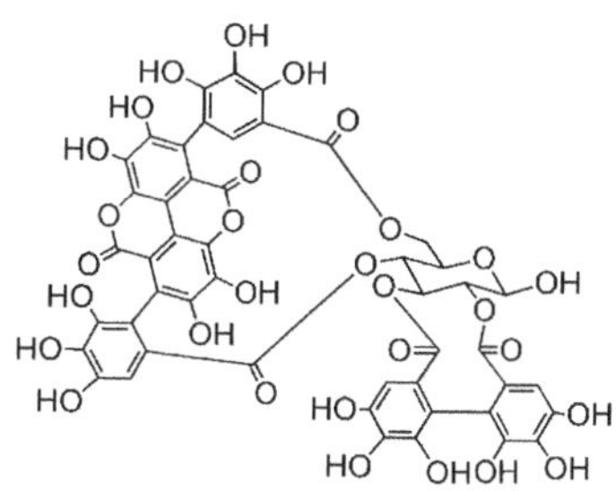

Punicalagin

Figure 1.27 Ellagitannins detected in pomegranate juice.

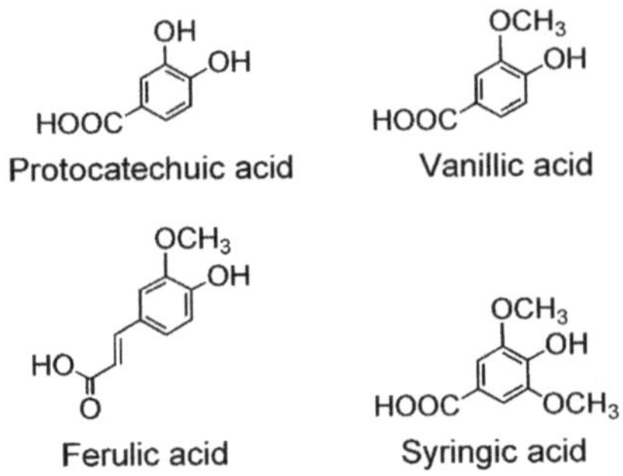

Figure 1.28 Phenolic acids in dates.

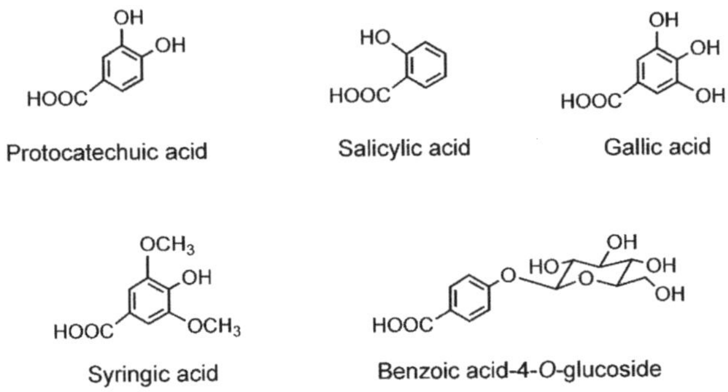

Figure 1.29 Some hydroxybenzoates derivative compounds found in herbs.

cinnamon bark accompanied by salicylic and syringic acid. Gallic acid occurs in clove buds (*Eugenia caryophyllata* Thunb.) along with protocatechuic and syringic acid (Fig. 1.29). Benzoic acid-4-*O*-glucoside is the common phenolic acid in many herbs such as in anise (*Pimpinella anisum*), star anise (*Illicium verum*), dill (*Anethum graveolens*), fennel (*Foeniculum vulgare*), caraway (*Carum carvi*), and parsley (*Petroselinum crispum*) (Fig. 1.25) [Crozier et al., 2006c].

Hydroxycinammates

The most common hydroxycinnamates, *p*-coumaric, caffeic, and ferulic acids, frequently accumulate as their respective tartrate esters, coutaric, caftaric, and fertaric acids. Quinic acid conjugates of caffeic acid, namely 3-, 4-, and 5-*O*-caffeoylquinic acid, which belong to a family of hydroxycinnmate-quinic acid conjugates known as chlorogenic acids, are commonly found in fruits and vegetables. Fruits such as apples and dates (*Phoenix dactylifera*) are a good source of diverse phenolic compounds. 5-*O*-Caffeoylquinic acid, 4-*O*-*p*-coumaroylquinic acid, and caffeic acid have been detected in apples [Clifford et al., 2003; Kahle et al., 2005], while dates contain ferulic acid [Crozier et al., 2006c].

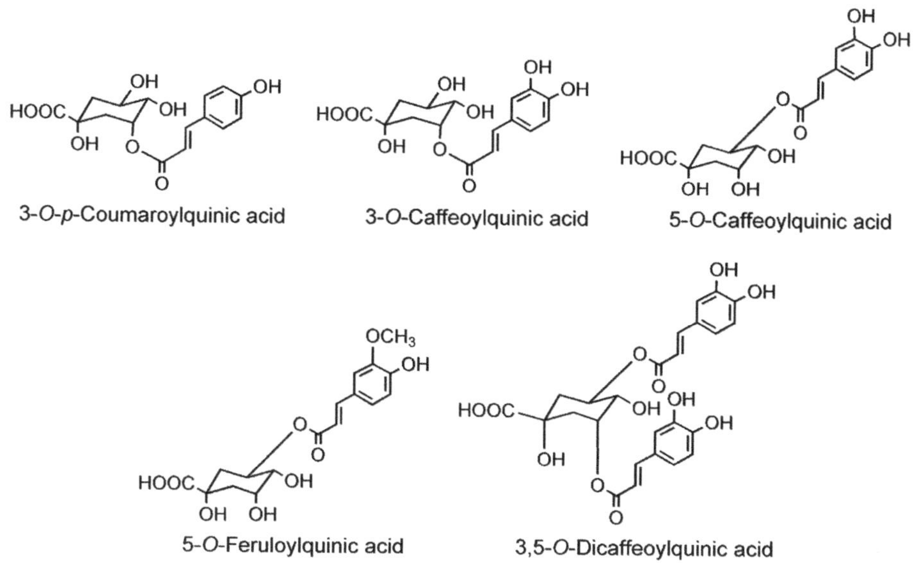

Figure 1.30 Chlorogenic acids in carrots.

Carrots *(Duacus carota)* contain a range of chlorogenic acids including 3-*O*- and 5-*O*-caffeoylquinic acids, 3-*O*-*p*-coumaroylquinic acid, 5-*O*-feruloylquinic acid, and 3,5-*O*-dicaffeoylquinic acids (Fig. 1.30). These chlorogenic acids are found in almost all varieties of carrot with a 10-fold higher level of 5-*O*-caffeoylquinic acid in purple carrots [Alasalvar et al., 2001].

The red-leaved lettuce Lollo Rosso contains the hydroxycinammates caffeoyltartaric acid, dicaffeoyltartaric acid, 5-*O*-caffeoylquinic acid, and 3,5-*O*-dicaffeoylquinic acid (Fig. 1.31) [Ferreres et al., 1997]. 5-*O*-Caffeoylquinic acid has also been detected in tomatoes [Paganga et al., 1999].

Green coffee beans *(Coffea arabica)* are one of the richest dietary sources of chlorogenic acids. 5-*O*-Caffeoylquinic acid is the dominant chlorogenic acid accounting for 50% of the total. This is followed by 3-*O*- and 4-*O*-caffeoylquinic acid, the three analogous feruloylquinic acids and 3,4-*O*-, 3,5-*O*- and 4,5-*O*-dicaffeoylquinic acids (Fig. 1.32) [Clifford, 1999]. Levels decline ca. 80% during the roasting of coffee beans, but sizable amounts with substantial antioxidant activity are still found in the typical cup of coffee.

There is also dietary interest in the curcuminoids, which are cinnamoylmethanes (diaryl-heptenoids), are characteristic of ginger (*Zingiber officinale*), cardamon (*Elettara cardamonum*), and turmeric (*Curcuma longa*). Curcumin is a diferuloylmethane. Three curcuminoids, curcumin, demethoxycurcumin, and bisdemethoxycurcumin (Fig. 1.33), are the principal components in tumeric, and all three impart the yellow pigmentation that is a hallmark of the spice [Jayaprakasha et al., 2005].

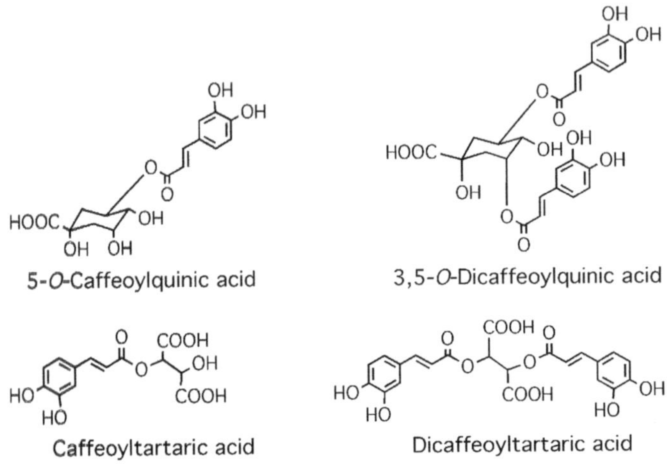

Figure 1.31 Hydroxycinammates in Lollo Rosso lettuce.

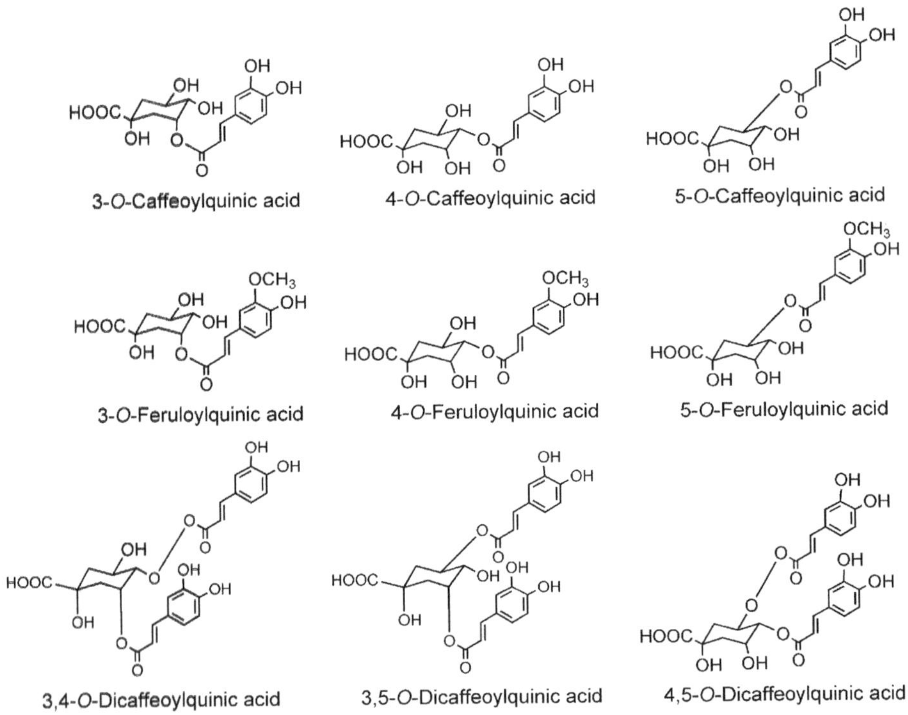

Figure 1.32 Main chlorogenic acids in green coffee beans.

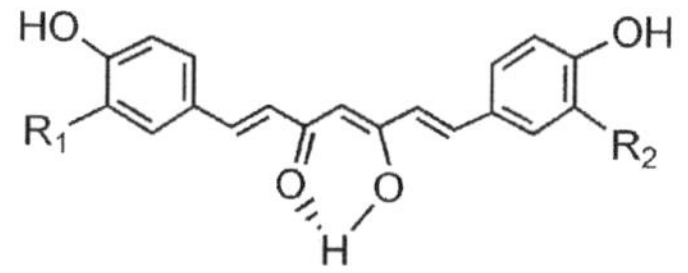

Figure 1.33 Curcuminoids accumulate in the rhizomes of turmeric.

Stilbenes

Members of the stilbene family have the C_6–C_2–C_6 structure and are phytoalexins produced by plants in response to disease, injury, and stress (Fig. 1.34) [Langcake and Pryce, 1977]. The main dietary source of stilbenes is resveratrol (3,5,4′-trihdroxystilbene) from red wine and peanuts (*Arachis hypogaea*) [Burns et al., 2002b] with lesser amounts found in berries, red cabbage (*Brassica oleraceae*), spinach, and certain herbs. Resveratrol occurs as *cis* and *trans* isomers and *trans*-resveratrol and *trans*-resveratrol-3-*O*-glucoside (*trans*-piceid) have recently been detected in pistachio nuts (*Pistacia vera* L.) [Grippi et al., 2008].

The woody root of the noxious weed *Polygonum cuspidatum* (Japanese knotweed or Mexican bamboo) has been shown to contain very high levels of *trans*-resveratrol and its glucosides with concentrations of up to 377 mg 100 g^{-1} dry weight [Vastano et al., 2000]. As well as resveratrol, Brazilian red wines have been shown to contain *trans*-piceatannol (3,3′,4,5′-tetrahydroxystilbene) and *trans*-astringin, its 3-*O*-glucoside [Vitrac et al., 2005] *trans*-resveratrol is transformed by *Botrytis cinerea*, a fungal grapevine pathogen, to pallidol and resveratrol *trans*-dehydrodimer, and both these compounds have been detected in grape cell cultures along with the 11-*O*- and 11′-*O*-glucosides of resveratrol *trans*-dehydrodimer [Waffo-Téguo et al., 2001]. Viniferins are another family of oxidized resveratrol dimers [Langcake and Pryce, 1977], and δ-viniferin and smaller amounts of its isomer δ-viniferin have been detected in *Vitis vinifera* leaves infected with *Plasmopara viticola* (downy mildew) [Pezet et al., 2003].

trans-Resveratrol that has gained significant worldwide attention because of its ability to inhibit or retard a wide variety of animal diseases [Baur and Sinclair, 2006] that include cardiovascular disease [Bradamante et al., 2004] and cancer [Jang et al., 1997]. It has also been reported to increase stress resistance and enhance longevity [Baur et al., 2006; Valenzano et al., 2006]. The protective effects of red wine consumption are regularly attributed to resveratrol [Kaeberlein and Rabinovitch, 2006]. However, this is highly unlikely as the levels of resveratrol in red wines are low, and for humans to ingest the quantity of resveratrol that affords protective effects in animals they would have to drink in excess of 100 L of red wine per day [Corder et al., 2003].

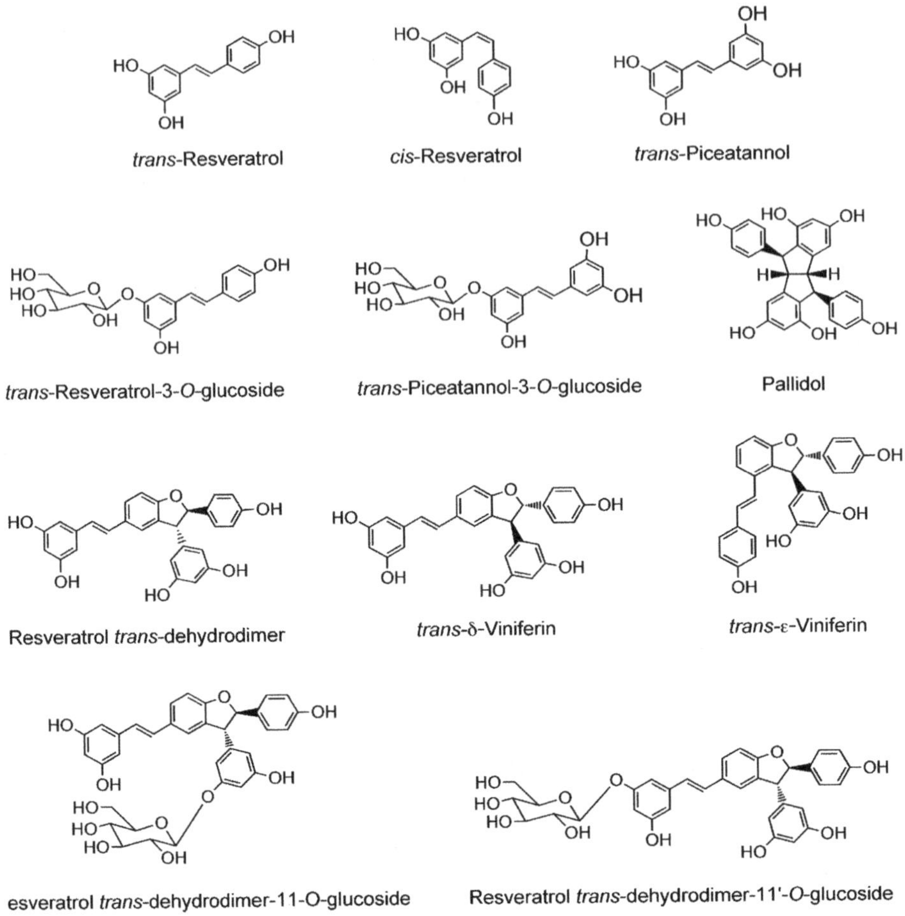

Figure 1.34 Structures of the *trans*- and *cis*-resveratrol and other stilbenes.

OVERVIEW OF FLAVONOID AND PHENOLIC BIOSYNTHETIC PATHWAYS

The biosynthesis of flavonoids, stilbenes, hydroxycinnamates, and phenolic acids involves a complex network of routes based principally on the shikimate, phenylpropanoid, and flavonoid pathways (Figs. 1.35 and 1.36). These biosynthetic pathways constitute a complex biological regulatory network that has evolved in vascular plants during their successful transition on land and that ultimately is essential for their growth, development, and survival [Costa et al., 2003].

From the 1970s to the 1990s, there was a rapid and substantial progress in the research on the phenylpropanoid pathway, focusing mainly on a broad understanding of the metabolic pathway [Hahlbrock and Grisebach, 1975; Ebel and Hahlbrock, 1982; Heller and Forkmann, 1988]. However, in more recent

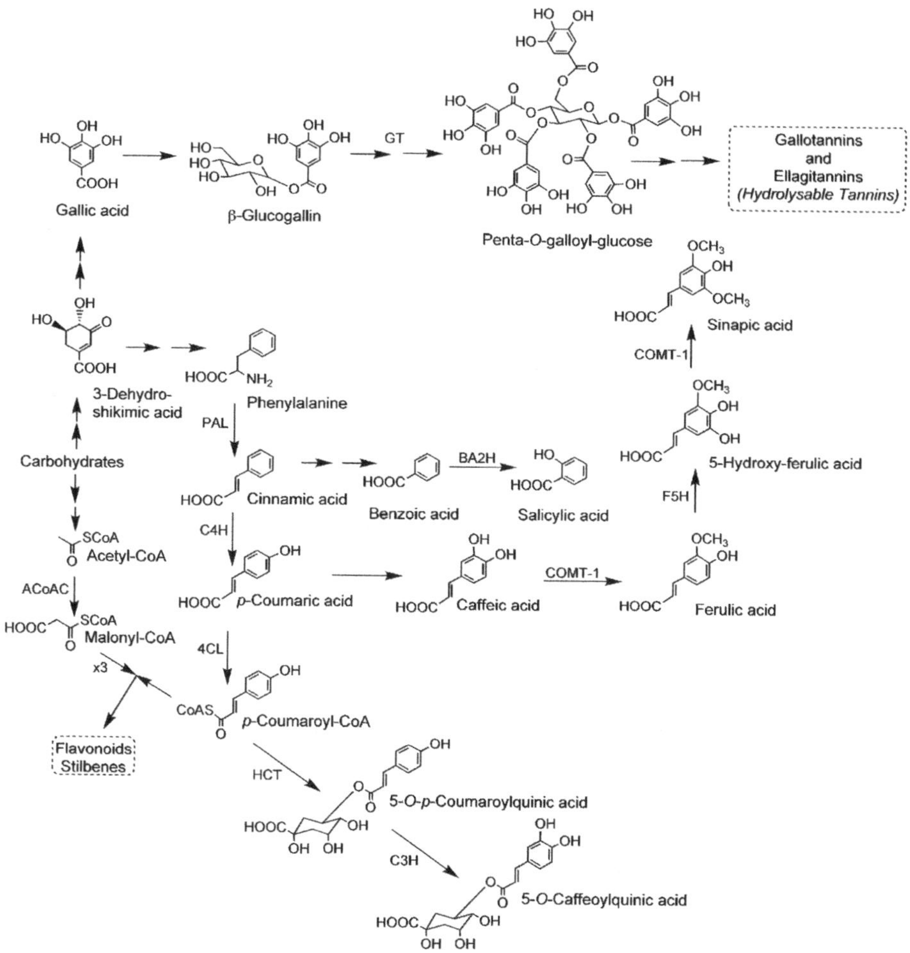

Figure 1.35 Schematic diagram of the phenolic biosynthetic pathway accompanied by the key enzymes involved. Enzyme abbreviations: PAL, phenylalanine ammonia-lyase; BA2H, benzoic acid 2-hydroxylase; C4H, cinnamate 4-hydroxylase; COMT-1, caffeic/5-hydroxyferulic acid *O*-methyltransferase; 4CL, *p*-coumarate:CoA ligase; F5H, ferulate 5-hydroxylase; GT, galloyltransferase; ACoAC, acetylCoA carboxylase.

years, much effort has been directed at elucidating the phenylpropanoid biosynthetic pathway from a biochemical and a molecular point of view by using approaches such as transposon tagging, positional cloning, co-immuno-precipitation, affinity chromatography, and two-hybrid experiments mainly utilizing *Arabidopsis thaliana* as a test system [Winkel-Shirley, 2001]. New information is also emerging regarding the regulation of the phenylpropanoid pathway. In the last few years, a great deal has been learned from studies in a variety of plant species, primarily about transcriptional regulation. A number of these studies were carried out using flavonoid mutants generated by

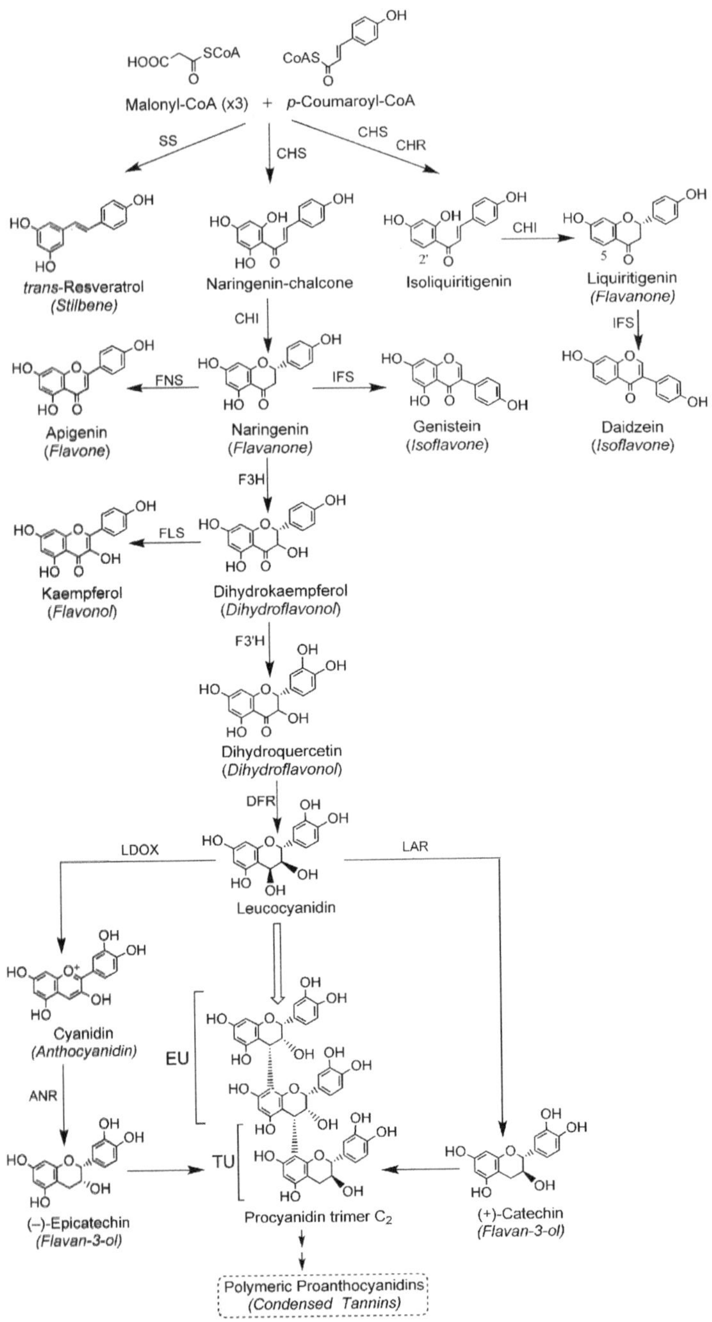

Figure 1.36 Schematic diagram of the stilbene and flavonoid biosynthetic pathway. Enzyme abbreviations: SS, stilbene synthase; CHS, chalcone synthase; CHR, chalcone reductase; CHI, chalcone isomerase; IFS, isoflavone synthase; FNS, flavone synthase; F3H, flavanone 3-hydroxylase; FLS, flavonol synthase; F3′H, flavonoid 3′-hydroxylase; DFR, dihydroflavonol 4-reductase; LAR, leucoanthocyanidin 4-reductase; LDOX, leucocyanidin deoxygenase; ANR, anthocyanidin reductase; EU, extension units; TU, terminal unit.

activation tagging [Borevitz et al., 2000; Mathews et al., 2003]. Characterization of flavonoid mutants in a variety of plant species has led to the identification of a number of novel regulatory proteins that are beginning to fill in the void between signals that induce the pathway and well-known flavonoid regulators such as the myb domain and basic helix–loop–helix transcription factors [Winkel-Shirley, 2001]. In addition, increasing evidence is being generated demonstrating that as well as inducing the phenylpropanoid pathway, these transcriptional regulators also influence the modification, transport, and deposition of metabolites in the vacuole [Broun, 2004].

In addition to the molecular techniques, technical advances both in chromatographic techniques and in identification tools, particularly the diverse forms of mass spectrometry, has allowed successful challenges to the separation and characterization of compounds of increasing complexity, poor stability, and low abundance [Whiting, 2001]. Information generated utilizing these techniques has resulted in characterization of a plethora of complex secondary metabolites that, in conjunction with the characterization of the enzymatic steps, has permitted the complete or partial elucidation of the flavonoid and the phenolic pathways present in many plants (Figs. 1.35 and 1.36).

Comprehensive information on the network of pathways responsible for the synthesis of numerous secondary metabolites can be found in Chapter 21. In addition, information on this aspect is also available in articles by Shimada et al. [2003], Toshiaki [2003], Tanner et al. [2003], Boatright et al. [2004], Hoffmann et al. [2004], Dixon et al. [2005], Niemetz and Gross [2005], Xie and Dixon [2005], and Ferrer et al. [2008]. Nonetheless, the complete dissection of phenolic metabolic pathway is far from being complete. For example, recent reports underline that important questions still remain to be answered in the field of protoanthocyanidins and tannins [Xie and Dixon, 2005], and that the exact nature of the biosynthetic pathway(s) leading to lignin monomers has not been fully elucidated [Boudet, 2007].

An example of the phenolic pathway, which produces secondary metabolites that have health benefiting effects, is the biosynthesis of curcuminoids. The initial investigations into the biosynthesis of curcuminoids were carried over 25 years ago [Denniff and Whiting, 1976; Macleod and Whiting, 1979; Denniff et al., 1980], although little has been done subsequently to elucidate fully the routes involved. The proposed biosynthetic pathway is presented in Figure 1.37. The curcuminoids are thought to be formed from condensation of two molecules of *p*-coumaroyl-CoA with one molecule of malonyl-CoA via the action of possibly a polyketide synthase. The resulting bisdemethoxycurcumin would then be transformed through demethoxycurcumin into curcumin via two sequential rounds of hydroxylation followed by *O*-methylation. Alternatively, it is possible that the curcuminoid synthase enzyme may utilize the CoA esters of both *p*-coumaric acid and ferulic acid as substrates [Ramirez-Ahumada et al., 2006].

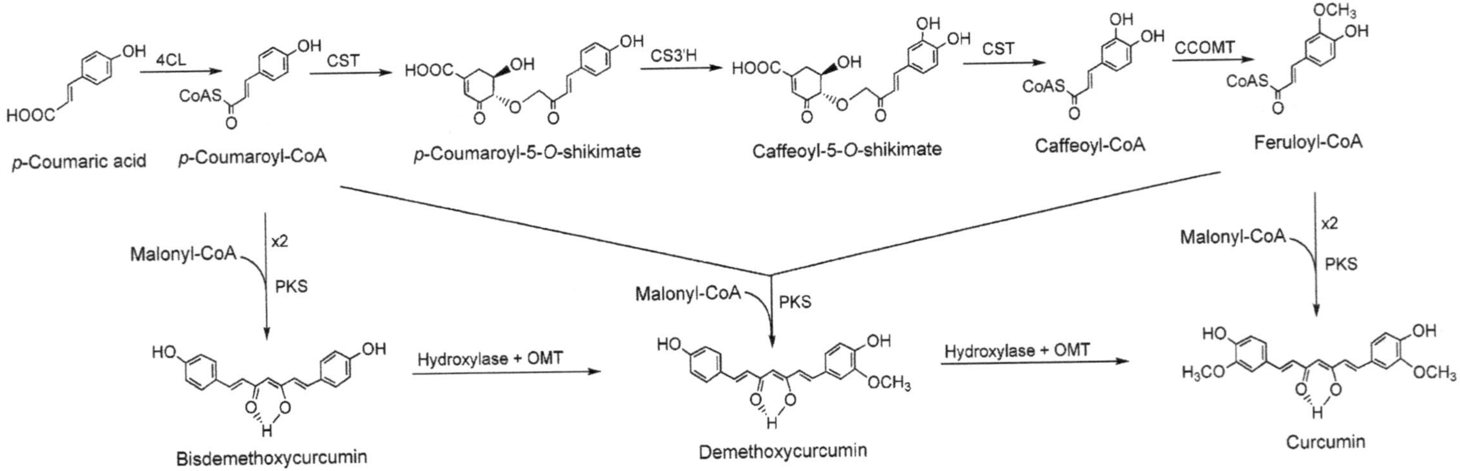

Figure 1.37 Proposed biosynthetic pathway of curcuminoids in tumeric. Enzyme abbreviations: CCOMT, caffeoyl-CoA *O*-methyltransferase; 4CL, 4-coumarate:CoA ligase; CST, shikimate transferase; CS3′H, *p*-coumaroyl 5-*O*-shikimate 3′-hydroxylase; OMT, *O*-methyltransferase; PKS, polyketide synthase. [Adapted from Ramirez-Ahumada et al. (2006)]

OPTIMIZATION OF THE FLAVONOID AND PHENOLIC PROFILES IN CROP PLANTS

The recent increase in consumer awareness on the health benefits of dietary phytochemicals accompanied by the rapid progress in the field of molecular biology have provided the means and incentive to enhance the functional value of plant material. This enhancement of health-promoting compounds is being tackled using a variety of approaches, which are discussed in the ensuing sections.

Agronomical and Physiological Modifications

Abiotic and biotic stresses are known to induce the accumulation of phenolic and flavonoid compounds in many higher plants. As sessile organisms, plants rely on the accumulation of such chemicals for defense, protection, cell-to-cell signaling, and other stress adaptations. As such, wild-type berries from harsh environmental growing regimes were found to be among the most biologically potent in terms of antioxidant content compared to their commercially grown counterparts [Deighton et al., 2000; Reyes-Carmona et al., 2005]. Hence agronomic manipulation by the application of mild stress at defined points during the growing season may have generic effects on phenolic and flavonoid accumulation. This is why deliberate stress on the target plants that are specifically cultivated for their health-benefiting compounds is now becoming a popular research strategy. Environmental deprivation, such as exposure of plants to low temperatures, as well as heavy metals, wounding, desiccation, and high irradiance are typical triggers that switch on a biochemical pathway cascade leading to increased secondary product accumulation [Lila, 2006].

High temperature is known to reduce and low temperature to enhance anthocyanin synthesis [Saure, 1990; Leng et al., 2000]. This was observed in Starkrimson and Golden Delicious apples where there was rapid anthocyanin accumulation in the skin in cooler habitats compared to warmer climates [Li et al., 2004]. Similarly, when grapes were grown under low night temperatures, anthocyanin synthesis, L-phenylalanine ammonia-lyase activity and *chalcone synthase 3* transcript levels were all markedly higher [Mori et al., 2005]. Apart from temperature, it has long been known that UV radiation, specifically the UV-B, can up-regulate key genes such as the *phenylalaline ammonia-lyase* [Kuhn et al., 1984] and *chalcone synthase* [Christie and Jenkins, 1996]. This up-regulation of the genes in the phenylpropanoid pathway is part of the plant's ability to offset the absorption of excessive UV radiation by accumulating UV-filtering secondary metabolites such as flavonols [Cuadra et al., 1997] and anthocyanins [Oelmüller and Mohr, 1985]. Similarly, investigations with apples have shown that covering the orchard floor with metallic foil in an effort to reflect increased light into the canopy resulted in an increase in anthocyanin concentration in the skin of the fruit [Ju et al., 1999].

Agronomical manipulation has also been employed to improve the phytochemical content in plants. In a number of investigations, anthocyanins were observed to accumulate in plants deficient in nutrients such as phosphorus and nitrogen [Cobbina and Miller, 1987; Hodges and Nozzolillo, 1996; Close et al., 2000]. In another study, it was found that production of the isoflavones, daidzein and genistein, could be modulated by changing the ammonia/nitrate ratio. In bean plants cultured on phosphorus-deficient media, higher concentrations of anthocyanins were found in the leaves, and this may play a role in protecting the plant against oxidative stress [Juszczuk et al., 2004]. Substantial variability in the levels of caffeoylquinic, sinapic, and ferulic acid derivates in eight broccoli *(Brassica oleracea)* cultivars grown under different agronomic conditions has also been reported by Vallejo et al. [2003]. In tomato, in addition to increasing flavonoid content, nitrogen stress also produces differential effects on expression of genes encoding anthocyanin biosynthetic enzymes [Bongue-Bartelsman and Phillips, 1995].

Another popular way to enhance the production of bioactive compounds is through elicitation. In this process, target plants are deliberately challenged with chemicals that trigger physiological responses that mimic the parallel environmental challenges. This in turn results in the accumulation of specific phytochemicals. This may include abiotic elicitors, such as metal ions and inorganic compounds, and biotic elicitors including fungi, bacteria, viruses or herbivores, plant cell wall components, as well as chemicals that are released by plants when they are subjected to pathogen or herbivore attack [Zhao et al., 2005]. Two well-known elicitors are salicylic acid and jasmonic acid, and these compounds have frequently been added to cell cultures to induce the accumulation of compounds with potential health benefits including flavonoids and phenylpropanoids [Zhao et al., 2005]. Other natural elicitors such as fish protein hydrolysates and lactoferrin have also been used. These elicitors stimulate the phenylpropanoid pathway in mung bean sprouts, probably through the pentose phosphate and shikimate pathways. This resulted in significant improvement of the phenolic content and antioxidant and antimicrobial properties of mung bean sprouts [Randhir et al., 2004]. In another study, preharvest treatment with benzothiadiazole increased *trans*-resveratol and anthocyanin levels in grapevine [Iriti et al., 2004]. Further investigations revealed that five monoglucosides of delphinidin, cyanidin, petunidin, peonidin, and malvidin, accompanied by the corresponding acetylated and *p*-coumaroyl derivatives, were enhanced by benzothiadiazole treatment [Fumagalli et al., 2006]. Plant hormones can also affect the phenolic and flavonoid content of plants as revealed in a number studies [Jeong et al., 2004; Peppi et al., 2007; Kondo and Inoue, 1997].

Genetic Manipulation—Conventional Breeding and Genetic Engineering

The roles that dietary flavonoids and phenolic compounds play in promoting human health have stimulated intense interest in genetically manipulating their accumulation in plants, through either conventional or molecular means. Genetic variability is one of the key factors in determining the amount of functional metabolites that accumulate in plants. Conventional breeding and cross-varietal screening tests have repeatedly revealed that genotypes within a plant species can have widely divergent levels of phytochemicals. For example, a number of different cultivars and species of blueberries exhibited varying levels of anthocyanins and proanthocyanidins, which were tightly correlated with the antioxidant capacity of fruit extracts [Kalt et al., 2001]. Enhanced lycopene and flavonoid levels have also been reported in some varieties of tomato and these lines are currently the preferred hosts for further genetic manipulation through conventional and molecular breeding [Long et al., 2006]. However, it is important to note that even though these genotypes are known to be capable of accumulating enhanced levels of specific phytochemicals, the final content is dependent on the selective pressure imposed by the environment. Genes are not always expressed, but instead can be triggered by environmental signals that may ultimately become the principal determinant for the accumulation of key secondary products. It is, therefore, important to note that gene–environment interactions are inherent as plants grow, which makes it difficult to predict phytochemical responses based on heritable traits and distinguish them from environmental influences [Lila, 2007]. The influences of the genome and the environment can be resolved through rigorous comparative tests of the identical plant genotypes in multiple environments followed by gene sequencing and phytochemical profiling of selected candidate plants [Taylor et al., 2002, Lim et al., 2005; Mpofu et al., 2006].

Due to the rapid speed at which knowledge of the genetic control of plant secondary metabolism has grown, it is hypothesized that over the next 25 years the most significant changes in the productivity and quality of crops will come about by applying genetic engineering tools. Normally, genetic engineering of a secondary metabolic pathway aims to increase the quantity of an individual or a group of specific compounds in the normal producing plant species or to transfer a pathway, or part of a pathway, to other plant species [Verpoorte and Memelink, 2002]. To increase the production of the compound(s) of interest, two general approaches have been followed. First, the structural genes encoding enzymes that participate directly in the formation of the compound of interest can be overexpressed. This is to enable the genetically modified plant to overcome specific rate-limiting steps in the pathway, to shut down competitive pathways, and to decrease catabolism of the product of interest. Secondly, attempts have been made to change the expression of regulatory genes that control the expression of the structural genes [Verpoorte and Memelink, 2002]. Regulatory genes control the expression of structural genes though the production of proteins called transcriptional factors. Transcriptional factors are believed to play an important role in regulating secondary metabolism pathways. Since transcriptional factors are able to control multiple steps within a pathway, they are potentially more powerful than structural genes that control only a single step, when attempting to manipulate metabolic pathways in plants [Broun, 2004].

The best studied route at the genetic level is the flavonoid biosynthesis pathway leading to the formation of anthocyanins. Most of the structural and several regulatory genes involved in this pathway have now been cloned. The use of structural genes in metabolic engineering was used by Jung et al. [2000] who introduced the *isofavone synthase* gene into the nonlegume arabidopsis (*Arabidopsis thaliana*) in order to convert naringenin, which is ubiquitous in higher plants, to the isoflavone genistein. In another study, chalcone isomerase (CHI), the key enzyme to increased flavonol production, was overexpressed in tomato. Results revealed a 78-fold increase of flavonol levels in the skin of tomatoes [Muir et al., 2001]. To date, several *leucoanthocyanidin reductase (LAR)* and/or *anthocyanidin reductase (ANR)* genes have been cloned and characterized from different plant species [Tanner et al., 2003; Xie et al., 2003; Bogs et al., 2005; Pang et al., 2007; Paolocci et al., 2007]. When the *ANR* gene was overexpressed in barrel clover (*Medicago truncatula*) and tobacco (*Nicotiana tabacum*), accumulation of proanthocyanidins was observed with a corresponding reduction of anthocyanin levels [Bogs et al., 2005; Xie et al., 2006]. Beyond the modified expression of one gene, more sophisticated strategies have been adopted such as the simultaneous introduction by cotransformation of a sense and an antisense construct to simultaneously up-regulate one enzyme and down-regulate another. For example, aspen trees (*Populus tremuloides*), expressing both antisense 4-coumarate-CoA ligase and sense coniferaldehyde 5-hydroxylase, had 52% reduced lignin content and a 64% higher syringyl/guaiacyl ratio [Li et al., 2003].

To control expression of structural genes, regulatory genes such as *LC*, *C1*, *MYB*, *HLH*, and the like are used. In an investigation where *LC* and *C1* genes were overexpressed in tomatoes, an increase in flavonols in the flesh of the fruit was observed. The total flavonol content of these overexpressed ripe transgenic tomatoes were ca. 20-fold higher than that of the controls where flavonol production occurred only in the skin [Bovy et al., 2002; Le Gall et al., 2003]. Similarly, when the *LC* gene was introduced into apple, both anthocyanin and proanthocyanidin accumulation was observed, and this was accompanied by induction of both the anthocyanin pathway genes and proanthocyanidin-specific pathway genes such as *LAR* and *ANR* [Li et al., 2007]. MYB and bHLH transcription factor is envisaged to be central to the control of proanthocyanidin biosynthesis. When two MYB transcription factors, AtTT2 and PAP1, together with one bHLH transcription factor, were introduced into Arabidopsis, the *ANR* gene was induced, which resulted in anthocyanin and proanthocyanidin accumulation [Sharma and Dixon, 2006].

Engineering of novel natural products by enzymatic modifications of core skeletons is another method where molecular tools are used to produce a range of novel products with enhanced/modified bioactivity. This is mainly carried out because the distribution of many of these compounds are either restricted or they accumulate at low levels, which is insufficient for large-scale extraction

[Tian et al., 2008]. Modifications to the ring structure and/or acyclic side structure of aglycones through the use of specific modification enzymes can be carried out to cause oxidation, *C*- or *O*-methylation, *C*- or *O*-glycosylation, and *C*- or *O*-prenylation to produce a range of phytochemical derivatives. Genes encoding some of the modification enzymes have been cloned and characterized in recent years using genetic, genomic, and biochemical approaches. The use of this method to produce a range of novel isoflavonoids has been reviewed in detail by Tian et al. [2008].

One of the major drawbacks of targeted induced modifications of key enzymes of phenolic and flavonoid metabolism, aiming to increase or decrease a specific phytochemical, is the observation of unexpected effects. This may be due partly to the effect of combined outcomes of a complex interplay of various metabolic pathways and variation between plant species. Other than the internetworking and regulation of endogenous pathways, the final result of metabolic engineering is also dependent on a number of factors such as the approach used, the encoded function of the introduced gene, and the type of promoter [Lessard et al., 2001; Broun, 2004]. Besseau et al. [2007] recently exhibited how network complexity and pathway interactions observed between different branches of phenolic biosynthesis resulted in an unexpected array of events. In arabidopsis plants silenced for hydroxycinnamoyl-CoA shikimate/quinate hydroxycinnamoyl transferase (HCT) expression lignin repression lead to an increase in chalcone synthase activity, which resulted in a metabolic flux into flavonoid pathway. Correlated with this was a prominant reduction in plant growth. When this process was reversed through the repression of *chalcone synthase* expression in HCT-silenced plants, the wild-type plant growth was restored. The results suggest that the dwarf phenotype may be due to an indirect effect of ectopic flavonoid accumulation altering auxin transport [Boudet, 2007].

In several cases, studies on overexpression of genes have resulted in the production of unexpected products, as revealed by Bovy et al. [2002] when C1 and R transcriptional factors were cloned into tomato. Although several flavonoid genes were induced, they were not sufficient to induce flavonoid-3′,5′-hydroxylase activity to enhance anthocyanin production by the fruit. Alternatively, the introduction of a new branch point into an existing pathway may interfere with endogenous flavonoid or phenolic biosynthesis and/or the transgenic enzyme may fail to compete with the native enzymes for the common substrate. This could, in part, be due to compartmentalization and metabolic channeling of substrates that may further complicate metabolic engineering strategies by limiting the access of substrates to introduced enzymes. This occurred when soybean-derived isoflavone synthase (IFS) was introduced into arabidopsis and tomato [Jaganath, 2005]. The nonleguminous species did not synthesize genistein despite expression of the IFS protein.

Based on these studies on biosynthetic pathways and metabolic engineering, it can be envisaged that once the plant cell factory has been assembled, the important determinants controlling the fluxes through the pathways are the posttranslational regulation of enzyme activity, enzyme, and metabolite compartmentation and transport [Verpoorte and Memelink, 2002].

FUTURE TRENDS AND PROSPECTS

Phenolic compounds and flavonoids are a unique category of plant phytochemicals especially in terms of their vast po ential health-benefiting properties. They represent the most abundant and the most widely represented class of plant natural products. A substantial amount of research has been carried out over the past two decades yet large information gaps still exist. For example, the inventory of these compounds is still incomplete, although there is continuous effort to provide new structures. In addition the dissection of the metabolic pathways for certain phenolic compounds remains to be resolved. Recent reports underline that important questions that still need to be answered in the field of proanthocyanidin and tannin biosynthesis [Xie and Dixon, 2005], and even the exact nature of the biosynthetic pathway(s) leading to lignin monomers is not fully elucidated.

Phenolic compounds and flavonoids are widely present in plant foods, and research in the last decade has increased dramatically. Two major objectives have been targeted: (i) to rationalize the potential health benefits of these phytochemicals and (ii) to redesign plants to enhance their production. The existing literature on biological activities suggests that polyphenol-rich products such as soya, teas, berries, red wine, and cocoa products may have positive effects on human health, especially by reducing the incidence of cardiovascular diseases and some types of cancer. Additional research is needed to substantiate whether it is a specific class of phenolic compounds and flavonoids present in plant foods that contributes to the observed bioactivities in man, or whether it is the consumption of a broad spectrum of phytochemicals that is more important. The exact mode of action of these phytochemicals still remains to be answered. Many earlier studies suggested that phenolic compounds and flavonoids protect cell constituents through direct scavenging of free radicals due to their antioxidant properties. However, recent data indicate that the protective effect of flavonoids and phenolic compounds may extend beyond their antioxidant activity. However, research in this field is still at its infancy as it has been carried out only on specific phytochemicals. Future research needs to focus on methods to better evaluate and optimize the in vivo effects of health-promoting compounds in biological system. Many promising results have been obtained to engineer or breed plants with enhanced levels of phenolic compounds and flavonoids. However, metabolomics and microarray analysis of global gene expression patterns have revealed that playing with a piece of the jigsaw may induce unfavorable changes in the fragile equilibrium of the interconnected pathways. Accurate controls should be envisaged in future studies to check for potential pitfalls. Once these setbacks are overcome there remains the possibility to develop super crop varieties containing enhanced health-promoting flavonoid and phenolic compounds.

REFERENCES

Alasalvar C, Grigorr JM, Zhang D. 2001. Comparison of volatiles, phenolic compounds, sugars, antioxidant vitamins and sensory quality of different coloured carrot varieties. *J Agric Food Chem* 49:1410–1416.

Al-Farsi M, Alasalvar C, Morris A, Baron M, Shahidi F. 2005. Comparison of antioxidant activity, anthocyanins, carotenoids, and phenolic compounds of three native fresh and sun dried date (Phoenix dactylifera L.) varieties grown in Oman. *J Agric Food Chem* 53:7592–7599.

Amakura Y, Okada M, Tsuji S, Tonagai Y. 2000. High performance liquid chromatographic determination with photo diode array detection of ellagic acid in fresh and processed fruits. *J Chromatogr A* 896:87–93.

Anderson JJ, Anthony MS, Cline JM, Washburn SA, Garner SC. 1999. Health potential of soy isoflavones for menopausal women. *Public Health Nutr* 2:489–504.

Aron PM, Kennedy JA. 2007. Compositional investigation of phenolic polymers isolated from Vitis vinifera L. cv. Pinot Noir during fermentation. *J Agric Food Chem* 55:5670–5680.

Auger C, Al Awwadi N, Bornet A, Rouanet JM, Gasc F, Cros G, Teissedre P L. 2004. Catechins and procyanidins in Mediterranean diets. *Food Res Int* 37:233–245.

Balentine DA, Wiseman SA, Bouwnes LC. 1997. The chemistry of tea flavonoids. *Crit Rev Food Sci Nutr* 37:693–704.

Barnes S. 2003. Phyto-oestrogens and osteoporosis: What is a safe dose? *Br J Nutr* 89:S101–S108.

Barnes S, Kirk M, Coward L. 1994. Isoflavones and their conjugates in soy foods: extraction conditions and analysis by HPLC–mass spectrometry. *J Agric Food Chem* 42:2466–2474.

Baur JA, Sinclair DA. 2006. Therapeutic potential of resveratrol: The in vivo evidence. *Nat Rev Drug Discov* 5:493–506.

Baur JA, Pearson KJ, Price NL, Jamieson HA, Lerin C, Kalra A, Prabhu VV, Allard JS, Lopez-Lluch G, Lewis K, Pistell PJ, Poosala S, Becker KG, Boss O, Gwinn D, Wang M, Ramaswamy S, Fishbein KW, Spencer RG, Lakatta EG, Couteur DL, Shaw RJ, Navas P, Puigserver P, Ingram DK, de Cabo R, Sinclair DA. 2006. Resveratrol improves health and survival of mice on a high-calorie diet. *Nature* 444:337–342.

Besseau S, Hoffmann L, Geoffroy P, Lapierre C, Pollet B, Legrand M. 2007. Flavonoid accumulation in arabidopsis repressed in lignin synthesis affects auxin transport and plant growth. *Plant Cell* 19:148–162.

Bogs J, Downey MO, Harvey JS, Ashton AR, Tanner GJ, Robinson SP. 2005 Proanthocyanidin synthesis and expression of genes encoding leucoanthocyanidin reductase and anthocyanidin reductase in developing grape berries and grapevine leaves. *Plant Physiol* 139:652–663.

Bongue-Bartelsman M, Phillips DA. 1995. Nitrogen stress regulates gene expression of enzymes in the flavonoid biosynthetic pathway of tomato. *Plant Physiol Biochem* 33:539–546.

Borevitz, JO, Xia YJ, Blount J, Dixon RA, Lamb C. 2000. Activation tagging identifies a conserved MYB regulator of phenylpropanoid biosynthesis. *Plant Cell* 12:2383–2393.

Boudet AM. 2007. Evolution and current status of research in phenolic compounds. *Phytochemistry* 68:2722–2735.

Bovy A, De Vos R, Kemper M, Schijlen E, Almenar Pertejo M, Muir SR, Collins J, Robinson S, Verhoeyen ME, Hughes SG, Van Tunen AJ. 2002. High-flavonol tomatoes resulting from heterologous expression of the maize transcription factor gene LC and C1. *Plant Cell* 14:2509–2526.

Bradamante S, Barenghi L, Villa A. 2004. Cardiovascular protective effects of resveratrol. *Cardiovasc Drug Rev* 22:169–188.

Bramati L, Aquilano F, Pietta P. 2003. Unfermented roooibos tea: Quantitative characteriztion of flavonoids by HPLC-UV and determination of total antioxidant activity. *J Agric Food Chem* 51:7472–7474.

Broun P. 2004. Transcription factors as tools for metabolic engineering in plants, *Current Opinion Plant Biol* 7:202–209.

Burns J, Mullen W, Landrault N, Teissedre P-L, Lean MEJ, Crozier A. 2002a. Variations in the profile and content of anthocyanins in wines made from Cabernet Sauvignon and hybrid grapes. *J Agric Food Chem* 50:4096–4102.

Burns J, Yokota T, Ashihara H, Lean MEJ, Crozier A. 2002b. Plant foods and herbal sources of resveratrol. *J Agric Food Chem* 50:3337–3340

Burns J, Gardner PT, Matthews D, Duthie, GG, Lean MEJ, Crozier A. 2001. Extraction of phenolic compounds and changes in antioxidant activity of red wines during vinification. *J Agric Food Chem* 49:5797–5808.

Chan MM, Mattiacci JA, Hwang HS, Shah A, Fong D. 2000. Synergy between ethanol and grape polyphenols, quercetin, andresveratrol, in the inhibition of the inducible nitric oxide synthase pathway. *Biochem Pharmacol* 60:1539–1548.

Cheel J, Theoduloz C, Rodríguez J, Schhmeda-Hirschmann G. 2005. Free radical scavengers and antioxidants from lemongrass (Cymbopogon citrata (DC.) Staf.). *J Agric Food Chem* 53:2511–2517.

Christie JM, Jenkins GI. 1996. Distinct UV-B and UV-A/blue light signal transduction pathways induce chalcone synthase gene expression in Arabidopsis cells. *Plant Cell* 8:1555–1567.

Clifford MN. 2000. Anthocyanins—nature, occurrence and dietary burden. *J Sci Food Agric* 80:1063–1072.

Clifford MN. 1999. Chlorogenic acids and other cinnamates—nature, occurrence and dietary burden. *J Sci Food Agric* 79:362–372.

Clifford MN, Johnston KL, Knight S, Kuhnert N. 2003. A hierarchical scheme for LC-MS identification of chlorogenic acids. *J Agric Food Chem* 51:2900–2911.

Close DC, Beadle CL, Brown PH, Holz GK. 2000. Cold-induced photoinhibition affects establishment of Eucalyptus nitens (Deane and Maiden) Maiden and Eucalyptus globules Labill. *Trees* 15:32–41.

Cobbina J, Miller JE. 1987. Purpling in maize hybrids as influenced by temperature and soil phosphorus. *Agron J* 79:576–582.

Corder R, Crozier A, Kroon PA. 2003. Drinking your health? It's too early to say. *Nature* 426:119.

Costa MA, Collins RE, Anterola AM, Cochrane FC, Davin LB. 2003. An in silico assessment of gene function and organization of the phenylpropanoid pathway metabolic networks in Arabidopsis thaliana and limitations thereof. *Phytochemistry* 64:1097–112

Coward L, Smith M, Kirk M, Barnes S. 1998. Chemical modification of isoflavones in soyfoods during cooking and processing. *Am J Clin Nutr* 68:14865–14915.

Croteau R, Kutchan TM, Lewis NG. 2000. Natural products (secondary metabolites). In: Buchanan B, Gruissem W, Jones R, Eds. *Biochemistry and Molecular Biology of Plants*. Rockville, MD: American Society of Plant Physiologists, PP. 1250–1318.

Crozier A, Ashihara H, Clifford MN (Eds). 2006a. *Plant Secondary Metabolites and the Human Diet*. Oxford: Blackwell Publishing.

Crozier A, Jaganath IB, Clifford MN. 2006b. Phenols, polyphenols and tannins: An overview. In: Crozier A, Ashihara H, Clifford MN, Eds. *Plant Secondary Metabolites and the Human Diet*. Oxford: Blackwell Publishing, PP. 1–31.

Crozier A, Yokota T, Jaganath IB, Marks SC, Saltmarsh M, Clifford MN. 2006c. Secondary metabolites in fruits, vegetables, beverages and other plant-based dietary components. In: Crozier A, Ashihara H, Clifford MN, Eds. Plant Secondary Metabolites and the Human Diet. Oxford: Blackwell Publishing, PP. 208–302.

Crozier A, McDonald MS, Lean MEJ, Black C. 1997. Quantitative analysis of the flavonoid content of tomatoes, onions, lettuce and celery. *J Agric Food Chem* 45:590–595.

Crozier A, Ashihara H, Clifford MN, Eds. *Plant Secondary Metabolites and the Human Diet*. Oxford: Blackwell Publishing, PP. 208–302.

Cuadra P, Harborne JB, Waterman PG, 1997. Increases in surface flavonols and photosynthetic pigments in Gnaphalium luteo-album in response to UV-B radiation. *Phytochemistry* 45:1377–1383.

Dégenéve A. 2004. *Antioxidants in Fruits and Vegetables*. MSc Thesis, University of Glasgow.

Deighton N, Brennan R, Finn C, Davies H. 2000. Antioxidant properties of domesticated and wild Rubus species. *J Sci Food Agric* 80:1307–1313.

Delmonte P, Perry J, Rader JI. 2006. Determination of isoflavones in dietary supplements containing soy, red clover and kudzu: Extraction followed by basic or acid hydrolysis. *J Chromatography A* 1107:59–69.

Del Rio D, Stewart AJ, Mullen W, Burns J, Lean MEJ, Brighenti F, Crozier A. 2004. HPLC-MSn analysis of phenolic compounds and purine alkaloids in green and black tea. *J Agric Food Chem* 52:2807–2815.

Dewick PM. 2002. *Medicinal Natural Products: A Biosynthetic Approach*, 2nd ed. Chichester: Wiley.

Dewick PM. 1993. Isoflavonoids. In: Harborne JB, Ed. *The Flavonoids: Advances in Research Since 1986*. London: Chapman and Hall, PP. 117–238.

Dixon RA, Steele CL. 1999. Flavonoids and isoflavonoids—a goldmine for metabolic engineering. *Trends Plant Sci* 4:394–400.

Dixon RA, Xie DY, Sharma SB. 2005. Proanthocyanidins—a final frontier in flavonoid research. *New Phytol* 165:9–28.

Donner H, Gao L, Mazza G. 1997. Separation and characterization of simple and malonylated anthocyanins in red onions. Allium cepa L. *Food Res Int* 30:637–643.

Dvoráková M, Hulín P, Karabín M, Dostálek P. 2007. Determination of polyphenols in beer by an effective method based on solid-phase extraction and high performance liquid chromatography with diode-array detection. *Czech J. Food Sci* 25:182–188.

Dykes L, Rooney LW. 2007. Phenolic compounds in cereal grains and their health benefits, *Cereal Foods World* 52:105–111.

Ebel J, Hahlbrock K. 1982. Biosynthesis. In: Harborne JB, Malory TJ, Eds. *The Flavonoids: Advances in Research.* London: Chapman and Hall, PP. 641–679.

Erlejman AG, Jaggers G, Fraga CG, Oteiza PI. 2008. TNFα-induced NF-κB activation and cell oxidant production are modulated by hexameric procyanidins in Caco-2 cells. *Arch Biochem Biophys* 476:186–195.

Ferrer J-L, Austin MB, Stewart C, Noel JP. 2008. Structure and function of enzymes involved in the biosynthesis of phenylpropanoids. *Plant Physiol Biochem* 46:356–370.

Ferreres F, Gil MI, Castañer M, Tomás-Barberán FA. 1997. Phenolic metabolites in red pigmented lettuce (Lactuca sativa). Changes with minimal processing and cold storage. *J Agric Food Chem* 45:4249–4254.

Fraga CG. 2007. Plant polyphenols: How to translate their in vitro antioxidant actions to in vivo conditions. *IUBMB Life* 59:308–315

Fumagalli F, Rossoni M, Iriti M, di Gennaro A, Faoro F, Borroni E, Borgo M, Scienza A, Sala A, Folco G. 2006. From field to health: A simple way to increase the nutraceutical content of grape as shown by NO-dependent vascular relaxation. *J Agric Food Chem* 54:5344–5349.

Gil MI, Tomás-Barberán FA, Hess-Pierce B, Holcroft DM, Kader AA. 2000. Antioxidant activity of pomegranate juice and its relationship with phenolic composition and processing. *J Agric Biol Chem* 48:4581–4589.

Gould K. 2004. Nature's Swiss army knife: The diverse protective roles of anthocyanins in leaves. *J Biomed Biotech* 5:314–320.

Graefe EU, Wittig J, Mueller S, Riethling A, Uehleke B, Drewelow B, Pforte H, Jacobasch G, Derendorf H, Veit M. 2001. Pharmacokinetics and bioavailability of quercetin glycosides in humans. *J Clin Pharmacol* 41:492–499.

Graham TL. 1991. Flavonoid and isoflavonoid distribution in developing soybean seedling tissues and in seed and root exudates. *Plant Physiol* 95:594–603.

Grippi F, Crosta L, Aiello G, Tolomeo M, Oliveri F, Gebbia N, Curione A. 2008. Determination of stilbenes in Sicilian pistachio by high-performance liquid chromatographic diode array (HPLC-DAD/FLD) and evaluation of eventual mycotoxin contamination. *Food Chem* 107:483–488.

Gu L, Kelm MA, Hammerstone JF, Beecher G, Holden J, Haytowitz D, Gebhardt S, Prior Rl. 2004. Concentration of proanthocyanidins in common foods and estimates of normal consumption. *J Nutr* 134:613–617.

Hahlbrock H, Grisebach H. 1975. Biosynthesis of flavonoids. In: Harborne JB, Mabry TJ, Mabry H, Eds. *The Flavonoids.* San Diego: Academic Press, PP. 866–915.

Harborne JB, Ed. 1993. *The Flavonoids; Advances in Research Since 1986*. London: Chapman and Hall.

Harborne JB, Williams CA. 2000. Advances in flavonoids research since 1992. *Phytochemistry* 55:481–504.

Heim K, Tagliaferro A, Bobilya D. 2002. Flavonoid antioxidants: Chemistry, metabolism and structure-activity relationships. *J Nutr Biochem* 13:572–584.

Heller W, Forkmann G. 1994. Biosynthesis of flavonoids. In: Harborne JB, Ed. *The Flavonoids*. London: Chapman and Hall, PP. 499–536.

Herrmann K. 1976. Flavonols and flavones in food plants: A review. *J Food Technol* 11:433–448.

Hertog MGL, Hollman P, Katan MP. 1992. Content of potentially anticarcenogenic flavonoids of 28 vegetables and 9 fruits commonly consumed in the Netherlands. *J Agric Food Chem* 40:2379–2383.

Hodges DM, Nozzolillo C. 1996. Anthocyanin and anthocyanoplast content of cruciferous seedlings subjected to mineral nutrient deficiencies. *J Plant Physiol* 14:749–754.

Hoffman L, Besseau S, Geoffroy P, Ritzenthaler C, Meyer D, Lapierre C, Pollet B, Legrand M. 2004. Silencing of hydroxycinnamoyl-coenzymeA shikimate/quinate hydroxycinnomyoltransferase affects phenylpropanoid biosynthesis. *Plant Cell* 16:1446–1465.

Hong YJ, Barrett DM, Mitchell AE. 2004. Liquid chromatography/mass spectrometry investigation of the impact of thermal processing and storage on peach procyanidins. *J Agric Food Chem* 52:2366–2371.

Horner JD, Gosz JR, Cates RG. 1988. The role of carbon-based plant secondary metabolites in decomposition in terrestrial ecosystems. *Am Nat* 132:869–883.

Iriti M, Rossoni M, Borgo M, Faoro F. 2004. Benzothiadiazole enhances resveratrol and anthocyanin biosynthesis in grapevine, meanwhile improving resistance to Botrytis cinerea. *J Agric Food Chem* 52:4406–4413.

Iwashina T. 2000. The structure and distribution of the flavonoids in plants. *J Plant Res* 113:287–299.

Jaganath IB. 2005. Dietary Flavonoids: Bioavailability and Biosynthesis. PhD thesis, University of Glasgow.

Jaganath IB, Mullen W, Edwards CA, Crozier A. 2006. The relative contribution of the small and large intestine to the absorption and metabolism of rutin in man. *Free Radical Res* 40:1035–1046.

Jang M, Cai L, Udeani GO, Slowing KV, Thomas CF, Beecher CWW, Fong HHS, Farnsworth NR, Kinghorn AD, Mehta RG, Moon RC, Pezzuto JM. 1997. Cancer chemopreventive activity of resveratrol: A natural product derived from grapes. *Science* 275:218–220.

Jayaprakasha GK, Jagan-Mohan-Rao L, Sakariah KK. 2005. Chemistry and biological activities of C. longa. *Trends Food Sci Technol* 16:533–548.

Jeong ST, Goto-Yamamoto N, Kobayashic S, Esaka M. 2004. Effects of plant hormones and shading on the accumulation of anthocyanins and the expression of anthocyanin biosynthetic genes in grape berry skins. *Plant Sci* 167:247–252.

Joubert E, Ferreira D. 1996. Antioxidants of rooibos tea—a possible explanation for its health promoting properties? *SA J Food Sci Nutr* 8:79–83.

Ju Z, Duana Y, Jub Z. 1999. Effects of covering the orchard floor with reflecting films on pigment accumulation and fruit coloration in "Fuji" apples. *Sci Hortic* 82:47–56.

Jung W, Yu O, Lau SM, O'Keefe DP, Odell J, Fader G, McGonigle B. 2000. Identification and expression of isofavone synthase, the key enzyme for biosynthesis of isofavones in legumes. *Nature Biotechnol* 18:208–212.

Juszczuk IM, Wiktorowska A, Malusa E, Rychter AM. 2004. Changes in the concentration of phenolic compounds and exudation induced by phosphate deficiency in bean plants (Phaseolus vulgaris L.). *Plant Soil* 267: 41–49.

Kaeberlein M, Rabinovitch PS. 2006. Grapes versus gluttony. *Nature* 444:280–281.

Kahle K, Kraus M, Richling E. 2005. Polyphenol profiles of apple juices. *Mol Nutr Food Res* 49:797–806.

Kalt W, Ryan D, Duy J, Prior R, Ehlenfeldt M, Koet S. 2001. Interspecific variation in anthocyanins, phenolic compounds, and antioxidant capacity among genotypes of highbush and lowbush blueberries (Vaccinium section cyanococcus spp.). *J Agric Food Chem* 49:4761–4767.

Kanazawa K, Sakakibara H. 2000. High content of dopamine, a strong antioxidant, in Cavendish banana. *J Agric Food Chem* 48:844–848.

Koes RE, Quattrocchio F, Mol JNM. 1994. The flavonoid biosynthetic pathway in plants: Function and evolution. *BioEssays* 16:123–132.

Kondo S, Inoue K. 1997. Abscisic acid (ABA) and 1-aminocyclopropane-1-carboxylic acid (ACC) content during growth of "Satohnishiki" cherry fruit, and the effect of ABA and ethephon application on fruit quality. *J Hortic Sci* 72:221–227.

Kong JM, Chia LS, Goh NK, Chia TF, Brouillard R. 2003. Analysis and biological activities of anthocyanins. *Phytochemistry* 64:923–933.

Krafczyk N, Glomb MA. 2008. Characterization of phenolic compounds in rooibos tea. *J Agric Food Chem* 56:3368–3376.

Krause M, Galensa G. 1992. Determination of naringenin and naringenin-chalcone in tomato skins by reversed phase HPLC after solid phase extraction. *Z Lebensmittel Forschung A* 194:29–32.

Kris-Etherton PM, Lefevre M, Beecher GR, Gross MD, Keen CL, Etherton TD. 2004. Bioactive compounds in nutrition and health-research methodologies for establishing biological function: The antioxidant and anti-inflammatory effects of flavonoids on atherosclerosis. *Ann Rev Nutr* 24:511–538.

Kroon PA, Clifford MN, Crozier A, Day AJ, Donovan JL, Manach C, Williamson G. 2004. How should we assess the effects of exposure to dietary polyphenols in vitro? *Am J Clin Nutr* 80:15–21.

Kuhn DN, Chappell J, Boudet A, Hahlbrock K. 1984. Induction of phenylalanine ammonia-lyase and 4-coumarate: CoA ligase mRNAs in cultured plant cells by UV light or fungal elicitor. *Proc Natl Acad Sci USA* 81:1102–1106.

Langcake P, Pryce RJ. 1977. The production of resveratrol and viniferins by grapevines in resposne to ultraviolet radiation. *Phytochemistry* 16:1193–1196.

Law MR, Morris JK. 1998. By how much does fruit and vegetable consumption reduce the risk of ischaemic heart disease? *Eur J Clin Nutr* 52:549–556.

Lee WK, Lee HJ. 2006. The roles of polyphenols in cancer chemoprevention. *Biofactors* 26:105–121.

Leng P, Itamura H, Yamamura H, Deng XM. 2000. Anthocyanin accumulation in apple and peach shoots during cold acclimation. *Sci Hortic* 83:43–50.

Lessard PA, Kulaveerasingam H, York GM, Strong A, Sinskey AJ. 2001. Manipulating gene expression for the metabolic engineering of plants. *Metab Engin* 4:67–79.

Li H, Flachowsky H, Fischer T, Hanke M-V, Forkmann G, Treutter D, Schwab W, Hoffmann T, Szankowski I. 2007. Maize Lc transcription factor enhances biosynthesis of anthocyanins, distinct proanthocyanidins and phenylpropanoids in apple (Malus domestica Borkh.) *Planta* 226:1243–1254.

Li XJ, Hou JH, Zhangb GL, Liu RS, Yang YG, Hu YX, Lin JX. 2004. Comparison of anthocyanin accumulation and morpho-anatomical features in apple skin during color formation at two habitats. *Sci Hortic* 99:41–53.

Li L, Zhou Y, Cheng X, Sun J, Marita JM, Ralph J, Chiang VL. 2003. Combinatorial modification of multiple lignin traits in trees through multigene cotransformation. *Proc Natl Acad Sci USA* 100:4939–4944.

Lila MA. 2007. The nature-versus-nurture debate on bioactive phytochemicals: The genome versus terroir. *J Sci Food Agric* 86:2510–2515.

Lim W, Mudge K, Vermeylen F. 2005. Effects of population, age, and cultivation methods on ginsenoside content of wild American ginseng (Panax uinquefolium). *J Agric Food Chem* 53:8498–8585.

Liu J, Xu K, Wen G, Guo H, Li S, Wu X, Dai R, Sheng Z, Liao E. 2008. Comparison of the effects of genistein and zoledronic acid on the bone loss in OPG-deficient mice. *Bone* 42:950–959.

Liu S, Manson JE, Lee IM, Cole SR, Hennekens CH, Willett WC, Buring JE. 2000. Fruit and vegetable intake and risk of cardiovascular heart disease: The Women's Health Study. *Am J Clin Nutr* 72:922–928.

Llorach R, Martinez-Sanchez A, Tomas-Barberan FA, Gil MI, Ferreres F. 2008 Characterisation of polyphenols and antioxidant properties of five lettuce varieties and escarole Food Chem 108:1028–1038.

Long M, Millar DJ, Kimura Y, Donovan G, Rees J, Fraser P, Bramley PM, Bolwell GP. 2006. Metabolite profiling of carotenoid and phenolic pathways in mutant and transgenic lines of tomato: Identification of a high antioxidant fruit line. *Phytochemistry* 67:1750–1757.

Mackenzie GG, Queisser N, Wolfson ML, Fraga CG, Adamo AM, Oteiza PI. 2008. Curcumin induces cell-arrest and apoptosis in association with the inhibition of constitutively active NF-kappaB and STAT3 pathways in Hodgkin's lymphoma cells. *Int J Cancer* 123:56–65.

Makris DP, Rossiter JT. 2001. Domestic processing of onion bulbs (Allium cepa) and asparagus spears (Asparagus officinalis): Effect on flavonol content and antioxidant status. *J Agric Food Chem* 49:3216–3222.

Makrisa DP, Kallithrakab S, Kefalas P. 2006. Flavonols in grapes, grape products and wines: Burden, profile and influential parameters. *J Food Comp Anal* 19:396–404.

Manach C, Scalbert A, Morand C, Rémésy C, Jimenez L. 2004. Polyphenols: Food sources and bioavailability. *Am J Clin Nutr* 79:727–747.

Marks SC, Mullen W, Crozier A. 2007. Flavonoid and chlorogenic acid profiles of English cider apples. *J Sci Food Agric* 87:719–728.

Mathews H, Clendennen SK, Caldwell CG, Liu XL, Connors K, Matheis N, Schuster DK, Menasco DJ, Wagoner W, Lightner J, Wagner DR. 2003. Activation tagging in tomato identifies a transcriptional regulator of anthocyanin biosynthesis, modification, and ransport. *Plant Cell* 15:1689–1703.

Mazza G, Miniati E. 1993. *Anthocyanins in Fruits, Vegetables and Grains*. Boca Raton, FL: CRC Press.

McColl Smith A. 1999. *The No.1 Ladies' Detective Agency*. Edinburgh: Polygon.

McColl Smith A. 2000. *Tears of the Giraffe*. Edinburgh: Polygon.

McKay DL, Blumberg JB. 2007. A review of the bioactivity of South African herbal teas: Rooibos (Aspalathus linearis) and honeybush (Cyclopia intermedia). *Phytotherap Res* 21:1–16.

Mertens-Talcott SU, Talcott ST, Percival SS. 2003. Low concentrations of quercetin and ellagic acid synergistically influence proliferation, cytotoxicity and apoptosis in MOLT-4 human leukemia cells. *J Nutr* 133:2669–2674.

Mori K, Sugaya S, Gemma H. 2005. Decreased anthocyanin biosynthesis in grape berries grown under elevated night temperature condition. *Sci Hort* 105:319–330.

Mouria M, Gukovskaya AS, Jung Y, Buechler P, Hines OJ, Reber HA, Pandol SJ. 2002. Food-derived polyphenols inhibit pancreatic cancer growth through mitochondrial cytochrome C release and apoptosis. *Int J Cancer* 98:761–769.

Mpofu A, Sapirstein H, Beta T. 2006. Genotype and environmental variation in phenolic content, phenolic acid composition, and antioxidant activity of hard spring wheat. *J Agric Food Chem* 54:1265–1270.

Muir SR, Collins GJ, Robinson S, Hughes S, Bovy A, Ric De Vos CH, Van Tunen AJ, Verhoeyen ME. 2001. Overexpression of petunia chalcone isomerase in tomato results in fruits containing increased levels of flavonols. *Nat Biotechnol* 19:470–474.

Mullen W, Marks SC, Crozier A. 2008. Evaluation of phenolic compounds in commercial fruit juices and fruit drinks. *J Agric Food Chem* 55:3148–3157.

Mullen W, Edwards CA, Crozier A. 2006. Absorption, excretion and metabolic profiling of methyl-, glucuronyl-, glucosyl and sulpho-conjugates of quercetin in human plasma and urine after ingestion of onions. *Brit J Nutr* 96:107–116.

Mullen W, Boitier A, Stewart AJ, Crozier A. 2004. Flavonoid metabolites in human plasma and urine after the consumption of red onions: Analysis by liquid chromatography with photodiode array and full scan tandem mass spectrometric detection. *J Chromatogr A* 1058:163–168.

Mullen W, McGinn J, Lean MEJ, Maclean MR, Gardner P, Duthie GG, Yokota T, Crozier A. 2002. Ellagitannins, flavonoids, and other phenolic compounds in red raspberries and their contribution to antioxidant capacity and vasorelaxation properties. *J Agric Food Chem* 50:6902–6909.

Niemetz R, Gross GG. 2005. Enzymology of gallotannin and ellagitannin biosynthesis. *Phytochemistry* 66:2001–2011.

Oelmüller R, Mohr H. 1985. Mode of co-action between blue/UV light and light absorbed by phytochrome in light-mediated anthocyanin formation in the milo (Sorghum vulgare Pers.) seedling, *Proc Natl Acad Sci USA* 82:6124–6128.

Paganga G, Miller NG, Rice-Evans CA. 1999. The phenolic content of fruits and vegetables and their antioxidant activities, What does a serving constitute? *FEBS Lett* 401:78–82.

Pang Y, Peel G, Wright E, Wang Z, Dixon RA. 2007. Early steps in proanthocyanidin biosynthesis in the model legume Medicago truncatula. *Plant Physiol* 145:601–615.

Paolocci F, Robbins MP, Madeo L, Arcioni S, Martens S, Damiani F. 2007. Ectopic expression of a basic helix-loop-helix gene transactivates parallel pathways of proanthocyanidin biosynthesis. Structure, expression, analysis, and genetic control of leucoanthocyanidin 4-reductase and anthocyanidin reductase genes in Lotus corniculatus. *Plant Physiol* 143:504–516.

Peppi MC, Fidelibus MW, Dokoozlian NK. 2007. Application timing and concentration of abscisic acid affect the quality of "Redglobe" grapes. *J Hort Sci Biotech* 82:304–310.

Peterson J, Dwyer MSJ. 1998. Flavonoids: Dietary occurrences and biochemical activity. *Nutr Res* 18:1995–2018.

Pezet R, Perret C, Jean-Denis JB, Tabacchi R, Gindro K, Viret O. 2003. δ-Viniferin, a resveratrol dihydrodimer: One of the major stilbenes synthesized by stressed grapevine leaves. *J Agric Food Chem* 51:5488–5492.

Pierpoint WS. 2000. Why do plants make medicines. *Biochemist* 22:37–40.

Prior RL, Wu X. 2006. Anthocyanins: Structural characteristics that result in unique metabolic patterns and biological activities. *Free Rad Res* 40:1014–1028.

Proteggente AR, Pannala AS, Paganga G, Van Buren L, Wagner E, Wiseman S, Van De Put F, Dacombe C, Rice-Evans CA. 2002. The antioxidant activity of regularly consumed fruit and vegetable reflects their phenolic and vitamin C composition. *Free Rad Res* 36:217–233.

Randhir R, Lin Y-T, Shetty K. 2004. Stimulation of phenolic compounds, antioxidant and antimicrobial activities in dark germinated mung bean sprouts in response to peptide and phytochemical elicitors. *Process Biochem* 39:637–646.

Reyes-Carmona J, Yousef G, Martinez-Peniche R, Lila M. 2005. Antioxidant capacity of fruit extracts of blackberry (Rubus sp.) produced in different climatic regions. *J Food Sci* 70:S497–S503.

Riboli E, Norat T. 2003. Epidemiologic evidence of the protective effect of fruit and vegetables on cancer risk. *Am J Clin Nutr* 78:559S–569S.

Sakamoto K. 2000. Synergistic effects of thearubigin and genistein on human prostate tumor cell (PC-3) growth via cell cycle arrest. *Cancer Lett* 151:103–109.

Santos-Buelga C, Scalbert A. 2000. Proanthocyanidins and tannin-like compounds—nature, occurrence, dietary intake and effects on nutrition and health. *J Sci Food Agric* 80:1094–1117.

Saure MC. 1990. External control of anthocyanin formation in apple. *Sci Hort* 42:181–218.

Schieber A, Ullrich W, Carle R. 2000. Characterization of polyphenols in mango puree concentrate by HPLC with diode array and mass spectrometric detection. *Innovative Food Sci Emerg Technol* 1:161–166.

Shan B, Cai YZ, Sun M, Corke H. 2005. Antioxidant capacity of 26 spice extracts and characterization of their phenolic constituents. *J Agric Food Chem* 53:7749–7759.

Sharma SB, Dixon RA. 2006. Metabolic engineering of proanthocyanidins by ectopic expression of transcription factors in Arabidopsis thaliana. *Plant J* 44:62–75.

Shimada N, Aoki T, Sato S, Nakamura Y, Tabata S, Ayabe S-I. 2003. A cluster of genes encodes the two types of chalcone isomerase involved in the biosynthesis of general flavonoids and legume-specific 5-deoxy(iso)flavonoids in Lotus japonicus. *Plant Physiol* 131:941–951.

Spencer JPE. 2007. The interactions of flavonoids within neuronal signalling pathways. *Genes Nutr* 2:257–273.

Steinmetz KA, Potter JD. 1996. Vegetables, fruit, and cancer prevention: a review. *J Am Diet Assoc* 96:1027–1039.

Stewart AJ, Mullen W, Crozier A. 2005. On-line high-performance liquid chromatography analysis of the antioxidant activity of phenolic compounds in green and black tea. *Mol Nutr Food Res* 49:19.

Sultana B, Anwar F (2008). Flavonols (kaempeferol, quercetin, myricetin) contents of selected fruits, vegetables and medicinal plants. *Food Chemi*, 108:879–884.

Szmitko PE, Verma S. 2005. Antiatherogenic potential of red wine: Clinician update. *Am J Physiol Heart Circ Physiol* 288:2023–2030.

Tanner GJ, Franki KT, Abrahams S, Watson JM, Larkin PJ, Ashton AR. 2003. Proanthocyanidin biosynthesis in plants: Purification of legume leucoanthocyanidin reductase and molecular cloning of its cDNA. *J Biol Chem* 278:31647–31656.

Taylor W, Zulyniak H, Richards K, Acharya S, Bittman S, Elder J. 2002. Variation in diosgenin levels among 10 accessions of fenugreek seeds produced in western Canada. *J Agric Food Chem* 50:5994–5997.

Temple NJ, Gladwin KK. 2003. Fruits, vegetables, and the prevention of cancer: Research challenges. *Nutrition* 19:467–470.

Thangapazham RL, Sharma A, Maheshwari RK. 2006. Multiple molecular targets in cancer chemoprevention by curcumin. *APPS J* 8:443–449.

Tian L, Pang Y, Dixon RA. 2008. Biosynthesis and genetic engineering of proanthocyanidins and (iso)flavonoids. *Phytochem Rev* 7:445–465.

Tomás-Barberán FA, Clifford MN. 2000. Dietary hydroxybenzoic acid derivatives—nature, occurrence and dietary burden. *J Sci Food Agric* 80:1024–1032.

Valenzano D, Terzibasi E, Genade T, Cattaneo A, Domenici L, Cellerino A. 2006. Resveratrol prolongs lifespan and retards the onset of age-related markers in a short-lived vertebrate. *Curr Biol* 16:296–300.

Vallejo F, Tomás-Barberán FA, Garcia-Viguera C. 2003. Effect of climatic and sulphur fertilisation conditions, on phenolic compounds and vitamin C, in the inflorescences of eight broccoli cultivars. *Eur Food Res Technol* 216:395–401.

Vastano BC, Chen Y, Zhu N, Ho C-T, Zhou Z, Rosen RT. 2000. Isolation and identification of stilbenes in two varieties of Polygonum cuspidatum. *J Agric Food Chem* 48:253–256.

Verpoorte R, Memelink J. 2002. Engineering secondary metabolite production in plants. *Curr Opin Biotechnol* 13:181–187.

Vitrac X, Bornet A, Vanderlinde R, Valls JM, Richard T, Delaunay J-C, Mérillon J-M, Teissédre P-L. 2005. Determination of stilbenes (δ-viniferin, trans-astringin, trans-piceid, cis- and trans-resveratrol, e-viniferin) in Brazilian wines. *J Agric Food Chem* 53:5664–5669.

Waffo-Téguo P, Lee D, Cuendet M, Mérillon J, Pezzuto JM, Kinghorn AD. 2001. Two new stilbene dimer glucosides from grape (Vitis vinifera) cell cultures. *J Nat Prod* 64:136–138.

Walle T. 2007. Methoxylated flavones, a superior cancer chemopreventive flavonoid subclass? *Semin Cancer Biol* 17:354–362.

Wang H, Race EJ, Shrikhande AJ. 2003a. Characterization of anthocyanins in grape juices by ion trap liquid chromatography–mass spectrometry. *J Agric Food Chem* 51:1839–1844.

Wang M, Simon JE, Aviles IF, Hirshberg J, Olmedilla B, Sandmann G, Southon S, Stahl W. 2003b. Analysis of antioxidative phenolic compounds in artichoke (Cynara scolymus L.). *J Agric Food Chem* 51:601–608.

Whiting DA. 2001. Natural phenolic compounds 1900–2000: A bird's eye view of a century's chemistry. *Nat Prod Rep* 18:583–606.

WHO. 2003. *Diet Nutrition and the Prevention of Chronic Diseases, Report of a Joint WHO/FAO Expert Consultation, WHO Technical Report Series 916*. Geneva, World Health Organization.

Williams RJ, Spencer JPE, Rice-Evans C. 2004. Flavonoids: Antioxidants or signalling molecules? *Free Radic Biol Med* 36:838–849.

Williamson G, Manach C. 2005. Bioavailability and bioefficacy of polyphenols in humans. II. Review of 93 intervention studies. *Am J Clin Nutr* 81:243S–255S.

Winkel-Shirley B. 2001. Flavonoid biosynthesis. A colorful model for genetics, biochemistry, cell biology, and biotechnology. *Plant Physiol* 126:485–493.

Wu X, Gu L, Prior RL, McKay S. 2004. Characterization of anthocyanins and proanthocyanidins in some cultivars of Ribes, Aronia, and Sambucus and their antioxidant capacity. *J Agric Food Chem* 52:7846–7856.

Xie DY, Dixon RA. 2005. Proanthocyanidin biosynthesis—still more questions than answers? *Phytochemistry* 66:2127–2144.

Xie DY, Sharma SB, Wright E, Wang ZY, Dixon RA. 2006. Metabolic engineering of proanthocyanidins through co-expression of anthocyanidin reductase and the PAP1 MYB transcription factor. *Plant J* 45:895–907.

Xie DY, Sharma SB, Paiva NL, Ferreira D, Dixon RA. 2003. Role of anthocyanidin reductase, encoded by BANYULS in plant flavonoid biosynthesis. *Science* 299:396–399.

Zhao J, Davis LC, Verpoorte R. 2005. Elicitor signal transduction leading to production of plant secondary metabolites. *Biotechnol Adv* 23:283–333.

2 Bioavailability of Flavanols and Phenolic Acids

LAURE POQUET[1], MICHAEL N CLIFFORD[2], AND GARY WILLIAMSON[3]

[1]Nestlé Research Centre, Lausanne, Switzerland
[2]Faculty of Health and Medical Sciences, The University of Surrey, Guildford, Surrey, United Kingdom
[3]School of Food Science and Nutrition, University of Leeds, Leeds, United Kingdom

CONTENTS

FLAVANOLS: STRUCTURE AND FOOD SOURCES

Flavanols consist of catechin and epicatechin monomers together with oligomeric tannins, which comprise the most abundant class of flavonoids in the human diet. The tannins are divided into three subclasses: the condensed

Plant Phenolics and Human Health: Biochemistry, Nutrition, and Pharmacology, Edited by Cesar G. Fraga.

tannins or proanthocyanidins, the derived tannins, and the hydrolysable tannins. Proanthocyanidins are present in fruits, bark, leaves, and seeds of many plants.

The procyanidins, the most abundant proanthocyanidins in plants, are composed of (epi)catechin units (Fig. 2.1) compared with the less common types of proanthocyanidins called propelargonidin and prodelphinidin, which contain, respectively, (epi)afzelechin or (epi)gallocatechin subunits [Rasmussen et al., 2005]. The three forms of proanthocyanidins differ in the degree of hydroxylation of the B-ring. Prodelphinidins possess a 3′, 4′, 5′ -trihydroxy substitution on the B-ring, procyanidins have hydroxyl groups on the 3′ - and 4′ - positions, and propelargonidin on the 4′ - position only [Dixon et al., 2005]. The monomeric units of proanthocyanidins are usually linked by C—C and occasionally by C—O—C bonds [Santos-Buelga and Scalbert, 2000]. The C—C bond is generally between the 4-position of the "upper" unit and the 8-position of the "lower" unit [Fine, 2000] (Fig. 2.1b), as in procyanidin B_1–B_4. However, an alternative linkage can be found between the C_4 of the "upper" unit and the

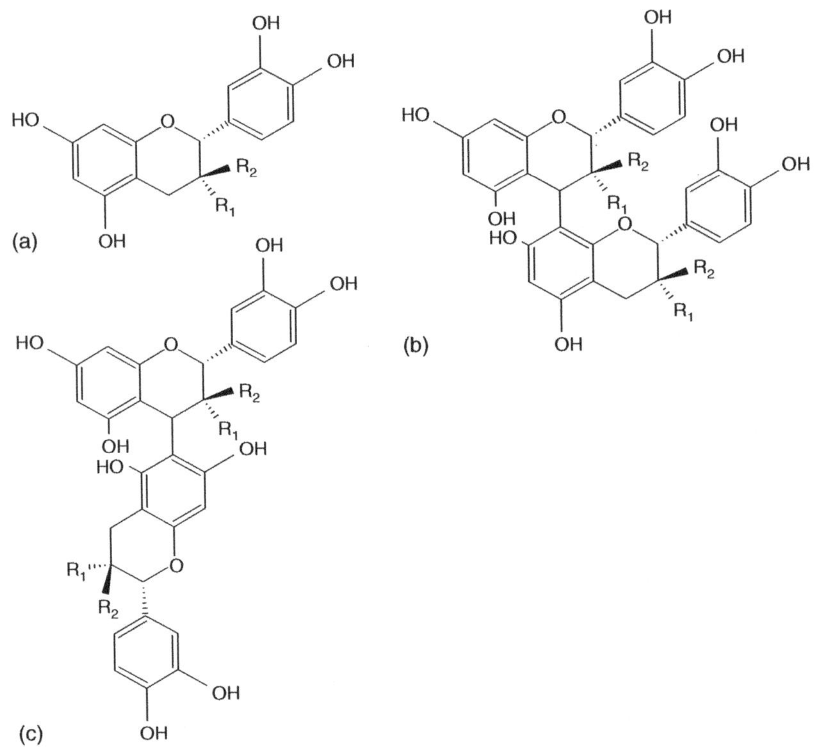

Figure 2.1 Structure of flavan-3-ol monomers and dimers. (A)(−)-Epicatechin with R_1=OH and R_2=H or (+)-catechin with R_1=H and R_2=OH;, (*B*) procyanidin (4β→8)-dimer; (*C*) procyanidin (4β→6)-dimer.

C_6 of the "lower" one (Fig. 2.1c), as for the procyanidins B_5–B_8. Both of these are called a B-type linkage. Compounds with doubly linked units, by a B-linkage and a C_2—C_7 bond, or A-linkage, have also been reported in some foods such as cranberry fruits [Prior et al., 2001] and cinnamon [Anderson et al., 2004]. Many foods have been shown to contain proanthocyanidins—fruits and berries, nuts, beans, some cereals (barley and sorghum), spices such as curry and cinnamon, but also beverages such as wine and beer [Gu et al., 2004]. However, attention has been focused on the most abundant proanthocyanidins, the procyanidins. A good example of a food source of procyanidins is cocoa, which, apart from the monomeric units, contains dimeric to decameric (or larger) forms [Hammerstone et al., 1999]. Commonly consumed food, like dark chocolate or different varieties of apples and beverages, such as wine or cranberry juice, contain large amounts of procyanidins [Hammerstone et al., 2000].

PHENOLIC ACIDS: STRUCTURE AND FOOD SOURCES

Phenolic acids are generally divided into two classes, hydroxybenzoic (C_6—C_1) and hydroxycinnamic (C_6—C_3) derivatives [Clifford, 2000; Tomás-Barberán and Clifford, 2000], but phenylacetic (C_6—C_2) and phenylpropionic (C_6—C_3) also occur. The related phenylvaleric (C_6—C_5), phenyl-lactic (α-hydroxy C_6—C_3), phenylmandelic (β-hydroxy C_6—C_2), and phenyl-hydracrylic (β-hydroxy C_6—C_3) acids are gut flora/mammalian metabolites but rarely occur in food. Most have one or two substitutions on the phenyl ring, on the 3- and 4-positions (Fig. 2.2a). However, some derivatives can have three substitutions, the additional one being on the 5-position (Fig. 2.2b). Phenylacetic acids in foods are less common than benzoic, phenylpropionic, and cinnamic acids. While the content of hydroxybenzoic acids in edible plants is low, hydroxycinnamic acids are more abundant. The most commonly found hydroxycinnamic acids are 4-coumaric, ferulic, sinapic, and caffeic acids, the latter occurring the most extensively [Shahidi and Wanasundara, 1992]. Hydroxycinnamic acids are present in a wide range of berries, fruits, and beverages [Mattila et al., 2006] but also in many vegetables [Clifford, 1999] and cereals [Kern et al., 2003a]. In addition, phenolic acids are found in traditional herbal Chinese medicines and supplements [Cai et al., 2006].

Phenolic acids are rarely present as free forms, except in processed food, but occur more frequently as soluble or insoluble esters. These esters are formed with polysaccharides or simple sugars, with quinic acid or other carboxylic acids such as tartaric or shikimic acids [Herrmann, 1989], with other phenolic acids, with lipids [Clifford, 2000], with sterols or glycerol [Clifford, 1999], or with amino acids [Clifford and Knight, 2004]. To quinic acid, they can be conjugated as mono-, di-, tri-, and tetra-esters [Clifford, 2000]. The multiple esters can contain the same or different hydroxycinnamic acids. Among the hydroxycinnamic conjugates, caffeoylquinic and di-caffeoylquinic acids

(a)

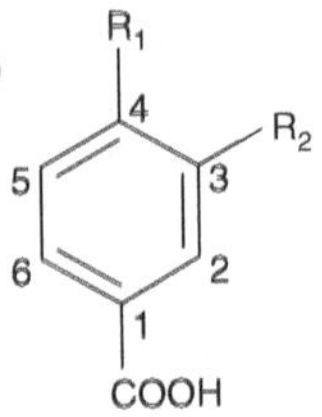

	R1	R2
Benzoic acid	H	H
4- or *p*-hydroxybenzoic acid	OH	H
3- or *m*-hydroxybenzoic acid	H	OH
protocatechuic acid	OH	OH
vanillic acid	OH	OCH_3
isovanillic acid	OCH_3	OH

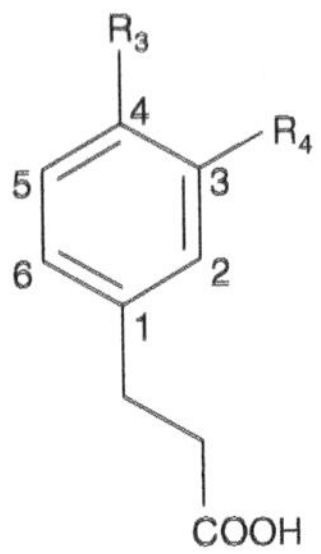

	R3	R4
Phenylpropionic acid	H	H
4- or *p*-hydroxyphenylpropionic acid	OH	H
3- or *m*-hydroxyphenylpropionic acid	H	OH
3,4-dihydroxyphenylpropionic acid (dihydrocaffeic acid)	OH	OH
dihydroferulic acid	OH	OCH_3
dihydroisoferulic acid	OCH_3	OH

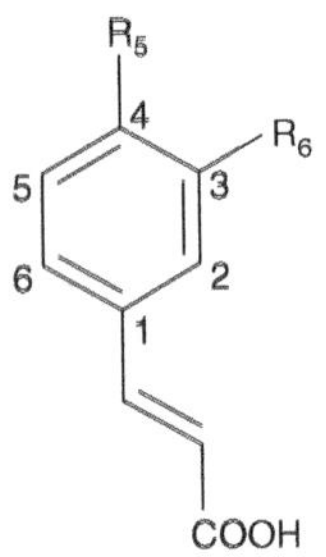

	R5	R6
Cinnamic acid	H	H
4- or *p*-hydroxycinnamic acid (*p*-coumaric)	OH	H
3- or *m*-hydroxycinnamic acid (*m*-coumaric)	H	OH
caffeic acid	OH	OH
ferulic acid	OH	OCH_3
isoferulic acid	OCH_3	OH

(b)

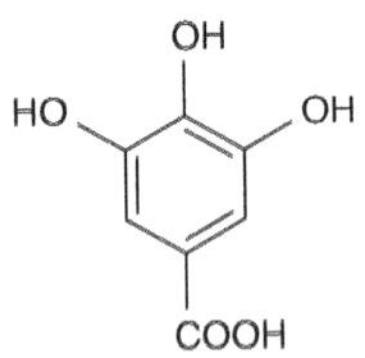

Gallic acid

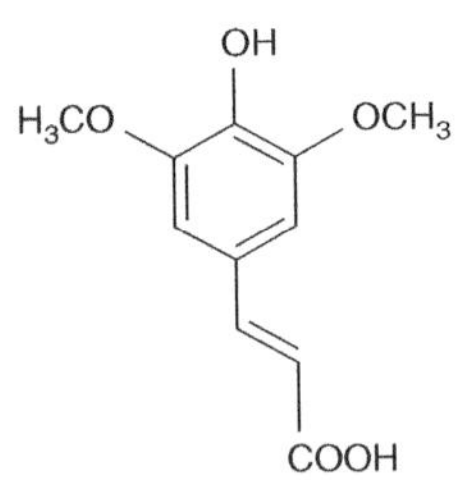

Sinapic acid

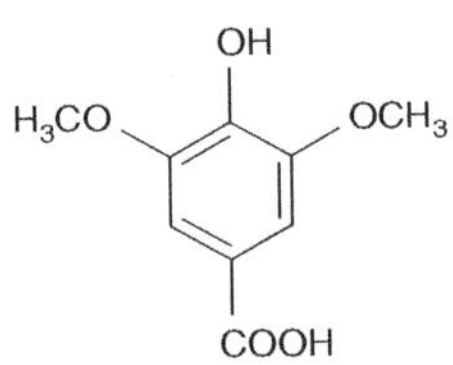

Syringic acid

Figure 2.2 Chemical structure of phenolic acids. Phenolic acids commonly found with two substitutions on the (*a*) phenyl ring on the 3- and 4-positions and (b) some less commonly found with triple substitutions on the 3-, 4-, and 5-positions.

are the most ubiquitous [Clifford, 2000]. The best known quinic acid ester or chlorogenic acid is the 5-*O*-caffeoylquinic acid, a caffeic acid esterified to position 5 of a quinic acid moiety (Fig. 2.3).

Phenolic acids, and especially ferulic acid, which is abundantly present in cereals, is found esterified to the polysaccharides present in primary and secondary cell walls of plants. Ferulic acid is the major phenolic acid occurring in the cell walls of monocotyledons and appears as *cis* and the more abundant *trans* isomers (reviewed in [Klepacka and Forna, 2006]). Ferulic acid is found in wheat, maize, rye, barley [Sun et al., 2001], oats, spinach, sugar beet, and water chesnuts [Clifford, 1999], generally esterified, and rarely as free form, such as in barley [Yu et al., 2001]. It is esterified in primary cell walls to arabinoxylans (Fig. 2.4) in the aleurone layer and pericarp [Clifford, 1999], as in spinach [Fry, 1982] or in wheat bran [Smith and Hartley, 1983]. Ferulic acid can also be found esterified to other hydroxycinnamic acids such as in Mongolian medicinal plants where it is found as feruloylpodospermic acid, which is

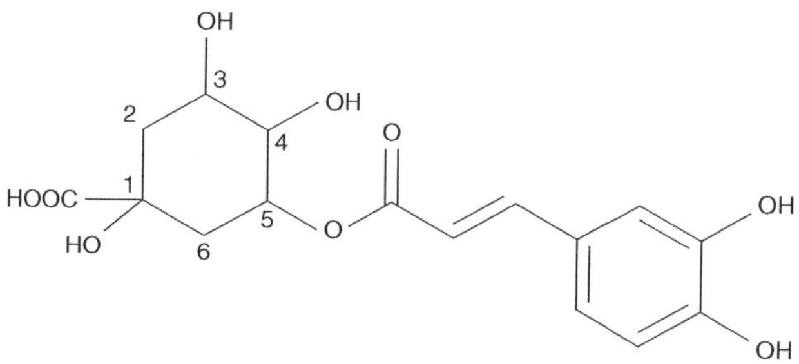

Figure 2.3 Example of chemical structure of a chlorogenic acid: 5-*O*-caffeoylquinic acid (with IUPAC numbering).

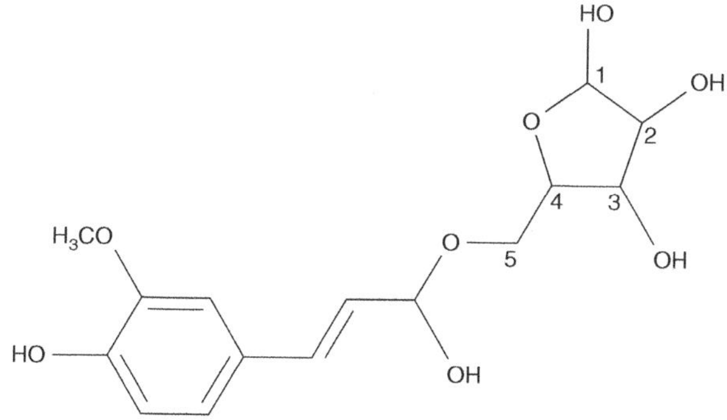

Figure 2.4 Example of chemical structure of 5-*O*-feruloyl-L-arabinofuranose a chlorogenic acid: 5-*O*-caffeoylquinic acid (with IUPAC numbering).

composed of two dihydrocaffeic acid units esterified at position 1 and 5 of a quinic acid moiety, with an additional feruloyl group being attached to position 3 [Tsevegsuren et al., 2007]. Dihydrocaffeic acid is also one of the major phenolic antioxidants abundantly present in black olive pericarp and in lower amounts in brined black olive, with traces in brined green olives [Owen et al., 2003].

ABSORPTION AND METABOLISM OF FLAVANOLS

Before reaching the systemic circulation, polyphenols orally administered will transit by the gastrointestinal tract (GIT) where they will be submitted to different biological and physiological parameters, which will participate to their bioavailability, such as the pH, the digestive enzymes, or the microflora.

Stability from the Mouth to the Small Intestine

The first fluid encountered by ingested procyanidins is saliva. Interaction between salivary proteins and flavan-3-ols has been investigated and demonstrated to be dependent on the chemical structure of the tested compounds, a gallate substitution on the position 3 on (−)-epicatechin or (−)-epigallocatechin, increasing their affinity to salivary proteins [Bacon and Rhodes, 1998]. Moreover, an extra hydroxyl group on the 5 -position of (−)-epicatechin gallate [(−)-epigallocatechin gallate] also favored interaction. However, epimerization of the hydroxyl group at position 3 seemed to have a less marked effect. The interactions between procyanidins and proline-rich proteins, optimum at two maxima at pH 3.5 and 5, were shown to lead to their precipitation [de Freitas and Mateus, 2001], the intraflavanol linkage and the structural monomeric units composing procyanidins influencing this precipitation, whereas the degree of polymerization had only a slight effect.

Different studies have investigated the effects of saliva and gastric fluid on the stability and, as a consequence, on the behavior of procyanidins during the first steps following their ingestion. Unpublished data, observed and mentioned by Spencer *et al.* in a review [Spencer et al., 2001a], showed that, incubated in human saliva for up to 30 min, procyanidins, from dimeric to hexameric forms, were not altered, 99% of the initial quantity being recovered at the end of the incubation period. Oligomeric procyanidins isolated from *Theobroma cacao* incubated in a phosphate buffer at pH 2 were degraded into monomeric and mainly into dimeric forms [Spencer et al., 2000]. After 2.5 hours at pH 2, 70% of the trimeric to hexameric forms were degraded compared with only 15% for the dimeric form during the same period of time. Since no difference was observed between the use of a phosphate buffer at pH 2 and a commercially available simulated gastric juice at the same pH without pepsin, the authors suggested that the pH was the only parameter responsible for the cleavage of oligomers. Another report mentioned that (−)-epicatechin and (+)-catechin are stable in simulated gastric juice (0.24% hydrochloric acid – 0.2% sodium chloride solution, pH 1.8), whereas dimers B_2 and B_5 are unstable, producing epicatechin units and their respective isomers B_5 and B_2 [Zhu et al., 2002]. These results are in contradiction with data obtained from an in vivo study investigating the stability of procyanidins from a chocolate beverage during stomach transit [Rios et al., 2002]. The pH of the stomach was at about 5.4 $\pm$ 0.2 after intake of the cocoa beverage and not at 2, the latter corresponding more to the pH of an empty stomach. The percentage of each oligomer among the total amount of flavonoids present in the beverage was the same after stomach transit and the components were not degraded during the stay in the stomach. This supports the idea that all the ingested procyanidins should reach the small intestine without any pre-depolymerization in the stomach. However, the pH of the GIT from stomach to jejunum varies from 2 to 8.5. It is known that polyphenols such as flavanols oxidize rapidly at alkaline pH due to the catechol moiety of the B-ring [Spencer et al., 2001a]. Thus, oxidation of procyanidins could occur in the first part of the small intestine. In simulated intestinal juice [1.5% (w/v) sodium hydrogen carbonate solution, pH 8.5], monomers, as well as dimers, degraded almost completely within several hours, but neither epicatechin/catechin nor isomers were found to be produced from the dimers [Zhu et al., 2002]. At higher pH, monomers and dimers degraded within minutes. Studies have also shown that (epi)catechins are much more stable than the corresponding gallocatechins under digestive conditions (gastric or small intestinal), whereas the gallate substitutes had an intermediate stability [Neilson et al., 2007]. However, catechins were more stable in the gastric phase than in the intestinal phase. This stability was more linked to pH than to enzymatic activity, the sensitivity being for rather neutral or basic pH. The instability of epigallocatechin gallate, epicatechin, and epigallocatechin resulted in the formation of homodimers.

Absorption from the Small Intestine

Unlike some classes of polyphenols such as flavonols and flavones, flavanols are almost always present in the nonglycosylated form. Removal of glycoside from flavonoids, usually necessary before the transport across the intestinal barrier, is not required in the case of flavanols [Scalbert and Williamson, 2000]. The absorption of procyanidins by the small intestine was investigated by studying ^{14}C-(+)-catechin, dimer, trimer, and procyanidin polymers permeation through Caco-2 cell cultures [Deprez et al., 2001]. There was little difference in permeability between monomer, dimmer, and trimer, based on the measurement of radioactivity present on the basal side of the cultures, whereas the permeability of the polymers was 10 times lower. The authors reported the absence of catechin metabolism but did not determine whether the radioactivity measured on the basal side of the cultures was from the parent dimers to polymers or from their products of degradation or metabolites, which could have resulted from instability of the parent compounds in the culture medium or from their metabolism by mammalian enzymes. From the results obtained, the authors suggested that procyanidins, from monomers to polymers, are poorly absorbed, transported via the paracellular pathway. The paracellular permeation of monomers does not fully support other studies reported in the literature, which rather suggested an absorption by intestinal cells of (epi)catechin in order to allow conjugation. For example, catechin glucuronide and 3′-*O*-methyl-catechin were found in the mesenteric vein after in situ perfusion of rat jejunum and ileum with catechin, indicating the intestinal origin of these conjugates [Donovan et al., 2001]. After oral administration of epicatechin by direct stomach intubation of rats, its free as well as its methylated forms were found in plasma of blood collected from the tail in addition to sulfated, glucuronidated, and sulfo-glucuronidated forms [Piskula and Terao, 1998]. Since the blood was taken from the tail 2 h after intake, it could not be concluded that conjugation took place in the intestinal cells. However, with isolated tissues, the authors reported a glucuronosyltransferase activity higher in the small and large intestinal mucosa than in the liver, kidney, and lung. In addition to global bioavailability of procyanidins, the mechanisms of transport have also been investigated, at least for the monomers. Indeed, an active transport, Na^+ independent but stimulated by an outwardly directed gradient of protons, could be used by the flavan-3-ols monomers (−)-epicatechin and (+)-catechin to cross the intestinal barrier [Starp et al., 2005]. Based on their results, the authors also proposed that free diffusion would be favored by high concentrations of flavan-3-ols, whereas carrier-mediated uptake would be important for lower amounts.

If the mechanisms of absorption of monomeric forms of flavanols are better understood, the results obtained from the studies investigating the absorption of dimers are contradictory and make understanding their permeation mechanism more difficult. The absorption of dimers has been observed from in vivo data showing procyanidin B_1 detected in human serum after intake of procyanidin-rich grape seed extract [Sano et al., 2003]. After oral administration of procyanidin dimer B_2 to rats, 0.34% of the provided dose was detected in urine [Baba et al., 2002]. Besides dimer B_2, the presence of epicatechin and its 3′-*O*-methylated product were also excreted in urine. All the compounds found in urine were present in plasma, peaking at 30–60 min following the intake of procyanidin dimer. The absorption of pure dimers B_2 and B_5 from the perfusion of rat jejunum was also shown, with six different components having flavanol-like spectral characteristics detected on the serosal side of the intestinal epithelium, which represented a transfer of flavanol of 23% [Spencer et al., 2001b]. In addition to catechin, dimer, and *O*-methylated dimer, the major compound (95.8%) obtained on the serosal side had the characteristics of epicatechin, with no evidence for the presence of its glucuronides or 3′-*O*-methyl and 4′-*O*-methyl conjugates. The authors also suggested that the cleavage of the dimer into the monomeric form would be energy dependent, taking place in the intestinal cells. More recently, tetramethylated dimeric procyanidins were detected in rat plasma and liver 1 h after intake of synthetic dimeric procyanidins [Garcia-Ramirez et al., 2006]. Appearance of procyanidin B_2 and B_3, purified from cinnamonomic cortex, has been measured in plasma, peaking, respectively, at 20 and 40 min after their oral administration to rats [Tanaka et al., 2003]. According to the chemical structure drawn and the mass reported in the study, the compound called dimer B_3 would be a trimer. This would be the first time that a trimer, found under the form of parent compound in plasma, is reported in the literature, which is unexpected due to its molecular weight.

Even if dimers and trimers are stable under GIT conditions and absorbed in their native form as seen for the monomers, the amount crossing the epithelial barrier of the small intestine and being further excreted in urine must be very low. The major part of the ingested procyanidins must reach the colon where an abundant microflora can metabolize them into smaller molecules, which are more readily absorbed. The general absorption and metabolism pathways of procyanidin monomers to oligomers in the GIT, from their ingestion to microbial metabolism, are summarized in Figure 2.5.

Formation of Phenolic Acids from Flavanols

Gut microflora account for a major part of the production of phenolic acids excreted in urine [Goodwin et al., 1994]. In addition, mammalian enzymes can also participate to the structural modification of these phenolic acids. Several studies have shown the presence of phenolic acids in plasma and/or urine after intake of different types of flavonoid-rich food, such as artichoke extract, apple cider, breakfast cereals, tomatoes, coffee, or beer, with 5-*O*-caffeoylquinic, caffeic, and ferulic acids being the most studied examples (as reviewed in [Manach et al., 2005]), but also after chocolate intake [Rios et al., 2003] or red wine consumption [Caccetta et al., 2000]. Several phenolic acids were identified in human urine after intake of a polyphenol-rich meal based on fruits and vegetables and comprising tomato and onion pasta sauce, pasta, cooked broccoli, cherry tomatoes, cucumber, continental leaf salad, pepper salad dressing, raspberries, red grape juice, and apple juice [Rechner et al., 2002]. Ferulic, isoferulic, sinapic, and 3-hydroxyphenylacetic acids were found glucuronidated, whereas vanillic and homovanillic acids were found as both free and glucuronidated forms. In plasma, dihydroferulic and 3-hydroxyphenylpropionic acids were present as free and glucuronidated forms with ferulic, isoferulic, 4-coumaric, hydroxybenzoic, and vanillic acids as glucuronide conjugates only. The authors proposed 3,4-dihydroxyphenylpropionic acid (dihydrocaffeic acid) as the first phenolic acid in the cascade of microbial metabolism of dietary flavonoids, based on their work and other reports from the literature. Monomeric units of procyanidins, which have not been absorbed through the small intestinal epithelium, can reach the colon to be metabolized by the intestinal microflora. 3-Hydroxyphenylpropionic acid has been proposed to be the major metabolite found in urine of rats after ingestion of (+)-catechin [Griffiths, 1962], in the free form but also conjugated [Griffiths, 1964],

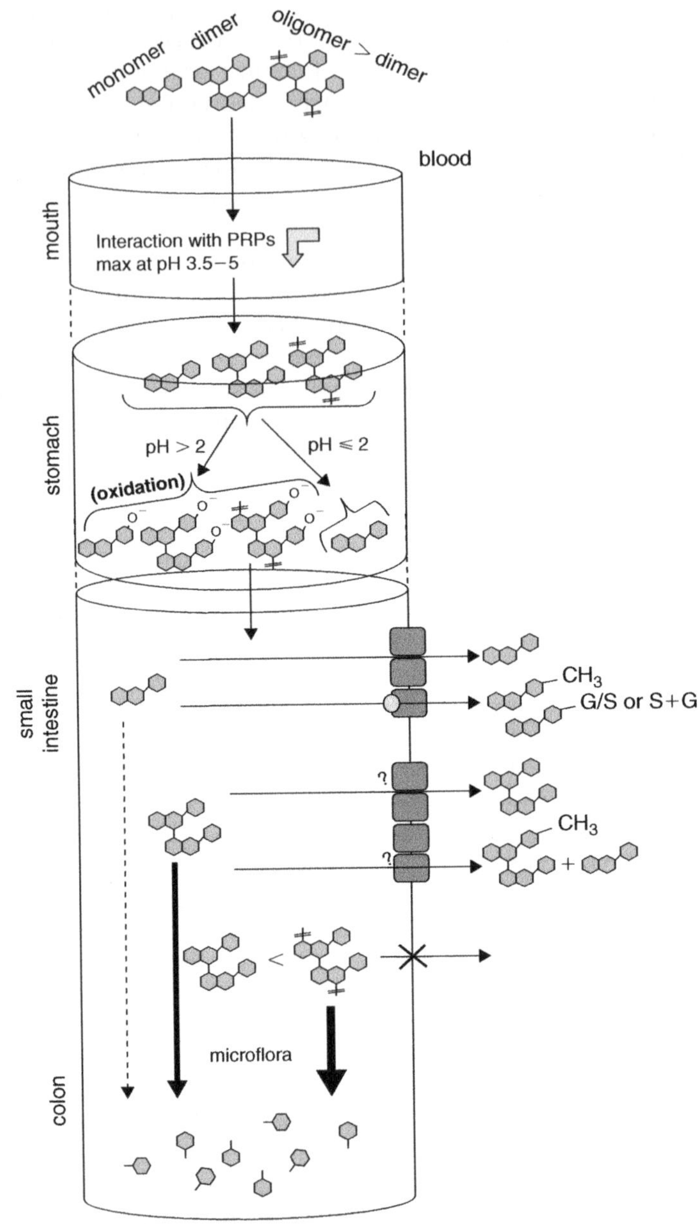

Figure 2.5 Summary of procyanidin behavior in the GIT. precipitation; phenolic acid; CH_3, methyl conjugate; G, glucuronide; O^{-}, oxidized hydroxyl group; S, sulfate; S + G, sulphoglucuronide conjugate.

together with 4-hydroxyphenylpropionic acid, 5-(3-hydroxyphenyl)-γ-valerolactone, 5-(3,4-dihydroxyphenyl)-γ-valerolactone, and hydroxyphenylvaleric acids, as minor metabolites [Scheline, 1970]. 5-(3,4-Dihydroxyphenyl)-γ-valerolactone was also detected in urine of rats from the ring fission of (−)-epicatechin by microflora [Meng et al., 2002; Unno et al., 2003] and found together with 3,5-dihydroxyphenyl-γ-valerolactone in human urine after ingestion of green tea, epigallocatechingallate, or epigallocatechin [Meng et al., 2002].

Benzoic, isomeric 2-, 3-, and 4-hydroxybenzoic, phenylacetic, 4-hydroxyphenylacetic, phenylpropionic, isomeric 2-, 3-, and 4-hydroxyphenylpropionic, 3,4-dihydroxyphenylpropionic acids, δ-(3-hydroxyphenyl)-γ-valerolactone, and 1-(3-hydroxyphenyl)-3-(2,4,6-trihydroxyphenyl)-propan-2-ol have been detected after 5 days incubation of dimer B_3 with rat-cecal microflora [Groenewoud and Hundt, 1986]. In a more recent study, also investigating the in vitro microbial metabolism of a ^{14}C-labeled procyanidins fraction, which did not contain monomer, dimer, and trimer, low-molecular-weight aromatic acids, differing by their hydroxylation status and the length of their aliphatic side chain, were detected [Deprez et al., 2000]. The six different compounds found were 3-hydroxyphenylacetic acid, and 3-hydroxyphenylpropionic acid, their isomers 4-hydroxyphenylacetic acid and 4-hydroxyphenylpropionic acid, 3-hydroxyphenylvaleric acid and phenylpropionic acid. Phenylacetic and benzoic acids were also detected, but their source was not clearly identified. In vivo work on rats or in human have also supported the results found in vitro, showing the presence of phenolic acids in plasma and urine after intake of isolated procyanidins, extracts, or food containing procyanidins. In vivo investigation of the metabolism of catechin monomer, dimer B_3, trimer C_2, and purified polymers using a rat model suggested that procyanidin degradation could start by a depolymerization into (epi)catechin monomers, due to similarities in the degradation products obtained from catechin and procyanidins, before further metabolism leading to the production of phenolic acids [Gonthier et al., 2003b]. Phenolic acids recovered in urine were also detected in humans after intake of a chocolate beverage [Rios et al., 2003] and in rats after intake of red wine powder [Gonthier et al., 2003a], food sources known to contain procyanidins. Phenolic acids can be expected to be found in plasma and urine after intake of procyanidin-rich foods such as grape seed extract [Ward et al., 2004], apples, cranberries, chocolate, and wine [Hammerstone et al., 2000]. The stability, as well as the microbial metabolism of monomeric and oligomeric/polypmeric procyanidins and the generation of the initial forms of hydroxycinnamic acids, are summarized in Figure 2.5. A nonexhaustive list of the phenolic acids found in plasma or urine in vivo after ingestion of flavonoid-rich foods or pure flavonoids is shown in Table 2.1.

Phenolic acids can come from the metabolism of polyphenols other than catechins and procyanidins. The degradation of flavonoid aglycones by microflora can be more or less rapid, the presence of a methoxyl group on the A- or B-ring of the flavonoid being protective against bacterial degradation by C-ring

Table 2.1 Summary of Phenolic Acids Found in Vivo and in Vitro from Different Food Sources and Flavonoids[a]

Diet	Model	Phenolic acids				References
(+)-Catechin	Rats (urine)	major metabolites: 3-HPPA, free and C				Griffiths, 1962 Griffiths, 1964
(+)-Catechin		• Major metabolite: 3-HPPA; minor metabolites: 4-HPPA, 5-(3-hydroxyphenyl)-γ-valerolactone, 5-(3,4-dihydroxyphenyl)-γ-valerolactone and HPVA				Scheline, 1970
(−)-Epicatechin		5-(3,4-Dihydroxyphenyl)-γ-valerolactone				Meng et al., 2002; Unno et al., 2003
• 8-d diet supp. with 0.12 g catechin /100 g • 0.25 or 0.50 g red wine powder /100 g	Rats (urine)	• Phenolic acids affected by the diet (represent 4.7 g/100 g catechin ingested): 3-HPPA (major), 3-HBA, 4-HBA, 3-HHA, 4-HBA, FA, 3,4-DHPAA, 3,4-DHPPA, 3-HPA • 3-HBA, 4-HBA, HA, 3-HHA, 4-HHA, VA, PAA, 3-HPA, 3,4-DHPAA, 3-HPPA, 3,4-DHPPA, *p*-CA, CA, FA.				Gonthier et al., 2003a
Catechin, dimer B3, trimer C2, and purified polymers	Rats (urine)	Phenolic Acids	From Dimer	Trimer	Polymer	Gonthier et al., 2003b
		3,4-DHPVA	nd	nd	nd	
		3-HPVA	×	×	nd	
		3,4-DHPPA	×	nd	nd	
		3-HPPA	×	×	×	
		m-CA	×	×	×	
		p-CA	Same as ctrl	Same as ctrl	Same as ctrl	
		FA	×	×	×	
		3,4-dHPAA	×	Same as ctrl	Same as ctrl	
		3-HPAA	Almost same as ctrl	Same as ctrl	Same as ctrl	
		PCA	×	Same as ctrl	Same as ctrl	
		3-HBA	×	nd	nd	
		4-HBA	×	Same as ctrl	Same as ctrl	

(Continued)

Table 2.1 *(Continued)*

Diet	Model	Phenolic acids				References
		VA	×	×	×	
		3-HHA	×	nd	nd	
		4-HHA	×	Same as ctrl	Same as ctrl	
		HA	×	Same as ctrl	×	
Green tea, epigallocatechingallate or epigallocatechin	Human (urine)	5-(3,4-dihydroxyphenyl)-γ-valerolactone and 3,5-dihydroxyphenyl-γ-valerolactone				Meng et al., 2002
Polyphenol-rich meal based on fruits and vegetables	Human	• urine: FA, isoFA, SA and 3-HPAA: G forms, VA and HVA: free and G forms • plasma: DHFA and 3-HPPA: free and G forms; FA, isoFA, *p*-CA, HBA and VA: G forms				Rechner et al., 2002
Chocolate intake	Human (urine)	3,4-DHPPA, 3-HPPA, FA, 3,4-dHPAA, 3-HPAA, PAA, VA, 3-HBA, 4-HBA, 4-HHA, HA				Rios et al., 2003
Proanthoyanidin dimer B3	5 d with rat-cecal microflora	BA, 2,3 and 4-HBA, PA, 4-HPAA, phenylpropionic, 2,3,4-HPPA, 3,4-diHPPA, -(3-hydroxyphenyl)-γ-valerolactone and 1-(3-hydroxyphenyl)-3-(2,4,6-trihydroxyphenyl) -propan-2-ol				Groenewoud and Hundt, 1986
^{14}C-labeled procyanidins fraction, without monomer, dimmer, and trimer	Human colonic microflora	3-HPAA and 3-HPPA acid, 4-HPAA and 4-HPPA, 3-HPVA, PPA, PA, BA				Deprez et al., 2000

[a] BA, benzoic acid; C, conjugated; CA, caffeic acid; *m* or *p*-CA, *m* or *p*-coumaric acid; DHFA, dihydroferulic acid; 3,4-DHPAA, 3,4-dihydroxyphenylacetic acid; 3,4-DHPPA, 3,4-dihydroxyphenylpropionic acid; 3,4-DHPVA, 3,4-dihydroxyphenylvaleric acid; FA, ferulic acid; G, glucuronide; HA, hippuric acid; 2-, 3- or 4-HBA, 2-, 3- or 4-hydroxybenzoic acid; 3- or 4-HHA, 3- or 4-hydroxyhippuric acid; 3- or 4-HPAA, 3- or 4-hydroxyphenylacetic acid; 3- or 4-HPPA, 3- or 4-hydroxyphenylpropionic acid; 3- or 4-HPVA, 3- or 4-hydroxyphenylvaleric acid; HVA, homovanillic acid; isoFA, isoferulic acid; nd, not detected; PAA, phenylacetic acid; PCA, protocatechuic acid; PPA, phenylpropionic acid; SA, sinapic acid; VA, vanillic acid.

cleavage [Lin et al., 2003]. 3-Hydroxyphenylpropionic acid results from ring fission followed by water elimination and dehydroxylation of quercetin [Rechner et al., 2004; Schoefer et al., 2003], which itself resulted from deglycosylation of rutin. However, quercetin has also been reported to be metabolized by C-ring fission into 3,4-dihydroxyphenylacetic acid (coming from the B-ring) and phloroglucinol (three hydroxyl groups on a phenyl ring, coming from the A-ring) [Winter et al., 1989]. 3-Hydroxyphenylpropionic acid also resulted from the ring fission of naringenin, itself the aglycone obtained by microbial deglycosylation of naringin, and could be further dehydroxylated to give phenylpropionic acid [Rechner et al., 2004]. Phenylhydracrylic acids are other metabolites that have been reported from the metabolism of flavanoids such as catechin [Baba et al., 1981] or rutin [Das, 1974].

Dihydrocaffeic acid, the metabolite of dihydroxyphenylvaleric acid in the cascade of microbial degradation of catechins and procyanidins (Fig. 2.6), is one of the major phenolic acids found in human fecal water [Jenner et al., 2005]. It was detected in the plasma of coffee drinkers [Goldstein et al., 1984], in urine as the free, form, and mainly conjugated in human plasma after ingestion of artichoke leaf extracts, which contain monocaffeoylquinic, dicaffeoylquinic, and caffeic acids as well as flavonoids [Wittemer et al., 2005], in human urine after chocolate intake [Rios et al., 2003], in rat urine after ingestion of polyphenol-rich red wine extract [Gonthier et al., 2003a], and produced by bacteria from luteolin and apigenin [Schoefer et al., 2003].

PHENOLIC ACIDS BIOAVAILABILITY

Release of Phenolic Acids from Esters

Phenolic acids, and most particularly ferulic acid, are mainly found in food as soluble or insoluble esters. For most of the insoluble esters, a preliminary hydrolysis and a release of the free form is required before absorption. In the gastrointestinal tract, the insoluble wheat bran fiber is first hydrolyzed by xylanase, releasing feruloylated oligosaccharides into the soluble phase, where cinnamoyl esterase can hydrolyze the ester bond and release ferulic acid [Kroon et al., 1997]. Human and rat colonic microflora contain an esterase activity able to release diferulic acids from dietary cereal brans [Andreasen et al., 2001b]. A small proportion of ferulic acid, as free form and esterified groups, has been shown to be released from wheat bran by in vitro gastric and small intestinal incubation, indicating the presence of enzymatic activities in the upper part of the GIT able to hydrolyze the xylan backbone and the ester bond of ferulic acid from the xylan backbone. However, 95% of the total release of feruloyl groups appear to take place in the colon, where xylanase and cinnamoyl esterase activities of microbial origin have been detected [Couteau et al., 2001]. Small amounts of cinnamoyl esterases of mammalian origin have been identified in the mucosa of the small intestine, which could possibly also come from

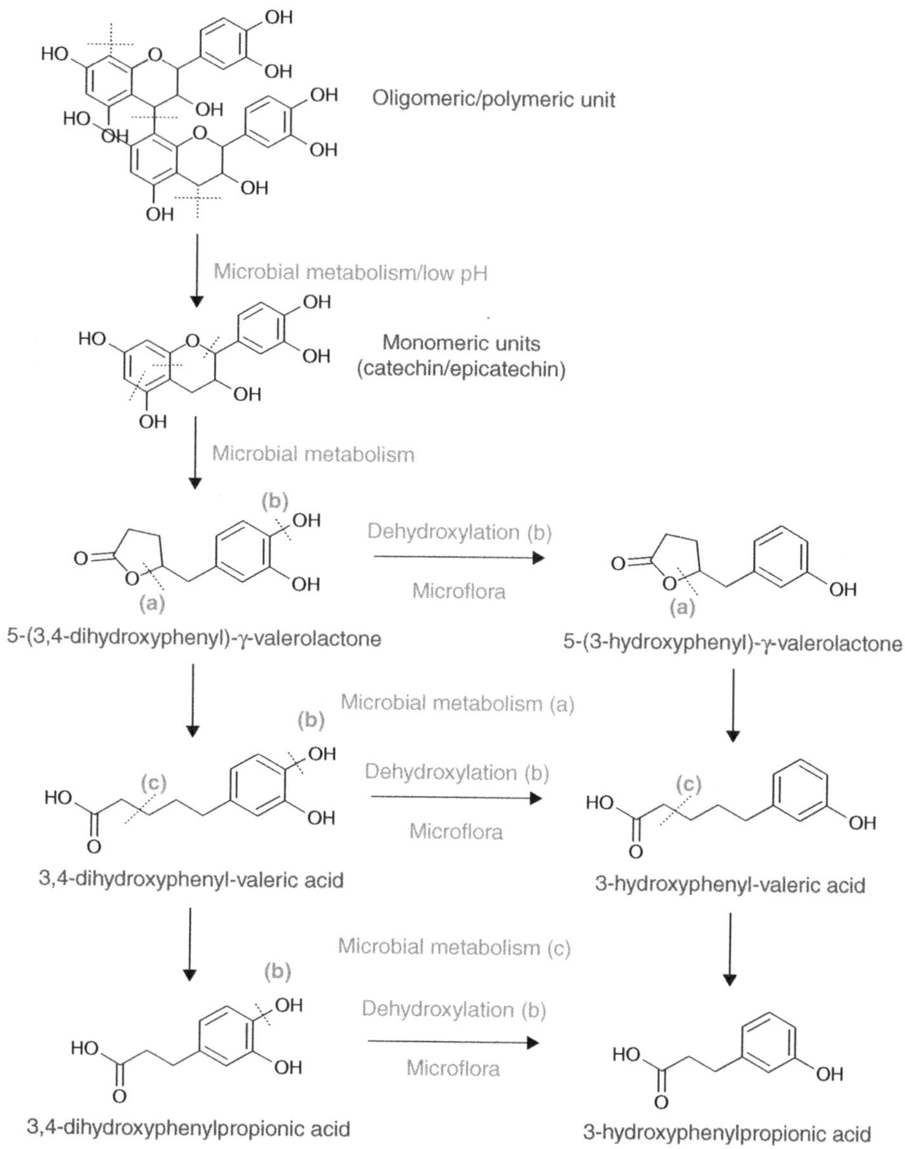

Figure 2.6 Summary of main possible interconversions between the different phenolic acids by microbial and mammalian enzymes, starting from dihydrocaffeic acid, the microbial metabolite of 3,4-dihydroxyphenylvaleric acid.

pancreatic secretions [Andreasen et al., 2001a, 2001b]. Recent findings of D-quinic acid in ileostomy bags of patients after consumption of apple juice containing 4- and 5-caffeoylquinic and 4-coumaroylquinic acids support this assumption of a possible small intestinal esterase activity [Kahle et al., 2007].

The enterocyte-like cells, Caco-2, also exhibited hydrolase activity toward esters of hydroxycinnamic acids and diferulates, part of the esterases produced by the cells being secreted in the culture medium [Kern et al., 2003b]. The hydrolysis of hydroxycinnamic acids by the intestinal esterases appeared, however, to be dependent on the substitution of the phenyl ring of the substrate, the activity decreasing with the number of substitutions and the substrate specificity being not as marked with microbial esterases [Andreasen et al., 2001a]. Interestingly, intestinal esterases are not active toward 5-*O*-caffeoylquinic acid, whereas microbial ones are able to produce caffeic acid [Plumb et al., 1999; Rechner et al., 2004]. Very low amounts of caffeic acid have been detected in gastric and small intestinal contents after ingestion by rats of a 5-*O*-caffeoylquinic-acid-supplemented diet, appearance of the free form being much more significant in the cecum [Lafay et al., 2006a]. From coffee consumption, the picture is uncertain. In one study, 5-*O*-caffeoylquinic acid was not found in human plasma but high amounts of caffeic acid were detected, suggesting no absorption of the ester and a requirement of precleavage of the ester bond before absorption in the gastrointestinal tract [Nardini et al., 2002]. However, another study showed high levels of chlorogenic acids in human plasma after intake of a large amount of coffee [Monteiro et al., 2007]. It can be extrapolated that ferulic acid, found esterified with carboxylic acids such as quinic acid in food [Herrmann, 1989] or with polysaccharides in cereals, for instance, could also reach the colon almost intact and be released by the action of xylanases and cinnamoyl esterases, mainly in the cecum and colon. Once ingested or released in the gastrointestinal tract, phenolic acids can be further metabolized by the microflora and/or by mammalian enzymes to produce new compounds.

Mammalian and Microbial Metabolism of Phenolic Acids

One of the first papers presenting microbial metabolism of phenolic acids was the report of Scheline who used rat intestinal microflora [Scheline, 1968]. This paper, covering most of the interconversions possible between phenolic acids by microflora, reported the reduction of the unsaturated side chain in 2-, 3- and 4-coumaric, caffeic, isoferulic and ferulic acids. Microflora were also capable of dehydroxylating position 4 of protocatechuic and dihydrocaffeic acids to form their respective 3-hydroxy acid forms. Dehydroxylation of hydroxycinnamic acids has also been demonstrated for human microflora, phenylpropionic acid obtained from 3-hydroxyphenylpropionic acid, itself resulting from dehydroxylation of dihydrocaffeic acid [Rechner et al., 2004]. Demethylation of the 3-hydoxy-4-methoxy- and 3-methoxy-4-hydroxy-phenylacetic, phenylpropionic, and cinnamic acids also occurred with rat intestinal microflora [Scheline, 1968]. In addition, the author showed decarboxylation of phenylacetic and cinnamic, but not phenylpropionic, acids that possessed a 4-hydroxyl group. Dihydrocaffeic and caffeic acids were later shown to be decarboxylated by α-oxidation to form, respectively, 4-vinyl-catechol and

4-ethylcatechol in vitro [Peppercorn and Goldman, 1971]. Vinyl-catechol was formed by *Lactobacillus* sp. 1 in vitro but not in vivo, in germ-free rats colonized by this strain [Peppercorn and Goldman, 1972]. The appearance of propionic and benzoic acids would be the result of progressive loss of carbons through β-oxidation from the phenylvaleric compounds [Deprez et al., 2000]. β-Oxidation, taking place in the liver by mammalian enzymes, was responsible for the production of vanillic acid from dihydroferulic acid, itself coming from the hydrogenation of ferulic acid [Rechner et al., 2002]. However, another report mentioned the production of vanillic acid from caffeic acid metabolism, with the detection of ferulic and isoferulic acid but not dihydroferulic acid, suggesting that vanillic acid could also directly come from ferulic acid instead of dihydroferulic acid [Jacobson et al., 1983]. This hydroxycinnamic metabolism seems to be similar to the β-oxidation of fatty acids. Indeed, vanillic acid has been proposed to result from ferulic acid via its reduction through the formation of feruloyl-SCoA, which is hydrogenated into 4-hydroxy-3-methoxyphenyl-β-hydroxypropionyl-SCoA further converted to vanillic acid and acetyl-SCoA by an enoyl-SCoA hydratase/lyase enzyme [Gasson et al., 1998]. This enzyme belongs to a superfamily, which includes bacterial, mitochondrial, peroxisomal, and glyoxysomal enzymes of the fatty acid oxidation pathway, suggesting that this enzymatic reaction can happen through bacterial or mammalian metabolism. The reduction, which has been shown to give dihydrocaffeic acid from caffeic acid by microflora [Peppercorn and Goldman, 1971; Rechner et al., 2004], could be an intermediate step in the β-oxidation. This reduction, which could also take place with mammalian enzymes due to the presence of the enzymes in mammalian cells, has never been clearly demonstrated. Indeed, gnotobiotic rats, germ-free rats harboring a strain of *Streptococcus*, were able to reduce caffeic into dihydrocaffeic acid, whereas germ-free rats were not [Peppercorn and Goldman, 1972]. It has also been shown more recently that hepatocytes reduced ferulic acid to dihydroferulic acid [Moridani et al., 2002].

Intestinal microflora is capable of demethlylation, but not methylation, of phenolic acids, since dihydroferulic and ferulic acids were excreted in urine of germ-free rats after ingestion of, respectively, dihydrocaffeic and caffeic acids [Peppercorn and Goldman, 1972]. The mammalian enzyme catechol-*O*-methyl transferase catalyses the methylation of hydroxyl groups. Ferulic acid was an in vivo metabolite of caffeic acid [Azuma et al., 2000], as was its isomer isoferulic acid [Rechner et al., 2002], secreted back into the lumen by intestinal cells after caffeic acid perfusion of rat small intestine [Lafay et al., 2006b]. Ferulic and isoferulic acids, absent from the plasma samples, were detected in urine of rats, together with caffeic and 5-*O*-caffeoylquinic acids, 1.5 h after the beginning of a meal containing 5-*O*-caffeoylquinic acid [Lafay et al., 2006a]. Usually, ferulic acid is found in nature as the *trans* isomer. However, *cis* ferulic acid, found in biological samples [Zhao et al., 2004], was attributed to an isomerization of the form taking place during the storage of the samples, as already hypothesized [Graf, 1992].

In addition to different hydroxyl derivatives of cinnamic, phenylacetic, and benzoic acids, hydroxyl derivatives of hippuric acid are found as metabolites of flavonoids and phenolic acids. Hippuric acids come from the glycination (conjugation with glycine moiety) of benzoic acid derivatives in the liver and kidneys. The first step of this conjugation is the activation of benzoic acid by benzoyl-CoA ligase to form benzoyl-CoA, similarly to the first step of β-oxidation [Schwab et al., 2001]. However, benzoic acid, because of its short side chain, is not degraded by *β*-oxidation, but CoA is rather substituted by the amino acid glycine to form the hippuric conjugate. Hippuric acid derivatives are the end products of metabolism of phenolic acids and consequently of flavonoids. Indeed, they were detected in human urine after intake of a polyphenol-rich meal [Rechner et al., 2002], 3-hydroxyhippuric, 4-hydroxyhippuric, and hippuric acids present as both free and glucuronidated forms. Hippuric acid and/or its hydroxy derivatives were also detected in rat urine after intake of 5-*O*-caffeoylquinic acid [Gonthier et al., 2003c], dimer, trimer, and polymers of procyanidins [Gonthier et al., 2003b], of a diet supplemented with catechin or red wine powder [Gonthier et al., 2003a] and in human urine after chocolate intake [Rios et al., 2003] or black tea consumption [Clifford et al., 2000]. Figure 2.7 summarizes most of the possible interconversions of the commonly found phenolic acids.

Behavior and Absorption in the Gastrointestinal Tract

It is thought that the soluble esters of phenolic acids with carboxylic acids or sugars will be stable during their passage in the stomach, since acidic gastric juice was not sufficient to cleave the ester linkage in 5-*O*-caffeoylquinic acid [Rechner et al., 2001], nor was an incubation in HCl-NaOH (9 g/L) aqueous solution (pH 2.5) of the 5-*O*-feruloyl-L-arabinofuranose [Zhao et al., 2004]. In a study involving healthy ileostomy subjects, it was shown that 5-*O*-caffeoylquinic acid is absorbed as such and further metabolized in the liver, rather than hydrolyzed into caffeic acid before absorption in the stomach or in the small intestine, since it was fairly stable when incubated in gastrointestinal fluids [Olthof et al., 2001]. However, the absorption of the ester was lower (33% of the ingested dose) than the free form (95%). Other reports have also mentioned the absorption of the soluble ester 5-*O*-caffeoylquinic acid [Lafay et al., 2006a, 2006b] without a prelysis of the ester bond. Indeed, 5-*O*-caffeoylquinic acid was detected in urine [Gonthier et al., 2003c] or plasma [Lafay et al., 2006a] of rats after intake of a semipurified diet containing 5-*O*-caffeoylquinic acid. In a more recent study performed in humans, 3-, 4-, and 5-caffeoylquinic acids, as well as 3,4-, 3,5-, and 4,5-dicaffeoylquinic acids were identified in plasma, peaking at 1 or 3 h following ingestion of brewed coffee, depending on the subjects [Monteiro et al., 2007], one subject showing the additional presence of 4- and 5-feruloylquinic acid. Results on 5-*O*-caffeoylquinic acid absorption (summarized in Table 2.2) are, however, contradictory since some reports also indicated no detection of this ester in plasma of rats after direct stomach administration

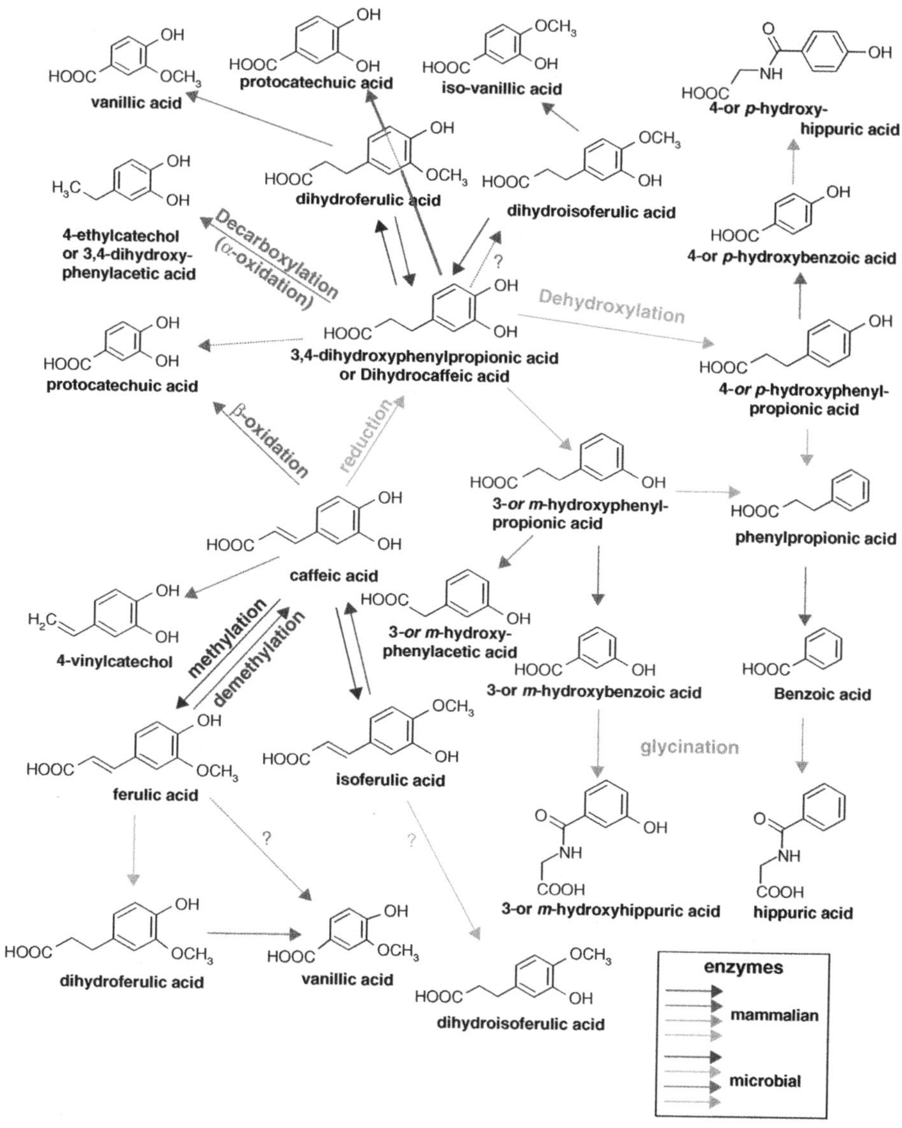

Figure 2.7 Summary of main possible interconversions between the different phenolic acids by microbial and mammalian enzymes, starting from dihydrocaffeic acid, the microbial metabolite of 3,4-dihydroxyphenylvaleric acid.

of a solution containing 5-*O*-caffeoylquinic acid [Azuma et al., 2000] or of humans after coffee consumption [Nardini et al., 2002]. These discrepancies could result from the methods used to detect chlorogenic acids in plasma, liquid chromatography–mass spectroscopy (LC–MS) with electrospray ionization appearing to be the most utilized [Matsui et al., 2007]. They could also result from the instability of chlorogenic acids toward the methods of purification used to process plasma or urine, leading to the lysis of the ester bond and to the release of the free form. Interestingly, a recent report also proposed that isomerization of caffeoylquinic acids could occur in the human gastrointestinal tract as 1- and 3-caffeoylquinic acids were detected in ileostomy bags of patients after consumption of apple juice, which contained only the 4- and 5-caffeoylquinic acid forms, isomerization hypothesized by the authors to result from enterocyte esterase activity [Kahle et al., 2007].

Ferulic acid has often been reported as a model compound for the investigation of bioavailability of phenolic acids, as free or esterified forms (Table 2.3), since it is abundantly present in food. Absorption via the gastrointestinal tract has also been found for dimers of ferulic acid, or diferulic acids, released from cereal brans [Andreasen et al., 2001b].

A scheme of digestion and absorption of ferulic acid has been proposed, based on the study performed on rats fed ferulic acid sugar esters, 5-*O*-feruloyl-L-arabinofuranose and feruloyl-arabinoxylan [Zhao et al., 2003a]. Released from the ester, it would preferably be more absorbed by the small intestine and the cecum. The authors also proposed that 5-*O*-feruloyl-L-arabinofuranose could be absorbed as such and further hydrolyzed in the mucosa or in the circulation. The absorption of free phenolic acids is mainly expected to occur in the small and large intestine. However, a possible permeation of phenolic acids already in the stomach has been proposed in addition to the small intestine, such as for ferulic acid and its sugar ester, based on its rapid absorption observed with rats [Zhao et al., 2003a]. This hypothesis was later supported by a study using a rat in situ gastric perfusion model in which the pylorus was ligatured, followed by an injection of a solution of ferulic acid free or 5-*O*-feruloyl-L-arabinofuranose in the stomach before the ligature of the cardia [Zhao et al., 2004]. The authors showed a high absorption rate of ferulic acid from the stomach ($74 \pm 11\%$ of the administered dose), much higher than the absorption of its sugar ester ($19 \pm 4\%$ of the administered dose). Uptake of ferulic acid and of 5-*O*-feruloyl-L-arabinofuranose by the gastric cells only as unconjugated forms was shown in vitro. They were present in the gastric mucosa, after incubation in an excised stomach, even though the experiment was performed on ice, suggesting that the uptake by the gastric cells did not require any energy and was probably carrier independent. If the uptake was by a transcellular route, the incubation on ice must have lowered the amount uptaken by the cells, since the fluidity of the phospholipidic membranes would be reduced by lowering the temperature. Whereas ferulic acid was found in the gastric mucosa only as free form, it was also present as glucuronide, sulfate, and sulfoglucuronide in plasma, bile, and urine, in the portal vein after its perfusion in the stomach in situ, even though free ferulic acid was the major form. A further in vivo investigation confirmed the results obtained by Zhao *et al.* (above) and allowed the detection of metabolites of ferulic acid in the portal vein and abdominal artery, by using a more sensitive method of detection [coulometric instead of ultraviolet (UV)] [Konishi et al., 2006]. Based

Table 2.2 Summary of Different in Vivo Studies Investigating the Bioavailability of Chlorogenic Acid or 5-*O*-Caffeoylquinic Acid[b]

Model	Vehicle/ Administration	Compound	Biological Fluid	Metabolites			Reference
Rat	Stomach intubation with 700 mol/kg	Chlorogenic acid	Plasma (t_{max} 0.5–1h)	CA-G, CA-S/G, FA-S/G (small amounts)			Azuma et al., 2000
		Caffeic acid	Plasma (t_{max} within the first 2 h)	FA, FA-G, FA-S, FA-S/G (50% total CA conjugates = 34% total metabolites); CA, CA-G (41% total metabolites), CA-S/G			
Rat	Supplemented diets with 250 μmol	With CLA or CA	Urine	**% of the intake**	**From CLA**	**From CA**	
				CLA, CA, FA, and isoFA	1.3	34	Gonthier et al., 2003c
				Microbial metabolites: *m*-CA, derivatives of PPA, BA and HA	57	28.1	
				From CLA	**From CA**		
			Plasma	FA, *m*-CA, *m*-HPPA, hippuric	CA, FA, isoFA, m-HPPA		
Rat	Duodeno-jejunal perfusion	50μM CA or CLA	45 min after perfusion collection of :	**Parent (CA, CLA) and metabolites (FA, isoFA)** (% ingested dose)			Lafay et al., 2006b
				From CLA	**From CA**		
			Bile	traces	0.4		
			Intestinal cells	9.2	20		
			Secretion from cells	1.2	0.5		
			Ileum effluent	90.8	80		
			Net absorption	8	19.5		
Rat	Supplemented diet	CLA (0.25%)	Content in % of total phenolic acids at 1.5–7 h	**Stomach**	**Small intestine**	**Cecum**	Lafay et al., 2006a
				Traces CA (1%), CLA	Traces CA (1%), CLA	15–32 % CA, CLA (peak at 4.5 h)	
	In situ gastric perfusion	CLA (35 mols)	Plasma	16% CLA absorption			

(Continued)

Table 2.2 *(Continued)*

Model	Vehicle/ Administration	Compound	Biological Fluid	Metabolites	Reference
Human	Artichoke extract in 9 capsules	CLA (= 123 mg)	24 h urine	• Estimation of CA absorption: 5.6% • No CLA but FA, isoFA, DHFA, and VA (and their conjugates)	Rechner et al., 2001
Healthy ileostomy subjects	Supplement	2.8 mmol	**24 h urine**	**24 h ileostomy bag**	Olthof et al., 2001
		CLA	Traces	67% ingested dose (estimated absorption: 33%)	
		CA	11% ingested dose	5% ingested dose (estimated absorption: 95%)	
Human	Brewed coffee	96 mg CLA, 166 mg CA, 3 mg *p*-CA, 29 mg FA	Plasma, at 1 and 2 h after intake; t_{max}: 1 h	No CLA, *p*-CA and FA; CA and its glucuronide or sulfate conjugates	Nardini et al., 2002
Human	Brewed coffee	Containing (in mmol/serving): total CQA (2.9), total diCQA (0.24)	**Plasma (%)**	5-CQA: 40; 3- and 4-CQA: 18; di-CQA: 28FQA: present in 1 subject over 6; CA: 14	Monteiro et al., 2007
			Urine (%)	dHCA and GA: 56; p-HBA and p-CA: not in all subjects samples	

Summary of the Different in vivo Studies Investigating Bioavailability of Ferulic Acid.[b]

Models	Vehicle Adminisration	Compounds	Biological-fluids				Metabolites	References
		Absorption (% of ingested dose)	**Foregut (small intestine/ stomach)**	**Cecum**	**Colon**	**Residues**	**FA in plasma (μM)**	
Rat	Supplemented diet with (in g/kg):FA (1.5), FAA (2.5), FAXn (59.5)	FA	~99	<1			11	Zhao et al., 2003b
		FAA	40 (6% as free FA)	57 (11% as free FA)		3	12	
		FAXn	Traces (0.3% as free FA)	44 (0.7% as free FA)	23	33 (2% as free FA)	~1	

(Continued)

Table 2.2 *(Continued)*

Model	Vehicle/ Administration	Compound	Biological Fluid			Metabolites			Reference
Rat	Supplemented water	FA, FAA, or FAXn at 70 mol/kg bw	Plasma (after 5, 15, 30, 60, 120 min)			Urine (over 24 h)		In % of total FA t_{max} (min) :FA-GFA-S/GFA-SFA	Zhao et al., 2003a
			FA	FAA	FA	FAA	FAXn		
			4–23 [60']	29–41 [60']	6	18	5		
			60–70 [120']	54–63 [15']	87	76	91		
			4–6 [15 & 120']	2–4 [15 & 120']	2	2	1		
			1–34 [5']	2–4 [15']	4	4	3		
						Total FA excreted (% dose)			
			Relative BV of FA from diets			Urine	Feces		
				FA	1	72	1		
				FAA	0.56	54	1		
				FAXn	0.21	20	20		
Rat	Supplemented diet	FA at 5.15 mg/kg	In % of the total FA (30 min after ingestion)						Rondini et al., 2002
			Plasma			Urine			
				24 (1.7 M)		11		FA	
				18 (0.9 M)		3		FA-G	
				58 (2.7 M)		86		FA-S and –S/G	
Rat	Small intestinal perfusion	Concentrations of FA from 10 to 50 µM	Absorption of FA		Biliary secretion of FA	Circulating conjugates	Total excretion	No FA free but FA-G and FA-G/S	Adam et al., 2002
			56 (for all doses)		~6	45–53	50		
	Supplemented diets	With FA (çmol/d):1050250		Plasma (µM)		Urine (% of intake)		No FA free but FA-G and FA-G/S	
				ND		42			
				1		52			
				7–8		39			
	Cereals with different amounts of FA	In mg/g: whole cereals (1.2) or bran (4.6)		0.25		Recovery quite limited (low absorption)			

(Continued)

Table 2.2 *(Continued)*

Model	Vehicle/ Administration	Compound	Biological Fluid	Metabolites					Reference
Rat				% of total FA (over 24h)					Rondini et al., 2004
				Plasma		Urine			
	26% bran-rich diet	Containing 95.8% bound and 4.2% of free FA (0.14 mg/g)		9		15		FA	
				19		11		FA-G	
				72		74		FA-S + FA-G/S	
	Supplemented diet	5.05 mg/kg free ferulic acid		25		14		FA	
				17		2		FA-G	
				58		84		FA-S + FA-G/S	
Rat	Supplemented saline solution	2.25 mol(8 μmol/kg BW) ofFA or FAA	25 min after gastric perfusion of	Gastric content		Gastric mucosa		FA in % of the ingested dose	Zhao et al., 2004
				in vitro	in situ	in vitro	in situ		
			FA	80	26	7	4		
			FAA	91	81	2	2		
			25 min after gastric perfusion ofFA in situ	Plasma				in % of total FAFAFA-SFA-GFA-G/S	
				Portal vein	Bile	Urine	Celiac arteria		
				49	6	8	5		
				5	18	6	11		
				3	3	24	15		
				43	72	62	70		

(Continued)

Table 2.2 *(Continued)*

Model	Vehicle/ Administration	Compound	Biological Fluid		Metabolites	Reference
Rat	Perfusion of jejunum + ileum	With 0.8 Ml/min of 15 or 120 M FA	**Mesenteric vein**	**Extrasplanchnic**	Conjugates (% of total)	Silberberg et al., 2006
			40 Intestinal secretion: 5	75	Conjugates (% of ingested dose)	
Human	Tomato ingestion	21–44 mg FA	**Urine (% of the ingested amount)**		FA and conjugates (the major one being FA-G)	Bourne and Rice-Evans, 1998
			7–9 h after intake	**Over 24 h**		
			11–25	4–5		

[a]Chlorogenic acid has been used as a name for 5-*O*-caffeoylquinic acid by the authors of the reports cited except [Monteiro et al., 2007]. 3-, 4-, or 5-CQA, 3-, 4- or 5-*O*-caffeoylquinic acid; BA, benzoic acid; CA, caffeic acid; CLA, chlorogenic acid or 5-*O*-caffeoylquinic acid; dHCA, dihydrocaffeic acid; diCQA, dicaffeoylquinic acid; FA, ferulic acid; FQA, feruloylquinic acid; GA, gallic acid; HA, hippuric; isoFA, isoferulic acid; *m*-CA, *m*-coumaric acid; *m*-HPPA, *m*-hydroxyphenylpropionic acid; *p*-CA, p-coumaric acid; p-HBA, *p*-hydroxybenzoic acid; PPA, phenylpropionic acid.
[b]bw, body weigh; FA: ferulic acid; FAA: 5-*O*-feruloyl-L-arabinofuranose; FA-G, feruloyl-glucuronide; FA-S, feruloyl-sulphate; FA-S/G, feruloyl-sulphoglucuronide; FAXn: feruloyl-arabinoxylan.

on their previous report [Zhao et al., 2004], the authors suggested that these metabolites would come from a further metabolism by the liver of the free form absorbed by the stomach, and/or reabsorption by the enterohepatic circulation rather than from a direct metabolism by the gastric cells. Ferulic acid and its sugar ester are not the only phenolic acids shown to be absorbed by the stomach. 5-*O*-Caffeoylquinic [Lafay et al., 2006a], but also gallic, caffeic, and 4-coumaric acids could also be absorbed at the gastric level, their absorption being dependent on their structure in the following order: 5-*O*-caffeoylquinic = gallic < caffeic < 4-coumaric acids [Konishi et al., 2006]. Indeed, caffeic acid has been shown to be three times better absorbed (95% of the ingested dose) than 5-*O*-caffeoylquinic acid (33%) [Olthof et al., 2001]. Figure 2.8 summarizes the behavior of phenolic acid esters and free forms in the GIT.

The Mechanisms of Uptake of Phenolic Acids

The partition coefficient (log *P*, or log octanol/water), which reflects the lipophilicity of a molecule, will determine its affinity for a biological membrane and its partitioning through and into a membrane. This coefficient has been shown to govern the capacity of different flavonoids to be absorbed by cells [Crespy et al., 2003], suggesting passive diffusion as a main mechanism of flavonoid transport. The partition coefficient determined for ferulic acid at 1.86 [Voisin-Chiret et al., 2007] rather suggests that the phenolic acid will have affinity for cytoplasmic membrane and permeate by transcellular diffusion. Moreover, the percentage of ferulic acid absorbed after perfusion of the rat small intestine was independent of the concentration perfused, indicating that there was no saturation of the permeation system involved [Silberberg et al., 2006], findings which do not support the existence of a specific transport system. However, some studies have attempted to discover other permeation mechanisms for phenolic acids.

Various drug transporters have been identified and characterized, among which are the monocarboxylate transporters (MCTs). The physiological role of these transporters is to co-transport the endogenous monocarboxylates such as pyruvate, lactate, acetoacetate, and β-hydroxybutyrate together with equimolar protons by a symport mechanism (as reviewed in [Enerson and Drewes, 2003]). MCTs may also have an important role in the transport of some exogenous compounds, which consist of weak organic acids that are monovalent, with the carboxyl group attached to a relatively small alkyl group having either hydrophobic or hydrophilic properties. These transporters are expressed by various tissues such as the intestine, heart, liver, kidney, and brain [Halestrap and Price, 1999]. Different isoforms of MCTs have recently been localized in the different regions of the human intestine and on the membrane of intestinal cells [Gill et al., 2005]. MCT1 expression is restricted to the luminal membranes of human intestinal cells, whereas MCT4 and MCT5 are present on the basolateral side. The isoform 3 is also expressed on the basolateral side but at a very low level compared with MCT4 and 5. However, no MCT6 is detected

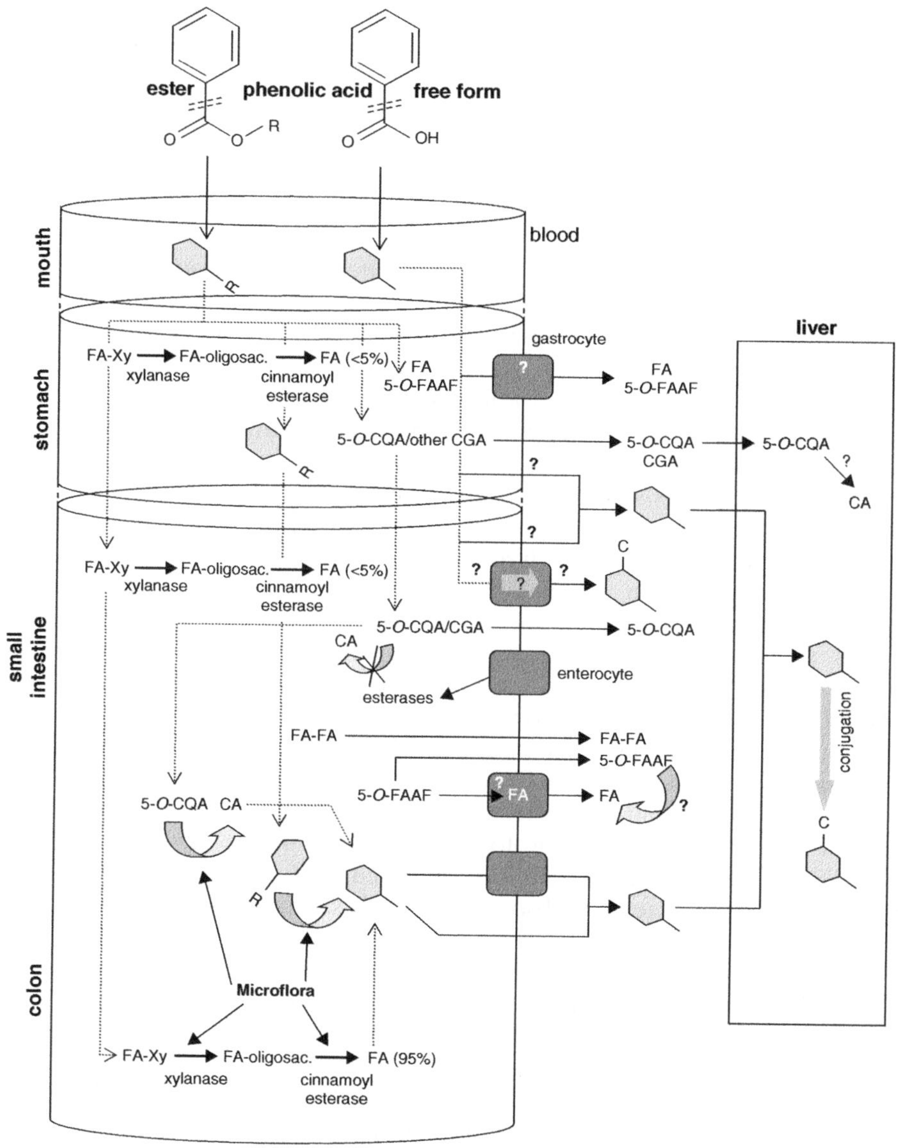

Figure 2.8 Summary of behavior of phenolic acid esters and free forms in the GIT and the liver. 5-*O*-CQA, 5-*O*-caffeoylquinic acid; 5-*O*-FAAF, 5-*O*-feruloyl-L-arabinofuranose; C, conjugate (sulfate, glucuronide or methyl); CA, caffeic acid; FA, ferulic acid; FA-FA, dimer of ferulic acid; FA-oligosac., ferulic acid esterified to an oligosaccharide; FA-Xy, ferulic acid esterified to xylan; R, alkyl group.

on either the basolateral or apical side of small intestinal or colonic cells. Expression of the transporters increases along the length of the intestine, being maximal in the distal colon and even absent in the jejunum for MCT4 and 5. MCT1, expressed on the luminal side of intestinal cells, is a potential candidate for the transport of phenolic acids.

An in vitro model of Caco-2 cells grown on semipermeable membranes was used to investigate the potential involvement of the MCTs in the transport of phenolic acids [Konishi et al., 2002]. The study was based on competition for transport between fluorescein, an MCT substrate [Kuwayama et al., 2002], and phenolic acids. Ferulic, cinnamic, 4-coumaric, and vanillic acids competitively inhibited the transport of fluorescein, with the following order of inhibition: cinnamic > ferulic ≥ vanillic > 4-coumaric acids, suggesting the transport of these phenolic acids via MCTs. Caffeic acid had no effect, whereas ferulic and vanillic acids had a very similar effect (respectively, 49 and 45% of inhibition). However, it must be mentioned that the concentrations of phenolic acids used as competitors in this study were high (10 mM) and one could wonder, even though not mentioned by the authors, if phenolic acids used at such concentrations could be not fully dissolved and/or be toxic for the cells, the resulting percentage of inhibition influenced by the toxicity. Further studies aimed to better understand the mechanisms of phenolic acid transport. Using the same model, transport of 4-coumaric acid was shown to be pH dependent, higher, but saturable, in the presence of a proton gradient (basal side, pH 6; apical side, pH 7.4), and 10 times higher from the apical to the basal side compared with the opposite direction [Konishi et al., 2003b]. According to these results, the authors suggested that the permeation of 4-coumaric acid in the presence of a proton gradient would be dependent on an active transport mechanism, whereas in the absence of a proton gradient, it must be restricted by tight junctions. The active transport was proposed to be MCT mediated, since benzoic and acetic acids, substrates of MCTs, inhibited 4-coumaric acid permeation, although lactic acid, a good substrate for MCT1–MCT4, had no effect, suggesting the involvement of a different isoform. However, no metabolite of 4-coumaric acid could be detected on the basal side of the Caco-2 cell model. This model also served to study the permeation of ferulic acid, shown to be Na^+-independent [Konishi and Shimizu, 2003], contrary to what has been proposed earlier for cinnamic acid [Ader et al., 1996; Wolffram et al., 1995]. As for 4-coumaric acid, ferulic acid was absorbed via an MCT system and was not conjugated by the in vitro model. The authors also proposed that tea polyphenols shared the same MCT transporter(s) with ferulic acid, since its permeation was inhibited by (+)-catechin, (–)-epicatechin, (–)-epicatechin-gallate, (–)-epigallocatechin, and (–)-epigallocatechingallate [Konishi et al., 2003a], the involvement of an MCT in the transport of electrocardiogram (ECG) also observed with Caco-2 cells as proposed by another report [Vaidyanathan and Walle, 2003]. Transport of caffeic, 3-coumaric, and 3-hydroxyphenylpropionic acids [Konishi and Kobayashi, 2004b], as well as dihydrocaffeic acid [Konishi and Kobayashi, 2004a], were also proposed to occur via MCTs, as shown using Caco-2 cells, later proposed to be an isoform different from MCT1 [Konishi et al., 2006]. More recently, the mechanism of transport of phenolic acids through the epithelium of the colon, the major site of release of phenolic acids from esters and flavanol precursors, was clarified [Poquet et al., 2008b]. The initial purpose of investigation was to determine the eventual role of the mucus layer, much thicker in the colon than in the small intestine, which could influence the absorption of phenolic acids due to its property of maintaining the pH close to the mucosa around neutrality [McNeil et al., 1987]. However, the mucus layer secreted by the mucus secreting HT29-MTX cells in a co-culture with Caco-2 had only a slight negative role on the transport of ferulic acid. However, using this model and more physiological concentrations of phenolic acids, compared with studies reported in the literature, ferulic acid was shown to mainly ($\sim$80%) permeate through the intestinal epithelium by transcellular diffusion, which was dependent on the hydrophobicity and on the pK_a of the phenolic acid, and on the pH of the donor compartment. Moreover, the remaining 20% of ferulic acid seemed to be transported by a Na^+-dependent carboxylic acid transporter (S-MCT) as well as by an MCT.

Circulating Metabolites and Conjugates

Free ferulic acid and its glucuronide were the forms excreted in human urine over the 7–9 h following ingestion of a meal based on fresh tomatoes, a source of ferulic, 5-*O*-caffeoylquinic, caffeic, and 4-coumaric acids [Bourne and Rice-Evans, 1998]. They represented 11–25% of the ingested amount, a considerable proportion as the glucuronide. Ferulic acid and its conjugates, glucuronide, sulfate, and sulfoglucuronide, peaked in rat plasma between 0 and 1 h, and lasted for at least 24 h after ingestion of a 26% bran-rich diet, which contained 95.8 and 4.2%, respectively, of bound and free ferulic acid [Rondini et al., 2004]. They also started to appear in urine only 30 min after the ingestion of bran, reaching a plateau after 4.5 h. In plasma, over the 24 h following intake of bran, the sulfate and sulfoglucuronide were the main metabolites (72% of total ferulic acid), followed by the glucuronide (19%) and the free form (9%). The free form was more abundant (25%) when animals received pure ferulic acid instead of the bran-based diet. Conjugated ferulic acid was also detected in the intestinal contents, suggesting a metabolism in the mucosa and a secretion back into the lumen via the bile duct. Free ferulic and its glucuronide, sulfate and sulfoglucuronide conjugates were detected in plasma after intake of ferulic acid or of 5-*O*-feruloyl-L-arabinofuranose [Zhao et al., 2003b]. Given as free form in a diet, the amount of ferulic acid in plasma peaked 30 min after ingestion of the meal and quickly decreased between 0.5 and 1.5 h and more slowly until 4.5 h [Rondini et al., 2002]. Ferulic acid was present in plasma and urine as the free form but also conjugated with glucuronide, sulfate, or both, the major form being the feruloyl-sulfate and feruloyl-sulfoglucuronide. Another study showed that, after oral administration of pure ferulic acid (1 mmol) to rats, three metabolites, in addition to the parent compound, which was the main form excreted, were detected in urine 0–12 h after the intake [Zhang et al., 2005]. Two of the metabolites were identified as monoglucuronides of ferulic acid, the conjugation taking place on the 4-OH group or on the carboxylic function, as suggested by the authors. The third conjugate was proposed to be a di-glucuronide, with one glucuronide on the free hydroxyl of the phenyl ring and the other one on the carboxylic function. This study also investigated the tissue distribution of ferulic acid in rats, found in the liver, spleen, kidney, and lungs, but not in the heart, 2 h after the oral administration. Free caffeic acid, together with ferulic acid, and their respective sulfate, glucuronide, and sulfoglucuronides were found in rat plasma following the oral administration of caffeic acid [Azuma et al., 2000; Plumb et al., 1999]. However, only caffeoyl-glucuronide, caffeoyl-sulfoglucuronide, and feruloyl-sulfoglucuronide were detected with a peak at 0.5–1 h in rat plasma after oral administration of 5-*O*-caffeoylquinic acid [Azuma et al., 2000]. In this study, interestingly, when caffeic acid was orally administered, caffeoyl-glucuronide, caffeoyl-sulfoglucuronide, and feruloyl-sulfoglucuronide peaked after 2 h instead of 0.5 h when 5-*O*-caffeoylquinic acid was administered. It is surprising to see that metabolites, which require microbial metabolism to be generated, appear quicker in plasma from 5-*O*-caffeoylquinic acid than when caffeic acid is absorbed. The results suggest that caffeic acid could be absorbed from the stomach but also in the upper part of the intestine, with a better absorption rate from the small intestine than from the stomach, explaining the delay of the peak of absorption. Even though not detected by the authors in plasma, 5-*O*-caffeoylquinic acid could be absorbed as such by the stomach, better than caffeic acid, and quickly hydrolyzed by gastric cells, intestinal cells, or liver, allowing the rapid presence of caffeic acid in plasma (0.5 h). Konishi et al. [2006] demonstrated that caffeic acid is better absorbed than 5-*O*-caffeoylquinic acid by the stomach, which does not support the above-mentioned hypothesis. 5-*O*-caffeoylquinic acid may be better absorbed than caffeic acid but, due to instability, its amount in biological fluids could be underestimated. The absorption by the stomach was also proposed for dihydrocaffeic acid, as the parent compound and its metabolites peaked in plasma before 30 min following its gastric administration to rats [Poquet et al., 2008a].

In situ perfusion of ferulic acid showed that ferulic acid could be absorbed from the small intestine and that the forms circulating in plasma were glucuronide and sulfate conjugates, ferulic acid being rapidly excreted in urine [Adam et al., 2002]. Dihydrocaffeic acid was rapidly excreted, detected in urine at 6 and 12 h, but not at 24 h, however, not detected in plasma at a time even as short as 30 min following its administration by gavage to rats, suggesting a quick absorption, before 30 min [Poquet et al., 2008a].

Caco-2 cells have also been used to study the metabolism of phenolic acids. Contrary to the transport studies, which usually use the cells grown on semipermeable filters (two compartments), the cultures used to study metabolism are prepared with cells grown and partially differentiated on dishes (one compartment). The Caco-2 cell model could perform sulfation, glucuronidation, and methylation of various hydroxycinnamic acids [Kern et al., 2003b]. Ferulic, sinapic and 4-coumaric acids could enter the cells, be sulfated, but not glucuronidated, and then be effluxed back into the medium. However, the corresponding methyl esters were both sulfated and glucuronidated. The authors proposed that methyl esters could be more hydrophobic and as a consequence better access to the UGTs localized in the membrane of the endoplasmic reticulum, whereas the more polar corresponding hydroxycinnamic acids would remain in the cytosol where the sulfotransferases are located. When a co-culture of HT29-MTX and Caco-2 cells, differentiated on semipermeable membranes to study the metabolism and efflux of ferulic acid [Poquet et al., 2008b], was used, the parent compound was the major form excreted to the basal side of the culture, followed by the glucuronide conjugate, the reduced form (dihydroferulic acid) and the sulfate conjugate. The reduced metabolite was equally distributed to the apical and basal sides, whereas the concentrations of the conjugates were higher on the serosal side. An MRP transporter could contribute to the major part of the glucuronide efflux and also of some dihydroferulic acid at the serosal side, the strongest candidate being MRP3, without excluding the possible involvement of MRP6. Moreover, P-gp protein could be responsible for some of the efflux of these metabolites to the luminal side of the cells. Since the amount of sulfate conjugate effluxed was much lower, its mechanism of efflux could not be investigated. The pattern of dihydrocaffeic acid metabolism by HT29-MTX and Caco-2 cells suggested that Caco-2 cells do not possess the isoenzyme responsible for the glucuronidation of phenolic acids, proposed to be UGT1A7 [Poquet et al., 2008a]. Methylation of dihydrocaffeic acid was also proposed to be regioselective, favored on the 3-OH and dependent on the pK_a of the 4-OH. Apart from sulfation, methylation, or glucuronidation, a report mentioned the possible formation of mono- or biglutathione conjugates of caffeic, 5-*O*-caffeoylquinic and dihydrocaffeic acids by rat hepatocyte microsomes, the reaction taking place during peroxidase/H_2O_2-catalyzed oxidation of the phenolic acids [Moridani et al., 2001].

All these studies suggest that the metabolic pathway is the same for free and bound forms of ferulic acid. Whereas the free form is rapidly absorbed, metabolized and excreted in urine, the bound form allows a longer time period (free and conjugated) in plasma, as already suggested by the absence of free ferulic acid in rat plasma 18 h after intake from an enriched diet, whereas its metabolites were still present [Adam et al., 2002]. Even though its bioavailability, in terms of half-life in plasma and low excretion in urine, was improved when esterified, the absorption of ferulic acid from a wheat source was lower than if present as free form in a diet [Rondini et al., 2004]. This indicates that the amounts of ferulic acid circulating in plasma as well as its circulating metabolites are dependent on its form present in the food source.

Depending on the food source, the model used, the time points, the method of analysis, and the reports, phenolic acids and, especially ferulic acid can be detected under different forms in the biological samples. However, there is a common tendency to suggest that phenolic acids can be absorbed through the gastrointestinal epithelium and be metabolized to circulate and be excreted as free and/or conjugated forms.

REFERENCES

Adam A, Crespy V, Levrat-Verny MA, Leenhardt F, Leuillet M, Demigne C, Remesy C. 2002. The bioavailability of ferulic acid is governed primarily by the food matrix rather than its metabolism in intestine and liver in rats. *J Nutr* 132:1962–1968.

Ader P, Grenacher B, Langguth P, Scharrer E, Wolffram S. 1996. Cinnamate uptake by rat small intestine: Transport kinetics and transepithelial transfer. *Exp Physiol* 81:943–955.

Anderson RA, Broadhurst CL, Polansky MM, Schmidt WF, Khan A, Flanagan VP, Schoene NW, Graves DJ. 2004. Isolation and characterization of polyphenol type-A polymers from cinnamon with insulin-like biological activity. *J Agric Food Chem* 52:65–70.

Andreasen MF, Kroon PA, Williamson G, Garcia-Conesa MT. 2001a. Esterase activity able to hydrolyze dietary antioxidant hydroxycinnamates is distributed along the intestine of mammals. *J Agric Food Chem* 49:5679–5684.

Andreasen MF, Kroon PA, Williamson G, Garcia-Conesa MT. 2001b. Intestinal release and uptake of phenolic antioxidant diferulic acids. *Free Radic Biol Med* 31:304–314.

Azuma K, Ippoushi K, Nakayama M, Ito H, Higashio H, Terao J. 2000. Absorption of chlorogenic acid and caffeic acid in rats after oral administration. *J Agric Food Chem* 48:5496–5500.

Baba S, Osakabe N, Natsume M, Terao J. 2002. Absorption and urinary excretion of procyanidin B2 [epicatechin-(4beta-8)-epicatechin] in rats. *Free Radic Biol Med* 33:142–148.

Baba S, Furuta T, Horie M, Nakagawa H. 1981. Studies on drug metabolism by use of isotopes XXVI: Determination of urinary metabolites of rutin in humans. *J Pharm Sci* 70:780–782.

Bacon JR, Rhodes MJ. 1998. Development of a competition assay for the evaluation of the binding of human parotid salivary proteins to dietary complex phenols and tannins using a peroxidase-labeled tannin. *J Agric Food Chem* 46:5083–5088.

Bourne LC, Rice-Evans C. 1998. Bioavailability of ferulic acid. *Biochem Biophys Res Commun* 253:222–227.

Caccetta RA, Croft KD, Beilin LJ, Puddey IB. 2000. Ingestion of red wine significantly increases plasma phenolic acid concentrations but does not acutely affect ex vivo lipoprotein oxidizability. *Am J Clin Nutr* 71:67–74.

Cai YZ, Mei S, Jie X, Luo Q, Corke H. 2006. Structure-radical scavenging activity relationships of phenolic compounds from traditional Chinese medicinal plants. *Life Sci* 78:2872–2888.

Clifford MN. 2000. Chlorogenic acids and other cinnamates—nature, occurrence, dietary burden, absorption and metabolism. *J Sci Food and Agric* 80:1033–1043.

Clifford MN. 1999. Chlorogenic acids and other cinnamates—nature, occurrence and dietary burden. *J Sci Food Agric* 79:362–372.

Clifford MN, Knight S. 2004. The cinnamoyl-amino acid conjugates of green robusta coffee beans. *Food Chem* 87:457–463.

Clifford MN, Copeland EL, Bloxsidge JP, Mitchell LA. 2000. Hippuric acid as a major excretion product associated with black tea consumption. *Xenobiotica* 30:317–326.

Couteau D, McCartney AL, Gibson GR, Williamson G, Faulds CB. 2001. Isolation and characterization of human colonic bacteria able to hydrolyse chlorogenic acid. *J Appl Microbiol* 90:873–881.

Crespy V, Morand C, Besson C, Cotelle N, Vezin H, Demigne C, Remesy C. 2003. The splanchnic metabolism of flavonoids highly differed according to the nature of the compound. Am J Physiol Gastrointest Liver Physiol 284:G980–G988.

Das NP. 1974. Studies on flavonoid metabolism. Excretion of m-hydroxyphenylhydracrylic acid from (plus)-catechin in the monkey (Macaca iris sp.). *Drug Metab Dispos* 2:209–213.

de Freitas V, Mateus N. 2001. Structural features of procyanidin interactions with salivary proteins. *J Agric Food Chem* 49:940–945.

Deprez S, Mila I, Huneau JF, Tome D, Scalbert A. 2001. Transport of proanthocyanidin dimer, trimer, and polymer across monolayers of human intestinal epithelial Caco-2 cells. *Antioxid Redox Signal* 3:957–967.

Deprez S, Brezillon C, Rabot S, Philippe C, Mila I, Lapierre C, Scalbert A. 2000. Polymeric proanthocyanidins are catabolized by human colonic microflora into low-molecular-weight phenolic acids. *J Nutr* 130:2733–2738.

Dixon RA, Xie DY, Sharma SB. 2005. Proanthocyanidins—a final frontier in flavonoid research? *New Phytol* 165:9–28.

Donovan JL, Crespy V, Manach C, Morand C, Besson C, Scalbert A, Remesy C. 2001. Catechin is metabolized by both the small intestine and liver of rats. *J Nutr* 131:1753–1757.

Enerson BE, Drewes LR. 2003. Molecular features, regulation, and function of monocarboxylate transporters: Implications for drug delivery. *J Pharm Sci* 92:1531–1544.

Fine AM. 2000. Oligomeric proanthocyanidin complexes: history, structure, and phytopharmaceutical applications. *Altern Med Rev* 5:144–151.

Fry SC. 1982. Phenolic components of the primary cell wall. Feruloylated disaccharides of D-galactose and L-arabinose from spinach polysaccharide. *Biochem J* 203:493–504.

Garcia-Ramirez B, Fernandez-Larrea J, Salvado MJ, Ardevol A, Arola L, Blade C. 2006. Tetramethylated dimeric procyanidins are detected in rat plasma and liver early after oral administration of synthetic oligomeric procyanidins. *J Agric Food Chem* 54:2543–2551.

Gasson MJ, Kitamura Y, McLauchlan WR, Narbad A, Parr AJ, Parsons EL, Payne J, Rhodes MJ, Walton NJ. 1998. Metabolism of ferulic acid to vanillin. A bacterial gene of the enoyl-SCoA hydratase/isomerase superfamily encodes an enzyme for the hydration and cleavage of a hydroxycinnamic acid SCoA thioester. *J Biol Chem* 273:4163–4170.

Gill RK, Saksena S, Alrefai WA, Sarwar Z, Goldstein JL, Carroll RE, Ramaswamy K, Dudeja PK. 2005. Expression and membrane localization of MCT isoforms along the length of the human intestine. *Am J Physiol Cell Physiol* 289:C846–C852.

Goldstein DS, Stull R, Markey SP, Marks ES, Keiser HR. 1984. Dihydrocaffeic acid: A common contaminant in the liquid chromatographic-electrochemical measurement of plasma catecholamines in man. *J Chromatogr* 311:148–153.

Gonthier MP, Cheynier V, Donovan JL, Manach C, Morand C, Mila I, Lapierre C, Remesy C, Scalbert A. 2003a. Microbial aromatic acid metabolites formed in the gut account for a major fraction of the polyphenols excreted in urine of rats fed red wine polyphenols. *J Nutr* 133:461–467.

Gonthier MP, Donovan JL, Texier O, Felgines C, Remesy C, Scalbert A. 2003b. Metabolism of dietary procyanidins in rats. *Free Radic Biol Med* 35:837–844.

Gonthier MP, Verny MA, Besson C, Remesy C, Scalbert A. 2003c. Chlorogenic acid bioavailability largely depends on its metabolism by the gut microflora in rats. *J Nutr* 133:1853–1859.

Goodwin BL, Ruthven CR, Sandler M. 1994. Gut flora and the origin of some urinary aromatic phenolic compounds. *Biochem Pharmacol* 47:2294–2297.

Graf E. 1992. Antioxidant potential of ferulic acid. *Free Radic Biol Med* 13:435–448.

Griffiths LA. 1964. Studies on flavonoid metabolism. Identification of the metabolities of (+)-catechin in rat urine. *Biochem J* 92:173–179.

Griffiths LA. 1962. m-Hydroxyphenylpropionic acid, a major urinary metabolite of catechin in the rat. *Nature* 194:869–870.

Groenewoud G, Hundt HK. 1986. The microbial metabolism of condensed (+)-catechins by rat-caecal microflora. *Xenobiotica* 16:99–107.

Gu L, Kelm MA, Hammerstone JF, Beecher G, Holden J, Haytowitz D, Gebhardt S, Prior RL. 2004. Concentrations of proanthocyanidins in common foods and estimations of normal consumption. *J Nutr* 134:613–617.

Halestrap AP, Price NT. 1999. The proton-linked monocarboxylate transporter (MCT) family: Structure, function and regulation. *Biochem J 343 Pt* 2:281–299.

Hammerstone JF, Lazarus SA, Schmitz HH. 2000. Procyanidin content and variation in some commonly consumed foods. *J Nutr* 130:2086S–2092S.

Hammerstone JF, Lazarus SA, Mitchell AE, Rucker R, Schmitz HH. 1999. Identification of procyanidins in cocoa (Theobroma cacao) and chocolate using high-performance liquid chromatography/mass spectrometry. *J Agric Food Chem* 47:490–496.

Herrmann K. 1989. Occurrence and content of hydroxycinnamic and hydroxybenzoic acid compounds in foods. *Crit Rev Food Sci Nutr* 28:315–347.

Jacobson EA, Newmark H, Baptista J, Bruce WR. 1983. A preliminray investigation of the metabolism of dietary phenolics in humans. *Nutr Rep Int* 28:1409–1417.

Jenner AM, Rafter J, Halliwell B. 2005. Human fecal water content of phenolics: The extent of colonic exposure to aromatic compounds. *Free Radic Biol Med* 38:763–772.

Kahle K, Huemmer W, Kempf M, Scheppach W, Erk T, Richling E. 2007. Polyphenols are intensively metabolized in the human gastrointestinal tract after apple juice consumption. *J Agric Food Chem* 55:10605–10614.

Kern SM, Bennett RN, Mellon FA, Kroon PA, Garcia-Conesa MT. 2003a. Absorption of hydroxycinnamates in humans after high-bran cereal consumption. *J Agric Food Chem* 51:6050–6055.

Kern SM, Bennett RN, Needs PW, Mellon FA, Kroon PA, Garcia-Conesa MT. 2003b. Characterization of metabolites of hydroxycinnamates in the in vitro model of human small intestinal epithelium caco-2 cells. *J Agric Food Chem* 51:7884–7891.

Klepacka J, Forna L. 2006. Ferulic acid and its position among the phenolic compounds of wheat. *Crit Rev Food Sci Nutr* 46:639–647.

Konishi Y, Kobayashi S. 2004a. Microbial metabolites of ingested caffeic acid are absorbed by the monocarboxylic acid transporter (MCT) in intestinal Caco-2 cell monolayers. *J Agric Food Chem* 52:6418–6424.

Konishi Y, Kobayashi S. 2004b. Transepithelial transport of chlorogenic acid, caffeicacid, and their colonic metabolites in intestinal caco-2 cell monolayers. *J Agric Food Chem* 52:2518–2526.

Konishi Y, Shimizu M. 2003. Transepithelial transport of ferulic acid by monocarboxylic acid transporter in Caco-2 cell monolayers. *Biosci Biotechnol Biochem* 67:856–862.

Konishi Y, Zhao Z, Shimizu M. 2006. Phenolic acids are absorbed from the rat stomach with different absorption rates. *J Agric Food Chem* 54:7539–7543.

Konishi Y, Kobayashi S, Shimizu M. 2003a. Tea polyphenols inhibit the transport of dietary phenolic acids mediated by the monocarboxylic acid transporter (MCT) in intestinal Caco-2 cell monolayers. *J Agric Food Chem* 51:7296–7302.

Konishi Y, Kobayashi S, Shimizu M. 2003b. Transepithelial transport of p-coumaric acid and gallic acid in Caco-2 cell monolayers. *Biosci Biotechnol Biochem* 67:2317–2324.

Konishi Y, Hagiwara K, Shimizu M. 2002. Transepithelial transport of fluorescein in Caco-2 cell monolayers and use of such transport in in vitro evaluation of phenolic acid availability. *Biosci Biotechnol Biochem* 66:2449–2457.

Kroon PA, Faulds CB, Ryden P, Robertson JA, Williamson G. 1997. Release of covalently bound ferulic acid from fiber in the human colon. *J Agric Food Chem* 45:661–667.

Kuwayama K, Miyauchi S, Tateoka R, Abe H, Kamo N. 2002. Fluorescein uptake by a monocarboxylic acid transporter in human intestinal Caco-2 cells. *Biochem Pharmacol* 63:81–88.

Lafay S, Gil-Izquierdo A, Manach C, Morand C, Besson C, Scalbert A. 2006a. Chlorogenic acid is absorbed in its intact form in the stomach of rats. *J Nutr* 136:1192–1197.

Lafay S, Morand C, Manach C, Besson C, Scalbert A. 2006b. Absorption and metabolism of caffeic acid and chlorogenic acid in the small intestine of rats. *Br J Nutr* 96:39–46.

Lin YT, Hsiu SL, Hou YC, Chen HY, Chao PD. 2003. *Degradation of flavonoid aglycones by rabbit, rat and human fecal flora. Biol Pharm Bull* 26:747–751.

Manach C, Williamson G, Morand C, Scalbert A, Remesy C. 2005. Bioavailability and bioefficacy of polyphenols in humans. I. Review of 97 bioavailability studies. Am J Clin Nutr 81:230S–242S.

Matsui Y, Nakamura S, Kondou N, Takasu Y, Ochiai R, Masukawa Y. 2007. Liquid chromatography-electrospray ionization-tandem mass spectrometry for simultaneous analysis of chlorogenic acids and their metabolites in human plasma. *J Chromatogr B Analyt Technol Biomed Life Sci* 858:96–105.

Mattila P, Hellstrom J, Torronen R. 2006. Phenolic acids in berries, fruits, and beverages. *J Agric Food Chem* 54:7193–7199.

McNeil NI, Ling KL, Wager J. 1987. Mucosal surface pH of the large intestine of the rat and of normal and inflamed large intestine in man. *Gut* 28:707–713.

Meng X, Sang S, Zhu N, Lu H, Sheng S, Lee MJ, Ho CT, Yang CS. 2002. Identification and characterization of methylated and ring-fission metabolites of tea catechins formed in humans, mice, and rats. *Chem Res Toxicol* 15:1042–1050.

Monteiro M, Farah A, Perrone D, Trugo LC, Donangelo C. 2007. Chlorogenic acid compounds from coffee are differentially absorbed and metabolized in humans. *J Nutr* 137:2196–2201.

Moridani MY, Scobie H, O'Brien PJ. 2002. Metabolism of caffeic acid by isolated rat hepatocytes and subcellular fractions. *Toxicol Lett* 133:141–151.

Moridani MY, Scobie H, Jamshidzadeh A, Salehi P, O'Brien PJ. 2001. Caffeic acid, chlorogenic acid, and dihydrocaffeic acid metabolism: Glutathione conjugate formation. *Drug Metab Dispos* 29:1432–1439.

Nardini M, Cirillo E, Natella F, Scaccini C. 2002. Absorption of phenolic acids in humans after coffee consumption. *J Agric Food Chem* 50:5735–5741.

Neilson AP, Hopf AS, Cooper BR, Pereira MA, Bomser JA, Ferruzzi MG. 2007. Catechin degradation with concurrent formation of homo- and heterocatechin dimers during in vitro digestion. *J Agric Food Chem* 55:8941–8949.

Olthof MR, Hollman PC, Katan MB. 2001. Chlorogenic acid and caffeic acid are absorbed in humans. *J Nutr* 131:66–71.

Owen RW, Haubner R, Mier W, Giacosa A, Hull WE, Spiegelhalder B, Bartsch H. 2003. Isolation, structure elucidation and antioxidant potential of the major phenolic and flavonoid compounds in brined olive drupes. *Food Chem Toxicol* 41:703–717.

Peppercorn MA, Goldman P. 1972. Caffeic acid metabolism by gnotobiotic rats and their intestinal bacteria. *Proc Natl Acad Sci USA* 69:1413–1415.

Peppercorn MA, Goldman P. 1971. Caffeic acid metabolism by bacteria of the human gastrointestinal tract. *J Bacteriol* 108:996–1000.

Piskula MK, Terao J. 1998. Accumulation of (−)-epicatechin metabolites in rat plasma after oral administration and distribution of conjugation enzymes in rat tissues. *J Nutr* 128:1172–1178.

Plumb GW, Garcia-Conesa MT, Kroon PA, Rhodes M, Ridley S, Williamson G. 1999. Metabolism of chlorogenic acid by human plasma, liver, intestine and gut microflora. *J Sci Food Agric* 79:390–392.

Poquet L, Clifford MN, Williamson G. 2008a. Investigation of the metabolic fate of dihydrocaffeic acid. *Biochem Pharmacol* 75:1218–1229.

Poquet L, Clifford MN, Williamson G. 2008b. Transport and metabolism of ferulic acid through the colonic epithelium. *Drug Metab Dispos* 36:190–197.

Prior RL, Lazarus SA, Cao G, Muccitelli H, Hammerstone JF. 2001. Identification of procyanidins and anthocyanins in blueberries and cranberries (Vaccinium spp.) using high-performance liquid chromatography/mass spectrometry. *J Agric Food Chem* 49:1270–1276.

Rasmussen SE, Frederiksen H, Struntze KK, Poulsen L. 2005. Dietary proanthocyanidins: Occurrence, dietary intake, bioavailability, and protection against cardiovascular disease. *Mol Nutr Food Res* 49:159–174.

Rechner AR, Smith MA, Kuhnle G, Gibson GR, Debnam ES, Srai SK, Moore KP, Rice-Evans CA. 2004. Colonic metabolism of dietary polyphenols: Influence of structure on microbial fermentation products. *Free Radic Biol Med* 36:212–225.

Rechner AR, Kuhnle G, Bremner P, Hubbard GP, Moore KP, Rice-Evans CA. 2002. The metabolic fate of dietary polyphenols in humans. *Free Radic Biol Med* 33:220–235.

Rechner AR, Pannala AS, Rice-Evans CA. 2001. Caffeic acid derivatives in artichoke extract are metabolised to phenolic acids in vivo. *Free Radic Res* 35:195–202.

Rios LY, Gonthier MP, Remesy C, Mila I, Lapierre C, Lazarus SA, Williamson G, Scalbert A. 2003. Chocolate intake increases urinary excretion of polyphenol-derived phenolic acids in healthy human subjects. *Am J Clin Nutr* 77:912–918.

Rios LY, Bennett RN, Lazarus SA, Remesy C, Scalbert A, Williamson G. 2002. Cocoa procyanidins are stable during gastric transit in humans. *Am J Clin Nutr* 76:1106–1110.

Rondini L, Peyrat-Maillard MN, Marsset-Baglieri A, Fromentin G, Durand P, Tome D, Prost M, Berset C. 2004. Bound ferulic acid from bran is more bioavailable than the free compound in rat. *J Agric Food Chem* 52:4338–4343.

Rondini L, Peyrat-Maillard MN, Marsset-Baglieri A, Berset C. 2002. Sulfated ferulic acid is the main in vivo metabolite found after short-term ingestion of free ferulic acid in rats. *J Agric Food Chem* 50:3037–3041.

Sano A, Yamakoshi J, Tokutake S, Tobe K, Kubota Y, Kikuchi M. 2003. Procyanidin B1 is detected in human serum after intake of proanthocyanidin-rich grape seed extract. *Biosci Biotechnol Biochem* 67:1140–1143.

Santos-Buelga C, Scalbert A. 2000. Proanthocyanidins and tannin-like compounds—nature, occurence, dietary intake and effects on nutrition and health. *J Sci Food Agric* 80:1094–1117.

Scalbert A, Williamson G. 2000. Dietary intake and bioavailability of polyphenols. *J Nutr* 130:2073S–2085S.

Scheline RR. 1970. The metabolism of (+)-catechin to hydroxyphenylvaleric acids by the intestinal microflora. *Biochim Biophys Acta* 222:228–230.

Scheline RR. 1968. Metabolism of phenolic acids by the rat intestinal microflora. *Acta Pharmacol Toxicol (Copenh)* 26:189–205.

Schoefer L, Mohan R, Schwiertz A, Braune A, Blaut M. 2003. Anaerobic degradation of flavonoids by Clostridium orbiscindens. *Appl Environ Microbiol* 69:5849–5854.

Schwab AJ, Tao L, Yoshimura T, Simard A, Barker F, Pang KS. 2001. Hepatic uptake and metabolism of benzoate: A multiple indicator dilution, perfused rat liver study. *Am J Physiol Gastrointest Liver Physiol* 280:G1124–G1136.

Shahidi F, Wanasundara PK. 1992. Phenolic antioxidants. *Crit Rev Food Sci Nutr* 32:67–103.

Silberberg M, Morand C, Mathevon T, Besson C, Manach C, Scalbert A, Remesy C. 2006. The bioavailability of polyphenols is highly governed by the capacity of the intestine and of the liver to secrete conjugated metabolites. *Eur J Nutr* 45:88–96.

Smith MM, Hartley RD. 1983. Occurrence and nature of ferulic acid substitution of cell wall polysaccharides in gramineous plants. *Carbohydrate Research* 118:65–80.

Spencer JP, Schroeter H, Rechner AR, Rice-Evans C. 2001a. Bioavailability of flavan-3-ols and procyanidins: gastrointestinal tract influences and their relevance to bioactive forms in vivo. *Antioxid Redox Signal* 3:1023–1039.

Spencer JP, Schroeter H, Shenoy B, Srai SK, Debnam ES, Rice-Evans C. 2001b. Epicatechin is the primary bioavailable form of the procyanidin dimers B2 and B5 after transfer across the small intestine. *Biochem Biophys Res Commun* 285:588–593.

Spencer JP, Chaudry F, Pannala AS, Srai SK, Debnam E, Rice-Evans C. 2000. Decomposition of cocoa procyanidins in the gastric milieu. *Biochem Biophys Res Commun* 272:236–241.

Starp C, Alteheld B, Stehle P. 2005. Characteristics of (+)-catechin and (–)-epicatechin transport across pig intestinal brush border membranes. *Ann Nutr Metab* 50:59–65.

Sun RC, Sun XF, Zhang SH. 2001. Quantitative determination of hydroxycinnamic acids in wheat, rice, rye, and barley straws, maize stems, oil palm frond fiber, and fast-growing poplar wood. *J Agric Food Chem* 49:5122–5129.

Tanaka N, Sekiya N, Hattori M, Goto H, Shibahara N, Shimada Y, Terasawa K. 2003. Measurement of plasma procyanidin B-2 and procyanidin B-3 levels after oral administration in rat. *Phytomedicine* 10:122–126.

Tomás-Barberán FA, Clifford MN. 2000. Dietary hydroxybenzoic acid derivatives-nature, occurrence and dietary burden. *J Sci Food Agric* 80:1024–1032.

Tsevegsuren N, Edrada R, Lin W, Ebel R, Torre C, Ortlepp S, Wray V, Proksch P. 2007. Biologically active natural products from mongolian medicinal plants Scorzonera divaricata and Scorzonera pseudodivaricata. *J Nat Prod* 70:962–967.

Unno T, Tamemoto K, Yayabe F, Kakuda T. 2003. Urinary excretion of 5-(3, 4 -dihydroxyphenyl)-gamma-valerolactone, a ring-fission metabolite of (–)-epicatechin, in rats and its in vitro antioxidant activity. *J Agric Food Chem* 51:6893–6898.

Vaidyanathan JB, Walle T. 2003. Cellular uptake and efflux of the tea flavonoid (–)epicatechin-3-gallate in the human intestinal cell line Caco-2. *J Pharmacol Exp Ther* 307:745–752.

Voisin-Chiret AS, Bazin MA, Lancelot JC, Rault S. 2007. Synthesis of new L-ascorbic ferulic acid hybrids. *Molecules* 12:2533–2545.

Ward NC, Croft KD, Puddey IB, Hodgson JM. 2004. Supplementation with grape seed polyphenols results in increased urinary excretion of 3-hydroxyphenylpropionic acid, an important metabolite of proanthocyanidins in humans. *J Agric Food Chem* 52:5545–5549.

Winter J, Moore LH, Dowell VR, Bokkenheuser VD. 1989. C-ring cleavage of flavonoids by human intestinal bacteria. *Appl Environ Microbiol* 55:1203–1208.

Wittemer SM, Ploch M, Windeck T, Muller SC, Drewelow B, Derendorf H, Veit M. 2005. Bioavailability and pharmacokinetics of caffeoylquinic acids and flavonoids after oral administration of artichoke leaf extracts in humans. *Phytomedicine* 12:28–38.

Wolffram S, Weber T, Grenacher B, Scharrer E. 1995. A Na(+)-dependent mechanism is involved in mucosal uptake of cinnamic acid across the jejunal brush border in rats. *J Nutr* 125:1300–1308.

Yu J, Vasanthan T, Temelli F. 2001. Analysis of phenolic acids in barley by high-performance liquid chromatography. *J Agric Food Chem* 49:4352–4358.

Zhang JL, Zhang GD, Zhou TH. 2005. Metabolism of ferulic acid in rats. *J Asian Nat Prod Res* 7:49–58.

Zhao Z, Egashira Y, Sanada H. 2004. Ferulic acid is quickly absorbed from rat stomach as the free form and then conjugated mainly in liver. *J Nutr* 134:3083–3088.

Zhao Z, Egashira Y, Sanada H. 2003a. Digestion and absorption of ferulic acid sugar esters in rat gastrointestinal tract. *J Agric Food Chem* 51:5534–5539.

Zhao Z, Egashira Y, Sanada H. 2003b. Ferulic acid sugar esters are recovered in rat plasma and urine mainly as the sulfoglucuronide of ferulic acid. *J Nutr* 133:1355–1361.

Zhu QY, Holt RR, Lazarus SA, Ensunsa JL, Hammerstone JF, Schmitz HH, Keen CL. 2002. Stability of the flavan-3-ols epicatechin and catechin and related dimeric procyanidins derived from cocoa. *J Agric Food Chem* 50:1700–1705.

3 Biochemical Actions of Plant Phenolics Compounds: Thermodynamic and Kinetic Aspects

CESAR G. FRAGA[1,2], GULCIN SAGDICOGLU CELEP[2,3], AND MONICA GALLEANO[1]

[1]Physical Chemistry-PRALIB, School of Pharmacy and Biochemistry, University of Buenos Aires–CONICET, Buenos Aires, Argentina
[2]Department of Nutrition, University of California, Davis, California
[3]Food and Nutrition Technology, Gazi University, Ankara, Turkey

CONTENTS

Plant Phenolics and Human Health: Biochemistry, Nutrition, and Pharmacology,
Edited by Cesar G. Fraga.

INTRODUCTION

Polyphenols are secondary metabolites of plants and include a myriad of chemical structures, from simple molecules such as phenolic acids to highly polymerized compounds such as condensed tannins. Considering this heterogeneity, we have focused this chapter on one of the most abundant family of polyphenols in human diets, the flavonoids. Flavonoids are a chemically defined family of polyphenols that have a basic structure of two aromatic rings (A and B) linked through three carbons that usually form an oxygenated heterocycle (ring C) and differ in the arrangements of hydroxy, methoxy, and glycosidic groups and in the conjugation between the B- and C-rings [Butkovic et al., 2004].

The benefits of polyphenols on human health are often ascribed to their potential ability to act as antioxidants. Such ability would be possible considering that polyphenols have chemical structures supporting antioxidant chemical reactions (i.e., scavenging of free radicals and chelation of redox-active metals). Importantly, many polyphenols retain key antioxidant features in their chemical structure after ingestion and metabolism by mammals. In parallel, it has been reported that certain polyphenols can provide benefits in pathological situations associated with high free-radical production, for example, hypertension and cardiovascular disease [Bazzano et al., 2002; Hung et al., 2004; Rimm, 2002]. However, the physiological mechanisms linking the antioxidant chemical characteristics of flavonoids with their health effects are still a subject of discussion.

Based on chemical, thermodynamic, and kinetic data, this chapter will discuss the mechanisms proposed for polyphenol biological actions, especially antioxidant reactions. Although most of the points addressed in this chapter are valid for an important number of plant polyphenols, we will focus the analysis on flavonoids and, in particular, on flavanols.

POLYPHENOLS AS ANTIOXIDANTS

Polyphenols as Chain-Breaking Antioxidants

The inhibition of lipid (LH) oxidation may be considered as one of the most important chemical reaction mechanisms that could explain the antioxidant function of flavonoids. In general terms, chain-breaking antioxidants (AH) inhibit or retard lipid oxidation (reactions 1–7) by interfering with initiation [generically represented by reaction 1] or with chain propagating reactions (reactions 2 and 3) by readily donating hydrogen atoms to lipid peroxyl radicals (LOO•) or lipid radicals (L•) (reactions 4 and 5) [Frankel, 1998]:

$$LH \rightarrow L\bullet \quad (1)$$

$$L\bullet + O_2 \rightarrow LOO\bullet \quad (2)$$

$$LOO\bullet + LH \rightarrow ROOH + L\bullet \quad (3)$$

$$LOO\bullet + AH \rightleftarrows LOOH + A\bullet \quad (4)$$

$$L\bullet + AH \rightarrow LH + A\bullet \quad (5)$$

$$A\bullet + LOO\bullet \rightarrow \text{nonradical products} \quad (6)$$

$$A\bullet + A\bullet \rightarrow \text{nonradical products} \quad (7)$$

The analysis of the thermodynamic and kinetic characteristics of these reactions will allow a better understanding of the biological importance of chain breaking in the antioxidant actions of polyphenols.

Polyphenols as Chain-Breaking Antioxidants: Thermodynamics

Polyphenols could act as effective chain-breaking antioxidants (AH = POH) through the one-electron transfer reactions 4 and 5 if they produce a stable and relatively nonreactive "antioxidant radical" (A• = PO•) [Jovanovic et al., 1994]. Reactions 4 and 5 can be represented by the reaction 8, where L• or LOO• represents the oxidant free radical. This reaction can be decomposed in two half-reactions: one reduction (reaction 9) and one oxidation (reaction 10):

$$L\bullet \text{ or } LOO\bullet + POH \rightarrow LH \text{ or } LOOH + PO\bullet \quad (8)$$

$$L\bullet \text{ or } LOO\bullet + e^- + H^+ \rightarrow LH \text{ or } LOOH \quad (9)$$

$$POH \rightarrow PO\bullet + e^- + H^+ \quad (10)$$

Thermodynamics is used to predict whether reactants have a spontaneous tendency to change into products. This tendency is associated with a decrease in the free energy or Gibbs energy of the system (G) to a minimum. As a consequence, the thermodynamic criterion for spontaneous change at constant temperature and pressure is $\Delta G < 0$. Under standard conditions (concentrations = 1 M, and P = 1 atm), the standard Gibbs energy variation (ΔG°) is related with the equilibrium constant (K) by equation 11:

$$\Delta G^\circ = -RT \ln K \quad (11)$$

To explore the spontaneity of electron/hydrogen atom transfer reactions ΔG° can be calculated by the following expression:

$$\Delta G^{\circ} = -nF\,\Delta E^{\circ} \qquad (12)$$

where n is the number of electrons that are transferred from the reducing agent (antioxidant) to the oxidizing agent per mole of reaction events; F represents the Faraday constant (96485 C/mol of electrons); and ΔE° is the difference between the standard reduction potential of the involved couples (E°, for the reduction half-reaction and E°, for the oxidation half-reaction). Table 3.1 shows the $E^{\circ\prime}$, that is, E° under biological conditions (see footnote a in Table 3.1) of well-known antioxidant compounds and of select flavonoids. Table 3.1 also depicts the $\Delta G^{\circ\prime}$ obtained after applying equation 14 for the reactions of these flavonoids with HO•, LO•, and LOO•. The negative value of $\Delta G^{\circ\prime}$ observed for the processes indicate that these reactions are thermodynamically (energetically) favored, pointing out the feasibility of flavonoids as free-radical scavengers.

The possibility that some flavonoids can be prooxidant has also been suggested [Halliwell, 2008]. Such prooxidant activity is based on a sum of complex reactions involving the reaction between POH and O_2 to generate $O_2^{\bullet -}$, and the reaction of POH with $O_2^{\bullet -}$ yielding H_2O_2. The former process, that is, POH autoxidation, is thermodynamically unfavored ($E^{\circ\prime}\,O_2/O_2^{\bullet -} = -330$ mV [Wardman, 1989] or -160 mV [Sawyer and Valentine, 1981]), although it has been claimed to occur under certain conditions [Miura et al., 1998; Nakayama et al., 1995]. By contrast, the latter reaction is energetically favored ($E^{\circ\prime}\,O_2^{\bullet -}/H_2O_2 = 940$ mV [Buettner, 1993]), but being dependent on the presence of a product of the first reaction has low possibilities to occur in a significant extent.

A strict approximation should take into consideration that in vivo conditions are usually nonstandard conditions, and the actual concentrations of the reactants (oxidants and antioxidants) are also important. This last situation will be considered in the subsequent kinetic analysis.

Polyphenols as Chain-Breaking Antioxidants: Kinetics

Thermodynamics only identify whether a particular reaction mixture has a tendency to form products, but do not indicate whether that tendency will ever occur in a biologically appropriate time scale. To have a real idea if that reaction occurs, it would be necessary to know the rate of the chemical reactions; that is, making a kinetic analysis. For a given reaction 13, the reaction rate (v) is proportional to the molar concentration of the reactants (A, B) raised to a simple power (α, β). These values are called partial orders with respect to each of the species participating in the reaction. The rate constant (k) is essentially defined by the thermodynamic characteristics of the species under reaction and

Table 3.1 Biological Standard Reduction Potentials ($E^{\circ\prime}$) for a Series of Antioxidant Compounds and Flavonoids and Biological Standard Free Energy Variation ($\Delta G^{\circ\prime}$) for the Reactions between These Compounds and Oxygen Free Radicals

		$\Delta G^{\circ\prime b}$(kJ)		
Compound (Redox Couple)	$(E^{\circ\prime})^{a}$(mV)	HO•/H_2O $E^{\circ\prime} = 2310$ mV[cc]	LO•/LOH $E^{\circ\prime} = 1600$ mV[d]	LOO•/LOOH $E^{\circ\prime} = 1000$ mV[c]
Glutathione–cysteine (RS•/RS^-)	920[e]	−134	−66	−8
α-Tocopherol (α-TO•/α-TOH)	500[f]	−175	−106	−48
Trolox C (TO•/TOH)	480[f]	−177	−108	−50
Ascorbate (A•/AH^-)	282[g]	−196	−127	−69
Catechol (Cat-O•/Cat-OH)	530[h]	−172	−103	−45
(−)-Epigallocatechin (EGC•/EGC)	430[i]	−181	−113	−55
(−)-Epigallocatechin gallate (EGCG•/EGCG)	430[i]	−181	−113	−55
(−)-Epicatechin (EC•/EC)	570[i]	−168	−99	−41
(−)-Epicatechin gallate (ECG•/ECG)	550[i]	−170	−101	−43
Quercetin (Q•/QH)	330[j]	−191	−123	−65
Kaempferol (K•/KH)	750[j]	−151	−82	−24
Rutin (R•/RH)	600[j]	−165	−96	−39

[a] Several considerations must be made about biological standard reduction potentials ($E^{\circ\prime}$) values. IUPAC convention of writing the involved couples always in the sense of the reduction is widely used. With this convention the potentials are presented in the form $E^{\circ\prime}$(A/B) where **B** is the one-electron reduction product of A. The standard reduction potentials (E°) are expressed in relation to the normal or standard hydrogen electrode (NHE or SHE) that has assigned the value zero [Atkins and de Paula, 2006]. Standard conditions refer to 1 M concentrations for all dissolved species [Atkins and de Paula, 2006]. As a consequence, E°corresponds to the reaction under these condtions, that includes pH = 0 when H^+ is involved. Considering that reactions in biological systems, $E^{\circ\prime}$ (biological standard reduction potentials) is calculated for all dissolved species at 1 M concentration and pH = 7.

[b] $\Delta G^{\circ\prime}$ for the reaction can be obtained by using equation 12 or by calculating the $\Delta G^{\circ\prime}$ of the overall reaction as the sum of the Gibbs energies of the reactions into which it can be divided ($\Delta G^{\circ\prime} = \Delta G^{\circ\prime}{}_1 + \Delta G^{\circ\prime}{}_2$) (where 1 corresponds to the reduction half-reaction and 2 corresponds to the oxidation half-reaction). In general, we cannot combine the $E^{\circ\prime}$ values directly because they depend on the value of n, which may be different for the two couples. In this case both of the approximations are valid because the $E^{\circ\prime}$ considered correspond to one electron transferences.

[c] Buettner, [1993].

[d] Koppenol, [1990].

[e] [Surdhar and Armstrong, 1986; Surdhar and Armstrong, 1987].

[f] [Simic, 1990].

[g] [Williams and Yandell, 1982].

[h] [Scurlock et al., 1990]

[i] [Frei and Higdon, 2003].

[j] [Jovanovic and Simic, 2000].

is independent of the concentration of these species. The reaction rates are experimentally determined.

$$A + B \xrightarrow{k} \text{Products} \tag{13}$$

$$v = k[A]^{\alpha}[B]^{\beta} \tag{14}$$

Reactions can be classified on the basis of their order, which is the sum of the powers to which the concentrations of the participating species are raised in the rate law. If $\alpha = \beta = 1$ in equation 14, the reaction is first-order in A, first-order in B, and is globally of second order. Reactions 4 and 5 respond to this kind of second-order reaction rate law, and the k values have been established for the reactions between several oxidant species and antioxidants.

Table 3.2 shows the relative reaction rates of two well-known antioxidants (ascorbic acid and α-tocopherol) and of the flavanol (−)-epicatechin, with different radicals, that is, hydroxyl radical (HO•), alcoxyl radical (LO•), and peroxyl radical (LOO•). These calculated values provide a relevant limitation for the "free-radical scavenging" hypothesis. The relatively low plasma concentrations of flavanols, which occur even after the consumption of foods rich in these compounds [Holt et al., 2002], leads to a situation kinetically unfavorable with respect to that of other compounds with similar thermodynamic free-radical scavenging capabilities. These compounds, that is, α-tocopherol and ascorbate, are present in blood and/or in the intracellular milieu in micromolar and even millimolar ranges. The bioavailability of flavonoids is generally low; the actual concentrations that can be reached in plasma of humans subjected to realistic polyphenol consumption are in the nanomolar range. These concentrations are transient in nature, peaking at 2–4 h after consumption [Holt et al., 2002; Rein et al., 2000; Schroeter et al., 2006]. Even assuming the highest flavanol concentration found in plasma (low micromolar values), the relative physiological reaction rate with the more common radicals does not support the in vivo free-radical scavenging effects proposed for polyphenols. Nevertheless, this kinetic limitation implies the existence of other mechanisms, compatible with the tissue levels reached by polyphenols, that may explain the observed changes in cell or tissue oxidation [Fraga, 2007]. Taking into consideration these limitations, a direct effect should be possible only in tissues directly exposed to polyphenols after their consumption, but unlikely to occur in other tissues in which polyphenols are present at significantly lower (nanomolar) concentrations.

It is possible to further expand this scenario considering that, in cells, flavonoids (or other polyphenols) should be competing with other relevant antioxidants, as glutathione. Glutathione is able to function as both a direct antioxidant (scavenging free radicals) and an indirect antioxidant, being the reducing cofactor of glutathione peroxidase, one of the main enzymatic antioxidant defenses. Additionally, most of biological systems are able to synthesize reduced glutathione and

Table 3.2 Relative Reaction Rates for the Reactions between Oxygen Free Radicals and Selected Antioxidant Compounds and (−)-Epicatechin

		Relative reaction rate (s^{-1})		
	Compound concentration (M)	v/[HO•][a]	v/[RO•][b]	v/[ROO•][c]
Plasma	Ascorbate = $[50 \times 10^{-6}]$[d]	500,000	80,000	50
	α-Tocopherol = $[28 \times 10^{-6}]$[e]	280,000	2,800	28
	(−)-Epicatechin ≈ $[0.3 \times 10^{-6}]$[f]	2,000	40	2
Cells	Ascorbate = $[5 \times 10^{-3}]$[d]	50,000,000	8,000,000	5,000
	(−)-Epicatechin ≈ $[2 \times 10^{-6}]$[f]	13,000	270	15

[a] Relative reaction rates between HO• and ascorbate (AH^-) are based on equation 3: v = k [HO•] M × [AH^-] M assuming $k = 1 \times 10^{10}\ M^{-1}\ s^{-1}$ [Buettner and Jurkiewicz, 1996]. Idem for α-tocopherol assuming k as $1 \times 10^{10}\ M^{-1}\ s^{-1}$ [Niki, 1996]. Idem for (−)-epicatechin assuming $k_{(-)\text{-epicatechin}} \sim k_{(+)\text{-catechin}} = 66 \times 10^8\ M^{-1}\ s^{-1}$ [Bors et al., 1990].

[b] Relative reaction rates between RO• and AH^- are based on equation 3: v = k [RO•] M × [AH^-] M assuming $k = 1.6 \times 10^9\ M^{-1}\ s^{-1}$ [Buettner and Jurkiewicz, 1996]. Idem for α-tocopherol assuming $k = 1 \times 10^8\ M^{-1}\ s^{-1}$ [Niki, 1996]. Idem for (−)-epicatechin assuming $k_{(-)\text{-epicatechin}} \sim k_{(+)\text{-catechin}}$ as $1.35 \times 10^8\ M^{-1}\ s^{-1}$ [Bors et al., 1990].

[c] Relative reaction rates between ROO• and AH^- are based on equation 3: v = k [ROO•] M × [AH^-] M assuming $k = 1 \times 10^6\ M^{-1}\ s^{-1}$ [Buettner and Jurkiewicz, 1996]. Idem for α-tocopherol assuming $k = 1 \times 10^6\ M^{-1}\ s^{-1}$ [Niki, 1996]. Idem for (−)-epicatechin assuming $k = 7.3 \times 10^6\ M^{-1}\ s^{-1}$ [Scott et al., 1993]

[d] [Frei and Higdon, 2003];

[e] [Fraga et al., 2005];

[f] [Rein et al., 2000].

reduce oxidized glutathione allowing the maintenance of an adequate steady-state concentration of reduced glutathione. By contrast, polyphenols are not synthesized, and there is no known recycling in animals. Furthermore, polyphenol levels in the organism have been reported to be transient, requiring a permanent consumption of polyphenol-rich foods to maintain a steady-state concentration. As a consequence, cells would have 10^7–10^{10} molecules of glutathione and less than 10^3 molecules of flavonoids as a protection from the oxidative injury. These considerations are not always taken into account in in vitro studies, where the number of molecules per cell present to exert a function is manipulated at concentrations that exceed biologically relevant ranges. The described considerations open the possibility for ascribing the well-documented beneficial health effects of flavonoids to other mechanisms, antioxidant or not, beyond the free-radical scavenger and metal chelation activity.

Other Physicochemical Aspects Associated to the Chain-Breaking Antioxidant Effect: The Stability of the Polyphenol-Derived Free Radicals

The reactions between free radicals and flavonoids (or polyphenols) are assumed to form aroxyl radicals (PO•) (reaction 8). The stability of these secondary radical species is an important element to be considered in their antioxidant actions. Flavonoids with similar reduction potentials can originate radicals with very different reactivity toward other molecules present in biological systems. While a stable and relatively nonreactive PO• is also nonreactive to propagate the chain reaction, a high reactive PO• would propagate rather than interrupt a chain reaction.

To compare the stability of aroxyl radicals derived from different flavonoids, the rate constants for the disproportionation reaction between two PO• to render nonradical species (reaction 15) can be used:

$$2\text{PO}\bullet \rightarrow \text{POH} + \text{P}=\text{O} \tag{15}$$

The rate of reaction 15 can be described by the following equations:

$$v = -\frac{1}{2}d[\text{PO}\bullet]/dt = k_D[\text{PO}\bullet]^2 \tag{16}$$

$$-d[\text{PO}\bullet]/dt = 2k_D[\text{PO}\bullet]^2 \tag{17}$$

When a series of flavonoids were compared through the analysis of the second-order rate constants (k_D), it was concluded that (i) k_D values differ considerably, ranging from $10^5\ \text{M}^{-1}\ \text{s}^{-1}$ (the most stable) to $10^8\ \text{M}^{-1}\ \text{s}^{-1}$ (the less stable), and (ii) these data, together with the spectral characteristics of transient aroxyl radicals, were used to establish the structural requirements for radical stabilization (the presence of the catechol moiety, the 2,3-double bond in conjugation with a 4-oxo function, and the additional presence of 3- and 5-hydroxyl groups) [Bors et al., 1990].

POLYPHENOLS AS METAL CHELATORS

Polyphenols as Metal Chelators: Thermodynamics

Redox active metals (M) such as iron, manganese, and cooper can promote free radical formation by catalyzing otherwise slow reactions in biological settings: (i) catalyzing the decomposition of hydrogen peroxide (H_2O_2) into hydroxyl radicals (•OH) (reaction 18), and (ii) catalyzing the decomposition of other hydroperoxides, mostly lipid hydroperoxides yielding peroxyl and alcoxyl radicals (reactions 19 and 20):

$$H_2O_2 + \text{Ligand} - M^{(n)+} \rightarrow \bullet\, OH + -OH + \text{Ligand} - M^{(n+1)+} \qquad (18)$$

$$LOOH + \text{Ligand} - M^{(n+1)+} \rightarrow LOO\bullet\; + H^+ + \text{Ligand} - M^{(n)+} \qquad (19)$$

$$LOOH + \text{Ligand} - M^{n+} \rightarrow LO\bullet\; + OH^- + \text{Ligand} - M^{(n+1)+} \qquad (20)$$

As a consequence, metal chelation is an alternative antioxidant strategy based on sequestering metal ions to prevent metal-catalyzed free radical formation [Brown et al., 1998; Guo et al., 1996; Morel et al., 1998]. To have an antioxidant action by metal chelation, a compound should be: (i) able to complex the metal and yield a stable product, and (ii) the resultant complex should be a poor catalyst for the free radical generating reactions where the metal is involved.

Some polyphenols, and specially flavonoids, have a strong affinity for metal ions due to the presence and position of several hydroxyl groups: the 3′- 4′-hydroxyl groups in the catechol unit, together with the hydroxyl groups at positions 3 and 5 and the 4-oxo function, have been postulated as metal-complexing sites within a flavonoid molecule [van Acker et al., 1996]. In addition, the electronic properties of the ring system make flavonoids good ligands for the *d* electrons of the elements of the fourth period, for example, iron [Havsteen, 1983]. Because of its biological relevance, there are numerous in vitro studies using iron as the transition metal. By defining the chelation strength as the ability to release Fe^{2+} from the EDTA–Fe^{2+} complex, a wide range of chelating activities for flavonoids were reported, including those that act as efficient chelators, for example, apigenin, and those with nonchelating possibilities, for example, naringinin [van Acker et al., 1996].

The used experimental designs very often do not allow discriminating if the inhibitory effecs on metal-catalyzed free radical reactions are mediated by metal chelation or by the scavenging of the resulting free radical. In this regard, Arora et al. [1998] analyzed the effects of a series of flavonoids on lipid oxidation comparing a metal-initiated process versus a metal-independent process. Results showed that flavonoids were less effective inhibiting metal-independent than metal-initiated lipid oxidation, suggesting that metal chelation would play a role in determining the antioxidant activities of these compounds. The large differences observed in the antioxidant action based on the iron-chelating capacity of different flavonoids can be explained by flavonoid–metal complex thermodynamic aspects, for example, stabilities and redox properties [Thompson et al., 1976] that involved the flavonoid chemical structure.

The stability of metal complex is also given by the number of chelate rings formed in the resultant ligand–metal complex. For example, desferrioxamine, the most widely used iron chelator, minimizes •OH production by acting as a hexadentate ligand [Liu and Hider, 2002]. Unfortunately, there is not enough information on the denticity of polyphenols as metal chelators to assess the relevance of the stability of the flavonoid–metal complex formed.

From the point of view of the redox properties of the formed complex, Merkofer et al. [2006] have studied the Fenton reaction 18 catalyzed by different iron chelators and concluded that a Fe-complex can participate in cell redox cycling only when two conditions are fulfilled: (i) the oxidized complex (ligand–$M^{(n+1)+}$) can be reduced by a physiologically relevant compound [e.g., NAD(P)H] [Pierre et al., 2002], then the $E^{\circ\prime}{}_{(\text{Ligand}-M^{(n+1)+}/\text{Ligand}-M^{n+})}$ is higher than -320 mV; and (ii) the reduced complex (ligand–M^{n+}) can transfer an electron to H_2O_2, thus the $E^{\circ\prime}{}_{(\text{Ligand}-M^{(n+1)+}\text{Ligand}-M^{n+})}$ should be smaller than 320–390 mV [Buettner, 1993; Merkofer et al., 2006]. Then, complexes $E^{\circ\prime}$ in the -320 mV/320–390 mV range could participate in the Fenton reaction promoting free radical generation. Complexes that are out of such range could act as good chelators and avoid free radical production. As was explained in a previous section, a strict approximation should take into consideration that in vivo conditions are not standard conditions as those used for defining the range of potentials.

Polyphenols as Metal Chelators: Kinetics

The actual biological concentrations of the reactants (ligands, iron, complexes, H_2O_2) are also important and need to be considered when assessing the redox properties of the formed complex.

When assessing the in vivo relevance of metal chelation to explain the antioxidant capacity of polyphenols, another physiological fact that should be considered is the very modest amount of metals available to catalyze free radical formation. A basic kinetic analysis of an in vivo Fenton reaction mediated by Fe^{2+} reaction 18, in which M is iron], would include a second-order rate equation (equation 21) [Halliwell and Gutteridge, 1989]:

$$v = k[H_2O_2][\text{ligand} - Fe^{2+}] \qquad (21)$$

The iron involved in this process, usually called "redox active iron" or "labile iron pool," has been estimated to be ≤ 5 μM in the cell cytosol and ≤ 1 μM in human plasma [Woodmansee and Imlay, 2002]. Polyphenols should be able to diminish the rate of H_2O_2 decomposition (or •OH generation) by chelating a fraction of the "redox active iron" pool generating a flavonoid–iron complex with a very low rate constant (k) for Fenton reaction (reaction 18, equation 21). Nevertheless, considering that the physiological rate of •OH generation is indeed estimated to range between 2.5×10^{-10} and 1×10^{-9} $M\,s^{-1}$; and that the polyphenol concentrations in human and animal tissues are not enough to displace physiological metal chelators, the antioxidant actions of flavonoids could be only relevant in conditions in which "redox active iron", and/or other redox active metal, are in excess.

Note: The •OH generation rate was calculated considering [citrate–Fe^{2+}] = 5 μM; $k_{citrate}$ with $H_2O_2 = 4.9 \times 10^3 M^{-1} s^{-1}$ [Rush et al., 1990]; $[H_2O_2]$ steady state = 10–100 nM [Boveris et al., 1999; Merkofer et al., 2006]

POLYPHENOL INTERACTIONS WITH MEMBRANES

Flavonoids and other polyphenols can interact with lipids and proteins. The interactions with proteins could be both unspecific or specific, meanwhile the interactions with lipids seems to be rather unspecific, based essentially on physical adsorption. This physical adsorption would mostly depend on the hydrophobic/hydrophilic characteristics of the flavonoid molecule, the number of hydroxyl substituents, and the polymerization degree [Erlejman et al., 2004; Verstraeten et al., 2005, 2003, 2004].

Adsorption is the process of accumulation at an interface and should be distinguished from absorption, which implies the penetration of one component, for example, the polyphenol molecule, throughout the mass of a second, for example, membrane [Florence and Attwood, 1998]. Additionally, it is possible to distinguish two types of adsorption: (i) physical adsorption: the adsorbate is bound to the surface through weak van der Waals forces; and (ii) chemical adsorption: involves stronger interactions, usually covalent bonds [Florence and Attwood, 1998]. The theoretical study of this type of process has been originally developed for the adsorption of gases on solids, for example, Langmuir equation. Experimentally, this equation can be applied to flavonoid adsorption on membranes, which based on such processes, occurs as described by reaction 22. In this reaction, S represents the available sites on the membrane surface; and k_a and k_d are the adsorption and desorption rate constants, respectively:

$$\text{POH} + \text{S} \underset{k_d}{\overset{k_a}{\rightleftarrows}} \text{POH-S} \tag{22}$$

Adsorption rates can be described by (equation 23) and depend on [POH] and the number of available sites in the membrane. Being N the total number of surface active sites and θ the fraction of surface sites occupied by a single molecule of the adsorbent, the number of empty active sites will be $N\,(1 - \theta)$ (equation 24). Desorption rate is directly proportional to the number of adsorbed species ($N.\theta$) [(Eq. (3.8)]:

$$V_{\text{adsorption}} = k_{\text{a}}[\text{POH}]N(1\text{-}\theta) \tag{23}$$

$$V_{\text{desorption}} = -k_{\text{d}}N\theta \tag{24}$$

At equilibrium, both rates are equal and it is possible to obtain the Langmuir equation (equation 25), in which K is the equilibrium constant ($K = \text{k}_\text{a}/\text{k}_\text{d}$):

$$1 + K[\text{POH}]_{\text{eq}}\ \theta = K[\text{POH}]_{\text{eq}} \tag{25}$$

The use of this type of equation to evaluate the interaction between tea catechins and lipid bilayers of dimyristoylphosphatidylcholine revealed that the affinities of (−)-epicatechin gallate and (−)-epigallocatechin gallate for the bilayers are similar, and 1000 times stronger than the affinity of (-)-epicatechin or (−)-epigallocatechin. These results indicate that the presence of the galloyl groups should be responsible for defining the extent of the interaction [Kamihira et al., 2008].

Polyphenol–protein interactions have been epitomized as similar to antigen–antibody interactions in that a binding agent and a ligand associate through single or multiple moieties to form a complex. For polyphenol–protein interactions, the chemical characteristics of such "multivalent" associations are mainly related to: (i) the hydrophobicity of the aromatic nuclei of polyphenols and (ii) the availability of multiple phenolic hydroxyl groups that allow hydrogen bonding [Hagerman and Butler, 1981]. Recently, it has been suggested that the inhibitory activity of flavonoids toward human α-amilase depends on the formation of hydrogen bonds between the OH groups in R7 and/or R4′ of the flavonoid ligand, and the side chains of Asp^{197} and Glu^{233}, and a possible conjugated π-system between either the AC or the B ring system and Trp^{59} [Lo Piparo et al., 2008]. By contrast, ligands without inhibitory activity (flavonols, flavanones, and isoflavones) do not interact at the same time with the catalytic residues as well as with Trp^{59}. Flavanols and flavanones posses a C2–C3 single bond that prevents electron delocalization over the AC condensate ring system, and the isoflavone genistein may form a π-bond to Trp^{59} but is unable to form a hydrogen bond to the catalytic residues because of the different position of the aromatic B-ring in the ligand structure. Proline-rich proteins are preferential targets for polyphenol interactions, that is, the precipitation of proline-rich proteins present in saliva by the tannins (polymers of flavanols) from wine, chocolate, tea, and other flavanol-containing foods [Luck et al., 1994]. It has also been reported that a significant number of enzyme activities are inhibited by polyphenols, flavonoids, and flavanols [Middleton et al., 2000].

CONCLUSIONS

The capacity of flavonoids, as well as other plant polyphenols, to scavenge free radicals and/or chelate metals has been often claimed as responsible for the antioxidant actions of these phytochemicals. This action is assumed based on some thermodynamic properties of polyphenols to react with free radicals and metals. Nevertheless, the physiological concentrations reached by polyphenols in animals and human tissues are incompatible with the kinetic requirements necessary to reach reaction rates of physiological relevance. Other biochemical mechanisms can explain the observed in vitro and in vivo antioxidant effects. These mechanisms are related to polyphenol–lipid and polyphenol–protein interactions and are based on the presence of OH groups that give polyphenols high possibilities for physical and chemical interactions. These mechanisms do not preclude direct free radical scavenging or metal chelating actions but are more consistent with the polyphenol levels observed in human and animal tissues. Furthermore, the actual occurrence of these potential actions in vivo should be considered when linking an antioxidant action to a polyphenol, or to any other molecule, whose metabolic fate and biological actions are not firmly established.

ACKNOWLEDGMENTS

This work was supported by NIH AT2966; UC Davis CHNR (State of California Vitamin Price Fixing Consumer Settlement fund); and UBACyT (B801 and B802). MG and CGF are members of the CIC, CONICET, Argentina. GSC has a postdoctoral fellowship from TUBITAK, Turkey.

REFERENCES

Arora A, Nair MG, Strasburg GM. 1998. Structure-activity relationships for antioxidant activities of a series of flavonoids in a liposomal system. *Free Radic Biol Med* 24:1355–1363.

Atkins PW, de Paula J. 2006. *Physical Chemistry for the Life Sciences*. New York: W.H. Freeman.

Bazzano LA, He J, Ogden LG, Loria CM, Vupputuri S, Myers L, Whelton PK. 2002. Fruit and vegetable intake and risk of cardiovascular disease in US adults: The first National Health and Nutrition Examination Survey Epidemiologic Follow-up Study. *Am J Clin Nutr* 76:93–99.

Bors W, Heller W, Michel C, Saran M. 1990. Flavonoids as antioxidants: Determination of radical-scavenging efficiencies. *Methods Enzymol* 186:343–355.

Boveris A, Costa LE, Cadenas E. 1999. The mitochondrial production of oxygen radicals and cellular aging. In: Cadenas E, Packer L, Eds. *Understanding the Process of Aging—The Role of Mitochondria, Free Radicals and Antioxidants*. New York: Marcel Dekker, Inc pp. 1–16.

Brown JE, Khodr H, Hider RC, Rice-Evans CA. 1998. Structural dependence of flavonoid interactions with Cu^{2+} ions: Implications for their antioxidant properties. *Biochem J* 330:1173–1178.

Buettner GR. 1993. The pecking order of free radicals and antioxidants: Lipid peroxidation, alpha-tocopherol, and ascorbate. *Arch Biochem Biophys* 300:535–543.

Buettner GR, Jurkiewicz BA. 1996. Chemistry and biochemistry of ascorbic Acid. In: Cadenas E, Packer L, Eds. *Handbook of Antioxidants*. New York: Marcel Dekker, pp. 91–115.

Butkovic V, Klasinc L, Bors W. 2004. Kinetic study of flavonoid reactions with stable radicals. *J Agric Food Chem* 52:2816–2820.

Erlejman AG, Verstraeten SV, Fraga CG, Oteiza PI. 2004. The interaction of flavonoids with membranes: Potential determinant of flavonoid antioxidant effects. *Free Radic Res* 38:1311–1320.

Florence AT, Attwood D. 1998. *Physicochemical Principles of Pharmacy*, 3rd ed. London: MacMillan Press.

Fraga CG. 2007. Plant polyphenols: How to translate their in vitro antioxidant actions to in vivo conditions. *IUBMB Life* 59:308–315.

Fraga CG, Actis-Goretta L, Ottaviani JI, Carrasquedo F, Lotito SB, Lazarus S, Schmitz HH, Keen CL. 2005. Regular consumption of a flavanol-rich chocolate can improve oxidant stress in young soccer players. *Clin Dev Immunol* 12:11–17.

Frei B, Higdon JV. 2003. Antioxidant activity of tea polyphenols in vivo: Evidence from animal studies. *J Nutr* 133:3275S–3284S.

Guo Q, Zhao B, Li M, Shen S, Xin W. 1996. Studies on protective mechanisms of four components of green tea polyphenols against lipid peroxidation in synaptosomes. *Biochim Biophys Acta* 1304:210–222.

Hagerman AE, Butler LG. 1981. The specificity of proanthocyanidin-protein interactions. *J Biol Chem* 256:4494–4497.

Halliwell B. 2008. Are polyphenols antioxidants or pro-oxidants? What do we learn from cell culture and in vivo studies? *Arch Biochem Biophys* 476:107–112.

Halliwell B, Gutteridge JMC. 1989. *Free Radicals in Biology and Medicine*, 2nd ed. New York: Oxford University Press.

Havsteen B. 1983. Flavonoids, a class of natural products of high pharmacological potency. *Biochem Pharmacol* 32:1141–1148.

Holt RR, Lazarus SA, Sullards MC, Zhu QY, Schramm DD, Hammerstone JF, Fraga CG, Schmitz HH, Keen CL. 2002. Procyanidin dimer B2 [epicatechin-(4beta-8)-epicatechin] in human plasma after the consumption of a flavanol-rich cocoa. *Am J Clin Nutr* 76:798–804.

Hung HC, Joshipura KJ, Jiang R, Hu FB, Hunter D, Smith-Warner SA, Colditz GA, Rosner B, Spiegelman D, Willett WC. 2004. Fruit and vegetable intake and risk of major chronic disease. *J Natl Cancer Inst* 96:1577–1584.

Jovanovic SV, Simic MG. 2000. Antioxidants in nutrition. *Ann N Y Acad Sci* 899:326–334.

Jovanovic SV, Steenken S, Tosic M, Marjanovic B, Simic MG. 1994. Flavonoids as antioxidants. *J Am Chem Soc* 116:4846–4851.

Kamihira M, Nakazawa H, Kira A, Mizutani Y, Nakamura M, Nakayama T. 2008. Interaction of tea catechins with lipid bilayers investigated by a quartz-crystal microbalance analysis. *Biosci Biotechnol Biochem* 72:1372–1375.

Liu ZD, Hider RC. 2002. Design of iron chelators with therapeutic application. *Coord Chem Rev* 232:151–171.

Lo Piparo E, Scheib H, Frei N, Williamson G, Grigorov M, Chou CJ. 2008. Flavonoids for controlling starch digestion: Structural requirements for inhibiting human alpha-amylase. *J Med Chem* 51:3555–3561.

Luck G, Liao H, Murray NJ, Grimmer HR, Warminski EE, Williamson MP, Lilley TH, Haslam E. 1994. Polyphenols, astringency and proline-rich proteins. *Phytochemistry* 37:357–371.

Merkofer M, Kissner R, Hider RC, Brunk UT, Koppenol WH. 2006. Fenton chemistry and iron chelation under physiologically relevant conditions: Electrochemistry and kinetics. *Chem Res Toxicol* 19:1263–1269.

Middleton E, Jr., Kandaswami C, Theoharides TC. 2000. The effects of plant flavonoids on mammalian cells: Implications for inflammation, heart disease, and cancer. *Pharmacol Rev* 52:673–751.

Miura YH, Tomita I, Watanabe T, Hirayama T, Fukui S. 1998. Active oxygens generation by flavonoids. *Biol Pharm Bull* 21:93–96.

Morel I, Cillard P, Cillard J. 1998. Flavonoid-metal interactions in biological systems. In: Rice-Evans CA, Packer L, Eds. *Flavonoids in Health and Disease*. New York: Marcel Dekker, pp. 163–197.

Nakayama T, Enoki I, Hashimoto K. 1995. Hydrogen peroxide formation during catechin oxidation is inhibited by superoxide dismutase. *Food Sci Technol Int* 1:65–69.

Niki E. 1996. alpha-Tocopherol. In: Cadenas E, Packer L, Eds. *Handbook of Antioxidants*. New York: Marcel Dekker, pp. 3–25.

Pierre JL, Fontecave M, Crichton RR. 2002. Chemistry for an essential biological process: the reduction of ferric iron. *BioMetals* 15:341–346.

Rein D, Lotito S, Holt RR, Keen CL, Schmitz HH, Fraga CG. 2000. Epicatechin in human plasma: In vivo determination and effect of chocolate consumption on plasma oxidation status. *J Nutr* 130:2109S–2114S.

Rimm EB. 2002. Fruit and vegetables—building a solid foundation. *Am J Clin Nutr* 76:1–2.

Rush JD, Maskos Z, Koppenol WH. 1990. Reactions of iron(II) nucleotide complexes with hydrogen peroxide. *FEBS Lett* 261:121–123.

Sawyer DT, Valentine JS. 1981. How super is superoxide? *Acc Chem Res* 14:393–400.

Schroeter H, Heiss C, Balzer J, Kleinbongard P, Keen CL, Hollenberg NK, Sies H, Kwik-Uribe C, Schmitz HH, Kelm M. 2006. (−)-Epicatechin mediates beneficial effects of flavanol-rich cocoa on vascular function in humans. *Proc Natl Acad Sci USA* 103:1024–1029.

Scott BC, Butler J, Halliwell B, Aruoma OI. 1993. Evaluation of the antioxidant actions of ferulic acid and catechins. *Free Radic Res Commun* 19:241–253.

Scurlock R, Rougee M, Bensasson RV. 1990. Redox properties of phenols, their relationships to singlet oxygen quenching and to their inhibitory effects on benzo(a)-pyrene-induced neoplasia. *Free Radic Res Commun* 8:251–258.

Simic MG. 1990. Pulse radiolysis in study of oxygen radicals. *Methods Enzymol* 186:89–100.

Surdhar PS, Armstrong DA. 1987. Reduction potentials and exchange reactions of thiyl radicals and disulfide anion radicals. *J Phys Chem* 91:6532–6537.

Surdhar PS, Armstrong DA. 1986. Redox potentials of some sulfur-containing radical. *J Phys Chem* 90:5915–5917.

Thompson M, Williams CR, Elliot GE. 1976. Stability of flavonoid complexes of copper(II) and flavonoid antioxidant activity. *Anal Chim Acta* 85:375–381.

van Acker SA, van den Berg DJ, Tromp MN, Griffioen DH, van Bennekom WP, van der Vijgh WJ, Bast A. 1996. Structural aspects of antioxidant activity of flavonoids. *Free Radic Biol Med* 20:331–342.

Verstraeten SV. Hammmerstone JF. Keen CL, Fraga CG, Oteiza PL. 2005. Antioxidant and membrane effects of procyabidin dimers and trmeis isolated from peanut and Cocoa. *J Agric Food Chem* 53:5041–5048.

Verstraeten SV, Oteiza PI, Fraga CG. 2004. Membrane effects of cocoa procyanidins in liposomes and Jurkat T cells. *Biol Res* 37:293–300.

Verstraeten SV, Keen CL, Schmitz HH, Fraga CG, Oteiza PI. 2003. Flavan-3-ols and procyanidins protect liposomes against lipid oxidation and disruption of the bilayer structure. *Free Radic Biol Med* 34:84–92.

Wardman P. 1989. Reduction potentials of one-electron couples involving free radicals in aqueous solution. *J Phys Chem Ref Data* 118:1637–1755.

Williams NH, Yandell JK. 1982. Outer-sphere electron-transfer reactions of ascorbate anions. *Aust J Chem* 35:1133–1144.

Woodmansee AN, Imlay JA. 2002. Quantitation of intracellular free iron by electron paramagnetic resonance spectroscopy. *Methods Enzymol* 349:3–9.

4 Flavonoid–Membrane Interactions: Consequences for Biological Actions

SANDRA V. VERSTRAETEN[1], CESAR G. FRAGA[2,3], AND PATRICIA I. OTEIZA[3,4]

[1]Department of Biological Chemistry, IIMHNO (UBA) and IQUIFIB, University of Buenos Aires–CONICET, Buenos Aires, Argentina
[2]Physical-Chemistry-PRALIB, School of Pharmacy and Biochemistry, University of Buenos Aires–CONICET, Buenos Aires, Argentina
[3]Department of Nutrition, University of California, Davis, California
[4]Department of Environmental Toxicology, University of California, Davis, California

CONTENTS

Plant Phenolics and Human Health: Biochemistry, Nutrition, and Pharmacology, Edited by Cesar G. Fraga.

INTRODUCTION

Flavonoids constitute one of the most common forms of phenolic compounds present in plants. Multiple health benefits have been proposed for flavonoids, although the biochemical and physiological mechanisms are not fully defined. Among others, the existence of a large number of flavonoids bearing different chemical substituent and the limited knowledge of their tissue distribution and concentration are factors that complicate the understanding of their biological effects.

The interactions of flavonoids with membranes emerge as an important mechanism involved in their biological effects. Importantly, certain flavonoids that, due to their chemical properties, cannot be absorbed by cells could still exert certain biological actions at the membrane level. Flavonoids can interact with different membrane components. As a consequence of a direct interaction with membrane lipids, flavonoids can alter membrane physical properties, modulating enzyme activity, ligand–receptor interactions, ion fluxes, signal transduction, transport, and other membrane-associated functions. These membrane–flavonoid interactions can affect the occurrence and extent of deleterious events, for example, lipid oxidation and membrane disruption. Flavonoids can also interact with membrane proteins and modify their biological function. While the interactions of flavonoids with proteins may have a specific requirement of flavonoid structure and conformation, the interactions with membrane lipids could be in general driven by the hydrophobic/hydrophilic character of the molecule, the number of hydroxyl substituent, and the degree of polymerization. This chapter will address current knowledge on the interactions of flavonoids with membrane lipids and proteins, the nature of these interactions, and their biological consequences.

INTERACTIONS OF FLAVONOIDS WITH MEMBRANE LIPIDS

Flavonoids bear different degrees of hydroxylation, polymerization, and methylation that define both specific and nonspecific interactions with membrane lipids. Molecule size, tridimensional structure, and hydrophilicity/hydrophobicity are chemical parameters that determine the nature and extent of flavonoid interactions with lipid bilayers. The hydrophilic character of certain flavonoids and their oligomers endows these molecules with the ability to bind to the polar headgroups of lipids localized at the water–lipid interface of membranes. On the other hand, flavonoids with hydrophobic character can reach and cross the lipid bilayer. In this section, we will discuss current experimental evidences on the consequences of flavonoid interactions with both the surface and the hydrophobic core of the lipid bilayer.

Interaction of Flavonoids and Procyanidins at Membrane Water–Lipid Interfaces

Flavonoids are a family of compounds that present subtle variations in the number and location of hydroxyl groups (Crozier et al., this volume). Considering that most flavonoids have $pK_a > 8$ [Jovanovic et al., 1994], these compounds will mainly make contact with polar surfaces through hydrogen bonds rather than ionic interactions [Singleton, 1987]. This is especially relevant in biological environments, where flavonoids could interact with the polar headgroup of phospholipids, e.g. phosphatidylcholine (PC) and adsorb to the cell surface. Using dipalmitoyl-PC vesicles as membrane models, Yoshioka et al. [2006] demonstrated that (+)-catechin, which is a highly hydrophilic molecule, adsorbs to the vesicle surface. As a result, (+)-catechin prevented membrane defects as it evolved from gel to liquid-crystalline phase transition and increased lipids transition temperature [Yoshioka et al., 2006]. The adsorption of catechin to dipalmitoyl-PC liposomes extended to the membrane core, decreasing the mobility of dipalmitoyl-PC acyl chains [Yoshioka et al., 2006]. It is noteworthy that not only the number of hydroxyl moieties but their distribution is determinant of the degree of flavonoid adsorption to membranes. Supporting this, it has been demonstrated that (−)-epicatechin, which is slightly more hydrophobic than its geometrical isomer (+)-catechin, has higher affinity for membrane lipids [Tsuchiya, 2001]. Additionally, the oligomerization degree of procyanidins built on catechin and epicatechin monomers (dimers A1, A2, B2, and trimers A, and C2) is also important in determining their adsorption to membranes, as the number of flavonoid–membrane contacts increases with the number of monomers present in these molecules [Verstraeten et al., 2005]. Similarly, the adsorption of procyanidins to PC liposomes affected the fluidity of the bilayer not only when evaluated with a probe located at the membrane water–lipid interface, but also in the hydrophobic region of the bilayer [Verstraeten et al., 2005]. Together, these findings demonstrate that upon their interaction with lipid polar headgroups, both, flavanol monomers and their oligomers cause the tightening of the liposome surface, which leads to the immobilization of phospholipids acyl chains and to the loss of membrane fluidity.

The interactions of flavonoids with polar surfaces can also be predicted using simpler membranes. Using a phosphatidyl choline-coated high-pressure liquid chromatography (HPLC) column, it has been observed that the surface interaction (measured as the retention delay) of eight different flavonoids with phospholipids was positively correlated with the number of hydroxyl groups in the flavonoids [Ollila et al., 2002]. To investigate the nature of the interactions between certain flavonoids and polar surfaces, Cartalade and Vernhet [2006] used three artificial membranes with a different degree of hydrophobicity. Interestingly, both, (+)-catechin and (−)-epicatechin mainly adsorbed to polar membranes, while (−)-epigallocatechin gallate (EGCG) adsorbed to both, polar and nonpolar membranes [Cartalade and Vernhet, 2006]. These flavanols were easily desorbed by dilution, indicating the reversibility of their adsorption. Larger procyanidins, with an oligomerization degree that ranged from 3 to 10, adsorbed to all the studied membranes, regardless of their polar character. In addition, the desorption of these oligomers showed large hysteresis, indicating that the strength of their interactions with membranes was higher than that of the monomeric units [Cartalade and Vernhet, 2006]. By contrast, hexameric procyanidins presented a reversible adsorption to Caco-2 cell monolayers [Erlejman et al., 2006]. Similar findings in Caco-2 cells were observed for high-molecular-weight procyanidins isolated from willow shoots [Deprez et al., 2001]. Large amounts of hydroxyl groups in small flavonoid molecules can prevent their adsorption by causing a large increase in their hydrophylicity. For example, two highly hydroxylated flavonoids, myricetin and rutin, did not interact with a probe located in the polar region of PC liposomes-indicating the lack of penetration into the bilayer even at the most superficial membrane level, the polar region [Ratty et al., 1988]. Accordingly, Kayika et al. [2001] reported that green tea flavanols, which can be adsorbed to the membrane surface, interact with probes shallowly located at the water–lipid interface whereas no interaction was found with probes deeply buried into the bilayer.

The relevance of hydroxyl groups in flavonoid binding was also studied by Yano et al. [2007], who investigated how methylation at different positions of the gallate ring in EGCG affects its binding to biological membranes. Methylation at 3′- and/or 4′-positions of EGCG markedly reduced its affinity for membranes, which was abolished when EGCG was simultaneously methylated at the 4′- and 4″-positions. These findings stress the importance of flavonoid hydroxylation degree and of hydroxyl group location to define flavonoid–membrane interactions [Yano et al., 2007]. Using nuclear magnetic resonance (NMR) in both, solid-state and in bicelle systems, it was recently demonstrated that the gallate moiety of EGCG inserts into lipid bilayers and interacts with the choline moiety of the PC [Uekusa et al., 2007; Kajiya et al., 2008]. Although methylation decreases the affinity of the flavonoids for lipid polar headgroups, it does not prelude the flavonoid capacity to interact with membrane proteins.

Another flavonoid characteristic that will determine the extent of flavonoid adsorption to the bilayer is their tridimensional structure. Van Dijk et al. [2000] evaluated the affinity of several flavonoids for liposomes, measuring their capacity to quench the fluorescence of a probe inserted in the lipid bilayer. The planar structure adopted by flavanols morin and quercetin confers to these molecules a markedly higher affinity for membranes than that of the flavanones pinocembrin, naringenin, eriodictyol, and hesperetin which adopt a tilted configuration. Accordingly, glycosylation at position 7 of naringenin and eriodictyol significantly increased their affinity for membranes when compared with their respective aglycons, effect that was associated with a molecule torsion due to glycosylation [van Dijk et al., 2000].

As a consequence of flavonoid–lipid interactions, flavonoids affect several physical properties of the bilayer. Studies from our laboratory demonstrated that procyanidins isolated from cocoa adsorb to the surface of liposomes, preventing the disruption of the bilayer caused by the incorporation of Triton X-100 [Verstraeten et al., 2003]. Interestingly, the effect of oligomerization was not the additive for the number of monomer units but synergistic [Verstraeten et al., 2003]. These observations were extended to the plasma membranes of Caco-2 cells, in which the incubation with micromolar concentrations of procyanidins (dimer to hexamer) prevented bile-induced cell damage [Erlejman et al., 2006]. Both in liposomes and in Caco-2 cells the protective effects of procyanidins on detergent-mediated membrane disruption depended on the oligomerization degree, being more effective as the number of monomers increased [Verstraeten et al., 2003; Erlejman et al., 2006]. On the contrary, in giant unilamellar vesicles EGCG increases per se membrane permeability and promoted the interaction between vesicles (aggregation) and structural changes in vesicles topology, generating small lumps [Tamba et al., 2007]. As previously discussed, this effect could be due to the fact that EGCG is a relatively small molecule that could partially penetrate the bilayer and induce the formation of membrane pores. Through these nonspecific pores, molecules present in the inner liposomes aqueous space could reach the external milieu. In this regard, a strong dependence between flavonoid–mediated release of the water-soluble probe calcein and flavonoid hydrophobicity was reported by Olilla et al. [2002]. Thus, flavonoids could exert different effects on membrane permeability strictly based on their hydrophobicity: while the adsorption of highly hydrophilic molecules onto the membrane surface has a protective effect on the permeabilization induced by external agents, more hydrophobic flavonoids can incorporate into the bilayer and increase permeability by favoring the creation of unspecific pores.

Isoflavones obtained from different *Leguminosae* species were demonstrated to promote dipalmitoyl-PC liposomes aggregation [Hendrich et al., 2002]. Similarly, purified daidzein promoted liposome aggregation, with a magnitude that relied on the chemical nature of the phospholipid polar headgroup and that followed the order: phosphatidylinositol > phosphatidylserine (PS) > phosphatidylglycerol > phosphatidic acid > PC [Lehtonen et al., 1996]. Huh et al. [1996] proposed that flavonoids could generate polyphenol bridges between adjacent lipid surfaces and, by decreasing surface hydration, reduce the repulsion among them, leading to membrane aggregation. The ability of EGCG to promote membrane aggregation was postulated to have beneficial health effects by inducing Gram-positive bacteria death [Ikigai et al., 1993]. This bactericidal effect was ascribed to EGCG interactions with negatively charged lipopolysacarides, whose absence in Gram-negative bacteria results in the resistance against the action of EGCG. By contrast, (−)-epicatechin, which has a poor capability to induce membrane aggregation, also has low bactericidal action [Ikigai et al., 1993]. However, the bactericidal properties of flavonoids should be carefully considered. Recently, the flavonol galangin, which has been proposed as a potent agent against *Staphylococcus aureus, Enterococcus* spp., and *Pseudomonas aeruginosa* [Pepeljnjak and Kosalec,

2004], was shown to cause bacteria clumping but not to decrease cell viability [Cushnie et al., 2007].

Another membrane property physically affected by flavonoids is the surface potential. Working with liposomes composed of PC and PS, we found that (−)-epicatechin and certain procyanidins (dimer to hexamer) decreased liposome surface potential. This effect relied on both, procyanidin concentration and number of monomeric units [Verstraeten et al., 2003]. On the other hand, when liposomes were composed exclusively of PC, it was found that (−)-epicatechin dimers A1 and B2, and the trimers A and C2 increased liposome surface potential in a concentration-dependent manner [Verstraeten et al., 2005]. In PC liposomes, the distance between two neighbor polar headgroups would allow a partial penetration of the oligomers into the polar region of the bilayer. Meanwhile, in PC:PS liposomes, the membrane is tightly packed due to ionic interactions between PC and PS headgroups, making the interactions with flavonoids shallower than in PC vesicles. As a consequence, differences in flavonoid penetration into the bilayer could lead to differences in their effects on membrane surface potential.

Partitioning of Flavonoids in the Hydrophobic Portion of Membranes

As previously mentioned, certain flavonoids can penetrate into the hydrophobic core of membranes, a feature that mainly relies on their hydrophobic character, which is dictated by flavanoid chemical structure and spatial conformation. When flavonoid hydrophobicity was assessed from the partition coefficient between *n*-octanol and an aqueous solution, the following order of hydrophobicity was observed: flavone, genistein > eriodictyol, myricetin, quercetin, kaempferol, hesperetin, daidzein >> galangin, morin, flavanone, naringenin, taxifolin (Table 4.1).

Interestingly, the partition of flavanols into *n*-octanol was almost negligible, with the exception of (−)-epicatechin gallate [Erlejman et al., 2004]. Using the partition system water:olive oil, van Dijk et al. [2000] found that flavonoids followed the hydrophobicity order: flavones > flavanones > flavanols. The authors concluded that at the same hydroxylation degree, flavones are more hydrophobic than flavanones, an effect that could be related to differences in their tridimensional structure. Accordingly, they observed that the capacity of flavonoids to affect transmembrane potential and to dissipate pH between inner and outer liposome aqueous spaces is restricted to the more hydrophobic flavones and flavanones, while the flavonols quercetin and morin do not affect these parameters [van Dijk et al., 2000]. Working with palmitoly-oleoyl-PC vesicles, Scheidt et al. [2004] demonstrated, by high-resolution magic angle spinning NMR spectroscopy, that flavonoids show a broad bilayer distribution determined by their hydrophobicity (flavone, chrysin > luteolin, myricetin). Furthermore, flavonoids deeply buried into the bilayer have similar motilities to that of fatty acids, by contrast more polar flavonoids are bound at the water–lipid interface having motilities similar to water molecules [Scheidt et al., 2004].

Table 4.1 Partition Coefficients of Selected Flavonoids[a]

Family	Compound	Partition Coefficient
Flavonols	Flavone	108.55 ± 7.07
	Flavonol	0.70 ± 0.21
	Galangin	4.62 ± 4.65
	Kaempferol	23.65 ± 2.04
	Morin	4.59 ± 0.14
	Quercetin	20.65 ± 2.84
	Myricetin	39.34 ± 1.94
Flavanols	(+)-Catechin	1.80 ± 0.40
	(−)-Epicatechin	1.53 ± 0.03
	(−)-Epigallocatechin	0.14 ± 0.01
	(−)-Epicatechin gallate	13.47 ± 0.13
	(−)-Epigallocatechin gallate	5.22 ± 0.49
	Dimer	0.42 ± 0.03
	Hexamer	0.01 ± 0.001
Flavanones	Flavanone	7.87 ± 0.66
	Naringenin	6.70 ± 0.30
	Eriodictyol	61.94 ± 6.27
	Hesperetin	20.60 ± 1.53
	Taxifolin	5.67 ± 0.05
Isoflavones	Daidzein	20.81 ± 0.20
	Genistein	109.25 ± 7.79

[a] Flavonoid partition coefficients were evaluated in the solvents mixture *n*-octanol and 20 mM Tris-HCl (pH 7.4), 140 mM NaCl buffer.

Source: Data taken from Erlejman et al. [2004].

Additionally, the authors did not observe flavonoid-induced alterations in membrane structures, indicating that bilayers can easily accommodate a large number of flavonoid molecules.

Similar to hydrophilic flavonoids, hydrophobic flavonoids can affect membrane permeability. Alterations in this biophysical property of liposome bilayers lead to the release of bulky molecules entrapped into the inner aqueous space. As mentioned in the previous section, a strong correlation was found between flavonoid retention to a hydrophobic matrix and their capacity to induce membrane leakage [Ollila et al., 2002]. Interestingly, hydrophilic flavonoids, such as (−)-epicatechin and related procyanidins (dimer to hexamer) prevented Fe^{2+}-mediated liposome permeabilization, although in this case the beneficial effect could be related to both their antioxidant and metal chelating capacities and their membrane stabilizing properties [Verstraeten et al., 2004].

In addition, some flavonoids can differentially interact with membrane polar surfaces or penetrate into the bilayer, depending on certain characteristics of the reaction milieu. This is the case of quercetin, a flavonoid that at acidic pH is deeply embedded into planar lipid bilayers [Movileanu et al., 2000], while at physiological pH it mainly interacts with lipid polar headgroups [Terao et al., 1994; Movileanu et al., 2000; Pawlikowska-Pawlega et al., 2003]. At low concentrations (10 μM), the isoflavone puerarin locates at the inner region of cetyl-trimethyl-ammonium bromide micelles, while it displaces toward the outer region of these micelles as its concentration increases, effect that is associated with puerarin deprotonation [Xi and Guo, 2007].

As previously mentioned, flavonoids can modulate membrane fluidity, with polar flavonoids binding to the membrane surface and propagating their effects into the hydrophobic region of the bilayer. Also, and using a set of fluorescent probes that sense lipid packing at different depths of the bilayer, it has been demonstrated that upon their incorporation, hydrophobic flavonoids, such as naringenin, rutin, and a series of isoflavones, reduce membrane fluidity especially in its deepest region [Arora et al., 2000]. The isoflavones daidzein and genistein have differential effects on membrane fluidity. Daidzein increases the fluidity of membranes when in gel phase or close to their transition temperature and do not affect membranes in the liquid-crystalline phase [Lania-Pietrzak et al., 2005]. On the contrary, genistein, which only has one hydroxyl group more than daidzein, decreases membrane fluidity in membranes in liquid-crystalline phase without affecting those in gel phase or close to their transition temperature [Lania-Pietrzak et al., 2005]. Other hydrophobic flavanoids, such as hesperetin and naringenin, which incorporate into the bilayer, also decreased dipalmitoyl-PC liposomes transition temperature [Saija et al., 1995].

In summary, current evidence supports the interaction of flavonoids with membrane lipids. The associated perturbations in membrane physical properties can have a significant impact on membrane-associated processes.

Interactions of Flavonoids with Membrane Proteins

Flavonoids can also interact with membrane proteins, such as those functioning as receptors, transporters, channels, and enzymes, and potentially affect their biological activities. A summary of recent advances on the study of flavonoid interactions with plasma membrane proteins is presented in Table 4.2.

Among plasma membrane transporters, the ABC [adenosine 5–triphosphate (ATP)-binding cassette] superfamily is especially important due to its involvement in the transmembrane movement of a myriad of organic and inorganic compounds, participating in the absorption of nutrients and the efflux of intracellular metabolites (for a review about ABC protein classifications and functions, see Pohl et al. [2005]). Several cancer cells present multidrug resistance through the overexpression of specific ABC proteins that extrude the chemotherapeutic agents. Since the use of specific ABC protein inhibitors that have adverse side effects on human health, flavonoids have risen as potential non-toxic inhibitors of these transporters [Brand et al., 2006].

Almost every known member of the ABC family can be inhibited by one or more flavonoids [Brand et al., 2006]. For example, in a human epidermal cell

Table 4.2 **Effects of Interactions of Flavonoids and Selected Plasma Membrane Proteins**

Function	Protein	Flavonoid	System	Effect	References
Broad range transporters	**ABC transporters**	Chrysin, quercetin, genistein	Caco-2 cells	Inhibition	Sergent et al., 2005
		Sylimarin	Caco-2 cells	Inhibition	Zhang and Morris, 2003
		Green tea catechins	Various tissues	Stimulation and inhibition	Zhou et al., 2004
		32 flavonoids	K562 cells	Inhibition	Katayama et al., 2007
		16 flavonoids	MCF-7 cells	Inhibition	Zhang et al., 2005
		Silymarin, hesperetin, quercetin, and daidzein	MCF/MR cells	Inhibition	Cooray et al., 2004
Cl^- transport	CFTR	Kaempferol	Mice trachea	Stimulation	Sousa et al., 2007
			IB3-1 cells	Stimulation	Lim et al., 2004
		Genistein	Baby hamster kidney cells	Stimulation and inhibition	Schmidt et al., 2008
			Rat aorta	Stimulation	Valero et al., 2006
			epithelial cells	Stimulation and inhibition	Moran and Zegarra-Moran, 2005
			Renal epithelia	Stimulation	Lim et al., 2004
			Mouse jejunum	Stimulation	Baker and Hamilton, 2004
			Airway epithelia	Stimulation	Andersson et al., 2003
			CHO cells	Stimulation	Bulteau-Pignoux et al., 2002
			Murine trachea and colon	Stimulation	Goddard et al., 2000
			Xenopus laevis oocytes	Stimulation	Suaud et al., 2002
		7,8-Benzoflavone	Mouse fibroblasts	Stimulation	Ferrera et al., 2007

(Continued)

Table 4.2 *(Continued)*

Function	Protein	Flavonoid	System	Effect	References
			Human bronchial epithelial cells	Stimulation	Caci et al., 2003
		Cocoa flavanols	Human colon epithelia	Inhibition	Schuier et al., 2005
		Apigenin	Renal epithelia	Stimulation	Lim et al., 2004
			Xenopus laevis oocytes	Inhibition	Lee et al., 2005
Ca^{2+} transport	CRAC channel	*Opuntia ficus* indica polyphenols	Human Jurkat T-cells	Stimulation	Aires et al., 2004
	Voltage-gated Ca^{2+} channels	Red wine polyphenols	Dorsal root ganglion neurons	Inhibition	Wu et al., 2005
		Quercetin	Rat pituitary GH3 cells	Stimulation	Wu et al., 2003
			NG108-15 neuronal cells	Inhibition	Wu et al., 2003
		Galangin	Rat urinary bladder	Inhibition	Capasso and Tavares, 2002
		Myricetin	Rat tail artery myocytes	Activation	Fusi et al., 2005
			Vascular smooth muscle	Activation	Fusi et al., 2003
	NMDA channels	Apigenin	HEK293 cells	Inhibition	Losi et al., 2004
K^{+} transport	K_{ATP} channels	Epigallocatechin gallate	*Xenopus laevis* oocytes	Uncoupling	Jin et al., 2007
				Inhibition	Baek et al., 2005
		Aspalathus linearis flavonoids	Rabbit jejunum	Activation	Gilani et al., 2006
	K_v channels	Chalcones	HEK293 cells	Inhibition	Yarishkin et al., 2008
		Red wine polyphenols	Dorsal root ganglion neurons	Inhibition	Wu et al., 2005
	BKCa channels	Isoliquiritigenin	Guinea-pig tracheal smooth muscle	Inhibition	Liu et al., 2008
		Naringenin	Rat thoracic aorta rings	Activation	Saponara et al., 2006
		Quercetin	Rat coronary artery	Activation	Cogolludo et al., 2007

(Continued)

Table 4.2 *(Continued)*

Function	Protein	Flavonoid	System	Effect	References
Na^+ transport	Voltage-gated Na^+ channels	Quercetin	Human endothelial cells	Activation	Kuhlmann et al., 2005
		Quercetin, catechin	TSA201 cells	Inhibition	Wallace et al., 2006
		Red wine polyphenols	Mouse dorsal root ganglionic neurons	Inhibition	Wu et al., 2005
	$Na^+/K^+/2Cl^-$ cotransporter	Quercetin	A6 cells	Activation	Fujimoto et al., 2005
Glucose transport	GLUT1	Myricetin, quercetin, and catechin gallate	Rat adipocytes	Inhibition	Strobel et al., 2005
		Cinnamon polyphenols	Mouse macrophages	Increased expression	Cao et al., 2008
		Genistein	Human erythrocytes	Inhibition	Afzal et al., 2002
	GLUT4	Genistein	3T3-L1 adipocytes	Inhibition	Bazuine et al., 2005
		Myricetin	Insulin-resistant rats muscle	Stimulation	Liu et al., 2007b
			Obese Zucker rats	stimulation	Liu et al., 2007a
	GLUT1 + GLUT4	Green tea polyphenols	Insulin-resistant rats muscle	Increased expression	Cao et al., 2007
	GLUT2 + GLUT5	Epicatechin gallate	Caco-2 cells	Inhibition	Shimizu et al., 2000
Phytoestrogen	Plasma membrane estrogen receptor	Daizein	Bovine adrenal medullary cells	Stimulation and inhibition	Liu et al., 2007c
			Embryonic rat and hippocampal neurons	Increased expression	Wang et al., 2008
		Naringenin	HepG2 cells	Inhibition	Galluzzo et al., 2008
		Naringenin, quercetin	HeLA cells	Inhibition	Virgili et al., 2004
		Ginkgo biloba polyphenols	MCF-7 cells	Stimulation	Oh and Chung, 2004
			MCF-7 cells	Inhibition	Oh and Chung, 2006

(Continued)

Table 4.2 ***(Continued)***

Function	Protein	Flavonoid	System	Effect	References
Neurotrans-mission	Glutamate receptor	Quercetin, genistein	Brain synaptic membranes	Inhibition	Martini et al., 2007
	Dopamine receptor	Epigallocatechin gallate	PC12 cells	Inhibition	Li et al., 2006
		Croton celtidifolius proanthocyanidins	Swiss mice	D_2- agonist	DalBo et al., 2006
	GABA receptor	Apigenin	Rat cerebellar granule cells	Inhibition	Avallone et al., 2000
		Apigenin, epigallocatechin gallate	*Xenopus laevis* oocytes	Stimulation	Campbell et al., 2004
	Glycine receptor	Quercetin	Rat hippocampal neurons	Desensitization	Sun et al., 2007
			Xenopus laevis oocytes	Inhibition	Lee et al., 2008
		Genistein	Rat neurons	Inhibition	Zhu et al., 2003
			Rat hypothalamic Neurons	Inhibition	Huang and Dillon, 2000
	5-HT receptor	Quercetin	*Xenopus laevis* oocytes	Inhibition	Lee et al., 2005
		Puerarin	Rat cerebral cortex	Inhibition	Overstreet et al., 2003
		Scutellaria lateriflora flavonoids extract	CHO cells	Inhibition	Gafner et al., 2003
	Adrenergic receptor	Quercetin	Rats	α_2- agonist	Kaur et al., 2007
		Genistein	C2C12 cells	α_1- agonist	Jou et al., 2004
			Canine ventricular myocardium	β- agonist	Chu et al., 2005
	Acetylcholine receptor	Genistein	Bovine adrenal chromaffin cells	Nicotinic antagonist	Matsumura et al., 2007
		Daidzein	Bovine adrenal medulla	Nicotinic antagonist	Liu et al., 2007c

line (KB-C2 cells) that overexpresses P-glycoprotein (also known as ABCB1 protein, or MDR1), it has been demonstrated that select flavonoids, inhibit P-glycoprotein-mediated rhodamine 123 efflux [Kitagawa et al., 2005]. Among the flavonols assessed, the magnitude of the inhibitory effect followed the order: kaempferol > quercetin, baicalein > myricetin > fisetin and morin, which is in accordance with their hydrophobicity [Kitagawa et al., 2005]. On the contrary, in MCF-7 cells, flavanols stimulated P-glycoprotein-mediated transport of the carcinogen 7,12–dimethylbenz(*a*)anthracene [Phang et al., 1993]. As hypothesized by Morris and Zhang [2006], this conflictive behavoir could be explained based on the relative affinity of flavonoids for either the R or H sites of the protein, responsible for rhodamine 123 or Hoescht 33342 transport, respectively. Binding to one of these sites stimulates the transport activity of the other one. Supporting that, quercetin stimulates rhodamine 123 efflux through its binding to the H site, an effect that strongly depends on quercetin concentration [Shapiro and Ling, 1997]. At low concentrations (5 μM) quercetin stimulates P-glycoprotein phosphorylation and activation, while at high concentrations (50 μM) it causes P-glycoprotein inhibition [Mitsunaga et al., 2000]. Even when not all the flavonoids biologically tested were investigated in silico for the assignment of their specific binding sites in ABC proteins, a recent study [Badhan and Penny, 2006] located flavones, flavonols, flavanones, and chalcones binding sites to the nucleotide binding domain of P-glycoprotein. It was demonstrated that in the flavonoid–P-glycoprotein complex, flavonoids' B ring is oriented in a manner similar to the spatial disposition adopted by the adenine base of ATP, and thus, they can compete with the ATP for its binding to the binding domain [Badhan and Penny, 2006].

Another member of the ABC superfamily is cystic fibrosis transmembrane conductance regulator (CFTR, ABCC7), a plasma membrane channel involved in chloride transport. Recently, Sousa et al. [2007] demonstrated that an extract from the medicinal plant *Phyllantus acidus* composed mainly by kaempferol, adenosine, and hypogallic acid, significantly stimulated the CFTR channel, increasing chloride transport. This effect was due to the activation of multiple events, including the modulation of intracellular Ca^{2+} concentration, cyclic adenosine 5–monophosphate (cAMP) generation, and CFTR redistribution. Individually, all three components of the extract activate the CFTR. Kaempferol has a direct effect on CFTR with a magnitude lower than that of adenosine [Sousa et al., 2007]. Accordingly, working with IB3-1 cells that bear the ΔF508 mutation that accounts for ~70% of the mutations found in patients with cystic fibrosis, Lim et al. [2004] demonstrated that kaempferol, apigenin, and genistein cause the redistribution of the mutant CFTR, restoring the usual location in the plasma membrane instead of the abnormal perinuclear location. These findings open a promising therapeutic strategy to ameliorate the progression of cystic fibrosis by using plant products.

Another key group of transporters present in the cell plasma membrane is integrated by Na^+, K^+, and Ca^{2+} channels. Among them, there are ligand- and voltage-gated ion channels. The latter comprises a group of channels, all of them being transmembrane proteins with a central ion-conducting pore. These channels are operationally relevant in excitable cells, such as neurons and muscle cells, in which changes in the surrounding electrical potential lead to transmembrane movement of ions responsible for neurotransmitter release and action potentials, among other events. These channels can be also found in nonexcitable cells, although their biological function in this kind of cells is not established. The activities of several voltage-gated ion channels can be modulated by flavonoids (Table 4.2). For these channels, flavonoids do not show a unique behavior; depending on the cell model and the flavonoid assessed, the reported results are dissimilar. Wu et al. [2005] demonstrated that red wine polyphenols inhibit voltage-gated Ca^{2+} currents in dorsal root ganglion neurons. Conversely, L-type Ca^{2+} channels are specifically activated by myricetin [Fusi et al., 2003, 2005]. Quercetin has a stimulatory effect on these channels, although its effects also depend on the cell type. For example, working with rat pituitary GH3 cells, Wu et al. [2003] found that quercetin stimulates L-type Ca^{2+} channels, while it has an opposed effect on the same channels from NG108–15 neuronal cells. These differential effects of quercetin could be related to the various isoforms of L-type channels present in different cells.

Several lines of investigation demonstrated that flavonoids can also affect glucose uptake (Table 4.2). This process occurs through two groups of transporters, the Na^{+}-coupled carrier system (SGLT) and the facilitative glucose transporters (GLUT) (reviewed in Scheepers et al. [2004]). Given the relevance of GLUT activities in the context of human pathologies such as obesity and insulin resistance, it is very important to elucidate if flavonoids could act as natural modulators of glucose uptake in muscle and adipose tissues. Most of the flavonoids studied have stimulatory properties on GLUT expression and/or activities (Table 4.2). However, the behavior of isoflavones should be analyzed separately. Genistein inhibit GLUT1 and GLUT4 activities in different experimental models [Afzal et al., 2002; Bazuine et al., 2005]. On one side, GLUT1 needs non-hydrolytic ATP binding to exert its biological effects [Carruthers and Helgerson, 1989; Cloherty et al., 2002,] which is prevented by genistein [Vera et al., 1996]. On the other hand, the signaling cascade that ends in GLUT4 translocation to the plasma membrane and to glucose uptake depends on the tyrosine kinase activity of the insulin receptor. Other non-insulin receptor-related tyrosine kinases can also stimulate GLUT4 translocation in adipocytes [Chen et al., 1997; Bazuine et al., 2003]. Genistein is a potent tyrosine kinase inhibitor [Akiyama et al., 1987] that raises a plausible explanation for its inhibitory effect on GLUT4 activation. In addition, Bazuine et al. [2005] found in 3T3-L1 adipocytes that genistein can also inhibit per se GLUT4 activity. Therefore, genistein has a dual inhibitory effects on GLUT4, impairing insulin-dependent GLUT4 translocation to the plasma membrane, and inhibiting GLUT4 once the transporter reaches the cell membrane, both events leading to a lower glucose uptake.

Other important effects of isoflavones is their capability to act as phytoestrogens. The finding that their regular consumption reduced osteoporosis and arteriosclerosis symptoms in postmenopausal women, and presumably, the incidence of certain cancers [Wuttke et al., 2007], fueled extensive research. Given that genistein structure and spatial topology highly resembles those of estradiol 17β [Vaya and Tamir, 2004], genistein could bind to the estrogen receptors α and β (ERα and ERβ) and mimic some biological effects of estradiol, although with weaker transcriptional potency [Wuttke et al., 2007]. Since human mammary tissue is highly enriched in ERα, and the stimulation of ERα enhances cell proliferation, a controversy exists on the advice of increasing the consumption of isoflavones and the potential risk of breast cancer development [Warner et al., 2000]. Besides isoflavones, other flavonoids have been recently reported as phytoestrogens, including naringenin, quercetin, and certain polyphenols present in *Ginkgo biloba* extracts (Table 4.2). The flavanone naringenin inhibited ERα through a different mechanism than that of isoflavones. Naringenin inhibitory effect takes place at the plasma membrane level where it induces a rapid ERα depalmitoylation event that results in ERα redistribution in the cell membrane and in the inability to activate extracellular signal-regulated kinase (ERK) and protein kinase B (Akt) phosphorylation [Galluzzo et al., 2008]. On the other hand, *Ginkgo biloba* polyphenols, which include kaempferol, quercetin, and isorhamnetin, could have both estrogenic effects through their binding to ERβ [Oh and Chung, 2004] and antiestrogenic effects by stimulating ERα catabolism and inhibiting its synthesis [Oh and Chung, 2006].

Finally, we will briefly focus on plasma membrane receptors involved in neurotransmission and on how flavonoids can affect their activity (Table 4.2). In general, the interactions of flavonoids with these receptors result in an inhibition of their biological response involving multiple mechanisms. As an example, genistein blocks glycine receptor-mediated electrical current but strictly requires glycine prebinding to the receptor, suggesting that genistein binds to the open state of the channel not competing with glycine for its binding site [Zhu et al., 2003]. Genistein can also act as a β-adrenergic agonist, which is due to its tyrosine kinase inhibitor properties [Chu et al., 2005]. Finally, genistein inhibits long-term potentiation in rat brain dentate gyrus, through the inhibition of the phosphorylation of presynaptic α-subunits of Ca^{2+} channels involved in glutamate release [Casey et al., 2002].

Flavonoids can also regulate the activity of postsynaptic neurotransmiter receptors. It has been demonstrated that apigenin has neuroprotective effects against glutamate-mediated neurotoxicity in cortical and cerebellar neurons antagonizing γ-aminobutyric acid (GABA) and *N*-methyl *D*-aspartate (NMDA) receptors, thus reducing the overall excitability of the neuronal network [Losi et al., 2004]. These effects of apigenin are closely related to its sedative effects, which have been recognized for a long time and used with therapeutic purposes against anxiety (for a review see Johnston et al. [2006]). Based on the capacity of apigenin, its dimer amentoflavone, and of other flavonoids to inhibit GABA receptors through their binding to benzodiazepine's binding site [Marder and Paladini, 2002], several compounds have been synthesized and assessed as sedatives, with the purpose to develop better anxiolytic and sleep-enhancing compounds.

The above evidence shows that flavonoids can interact and modulate the activity of several membrane proteins. These effects can in part explain the beneficial health properties of flavonoids and could also help develop new therapeutic approaches for major human diseases.

CONSEQUENCES OF FLAVONOID INTERACTIONS ON MEMBRANE-ASSOCIATED PROCESSES

Antioxidant Actions of Flavonoids at the Membrane Level

Flavonoids are well-recognized antioxidant molecules with proven antioxidant activity in numerous biological systems [Lotito and Fraga, 1998; Lotito et al., 2000; Rice-Evans, 2001]. Chemically, flavonoid antioxidant ability relies mostly on three characteristics: (a) the number and location of hydroxyl groups; (b) the presence of a 2,3 double bond conjugated with a 4-oxo group in ring C; and (c) the presence of 3- and 5-OH groups [Bors et al., 1990]. The presence of an *o*-catechol group in the B-ring [Burda and Oleszek, 2001; Silva et al., 2002] could also be important, as it is the degree of oligomerization when flavonoids form oligomers [Verstraeten et al., 2003, 2004, 2005; Erlejman et al., 2004, 2006].

Several structure–activity studies have been performed to find the best antioxidant at the membrane level [Amic et al., 2007; Om and Kim, 2008; Ray et al., 2008]. Analyzing a set of 12 different flavonoids, Rackova et al. [2005] found a close association between flavonoid-mediated inhibition of dioleoyl-PC liposome oxidation by the azo-compound 2,2′-azobis-(2-amidino-propane)hydrochloride (AAPH) and flavonoid hydration energy, a parameter that evaluates their hydrophilicity. Given the fact that the initiator used for the experiments is a water-soluble compound, the observed relationship between flavonoids hydration energy and their antioxidant activity could be due mainly to the free-radical scavenger properties of the flavonoids OH groups. Thus, even when these findings are valuable in the prediction of certain flavonoid antioxidant ability, the assay leaves aside the flavonoids with hydrophobic character.

Besides their free-radical trapping properties, flavonoids can interfere with the capacity of oxidants to reach the bilayer. A study from our laboratory demonstrated that the adsorption of water-soluble (−)-epicatechin oligomers (dimer to hexamer) to membranes prevents lipid oxidation initiated by the azo-compound 2,2′-azobis (2,4-dimethylvaleronitrile), (AMVN), a hydrophobic molecule that upon its incorporation into the bilayer decomposes yielding peroxyl radicals [Verstraeten et al., 2003]. In this case, given that the oxidant species are generated in a membrane region relatively far from the adsorption sites of flavonoids, a direct scavenging action of these molecules should not be expected. These effect of (−)-epicatechin oligomers can be explained in terms of a protection to membranes by large oligomers adsorbed to the membrane surface limiting the accessibility of AMVN to the membrane interior. Supporting this hypothesis, a significant correlation between oligomers antioxidant capacity and the prevention of membrane disruption by a nonionic detergent was observed [Verstraeten et al., 2003]. Similarly, Aldini and co-workers [2003] observed that procyanidins isolated from *Vitis vinifera* bind to endothelial cell membrane and prevented peroxynitrite-mediated oxidative damage, an effect that the authors partially ascribed to procyanidin-mediated prevention of peroxynitrite diffusion into the cells.

The procyanidin structure further determines the extent of the protection to membranes. In this regard, dimers A1 and A2 had higher protective effects than dimer B2 against both, liposome membrane damage induced by the detergent Triton X-100 and AMVN-initiated lipid oxidation [Verstraeten et al., 2005]. This differential effect of the dimers could be due to a different molecule folding. Mackenzie et al. [2004] showed by using computational modelling that dimer B2 can establish internal hydrogen bonds, allowing epicatechin monomers the possibility to be stacked over each other. On the contrary, the monomeric units in the dimers A1 and A2 are connected by two covalent bonds that bring higher rigidity to the dimer and that could impede their folding and confer an elongated shape [Mackenzie et al., 2009]. Therefore, the molecules of dimers A1 and A2 will have higher membrane surface coverage per molecule compared to dimer B2.

The finding that water-soluble flavonoids could exert their beneficial properties at the hydrophobic portion of the membrane was also observed in in vivo studies and in cells in culture. For example, erythrocytes obtained from animals fed a flavanol- and procyanidin-rich meal showed reduced susceptibility to free-radical-mediated hemolysis [Zhu et al., 2002]. Consistently, we demonstrated that procyanidin hexamers, which interact with membranes but would not be internalized, protected Caco-2 cells from AMVN- and bile-induced oxidation [Erlejman et al., 2006]. When liposomes were preincubated with a series of flavonoids with diverse hydrophobicity, not only hydrophobic flavonoids prevented AMVN-mediated lipid oxidation but also the more hydrophilic ones [Erlejman et al., 2004]. Similarly to what was previously found in liposomes, the protective effects of flavonoids against AMVN-supported oxidation was strongly associated with their capacity to prevent membrane disruption by detergents, supporting the hypothesis of a physical protection of membranes by preventing oxidants to reach fatty acids.

A different mechanism should account for the antioxidant ability of hydrophobic flavonoids, which can be incorporated into the bilayer where they can affect certain membrane physical properties. It is well-known that alterations in membrane rheology will affect the extension and rate of lipid oxidation. For example, increased lipid oxidation rates have been observed when membranes are incubated at temperatures below the transition temperature of their lipids [Mowri et al., 1984; Cervato et al., 1988]. In this situation, lipids are tightly packed with their acyl chains located in close vicinity, hence creating an environment where lipid oxidation will easily propagate [McLean and Hagaman, 1992]. As mentioned in a previous section, the isoflavones daidzein and genistein have differential effects on membrane fluidity; however, both genistein and daidzein act like antioxidants in membranes above lipid transition temperature [Erlejman et al., 2004]. These findings indicate that daidzein and genistein antioxidant properties reside in both their capacity to modulate membrane physical properties and their ability to trap oxidant species.

Modulation of Cell Signals by Flavonoids at the Membrane Level

Flavonoids could play a key role in the regulation of certain cell signals triggered at the plasma membrane level. As mentioned before, the flavanone naringenin inhibited ERα by inducing its rapid depalmitoylation, causing ERα detachment from the plasma membrane [Galluzzo et al., 2008]. It has been demonstrated that through cycles of palmitoylation–depalmitoylation catalized by protein acyltransferases and thioesterases [Linder and Deschenes, 2006], cells regulate the amount of ERα protein in the membrane. ERα Cys_{447} palmitoylation leads to a strong protein association with caveolin-1 followed by its specific translocation to lipid rafts and to the triggering of numerous cell signals [Levin, 2005]. Even when the molecular mechanisms involved in naringenin-mediated ERα depalmitoylation is still under elucidation, it was hypothesized that narigerin could directly bind to ERα and promote changes in protein conformation that makes it prone to deacylation [Galluzzo et al., 2008].

(−)-Epicatechin hexamer (Hex) reversibly adsorbs to Caco-2 cell surface [Erlejman et al., 2006] and prevents the deleterious effects of the biliary salt deoxycholate. Hex protects cells from deoxycholate-induced citotoxicity, oxidant production, and permeabilization of Caco-2 cell monolayer surfaces [Erlejman et al., 2008]. Hex also prevents tumor necrosis alpha (TNFα) activation of the transcription factor NF-κB [Erlejman et al., 2008]. Hex is not incorporated into cells and the inhibition of TNFα-mediated NF-κB activation could be due to either a direct interaction of this procyanidin with the TNFα receptor or secondary to alterations of the physical properties of the lipids in the receptor's immediate surroundings that impede the normal binding of its specific ligand. In the same experimental conditions, 30 min of preincubation with the (−)-epicatechin or dimer B2 did not inhibit TNFα-induced NF-κB activation. These findings, together with the observation that after 24 h incubation, epicatechin and dimer B2 inhibited phorbol myristate acetate-mediated NF-κB activation and binding to DNA in Jurkat T cells [Mackenzie et al., 2004, 2009] suggest that epicatechin and dimer B2 internalization is necessary to inhibit NF-κB activation.

Lipids such as phosphoinositides, sphingolipids, and eicosanoids have been recognized as signaling molecules that control important cellular events such as cell proliferation, metabolism, apoptosis, and others. Alterations in normal lipid-mediated cell signaling have been related to the development of cancer, and inflammatory and metabolic diseases (recently reviewed by Wymann and Scheiter [2008]). The flavonoid apigenin-7-glucoside and those from German chamomile extracts prevent carbon tetrachloride (CCl_{4-})-mediated activation of neutral sphingomyelinase [Babenko and Shakhova, 2008]. This plasma membrane protein is responsible for sphingomyelin hydrolysis with formation of a key signaling molecule, ceramide. By preventing sphingomyelinase activation, those flavonoids helped to maintain rat hepatocytes membrane integrity upon CCl_{4-} administration, and from the physiological membrane alterations caused by ageing [Babenko and Shakhova, 2006, 2008]. The modulatory effects of flavonoids on signaling lipids formation could also occur through indirect mechanisms. For example, luteolin and luteolin-7-glucoside inhibited platelet-derived growth factor beta (PDGFβ) receptor phosphorylation in vascular smooth muscle cells [Kim et al., 2005]. The activation of this receptor causes phospholipase C gamma (PLCγ) activation with the subsequent hydrolysis of phosphoinositides and Ca^{2+} mobilization. In addition, PDGFβ receptor activation leads to Akt and ERK 1/2 phosphorylation. All these events were prevented by luteolin and luteolin-7-glucoside and therefore, the mitogenic effects of PDGFβ were abolished by these flavonoids [Kim et al., 2005, 2006].

Finally, another possible mechanism that could explain the regulatory effects of certain flavonoids on membrane-associated events is the creation of a lipid environment that ultimately affects membrane protein functionality. Recently, Tarahovsky et al. [2008] proposed that quercetin and taxifolin may have a preferential affinity for some membrane proteins and lipids, and then accumulate and promote lipid segregation with formation of raftlike domains. Among these interactions, flavonoids carboxyl and hydroxyl residues could interact with sphingosine moiety in sphingolipids, in a similar fashion as the cholesterolsphingomyelin interactions observed in lipid rafts [McConnell and Radhakrishnan, 2003; McConnell and Vrljic, 2003]. Also, as flavonoids can interact with annular lipids in proteins, it has been hypothesized that flavonoids "may serve as adaptors mediating the interactions between membrane proteins and surrounding annular lipids" [Tarahovsky et al., 2008]. If true, flavonoids possibly could modulate the activity of many different membrane proteins. However, further research is required to assess this hypothesis.

CONCLUSIONS

As flavonoids are ubiquitous components of fruits and vegetables, they are part of almost any human diet. The identification of potential targets for the beneficial health effects of flavonoids exponentially increased over the years. The physiological relevance of the interactions of flavonoids with cell membranes is becoming of interest to explain at least part of their beneficial health effects. As presented in this chapter, the interactions of flavonoids with major membrane components, lipids and proteins, modulate a number of cellular events involved in both physiological and pathological conditions. The knowledge of the nature and biological consequences of flavonoid–membrane interactions will certainly contribute to develop nutritional and pharmacological strategies oriented to prevent major human diseases.

ACKNOWLEDGMENTS

This work was supported by grants from the University of Buenos Aires, Argentina (UBACYT 081); NIH, USA (R21AT2966); UC Davis CHNR-which was established with funding from the State of California Vitamin Price Fixing Consumer Settlement Fund, and Mars Incorporated.

REFERENCES

Afzal I, Cunningham P, Naftalin RJ. 2002. Interactions of ATP, oestradiol, genistein and the anti-oestrogens, faslodex (ICI 182780) and tamoxifen, with the human erythrocyte glucose transporter, GLUT1. *Biochem J* 365:707–719.

Aires V, Adote S, Hichami A, Moutairou K, Boustani ES, Khan NA. 2004. Modulation of intracellular calcium concentrations and T cell activation by prickly pear polyphenols. *Mol Cell Biochem* 260:103–110.

Akiyama T, Ishida J, Nakagawa S, Ogawara H, Watanabe S, Itoh N, Shibuya M Fukami Y. 1987. Genistein, a specific inhibitor of tyrosine-specific protein kinases. *J Biol Chem* 262:5592–5595.

Aldini G, Carini M, Piccoli A, Rossoni G Facino RM. 2003. Procyanidins from grape seeds protect endothelial cells from peroxynitrite damage and enhance endothelium-dependent relaxation in human artery: New evidences for cardio-protection. *Life Sci* 73:2883–2898.

Amic D, Davidovic-Amic D, Beslo D, Rastija V, Lucic B, Trinajstic N. 2007. SAR and QSAR of the antioxidant activity of flavonoids. *Curr Med Chem* 14:827–845.

Andersson C, Servetnyk Z, Roomans GM. 2003. Activation of CFTR by genistein in human airway epithelial cell lines. *Biochem Biophys Res Commun* 308:518–522.

Arora A, Byrem TM, Nair MG, Strasburg GM. 2000. Modulation of liposomal membrane fluidity by flavonoids and isoflavonoids. *Arch Biochem Biophys* 373:102–109.

Avallone R, Zanoli P, Puia G, Kleinschnitz M, Schreier P, Baraldi M. 2000. Pharmacological profile of apigenin, a flavonoid isolated from Matricaria chamomilla. *Biochem Pharmacol* 59:1387–1394.

Babenko NA, Shakhova EG. 2008. Effects of flavonoids on sphingolipid turnover in the toxin-damaged liver and liver cells. *Lipids Health Dis* 7:1.

Babenko NA, Shakhova EG. 2006. Effects of Chamomilla recutita flavonoids on age-related liver sphingolipid turnover in rats. *Exp Gerontol* 41:32–39.

Badhan R, Penny J. 2006. In silico modelling of the interaction of flavonoids with human P-glycoprotein nucleotide-binding domain. *Eur J Med Chem* 41:285–295.

Baek WK, Jang BC, Lim JH, Kwon TK, Lee HY, Cho CH, Kim DK, Shin DH, Park JG, Lim JG, Bae JH, Yoo SK, Park WK, Song DK. 2005. Inhibitory modulation of ATP-sensitive potassium channels by gallate-ester moiety of (−)-epigallocatechin-3-gallate. *Biochem Pharmacol* 70:1560–1567.

Baker MJ, Hamilton KL. 2004. Genistein stimulates electrogenic Cl(−) secretion in mouse jejunum. *Am J Physiol Cell Physiol* 287:C1636–1645.

Bazuine M, van den Broek PJ, Maassen JA. 2005. Genistein directly inhibits GLUT4-mediated glucose uptake in 3T3-L1 adipocytes. *Biochem Biophys Res Commun* 326:511–514.

Bazuine M, Ouwens DM Gomes de Mesquita DS, Maassen JA. 2003. Arsenite stimulated glucose transport in 3T3-L1 adipocytes involves both Glut4 translocation and p38 MAPK activity. *Eur J Biochem* 270:3891–3903.

Bors W, Heller W, Michel C, Saran M. 1990. Flavonoids as antioxidants: Determination of radical-scavenging efficiencies. *Methods Enzymol* 186:343–355.

Brand W, Schutte ME, Williamson G, van Zanden JJ, Cnubben NH, Groten JP, van Bladeren PJ, Rietjens IM. 2006. Flavonoid-mediated inhibition of intestinal ABC transporters may affect the oral bioavailability of drugs, food-borne toxic compounds and bioactive ingredients. *Biomed Pharmacother* 60:508–519.

Bulteau-Pignoux L, Derand R, Metaye T, Joffre M, Becq F. 2002. Genistein modifies the activation kinetics and magnitude of phosphorylated wild-type and G551D-CFTR chloride currents. *J Membr Biol* 188:175–182.

Burda S, Oleszek W. 2001. Antioxidant and antiradical activities of flavonoids. *J Agric Food Chem* 49:2774–2779.

Caci E, Folli C, Zegarra-Moran O, Ma T, Springsteel MF, Sammelson RE, Nantz MH, Kurth MJ, Verkman AS, Galietta LJ. 2003. CFTR activation in human bronchial epithelial cells by novel benzoflavone and benzimidazolone compounds. *Am J Physiol Lung Cell Mol Physiol* 285:L180–188.

Campbell EL, Chebib M, Johnston GA. 2004. The dietary flavonoids apigenin and (−)-epigallocatechin gallate enhance the positive modulation by diazepam of the activation by GABA of recombinant GABA(A) receptors. *Biochem Pharmacol* 68:1631–1638.

Cao H, Urban JF. Jr., Anderson RA. 2008. Cinnamon polyphenol extract affects immune responses by regulating anti- and proinflammatory and glucose transporter gene expression in mouse macrophages. *J Nutr* 138:833–840.

Cao H, Hininger-Favier I, Kelly MA, Benaraba R, Dawson HD, Coves S, Roussel AM, Anderson RA. 2007. Green tea polyphenol extract regulates the expression of genes involved in glucose uptake and insulin signaling in rats fed a high fructose diet. *J Agric Food Chem* 55:6372–6378.

Capasso R, Tavares IA. 2002. Effect of the flavonoid galangin on urinary bladder rat contractility in-vitro. *J Pharm Pharmacol* 54:1147–1150.

Carruthers A, Helgerson AL. 1989. The human erythrocyte sugar transporter is also a nucleotide binding protein. *Biochemistry* 28:8337–8346.

Cartalade D, Vernhet A. 2006. Polar interactions in flavan-3-ol adsorption on solid surfaces. *J Agric Food Chem* 54:3086–3094.

Casey M, Maguire C, Kelly A, Gooney MA, Lynch MA. 2002. Analysis of the presynaptic signaling mechanisms underlying the inhibition of LTP in rat dentate gyrus by the tyrosine kinase inhibitor, genistein. *Hippocampus* 12:377–385.

Cervato G, Viani P, Masserini M, Di Iorio C, Cestaro B. 1988. Studies on peroxidation of arachidonic acid in different liposomes below and above phase transition temperature. *Chem Phys Lipids* 49:135–139.

Chen D, Elmendorf JS, Olson AL, Li X, Earp HS, Pessin JE. 1997. Osmotic shock stimulates GLUT4 translocation in 3T3L1 adipocytes by a novel tyrosine kinase pathway. *J Biol Chem* 272:27401–27410.

Chu L, Zhang JX, Norota I, Endoh M. 2005. Differential action of a protein tyrosine kinase, inhibitor genistein, on the positive inotropic effect of endothelin-1 and norepinephrine in canine ventricular myocardium. *Br J Pharmacol* 144:430–442.

Cloherty EK, Levine KB, Graybill C, Carruthers A. 2002. Cooperative nucleotide binding to the human erythrocyte sugar transporter. *Biochemistry* 41:12639–12651.

Cogolludo A, Frazziano G, Briones AM, Cobeno L, Moreno L, Lodi F, Salaices M, Tamargo J, Perez-Vizcaino F. 2007. The dietary flavonoid quercetin activates BKCa currents in coronary arteries via production of H_2O_2. Role in vasodilatation. *Cardiovasc Res* 73:424–431.

Cooray HC, Janvilisri T, van Veen HW, Hladky SB, Barrand MA. 2004. Interaction of the breast cancer resistance protein with plant polyphenols. *Biochem Biophys Res Commun* 317:269–275.

Cushnie TP, Hamilton VE, Chapman DG, Taylor PW, Lamb AJ. 2007. Aggregation of Staphylococcus aureus following treatment with the antibacterial flavonol galangin. *J Appl Microbiol* 103:1562–1567.

DalBo S, Jurgensen S, Horst H, Soethe DN, Santos AR, Pizzolatti MG, Ribeiro-do-Valle RM. 2006. Analysis of the antinociceptive effect of the proanthocyanidin-rich fraction obtained from Croton celtidifolius barks: evidence for a role of the dopaminergic system. *Pharmacol Biochem Behav* 85:317–323.

Deprez S, Mila I, Huneau JF, Tome D, Scalbert A. 2001. Transport of proanthocyanidin dimer, trimer, and polymer across monolayers of human intestinal epithelial Caco-2 cells. *Antioxid Redox Signal* 3:957–967.

Erlejman AG, Jaggers G, Fraga CG, Oteiza PI. 2008. TNFα-induced NF-κB activation and cell oxidant production are modulated by hexameric procyanidins in Caco-2 cells. *Arch Biochem Biophys* 476:186–195.

Erlejman AG, Fraga CG, Oteiza PI. 2006. Procyanidins protect Caco-2 cells from bile acid- and oxidant-induced damage. *Free Radic Biol Med* 41:1247–1256.

Erlejman AG, Verstraeten SV, Fraga CG, Oteiza PI. 2004. The interaction of flavonoids with membranes: Potential determinant of flavonoid antioxidant effects. *Free Radic Res* 38:1311–1320.

Ferrera L, Pincin C, Moran O. 2007. Characterization of a 7,8-benzoflavone double effect on CFTR Cl(−) channel activity. *J Membr Biol* 220:1–9.

Fujimoto S, Niisato N, Sugimoto T, Marunaka Y. 2005. Quercetin and NPPB-induced diminution of aldosterone action on Na + absorption and ENaC expression in renal epithelium. *Biochem Biophys Res Commun* 336:401–407.

Fusi F, Sgaragli G, Saponara S. 2005. Mechanism of myricetin stimulation of vascular L-type Ca^{2+} current. *J Pharmacol Exp Ther* 313:790–797.

Fusi F, Saponara S, Frosini M, Gorelli B, Sgaragli G. 2003. L-type Ca^{2+} channels activation and contraction elicited by myricetin on vascular smooth muscles. *Naunyn Schmiedebergs Arch Pharmacol* 368:470–478.

Gafner S, Bergeron C, Batcha LL, Reich J, Arnason JT, Burdette JE, Pezzuto JM, Angerhofer CK. 2003. Inhibition of [3H]-LSD binding to 5-HT7 receptors by flavonoids from Scutellaria lateriflora. *J Nat Prod* 66:535–537.

Galluzzo P, Ascenzi P, Bulzomi P, Marino M. 2008. The nutritional flavanone naringenin triggers antiestrogenic effects by regulating estrogen receptor alpha-palmitoylation. *Endocrinology* 149:2567–2575.

Gilani AH, Khan AU, Ghayur MN, Ali SF, Herzig JW. 2006. Antispasmodic effects of Rooibos tea (Aspalathus linearis) is mediated predominantly through K + -channel activation. *Basic Clin Pharmacol Toxicol* 99:365–373.

Goddard CA, Evans MJ, Colledge WH. 2000. Genistein activates CFTR-mediated Cl(−) secretion in the murine trachea and colon. *Am J Physiol Cell Physiol* 279:C383–392.

Hendrich AB, Malon R, Pola A, Shirataki Y, Motohashi N, Michalak K. 2002. Differential interaction of Sophora isoflavonoids with lipid bilayers. *Eur J Pharm Sci* 16:201–208.

Huang RQ, Dillon GH. 2000. Direct inhibition of glycine receptors by genistein, a tyrosine kinase inhibitor. *Neuropharmacology* 39:2195–2204.

Huh NW, Porter NA, McIntosh TJ, Simon SA. 1996. The interaction of polyphenols with bilayers: Conditions for increasing bilayer adhesion. *Biophys J* 71:3261–3277.

Ikigai H, Nakae T, Hara Y, Shimamura T. 1993. Bactericidal catechins damage the lipid bilayer. *Biochim Biophys Acta* 1147:132–136.

Jin JY, Park SH, Bae JH, Cho HC, Lim JG, Park WS, Han J, Lee JH, Song DK. 2007. Uncoupling by (−)-epigallocatechin-3-gallate of ATP-sensitive potassium channels from phosphatidylinositol polyphosphates and ATP. *Pharmacol Res* 56:237–247.

Johnston GA, Hanrahan JR, Chebib M, Duke RK, Mewett KN. 2006. Modulation of ionotropic GABA receptors by natural products of plant origin. *Adv Pharmacol* 54:285–316.

Jou SB, Huang CC, Liu IM, Cheng JT. 2004. Activation of alpha1A-adrenoceptors by genistein at concentrations lower than that to inhibit tyrosine kinase in cultured C2C12 cells. *Planta Med* 70:610–614.

Jovanovic SV, Steenken S, Tosic M, Marjanovic B, Simic MG. 1994. Flavonoids as Antioxidants. *J Am Chem Soc* 116:4846–4851.

Kajiya K, Kumazawa S, Naito A, Nakayama T. 2008. Solid-state NMR analysis of the orientation and dynamics of epigallocatechin gallate, a green tea polyphenol, incorporated into lipid bilayers. *Magn Reson Chem* 46:174–177.

Kajiya K, Kumazawa S, Nakayama T. 2001. Steric effects on interaction of tea catechins with lipid bilayers. *Biosci Biotechnol Biochem* 65:2638–2643.

Katayama K, Masuyama K, Yoshioka S, Hasegawa H, Mitsuhashi J, Sugimoto Y. 2007. Flavonoids inhibit breast cancer resistance protein-mediated drug resistance: Transporter specificity and structure-activity relationship. *Cancer Chemother Pharmacol* 60:789–797.

Kaur R, Chopra K, Singh D. 2007. Role of alpha2 receptors in quercetin-induced behavioral despair in mice. *J Med Food* 10:165–168.

Kim TJ, Kim JH, Jin YR, Yun YP. 2006. The inhibitory effect and mechanism of luteolin 7-glucoside on rat aortic vascular smooth muscle cell proliferation. *Arch Pharm Res* 29:67–72.

Kim JH, Jin YR, Park BS, Kim TJ, Kim SY, Lim Y, Hong JT, Yoo HS, Yun YP. 2005. Luteolin prevents PDGF-BB-induced proliferation of vascular smooth muscle cells by inhibition of PDGF beta-receptor phosphorylation. *Biochem Pharmacol* 69:1715–1721.

Kitagawa S, Nabekura T, Takahashi T, Nakamura Y, Sakamoto H, Tano H, Hirai M, Tsukahara G. 2005. Structure-activity relationships of the inhibitory effects of flavonoids on P-glycoprotein-mediated transport in KB-C2 cells. *Biol Pharm Bull* 28:2274–2278.

Kuhlmann CR, Schaefer CA, Kosok C, Abdallah Y, Walther S, Ludders DW, Neumann T, Tillmanns H, Schafer C, Piper HM, Erdogan A. 2005. Quercetin-induced induction of the NO/cGMP pathway depends on \$\$$Ca^{2+}$-activated K + channel-induced hyperpolarization-mediated Ca^{2+}-entry into cultured human endothelial cells. *Planta Med* 71:520–524.

Lania-Pietrzak B, Hendrich AB, Zugaj J, Michalak K. 2005. Metabolic O-demethylation does not alter the influence of isoflavones on the biophysical properties of membranes and MRP1-like protein transport activity. *Arch Biochem Biophys* 433:428–434.

Lee BH, Pyo MK, Lee JH, Choi SH, Shin TJ, Lee SM, Lim Y, Han YS, Paik HD, Cho SG, Yun-Choi HS, Rhim H, Nah SY. 2008. Differential regulations of quercetin and its glycosides on ligand-gated ion channels. *Biol Pharm Bull* 31:611–617.

Lee BH, Jeong SM, Lee JH, Kim JH, Yoon IS, Choi SH, Lee SM, Chang CG, Kim HC, Han Y, Paik HD, Kim Y, Nah SY. 2005. Quercetin inhibits the 5-hydroxytryptamine type 3 receptor-mediated ion current by interacting with pre-transmembrane domain I. *Mol Cells* 20:69–73.

Lehtonen JY, Adlercreutz H, Kinnunen PK. 1996. Binding of daidzein to liposomes. *Biochim Biophys Acta* 1285:91–100.

Levin ER. 2005. Integration of the extranuclear and nuclear actions of estrogen. *Mol Endocrinol* 19:1951–1959.

Li R, Peng N, Li XP, Le WD. 2006. (−)-Epigallocatechin gallate regulates dopamine transporter internalization via protein kinase C-dependent pathway. *Brain Res* 1097:85–89.

Lim M, McKenzie K, Floyd AD, Kwon E, Zeitlin PL. 2004. Modulation of deltaF508 cystic fibrosis transmembrane regulator trafficking and function with 4-phenylbutyrate and flavonoids. *Am J Respir Cell Mol Biol* 31:351–357.

Linder M, Deschenes R. 2006. Protein palmitoylation. *Methods* 40:125–126.

Liu B, Yang J, Wen Q, Li Y. 2008. Isoliquiritigenin, a flavonoid from licorice, relaxes guinea-pig tracheal smooth muscle in vitro and in vivo: Role of cGMP/PKG pathway. *Eur J Pharmacol* 587:257–266.

Liu IM, Tzeng TF, Liou SS, Lan TW. 2007a. Improvement of insulin sensitivity in obese Zucker rats by myricetin extracted from Abelmoschus moschatus. *Planta Med* 73:1054–1060.

Liu IM, Tzeng TF, Liou SS, Lan TW. 2007b. Myricetin, a naturally occurring flavonol, ameliorates insulin resistance induced by a high-fructose diet in rats. *Life Sci* 81:1479–1488.

Liu M, Yanagihara N, Toyohira Y, Tsutsui M, Ueno S, Shinohara Y. 2007c. Dual effects of daidzein, a soy isoflavone, on catecholamine synthesis and secretion in cultured bovine adrenal medullary cells. *Endocrinology* 148:5348–5354.

Losi G, Puia G, Garzon G, de Vuono MC, Baraldi M. 2004. Apigenin modulates GABAergic and glutamatergic transmission in cultured cortical neurons. *Eur J Pharmacol* 502:41–46.

Lotito SB, Fraga CG. 1998. (+)-Catechin prevents human plasma oxidation. *Free Radic Biol Med* 24:435–441.

Lotito SB, Actis-Goretta L, Renart ML, Caligiuri M, Rein D, Schmitz HH, Steinberg FM, Keen CL, Fraga CG. 2000. Influence of oligomer chain length on the antioxidant activity of procyanidins. *Biochem Biophys Res Commun* 276:945–951.

Mackenzie GG, Delfino JM, Keen CL, Fraga CG, Oteiza PI. 2009. Dimeric procyanidins are inhibitors of NF-κB-DNA binding. *Biochem Pharmacol,* in press (doi:10.1016/j.bcp.2009.06.111).

Mackenzie GG, Carrasquedo F, Delfino JM, Keen CL, Fraga CG, Oteiza PI. 2004. Epicatechin, catechin, and dimeric procyanidins inhibit PMA-induced NF-kappaB activation at multiple steps in Jurkat T cells. *FASEB J* 18:167–169.

Marder M, Paladini AC. 2002. GABA(A)-receptor ligands of flavonoid structure. *Curr Top Med Chem* 2:853–867.

Martini LH, Jung F, Soares FA, Rotta LN, Vendite DA, Frizzo ME, Yunes RA, Calixto JB, Wofchuk S, Souza DO. 2007. Naturally occurring compounds affect glutamatergic neurotransmission in rat brain. *Neurochem Res* 32:1950–1956.

Matsumura C, Kuwashima H, Soma S, Kimura T. 2007. Effects of genistein and herbimycin, tyrosine kinase inhibitors, on catecholamine release in bovine adrenal chromaffin cells. *Auton Autacoid Pharmacol* 27:181–187.

McConnell HM, Radhakrishnan A. 2003. Condensed complexes of cholesterol and phospholipids. *Biochim Biophys Acta* 1610:159–173.

McConnell HM, Vrljic M. 2003. Liquid-liquid immiscibility in membranes. *Annu Rev Biophys Biomol Struct* 32:469–492.

McLean LR, Hagaman KA. 1992. Effect of lipid physical state on the rate of peroxidation of liposomes. *Free Radic Biol Med* 12:113–119.

Mitsunaga Y, Takanaga H, Matsuo H, Naito M, Tsuruo T, Ohtani H, Sawada Y. 2000. Effect of bioflavonoids on vincristine transport across blood-brain barrier. *Eur J Pharmacol* 395:193–201.

Moran O, Zegarra-Moran O. 2005. A quantitative description of the activation and inhibition of CFTR by potentiators: Genistein. *FEBS Lett* 579:3979–3983.

Morris ME, Zhang S. 2006. Flavonoid-drug interactions: effects of flavonoids on ABC transporters. *Life Sci* 78:2116–2130.

Movileanu L, Neagoe I, Flonta ML. 2000. Interaction of the antioxidant flavonoid quercetin with planar lipid bilayers. *Int J Pharm* 205:135–146.

Mowri H, Nojima S, Inoue K. 1984. Effect of lipid composition of liposomes on their sensitivity to peroxidation. *J Biochem (Tokyo)* 95:551–558.

Oh SM, Chung KH. 2006. Antiestrogenic activities of Ginkgo biloba extracts. *J Steroid Biochem Mol Biol* 100:167–176.

Oh SM, Chung KH. 2004. Estrogenic activities of Ginkgo biloba extracts. *Life Sci* 74:1325–1335.

Ollila F, Halling K, Vuorela P, Vuorela H, Slotte JP. 2002. Characterization of flavonoid–biomembrane interactions. *Arch Biochem Biophys* 399:103–108.

Om A, Kim JH. 2008. A quantitative structure-activity relationship model for radical scavenging activity of flavonoids. *J Med Food* 11:29–37.

Overstreet DH, Kralic JE, Morrow AL, Ma ZZ, Zhang YW, Lee DY. 2003. NPI-031G (puerarin) reduces anxiogenic effects of alcohol withdrawal or benzodiazepine inverse or 5-HT2C agonists. *Pharmacol Biochem Behav* 75:619–625.

Pawlikowska-Pawlega B, Gruszeckib WI, Misiakb LE, Gawron A. 2003. The study of the quercetin action on human erythrocyte membranes. *Biochem Pharmacol* 66:605–612.

Pepeljnjak S, Kosalec I. 2004. Galangin expresses bactericidal activity against multiple-resistant bacteria: MRSA, Enterococcus spp. and Pseudomonas aeruginosa. *FEMS Microbiol Lett* 240:111–116.

Phang JM, Poore CM, Lopaczynska J, Yeh GC. 1993. Flavonol-stimulated efflux of 7,12-dimethylbenz(a)anthracene in multidrug-resistant breast cancer cells. *Cancer Res* 53:5977–5981.

Pohl A, Devaux PF, Herrmann A. 2005. Function of prokaryotic and eukaryotic ABC proteins in lipid transport. *Biochim Biophys Acta* 1733:29–52.

Rackova L, Firakova S, Kostalova D, Stefek M, Sturdik E, Majekova M. 2005. Oxidation of liposomal membrane suppressed by flavonoids: Quantitative structure-activity relationship. *Bioorg Med Chem* 13:6477–6484.

Ratty AK, Sunamoto J, Das NP. 1988. Interaction of flavonoids with 1,1-diphenyl-2-picrylhydrazyl free radical, liposomal membranes and soybean lipoxygenase-1. *Biochem Pharmacol* 37:989–995.

Ray S, Sengupta C, Roy K. 2008. QSAR modeling for lipid peroxidation inhibition potential of flavonoids using topological and structural parameters. *Cent Eur J Chem* 6:267–276.

Rice-Evans C. 2001. Flavonoid antioxidants. *Curr Med Chem* 8:797–807.

Saija A, Scalese M, Lanza M, Marzullo D, Castelli F. 1995. Flavonoids as antioxidant agents: Importance of their interaction with biomembranes. *Free Radic Biol Med* 19:481–486.

Saponara S, Testai L, Iozzi D, Martinotti E, Martelli A, Chericoni S, Sgaragli G, Fusi F, Calderone V. 2006. (+ /-)-Naringenin as large conductance Ca(2 +)-activated K + (BKCa) channel opener in vascular smooth muscle cells. *Br J Pharmacol* 149:1013–1021.

Scheepers A, Joost HG, Schurmann A. 2004. The glucose transporter families SGLT and GLUT: Molecular basis of normal and aberrant function. *J Parenter Enteral Nutr* 28:364–371.

Scheidt HA, Pampel A, Nissler L, Gebhardt R, Huster D. 2004. Investigation of the membrane localization and distribution of flavonoids by high-resolution magic angle spinning NMR spectroscopy. *Biochim Biophys Acta* 1663:97–107.

Schmidt A, Hughes LK, Cai Z, Mendes F, Li H, Sheppard DN, Amaral MD. 2008. Prolonged treatment of cells with genistein modulates the expression and function of the cystic fibrosis transmembrane conductance regulator. *Br J Pharmacol* 153:1311–1323.

Schuier M, Sies H, Illek B, Fischer H. 2005. Cocoa-related flavonoids inhibit CFTR-mediated chloride transport across T84 human colon epithelia. *J Nutr* 135:2320–2325.

Sergent T, Garsou S, Schaut A, De Saeger S, Pussemier L, Van Peteghem C, Larondelle Y, Schneider YJ. 2005. Differential modulation of ochratoxin A absorption across Caco-2 cells by dietary polyphenols, used at realistic intestinal concentrations. *Toxicol Lett* 159:60–70.

Shapiro AB, Ling V. 1997. Positively cooperative sites for drug transport by P-glycoprotein with distinct drug specificities. *Eur J Biochem* 250:130–137.

Shimizu M, Kobayashi Y, Suzuki M, Satsu H, Miyamoto Y. 2000. Regulation of intestinal glucose transport by tea catechins. *Biofactors* 13:61–65.

Silva MM, Santos MR, Caroco G, Rocha R, Justino G, Mira L. 2002. Structure-antioxidant activity relationships of flavonoids: A re-examination. *Free Radic Res* 36:1219–1227.

Singleton VL. 1987. Oxygen with phenols and related reactions in musts, wines, and model systems: observations and practical iImplications. *Am J Enol Vitic* 38:69–77.

Sousa M, Ousingsawat J, Seitz R, Puntheeranurak S, Regalado A, Schmidt A, Grego T, Jansakul C, Amaral MD, Schreiber R, Kunzelmann K. 2007. An extract from the medicinal plant Phyllanthus acidus and its isolated compounds induce airway chloride secretion: A potential treatment for cystic fibrosis. *Mol Pharmacol* 71:366–376.

Strobel P, Allard C, Perez-Acle T, Calderon R, Aldunate R, Leighton F. 2005. Myricetin, quercetin and catechin-gallate inhibit glucose uptake in isolated rat adipocytes. *Biochem J* 386:471–478.

Suaud L, Li J, Jiang Q, Rubenstein RC, Kleyman TR. 2002. Genistein restores functional interactions between Delta F508-CFTR and ENaC in Xenopus oocytes. *J Biol Chem* 277:8928–8933.

Sun H, Cheng XP, You-Ye Z, Jiang P, Zhou JN. 2007. Quercetin subunit specifically reduces GlyR-mediated current in rat hippocampal neurons. *Neuroscience* 148:548–559.

Tamba Y, Ohba S, Kubota M, Yoshioka H, Yamazaki M. 2007. Single GUV method reveals interaction of tea catechin (−)-epigallocatechin gallate with lipid membranes. *Biophys J* 92:3178–3194.

Tarahovsky YS, Muzafarov EN, Kim YA. 2008. Rafts making and rafts braking: how plant flavonoids may control membrane heterogeneity. *Mol Cell Biochem* 314:65–71.

Terao J, Piskula M, Yao Q. 1994. Protective effect of epicatechin, epicatechin gallate, and quercetin on lipid peroxidation in phospholipid bilayers. *Arch Biochem Biophys* 308:278–284.

Tsuchiya H. 2001. Stereospecificity in membrane effects of catechins. *Chem Biol Int* 134:41–54.

Uekusa Y, Kamihira M, Nakayama T. 2007. Dynamic behavior of tea catechins interacting with lipid membranes as determined by NMR spectroscopy. *J Agric Food Chem* 55:9986–9992.

Valero MS, Garay RP, Gros P, Alda JO. 2006. Cystic fibrosis transmembrane conductance regulator (CFTR) chloride channel and Na-K-Cl cotransporter NKCC1 isoform mediate the vasorelaxant action of genistein in isolated rat aorta. *Eur J Pharmacol* 544:126–131.

van Dijk C, Driessen AJ, Recourt K. 2000. The uncoupling efficiency and affinity of flavonoids for vesicles. *Biochem Pharmacol* 60:1593–1600.

Vaya J, Tamir S. 2004. The relation between the chemical structure of flavonoids and their estrogen-like activities. *Curr Med Chem* 11:1333–1343.

Vera JC, Reyes AM, Carcamo JG, Velasquez FV, Rivas CI, Zhang RH, Strobel P, Iribarren R, Scher HI, Slebe JC, et al. 1996. Genistein is a natural inhibitor of hexose and dehydroascorbic acid transport through the glucose transporter, GLUT1. *J Biol Chem* 271:8719–8724.

Verstraeten SV, Hammerstone JF, Keen CL, Fraga CG, Oteiza PI. 2005. Antioxidant and membrane effects of procyanidin dimers and trimers isolated from peanut and cocoa. *J Agric Food Chem* 53:5041–5048.

Verstraeten SV, Oteiza PI, Fraga CG. 2004. Membrane effects of cocoa procyanidins in liposomes and Jurkat T cells. *Biol Res* 37:293–300.

Verstraeten SV, Keen CL, Schmitz HH, Fraga CG, Oteiza PI. 2003. Flavan-3-ols and procyanidins protect liposomes against lipid oxidation and disruption of the bilayer structure. *Free Radic Biol Med* 34:84–92.

Virgili F, Acconcia F, Ambra R, Rinna A, Totta P, Marino M. 2004. Nutritional flavonoids modulate estrogen receptor alpha signaling. *IUBMB Life* 56:145–151.

Wallace CH, Baczko I, Jones L, Fercho M, Light PE. 2006. Inhibition of cardiac voltage-gated sodium channels by grape polyphenols. *Br J Pharmacol* 149:657–665.

Wang P, Jeng CJ, Chien CL, Wang SM. 2008. Signaling mechanisms of daidzein-induced axonal outgrowth in hippocampal neurons. *Biochem Biophys Res Commun* 366:393–400.

Warner M, Saji S, Gustafsson JA. 2000. The normal and malignant mammary gland: A fresh look with ER beta onboard. *J Mammary Gland Biol Neoplasia* 5:289–294.

Wu YL, Ohsaga A, Oshiro T, Iinuma K, Kondo Y, Ebihara S, Sasaki H, Maruyama Y. 2005. Suppressive effects of red wine polyphenols on voltage-gated ion channels in dorsal root ganglionic neuronal cells. *Tohoku J Exp Med* 206:141–150.

Wu SN, Chiang HT, Shen AY, Lo YK. 2003. Differential effects of quercetin, a natural polyphenolic flavonoid, on L-type calcium current in pituitary tumor (GH3) cells and neuronal NG108–15 cells. *J Cell Physiol* 195:298–308.

Wuttke W, Jarry H, Seidlova-Wuttke D. 2007. Isoflavones–safe food additives or dangerous drugs? *Ageing Res Rev* 6:150–188.

Wymann MP, Schneiter R. 2008. Lipid signalling in disease. *Nat Rev Mol Cell Biol* 9:162–176.

Xi J, Guo R. 2007. Acid-base equilibrium of puerarin in CTAB micelles. *J Pharm Biomed Anal* 43: 111–118.

Yano S, Fujimura Y, Umeda D, Miyase T, Yamada K, Tachibana H. 2007. Relationship between the biological activities of methylated derivatives of (–)-epigallocatechin-3-O-gallate (EGCG) and their cell surface binding activities. *J Agric Food Chem* 55:7144–7148.

Yarishkin OV, Ryu HW, Park JY, Yang MS, Hong SG, Park KH. 2008. Sulfonate chalcone as new class voltage-dependent K + channel blocker. *Bioorg Med Chem Lett* 18:137–140.

Yoshioka H, Haga H, Kubota M, Sakai Y. 2006. Interaction of (+)-catechin with a lipid bilayer studied by the spin probe method. *Biosci Biotechnol Biochem* 70:395–400.

Zhang S, Morris ME. 2003. Effect of the flavonoids biochanin A and silymarin on the P-glycoprotein-mediated transport of digoxin and vinblastine in human intestinal Caco-2 cells. *Pharm Res* 20:1184–1191.

Zhang S, Yang X, Coburn RA, Morris ME. 2005. Structure activity relationships and quantitative structure activity relationships for the flavonoid-mediated inhibition of breast cancer resistance protein. *Biochem Pharmacol* 70:627–639.

Zhou S, Lim LY, Chowbay B. 2004. Herbal modulation of P-glycoprotein. *Drug Metab Rev* 36:57–104.

Zhu L, Jiang ZL, Krnjevic K, Wang FS, Ye JH. 2003. Genistein directly blocks glycine receptors of rat neurons freshly isolated from the ventral tegmental area. *Neuropharmacology* 45:270–280.

Zhu QY, Holt RR, Lazarus SA, Orozco TJ, Keen CL. 2002. Inhibitory effects of cocoa flavanols and procyanidin oligomers on free radical-induced erythrocyte hemolysis. *Exp Biol Med (Maywood)* 227:321–329.

5 The Biochemistry Behind the Potential Cardiovascular Protection by Dietary Flavonoids

WAI MUN LOKE[1,2], JONATHAN M. HODGSON[1], AND KEVIN D. CROFT[1]

[1]School of Medicine and Pharmacology
[2]School of Biomedical, Biomolecular and Chemical Sciences, University of Western Australia, Perth, Western Australia

CONTENTS

INTRODUCTION

Flavonoids are natural dietary phytochemicals and are currently the focus of much nutritional and therapeutic interest. However, the benefit of dietary flavonoids as part of a diet or as supplements is unclear. Results of population studies suggest that adopting flavonoid-rich diets may protect against cardiovascular disease [Hertog et al., 1995; Keli et al., 1996; Knekt et al., 1996; Huxley et al., 2003; Sesso et al., 1999; Yochum et al., 1999; Hirvonen et al., 2001;

Plant Phenolics and Human Health: Biochemistry, Nutrition, and Pharmacology,
Edited by Cesar G. Fraga.

Geleijnse et al., 2002; Mukamal et al., 2002]. Mechanisms by which these compounds exert their cardiovascular protective effects are still unknown. It is widely hypothesized that dietary flavonoids improve cardiovascular health and may help to prevent cardiovascular diseases/events by inhibiting pathogenic processes such as oxidative stress (lipid and protein peroxidation), inflammation, endothelial dysfunction, and platelet activation [Nijveldt et al., 2001].

Daily intake of flavonoids such as quercetin and (−)-epicatechin has been estimated at between 20 and 35 mg/day, in the form of various glycosides, although intact glycosides are not found in plasma [Scalbert and Williamson, 2000]. Following supplementation with flavonoid-rich foods (such as onions or apples), or various glucosides of the pure flavonoids, at doses of 50–200 mg equivalents, their plasma concentrations can reach between 0.5 and 7 μM [Manach et al., 2005 and references therein]. However, due to metabolic transformation, many flavonoids are not present as the aglycone form in plasma but in conjugated forms. For example, 20–40% of circulating quercetin is methylated at the 3′-position with other identified metabolites being the 3-*O*-glucuronide, 3′-*O*-sulfate or the 3′-*O*-methylquercetin-3-*O*-glucuronide [Manach et al., 2005] (see Fig. 5.1). In this chapter, we will discuss the results from recent studies regarding the possible protective roles of dietary flavonoids against cardiovascular diseases.

ANTIOXIDANT ACTIVITY

The biological activity of dietary flavonoids was thought to be due to their antioxidant activity. Lipid peroxidative damage may be a critical step in the pathogenesis of atherosclerosis [Stocker and Keaney, 2004] and the well recognized in vitro antioxidant activity of many flavonoids, has led to the proposal that the mechanism for this benefit may involve antioxidant effects [Duthie and Crozier, 2000]. Flavonoids can potentially prevent free-radical–related injury as they exhibit powerful antioxidant activities in vitro. They are able to scavenge a wide range of reactive oxygen, nitrogen, and chlorine species [Hanasaki et al., 1994; Santos and Mira, 2004; Sadeghipour et al., 2005] as well as being able to inhibit the production of such reactive species [Selloum et al., 2001; Pincemail et al., 1988]. Flavonoids can also chelate metal ions and often decrease metal ion pro-oxidant activity [Mira et al., 2002]. Ishige et al. [2001] proposed that flavonoids may protect against oxidative stress by increasing intracellular glutathione, directly reducing reactive oxygen species and preventing the influx of Ca^{2+}. However, the studies carried out in this area should be interpreted with caution as the native unmodified forms of flavonoids found in the diet were utilized in in vitro experiments instead of the metabolites found in vivo.

Recent evidence on the bioavailability and metabolism of these compounds in vivo suggests that dietary flavonoids are less likely to act as antioxidants. Most flavonoids may not have significant antioxidant activity in vivo for two reasons: (1) bioavailability of dietary flavonoids is very poor (their concentration in vivo

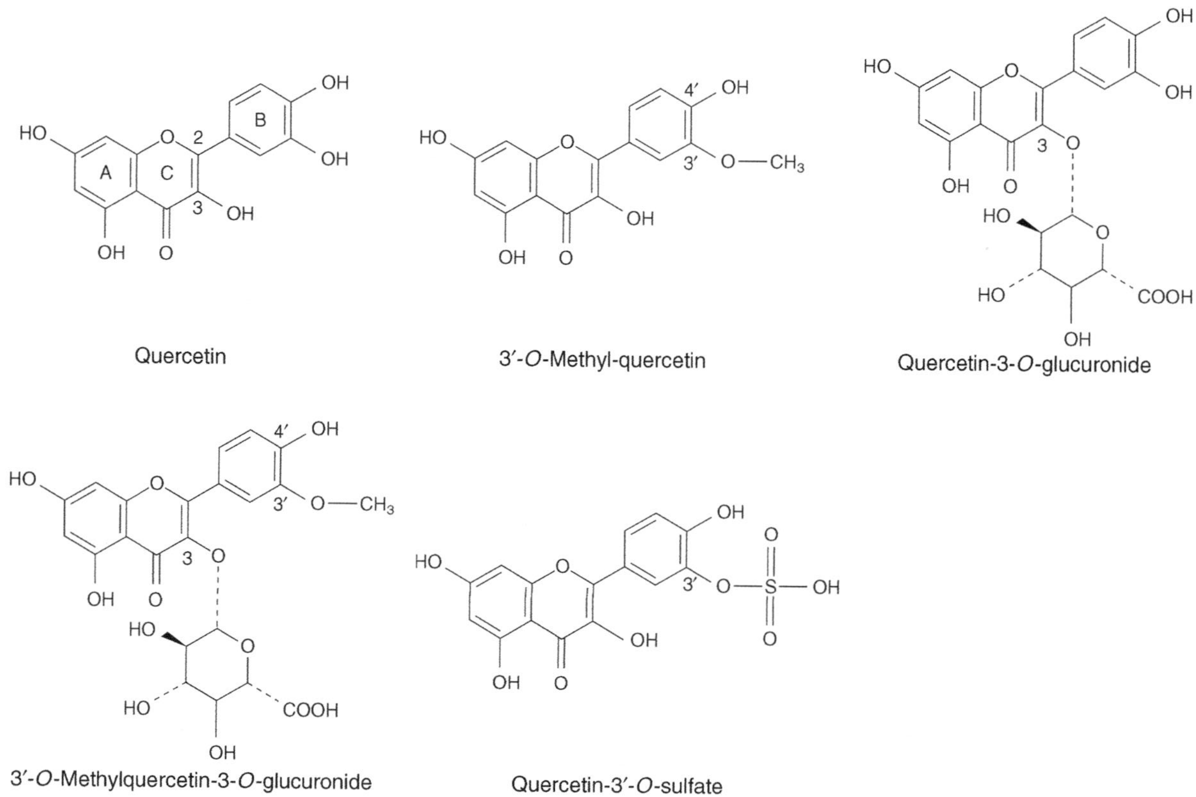

Figure 5.1 Backbone structure of dietary flavonoids.

is likely to be much lower than vitamin C or vitamin E, although they may act as co-antioxidants), and (2) biotransformation will lead to diminished antioxidant activity. Methylation, sulfation, and glucuronidation block the radical scavenging phenolic hydroxyl groups and may decrease antioxidant activity. It may be more useful to examine the antioxidant activity of the metabolites instead of the parent compounds. However, it has been suggested by some investigators that conjugated flavonoids may be hydrolyzed back to the free aglycones in certain cells [Kawai et al., 2008]. One pathway for low-density lipoprotein (LDL) modification in vivo is through reactive species generated by myeloperoxidase (MPO) [Zhang et al., 2002a; Podrez et al., 2000; Wang et al., 2007]. MPO has been suggested as a physiological catalyst for in vivo LDL modification in studies using monocytes and neutrophils isolated from humans [Hazen et al., 1999; Zhang et al., 2002b]. Previous reports of elevated plasma MPO in early adverse cardiac events [Brennan et al., 2003], in acute coronary syndrome [Baldus et al., 2003], and after acute myocardial infarction [Mocatta et al., 2007] provide further evidence that MPO may contribute to cardiovascular disease. More recently MPO has been implicated in the production of cyanate which can cause carbamylation of proteins [Wang et al., 2007].

We have recently concluded an in vitro study investigating and comparing the antioxidant activity of various major in vivo metabolites of quercetin against neutrophil–mediated LDL modification. The study showed that quercetin and its major metabolites at physiological realistic concentrations (1 μM) significantly protect LDL against neutrophil–inflicted oxidative damage as well as peroxyl radical-initiated oxidation, both measured by the formation of F_2-isoprostanes (a stable marker of lipid peroxidation) compared to the untreated control experiments [Loke et al., 2008a]. Quercetin acts predominantly through the inhibition of MPO as evidenced by reduction in both the MPO enzyme activity and 3-chlorotyrosine formation from LDL by freshly isolated human neutrophils in the presence of quercetin [Loke et al., 2008a]. Of the metabolites, only one (quercetin-3-*O*-glucuronide) showed comparable antioxidant activity as the parent molecule. Most structural activity relationship studies of flavonoids antioxidant action concluded that flavonoids owe their powerful antioxidant actions to the 2-phenyl-1,4-benzopyrone structure with multiple H-donating phenolic hydroxyl groups. Choi et al. [2002] reported that all flavonoids possessing the ortho dihydroxy functional group, whether on the A-ring or B-ring, and the OH group at position 3, containing either saturated or unsaturated C2–C3 double bonds, appear to have radical scavenging activities. While the 5-OH and 7-OH groups on the A-ring appear to have negligible radical scavenging abilities [Choi et al., 2002], they may enhance the radical scavenging activity of the 3-OH group on the C-ring [Heijnen et al., 2001]. Kettle et al. [2008] showed that myricitrin could irreversibly inactivate MPO and that the OH groups on the B-ring as well as the presence of a free *m*-phenol in the A-ring were important for the inhibitory effects. The catechol structure on the B-ring is widely reported to play a predominant role in scavenging free radicals [Santos and Mira, 2004; Justino et al. 2004].

Experiments carried out at physiological concentrations [Manach et al., 2005] showed that the 3′-OH and 3-OH groups on the B- and C-rings, respectively, played similar roles in both radical scavenging and MPO inhibiting activities [Loke et al., 2008a]. The 2,3 double bond on the B-ring played a negligible part in protecting LDL against oxidative damage because taxifolin (which lacks the 2,3 double bond) retained similar activity to quercetin (see Fig. 5.2). Similar observations were reported for cell-free Cu^{2+} and AAPH-mediated LDL oxidation under nonphysiological concentrations [Vaya et al., 2003; Hou et al., 2004]. Interestingly, conjugation at either 3- or 3′- positions did not significantly reduce quercetin activity against MPO-catalyzed events unlike that against radical scavenging activities. Therefore, metabolic transformation of quercetin may not render the molecule significantly less potent as a MPO inhibitor. These results are particularly interesting considering recent evidence that associates MPO with cardiovascular disease development and progression [Brennan et al., 2003; Baldus et al., 2003; Mocatta et al., 2007].

Studies examining whether dietary flavonoids exert antioxidant effects in vivo produced confusing and self-contradicting data. O'Reilly et al. [2001] reported that F_2-isoprostane concentrations and plasma levels of oxidized LDL did not differ whether healthy subjects were on the flavonoid-rich diet or flavonoid-poor diet. A similar lack of effect was observed when rutin supplementation did not affect urinary concentrations of 8-hydroxy-2′-deoxyguanosine, F_2-isoprostanes, or malondialdehyde in human volunteers [Boyle et al., 2000]. While the consumption of fruits and vegetables failed to decrease markers of oxidative damage, including F_2-isoprostanes and DNA damage markers in some studies [Berg et al., 2001; M ½ller et al., 2003], Thompson et al. (1999) reported decreased lipid peroxidation markers (F_2-isoprostanes and malondialdehyde) among subjects who consumed more fruits and vegetables. Dietary intervention studies involving green tea [Freese et al., 1999; Hakim et al., 2003], black tea [Hodgson et al., 2002; Klaunig et al., 1999] and red wines [Young et al., 2000; Caccetta et al., 2000], which represent rich sources of flavonoids in human diets, produced similar contradicting results. These confusing data may result, in part, from the choice of the biomarkers of lipid and protein peroxidation, as they do not fulfill all of the criteria for ideal biomarkers [Halliwell et al., 2004]. Much more work must be done to establish more accurate in vivo biomarkers of oxidative damage in order to draw conclusions on the antioxidant effects of dietary flavonoids.

LIPIDS AND LIPOPROTEINS

Hypercholesterolemia is well established as a risk factor for atherosclerosis. Dietary flavonoids are thought to protect against the disease by exerting hypocholesterolemic effects. However, clinical studies are not conclusive enough to provide the supporting evidence. Recent clinical studies on both normo- and hypercholesterolemic subjects with different flavonoids sources lasting 1–13 weeks showed improvement or no change in lipid profiles. Consumption of black tea (5 servings/day) in hypercholesterolemic subjects over a period of 3 weeks resulted in significant reductions in total cholesterol, LDL cholesterol, and apolipoprotein B levels [Davies et al., 2003] (HDL) cholesterol levels in healthy volunteers were elevated, but LDL/HDL ratio remained unaffected after 4 weeks of intervention with cocoa powder and dark chocolate [Wan et al., 2001]. Two independent studies on healthy subjects and smokers showed no change in the levels of triglyceride, LDL, HDL, and total cholesterol after red wine intervention [Cartron et al., 2003; Caccetta et al., 2001]. Lyophilized grape powder significantly decreased triglyceride and LDL cholesterol concentrations in both pre- and postmenstrual women [Zern et al., 2005].

Mechanisms by which dietary flavonoids may influence plasma lipids are yet to be defined. Flavonoids may reduce cholesterol absorption via interaction with cholesterol carriers and transporters across the brush border membrane [Loest et al., 2002]. This reduction in cholesterol absorption decreases the delivery of cholesterol to the liver, which in turn up-regulates the expression of LDL receptor to compensate for less substrate availability and induces reductions in plasma cholesterol [Zern et al., 2003]. Dietary flavonoids were also shown to affect hepatic production of lipoproteins and inhibit cholesterol esterification through their binding with the plasma membrane transport P-glycoprotein [Debry et al., 1997]. Reductions in plasma triglyceride by the flavonoids may occur as a result of lower microsomal transfer protein activity and increased lipoprotein lipase activity, which may further alter the delipidation cascade, yielding less LDL in circulation [Zern et al., 2003]. The citrus flavonoid, naringenin has been shown to improve dyslipidemia and insulin resistance in a high-fat-fed animal model of LDL receptor knockout mouse [Borradaile et al., 2003]. The mechanism is via the inhibition of apoB100 secretion in the liver. Whether these benefits are observed in humans is yet to be established.

DIETARY FLAVONOIDS AND INFLAMMATION

Inflammation is now recognized as a key process in atherogenesis [Libby, 2002]. The potential for dietary flavonoids to inhibit inflammatory activities is of particular interest. A potential anti-inflammatory feature of the flavonoids is the ability to inhibit the biosynthesis of eicosanoids. Selected phenolic acids and some flavonoids have been shown to inhibit both cyclooxygenase (COX) and 5-lipoxygenase (5-LO) pathways [Nijveldt et al., 2001; Takano-Ishikawa et al., 2006]. Epicatechin and related flavonoids have been shown to inhibit the synthesis of pro-inflammatory cytokines in vitro [Sanbongi et al., 1997], and plasma metabolites of catechin and quercetin inhibit the adhesion of monocytes to cultured endothelial cells [Koga and Meydani, 2001]. Silymarin has been shown to inhibit the production of inflammatory cytokines, such as interleukin-1, interferon-, and tumor necrosis factor-α (TNFα), from macrophages and T-cells [Matsuda et al., 2005]. Some flavonoids can inhibit neutrophil degranulation, diminishing the release of free arachidonic acid [Tordera et al., 1994]. These activities are of interest since the possible roles of COX-2 and leukotriene pathways in atherosclerosis are becoming more appreciated [Jala and Haribabu, 2004; Vila, 2004].

We have recently reported that although quercetin is a very potent inhibitor of LTB_4 production in human peripheral monocytes and neutrophils at a realistic physiological concentration ($\sim$2 μM), some of its major metabolites show significantly diminished activity [Loke et al., 2008b]. Metabolism via conjugation at 3′-OH of quercetin's phenyl *benzopyrone* structure (3′-*O*-methylquercetin and quercetin-3′-*O*-sulfate) decreased LTB_4 inhibitory activity by up to 50% while conjugation at the 3-OH (quercetin-3-*O*-glucuronide and 3′-*O*-methylquercetin-3-*O*-glucuronide) greatly diminished LTB_4 inhibiting activity within the physiological concentration range (2–10 μM) tested. When quercetin was compared with structural analogs (luteolin, kaempferol, and taxifolin), it became apparent that the 3′-OH of the B-ring played a more critical role in LTB_4 inhibition than the 3-OH of the C-ring [Loke et al., 2008b]. This result is in agreement with a previous observation that formation of the 3-*O*-glucuronide by incubation of quercetin with liver cell-free extracts substantially reduces lipoxygenase inhibitory activity [Day et al., 2000]. Interestingly, glucuronide formation at other sites (3′, 4′, and 7) had little effect on soybean lipoxygenase inhibition in a cell-free preparation [Day et al., 2000]. These results also highlighted that the 2,3 double bond within the C-ring in quercetin is an essential structural requirement for inhibition of LTB_4 production in neutrophils, as its absence (as in taxifolin) totally diminished the inhibiting action. The results are consistent with previous reports that the 2,3 double bond of the C-ring in flavonols is essential for inhibition of adhesion molecule expression in endothelial cells [Lotito et al., 2006] and inflammatory cytokine production in mouse macrophages [Comalada et al., 2006; Takano-Ishikawa et al., 2006]. In addition, we have observed a remarkable dissociation between structural features that determine anti-inflammatory activity and antioxidant activity, which are illustrated in Figure 5.2 [Loke et al., 2008b]. Another factor that may influence the activity of quercetin metabolites is change in polarity and reduced cellular uptake. This is particularly noticeable with polar moieties such as glucuronide and sulfate, while methylation has little effect on cellular uptake. However, the actual mechanisms by which neutrophils sequester quercetin remain unknown at this time.

There is evidence that quercetin and other flavonols such as (−)-epicatechin can directly inhibit human 5-lipoxygenase. It has been established that quercetin does not suppress the hydrolysis of LTA_4 to LTB_4, indicating that quercetin acts directly on 5-lipoxygenase or possibly FLAP [Loke et al., 2008b]. Using a recombinant enzyme, Schewe et al. [2002] showed direct inhibition in a cell-free system with quercetin having IC_{50} = 0.6 μM for 5-lipoxygenase and 4 μM for inhibition of 15-lipoxygenase. Since these were cell-membrane-free systems, the involvement of FLAP appears to be ruled out. Moreover, the flavonols appear to be nonspecific lipoxygenase inhibitors [Schewe et al., 2002].

Whatever the mechanism of action for the inhibition of 5-lipoxygenase by flavonoids, it appears to be distinct from the antioxidant properties of these compounds. The results comparing antioxidant activity with leukotriene inhibitory activity clearly demonstrate this distinction. The profound effects of metabolic transformation on the anti-inflammation activity of dietary flavonoids such as quercetin must also be considered in relation to in vitro studies, and further highlights the need to use actual metabolic forms of flavonoids rather than the free aglycone or glycosides occurring in the diet.

While in vitro studies have provided extensive augments for anti-inflammatory effects of various dietary flavonoids, human studies provide only a few tangible results. Flavonoids may exert anti-inflammatory effects through modulation of immune processes involving cytokines, inflammatory mediators, as well as circulating adhesion molecules. TNF-induced adhesion of monocytes to endothelial cells was virtually abolished after red wine consumption in human volunteers but was only partially reduced after gin consumption [Badia et al., 2004]. Similar observations were made in another 11-week study with a crossover design comparing red wine and gin consumption in healthy men where adhesion molecules and monocyte adhesion to endothelial cells were significantly reduced due to red wine [Estruch et al., 2004]. Lyophilized grape powder treatment was shown to significantly decrease TNF and IL-6 concentrations in both pre- and postmenopausal women [Zern et al., 2005]. Circulating soluble P-selectin in healthy human volunteers was significantly reduced after black tea consumption [Hodgson et al., 2001]. However, supplementation with cocoa did not affect circulating concentrations of cytokines [Mathur et al., 2002] as well as urinary thromboxane B_2 and 6-keto-prostaglandin $F_{1\alpha}$ (a metabolite of prostacyclin) excretions [Wan et al., 2001] in healthy human volunteers. Circulating VCAM-1 concentrations were significantly reduced after 6 weeks administration of formononetin-enriched isoflavones [Teede et al., 2003]. This effect was totally absent in another human intervention study involving 6 weeks of soy isoflavones [Steinberg et al., 2003]. As the transcription factor, nuclear factor-kB (NF-kB) is responsible for activating cytokines, adhesion molecules and other pro-inflammatory mediators, flavonoids may act by inhibiting NF-kB [Murase et al., 1999]. The limited anti-inflammatory effects of flavonoids in these human studies may be due to the lack of inflammatory immune responses in healthy subjects.

In summary, while in vitro studies indicate that several flavonoids may have profound anti-inflammatory activity, results from human intervention studies are less consistent.

ENDOTHELIAL FUNCTION

Endothelial dysfunction is a critical event in the pathogenesis of atherosclerosis and its clinical manifestations [Mano et al., 1996; Cayatte et al., 1994]. It accelerates the development of atherosclerosis and may be one of the earliest

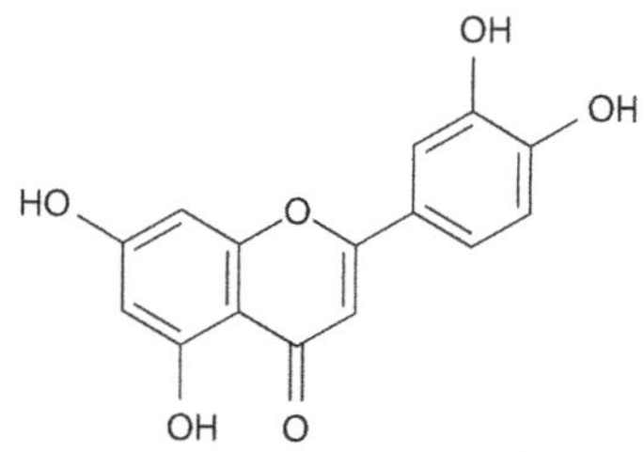

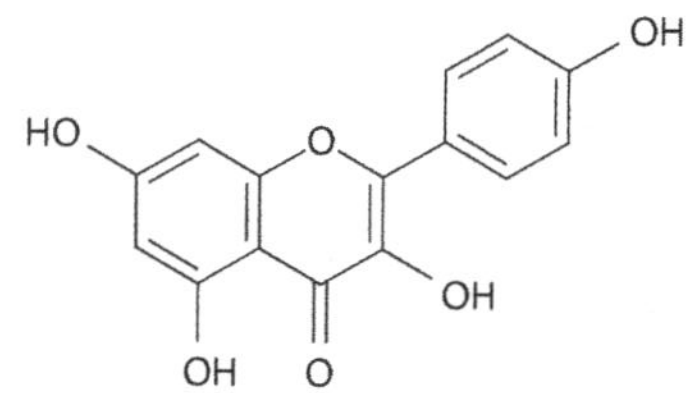

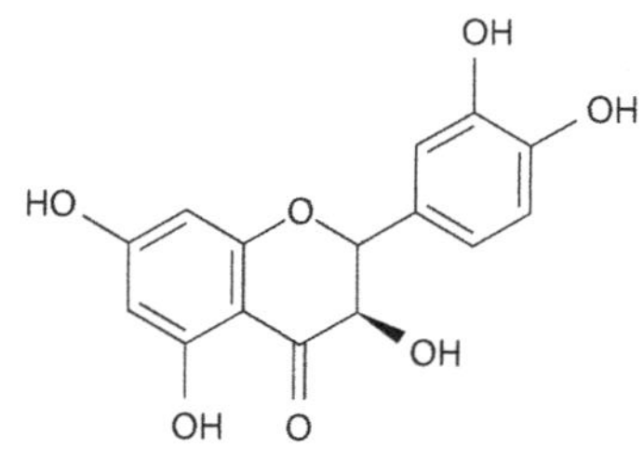

	Luteolin (L)	Kaempferol (K)	Taxifolin (T)
Structural difference compared to Q	Absence of 3-OH group	Absence of 3′-OH group	Absence of 2,3 double bond
MPO inhibiting activity	Diminished activity compared to Q, similar to K	Diminished activity compared to Q, similar to L	Strong activity, similar to Q
Radical scavenging activity	Diminished activity compared to Q, similar to K	Diminished activity compared to Q, similar to L	Strong activity, similar to Q
Antiinflammatory activity	Strong activity, similar to Q	Some activity	No activity

Figure 5.2 Structures of luteolin (L), kaempferol (K), and taxifolin (T). A comparison of the antioxidant (inhibition of MPO and radical scavenging activity) and anti-inflammatory (inhibition of LTB_4) activity with that of quercetin (Q) is summarized.

manifestations of this disease [Celermajer, 1997; Vogel, 1997]. The endothelium regulates vascular tone by balancing the production of vasodilators, most importantly nitric oxide (NO) [Rassaf et al., 2002] and vasoconstrictors, such as endothelin-1 (ET-1) [Strachan et al., 2002]. It maintains vascular homeostasis through multiple complex interactions with cells in the vessel wall. Therefore, endothelial function may serve as an indication for cardiovascular health and be used for evaluation of new therapeutic strategies [Ganz et al., 2003].

Flavonoid rich tea [Hirata et al., 2004], purple grape juice [Stein et al., 1999], and cocoa [Fisher et al., 2003] have all been found to improve endothelial function in acute and short-term intervention trials in humans [Hodgson et al., 2006]. Improving endothelium-dependent vasodilation is believed to be one possible mechanism by which flavonoids may improve cardiovascular risk [Heiss et al., 2006]. There are many hundreds of flavonoids in the human diet. However, it is likely that bioactivity relevant to endothelial function is limited to fewer compounds. Isolated (−)-epicatechin [Schroeter et al., 2006] and epigallocatechin gallate (EGCG) [Widlansky et al., 2005] have been found to acutely improve endothelial function in humans. Isolated (−)-epicatechin has also been shown to augment NO status in humans [Schroeter et al., 2006]. Resveratrol was able to up-regulate the expression of endothelial nitric oxide synthase (eNOS) messenger ribonucleic acid (mRNA) in a time- and dose-dependent manner and thereby increasing NO production in the cultured human endothelial cells [Wallerath et al., 2002]. Similar results were obtained with red wine polyphenol extract [Leikert et al., 2002]. Epicatechin was shown to protect endothelial cells against oxidized LDL and maintain NO synthase [Steffen et al., 2005]. It is therefore possible that flavonoids may improve endothelial function in vivo via NO production, explaining beneficial effects of dietary flavonoids in reducing atherosclerosis.

We have recently conducted a randomized, placebo controlled, crossover trial in healthy men to compare the acute effects on nitric oxide, ET-1, and oxidative stress after oral administration of 200 mg of quercetin, (−)-epicatechin, and EGCG [Loke et al., 2008c]. NO production was assessed by measuring plasma *S*-nitrosothiols, plasma and urinary nitrite and nitrate concentrations, while their effects on oxidative stress by measuring plasma and urinary F_2-isoprostanes. Plasma and urinary concentrations of quercetin, (−)-epicatechin, and EGCG were measured to establish the absorption of these flavonoids. Relative to water (control), quercetin and (−)-epicatechin resulted in a significant increase in plasma *S*-nitrosothiols and nitrite and urinary nitrate concentrations, but not plasma nitrate or urinary nitrite. EGCG did not alter any of the measures of NO production. Only treatment with quercetin and epicatechin resulted in a significant reduction in plasma ET-1 concentration, and quercetin alone decreasing urinary ET-1. All three treatments did not significantly change plasma or urinary F_2-isoprostane concentrations. Significant increases in the circulating levels of the three flavonoids were observed after each treatment, however, plasma levels of EGCG were very low (approx. 0.1 μM) even after supplementation. Our results showed that dietary flavonoids, such as quercetin and (−)-epicatechin augment NO status and may thereby improve endothelial function [Loke et al., 2008c]. The changes in plasma *S*-nitrosothiols concentrations were also shown to be significantly correlated to the changes in plasma quercetin and (−)-epicatechin concentrations [Loke et al., 2008c].

Quercetin may have increased NO production by enhancing the bioavailability of endothelium-derived NO [Machha et al., 2007] or increasing eNOS activity [Sanchez et al., 2006; Benito et al., 2002]. (−)-Epicatechin has been shown to elevate NO in endothelial cells in vitro via the inhibition of NADPH oxidase [Steffen et al., 2007]. Oral administration of pure (−)-epicatechin to humans closely emulated the acute vascular effects of the flavonol-rich cocoa [Schroeter et al., 2006]. Quercetin and (−)-epicatechin may also act as antioxidants, reducing nitrites and nitrates into free NO in the stomach [Gago et al., 2007]. ET-1 has been demonstrated to be associated with increased oxidative stress and endothelial dysfunction in humans. ET-1 stimulates superoxide production and vasoconstriction through activation of NADPH oxidase and uncoupled NOS in the rat aorta [Loomis et al., 2005]. It also reduces NO bioavailability via interference with the expression and activity of eNOS [Ramzy et al., 2006], indicating that diminished ET-1 level may be accompanied by elevated NO bioavailability. It has been reported that NO inhibits ET-1 production through the suppression of NF- B [Ohkita et al., 2002]. There seems to be an inverse relationship between NO and ET-1, which may serve to modulate the endothelial function in the vasculature. Quercetin was shown to decrease ET-1 production in thrombin-stimulated cultured human umbilical vein endothelial cells in a dose-dependent manner with an IC_{50} of 1.54 μM [Zhao et al., 1999]. Red wine polyphenols have also been shown to prevent vascular oxidative stress by inhibiting NADPH oxidase activity and/or by reducing ET-1 release [Jimenez et al., 2007]. The potential mechanistic pathways by which dietary flavonoids may improve NO status and endothelial function are summarized in Figure 5.3.

Surprisingly, EGCG did not show the same augmentation of NO products as quercetin and (−)-epicatechin [Loke et al., 2008c]. It has been widely assumed to be the vasoactive flavonoid present in green tea that offers vascular protection against cardiovascular diseases [Demeule et al., 2002]. EGCG was shown to mediate NO-dependent vasodilation in rat aortic rings [Kim et al., 2007], and it works primarily by the rapid activation of eNOS and increase in eNOS activity, independent of an altered eNOS protein content [Lorenz et al., 2004]. However, it must be noted that these studies reported the effects of EGCG at nonphysiological concentrations. EGCG was present at much lower concentrations (0.1 ± 0.01 μM) in the circulation than quercetin (3.54 ± 1.57 μM) and (−)-epicatechin (3.57 ± 1.21 μM) after acute treatment. Similar circulating concentrations of EGCG was reported in a recent study in which 300 mg dose of EGCG acutely improves brachial artery flow-mediated dilation by vascular ultrasound in humans with coronary artery disease [Widlansky et al., 2005]. If improved endothelial function is brought about by EGCG, the

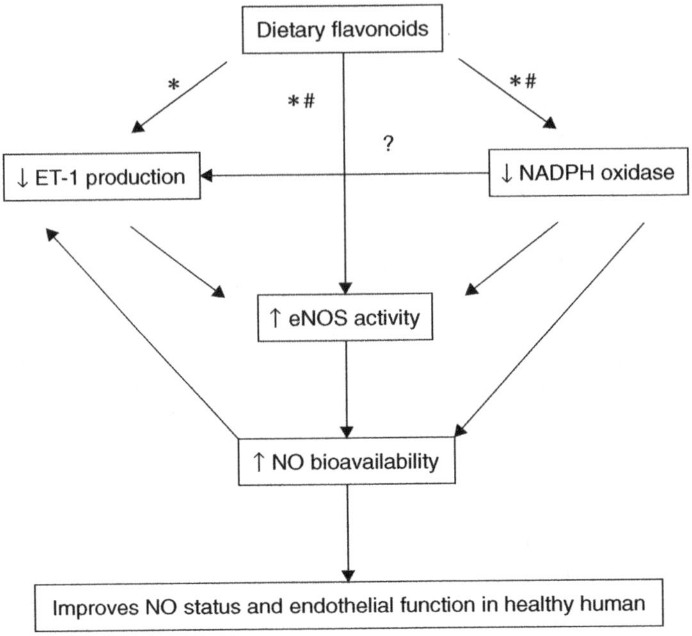

Figure 5.3 Potential mechanistic pathways by which dietary flavonoids may improve NO status and endothelial function. (*) Based on in vitro studies. (#) Evidence from animal studies.(?) Possible proposed action.

compound is likely to have exerted its effects through mechanisms other than those of NO or ET-1.

There is now emerging data that specific dietary flavonoids may improve endothelial function by increasing NO bioavailability. Longer term human intervention studies are required to establish a proven benefit on vascular function in healthy subjects as well as those who may be at increased risk of cardiovascular disease.

DIETARY FLAVONOIDS AND PLATELET REACTIVITY

Platelet aggregation plays a critical role in the pathogenesis of acute coronary syndromes with increasing evidence that antiplatelet therapy reduces cardiovascular disease risk [Awtry and Loscalzo, 2000]. The effect of dietary flavonoids to reduce platelet activity may provide an important mechanistic explanation for the available epidemiological data regarding flavonoids and cardiovascular disease.

Demrow et al. (1995) used the Folts model of unstable coronary stenosis, which closely mimics ruptured atherosclerotic plaque, causing unstable angina, to examine the effects of grape juice on platelet function in vivo. In this model, transient platelet aggregation and release are reflected in cyclic variations in coronary blood flow. Acute intragastric administration of red wine or grape juice was associated with marked reductions in cyclic flow variations, which was indicative of an antiplatelet effect. Tea consumption reduced plasma concentrations of P-selectin (a marker of in vivo platelet activation) [Hodgson et al., 2001]. However, other studies that examined the effects of short-term and/or long-term tea consumption on ex vivo platelet aggregation did not demonstrate any effect on platelet function [Duffy et al., 2001; Hodgson et al., 2002]. More studies are required to define the effects of tea consumption on platelet function. Cocoa decreased epinephrine-stimulated or adenosine diphosphate-stimulated glycoprotein IIb/IIIa and P-selectin expression in human subjects 2 and 6 h after consumption [Rein et al., 2000]. When platelet-related primary homeostasis was measured as the time taken for epinephrine-stimulated or adenosine diphosphate-stimulated whole blood to occlude an aperture in a collagen membrane under stimulated small vessel shear condition, the respective closure time increased 6 h and 2 h after flavonoid-rich cocoa ingestion. Quercetin has been implicated as a dietary inhibitor of platelet cell signaling when a human acute intervention study showed that platelet aggregation was inhibited 30 and 120 min after ingestion of 150 and 300 mg of quercetin-4′-*O*-β-glucoside, accompanied by reduced tyrosine phosphorylation of the tyrosine kinase Syk and phospholipase Cγ2 components of the platelet glycoprotein VI collagen receptor signaling pathway [Hubbard et al., 2004].

The exact mechanisms by which flavonoids inhibit platelet activity are not yet fully understood, but it is possible that flavonoids effect changes in membrane fluidity, ligand–receptor affinity, and intracellular signaling pathways. Flavonoids may mediate their effects through antioxidant and NO-related pathways. Catechin was shown to reduce platelet aggregation and hydrogen peroxide in vitro [Pignatelli et al., 2000] and attenuate oxidant-induced platelet activation in vivo [Blache et al., 2002]. The addition of grape juice to platelets ex vivo reduced platelet aggregation, decreased platelet production of superoxide anion, and increased the platelet production of nitric oxide [Freedman et al., 2001]. Flavonoids or flavonoid-rich food can modulate different aspects of eicosanoid metabolism in vivo [Schramm et al., 2001] and were shown to inhibit platelet 12-lipoxygenase and 5-lipoxygenase in vitro [Sies et al., 2005]. Flavonoids can also modulate membrane fluidity by interacting with the lipid bilayer [Verstraeten et al., 2003], which in turn can result in changes in membrane receptor function and enzymatic activity [Oteiza et al., 2005]. Lipid rafts may be important membrane components involved in cell signaling [Simons et al., 1997; Bodin et al., 2003] and platelet activation [Gousset et al., 2002]. Recently, epigallocatechin-3-*O*-gallate was found to associate strongly with the cholesterol component of the lipid raft in human basophilic KU812 cells [Fujimura et al., 2004], suggesting that certain flavonoids may modulate cellular activation through interaction with lipid rafts.

CONCLUSIONS

If flavonoids are cardiovascular protective nutrients, the bioactivity must originate from their metabolites rather than from the native forms present in our food and beverages. These metabolites may be present in the circulation at sufficient concentrations to exert their protective actions. These include: (a) inhibition of MPO enzyme activity, which protects LDL against lipid peroxidation and protein modification; (b) inhibition of eicosanoid biosyntheses, especially from the COX and lipoxygenase pathways; (c) regulation of endothelial function by modulating the production of NO; and (d) platelet activity and thrombosis inhibition. Since NF- B is implicated in the transcriptional activation of numerous genes, including those relevant to atherosclerosis and inflammatory responses, its inhibition by these compounds may explain some of their potential properties. Flavonoids and their metabolites may also act in cells as signaling molecules through actions at protein kinase and lipid kinase signaling pathways [Williams et al., 2004].

To fully understand the actual mechanism of action of dietary flavonoids either as antioxidants, modulators of cell signalling, or inflammatory pathways, it is important to detect and identify their metabolites in vivo as well as to study the consequences of interaction of these circulating metabolites with cells. Animal models and human studies should also be conducted to identify the in vivo mechanism of action of dietary flavonoids. Results of such studies are crucial in the evaluation of their potential as cardiovascular protective agents and may eventually lead to specific advice regarding intake of foods and beverages rich in flavonoids and attempts to alter and enhance flavonoid content of a range of foods.

REFERENCES

Awtry EH, Loscalzo J. 2000. *Aspirin. Circulation* 101:1206–1218.

Badia E, Sacanella E, Fernandez-Sola J, Nicolas JM, Antunez E, Rotilio D, de Gaetano G, Urbano-Marquez A, Estruch R. 2004. Decreased tumor necrosis factor-induced adhesion of human monocytes to endothelial cells after moderate alcohol consumption. *Am J Clin Nutr* 80:225–230.

Baldus S, Heeschen C, Meinertz T, Zeiher AM, Eiserich JP, Mu?nzel T, Simoons ML, Hamm CW. 2003. Myeloperoxidase serum levels predict risk in patients with acute coronary syndromes. *Circulation* 108:1440–1445.

Benito S, Lopez D, Saiz MP, Buxaderas S, Sa?nchez J, Puig-Parellada P, Mitjavila MT. 2002. A flavonoid-rich diet increases nitric oxide production in rat aorta. *Br J Pharmacol* 135:910–916.

Blache D, Durand P, Prost M, Loreau N. 2002. (+)-Catechin inhibits platelet hyperactivity induced by an acute iron load in vivo. *Free Radic Biol Med* 33:1670–1680.

Bodin S, Tronchere H, Payrastre B. 2003. Lipid rafts are critical membrane domains in blood platelet activation processes. *Biochim Biophys Acta* 1610:247–257.

Borradaile NM, de Dreu LE, Huff MW. 2003. Inhibition of net HepG2 cell apolipoprotein B secretion by the citrus flavonoid naringenin involves activation of phosphatidylinositol 3-kinase, independent of insulin receptor substrate-1 phosphorylation. *Diabetes* 52:2554–2561.

Boyle S, Dobson V, Duthie S, Hinselwood D, Kyle JA, Collins AR. 2000. Bioavailability and efficiency of rutin as an antioxidant: A human supplementation study. *Eur J Clin Nutr* 54:774–782.

Brennan M-L, Penn MS, Van Lente F, Nambi V, Shishehbor MH, Aviles RJ, Goormastic M, Pepoy ML, McErlean ES, Topol EJ, Nissen SE, Hazen SL. 2003. Prognostic value of myeloperoxidase in patients with chest pain. *N Engl J Med* 349:1595–1604.

Caccetta RA, Burke V, Mori TA, Beilin LJ, Puddey IB, Croft KD. 2001. Red wine polyphenols, in the absence of alcohol, reduce lipid peroxidative stress in smoking subjects. *Free Radic Biol Med* 30:636–642.

Cartron E, Fouret G, Carbonneau M-A, Lauret Cel, Michel Fco, Monnier L, Descomps B, Léger CL. 2003. Red-wine beneficial long-term effect on lipids but not on antioxidant characteristics in plasma in a study comparing three types of wine—description of two *O*-methylated derivatives of gallic acid in humans. *Free Radic Res* 37:1021–1035.

Cayatte AJ, Palacino JJ, Horten K, Cohen RA. 1994. Chronic inhibition of nitric oxide production accelerates neointima formation and impairs endothelial function in hypercholesterolemic rabbits. *Arterioscler Thromb Vasc Biol* 14:753–759.

Celermajer DS. 1997. Endothelial dysfunction: Does it matter? Is it reversible? *J Am Coll Cardiol* 30:325–333.

Choi JS, Chung HY, Kang SS, Jung MJ, Kim JW, No JK, Jung HA. 2002. The structure-activity relationship of flavonoids as scavengers of peroxynitrite. *Phytother Res* 16:232–235.

Comalada M, Ballester I, Bailon E, Sierra S, Xaus J, Galvez J, Medina FSD, Zarzuelo A. 2006. Inhibition of pro-inflammatory markers in primary bone marrow-derived mouse macrophages by naturally occurring flavonoids: Analysis of the structure-activity relationship. *Biochem Pharmacol* 72:1010–1021.

Davies MJ, Judd JT, Baer DJ, Clevidence BA, Paul DR, Edwards AJ, Wiseman SA, Muesing RA, Chen SC. 2003. Black tea consumption reduces total and ldl cholesterol in mildly hypercholesterolemic adults. *J Nutr* 133:3298S–3302.

Day AJ, Bao Y, Morgan MRA, Williamson G. 2000. Conjugation position of quercetin glucuronides and effect on biological activity. *Free Radic Biol Med* 29:1234–1243.

Debry P, Nash EA, Neklason DW, Metherall JE. 1997. Role of multidrug resistance p-glycoproteins in cholesterol esterification. *J Biol Chem* 272:1026–1031.

Demeule M, Michaud-Levesque J, Annabi B, Gingras D, Boivin D, Jodoin J, Lamy S, Bertrand Y, Beliveau R. 2002. Green tea catechins as novel antitumor and antiangiogenic compounds. *Curr Med Chem* 2:441–463.

Demrow HS, Slane PR, Folts JD. 1995. Administration of wine and grape juice inhibits in-vivo platelet activity and thrombosis in stenosed canine coronary arteries. *Circulation* 91:1182–1188.

Duffy SJ, Vita JA, Holbrook M, Swerdloff PL, Keaney JFJr. 2001. Effect of acute and chronic tea consumption on platelet aggregation in patients with coronary artery disease. *Arterioscler Thromb Vasc Biol* 21:1084–1089.

Duthie G, Crozier A. 2000. Plant-derived phenolic antioxidants. *Curr Opin Lipidol* 11:43–47.

Estruch R, Sacanella E, Badia E, Antunez E, Nicolas JM, Fernandez-Sola J, Rotilio D, de Gaetano G, Rubin E, Urbano-Marquez A. 2004. Different effects of red wine and gin consumption on inflammatory biomarkers of atherosclerosis: A prospective randomized crossover trial: Effects of wine on inflammatory markers. *Atherosclerosis* 175:117–123.

Fisher NDL, Hughes M, Gerhard-Herman M, Hollenberg NK. 2003. Flavanol-rich cocoa induces nitric-oxide-dependent vasodilation in healthy humans. *J Hypertens* 21:2281–2286.

Ford-Hutchinson AW. 1991. FLAP: A novel drug target for inhibiting the synthesis of leukotrienes. *Trends Pharmacol Sci* 12:68–70.

Freedman JE, Parker CIII, Li L, Perlman JA, Frei B, Ivanov V, Deak LRBS, Iafrati MD, Folts JD. 2001. Select flavonoids and whole juice from purple grapes inhibit platelet function and enhance nitric oxide release. *Circulation* 103:2792–2798.

Freese R, Basu S, Hietanen E, Nair J, Nakachi K, Bartsch H, Mutanen M. 1999. Green tea extract decreases plasma malondialdehyde concentration but does not affect other indicators of oxidative stress, nitric oxide production, or hemostatic factors during a high-linoleic acid diet in healthy females. *Eur J Nutr* 38:149–157.

Fujimura Y, Tachibana H, Yamada K. 2004. Lipid raft-associated catechin suppresses the FcεRI expression by inhibiting phosphorylation of the extracellular signal-regulated kinase1/2. *FEBS Lett* 556:204–210.

Gago B, Lundberg JO, Barbosa RM, Laranjinha J. 2007. Red wine-dependent reduction of nitrite to nitric oxide in the stomach. *Free Radic Biol Med* 43:1233–1242.

Ganz P, Vita JA. 2003. Testing endothelial vasomotor function: Nitric oxide, a multipotent molecule. *Circulation* 108:2049–2053.

Geleijnse JM, Launer LJ, van der Kuip DA, Hofman A, Witteman JC. 2002. Inverse association of tea and flavonoid intakes with incident myocardial infarction: The Rotterdam Study. *Am J Clin Nutr* 75:880–886.

Gousset K, Wolkers WF, Tsvetkova NM, Oliver AE, Field CL, Walker NJ, Crowe JH, Tablin F. 2002. Evidence for a physiological role for membrane rafts in human platelets. *J Cell Physiol* 190:117–128.

Hakim IA, Harris RB, Brown S, Chow H-HS, Wiseman S, Agarwal S, Talbot W. 2003. Effect of increased tea consumption on oxidative DNA damage among smokers: A randomized controlled study. *J Nutr* 33:3303S–3309.

Halliwell B, Long L, Lee T, Lim S, Kelly R. 2004. Establishing biomarkers of oxidative stress: The measurement of hydrogen peroxide in human urine. *Curr Med Chem* 11:1085–10392.

Hanasaki Y, Ogawa S, Fukui S. 1994. The correlation between active oxygens scavenging and antioxidative effects of flavonoids. *Free Radic Biol Med* 16:845–850.

Hazen SL, Zhang R, Shen Z, Wu W, Podrez EA, MacPherson JC, Schmitt D, Mitra SN, Mukhopadhyay C, Chen Y, Cohen PA, Hoff HF, Abu-Soud HM. 1999. Formation of nitric oxide-derived oxidants by myeloperoxidase in monocytes: Pathways for monocyte-mediated protein nitration and lipid peroxidation in vivo. *Circ Res* 85:950–958.

Heijnen CGM, Haenen GRMM, Vekemans JAJM, Bast A. 2001. Peroxynitrite scavenging of flavonoids: Structure activity relationship. *Environ Toxicol Pharmacol* 10:199–206.

Heiss C, Schroeter H, Balzer J, Kleinbongard P, Matern S, Sies H, Kelm M. 2006. Endothelial function, nitric oxide, and cocoa flavanols. *J Cardiovasc Pharmacol* 47:S128–S135.

Hertog MG, Kromhout D, Aravanis C, Blackburn H, Buzina R, Fidanza F, Giampaoli S, Jansen A, Menotti A, Nedeljkovic S. 1995. Flavonoids intake and long-term risk of coronary heart disease and cancer in the seven countries study. *Arch Intern Med* 155:381–386.

Hirata K, Shimada K, Watanabe H, Otsuka R, Tokai K, Yoshiyama M, Homma S, Yoshikawa J. 2004. Black tea increases coronary flow velocity reserve in healthy male subjects. *Am J Cardiol* 93:1384–1388.

Hirvonen T, Pietinen P, Virtanen M, Ovaskainen M-L, Ha?kkinen S, Albanes D, Virtamo J. 2001. Intake of flavonols and flavones and risk of coronary heart disease in male smokers. *Epidemiology* 12:62–67.

Hodgson JM, Croft KD. 2006. Dietary flavonoids: Effects on endothelial function and blood pressure. *J Sci Food Agric* 86:2492–2498.

Hodgson JM, Croft KD, Mori TA, Burke V, Beilin LJ, Puddey IB. 2002a. Regular ingestion of tea does not inhibit in vivo lipid peroxidation in humans. *J Nutr* 132:55–58.

Hodgson JM, Puddey IB, Burke V, Beilin LJ, Mori TA, Chan SY. 2002b. Acute effects of ingestion of black tea on postprandial platelet aggregation in humans. *Br J Nutr* 87:141–145.

Hodgson JM, Puddey IB, Mori TA, Burke V, Baker RI, Beilin LJ. 2001. Effects of regular ingestion of black tea on haemostasis and cell adhesion molecules in humans. *Eur J Clin Nutr* 55:881–886.

Hou L, Zhou B, Yang L, Liu Z-L. 2004. Inhibition of human low density lipoprotein oxidation by flavonols and their glycosides. *Chem Phys Lipids* 129:209–219.

Huxley RR, Neil HAW. 2003. The relation between dietary flavonol intake and coronary heart disease mortality: A meta-analysis of prospective cohort studies. *Eur J Clin Nutr* 57:904–908.

Ishige K, Schubert D, Sagara Y. 2001. Flavonoids protect neuronal cells from oxidative stress by three distinct mechanisms. *Free Radic Biol Med* 30:433–446.

Jala VR, Haribabu B. 2004. Leukotrienes and atherosclerosis: New roles for old mediators. *Trends Immunol* 25:315–322.

Jimenez R, Lopez-Sepulveda R, Kadmiri M, Romero M, Vera R, Sanchez M, Vargas F, O'Valle F, Zarzuelo A, Duenas M, Santos-Buelga C, Duarte J. 2007. Polyphenols restore endothelial function in DOCA-salt hypertension: Role of endothelin-1 and NADPH oxidase. *Free Radic Biol Med* 43:462–473.

Justino GC, Santos MR, Canario S, Borges C, Florencio MH, Mira L. 2004. Plasma quercetin metabolites: Structure-antioxidant activity relationships. *Arch Biochem Biophys* 432:109–121.

Kawai Y, Nishikawa T, Shiba Y, Saito S, Murota K, Shibata N, Kobayashi M, Kanayama M, Uchida K, Terao J. 2008. Macrophage as a target of quercetin glucuronides in human atherosclerotic arteries: Implication in the anti-atherosclerotic mechanism of dietary flavonoids. *J Biol Chem* 283:9424–9434.

Keli SO, Hertog MG, Feskens EJ, Kromhout D. 1996. Dietary flavonoids, antioxidant vitamins, and incidence of stroke: the Zutphen study. *Arch Intern Med* 156:637–642.

Kim JA, Formoso G, Li Y, Potenza MA, Marasciulo FL, Montagnani M, Quon MJ. 2007. Epigallocatechin gallate, a green tea polyphenol, mediates NO-dependent vasodilation using signaling pathways in vascular endothelium requiring reactive oxygen species and Fyn. *J Biol Chem* 282:13736–13745.

Klaunig JE, Xu Y, Han C, Kamendulis LM, Chen J, Heiser C, Gordon MS, Mohler III ER. 1999. The effect of tea consumption on oxidative stress in smokers and nonsmokers. *Proc Soc Exp Biol Med* 220:249–254.

Knekt P, Jarvinen R, Reunanen A, Maatela J. 1996. Flavonoids intake and coronary mortality in Finland: A cohort study. *Br Med J* 312:478–481.

Koga T, Meydani M. 2001. Effect of plasma metabolites of (+)-catechin and quercetin on monocyte adhesion to human aortic endothelial cells. *Am J Clin Nutr* 73:941–948.

Leikert JF, Ra?thel TR, Wohlfart P, Cheynier V, Vollmar AM, Dirsch VM. 2002. Red wine polyphenols enhance endothelial nitric oxide synthase expression and subsequent nitric oxide release from endothelial cells. *Circulation* 106:1614–1617.

Libby P. 2002. Inflammation in atherosclerosis. *Nature* 420:868–874.

Loest HB, Noh SK, Koo SI. 2002. Green tea extract inhibits the lymphatic absorption of cholesterol and -tocopherol in ovariectomized rats. *J Nutr* 132:1282–1288.

Loke WM, Hodgson JM, Proudfoot JM, McKinley AJ, Puddey IB, Croft KD. 2008a. Pure dietary flavonoids quercetin and (–)-epicatechin augment nitric oxide products and reduce endothelin-1 acutely in healthy men. *Am J Clin Nutr* 88:1018–1025.

Loke WM, Proudfoot JM, McKinley AJ, Needs PW, Kroon PA, Hodgson JM, Croft KD. 2008b. Quercetin and its in vivo metabolites inhibit neutrophil-mediated LDL oxidation. *J Agric Food Chem* 56:3609–3615.

Loke WM, Proudfoot JM, Stewart S, McKinley AJ, Needs PW, Kroon PA, Hodgson JM, Croft KD. 2008c. Metabolic transformation has a profound effect on anti-inflammatory activity of flavonoids such as quercetin: Lack of association between antioxidant and lipoxygenase inhibitory activity. *Biochem Pharmacol* 75:1045–1053.

Loomis ED, Sullivan JC, Osmond DA, Pollock DM, Pollock JS. 2005. Endothelin mediates superoxide production and vasoconstriction through activation of NADPH oxidase and uncoupled nitric-oxide synthase in the rat aorta. *J Pharmacol Exp Ther* 315:1058–1064.

Lorenz M, Wessler S, Follmann E, Michaelis W, Dusterhoft T, Baumann G, Stangl K, Stangl V. 2004. A constituent of green tea, epigallocatechin-3-gallate, activates endothelial nitric oxide synthase by a phosphatidylinositol-3-OH-kinase-, cAMP-dependent protein kinase-, and Akt-dependent pathway and leads to endothelial-dependent vasorelaxation. *J Biol Chem* 279:6190–6195.

Lotito SB, Frei B. 2006. Dietary flavonoids attenuate tumor necrosis factor -induced adhesion molecule expression in human aortic endothelial cells: Structure-function relationships and activity after first pass metabolism. *J Biol Chem* 281:37102–37110.

Machha A, Achike FI, Mustafa AM, Mustafa MR. 2007. Quercetin, a flavonoid antioxidant, modulates endothelium-derived nitric oxide bioavailability in diabetic rat aortas. *Nitric Oxide* 16:442–447.

Manach C, Williamson G, Morand C, Scalbert A, Rémésy C. 2005. Bioavailability and bioefficacy of polyphenols in humans. I. Review of 97 bioavailability studies. *Am J Clin Nutr* 81:230S–242S.

Mano T, Masuyama T, Yamamoto K, Naito J, Kondo H, Nagano R, Tanouchi J, Hori M, Inoue M, Kamada T. 1996. Endothelial dysfunction in the early stage of atherosclerosis precedes appearance of intimal lesions assessable with intravascular ultrasound. *Am Heart J* 131:231–238.

Mathur S, Devaraj S, Grundy SM, Jialal I. 2002. Cocoa products decrease low density lipoprotein oxidative susceptibility but do not affect biomarkers of inflammation in humans. *J Nutr* 132:3663–3667.

Matsuda T, Ferreri K, Todorov I, Kuroda Y, Smith CV, Kandeel F, Mullen Y. 2005. Silymarin protects pancreatic -cells against cytokine-mediated toxicity: Implication of c-Jun NH2-terminal kinase and Janus kinase/signal transducer and activator of transcription pathways. *Endocrinology* 146: 175–185.

Meotti FC, Senthilmohan R, Harwood DT, Missau FC, Pizzolatti MG, Kettle AJ. 2008. Myricitrin as a substrate and inhibitor of myeloperoxidase: Implications for the pharmacological effects of flavonoids. *Free Radic Biol Med* 44:109–120.

Mira L, Tereza Fernandez M, Santos M, Rocha R, Helena Flore?ncio M, Jennings KR. 2002. Interactions of flavonoids with iron and copper ions: A mechanism for their antioxidant activity. *Free Radic Res* 36:1199–1208.

Mocatta TJ, Pilbrow AP, Cameron VA, Senthilmohan R, Frampton CM, Richards AM, Winterbourn CC. 2007. Plasma concentrations of myeloperoxidase predict mortality after myocardial infarction. *J Am Coll Cardiol* 49:1993–2000.

Moller P, Vogel U, Pedersen A, Dragsted LO, Sandstrom B, Loft S. 2003. No effect of 600 grams fruit and vegetables per day on oxidative DNA damage and repair in healthy nonsmokers. *Cancer Epidemiol Biomarkers Prev* 12:1016–1022.

Mukamal KJ, Maclure M, Muller JE, Sherwood JB, Mittleman MA. 2002. Tea consumption and mortality after acute myocardial infarction. *Circulation* 105:2476–2481.

Murase T, Kume N, Hase T, Shibuya Y, Nishizawa Y, Tokimitsu I, Kita T. 1999. Gallates inhibit cytokine-induced nuclear translocation of NF-kB and expression of leukocyte adhesion molecules in vascular endothelial cells. *Arterioscler Thromb Vasc Biol* 19:1412–1420.

Nijveldt RJ, van Nood E, van Hoorn DEC, Boelens PG, van Norren K, van Leeuwen PAM. 2001. Flavonoids: A review of probable mechanisms of action and potential applications. *Am J Clin Nutr* 74:418–425.

Ohkita M, Takaoka M, Shiota Y, Nojiri R, Matsumura Y. 2002. Nitric oxide inhibits endothelin-1 production through the suppression of nuclear factor kappa B. *Clin Sci* 103:68S–71S.

Oteiza PI, Erlejman AG, Verstraeten SV, Keen CL, Fraga CG. 2005. Flavonoid-membrane interactions: A protective role of flavonoids at the membrane surface? *Clin Dev Immunol* 12:19–25.

Pignatelli P, Pulcinelli FM, Celestini A, Lenti L, Ghiselli A, Gazzaniga PP, Violi F. 2000. The flavonoids quercetin and catechin synergistically inhibit platelet function by antagonizing the intracellular production of hydrogen peroxide. *Am J Clin Nutr* 72:1150–1155.

Pincemail J, Deby C, Thirion A, de Bruyn-Dister M, Goutier R. 1988. Human myeloperoxidase activity is inhibited in vitro by quercetin. Comparison with three related compounds. *Cell Mol Life Sci* 44:450–453.

Podrez EA, Abu-Soud HM, Hazen SL. 2000. Myeloperoxidase-generated oxidants and atherosclerosis. *Free Radic Biol Med* 28:1717–1725.

Ramzy D, Rao V, Tumiati LC, Xu N, Sheshgiri R, Miriuka S. 2006. Elevated endothelin-1 levels impair nitric oxide homeostasis through a PKC-dependent pathway. *Circulation* 114:I319–I326.

Rassaf T, Preik M, Kleinbongard P, Lauer T, Heiß C, Strauer BE, Feelisch M, Kelm M. 2002. Evidence for in vivo transport of bioactive nitric oxide in human plasma. *J Clin Invest* 109:1241–1248.

Rein D, Paglieroni TG, Wun T, Pearson DA, Schmitz HH, Gosselin R, Keen CL. 2000. Cocoa inhibits platelet activation and function. *Am J Clin Nutr* 72:30–35.

Sadeghipour M, Terreux R, Phipps J. 2005. Flavonoids and tyrosine nitration: Structure-activity relationship correlation with enthalpy of formation. *Toxicol in Vitro* 19:155–165.

Sanbongi C, Suzuki N, Sakane T. 1997. Polyphenols in chocolate, which have antioxidant activity, modulate immune functions in humans in-vitro. *Cell Immunol* 177:129–136.

Sanchez M, Galisteo M, Vera R, Villar IC, Zarzuelo A, Tamargo J, Perez-Vizcaino F, Duarte J. 2006. Quercetin downregulates NADPH oxidase, increases eNOS activity and prevents endothelial dysfunction in spontaneously hypertensive rats. *J Hypertens* 24:75–84.

Santos MR, Mira L. 2004. Protection by flavonoids against the peroxynitrite-mediated oxidation of dihydrorhodamine. *Free Radic Res* 38:1011–1018.

Scalbert A, Williamson G. 2000. Dietary intake and bioavailability of polyphenols. *J Nutr* 130:2073S–2085S.

Schewe T, Kuhn H, Sies H. 2002. Flavonoids of cocoa inhibit recombinant human 5-lipoxygenase. *J Nutr* 132:1825–1829.

Schramm DD, Wang JF, Holt RR, Ensunsa JL, Gonsalves JL, Lazarus SA, Schmitz HH, German JB, Keen CL. 2001. Chocolate procyanidins decrease the leukotriene-prostacyclin ratio in humans and human aortic endothelial cells. *Am J Clin Nutr* 73:36–40.

Schroeter H, Heiss C, Balzer J, Kleinbongard P, Keen CL, Hollenberg NK, Sies H, Kwik-Uribe C, Schmitz HH, Kelm M. 2006. (–)-Epicatechin mediates beneficial effects of flavanol-rich cocoa on vascular function in humans. *Proc Natl Acad Sci USA* 103:1024–1029.

Selloum L, Reichl S, Muller M, Sebihi L, Arnhold J. 2001. Effects of flavonols on the generation of superoxide anion radicals by xanthine oxidase and stimulated neutrophils. *Arch Biochem Biophys* 395:49–56.

Sesso HD, Gaziano JM, Buring JE, Hennekens CH. 1999. Coffee and tea intake and the risk of myocardial infarction. *Am J Epidemiol* 149:162–167.

Sies H, Schewe T, Heiss C, Kelm M. 2005. Cocoa polyphenols and inflammatory mediators. *Am J Clin Nutr* 81:304S–312S.

Simons K, Ikonen E. 1997. Functional rafts in cell membranes. *Nature* 387:569–572.

Steffen Y, Jung T, Klotz L-O, Schewe T, Grune T, Sies H. 2007. Protein modification elicited by oxidized low-density lipoprotein (LDL) in endothelial cells: Protection by (−)-epicatechin. *Free Radic Biol Med* 42:955–970.

Steffen Y, Schewe T, Sies H. 2005. Epicatechin protects endothelial cells against oxidized LDL and maintains NO synthase. *Biochem Biophys Res Commun* 331:1277–1283.

Stein JH, Keevil JG, Wiebe DA, Aeschlimann S, Folts JD. 1999. Purple grape juice improves endothelial function and reduces the susceptibility of LDL cholesterol to oxidation in patients with coronary artery disease. *Circulation* 100:1050–1055.

Steinberg FM, Guthrie NL, Villablanca AC, Kumar K, Murray MJ. 2003. Soy protein with isoflavones has favorable effects on endothelial function that are independent of lipid and antioxidant effects in healthy postmenopausal women. *Am J Clin Nutr* 78:123–130.

Stocker R, Keaney JFJr. 2004. Role of oxidative modifications in atherosclerosis. *Physiol Rev* 84:1381–1478.

Strachan FE, Newby DE, Sciberras DG, McCrea JB, Goldberg MR, Webb DJ. 2002. Repeatability of local forearm vasoconstriction to endothelin-1 measured by venous occlusion plethysmography. *Br J Clin Pharmacol* 54:386–394.

Takano-Ishikawa Y, Goto M, Yamaki K. 2006. Structure-activity relations of inhibitory effects of various flavonoids on lipopolysaccharide-induced prostaglandin E2 production in rat peritoneal macrophages: Comparison between subclasses of flavonoids. *Phytomedicine* 13:310–317.

Teede HJ, McGrath BP, DeSilva L, Cehun M, Fassoulakis A, Nestel PJ. 2003. Isoflavones reduce arterial stiffness: A placebo-controlled study in men and postmenopausal women. *Arterioscler Thromb Vasc Biol* 23:1066–1071.

Thompson HJ, Heimendinger J, Haegele A, Sedlacek SM, Gillette C, O'Neill C, Wolfe P, Conry C. 1999. Effect of increased vegetable and fruit consumption on markers of oxidative cellular damage. *Carcinogenesis* 20:2261–2266.

Tordera M, Ferrandiz ML, Alcaraz MJ. 1994. Influence of anti-inflammatory flavonoids on degradation and arachidonic acid release in rat neutrophils. *Z Naturforsch [C]* 49:235–240.

van den Berg R, van Vliet T, Broekmans W, Cnubben N, Vaes W, Roza L, Haenen G, Bast A, van den Berg H. 2001. A vegetable/fruit concentrate with high antioxidant capacity has no effect on biomarkers of antioxidant status in male smokers. *J Nutr* 131:1714–1722.

Vaya J, Mahmood S, Goldblum A, Aviram M, Volkova N, Shaalan A, Musa R, Tamir S. 2003. Inhibition of LDL oxidation by flavonoids in relation to their structure and calculated enthalpy. *Phytochemistry* 62:89–99.

Verstraeten SV, Keen CL, Schmitz HH, Fraga CG, Oteiza PI. 2003. Flavan-3-ols and procyanidins protect liposomes against lipid oxidation and disruption of the bilayer structure. *Free Radic Biol Med* 34:84–92.

Vieira O, Escargueil-Blanc I, Meilhac O, Basile JP, Laranjinha J, Alemeida L, Salvayre R, Negre-Salvayre A. 1998. Effect of dietary phenolic compounds on apoptosis of human cultured endothelial cells induced by oxidised LD L. *Br J Pharmacol* 123: 565–573.

Vila L. 2004. Cyclooxygenase and 5-lipoxygenase pathways in the vessel wall: Role in atherosclerosis. *Med Res Rev* 24:399–424.

Vogel RA. 1997. Coronary risk factors, endothelial function, and atherosclerosis: A review. *Clin Cardiol* 20:426–432.

Wallerath T, Deckert G, Ternes T, Anderson H, Li H, Witte K, Fo?rstermann U. 2002. Resveratrol, a polyphenolic phytoalexin present in red wine, enhances expression and activity of endothelial nitric oxide synthase. *Circulation* 106:1652–1658.

Wan Y, Vinson JA, Etherton TD, Proch J, Lazarus SA, Kris-Etherton PM. 2001. Effects of cocoa powder and dark chocolate on LDL oxidative susceptibility and prostaglandin concentrations in humans. *Am J Clin Nutr* 74:596–602.

Wang Z, Nicholls SJ, Rodriguez ER, Kummu O, Horkko S, Barnard J, Reynolds WF, Topol EJ, DiDonato JA, Hazen SL. 2007. Protein carbamylation links inflammation, smoking, uremia and atherogenesis. *Nat Med* 13:1176–1184.

Widlansky ME, Duffy SJ, Hamburg NM, Gokce N, Warden BA, Wiseman S, Keaney J, John F, Frei B, Vita JA. 2005. Effects of black tea consumption on plasma catechins and markers of oxidative stress and inflammation in patients with coronary artery disease. *Free Radic Biol Med* 38:499–506.

Williams RJ, Spencer JPE, Rice-Evans C. 2004. Flavonoids: Antioxidants or signalling molecules? *Free Radic Biol Med* 36:838–849.

Yochum L, Kushi LH, Meyer K, Folsom AR. 1999. Dietary flavonoid intake and risk of cardiovascular disease in postmenopausal women. *Am J Epidemiol* 149:943–949.

Young J, Dragsted L, Daneshvar B, Lauridsen S, Hansen M, Sandstrom B. 2000. The effect of grape-skin extract on oxidative status. *Br J Nutr* 84:505–13.

Zern TL, Wood RJ, Greene C, West KL, Liu Y, Aggarwal D, Shachter NS, Fernandez ML. 2005. Grape polyphenols exert a cardioprotective effect in pre- and postmenopausal women by lowering plasma lipids and reducing oxidative stress. *J Nutr* 135:1911–1917.

Zern TL, West KL, Fernandez ML. 2003. Grape polyphenols decrease plasma triglycerides and cholesterol accumulation in the aorta of ovariectomized guinea pigs. *J Nutr* 133:2268–2272.

Zhang R, Brennan M-L, Shen Z, MacPherson JC, Schmitt D, Molenda CE, Hazen SL. 2002a Myeloperoxidase functions as a major enzymatic catalyst for initiation of lipid peroxidation at sites of inflammation. *J Biol Chem* 277:46116–46122.

Zhang R, Shen Z, Nauseef WM, Hazen SL. 2002b Defects in leukocyte-mediated initiation of lipid peroxidation in plasma as studied in myeloperoxidase-deficient subjects: Systematic identification of multiple endogenous diffusible substrates for myeloperoxidase in plasma. *Blood* 99:1802–810.

Zhao X, Gu Z, Attele AS, Yuan C-S. 1999. Effects of quercetin on the release of endothelin, prostacyclin and tissue plasminogen activator from human endothelial cells in culture. *J Ethnopharmacol* 67:279–285.

6 Dietary Flavanols: Biochemical Basis of Short-Term and Longer-Term Vascular Responses

TANKRED SCHEWE, YVONNE STEFFEN, ELISABETH KRAVETS, AND HELMUT SIES*

Institute for Biochemistry and Molecular Biology I, Heinrich Heine University Düsseldorf, Düsseldorf, Germany

CONTENTS

INTRODUCTION

The literature on vascular responses to dietary components [e.g., Fitzpatrick et al., 1995] has increased substantially in recent years. Short-term and long-term consumption of black tea was shown to reverse endothelial dysfunction in patients with coronary artery disease [Duffy et al., 2001]. On the acute short-term time frame, which is taken here as the responses observed between 0 and 6 h, we demonstrated increases in flow-mediated dilation of the brachial artery upon intake of high-flavanol cocoa drink, but not of low-flavanol cocoa drink, in a crossover study [Heiss et al., 2003]. The effect of the flavanol-rich cocoa drink on flow-mediated dilation as well as the increase in circulating

* H.S. is a Fellow of the National Foundation for Cancer Research (NFCR), Bethesda, Maryland.

Plant Phenolics and Human Health: Biochemistry, Nutrition, and Pharmacology,
Edited by Cesar G. Fraga.

RNO [Heiss et al., 2003] was mimicked by isolated (−)-epicatechin administered to humans [Schroeter et al., 2006].

As in the studies on tea [Duffy et al., 2001], a chronic longer-term vascular response was identified with high-flavanol cocoa, when the flavanol-containing beverage was administered repetitively over several days [Heiss et al., 2007]. This effect was clinically characterized as continuously elevated baseline of flow-mediated dilation, on top of which the short-term effect was added on with every new dose; this additive effect was at about the same magnitude as without the pretreatment.

This chapter will summarize, in brief, results on the short-term and the longer-term responses, with emphasis on the former, where recent work led to some clarification and new perspectives [Schewe et al., 2008].

SHORT-TERM EFFECTS

Subsequent to ingestion of flavanol-rich cocoa products, the plasma levels of flavanols on their metabolites show a time course with increasing levels up to about 2 h, followed by a decrease toward the initial baseline, reached at about 4–6 h [Baba et al., 2000; Schroeter et al., 2006]. The vascular response, determined as an increase in flow-mediated dilation, follows a similar time course, as does the time course of the concentration of circulating nitric oxide (NO), assayed as RNO (RSNO and RNNO) [Heiss et al., 2003; Schroeter et al., 2006]. Several other studies also addressed these issues [Fisher et al., 2003; Engler et al., 2004; Heiss et al., 2005]. The responses on a time scale of only a few hours are not attributable to changes in gene expression followed by protein synthesis, but rather to more swift mechanisms like enhancement of flux through an enzyme reaction or to inhibitory effects directly influencing cellular and intercellular signaling.

Whereas enhancement of flux, that is, NO production, through the reaction catalyzed by endothelial nitric oxide synthase (eNOS) would be a straightforward explanation, our attempts to demonstrate such an activated flux through eNOS yielded no positive results, and we did not detect a potential short-term change in phosphorylation of serine-1177 of the enzyme, which would have led to an activation of the enzyme. However, we showed in cell culture experiments that indeed the steady-state level of cellular NO was increased in the presence of (-)-epicatechin, as visualized by DAF-2DA [Steffen et al., 2007]. A similar response was obtained with apocynin, 3-methoxy-4-hydroxyacetophenone, an efficient inhibitor of NADPH oxidase. When (−)-epicatechin was added in the presence of an inhibitor of the methylating enzyme, catechol-*O*-methyltransferase (COMT), the effect was blocked. Conversely, the methyl ether of (−)-epicatechin, 3′-*O*-methyl-epicatechin, was capable of elevating the fluorescence signal, indicating increased NO levels [Steffen et al., 2007]. Thus, it can be suggested that the effect of (−)-epicatechin is an indirect one, namely serving as precursor for the biologically active methyl ether, which acts as an inhibitor of

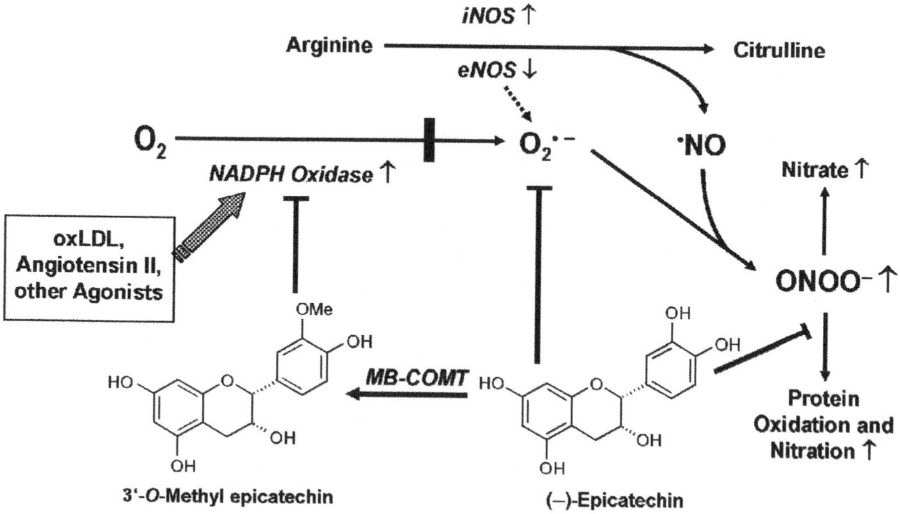

Figure 6.1 (−)-Epicatechin and 3′-*O*-methyl epicatechin increase NO levels in vascular endothelial cells via inhibition of NADPH-dependent formation of superoxide anions. Oxidized low-density lipoprotein (LDL), angiotensin II or other agonists evoke oxidative stress in vascular endothelial cells through activation of NADPH oxidase leading to elevated generation of $O_2^{\cdot -}$. Some $O_2^{\cdot -}$ may also arise from uncoupled eNOS (occurring, e.g. under conditions of tetrahydrobiopterin deficiency) (dotted arrow). $O_2^{\cdot -}$ reacts with NO generated by either eNOS or iNOS (the latter under proinflammatory conditions), forming peroxynitrite that, in turn, causes protein modifications or isomerizes to nitrate. (−)-Epicatechin is capable of scavenging $O_2^{\cdot -}$ and suppressing peroxynitrite-dependent reactions, but, more efficiently, serves as precursor of 3′-*O*-methyl epicatechin, which acts as inhibitor of NADPH oxidase. The methyl ether, generated by membrane-bound *O*-methyltransferase (MB-COMT), blocks NADPH oxidase-dependent $O_2^{\cdot -}$ generation, thus sparing NO.

NADPH oxidase in an apocynin-like fashion [Steffen et al., 2007, 2008; Schewe et al., 2008; Ximenes et al., 2007]. This is depicted schematically in Figure 6.1.

The occurrence of COMT in vascular cells has not been described so far. As shown in Figure 6.2, we confirm the presence of COMT by immunostaining of human umbilical vein endothelial cells. The localization is with intracellular membranes, and Western blots showed that the membrane-bound form, MB-COMT, is predominant by far over the soluble enzyme, S-COMT [Kravets, 2008]. Methylation of the B-ring of (−)-epicatechin exhibits some regioselectivity; the 3′-methyl ether was the preferred product, with the ratio of 3′- over 4′-*O*-methylepicatechin being about 2.0 [Kravets, 2008].

Among the metabolites circulating in human blood plasma after intake of (−)-epicatechin, the major one is the glucuronide [Natsume et al., 2003]. We suggest that the epicatechin glucuronide serves as transport metabolite, taken up into the cells and after deglucuronidation delivering the flavanol

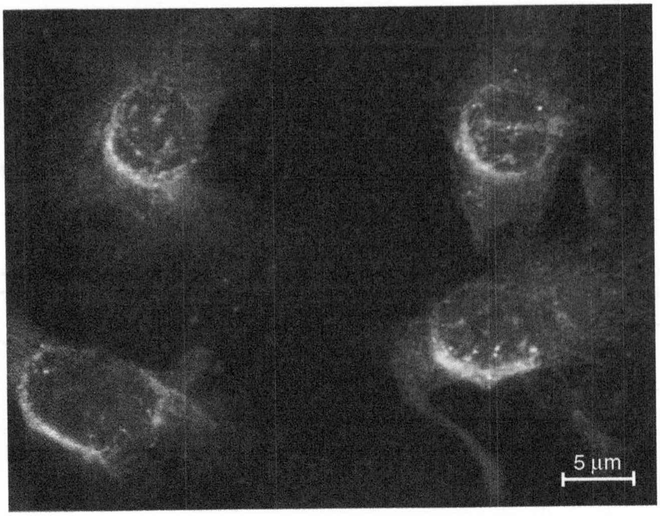

Figure 6.2 Immunocytochemical detection of catechol-*O*-methyltransferase (COMT) in human umbilical vein endothelial cells (HUVEC). Polyclonal COMT antibody from Chemicon Millipore, USA. (From Kravets E. [2008].)

intracellularly [Schewe et al., 2008]. These pathways will have to be worked out in more detail in the future.

The observations mentioned here suggest that short-term effects are correlated to metabolism of the flavanol moiety, potentially including a metabolic formation of a dimer, the procyanidin, and a main result being a modulation of the availability of circulating NO equivalents. In other words, there is an alternative pathway to induce a change in RNO levels without necessarily postulating an increase rate of production of NO through the NO synthase reaction. Abolition of the flavanol effects by inhibitors of eNOS, for example, L-NMMA, would not be in contradistinction, as the NO that is available is generated from eNOS, and only the disposition of it would be addressed.

LONGER-TERM EFFECTS

As mentioned above, repetitive ingestion over several days augments the baseline of flow-mediated dilation [Heiss et al., 2007]. The daily consumption of a flavanol-rich cocoa drink (3 × 306 mg flavanols/d) over 7 days ($n = 6$) resulted in continual FMD increases at baseline (after overnight fast and before flavanol ingestion) and in sustained FMD augmentation at 2 h after ingestion. Fasted FMD responses increased from 3.7 ± 0.4% on day 1 to 5.2 ± 0.6%, 6.1 ± 0.6%, and 6.6 ± 0.5% (each $P < 0.05$) on days 3, 5, and 8, respectively. FMD returned to 3.3 ± 0.3% after a washout week of cocoa-free diet (day 15). Increases observed in circulating nitrite, but not in circulating nitrate, paralleled the observed FMD augmentations. The short-term response assayed each morning following application of the daily morning dose was unaffected; it remained at similar magnitude throughout the study [Heiss et al., 2007]. This established for the cocoa-derived responses what was also observed in the 30-day black tea study [Duffy et al., 2001]. More recently, a similar result was obtained on a group of diabetic patients [Balzer et al., 2008].

Thus, there are responses on the longer term that change the setpoints of responses, most likely involving changes in gene expression, enzyme patterns, or activity state of enzymes. Recently, we observed a decrease in arginase following treatment with cocoa [Schnorr et al., 2008], and there are numerous other observations relating to changes in eNOS level and activity. This longer-term time frame may be of importance for assessing health effects, which may have more of an impact on long-term degenerative disease prevention [see Ding et al., 2006] than the short-term responses on flow-mediated dilation. The currently available time range on longer term is 8 up to 30 days, and long-term studies over several months should be encouraged. A recent compilation of studies in this research area is available [Shenouda and Vita, 2007].

REFERENCES

Baba S, Osakabe N, Yasuda A, Natsume M, Takizawa T, Nakamura T, Terao J. 2000. Bioavailability of (−)-epicatechin upon intake of chocolate and cocoa in human volunteers. *Free Radic Res* 33:635–641.

Balzer J, Rassaf T, Heiss C, Kleinbongard P, Lauer T, Merx M, Heussen N, Gross HB, Keen CL, Schroeter H, Kelm M. 2008. Benefits in vascular function through flavanol-containing cocoa in medicated diabetic patients a double-masked, randomized, controlled trial. *J Am Coll Cardiol* 51:2141–2149.

Ding EL, Hutfless SM, Ding X, Girotra S. 2006. Chocolate and prevention of cardiovascular disease: A systematic review. *Nutr Metab* (Lond) 3:2.

Duffy SJ, Keaney JF Jr, Holbrook M, Gokce N, Swerdloff PL, Frei B, Vita JA. (2001) Short- and long-term black tea consumption reverses endothelial dysfunction in patients with coronary artery disease. *Circulation* 104:151–156.

Engler MB, Engler MM, Chen CY, Malloy MJ, Browne A, Chiu EY, Kwak HK, Milbury P, Paul SM, Blumberg J, Mietus-Snyder ML. 2004. Flavonoid-rich dark chocolate improves endothelial function and increases plasma epicatechin concentrations in healthy adults. *J Am Coll Nutr* 23:197–204.

Fisher ND, Hughes M, Gerhard-Herman M, Hollenberg NK. 2003. Flavanol-rich cocoa induces nitric-oxide-dependent vasodilation in healthy humans. *J Hypertens* 21:2281–2286.

Fitzpatrick DF, Hirschfield SL, Ricci T, Jantzen P, Coffey RG. 1995. Endothelium-dependent vasorelaxation caused by various plant extracts. *J Cardiovasc Pharmacol.* 26:90–95.

Heiss C, Finis D, Kleinbongard P, Hoffmann A, Rassaf T, Kelm M, Sies H. 2007. Sustained increase in flow-mediated dilation after daily intake of high-flavanol cocoa drink over 1 week. *J Cardiovasc Pharmacol* 49:74–80.

Heiss C, Kleinbongard P, Dejam A, Perré S, Schroeter H, Sies H, Kelm M. 2005. Acute consumption of flavanol-rich cocoa and the reversal of endothelial dysfunction in smokers. *J Am Coll Cardiol* 46:1276–1283.

Heiss C, Dejam A, Kleinbongard P, Schewe T, Sies H, Kelm M. 2003. Vascular effects of cocoa rich in flavan-3-ols. *J Am Med Assoc* 290:1030–1031.

Kravets K. 2008. *Untersuchung zur O*-Methylierung von (–)-Epicatechin in Gefäß-Endothelzellen. Master Thesis, Faculty of Natural Sciences, Heinrich-Heine-Universität Düsseldorf. 1–100.

Natsume M, Osakabe N, Oyama M, Sasaki M, Baba S, Nakamura Y, Osawa T, Terao J. 2003. Structures of (–)-epicatechin glucuronide identified from plasma and urine after oral ingestion of (–)-epicatechin: differences between human and rat. *Free Radic Biol Med* 34:840–849.

Schewe T, Steffen Y, Sies H. 2008. How do dietary flavanols improve vascular function? A position paper. *Arch Biochem Biophys* 476:102–106.

Schnorr O, Brossette T, Momma TY, Kleinbongard P, Keen CL, Schroeter H, Sies H. 2008. Cocoa flavanols lower vascular arginase activity in human endothelial cells in vitro and in erythrocytes in vivo. *Arch Biochem Biophys* 476:211–215.

Schroeter H, Heiss C, Balzer J, Kleinbongard P, Keen CL, Hollenberg NK, Sies H, Kwik-Uribe C, Schmitz HH, Kelm M. 2006. (–)-Epicatechin mediates beneficial effects of flavanol-rich cocoa on vascular function in humans. *Proc Natl Acad Sci* 103:1024–1029.

Shenouda SM, Vita JA. 2007. Effects of flavonoid-containing beverages and EGCG on endothelial function. *J Am Coll Nutr* 26:366S–372S.

Steffen Y, Gruber C, Schewe T, Sies H. 2008. Mono-*O*-methylated flavanols and other flavonoids as inhibitors of endothelial NADPH oxidase. *Arch Biochem Biophys* 469:209–219.

Steffen Y, Schewe T, Sies H. 2007. (–)-Epicatechin elevates nitric oxide in endothelial cells via inhibition of NADPH oxidase. *Biochem Biophys Res Commun* 359:828–833.

Ximenes VF, Kanegae MP, Rissato SR, Galhiane MS. 2007. The oxidation of apocynin catalyzed by myeloperoxidase: proposal for NADPH oxidase inhibition. *Arch Biochem Biophys* 457:134–141.

7 Green Tea Catechins: Anticancer Effects and Molecular Targets

NAGHMA KHAN AND HASAN MUKHTAR
Department of Dermatology, University of Wisconsin, Madison, Wisconsin

CONTENTS

Plant Phenolics and Human Health: Biochemistry, Nutrition, and Pharmacology,
Edited by Cesar G. Fraga.

INTRODUCTION

Tea is a popular beverage, derived from the plant *Camellia sinensis* and has been brewed for nearly 5000 years. There are three major commercial varieties of tea: green (20% of consumption), oolong (2% of consumption), and black tea (78% of consumption). In the production of black tea, the plant leaves are picked and then allowed to wither indoors, ferment, and oxidize. For green tea, the plant leaves are steamed and parched after picking to prevent oxidation of the catechins present in the leaf. Oolong tea is produced by "semifermenting" the green leaves, resulting in a tea that is chemically a mixture of green and black teas [Khan et al., 2007]. Although green tea is becoming popular worldwide, traditionally it is most commonly consumed in Asia, whereas black tea is more popular in the United States and Europe. Since ancient times green tea is a widely consumed beverage and, for centuries, has been regarded to possess significant health-promoting effects. The health-promoting effects of green tea are mainly attributed to its polyphenol content. Green tea is a rich source of polyphenols, especially flavanols, which represent approximately 30% dry weight of the fresh leaf. Catechins are the predominant flavanols in green tea and are mainly comprised of (−)-epigallocatechin gallate (EGCG), (−)-epigallocatechin (EGC), (−)-epicatechin gallate (ECG), and (−)-epicatechin (EC) (Fig. 7.1).

Catechins contain a benzopyran skeleton with a phenyl group substituted at the 2-position and a hydroxyl (or ester) function at the 3-position. Variations to the catechin structure include the stereochemistry of the 2,3-substituents and the number of hydroxyl groups in the B- and D-ring. More than 50% of the mass of this catechin combination is composed of EGCG, and a vast body of scientific research suggests that EGCG is responsible for the majority of the potential health benefits attributed to green tea consumption. Theaflavins, which include theaflavin, theaflavin-3-gallate, theaflavin-3′ gallate, and theaflavin-3,3′-digallate, are key to the characteristic color and taste of black tea, and account for 2–6% of the solids in brewed black tea. The major fractions of black tea polyphenols, accounting for >20% of the solids in brewed black tea, are known as thearubigens [Khan et al., 2008]. In recent years, scientists throughout the world have investigated the potential benefits of green tea and its most abundant catechin, EGCG. The potential for use of EGCG in human cancer prevention and treatment seems very promising because EGCG acts against cancer through a variety of mechanisms. This potential is reflected by the growing number of in vivo and in vitro research studies on this topic. Tea is an important source of flavonoids in the diet with levels approaching 200 mg/cup for a typical brew of black tea. The flavonoids found in green and black tea are very effective antioxidants in vitro and may therefore be active as antioxidants in the body [Adhami and Mukhtar, 2007]. The mechanisms responsible for the anticancer actions of tea are not well understood but are being intensively investigated. Recent research findings indicate that tea polyphenols can protect against the multistages of carcinogenesis (Fig.7.2).

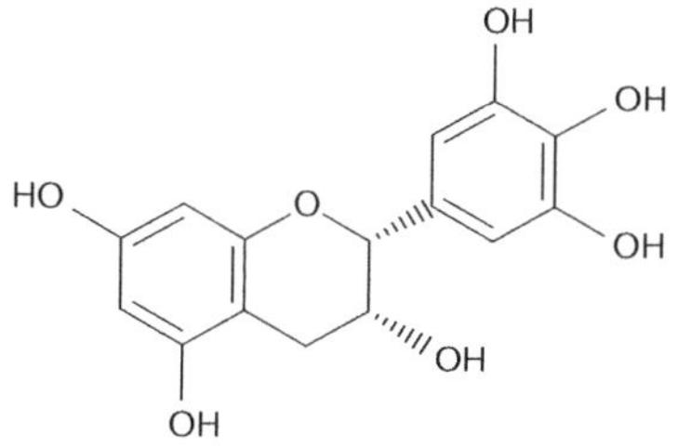

(−)-Epigallocatechin

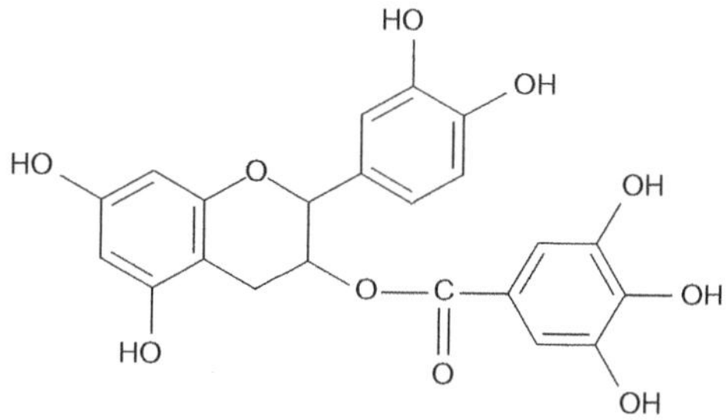

(−)-Epicatechin-3-gallate

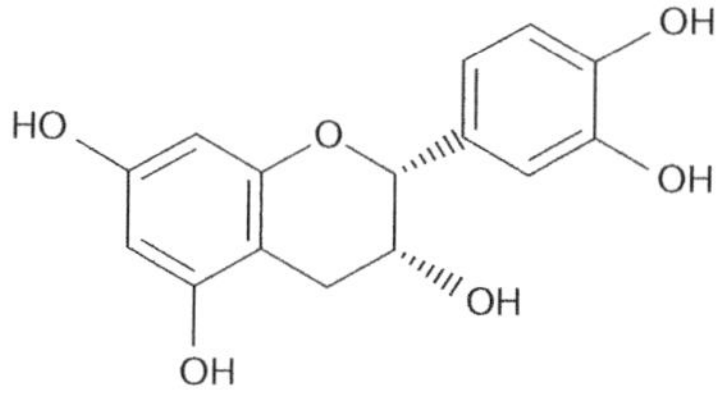

(−)-Epicatechin

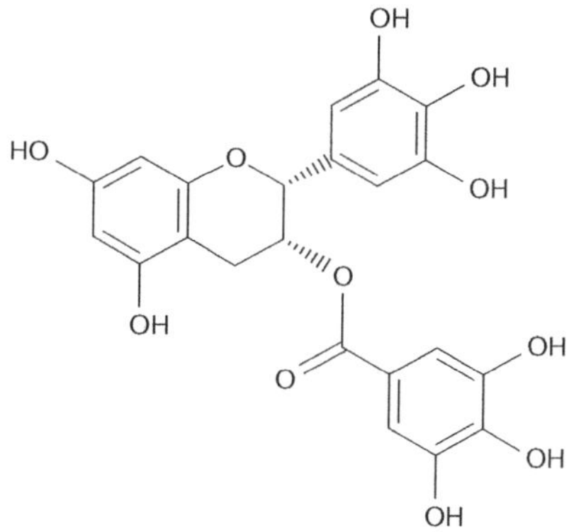

(−)-Epigallocatechin-3-gallate

Figure 7.1 Chemical structures of major green tea catechins.

This chapter discusses the anticarcinogenic effects of green tea catechins, especially EGCG in different organ sites and effects on cellular signaling events.

BIOTRANSFORMATION AND BIOAVAILABILITY OF TEA POLYPHENOLS

Yang et al. [1998] have reported that administration of 1.5, 3.0, and 4.5 g of decaffeinated green tea solids to human volunteers resulted in maximal plasma concentrations (C_{max}) of 326, 550, and 190 ng/L for EGCG, EGC, and EC, respectively. These C_{max} values were observed 1.4–2.4 h after the ingestion of the tea preparation. The elimination half-life ($t_{1/2}$) of EGCG (5.0–5.5 h) was found to be higher than those of EGC and EC (2.5–3.4 h). EGC and EC, but not EGCG, were excreted in the urine. Within 8 h, over 90% of the total urinary EGC and EC was excreted [Yang et al., 1998].

The major metabolic pathways for tea catechins include glucuronidation, sulfation, and methylation. There are species and tissue-specific differences in

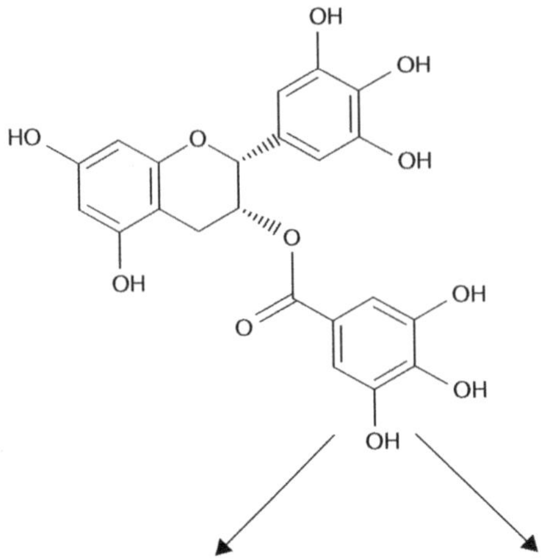

Inhibition of tumor initiation

- Antioxidant and reactive oxygen species scavenging effects
- Induction of detoxification pathways
- Inhibition of cytochrome P450
- Inhibition of histone deacetylation
- Inhibition of telomerase activity
- Inhibition of oncogene expression
- Inhibition of target genes

Inhibition of tumor promotion

- Inhibition of cell-proliferation
- Induction of apoptosis and cell-cycle arrest
- Modulation of multiple signaling pathways
- Inhibition of cyclooxygenases and lipooxygenases
- Inhibition of inducible nitric oxide synthase
- Inhibition of proteases
- Inhibition of ornithine decarboxylase activity
- Inhibition of genotoxic effects

Figure 7.2 Mechanisms of inhibition of carcinogenesis by EGCG.

EGCG and EGC glucuronidation, with humans and mice being more similar than humans and rats. Methylated catechins have been observed in the rat including 3′ and 4′-*O*-methyl EC, 4′-*O*-methyl EGC, and 4″-*O*-methyl EGC and EGCG. Following oral EGCG administration, the major metabolite detected in the bile of the rat was 4′, 4″-di-*O*-methyl-EGCG [Lambert and Yang, 2003]. The greatest catalytic efficiency for glucuronidation is in mouse intestinal microsomes followed in decreasing order by mouse liver, human liver, rat liver, and rat small intestine. EC undergoes sulfation catalyzed by human and rat intestinal and liver cytosol with the human liver being the most efficient [Vaidyanathan and Walle, 2002]. EGC is detected mainly as the glucuronidated form or sulfated form with only a small amount present as the free form in humans [Lee et al., 1995]. Tea catechins undergo metabolism in the gut to form the ring fission products 5-(3′,4′,5′-trihydroxyphenyl)- -valerolactone (M4), 5-(3′,4′-dihydroxyphenyl)-γ-valerolactone (M6), and 5-(3′,5′-dihydroxyphenyl)-γ-valerolactone (M6′) [Li et al., 2000]. These metabolic intermediates are further broken down by gut flora to phenylacetic and phenylpropionoic acids. M6 was previously shown to form during anaerobic incubation of ECG and EC with human intestinal bacteria [Meselhy et al., 1997].

Detailed pharmacokinetic and biotransformation studies of the tea catechins have been conducted in rats [Lambert and Yang, 2003]. It has been shown that treatment of rats with a green tea polyphenol (GTP) preparation in the drinking fluid result in increasing plasma levels over a 14-day period with levels of EGC and EC being higher than those of EGCG. EGCG levels are lower in the bladder, kidney, colon, lung, and prostate, whereas the levels of EGCG were found to be highest in the rat esophagus, intestine, and colon, which have direct contact with tea catechins. The EGCG levels in the plasma, lung, and liver are much higher than in rats when the same polyphenol preparation is given to mice. These levels appear to peak on day 4 and then decrease to $\leq$20% of the peak values on days 8–10 [Kim et al., 2000]. The absolute bioavailability of EGCG, EGC, and EC, after intragastric administration of decaffeinated green tea is 0.1, 14, and 31%, respectively. After oral administration of 100 mg of EGCG, in studies with bile-duct-cannulated rats, 3.28% of the dose is recovered in the bile as EGCG (2.65%), 4″-*O*-methyl-EGCG (0.25%), 3″-*O*-methyl-EGCG (0.11%), 4′-*O*-methyl-EGCG (0.11%), 3″-*O*-methyl-EGCG (0.10%), and 4′,4″-di-*O*-methyl-EGCG (0.06%) [Okushio et al., 1999].

ANTICARCINOGENIC EFFECTS OF GREEN TEA POLYPHENOLS

Green Tea Polyphenols and Lung Cancer

Green tea was also found to significantly inhibit benzo(*a*)pyrene-induced lung tumorigenesis in A/J mice [Wang et al., 1992]. There was reduction in lung tumor incidence and multiplicity in *N*-nitrosodiethylamine-treated A/J mice by green tea and decaffeinated green tea [Wang et al., 1992]. In LACA mice, green tea greatly reduced tumor incidence and multiplicity in *N*-methyl-N -nitro-*N*-nitrosoguanidine (MNNG)-induced lung cancers and precancerous lesions [Luo and Li, 1992]. Decaffeinated green and black tea showed a dose-dependent chemoprevention of lung tumors in dimethylnitrosamine-treated C3H mice [Cao et al., 1996]. Administration of polyphenon E, a standardized GTP preparation significantly reduced the NNK-induced lung tumor incidence and multiplicity in female A/J mice. Polyphenon E treatment inhibited cell proliferation and enhanced apoptosis in adenocarcinomas and adenomas and lowered levels of c-Jun and extracellular signal-regulated kinase (ERK) 1/2 phosphorylation [Lu et al., 2006]. There was significant reduction in tumor multiplicity in both $p53^{wt/wt}$ and $p53^{val135/wt}$ mice after the administration of green tea as the sole drinking source beginning 1 week after NNK administration [Zhang et al., 2000]. It has been reported that EGCG caused suppression of NF-κB/PI3K/AKT/mTOR and MAPKs in normal human bronchial epithelial (NHBE) cells, which may contribute to its ability to suppress inflammation, proliferation, and angiogenesis induced by cigarette smoke [Syed et al., 2007]. Lu et al [2006]. analyzed the gene expression changes caused by the administration of green tea or polyphenon E to chemically induced mouse model for lung tumorigenesis. They found that 88 genes that were differentially expressed in tumors (from the normal tissues) were reversed by the treatment and suggested that these genes may be used as markers for tea exposure [Lu et al., 2006].

Green Tea Polyphenols and Liver Cancer

It has been reported that EGCG inhibited the growth of hepatocellular carcinoma cell lines, induced apoptosis, and down-regulated Bcl-2 and Bcl-xL by inactivation of NF-κB. Oral administration of EGCG showed similar effects in HLE xenograft tumors. Co-treatment with EGCG and necrosis factor-related apoptosis-inducing ligand (TRAIL) synergistically induced apoptosis in HLE cells [Nishikawa et al., 2006]. All four tea catechins (EC, EGC, ECG, EGCG), black tea extract, and oolong tea extract, significantly decreased the diethylnitrosamine and phenobarbital-induced number and area of preneoplastic glutathione *S*-transferase placental form-positive foci in the liver [Matsumoto et al., 1996]. EGCG inhibited the platelet-derived rowth factor (PDGF)-BB-induced cell-proliferation and collagen α1(I) and (IV) messenger ribonucleic acid (mRNA) expressions. EGCG also reduced the autophosphorylation of the PDGF receptor and blocked PDGF-BB binding to its receptor in a noncompetitive manner [Sakata et al., 2004]. In C3H mice, green and black tea treatment caused a significant decrease in the diethylnitrosamine-induced liver tumors [Cao et al., 1996]. There was inhibition of hepatic glucose-6-phophatase system mediated through an elevated luminal glucose level by EGCG [Csala et al., 2007]. In mice, administration of green tea was able to prevent the increase in incidences and multiplicities of diethylnitrosamine-induced hepatocellular tumors and also arrest the progression of cholangiocellular tumors [Umemura et al., 2003].

Green Tea Polyphenols and Gastrointestinal Tract Cancer

The inhibitory effect of EGCG on MNNG-induced glandular stomach carcinogenesis in rats and on *N*-ethyl-*N* -nitro-*N*-nitrosoguanidine (ENNG)-induced duodenal carcinogenesis in mice has been reported [Yamane et al., 1996]. EGCG was found to inhibit growth and activation of the epidermal growth factor receptor (EGFR) and human EGFR-2 signaling pathways in human colon cancer cells [Shimizu et al., 2005]. Administration of EGCG in drinking fluid dose-dependently inhibited small intestinal tumorigenesis in ApcMin/+ mice. EGCG administration resulted in increased levels of E-cadherin as well as decreased levels of β-catechin in the nucleus, c-Myc, phospho-Akt, and phospho-ERK in the tumors [Ju et al., 2005]. IHC analysis showed that PPE or EGCG treatment increased apoptosis but decreased cell proliferation as well as levels of phospho-Akt and nuclear β-catenin. Green tea administration (0.6% in drinking fluid) inhibited the formation of azoxymethane (AOM)-induced aberrant crypt foci in CF-1 mice on a high-fat diet [Ju et al., 2003]. Green tea extract administration also inhibited intestinal tumor formation in the Apc mutant Min mice, and the effect was enhanced synergistically when combined with sulindac [Suganuma et al., 2001].

Green Tea Polyphenols and Skin Cancer

It was reported that GTP acts as anti-initiating agent against the skin tumorigenicity induced by polycyclic aromatic hydrocarbons in mice. In a complete skin tumorigenesis protocol using 3-methylcholanthrene, the topical application of GTP to female BALB/c mice resulted in substantial protection against the onset and subsequent development of tumors. In the two-stage DMBA-TPA skin tumorigenesis protocol, topical application of GTP to female SENCAR mice afforded significant protection against skin tumorigenicity. Oral feeding of GTP in drinking water to female SENCAR mice also protected against skin tumorigenesis [Wang et al., 1989]. In SKH-1 mice, oral administration of GTP reduced ultraviolet-B (UVB)-induced skin tumor incidence, tumor multiplicity, and tumor growth. There was also reduced expression of the matrix metalloproteinases (MMP)-2 and MMP-9, CD31, vascular endothelial growth factor (VEGF), and proliferating cell nuclear antigen (PCNA) in the GTP-treated group. There were also more cytotoxic CD8(+) T cells and greater activation of caspase-3 in the tumors of the orally administered GTP group indicating the apoptotic death of the tumor cells [Mantena et al., 2005]. We have shown that topical application of a GTP fraction to the skin of DMBA-initiated mice, prior to that of TPA or mezerein, resulted in protection against skin tumor promotion as indicated by the decrease in tumor incidence, tumor multiplicity, and tumor volume. GTP also protected against the malignant conversion of papillomas to squamous cell carcinomas [Katiyar et al., 1997]. We have also shown that topical application of GTP resulted in significant decrease in UVB-induced bifold-skin thickness, skin edema, infiltration of leukocytes, and inhibition of MAPKs and NF-κB pathways in SKH-1 hairless mice [Afaq et al., 2003]. Green tea, black tea, or EGCG when given orally inhibited the growth of well-established skin tumors and, in some cases, tumor regression was also observed. Complete regression was observed in papilloma-bearing mice [Conney et al., 1999]. In SKH-1 mice, there was enhancement in the rate and extent of disappearance of the mutant p53-positive patches by the oral administration of green tea as the sole source of drinking fluid starting immediately after discontinuation of UV-B treatment [Lu et al., 2005].

Green Tea Polyphenols and Prostate Cancer

It has been shown that EGCG-induced apoptosis, cell-growth inhibition, and cell-cycle dysregulation in human prostate cancer DU145 and LNCaP cells without producing similar effects in normal cells [Ahmad et al., 1997;

Adhami et al., 2003]. We have earlier reported that GTP consumption was found to significantly inhibit prostate cancer development and metastasis in transgenic adenocarcinoma of the mouse prostate (TRAMP) model. This was achieved by an oral infusion of GTP equivalent to six cups of green tea a day, that is, at an achievable dose in humans [Gupta et al., 2001]. Moreover, elevated levels of insulin-like growth factor (IGF)-1, coupled with lowering of IGF-binding protein-3 (IGFBP-3) are associated with a greater risk of developing prostate cancer. GTP infusion has been shown to reduce the levels of IGF-1 and concomitantly increase IGFBP-3 with marked inhibition of markers of angiogenesis and metastasis. This leads to a reduction in the downstream signaling and inhibition of protein expression, hence inhibiting prostate cancer development and progression [Adhami et al., 2004]. GTP feeding to TRAMP mice also resulted in marked inhibition of prostate cancer progression, which was associated with reduction of S100A4 and restoration of E-cadherin [Saleem et al., 2005]. It was reported recently that green and black tea inhibits CWR22R*ν*1 tumor growth and prostate-specific antigen (PSA) secretion in athymic nude mice [Siddiqui et al., 2006]. EGCG induced apoptosis in human prostate carcinoma LNCaP cells via stabilization of p53 by phosphorylation on critical serine residues and p14ARF-mediated down-regulation of murine double minute 2(MDM2) protein, negative regulation of NF-κB activity, activation of caspases, causing a change in the ratio of Bax/Bcl-2 in a manner that favors apoptosis [Hastak et al., 2003]. In LNCaP and DU145 cells, EGCG treatment resulted in induction of G1 phase ckis, which inhibits the cyclin-cdk complexes operative in the G0/G1 phase of the cell cycle, thereby causing an arrest that ultimately leads to apoptotic cell death [Gupta et al., 2003]. We have recently reported that treatment of human prostate cancer cells LNCaP, PC-3, and CWR22R*ν*1 with a combination of EGCG and Cox-2 inhibitor resulted in enhanced cell growth inhibition, apoptosis induction, and inhibition of NF-κB. In athymic nude mice, implanted with CWR22R*ν*1 cells, combination treatment with GTP and celecoxib resulted in enhanced tumor growth inhibition, lowering of PSA and IGF-1 levels, and increase in IGFBP-3 levels [Adhami et al., 2007]. It was demonstrated using isogenic cell lines that EGCG activates growth arrest and apoptosis in prostate carcinoma cells primarily via p53-dependent pathway that involves the function of both p21 and Bax such that down-regulation of either molecule confers a growth advantage to the cells [Hastak et al., 2005]. Recently, the stage(s) of prostate cancer development that is most vulnerable to chemopreventive intervention by GTP were identified. We selected animals and initiated oral feeding of GTP (0.2% in drinking water ad libitum) at ages representing different stages of the disease: (i) 6 weeks, mice with normal prostate pathology, (ii) 12 weeks, when mice histologically display hyperplasia, (iii) 18 weeks, when severe hyperplasia and adenocarcinoma are observed, and (iv) 28 weeks when mice display primary tumors and metastasis is common. It was found that the chemopreventive potential of GTP decreases with increasing tumor grade [Adhami et al., 2008].

EFFECTS ON SIGNAL TRANSDUCTION PATHWAYS

As shown in Figure 7.3, EGCG appears to afford protection against cancer through multiple biological mechanisms, although the specific molecular targets and intracellular pathways modulated by it remain incompletely defined as discussed in the following sections.

Inhibition of NF-κB Signaling Pathway

Nuclear factor-κB (NF-κB) is a transcription factor important for the expression of numerous genes contributing to inflammation and innate and adaptive immune responses. Regulation of NF-κB occurs at several different hierarchical levels. In most resting cells, NF-κB is retained in the cytoplasm by binding to the inhibitory IκB proteins, which blocks its nuclear localization sequences. NF-κB is activated in response to a wide variety of stimuli that promote the dissociation of the IκBα through phosphorylation, ubiquitination, and degradation, thus unmasking of the nuclear localization sequence of NF-κB thereby allowing NF-κB to enter the nucleus and bind κB-regulatory elements [Richmond, 2002]. The activation of this transcription factor is affected by a number of parallel signaling pathways that cross-talk with NF-κB activating signal cascades either positively or negatively. Because of the critical role of NF-κB in cell survival, cell adhesion, inflammation, differentiation, and cell growth, it has been implicated in carcinogenesis. EGCG has been shown to

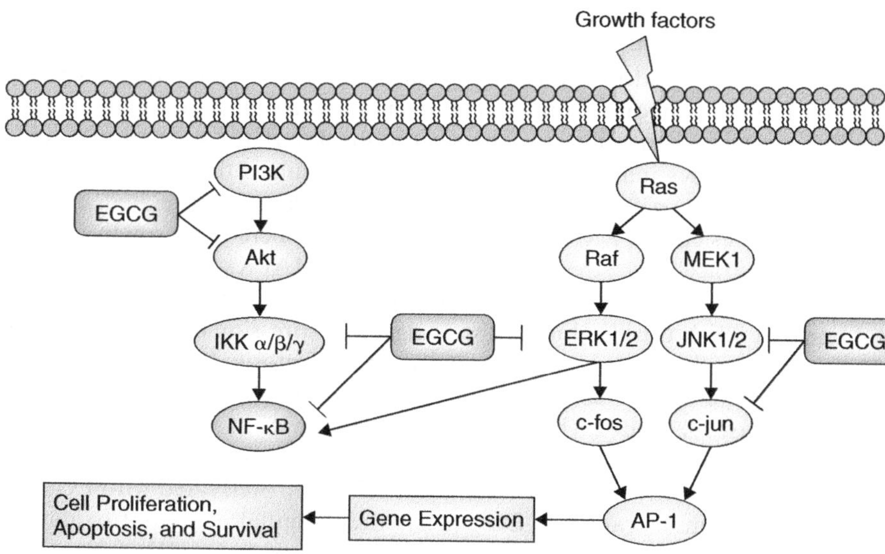

Figure 7.3 Effect of EGCG on the modulation of various signal transduction pathways.

inhibit TPA-induced DNA binding of NF-κB and CREB in mouse skin in vivo and also suppressed TPA-induced phosphorylation and subsequent degradation of IκBα, and prevented nuclear translocation of p65 [Kundu and Surh, 2007]. It has been reported that EGCG induced apoptosis in human prostate carcinoma LNCaP cells by negative regulation of NF-κB activity, thereby decreasing the expression of the proapoptotic protein Bcl-2 [Hastak et al., 2003]. Study from our lab has shown that treatment of EGCG dose and time-dependently increased IκB level, and inhibited NF-κB nuclear translocation in A431 epidermoid carcinoma cells [Gupta et al., 2004]. It was shown that UVB irradiation-induced NF-κB activation in NHEK was associated with increased IκB phosphorylation and degradation and EGCG was shown to block NF-κB activation and nuclear translocation [Afaq et al., 2003b]. Topical application of GTP to UV-B-irradiated SKH-1 hairless skin decreased phosphorylation and degradation of IκB and the subsequent activation of NF-κB [(Afaq et al., 2003a]. It has also been reported that pretreatment of NHBE cells with EGCG suppressed cigarette smoke condensate (CSC)-induced phosphorylation of IκBα and activation and nuclear translocation of NF-κB/p65. NHBE cells transfected with a luciferase reporter plasmid containing an NF-κB-inducible promoter sequence showed an increased reporter activity after CSC exposure that was specifically inhibited by EGCG pretreatment [Syed et al., 2007].

Inhibition of Mitogen-Activated Protein (MAP) Kinases and Activator Protein-1 (AP-1)

Mitogen-activated protein kinases (MAPKs) are major signaling transduction molecules that play an important role in regulating a variety of cellular responses, including cell proliferation, differentiation, and apoptosis. MAPKs pathways are comprised of a three-tier kinase module in which are MAPK is activated upon phosphorylation by a mitogen-activated protein kinase kinase (MAPKK), which in turn is activated when phosphorylated by a MAPKKK. Mammalian MAPKs mainly consist of three subfamilies; the Jun N-terminal kinases (JNKs), the p38 kinases, and the extracellular signal-related kinases (ERKs). MAPKs can be activated by diverse stimuli including growth factors and cellular/extracellular stresses. Once activated, MAPKs (ERK, JNK, and p38) can activate a variety of transcription factors, including ELK and c-Jun, a component of AP-1, thus leading to changes in the expression of genes that play critical roles in cell proliferation, migration, and apoptosis. It has been shown that treatment of H_2O_2 resulted in phosphorylation of extracellular signal-regulated kinases 1 and 2 (ERK1/2), JNK, and p38 in human epidermal keratinocytes. H_2O_2-induced phosphorylation of ERK1/2, JNK, and p38 was found to be significantly inhibited when these cells were pretreated with EGCG. These findings demonstrate that EGCG has the potential to inhibit oxidative stress-mediated phosphorylation of MAPK signaling pathways [Katiyar et al., 2001]. Recently, it has been reported that EGCG rapidly and substantially hampered UV-B irradiation-induced activation of ASK-1 and phosphorylation of ERK1/2, JNK, and p38 in dermal fibroblasts. These results demonstrate that EGCG has abilities to hamper UV-B-induced collagenolytic MMP production via interfering with the MAPK-responsive pathways [Bae et al., 2008]. A study from our lab has reported that topically application of GTP decreased UV-B-induced phosphorylation of ERK1/2 and JNK and p38 proteins in SKH-1 mice [Afaq et al., 2003a]. It has been shown that tea polyphenols, EGCG, and theaflavins can effectively block arsenite-induced apoptosis of JB6 cells and inhibited arsenite-induced AP-1 transcription activity and AP-1 DNA binding activity. EGCG and theaflavins also potently inhibited arsenite-induced ERK1/2 activity but not p38 kinase activity [Chen et al., 2000].

Inhibition of Epidermal Growth Factor Receptor (EGFR)-Mediated Pathways

The epidermal growth factor receptor (EGFR) family of receptor tyrosine kinases (TKs) includes four members: EGFR (HER-1), HER-2/neu, HER-3, and HER-4. This family has been shown to be important for proper regulation of many developmental, metabolic, and physiologic processes mediated by EGF, transforming growth factor-α and multiple other ligands. In numerous cancers, including glioblastomas, breast cancer, and non-small-cell lung cancer (NSCLC), there is often a transforming deregulation of EGFR family kinase activity. EGFR is also expressed by various epithelial tumors; high levels of EGFR expression in these cells indicate a poor prognosis or a late stage of disease. In this context, EGFR has become a novel molecular target of cancer therapies [Yarden and Sliwkowski, 2001]. EGCG was found to inhibit the activation of the EGFR, HER2, and multiple downstream signaling pathways in colon cancer cell lines [Shimizu et al., 2005]. It was shown that GTP resulted in substantial reduction in the levels of IGF-I and significant increase in the levels of IGFBP-3 in TRAMP mice [Adhami et al., 2004]. It has also been shown that EGCG inhibits the binding of EGF to the EGFR and the subsequent dimerization and activation of the EGFR by altering membrane organization (Adachi et al., 2007). It has been reported that EGCG binds to a specific metastasis associated 67-kDa laminin receptor that is expressed on a variety of tumor cells. It was shown using a subtraction cloning strategy involving cDNA libraries constructed from cells treated or untreated with all trans-retinoic acid that the anticancer action of EGCG is mediated by LR and it allows EGCG to bind to the cell surface [Tachibana et al., 2004]. EGCG has also been reported to suppress gene expression of EGFR in rat-activated hepatic stellate cells in vitro mediated by reducing the trans-activation activity of Egr-1 [Fu et al., 2006]

Inhibition of Insulin-like Growth Factor (IGF)-I Mediated Signal Transduction Pathway

The IGF system involves complex regulatory networks that operate at the whole organism, cellular and subcellular levels. Key molecules involved are the ligands IGF1 and IGF2, the type 1 and type 2 IGF receptors (IGF1R and IGF2R, respectively), the IGF-binding proteins (IGFBPs), and the proteins involved in intracellular signalling distal to IGF1R, which include members of the insulin-receptor substrate (IRS) family, AKT, target of rapamycin (TOR), and S6 kinase. A final important determinant of IGF activity is through a family of at least six distinct IGFBPs that modulate bioavailability of IGFs in the circulation. Low circulating levels of IGFBPs favor an increased IGF mitogenic activity. IGF could be an appropriate target for cancer chemoprevention as increasing levels of IGF-I are associated with an increased risk of cancer [Adhami et al., 2006]. We have earlier reported that 0.1% GTP (wt/vol) provided as the sole source of drinking fluid to TRAMP mice from 8 to 32 weeks of age resulted in significant inhibition in serum IGF-1 and restoration of IGFBP-3 levels in the prostate compared with water-fed TRAMP mice [Gupta et al., 2001]. EGCG has been reported to abrogate anchorage-eindependent growth induced by IGF-IR overexpression and also prevented human breast and cervical cancer cell phenotype expression through inhibition of IGF-IR downstream signaling [Li et al., 2007]. We have reported that GTP resulted in substantial reduction in the levels of IGF-I and significant increase in the levels of IGFBP-3 in TRAMP mice [Adhami et al., 2004]. Recently, we have also shown that combination treatment with GTP and celecoxib resulted in enhanced tumor growth inhibition by lowering of IGF-1 levels and increase in circulating levels of serum IGFBP-3 compared with results of single-agent treatment [Adhami et al., 2007].

Induction of Apoptosis and Cell-Cycle Arrest

Apoptosis is a preferential way of elimination of damaged cells. Apoptosis is characterized by a set of morphologic changes including chromatin condensation, nuclear fragmentation, membrane blebbing, and cell shrinkage. Generally, the growth rates of preneoplastic or neoplastic cells outpace that of normal cells because of malfunctioning or dysregulation of their cell growth and cell death machineries. Therefore, induction of apoptosis or cell-cycle arrest by dietary chemopreventive compounds can be an excellent approach to inhibit the promotion and progression of carcinogenesis and to remove genetically damaged, preinitiated, or neoplastic cells from the body [Khan et al., 2007]. For the very first time, we have shown that EGCG induces apoptosis and cell-cycle arrest in many cancer cells without affecting normal cells [Ahmad et al., 1997]. This observation has been verified by many subsequent studies in several cell types including lung, colon, pancreas, skin, and prostate [Khan et al., 2006]. We have earlier reported that in EGCG-treated LNCaP cells, p53 protein was stabilized, and EGCG inhibited NF-κB transcription activity. The balance between pro- and antiapoptotic Bcl-2 family proteins favored apoptosis, and LNCaP cells were arrested in the G0/G1 phase [Hastak et al., 2003]. EGCG induced expression of p21 and p27, inhibited the activity of CDK2 and CDK4, and caused Rb hypophosphorylation. EGCG increased the expression of p16, p18, p21, and p53, which are associated with negative regulation of cell-cycle progression in prostate cancer cells [Gupta et al., 2003; 2000]. In the A/J mouse model, green tea in drinking fluid increased the apoptosis index in lung adenoma in chemically induced lung tumors [Liao et al., 2004].

Inhibition of Urokinase-Plasminogen Activator (uPA)

The urokinase plasminogen activator (uPA) system is believed to play a key role in tissue degradation, cell migration, angiogenesis, cancer invasion, and metastasis. uPA is a member of the serine protease family and is strongly implicated as a promoter of tumor progression in various human malignancies. It is synthesized and secreted as a pro-enzyme, whose activation is markedly accelerated upon binding with high affinity to specific membrane-bounded or soluble cell surface uPA receptors (uPAR). Binding to uPAR, uPA efficiently converts the inactive zymogen, plasminogen, into the active serine protease, plasmin, which then directly or indirectly cleaves extracellular matrix components including laminin, fibronectin, fibrin, vitronectin, and collagen. Plasmin can activate latent elastase and MMPs, potent enzymes that can also digest a variety of ECM components. uPAR plays a key role not only in localizing uPA activity but also in mediating various signaling events essential for the differentiation and migration of cells within the tumor environment [Duffy et al., 2004]. EGCG has been reported to inhibit the activity of uPA. With the use of molecular modeling, it was shown that EGCG binds to urokinase, blocking His 57 and Ser 195 of the urokinase catalytic triad and extending toward Arg 35 from a positively charged loop of urokinase [Jankun et al., 1997]. We have reported that GTP infusion to TRAMP mice resulted in marked inhibition of uPA in the dorso-lateral prostate [Adhami et al., 2004]. Recently, we have shown that EGCG sensitizes TRAIL-resistant LNCaP cells to TRAIL-mediated apoptosis in part through inhibition in the protein expression of VEGF, uPA, and angiopoietin 1 and 2 in prostate cancer cells [Siddiqui et al., 2008]. EGCG was found to decrease the expressions of MMP-2, MMP-9, and uPA in a concentration-dependent manner in human oral cancer cell line [Ho et al., 2007]. It has also been reported that GTP inhibited constitutively active transcription factors AP-1 and NF-κB, which further suppressed secretion of uPA from breast cancer cells [Silvova et al., 2005].

INHIBITION OF PROTEASOME ACTIVITIES

The proteasome is a ubiquitous enzyme complex that plays a critical role in the degradation of many proteins involved in cell-cycle regulation, apoptosis, and angiogenesis. Since these pathways are fundamental for cell survival and proliferation, particularly in cancer cells, the inhibition of proteasome is an attractive potential anticancer therapy. The ubiquitin/proteasome pathway is the main nonlysosomal route for degradation and is responsible for the turnover of >80% of cellular proteins. This pathway involves the conjugation of multiple ubiquitin moieties to a substrate followed by the downstream degradation of the tagged protein by a multicatalytic proteasome complex [Montagut et al., 2006]. EGCG analogs with acetyl protected -OH groups were reported to be much more potent than natural EGCG in inhibiting the proteasome in cultured tumor cells. Consistently, these protected analogs showed much higher potency than EGCG to inhibit proliferation and transforming activity and to induce apoptosis in human leukemic, prostate, breast, and simian virus 40-transformed cells [Kuhn et al., 2005]. EGCG and ECG, but not EGC and EC, potently inhibited the chymotryptic activity of the 20s proteasome both in cell-free systems and in tumor cell lines at the concentrations found in the serum of green tea drinkers. This inhibition of the proteasome by EGCG in several tumor and transformed cell lines results in the accumulation of two natural proteasome substrates, p27(Kip1) and IκBα, followed by growth arrest in the G(1) phase of the cell cycle [Nam et al., 2001].

CONCLUSION AND FUTURE PROSPECTS

Green tea is an extremely popular beverage worldwide. Derivatives of green tea, particularly its major polyphenolic catechin, EGCG, have been proposed to have anticarcinogenic properties based on preclinical, observational, and clinical trial data. Furthermore, subsequent studies have shown that green tea can also prevent lifestyle-related diseases and that it has life-prolonging effects. Additional studies are required to determine how the active ingredients in green tea interact with environmental and genetic factors, as well as to identify mechanisms effective against given cancer types, so that these mechanisms can be fine-tuned or supplemented to increase the desired effects. The elucidation of the molecular mechanisms/targets associated with the antitumor effects exhibited by tea catechins will contribute greatly to the development and design of chemopreventive agents and more effective cancer chemoprevention trials.

ACKNOWLEDGMENT

The original work from the author's (HM) laboratory outlined in this review was supported by U.S. Public Health Service Grants RO1 CA 78809, RO1 CA 101039, RO1 CA 120451, and P50 DK065303.

REFERENCES

Adachi S, Nagao T, Ingolfsson HI, Maxfield FR, Andersen OS, Kopelovich L, Weinstein IB. 2007. The inhibitory effect of (−)-epigallocatechin gallate on activation of the epidermal growth factor receptor is associated with altered lipid order in HT29 colon cancer cells. *Cancer Res* 67:6493–6501.

Adhami VM, Mukhtar H. 2007. Anti-oxidants from green tea and pomegranate for chemoprevention of prostate cancer. *Mol Biotechnol* 37:52–57.

Adhami VM, Siddiqui IA. Sarfaraz S, Khwaja SI, Mukhtar H. 2008. Efficacious prostate cancer chemopreventive intervention with green tea polyphenols decreases with increasing tumor grade in the autochthonous transgenic TRAMP model. In: Proceedings of the 99th Annual Meeting of the American Association for Cancer Research, 2008 Apr 12–16; San Diego, CA, Philadelphia: AACR, 2008. (Abstract number: 3818).

Adhami VM, Malik A, Zaman N, Sarfaraz S, Siddiqui IA, Syed DN, Afaq F, Pasha FS, Saleem M, Mukhtar H. 2007. Combined inhibitory effects of green tea polyphenols and selective cyclooxygenase-2 inhibitors on the growth of human prostate cancer cells both in vitro and in vivo. *Clin Cancer Res* 13:1611–1619.

Adhami VM, Afaq F, Mukhtar H. 2006. Insulin-like growth factor-I axis as a pathway for cancer chemoprevention. *Clin Cancer Res* 12:5611–5614.

Adhami VM, Siddiqui IA, Ahmad N, Gupta S, Mukhtar H. 2004. Oral consumption of green tea polyphenols inhibits insulin-like growth factor-I-induced signaling in an autochthonous mouse model of prostate cancer. *Cancer Res* 64:8715–8722.

Adhami VM, Ahmad N, Mukhtar H. 2003. Molecular targets for green tea in prostate cancer prevention. *J Nutr* 133:2417S–2424S.

Afaq F, Adhami VM, Ahmad N, Mukhtar H. 2003a. Inhibition of ultraviolet B-mediated activation of nuclear factor kappaB in normal human epidermal keratinocytes by green tea constituent (−)-epigallocatechin-3-gallate. *Oncogene* 22:1035–1044.

Afaq F, Ahmad N, Mukhtar H. 2003b. Suppression of UVB-induced phosphorylation of mitogen-activated protein kinases and nuclear factor kappa B by green tea polyphenol in SKH-1 hairless mice. *Oncogene* 22:9254–9264.

Ahmad N, Feyes DK, Nieminen AL, Agarwal R, Mukhtar H. 1997. Green tea constituent epigallocatechin-3-gallate and induction of apoptosis and cell cycle arrest in human carcinoma cells. *J Natl Cancer Inst* 89:1881–1886.

Bae JY, Choi JS, Choi YJ, Shin SY, Kang SW, Han SJ, Kang YH. 2008. (−)Epigallocatechin gallate hampers collagen destruction and collagenase activation in ultraviolet-B-irradiated human dermal fibroblasts: Involvement of mitogen-activated protein kinase. *Food Chem Toxicol* 46:1298–1307.

Cao J, Xu Y, Chen J, Klaunig JE. 1996. Chemopreventive effects of green and black tea on pulmonary and hepatic carcinogenesis. *Fundam Appl Toxicol* 29:244–250.

Conney AH, Lu Y, Lou Y, Xie J, Huang M. 1999. Inhibitory effect of green and black tea on tumor growth. *Proc Soc Exp Biol Med* 220:229–233.

Csala M, Margittai E, Senesi S, Gamberucci A, Banhegyi G, Mandl J, Benedetti A. 2007. Inhibition of hepatic glucose 6-phosphatase system by the green tea flavanol epigallocatechin gallate. *FEBS Lett* 581:1693–1698.

Duffy MJ. 2004. The urokinase plasminogen activator system: Role in malignancy. *Curr Pharm Des* 10:39–49.

Fu Y, Chen A. 2006. The phyto-chemical (−)-epigallocatechin gallate suppresses gene expression of epidermal growth factor receptor in rat hepatic stellate cells in vitro by reducing the activity of Egr-1. *Biochem Pharmacol* 72:227–238.

Gupta S, Hastak K, Afaq F, Ahmad N, Mukhtar H. 2004. Essential role of caspases in epigallocatechin-3-gallate-mediated inhibition of nuclear factor kappa B and induction of apoptosis. *Oncogene* 23:2507–2522.

Gupta S, Hussain T, Mukhtar H. 2003. Molecular pathway for (−)-epigallocatechin-3-gallate-induced cell cycle arrest and apoptosis of human prostate carcinoma cells. *Arch Biochem Biophys* 410:177–185.

Gupta S, Hastak K, Ahmad N, Lewin JS, Mukhtar H. 2001. Inhibition of prostate carcinogenesis in TRAMP mice by oral infusion of green tea polyphenols. *Proc Natl Acad Sci U S A* 98:10350–10355.

Hastak K, Agarwal MK, Mukhtar H, Agarwal ML. 2005. Ablation of either p21 or Bax prevents p53-dependent apoptosis induced by green tea polyphenol epigallocatechin-3-gallate. *FASEB J* 19:789–791.

Hastak K, Gupta S, Ahmad N, Agarwal MK, Agarwal ML, Mukhtar H. 2003. Role of p53 and NF-kappaB in epigallocatechin-3-gallate-induced apoptosis of LNCaP cells. *Oncogene* 22:4851–4859.

Ho YC, Yang SF, Peng CY, Chou MY, Chang YC. 2007. Epigallocatechin-3-gallate inhibits the invasion of human oral cancer cells and decreases the productions of matrix metalloproteinases and urokinase-plasminogen activator. *J Oral Pathol Med* 36:588–593.

Jankun J, Selman SH, Swiercz R, Skrzypczak-Jankun E. 1997. Why drinking green tea could prevent cancer. *Nature* 387:561.

Ju J, Liu Y, Hong J, Huang MT, Conney AH, Yang CS. 2003. Effects of green tea and high-fat diet on arachidonic acid metabolism and aberrant crypt foci formation in an azoxymethane-induced colon carcinogenesis mouse model. *Nutr Cancer* 46:172–178.

Ju J, Hong J, Zhou JN, Pan Z, Bose M, Liao J, Yang GY, Liu YY, Hou Z, Lin Y, Ma J, Shih WJ, Carothers AM, Yang CS. 2005. Inhibition of intestinal tumorigenesis in Apcmin/ + mice by (−)-epigallocatechin-3-gallate, the major catechin in green tea. *Cancer Res* 65:10623–10631.

Katiyar SK, Afaq F, Azizuddin K, Mukhtar H. 2001. Inhibition of UVB-induced oxidative stress-mediated phosphorylation of mitogen-activated protein kinase signaling pathways in cultured human epidermal keratinocytes by green tea polyphenol (−)-epigallocatechin-3-gallate. *Toxicol Appl Pharmacol* 176:110–117.

Katiyar SK, Mohan RR, Agarwal R, Mukhtar H. 1997. Protection against induction of mouse skin papillomas with low and high risk of conversion to malignancy by green tea polyphenols. *Carcinogenesis* 18:497–502.

Khan N, Mukhtar H. 2007. Tea polyphenols for health promotion. *Life Sci* 81:519–533.

Khan N, Afaq F, Mukhtar H. 2008. Cancer chemoprevention through dietary antioxidants: Progress and promise. *Antioxid Redox Signal* 10:475–510.

Khan N, Afaq F, Mukhtar H. 2007. Apoptosis by dietary factors: the suicide solution for delaying cancer growth. *Carcinogenesis* 28:233–239.

Khan N, Afaq F, Saleem M, Ahmad N, Mukhtar H. 2006. Targeting multiple signaling pathways by green tea polyphenol (−)-epigallocatechin-3-gallate. *Cancer Res* 66:2500–2505.

Kim S, Lee MJ, Hong J, Li C, Smith TJ, Yang GY, Seril DN, Yang CS. 2000. Plasma and tissue levels of tea catechins in rats and mice during chronic consumption of green tea polyphenols. *Nutr Cancer* 37:41–48.

Kuhn D, Lam WH, Kazi A, Daniel KG, Song S, Chow LM, Chan TH, Dou QP. 2005. Synthetic peracetate tea polyphenols as potent proteasome inhibitors and apoptosis inducers in human cancer cells. *Front Biosci* 10:1010–1023.

Kundu JK, Surh YJ. 2007. Epigallocatechin gallate inhibits phorbol ester-induced activation of NF-kappa B and CREB in mouse skin: Role of p38 MAP K. *Ann N Y Acad Sci* 1095:504–512.

Lambert JD, Yang CS. 2003a. Cancer chemopreventive activity and bioavailability of tea and tea polyphenols. *Mutat Res* 523–524:201–208.

Lambert JD, Yang CS. 2003b. Mechanisms of cancer prevention by tea constituents. *J Nutr* 133:3262S–3267S.

Lee MJ, Wang ZY, Li H, Chen L, Sun Y, Gobbo S, Balentine DA, Yang CS. 1995. Analysis of plasma and urinary tea polyphenols in human subjects. *Cancer Epidemiol Biomarkers Prev* 4:393–399.

Li M, He Z, Ermakova S, Zheng D, Tang F, Cho YY, Zhu F, Ma WY, Sham Y, Rogozin EA, Bode AM, Cao Y, Dong Z. 2007. Direct inhibition of insulin-like growth factor-I receptor kinase activity by (−)-epigallocatechin-3-gallate regulates cell transformation. *Cancer Epidemiol Biomarkers Prev* 16:598–605.

Li C, Lee MJ, Sheng S, Meng X, Prabhu S, Winnik B, Huang B, Chung JY, Yan S, Ho CT, Yang CS. 2000. Structural identification of two metabolites of catechins and their kinetics in human urine and blood after tea ingestion. *Chem Res Toxicol* 13:177–184.

Liao J, Yang GY, Park ES, Meng X, Sun Y, Jia D, Seril DN, Yang CS. 2004. Inhibition of lung carcinogenesis and effects on angiogenesis and apoptosis in A/J mice by oral administration of green tea. *Nutr Cancer* 48:44–53.

Lu G, Liao J, Yang G, Reuhl KR, Hao X, Yang CS. 2006a. Inhibition of adenoma progression to adenocarcinoma in a 4-(methylnitrosamino)-1-(3-pyridyl)-1-butanone-induced lung tumorigenesis model in A/J mice by tea polyphenols and caffeine. *Cancer Res* 66:11494–11501.

Lu Y, Yao R, Yan Y, Wang Y, Hara Y, Lubet RA, You M. 2006b. A gene expression signature that can predict green tea exposure and chemopreventive efficacy of lung cancer in mice. *Cancer Res* 66:1956–1963.

Lu YP, Lou YR, Liao J, Xie JG, Peng QY, Yang CS, Conney AH. 2005. Administration of green tea or caffeine enhances the disappearance of UVB-induced patches of mutant p53 positive epidermal cells in SKH-1 mice. *Carcinogenesis* 26:1465–1472.

Luo D, Li Y. 1992. [Preventive effect of green tea on MNNG-induced lung cancers and precancerous lesions in LACA mice]. Hua Xi Yi Ke Da Xue Xue Bao 23:433–437.

Mantena SK, Meeran SM, Elmets CA, Katiyar SK. 2005. Orally administered green tea polyphenols prevent ultraviolet radiation-induced skin cancer in mice through activation of cytotoxic T cells and inhibition of angiogenesis in tumors. *J Nutr* 135:2871–2877.

Matsumoto N, Kohri T, Okushio K, Hara Y. 1996. Inhibitory effects of tea catechins, black tea extract and Oolong tea extract on hepatocarcinogenesis in rat. *Jpn J Cancer Res* 87:1034–1038.

Meselhy MR, Nakamura N, Hattori M. 1997. Biotransformation of (–)-epicatechin 3-O-gallate by human intestinal bacteria. *Chem Pharm Bull (Tokyo)* 45:888–893.

Montagut C, Rovira A, Albanell J. 2006. The proteasome: a novel target for anticancer therapy. *Clin Transl Oncol* 8:313–317.

Nam S, Smith DM, Dou QP. 2001. Ester bond-containing tea polyphenols potently inhibit proteasome activity in vitro and in vivo. *J Biol Chem* 276:13322–13330.

Nishikawa T, Nakajima T, Moriguchi M, Jo M, Sekoguchi S, Ishii M, Takashima H, Katagishi T, Kimura H, Minami M, Itoh Y, Kagawa K, Okanoue T. 2006. A green tea polyphenol, epigalocatechin-3-gallate, induces apoptosis of human hepatocellular carcinoma, possibly through inhibition of Bcl-2 family proteins. *J Hepatol* 44:1074–1082.

Okushio K, Suzuki M, Matsumoto N, Nanjo F, Hara Y. 1999. Identification of (–)-epicatechin metabolites and their metabolic fate in the rat. *Drug Metab Dispos* 27:309–316.

Richmond A. 2002. Nf-kappa B, chemokine gene transcription and tumor growth. *Nat Rev Immunol* 2:664–674.

Sakata R, Ueno T, Nakamura T, Sakamoto M, Torimura T, Sata M. 2004. Green tea polyphenol epigallocatechin-3-gallate inhibits platelet-derived growth factor-induced proliferation of human hepatic stellate cell line LI90. *J Hepatol* 40:52–59.

Saleem M, Adhami VM, Ahmad N, Gupta S, Mukhtar H. 2005. Prognostic significance of metastasis-associated protein S100A4 (Mtsl) in prostate cancer progression and chemoprevention regimens in an autochthonous mouse model. *Clin Cancer Res* 11:147–153.

Shimizu M, Deguchi A, Lim JT, Moriwaki H, Kopelovich L, Weinstein IB. 2005. (–)-Epigallocatechin gallate and polyphenon E inhibit growth and activation of the epidermal growth factor receptor and human epidermal growth factor receptor-2 signaling pathways in human colon cancer cells. *Clin Cancer Res* 11:2735–2746.

Siddiqui IA, Malik A, Adhami VM, Asim M, Hafeez BB, Sarfaraz S, Mukhtar H. 2008. Green tea polyphenol EGCG sensitizes human prostate carcinoma LNCaP cells to TRAIL-mediated apoptosis and synergistically inhibits biomarkers associated with angiogenesis and metastasis. *Oncogene* 27:2055–2063.

Siddiqui IA, Zaman N, Aziz MH, Reagan-Shaw SR, Sarfaraz S, Adhami VM, Ahmad N, Raisuddin S, Mukhtar H. 2006. Inhibition of CWR22Rnul tumor growth and PSA secretion in athymic nude mice by green and black teas. *Carcinogenesis* 27:833–839.

Slivova V, Zaloga G, DeMichele SJ, Mukerji P, Huang YS, Siddiqui R, Harvey K, Valachovicova T, Sliva D. 2005. Green tea polyphenols modulate secretion of urokinase plasminogen activator (uPA) and inhibit invasive behavior of breast cancer cells. *Nutr Cancer* 52:66–73.

Suganuma M, Ohkura Y, Okabe S, Fujiki H. 2001. Combination cancer chemoprevention with green tea extract and sulindac shown in intestinal tumor formation in Min mice. *J Cancer Res Clin Oncol* 127:69–72.

Syed DN, Afaq F, Kweon MH, Hadi N, Bhatia N, Spiegelman VS, Mukhtar H. 2007. Green tea polyphenol EGCG suppresses cigarette smoke condensate-induced NF-kappaB activation in normal human bronchial epithelial cells. *Oncogene* 26:673–682.

Tachibana H, Koga K, Fujimura Y, Yamada K. 2004. A receptor for green tea polyphenol EGC G. *Nat Struct Mol Biol* 11:380–381.

Umemura T, Kai S, Hasegawa R, Kanki K, Kitamura Y, Nishikawa A, Hirose M. 2003. Prevention of dual promoting effects of pentachlorophenol, an environmental pollutant, on diethylnitrosamine-induced hepato- and cholangiocarcinogenesis in mice by green tea infusion. *Carcinogenesis* 24:1105–1109.

Vaidyanathan JB, Walle T. 2002. Glucuronidation and sulfation of the tea flavonoid (−)-epicatechin by the human and rat enzymes. *Drug Metab Dispos* 30:897–903.

Wang ZY, Agarwal R, Khan WA, Mukhtar H. 1992a. Protection against benzo pyrene- and N-nitrosodiethylamine-induced lung and forestomach tumorigenesis in A/J mice by water extracts of green tea and licorice. *Carcinogenesis* 13:1491–1494.

Wang ZY, Hong JY, Huang MT, Reuhl KR, Conney AH, Yang CS. 1992b. Inhibition of N-nitrosodiethylamine- and 4-(methylnitrosamino)-1-(3-pyridyl)-1-butanone-induced tumorigenesis in A/J mice by green tea and black tea. *Cancer Res* 52:1943–1947.

Wang ZY, Khan WA, Bickers DR, Mukhtar H. 1989. Protection against polycyclic aromatic hydrocarbon-induced skin tumor initiation in mice by green tea polyphenols. *Carcinogenesis* 10:411–415.

Yamane T, Nakatani H, Kikuoka N, Matsumoto H, Iwata Y, Kitao Y, Oya K, Takahashi T. 1996. Inhibitory effects and toxicity of green tea polyphenols for gastrointestinal carcinogenesis. *Cancer* 77:1662–1667.

Yang CS, Chen L, Lee MJ, Balentine D, Kuo MC, Schantz SP. 1998. Blood and urine levels of tea catechins after ingestion of different amounts of green tea by human volunteers. *Cancer Epidemiol Biomarkers Prev* 7:351–354.

Yarden Y, Sliwkowski MX. 2001. Untangling the ErbB signalling network. *Nat Rev Mol Cell Biol* 2:127–137.

Zhang Z, Liu Q, Lantry LE, Wang Y, Kelloff GJ, Anderson MW, Wiseman RW, Lubet RA, You M. 2000. A germ-line p53 mutation accelerates pulmonary tumorigenesis: p53-independent efficacy of chemopreventive agents green tea or dexamethasone/myo-inositol and chemotherapeutic agents taxol or adriamycin. *Cancer Res* 60:901–907.

8 Flavonols: Metabolism, Bioavailability, and Health Impacts

JUNJI TERAO

Department of Food Science, Graduate School of Nutrition and Bioscience, The University of Tokushima, Tokushima, Japan

CONTENTS

INTRODUCTION

Flavonols are a subclass of flavonoids, which are distributed ubiquitously in the plant kingdom. The most common flavonols are kaempferol, quercetin, isorhamnetin, and myricetin (Fig. 8.1). They are frequently found as *O*-glycosides, in which glycosidation occurs mainly at the 3-position of the C-ring, but substitution can also occur at the 5′, 7′, 4′, 3′, and 5′-positions [Crozier et al., 2006]. Many types of glycosides are derived from flavonol aglycones because various sugar groups can conjugate to the hydroxyl groups of flavonols at different positions. Flavonol glycosides are often found in commonly consumed plants. Figure 8.1 shows typical examples of quercetin glycosides. In 1936, Szent-Györgyi found that rutin (quercetin 3-*O*-β-rutinoside) had

Plant Phenolics and Human Health: Biochemistry, Nutrition, and Pharmacology, Edited by Cesar G. Fraga.

vitamin properties and termed this flavonol "vitamin P" [Rusznyak and Szent-Györgyi, 1936]. Studies investigating deficiencies of dietary flavonols in humans were not reported. Dietary flavonols were believed to be little absorbed in the body due to their poor bioavailability. Nutritional scientists, therefore, excluded vitamin P from the category of vitamins at that time. In the 1970s, quercetin and other plant flavonols were suspected to be potential carcinogens because these polyphenolic compounds were found to possess mutagenic activity [Bjeldans and Chang, 1977; Sugimura et al., 1977]. A diet containing quercetin did not show a carcinogenic effect in animals. Recent studies strongly suggest that quercetin acts as a potential anticarcinogen in animal models of carcinogenesis [Murakami et al., 2008]. Epidemiological studies revealed that daily intake of flavonols from food is closely related to the prevention of atherosclerosis and other vascular diseases [Hertog et al., 1994]. The protective role of flavonols against oxidative damage in the central nervous system is thought to help prevent neurodegenerative diseases [Youdin et al., 2004]. Excessive intake of these polyphenolic compounds seems to exert an adverse effect on the body [Galati and O'Brien, 2004]. Studies on the bioavailability of plant flavonols, including intestinal absorption, metabolic conversion, and transport to target sites, are essential. Plant flavonols are recognized as xenobiotics by the body; they are subject to metabolic conversion, resulting in nontoxic compounds due to phase II enzymes [Spencer et al., 2004; Walle, 2004]. The mucosa of the small intestine and large intestine is the site for metabolic conversion of flavonols, as well as their absorption into the body [Scalbert and Williamson, 2000], that is, dietary flavonols are mostly converted into their conjugated metabolites before influx into the circulation. This conversion may be an effective tool for the attenuation of their toxicity and the regulation of their physiological function because some metabolites possess various biological activities [Williamson et al., 2005]. In this chapter, we focus on quercetin and its glycosides as typical flavonols originating from vegetables.

ANTIOXIDANT AND PRO-OXIDANT ACTIVITY

Flavonols are characterized by the substitution of a hydroxyl group at the 3-position and a double bond between the 2- and 3-position in diphenylpropane. These two structural characteristics contribute to the enhancement of the activity of flavonoids to scavenge free radicals. Flavonols are recognized as a subclass with strong antioxidant activity among flavonoids [Bors et al., 1990]. The essential part of flavonol structure for exerting such activity is the *o*-dihydroxyl structure at the 3′- and 4′-position of the B-ring [Bors, 1990; Terao and Piskula, 1997]. Hydroxyl groups at the 3-′ and 4′-position scavenge free radicals by donation of hydrogen or an electron. Quercetin and myricetin can exert strong antioxidant activity because they possess an *o*-dihydroxyl structure, so-called

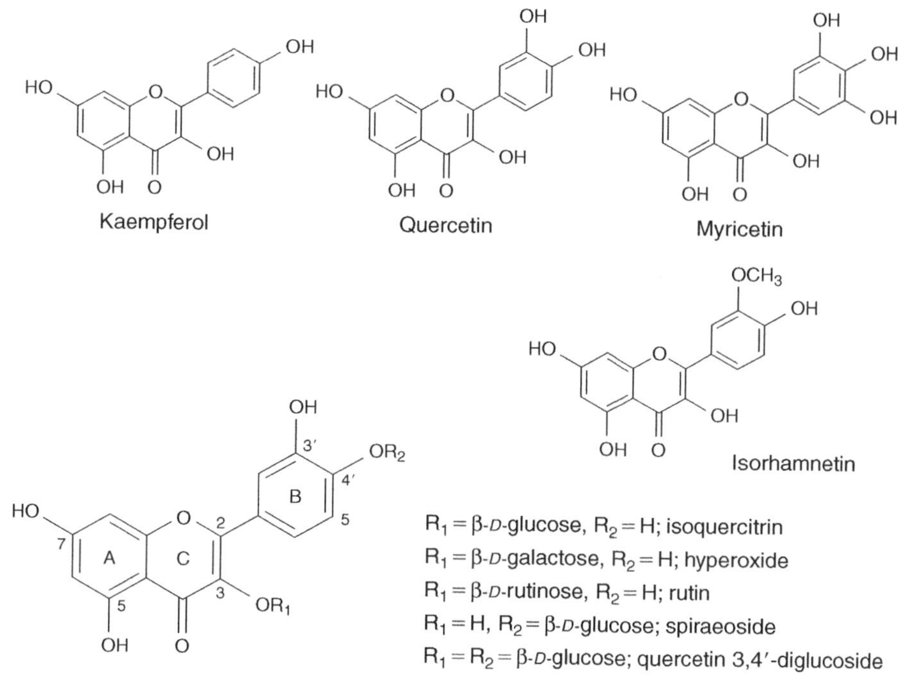

Figure 8.1 Structures of main flavonol aglycones and quercetin glycosides present in plant foods.

catechol and pyrogallol, respectively. Kaempferol is a weak antioxidant because of its monophenol structure in the B-ring [Terao and Piskula,1997].

Figure 8.2 illustrates the potential reaction pathways for catechol. One catechol can scavenge two peroxyl radicals, resulting in the formation of an *o*-semiquinone radical when receiving the first electron, and the formation of an *o*-quinone after receiving the second electron. Then, the *o*-dihydroxyl moiety possesses pro-oxidant properties [Gouphua et al., 1997] supported by the formation of an *o*-semiquinone radical that can react with an oxygen molecule to generate $O_2^{\bullet -}$, and also convert $O_2^{\bullet -}$ to H_2O_2 [Metodiewa et al., 1999; Kawanishi et al., 2005]. These reactive oxygen species (ROS) can act as both cellular redox signals for the expression of antioxidant enzymes and inducers of oxidative damage to cell components. Then, catechol/pyrogallol can yield antioxidant and pro-oxidant activity by scavenging or generating ROS depending on the environment of the reaction, and the relative concentration of other antioxidant or pro-oxidant molecules. Monophenol moieties in the B-ring do not show such a pro-oxidant effect. Loss of the *o*-dihydroxyl structure by conjugation of the hydroxyl group seems to help avoid the unfavorable pro-oxidant effect of flavonols on health. Physiologically, conjugation of the hydroxyl group in catechol/pyrogallol happens during intestinal absorption of flavonols by phase II enzymes.

ABSORPTION AND METABOLIC CONVERSION IN THE DIGESTIVE SYSTEM

Human saliva can hydrolyze glucose-bound quercetin glycosides, resulting in active aglycone within a few minutes [Walle et al., 2005]. This hydrolytic activity is suggested to be derived from bacterial colonies and epithelial cells in the mouth. Large interindividual variability was found in the hydrolytic activity of the oral cavity. Contribution of this activity is not obviously referred to the physiological function of dietary flavonols. Quercetin monoglucosides such as isoquercitrin (quercetin 3-*O*-β-D-glucoside) and spiraeoside (quercetin 4′-β-D-glucoside) are efficiently absorbed by the small intestine (Fig. 8.3) [Murota and Terao, 2003], that is, they are incorporated into the epithelial cells of the small intestine through independent pathways [Day et al., 2000a]. One is a sodium-dependent glucose transporter-1 (sGLT-1)-mediated cellular transport pathway in which glucosides are converted into aglycones by the activity of cellular β-glucosidase. In the other pathway, glucosides are first hydrolyzed by lactose-phlorizin hydrolase (LPH) on the surface of brush-border membranes, and the resulting lipophilic aglycone is readily transported into cells by passive absorption. LPH and cellular β-glucosidase prefer the glucoside bond at the 4′-position to that at the 3-position [Walle et al., 2000; Ioku et al., 1998]. Spiraeoside may therefore be more favorable than isoquercitrin in intestinal absorption.

Quercetin and other flavonols are subject to metabolic conversion during the absorption process in the small intestine. Phase II enzymes including uridine 5′-diphosphate glucuronosyl transferase (UGT), phenol-sulfotransferase (PST),

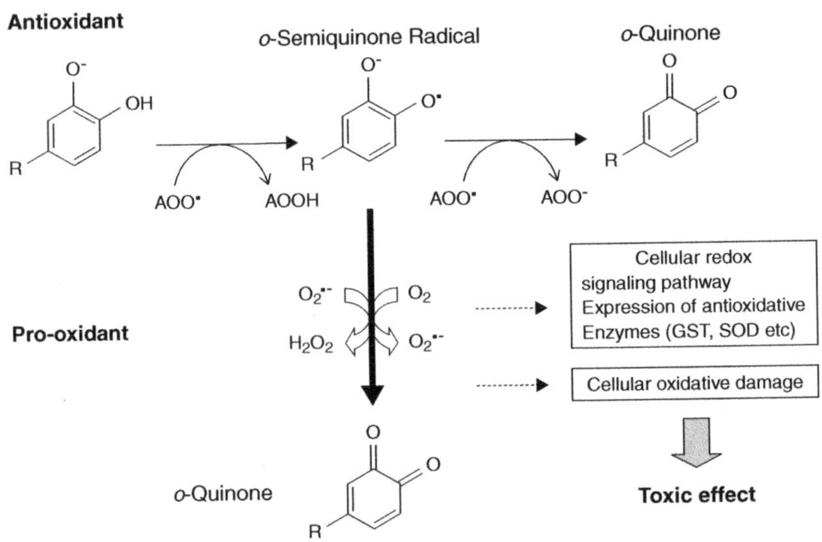

Figure 8.2 Free-radical scavenging and ROS-generating pathway of catechol-type flavonols (GST: glutathione-*S*-transferase: SOD: superoxidie dismutase).

and catechol-*O*-methyltransferase (COMT) are distributed in intestinal epithelial cells (Fig. 8.4) [Murota and Terao, 2003]. Quercetin aglycone is inevitably converted to its conjugated metabolites with or without *O*-methylation and then transferred to the liver through the portal vein. Some conjugated metabolites are likely to be transferred into the systemic circulation via lymph [Murota and Terao, 2005]. Most conjugated metabolites do not enter the body but return to the intestinal tract by the action of multidrug resistance-associated protein-2 (MRP-2).

Quercetin glycosides other than monoglucosides (e.g., rutin, hyperoside) reach the large intestine without absorption (Fig. 8.3). Quercetin glycosides from foods are mostly excreted into the feces without absorption. They are subject to hydrolysis by enterobacteria, resulting in their aglycone. Anaerobic bactera such as *Bacteroides distasonis, Bacteroides uniforms*, and *Bacteroides ovatus* are known to possess β-glucosidase activity [Bokkenheuse et al., 1987]. Enterobacteria also decompose aglycone to produce its ring scission products such as 3,4-dihydroxyphenylacetic acid, *m*-hydroxylphenylacetic acid, and *m*-homovanilic acid [Terao, 1999]. These products can be absorbed into the body by the monocarboxylic acid transporter and/or paracellular diffusion pathway [Konishi, 2005]. Aglycone is partly absorbed by large intestinal

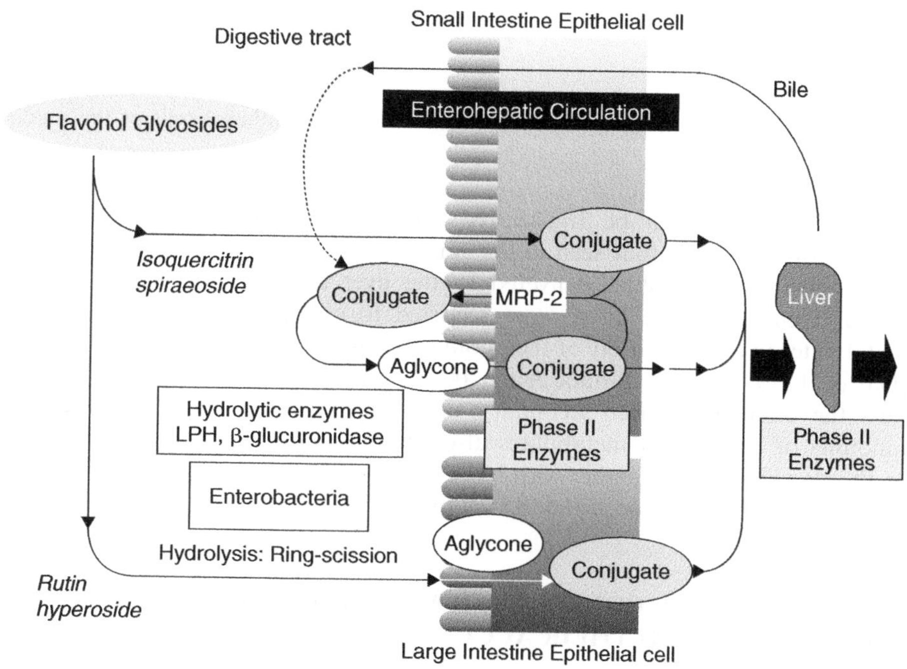

Figure 8.3 Pathway of absorption and metabolism of flavonols in the digestive system (MRP-2: multidrug resistance-associated protein-2; LPH: lactose phlorizin hydrolaze).

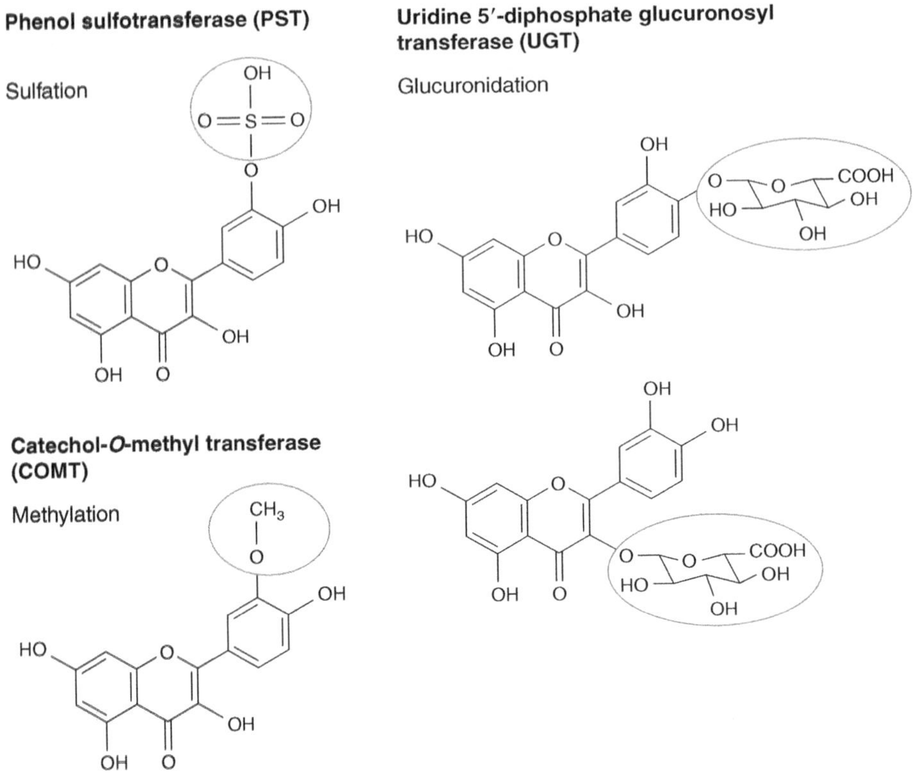

Figure 8.4 Metabolic conversion of quercetin in human epithelial cells of the small intestine.

mucosa and likely to be transferred into the circulation after metabolic conversion similar to that seen in the small intestine.

Conjugated metabolites delivered to the liver are metabolized by hepatic phase II enzymes. The latter were reported to convert conjugated quercetin metabolites to a wide variety of their conjugates with and without deconjugation [Day et al., 1998; van de Woude et al., 2004]. Conjugated metabolites are released into the circulation after secondary metabolism in the liver. Considerable amounts of conjugated metabolites also simultaneously enter bile and return to the intestinal tract together with bile components. This enterohepatic circulation should be considered if the overall bioavailability of flavonols is discussed.

DISTRIBUTION OF CONJUGATED METABOLITES IN THE CIRCULATION AND THEIR ACCUMULATION IN ORGANS

Day et al. [2001] detected >20 species of metabolites, including quercetin 3-*O*-β-D-glucuronide (Q3GA), quercetin 3′-*O*-β-D-glucuronide (Q3′GA), quercetin

4′-*O*-β-D-glucuronide (Q4′GA), quercetin 3′-*O*-β-sulphate (Q3′S), quercetin 3′-*O*-methyl 4′-*O*-β-D-glucuronide, and quercetin 3′-*O*-methyl 3-*O*-β-D-glucuronide in human plasma fractions, but aglycone was not detected. These metabolites were found in human urine after the intake of flavonol-rich foods [Mullen et al., 2004; Hong and Mitchell, 2005]. Quercetin metabolites are suggested to be located in the bloodstream in an albumin-bound form [McAnlis et al., 1999]. Quercetin aglycone is tightly bound to plasma albumin by noncovalent bonds [Papadopoulou et al., 2005]. It was reported that the binding strength of its conjugated metabolites to albumin is much lower than that of aglycone [Dufour and Dangles, 2005]. This implies that conjugated metabolites can be readily delivered to tissues by release from the albumin fraction.

In human plasma, quercetin metabolites transiently increased at ~ 1 μM after the intake of a flavonol-rich diet [Hollman et al., 1995], after which concentration gradually decreased to nearly zero in <8 h [Manach et al., 1997]. Continuous ingestion of flavonol-rich foods may maintain a low level of quercetin metabolites in plasma, but this baseline level may be greatly influenced by individual differences [Moon et al., 2000].

Whether or not dietary quercetin and other flavonols can accumulate in internal organs is controversial. It is unclear if they accumulate in the human body, but dietary quercetin was found to be distributed in each organ in experimental animals [de Boer et al., 2005]. In rats fed 0.1% quercetin for 11 weeks, the highest accumulation was in the lungs. Low amounts of quercetin metabolites were detected in the brain. This indicates that quercetin metabolites can access the central nervous system by passing the blood–brain barrier, although the mechanism of this passage is incompletely understood.

POTENTIAL HEALTH EFFECTS OF CONJUGATED METABOLITES AT TARGET SITES

Whether or not the intake of plant flavonols increases the antioxidant activity of human blood plasma is interesting. It was found that Q3GA (one of the major quercetin metabolites detected in human plasma) possesses substantial activity for scavenging free radicals and inhibits in vitro low-density lipoprotein (LDL) oxidation at higher concentration [Moon et al., 2001]. Various small-molecule antioxidants, including uric acid (160–450 μM), ascorbic acid (30–150 μM), bilirubin (5–20 μM), α-tocopherol (15–40 μM), and carotenoids (1.0–2.0 μM) contribute to the antioxidant activity of plasma [Wayner et al., 1987; Frei et al., 1988]. Plasma albumin is known to be an effective scavenger of free radicals and to contribute to plasma antioxidant activity [Roche et al., 2008]. It is therefore unlikely that albumin-bound flavonol metabolites affect plasma antioxidant activity if flavonols are ingested from an ordinary diet. We recently carried out a study in which cooked onion was consumed by human volunteers [Murota et al., 2007]. Onion consumption failed to enhance antioxidant activity of the albumin fraction against LDL oxidation, although about 80% of quercetin metabolites were in this fraction. Other major metabolites such as Q4′GA and Q3′S in human plasma seem to be devoid of antioxidant and pro-oxidant activity because they lack the catechol structure.

These metabolites are known to possess specific biological activities other than antioxidant activity in vitro (Table 8.1) [Yoshizumi et al., 2002; de Pascual-Teresa et al., 2004; Day et al., 2000b; O'leary et al., 2004; Mochizuki et al., 2004; Kim et al., 2007; Kawai et al., 2008; Shiba et al., 2008]. Steffen et al. [2008] recently demonstrated that isorhamnetin and its glucuronide conjugates, not quercetin aglycone, inhibit the activity of NADPH-oxidase of endothelial cells. Quercetin metabolites are likely to exert their physiological function at the target sites where they concentrate. Shimoi et al. [2005] demonstrated that conjugated metabolites of luteolin, a flavone-type flavonoid, were converted into their aglycone by the enhanced activity of plasma β-glucuronidase during inflammation when luteolin was orally administered to lipopolysaccharide-injected mice. This phenomenon implies that conjugated metabolites circulate as nontoxic forms in the bloodstream but can be converted to active aglycones to exert their function immediately. This effect was clarified using newly synthesized monoclonal antibody: Q3GA accumulated in atherosclerotic lesions but not in normal aorta (Fig. 8.5) [Kawai et al., 2008]. This finding indicates that quercetin metabolites translocate to the vascular target selectively when blood

Table 8.1 Biological Activities of Conjugated Quercetin Metabolites

Quercetin 3-glucuronide (Q3GA)	• Prevention of vascular smooth muscle cell hypertrophy by angiotensin II via effects on JNK and AP-1	Yoshizumi et al., 2002
	• Decreased expression of COX-2 in human lymphocytes	de Pascual-Teresa et al., 2004
	• Increase expression of bone sialoprotein mRNA expression	Kim et al., 2007
	• Inhibition of scavenger receptor mRNA expression in murine macrophages	Kawai et al., 2008
	• Potent inhibition of myeloperoxidase	Shiba et al., 2008
Quercetin 3′-glucuronide (Q3′GA)	• Potent inhibition of bovine xanthine oxidase	Day et al., 2000b
	• Inhibition of soybean lipoxygenase	Day et al., 2000b
Quercetin 4′-glucuronide (Q4′GA)	• Potent inhibition of bovine xanthine oxidase	Day et al., 2000b
	• Inhibition of soybean lipoxygenase	Day et al., 2000
Quercetin 3′-sulfate (Q3′S)	• Inhibition of COX-2 mRNA expression in Caco-2 Cell	O'Leary et al., 2004
Quercetin 3-sulfate	• Inhibition of ICAM-1 expression in human aortic endothelial cells	Mochizuki et al., 2004
Isorhamnetin glucuronide	• Inihibiton of endothelial NADPH oxidase	Steffen et al., 2008

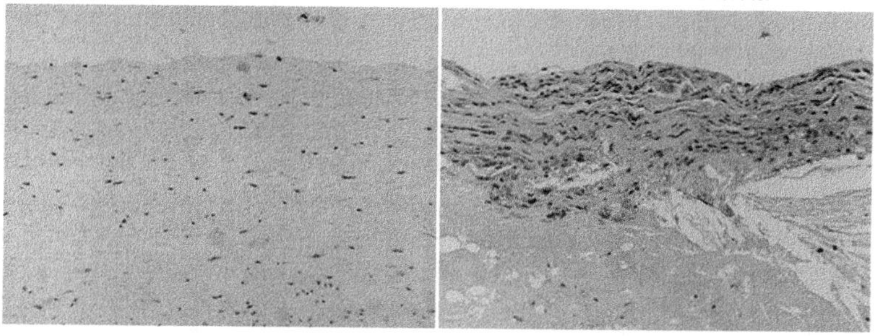

Figure 8.5 Detection of quercetin metabolites in human atherosclerotic aorta using Q3GA-specific monoclonal antibody mAb14A2. (Data from Kawai et al. [2008].)

vessels are injured by oxidation. It can be hypothesized that conjugated metabolites may exert specific activity by themselves or by conversion into their active aglycone under intolerable oxidative stress, for example, accelerated inflammation.

CONCLUSION

It is generally thought that plant flavonols have beneficial effects on human health by enhancing the antioxidant defense system. In recent years, their plasma metabolites were also found to possess various other biological activities, but excessive intake of plant flavonols may induce an adverse effect by generating ROS. Metabolic conversion of flavonols to their sulfate/glucuronide conjugates in the digestive system eliminates their toxic effects and, more importantly, regulates their biological activities. These conjugates may concentrate at the target site and exert their functions efficiently under oxidative stress.

REFERENCES

Bjeldans LF, Chang GW. 1977. Mutagenic activity of quercetin and related compounds. *Science* 197:577–578.

Bokkenheuse VD, Shacketon CHL, Winter J. 1987. Hydrolysis of dietaryflavonoide glycosides by strains of intestinal bacteroides from humans. *Biochem J* 248:953–956.

Bors W, Heller W, Michel C, Saran M. 1990. Flavonoids as antioxidants: Determination of radical-scavenging efficiencies. In: Packer L, Glazer AN, Eds. *Methods in Enzymology 186*, San Diego: Academic Press, pp. 343–355.

Crozier A, Jaganath IB, Cliford MN. 2006. Phenols, polyphenols and tannins: An overview. In: Crozier A, Clifford MN, Ashihaea H, Ed. *Plant Secondary Metabolites*. Oxford: Blackwell, pp. 1–24.

Day AJ, Mellon F, Barron D, Sarrazin G, Morgan MRA, Williamson G. 2001. Human metabolism of dietary flavonoids: identification of plasma metabolites of quercetin. *Free Radic Res* 35:941–952.

Day AJ, Canada FJ, Diaz JC, Kroon PA, Mclauchlan R, Faulds CB, Plumb GW, Morgan MR, Williamson G. 2000a. Dietary flavonoid and isoflavone glycosides are hydrolyzed by the lactase site of lactase phlorizin hydrolase. *FEBS Lett* 468:166–170.

Day AJ, Bao Y, Morgan MRA, Williamson G. 2000b. Conjugation position of quercetin glucuronide and effect on biological activity. *Free Radic Biol Med* 29:1234–1243.

Day AJ, Dupont MS, Ridley S, Rhodes M, Rhodes MJC, Morgan MRA, Williamson G. 1998. Deglucosidation of flavonoid and isoflavonoid glycosides by human small intestine and liver β-glucosidase activity. *FEBS Lett* 436:71–75.

de Boer VC, Dihal AA, Arts IC, Wolffram S, Alink GM, Rietjens IM, Keijer J, Hollman PC. 2005. Tissue distribution of quercetin in rats and pigs. *J Nutr* 135:1718–1725.

de Pascual-Teresa S, Johnston KL. Dupont MS, O'leary KA, Needs PW, Morgan LM, Clifford MN, Bao Y, Williamson G. 2004. Quercetin metabolism downregulate cyclooxygenase-2 transcription in human lymphocytes ex vivo but not in vivo. *J Nutr* 134:552–557.

Dufour C, Dangles O. 2005. Flavonoid-serum albumin complexation: determination of binding constants and binding sites by fluorescence spectroscopy. *Biochim Biophys Acta* 1721:164–173.

Frei B, Stocker R, Ames, BN. 1988. Antioxidant defenses and lipid peroxidation in human blood plasma. *Proc Natl Acad Sci USA* 85:9748–9752.

Galati G, O'Brien PJ. 2004. Potential toxicity of flavonoids and other dietary phenolics: significance for their chemopreventive and anticancer properties. *Free Radic Biol Med* 37:287–303.

Gouphua C, Sofic E, Prior RL. 1997. Antioxidant and prooxidant behavior of flavonoids: Structure-activity relationships. *Free Radic Biol Med* 22:749–760.

Hertog, MG, Feskens EJ, Hollman PC, Katan MB, Kromhout D. 1993. Antioxidant flavonols and coronary heart disease risk. *Lancet* 342:1007–1011.

Hollman PC, de Vries JH, van Leeuwen SD, Mengelers MJ, Katan MS. 1995. Absorption of dietary quercetin glycosides and quercetin in healthy ileostomy volunteers. *Am J Clin Nutr* 62:1276–1282.

Hong, Y-J, Mitchell E. 2005. Metabolic profiling of flavonol metabolites in human urine by liquid chromatography and tandem mass spectrometry. *J Agric Food Chem* 52:6794–6801.

Ioku K, Pongpiriyadacya Y, Konishi Y, Takei Y, Nakatani N, Terao J. 1998. β-glucosidase activity in the rat small intestine toward quercetin monoglucosides. *Biosci Biotecnol Biochem* 63:1428–1431.

Kawai Y, Nishikawa T, Shiba T, Saito S, Murota K, Shibata N, Kobayashi M, Kanayama M, Uchida K, Terao J. 2008. Macrophages as a target of quercetin glucuronides in human atherosclerotic arteries. *J Biol Chem* 283:9424–9434.

Kawanishi S, Oikawa S, Murata M. 2005. Evaluation for safety of antioxidant chemopreventive agents. *Antioxid Redox Signal* 7:1728–1739.

Kim DS, Takai H, Arai M, Araki S, Mezawa M, Kawai Y, Murota Y, Terao J, Ogata Y. 2007. Effect of quercetin and quercetin 3-glucuronide on the expression of bone sialoprotein gene. *J Cell Biochem* 101:790–800.

Konishi Y. 2005. Transepithelial transport of microbial metabolites of quercetin in intestinal Caco-2 cell monolayers. *J Agric Food Chem* 53:601–607.

Manach C, Morand C, Demigne C, Texier O, Regerat F, Rémésy, C. 1997. Bioavailability of rutin and quercetin in rats. *FEBS Lett* 409:12–16.

McAnlis GT, McEneny J, Pearce J, Young IS. Absorption and antioxidant effects of quercetin from onions in man. 1999. *Eur J Clin Nutr* 53:92–96.

Metodiewa D, Jaiswal AK, Cenas N, Dickancaite E, Segura-Aguilar J. 1999. Quinone may act as a cytotoxic prooxidant after its metabolic activation to semiquinone and quinoidal products. *Free Radic Biol Med* 26:107–116.

Mochizuki M, Kajiya K, Terao J, Kaji K, Kumazawa S, Nakayama T, Shimoi K. 2004. Effect of quercetin conjugation on vascular permeability and expression of adhesion molecules. *Biofactors* 22:201–204.

Moon J-H, Tsushida T, Nakahara K, Terao J. 2001. Identification of quercetin 3-*O*-β-glucuronide as an antioxidative metabolite in rat plasma after oral administration of quercetin. *Free Radic Biol Med* 30:1274–1285.

Moon J-H, Nakata R, Oshima S, Inakuma T, Terao J. 2000. Accumulation of quercetin conjugates in blood plasma after the short-term ingestion of onion by women. *Am J Physiol* 279: R461–R467.

Mullen W, Boitier A, Stewart AJ, Crozier A. 2004. Flavonoid metabolites in human plasma and urine after the consumption of red onions: analysis by liquid chromatography with photodiode array and full scan tandem mass spectrometric detection. *J Chromatogr A* 1058:163–168.

Murakami A, Ashida H, Terao J. 2008. Multitargeted cancer prevention by quercetin. Cancer Lett. 269:315–325.

Murota K and Terao J. 2005. Quercetin appears in the lymph of unanesthetized rats as its phase II metabolites after administered into the stomach. *FEBS Lett* 579:5343–5346.

Murota K, Terao J. 2003. Antioxidative flavonoid quercetin: implication of its intestinal absorption and metabolism. *Arch Biochem Biophys* 417:12–17.

Murota K, Hotta A, Ido H, Kawai Y, Moon J-H, Sekido K, Hayashi H, Inakuma T, Terao J. 2007. Antioxidant capacity of albumin-bound quercetin metabolites after onion consumption in humans. *J Med Invest* 54:37–374.

O'Leary KA, de Pascual-Teresa S, Needs PW, Bao YP, O'Brien NM, Williamson G. 2004. Effect of flavonoids and vitamin E on cyclooxygenase-2(COX-2) transcription. *Mutation Res* 38:877–884.

Papadopoulou A, Green RJ, Frazier RA. 2005. Interaction of flavonoids with bovine serum albumin: A fluorescence quenching study. *J Agric Food Chem* 53:158–163.

Roche M, Rondeau P, Singh NR, Tarnus E, Bourdon E. 2008. The antioxidant properties of serum albumin. FEBS Lett. 582:1783–1787.

Rusznyak S, Szent-Györgyi A. 1936. Vitamin nature of flavones. *Nature* 138:798.

Scalbert A, Williamson G. 2000. Dietary intake and bioavailability of polyphenols. *J Nutr* 130:2073S–2085S.

Shiba Y, Kinoshita T, Chuman H, Taketani Y, Takeda E, Kato Y, Naito M, Kawabata K, Ishisaka A, Terao J, Kawai Y. 2008. Flavonoids as substrates and inhibitors of myeloperoxidase: Molecular actions of aglycone and the metabolites. Chem Res Toxicol. 21:16–1609.

Shimoi K, Saka N, Nozawa R, Sato M, Amano I, Nakayama T, Kinae N 2001. Deglucuronidation of a flavonoid, luteolin monoglcucuronide, during inflammation. *Drug Metab Disposition* 29:1521–1524.

Spencer JP, Abd-el-Mohsen, Rice-Evans MC. 2004. Cellular uptake and metabolism of flavonoids and their metabolites: Implications for their bioactivity. *Arch Biochem Biophys* 423:148–161.

Steffen, Y, Gruber C, Schewe T, Sies H. 2008. Mono-O-*methylated* flavanols and other flavonoids as inhibitors of endothelial NADPH oxidase. *Arch Biochem Biophys* 469:209–219.

Sugimura T, Nagao MT, Matsushima T. 1977. Mutagenicity of flavone derivatives. *Proc Japan Acad* B53:194–197.

Terao J. 1999. Dietary flavonoids as plasma antioxidants on lipid peroxidation: Significance of metabolic conversion In: Packer L, Hiramatsu M, Yoshikawa T, Eds. *Antioxidant Food Supplements in Human Health*. San Diego: Academic Press, pp. 255–269.

Terao J, Piskula M. 1997. Flavonoid as inhibitors of lipid peroxidation in membranes. In: Rice-Evans CA, Packer L, Eds. *Flavonoids in Health and Disease*. New York: Marcel Dekker, pp. 277–294.

van de Woude H, Boersma MG, Vervoort J, Rietjens IMCM. 2004. Identification of 14 quercetin phase II mono and mixed conjugates and their formation by rat and human phase II in vitro model systems. *Chem Res Toxicol* 17:1520–1530.

Walle T. 2004. Absorption and metabolism of flavonoids. *Free Radic Biol Med* 36: 829–837.

Walle, T, Browning AM, Steed LL, Reed SG, Walle UK. 2005. Flavonoid glucosides are hydrolyzed and activated in the oral cavity in humans. *J Nutr* 135:48–52.

Walle T, Otake Y, Walle UK, Wilson FA. 2000. Quercetin glucosides are completely hydrolyzed in ileostomy patients before absorption. *J Nutr* 130:2658–2661.

Wayner DD, Burton GW, Ingold KU Barclay LR Locke SJ. 1987. The relative contributions to the total peroxyl radical-trapping antioxidant activity of human blood plasma. *Biochim Biophys Acta* 924:408–419.

Williamson G, Barron D, Shimoi K, Terao J. 2005. In vitro biological properties of flavonoid conjugates found in vivo. *Free Radic Res* 39:457–459.

Yoshizumi M, Tsuchiya K, Suzaki Y, Kirima K, Kya M, Moon JH, Terao J, Tamaki T. 2002. Quercetin glucuronide prevents VSMC hypertrophy by angiotensin II via the inhibition of JNK and AP-1 signaling pathway. *Biochem Biophys Res Commun* 293:1458–1465.

Youdin KA, Shukitt-Hale B, Joseph J. 2004. Flavonoids and the brain: interactions at the blood-brain barrier and their physiological effects on the central nervous system. *Free Radic Biol Med* 37:1683–1693.

9 Flavonols: Biochemistry Behind Cardiovascular Effects

FRANCISCO PÉREZ-VIZCAÍNO[1] AND JUAN DUARTE[2]
[1]Department of Pharmacology. School of Medicine. Universidad Complutense, Madrid, Spain
[2]Department of Pharmacology. School of Pharmacy. Universidad de Granada, Granada, Spain

CONTENTS

Plant Phenolics and Human Health: Biochemistry, Nutrition, and Pharmacology,
Edited by Cesar G. Fraga.

INTRODUCTION

Flavonoids constitute a large class of polyphenols found in fruits and vegetables that share a common skeleton of phenylchromane. This basic structure allows a large number of substitution patterns leading to several subclasses of flavonoids, such as flavonols, flavones, flavanones, flavanols, anthocyanidins, isoflavones, dihydroflavonols, and chalcones. Among the diverse flavonoid subclasses, flavonols (especially quercetin) and flavanols (catechins) are the most abundant in our food. Flavonols are present in foods as diverse glycosides, whereas flavanols are usually found as aglycones.

Flavonols are present in considerable amounts in our normal diet and show a wide range of biological activities. Therefore, they are candidates to explain the advantageous effects of fruits and vegetables on cardiovascular health. Formerly considered vitamins, the term "vitamin P," coined for their effect on capillary Permeability [Rusznyak and Szent-Györgyi, 1936] was discontinued in the 1950s [Joint Committee on Nomenclature of the American Society of Biological Chemists, 1950]. Some specific flavonols (e.g., quercetin) or flavonoid mixtures or extracts are commercialized as dietary supplements and recommended at doses that exceed by far the common daily intake. The health claims of these supplements are not clearly supported by clinical trials. As a rebound effect, for many within the academic community, the use of flavonoids, among other natural antioxidants used at high doses, have a certain whiff of the nonscientific. Within academic medicine they are only used as "venotonic" or "phlebotropic" drugs with limited results on several venous diseases [Lyseng-Williamson and Perry, 2003]. However, a glance at their pharmacology reveals that these compounds show a much wider therapeutic potential.

In this chapter, we will review the available evidence supporting a beneficial role of flavonols on cardiovascular disease and the potential molecular targets involved.

ARE FLAVONOLS REALLY EFFECTIVE IN CARDIOVASCULAR DISEASE?

Epidemiological Evidence

The interest in dietary flavonoids has grown in the last 15 years after the publication of the study of Hertog et al. [1993] showing an inverse correlation between dietary consumption of flavonols and flavones and reduced incidence and mortality from ischemic heart disease. Most prospective epidemiological studies carried out have found a similar relationship regarding ischemic heart disease while a possible inverse relationship with stroke is less clear [Hertog et al., 1997; Knekt et al., 1996; Rimm et al., 1996; Yochum et al., 1999; Hirvonen et al., 2001; Arai et al., 2000; Mursu et al., 2008]. The meta-analysis of seven prospective cohort studies concluded that the individuals in the top third of dietary flavonol intake are associated with a reduced risk of mortality from coronary heart disease as compared with those in the bottom third, after adjustment for known risk factors and other dietary components [Huxley and Neil, 2003].

Intervention Studies

No intervention studies published so far have analyzed the effects of flavonols or flavonol-rich food on cardiovascular morbidity or mortality, which would require a large number of patients followed for several years. However, the above-mentioned epidemiological studies and the basic research has stimulated several intervention trials analyzing the short-term effects on selected cardiovascular risk factors, mainly hypertension and endothelial function. Endothelial dysfunction is an early manifestation common to several cardiovascular risk factors such as arterial hypertension, arteriosclerosis, smoking, diabetes, or aging [Endemann and Schiffrin, 2004; Widlansky et al., 2003]. The presence of endothelial dysfunction has prognostic value independently of the traditional cardiovascular risk factors. Recently, it has been reported that high doses of quercetin reduced blood pressure in patients with stage 1 hypertension [Edwards et al., 2007]. However, high doses of quercetin have no effect on blood pressure or other cardiovascular risk factors in normotensive healthy individuals [Conquer et al., 1998]. A number of studies have analyzed the effects of flavonoid-rich food and beverages such as cocoa, tea, or fruit juices. Consumption of flavanol-rich cocoa was found to lower blood pressure and improve endothelial function in short-term human intervention studies [Taubert et al., 2003; Grassi et al. 2005; Fraga et al. 2005]. A recent study has also shown reduction in blood pressure and improvement of endothelial function with low doses of dark chocolate in untreated upper-range prehypertension or stage 1 hypertension followed for 18 weeks [Taubert et al., 2007]. Another double-blind, crossover study conducted in 12 hypertensive patients who received alternately high-flavonoid sweetie juice and low-flavonoid sweetie juice for 5 weeks showed that the flavonoid-rich juice significantly reduced blood pressure [Reshef et al, 2005]. These results, interesting from a nutritional point of view, have an important limitation because the effects, which are normally attributed to the flavonoids present, could be due to any of the several hundred other substances present in these food and beverages or due to interactions among them.

Animal Studies

Studies from several laboratories including ours have revealed that quercetin is able to prevent or revert hypertension and endothelial dysfunction, and we believe that this is a fundamental mechanism involved in the protective effects of flavonoids on cardiovascular health. Oral administration of quercetin (5 or

10 mg/kg/day) prevents the development of hypertension or produces a gradual decrease in blood pressure in various experimental models in rats, such as genetic hypertension in spontaneously hypertension rats (SHRs) [Duarte et al., 2001; Sanchez et al., 2006], L-NAME-induced hypertension [Duarte et al., 2002], hypertension induced by the administration of mineralcorticoids plus NaCl (DOCA-salt) [Galisteo et al., 2004], hypertension induced by unilateral nephrectomy and estenosis of the contralateral renal artery [Goldblatt model; García-Saura et al., 2005], genetic salt-sensitive rats (Dahl rats) [Aoi et al., 2004], high-fat high-sucrose diet-induced hypertension [Yamamoto and Oue, 2006], and in rats with aortic constriction [Jalili et al., 2006]. Therefore, regardless of the status of the renin–angiotensin system, volume expansion, NO, renal injury, or oxidative stress, quercetin produced consistent antihypertensive effects in vivo. These antihypertensive effects are accompanied by a reduction of associated end-organ damage: cardiac hypertrophy, kidney histological alterations and proteinuria, vascular remodelling, and endothelial dysfunction, as well as a decrease in the markers of oxidative stress in plasma, liver, and urine. The effects on endothelial dysfunction has been recently reviewed [Perez-Vizcaino et al., 2006]. In a recent puzzling report, Carlstrom et al. [2007] observed that when quercetin is added to the diet, it did not prevent cardiovascular complications in SHRs while it was effective when given orally by gavage. In principle, this could not be attributed to different bioavalability since the levels of quercetin metabolites were higher in the former. In contrast, in another study by the same group, quercetin-supplemented diets lowered blood pressure and attenuated cardiac hypertrophy in rats with aortic constriction [Jalili et al., 2006].

Consumption of low doses of quercetin reduced the progression of atherosclerosis in apolipoprotein E-deficient mice [Hayek et al., 1997]. The reduction in the atherosclerotic lesion area was not associated with changes in plasma low-density or high-density lipoprotein (LDL or HDL) cholesterol levels in this study but with a reduced susceptibility to oxidation (induced by different modes such as copper ions, free-radical generator, or macrophages) of LDL. Quercetin also efficiently protected hypercholesterolemic hamsters against aortic fatty streak accumulation, an early manifestation of the atherosclerotic process [Auger et al., 2005]. In addition, quercetin attenuated the hyperlipidemia and the lipid peroxidation in the aorta in high-cholesterol-fed rabbits [Kamada et al., 2005].

ARE ANTIOXIDANTS USEFUL IN CARDIOVASCULAR DISEASE?

As discussed below, the biological effects of flavonoids include the interaction with many biochemical targets in addition to their effects on the antioxidant status. However, as the cardiovascular effects of flavonoids are commonly attributed to their antioxidant effects, it seems pertinent to address herein the available evidence of the role of oxidative stress and the effectiveness of antioxidants in cardiovascular disease.

In the 1980s, when enthusiastic lecturers initiated us into the pathophysiological role of reactive oxygen species (ROS), the scientific community naively took for granted that, in a few years, the so-called antioxidants would be to cardiovascular diseases as antibiotics were to bacterial infections. Some years later, we are still in need of concrete evidence that allow the incorporation of antioxidants as pharmacologically accepted therapies. Nevertheless, the evidence from basic research studies in favor of the role of ROS in diseases such as atherosclerosis, ischemic heart disease, hypertension, heart failure, and diabetes has been continuously growing over the last three decades. Multiple epidemiological studies have suggested that increased intake of dietary antioxidant vitamins lowers the risk of atherosclerosis [Stampfer et al., 1993]. However, the scarce positive results with antioxidants—mainly with vitamin E, β-carotene, vitamin C, and cocktails—for the prevention of coronary heart disease are far outweighed by negative results in prospective controlled large-scale trials [Steinberg and Witztum, 2002]. A recent meta-analysis indicates that treatment with high doses of β-carotene, vitamin A, and vitamin E may even increase mortality while vitamin C and selenium had no effect [Bjelakovic et al., 2007].

On the other hand, it is becoming apparent that truly effective drugs such as statins, angiotensin-converting enzyme (ACE) inhibitors, angiotensin receptor antagonists, and possibly aspirin owe their clinical value, at least partly, to a direct or indirect reduction of oxidative stress. Besides the cholesterol-lowering effect, statins produce several alternative or pleiotropic beneficial effects. Among these pleiotropic effects they share the property of inhibiting LDL oxidation [Gotto, 2003]. ACE inhibitors and angiotensin receptor antagonists inhibit one of the most powerful triggers of oxidative status in cardiovascular disease, that is, angiotensin II [Reckelhoff and Romero, 2003]. Furthermore, these drugs have the advantage over classic antioxidant compounds that they target the signal transduction pathway leading to ROS formation in cells—rather than scavenging ROS already formed—which seems to be a more reasonable approach.

Several arguments have been used to answer why the ROS/antioxidant hypothesis is not translated into daily therapeutics [Heinecke, 2001]. First, ROS and antioxidants exhibit a complex chemistry, and there is an intricate interplay among the various ROS and antioxidant systems to produce or prevent damage and disease. A scavenger reacting with a radical species is converted into a radical itself, albeit less reactive. This radical could, under certain conditions, propagate lipid oxidation by attacking polyunsaturated fatty acids, or oxidize proteins or DNA. Thus, the behavior of a compound as antioxidant or pro-oxidant depends on the reaction conditions and available substrates. Second, claims of wrong antioxidant selection, dosages, and/or timing in clinical trials have been put forward. Notably, most clinical trials have analyzed the effects of antioxidants in established coronary artery disease while their effectiveness in primary prevention is unknown. Still, there is a general belief that some appropriate antioxidant intervention has the potential to improve prognosis. It should be emphasized herein that each antioxidant shows a differential pharmacological profile, and its effects may go beyond the changes in cellular antioxidant status. This is specially the case of flavonoids.

In summary, the evidence for the preventive effect of flavonols in cardiovascular disease is growing based on mechanistic, animal, epidemiological, and short-term intervention studies. However, we must learn from the lessons with dietary antioxidants, the last word awaits large-scale intervention clinical trials.

MOLECULAR TARGETS

Complex Pharmacology of Flavonoids

Flavonoids constitute a very large class, which includes several thousand compounds as found in nature, plus the hemisynthetic or synthetic derivatives. Second, as a group, they are far from being specific. Flavonoids are regarded as antioxidants. However, the interference of flavonoids with the oxidant species seems to be a tiny part of the whole range of their biological activities. The reader is invited to have a look to the classic review of Middleton and colleagues [2000] showing that there is an impressive number of enzymes whose activity is modulated (mostly inhibited) by flavonoids. Thus, it might be predicted that a huge number of signaling pathways can be affected by flavonoids. Moreover, flavonoids can modify the expression of multiple genes. One may think that, as there are so many different flavonoids, these effects would not be produced by all of them or, at least, in the same range of concentrations. While this is undoubtedly true, it is still clear that within the low micromolar range, a prototypical flavonol such as quercetin inhibits numerous enzymes and scavenges several ROS. Thus, flavonoids should be regarded as "dirty drugs," or, in other words, they present "pleiotropic effects." When any given cellular function is modified by quercetin, it is relatively easy to find out a mechanism potentially implicated, but difficult to ascertain whether this proposed target is a true biochemical mechanism involved. This promiscuous behavior of quercetin has been attributed to nonspecific interactions, for example, the formation of aggregates [McGovern and Soichet, 2003]. In recent years it has also become clear that most flavonoids, especially flavonols, are undetected in plasma or are present at extremely low concentrations. Only glucuronosylated and sulfated conjugates of the parent aglycone (or methylated derivatives) are present in plasma following ingestion of quercetin or quercetin glycosides, and the plasma concentrations reached are in the low micromolar range [Kroon et al., 2004]. Unfortunately, the literature on the vascular effects of these metabolites is scarce, and it is presently unknown which of the effects obtained in vitro reported for the parent compounds are also pertinent to the metabolites. The biological activities of these metabolites have been only recently started to be analyzed, which is complicated by the lack of commercial sources of these compounds. This issue is further complicated because the glucuronyl metabolites can be deglucuronidated at the target tissues rendering the original aglycone. There is some evidence that the activity of quercetin glucuronides depends on their deconjugation [Lee-Hilz et al., 2008].

In the following sections we will review some of the targets of flavonols that, in our opinion, are the most relevant to explain their cardiovascular effects.

Antioxidant Activity

The antioxidant activity of flavonols has been reviewed in detail elsewhere. Briefly, flavonols are good antioxidants in vitro, sometimes providing better protection than vitamin C or E. They are effective scavengers of superoxide, singlet oxygen, hydroxyl radicals, and peroxynitite, and they are powerful inhibitors of in vitro LDL oxidation. [Naderi et al., 2003; Ozgova et al., 2003; Pollard et al., 2006]. The copper chelating properties of flavonols may also contribute to prevent LDL oxidation. In addition, they prevent cytotoxicity of oxidized LDL. Flavonoids can prevent vitamin E consumption in phospholipid bilayers and can regenerate it once consumed in a similar way to what ascorbic acid does. The hydrophilic properties of flavonoids facilitate their localization at the water phase–lipid bilayer interface and thereby protect lipids and vitamin E from the initial attack by aqueous radicals. By scavenging superoxide anions, quercetin may protect nitric oxide (NO) [Lopez-Lopez et al., 2004]. Interestingly, the glucuronidated metabolites of quercetin retain its antioxidant properties [Moon et al., 2001].

Pro-oxidant Effects

It is often ignored that flavonols can also autoxidize to generate free radicals, behaving as pro-oxidants. At physiological conditions of pH and O_2, quercetin effectively scavenged NO in the low micromolar range [Lopez-Lopez et al., 2004]. This reaction involves the autoxidation of quercetin and the subsequent production of O_2^-, which rapidly reacts with NO to produce peroxynitrite. However, quercetin was apparently a more effective scavenger of O_2^- than of NO under conditions of increased O_2^- [Lopez-Lopez et al., 2004]. In addition, the glucuronidated metabolites lack a pro-oxidant effect at physiological concentrations [Lodi et al., 2008]. Therefore, even when the relevance of the pro-oxidant effect of flavonols in vivo may be limited, it should be taken into account when designing and analyzing in vitro studies.

NADPH Oxidase

Reduced nicotinamide adenine dinucleotide phosphate (NADPH) oxidase is an enzymatic complex considered the most important source of superoxide production in the vascular wall [Azumi et al., 1999] as well as in other tissues. Several lines of evidence indicate that NADPH oxidase plays a pathophysiological role in patients with endothelial dysfunction, atherosclerosis, and other cardiovascular risk factors [Guzik et al., 2000]. Quercetin was originally shown to inhibit NADPH oxidase in neutrophils [Tauber et al., 1984]. More recently, it was found that quercetin inhibited Protein kinase C (PKC)-dependent NADPH oxidase activation in platelets [Pignatelli et al., 2006]. In the vessel wall, quercetin inhibited NADPH oxidase activity and the associated spontaneous tone in SHR [Lodi et al., 2006]. The activity of NADPH oxidase is often measured by the production of superoxide stimulated by addition of NADPH. The analysis in intact cells or tissues is commonly criticized because NADPH added exogenously may not reach the intracellular target in the enzyme. However, we have found that quercetin exerts a similar inhibitory effect in intact aorta or in aortic homogenates [Lodi et al., 2006]. Moreover, these methods are also criticized because they do not allow distinguishing between a true inhibitory effect on the enzyme activity and the scavenging of the product of the enzyme, that is, superoxide. Recently, Steffen et al. [2008] have suggested that only *O*-methylated flavonoids do inhibit the enzyme activity while nonmethylated ones behave as superoxide scavengers. They found a structural analogy between methylated compounds and the standard inhibitor of NADPH oxidase apocynin. More intriguingly is the recent report suggesting that even apocynin predominantly acts as a superoxide scavenger rather than an enzyme inhibitor in endothelial cells and vascular smooth muscle cells and should not be used as a NADPH oxidase inhibitor in vascular systems [Heumüller et al., 2008]. Interestingly, besides the direct scavenging or the inhibitory effect on the enzyme, quercetin down-regulated the expression of NADPH oxidase subunit $p47^{phox}$ in SHR [Sanchez et al., 2006] and also prevented the overexpression of this subunit induced by angiotensin II [Sanchez et al., 2007] and endothelin-1 [Romero et al., 2008] in aortic rings. This mechanism of NADPH down-regulation is exerted at lower concentrations of quercetin, and it is also observed to occur with the glucuronosylated metabolites. Therefore, it seems that down-regulation of NADPH oxidase subunit expression is more important than the inhibitory effect on the enzyme or the superoxide scavenging effect.

Nitric Oxide

Nitric Oxide Synthase Expression and Activity; Nitric Oxide Synthase Uncoupling. Inhibitory effects of quercetin on the activity or expression of endothelial NO synthases (eNOS and iNOS) has been reported but only at high concentrations ($< 50\ \mu M$) [Camuesco et al., 2004; Chiesi and Schwaller 1995]. Therefore, direct changes in endothelial NO expression or activity, even when relevant to other polyphenolics, do not appear to be physiologically relevant for quercetin. However, quercetin may restore NOS expression or activity when unpaired by pathological stimuli. In fact, eNOS expression, which is up-regulated in SHR, can be corrected by chronic quercetin administration

(Sanchez et al., 2006). Under conditions of oxidative stress, substrate or cofactor deficiency eNOS is uncoupled, that is, it switches from an NO producing enzyme to a superoxide producing enzyme. Quercetin, by reducing superoxide levels, can prevent the uncoupling of eNOS [Romero et al., 2008].

Nitric Oxide and cGMP Metabolism During the autoxidation process of quercetin in aqueous buffers, superoxide anions are generated and the half-life of NO is reduced [Lopez-Lopez et al., 2004]. However, the rate of NO scavenging by quercetin is lower than that of the physiological NO scavenger hemoglobin. On the other hand, by scavenging superoxide anions or by inhibiting their synthesis (see above), quercetin protects NO from superoxide-driven inactivation and increases its half-life and its biological activity under conditions of high oxidative stress [Lopez-Lopez et al., 2004; Shutenko et al., 1999]. This property plays a key role in the protective effects on endothelial dysfunction. Quercetin and related flavonoids also inhibit several cyclic nucleotide phosphodiesterases (PDEs) [Picq et al., 1989], and thus reduce the degradation of cyclic (guanosine 5-munophosphate (eGMP), the main downstream effector of NO. However, these effects are produced only at concentrations above 10 μM.

Protein Kinases

Numerous signaling pathways involve the activation of serine-threonine protein kinases in both physiological and pathological conditions. Quercetin and related flavonoids, in the micromolar range, inhibit multiple serine-threonine kinases. Inhibition of protein kinases and the subsequent reduction in target phosphorylation is expected to affect multiple signaling pathways and, hence, numerous cellular functions. The inhibitory effects of quercetin on the activity of PKC and phosphatidylinositol-3 kinase (PI-3 kinase) are well known [Ferriola et al., 1989; Walker et al., 2000]. Quercetin also interacts with members of the family of mitogen-activated protein kinases [MAP kinases, ERK p38 and JNK) [Yoshizumi et al., 2002; Moon et al., 2003; Perez-Vizcaino et al., 2006a]. Akt/PKB, cyclin-dependent kinases (CDKs) and myosin light-chain kinase (MLCK) are also targets of quercetin. Quercetin has been crystallographically observed to specifically form an enzyme-inhibitor complex with PI-3 kinase [Walker et al., 2000]. The glucuronosylated metabolites of quercetin also inhibit the activity of JNK kinase [Yoshizumi et al., 2002]. In contrast to serine/threonine kinases, the inhibitory activity of quercetin on tyrosine kinases is generally weak and can only be observed at high concentrations of the flavonoid [Palmieri et al., 1999; Agullo et al., 1997].

Endothelin-1

A reduction in the gene expression of enodthelin-1 (ET-1) has been observed with several polyphenols. The influence of quercetin on ET-1 production by endothelial cells is controversial. Corder et al. [2001] found no effect of quercetin itself (but of other wine polyphenolics), while Zhao et al. [1999] and Nicholson et al. [2008] reported that quercetin is able to inhibit ET-1 release in human endothelial umbilical vein. The effects of quercetin or other flavonols on ET-1 in vivo has not been analyzed so far.

Ion Channels

The possible modulation of ion channels by flavonoids has been investigated by several groups. In rat tail artery smooth muscle, quercetin has been shown to activate L-type Ca^{2+} channels [Saponara et al., 2002]. This would lead to vascular smooth muscle contraction. However, mechanisms inducing vasodilation are more effective, and quercetin-activated Ca^{2+} entry does not appear to have a functional correlate [Fusi et al., 2003]. Much more attention has been paid on the possible modulation of potassium channels by quercetin and related flavonoids. Potassium channels do not seem to contribute to quercetin-induced relaxation in rat aorta and iliac arteries [Perez-Vizcaino et al., 2002]. However, in rat coronary arteries, quercetin has been shown to increase the current through large conductance calcium-activated potassium channels (BK_{Ca}), which contributes to its relaxant effect [Cogolludo et al., 2007]. Similar results have been reported with (+/−)-naringenin in rat tail arteries [Saponara et al., 2006]. Quercetin also increases the frequency of spontaneous transient outward currents (STOCs), which are carried by BK_{Ca} channels and triggered by Ca^{2+} sparks [Cogolludo et al., 2007]. Thus, quercetin potentiates a fundamental feedback mechanism in vascular function, limiting membrane depolarization and vasoconstriction. Notably, the modulation of BK_{Ca} channels by quercetin could be involved in its higher potency to relax coronary arteries as compared to aorta [Ibarra et al., 2002]. Interestingly, quercetin also increases endothelial NO production through the activation of endothelial BK_{Ca} channels, leading to hyperpolarization and subsequent Ca^{2+} influx [Kuhlmann et al., 2005].

Sirtuins

The sirtuins Sir2 and its mammalian analog SIRT1 have been implicated in the extended life span induced by caloric restriction [Cohen et al., 2004]. Regarding the cardiovascular system, activation of SIRT1 has been shown to promote endothelium-dependent vasodilation [Mattagajasingh et al., 2007] and to down-regulate angiotensin II receptor expression [Miyazaki et al., 2008]. The discovery of the role of sirtuins in oxidative stress and life span has led to a new and fascinating area of research in the field of polyphenols since several polyphenols can activate sirtuins [Howitz et al., 2003]. Resveratrol was the most potent sirtuin activator, but quercetin and other flavonoids were also active. Based on these ideas, it was proposed that sirtuin activators increase life span in several different organisms, mimicking the effects of caloric restriction.

In fact, resveratrol and quercetin extend the life spans of the yeast *Saccharomyces cerevisiae*, the worm *Caenorhabditis elegans*, and the fly *Drosophila melanogaster*, but only if the gene that encodes Sir2 is present in these organisms [Belinha et al., 2007]. In addition, resveratrol has been shown to improve health and survival of mice on a high-calorie diet [Baur et al., 2006].

Adhesion Molecules and Matrix Metalloproteinases

Adhesion molecules and matrix metalloproteinases are key proteins for several processes involved in atherosclerotic plaque formation such as infiltration of inflammatory cells. Furthermore, matrix degradation by matrix metalloproteinases may cause the plaque instability and rupture that leads to the clinical symptoms of atherosclerosis: unstable angina, myocardial infarction, and stroke. Quercetin inhibits the expression of the adhesion molecules such as monocyte chemoattractant protein-1 (MCP-1), intercellular adhesion molecule-1 (ICAM-1), vascular cell adhesion molecule-1 (VCAM-1), and E-selectin and also the expression of matrix metalloproteinase-2 and matrix metalloproteinase-9, which are induced by proliferative/inflammatory stimuli (e.g., angiotensin II, epidermal growth factor, or TNFα) [Huang et al., 1999]. The inhibitory effect on ICAM-1 expression occurs through a down-regulation of the JNK/AP-1 pathway [Kobuchi et al., 1999]. In addition, quercetin metabolites inhibited VCAM-1 endothelial cell surface expression at low concentrations [Tribolo et al., 2008].

Other Inflammatory Signaling Pathways

Quercetin has been reported to affect a number of enzymes and transcription factors involved in inflammation at the vascular level. Thus, quercetin inhibits several enzymes involved in cellular signaling through lipid mediators such as phospholipase A_2, phospholipase C, cyclooxygenases (COXs), lipoxygenases (LOXs), and cytochrome P450 oxygenases [Middleton et al., 2000; Sadik et al., 2003]. In addition, it also inhibits the expression of inducible cyclooxygenase (COX-2) and iNOS (as mentioned above), which are up-regulated during inflammation [Garcia-Mediavilla et al., 2007]. These later effects are supposed to be due to the inhibition of the transcription factor NF-κB. Quercetin may also decrease DNA binding of the transcription factor activator protein-1 (AP-1) [Yoshizumi et al., 2002]. On the other hand, quercetin up-regulates the expression of the anti-inflammatory enzyme heme oxygenase-1 [Chow et al., 2005].

Conclusions

Two fundamental questions regarding flavonols and cardiovascular disease remain open: (1) Are flavonols protective against human cardiovascular disease? and if so, (2) what is/are the molecular target(s) responsible of the protective effect? Concerning the first question, evidence from animal and clinical studies continues growing to support a beneficial effect on vascular homeostasis preventing the development of cardiovascular complications. Yet, long-term clinical trials have to confirm or refute it. Concerning the second, flavonols can potentially interact with many of the molecular targets known to be involved in the pathophysiology of ischemic heart disease and stroke. Thus, they might act by multiple mechanisms operating simultaneously.

ACKNOWLEDGMENTS

The authors research is funded by a grant from CICYT (SAF AGL2007-66108).

REFERENCES

Aoi W, Niisato N, Miyazaki H, Marunaka Y. 2004. Flavonoid-induced reduction of ENaC expression in the kidney of Dahl salt-sensitive hypertensive rat. *Biochem Biophys Res Commun* 315:892–896.

Arai Y, Watanabe S, Kimira M, Shimoi K, Mochizuki R, Kinae N. 2000. Dietary intakes of flavonols, flavones and isoflavones by Japanese women and the inverse correlation between quercetin intake and plasma LDL cholesterol concentration. *J Nutr* 130:2243–2250.

Auger C, Teissedre PL, Gerain P, Lequeux N, Bornet A, Serisier S, Besancon P, Caporiccio B, Cristol JP, Rouanet JM. 2005. Dietary wine phenolics catechin, quercetin, and resveratrol efficiently protect hypercholesterolemic hamsters against aortic fatty streak accumulation. *J Agric Food Chem* 53:2015–2021.

Azumi H, Inoue N, Takeshita S, Rikitake Y, Kawashima S, Hayashi Y, Itoh H, Yokoyama M. 1999. Expression of NADH/NADPH oxidase p22phox in human coronary arteries. *Circulation* 100:1494–1498.

Baur JA, Pearson KJ, Price NL, Jamieson HA, Lerin C, Kalra A, Prabhu VV, Allard JS, Lopez-Lluch G, Lewis K, Pistell PJ, Poosala S, Becker KG, Boss O, Gwinn D, Wang M, Ramaswamy S, Fishbein KW, Spencer RG, Lakatta EG, Le Couteur D, Shaw RJ, Navas P, Puigserver P, Ingram DK, de Cabo R, Sinclair DA. 2006. Resveratrol improves health and survival of mice on a high-calorie diet. *Nature* 444:337–342.

Belinha I, Amorim MA, Rodrigues P, de Freitas V, Moradas-Ferreira P, Mateus N, Costa V. 2007. Quercetin increases oxidative stress resistance and longevity in Saccharomyces cerevisiae. *J Agric Food Chem* 55:2446–2451.

Bjelakovic G, Nikolova D, Gluud LL, Simonetti RG, Gluud C. 2007. Mortality in randomized trials of antioxidant supplements for primary and secondary prevention: systematic review and meta-analysis. *JAMA* 297:842–857.

Camuesco D, Comalada M, Rodriguez-Cabezas ME, Nieto A, Lorente MD, Concha A, Zarzuelo A, Galvez J. 2004. The intestinal anti-inflammatory effect of quercitrin is associated with an inhibition in iNOS expression. *Br J Pharmacol* 143:908–918.

Carlstrom J, Symons JD, Wu TC, Bruno RS, Litwin SE, Jalili T. 2007. A quercetin supplemented diet does not prevent cardiovascular complications in spontaneously hypertensive rats. *J Nutr* 137:628–633.

Chiesi M, Schwaller R. 1995. Inhibition of constitutive endothelial NO-synthase activity by tannin and quercetin. *Biochem Pharmacol* 49:495–501.

Chow JM, Shen SC, Huan SK, Lin HY, Chen YC. 2005. Quercetin, but not rutin and quercitrin, prevention of H_2O_2-induced apoptosis via anti-oxidant activity and heme oxygenase 1 gene expression in macrophages. *Biochem Pharmacol* 69:1839–1851.

Cogolludo A, Frazziano G, Briones AM, Cobeno L, Moreno L, Lodi F, Salaices M, Tamargo J, Perez-Vizcaino F. 2007. The dietary flavonoid quercetin activates BKCa currents in coronary arteries via production of H2O2. Role in vasodilatation. *Cardiovasc Res* 73:424–431.

Cohen HY, Miller C, Bitterman KJ, Wall NR, Hekking B, Kessler B, Howitz KT, Gorospe M, de Cabo R, Sinclair DA. 2004. Calorie restriction promotes mammalian cell survival by inducing the SIRT1 deacetylase. *Science* 305:390–392.

Conquer JA, Maiani G, Azzini E, Raguzzini A, Holub BJ. 1998. Supplementation with quercetin markedly increases plasma quercetin concentration without effect on selected risk factors for heart disease in healthy subjects. *J Nutr* 128:593–597.

Corder R, Douthwaite JA, Lees DM, Khan NQ, Viseu Dos Santos AC, Wood EG, Carrier MJ. 2001. Endothelin-1 synthesis reduced by red wine. *Nature* 414:863–864.

Duarte J, Jimenez R, O'Valle F, Galisteo M, Perez-Palencia R, Vargas F, Perez-Vizcaino F, Tamargo J. 2002. Protective effects of the flavonoid quercetin in chronic nitric oxide deficient rats. *J Hypertens* 20:1843–1854.

Duarte J, Perez-Palencia R, Vargas F, Ocete MA, Perez-Vizcaino F, Zarzuelo A, Tamargo J. 2001. Antihypertensive effects of the flavonoid quercetin in spontaneously hypertensive rats. *Br J Pharmacol* 133:117–124.

Edwards RL, Lyon T, Litwin SE, Rabovsky A, Symons JD, Jalili T. 2007. Quercetin reduces blood pressure in hypertensive subjects. *J Nutr* 137:2405–2411.

Endemann DH, Schiffrin EL. 2004. Endothelial dysfunction. *J Am Soc Nephrol* 15:1983–1992.

Ferriola PC, Cody V, Middleton EJr. 1989. Protein kinase C inhibition by plant flavonoids. Kinetic mechanisms and structure-activity relationships. *Biochem Pharmacol* 38:1617–1624.

Fraga CG, Actis-Goretta L, Ottaviani JI, Carrasquedo F, Lotito SB, Lazarus S, Schmitz HH, Keen CL. 2005. Regular consumption of a flavanol-rich chocolate can improve oxidant stress in young soccer players. *Clin Dev Immunol* 12:11–17.

Fusi F, Saponara S, Pessina F, Gorelli B, Sgaragli G. 2003. Effects of quercetin and rutin on vascular preparations: A comparison between mechanical and electrophysiological phenomena. *Eur J Nutr* 42:10–17.

Galisteo M, Garcia-Saura MF, Jimenez R, Villar IC, Zarzuelo A, Vargas F, Duarte J. 2004. Effects of chronic quercetin treatment on antioxidant defence system and oxidative status of deoxycorticosterone acetate-salt-hypertensive rats. *Mol Cell Biochem* 259:91–99.

Garcia-Mediavilla V, Crespo I, Collado PS, Esteller A, Sanchez-Campos S, Tunon MJ, Gonzalez-Gallego J. 2007. The anti-inflammatory flavones quercetin and kaempferol cause inhibition of inducible nitric oxide synthase, cyclooxygenase-2 and reactive

C-protein, and down-regulation of the nuclear factor kappaB pathway in Chang Liver cells. *Eur J Pharmacol* 557:221–229.

García-Saura MF, Galisteo M, Villar IC, Bermejo A, Zarzuelo A, Vargas F. 2005. Effects of chronic quercetin treatment in experimental renovascular hypertension. *Mol Cell Biochem* 270:147–155.

Gotto AM. 2003. Antioxidants, statins and atherosclerosis. *J Am Coll Cardiol* 41:1205–1210.

Grassi D, Lippi C, Necozione S, Desideri G, Ferri C. 2005. Short-term administration of dark chocolate is followed by a significant increase in insulin sensitivity and a decrease in blood pressure in healthy persons. *Am J Clin Nutr* 81:611–614.

Guzik TJ, West NE, Black E, McDonald D, Ratnatunga C, Pillai R, Channon KM. 2000. Vascular superoxide production by NAD(P)H oxidase: Association with endothelial dysfunction and clinical risk factors. *Circ Res* 86:E85–90.

Hayek T, Fuhrman B, Vaya J, Rosenblat M, Belinky P, Coleman R. 1997. Reduced progression of atherosclerosis in apolipoprotein E-deficient mice following consumption of red wine, or its polyphenols quercetin or catechin, is associated with reduced susceptibility of LDL to oxidation and aggregation. *Arterioscler Thromb Vasc Biol* 17:2744–2752.

Heinecke JW. 2001. Is the emperor wearing clothes? Clinical trials of vitamin E and the LDL oxidation hypothesis. *Arterioscler Thromb Vasc Biol* 21:1261–2364.

Hertog MG, Feskens EJ, Kromhout D. 1997. Antioxidant flavonols and coronary heart disease risk. *Lancet* 349:699.

Hertog MG, Feskens EJ, Hollman PC, Katan MB, Kromhout D. 1993. Dietary antioxidant flavonoids and risk of coronary heart disease: The Zutphen Elderly Study. *Lancet* 342:1007–1011.

Heumüller S, Wind S, Barbosa-Sicard E, Schmidt HH, Busse R, Schröder K, Brandes RP. 2008. Apocynin is not an inhibitor of vascular NADPH oxidases but an antioxidant. *Hypertension* 51:211–217.

Hirvonen T, Pietinen P, Virtanen M, Ovaskainen ML, Hakkinen S, Albanes D, Virtamo J. 2001. Intake of flavonols and flavones and risk of coronary heart disease in male smokers. *Epidemiology* 12:62–67.

Howitz KT, Bitterman KJ, Cohen HM, Lamming DH, Lavu S, Wood JG, Zipkin RE, Chung P, Kisielewski A, Zhang L-L, Scherer B, Sinclair DA. 2003. Small molecule activators of sirtuins extend Saccharomyces cerevisiae lifespan. *Nature* 425:191–196.

Huang YT, Hwang JJ, Lee PP, Ke FC, Huang JH, Huang CJ, Kandaswami C, Middleton EJr, Lee MT. 1999. Effects of luteolin and quercetin, inhibitors of tyrosine kinase, on cell growth and metastasis-associated properties in A431 cells overexpressing epidermal growth factor receptor. *Br J Pharmacol* 128:999–1010.

Huxley RR, Neil HA. 2003. The relation between dietary flavonol intake and coronary heart disease mortality: A meta-analysis of prospective cohort studies. *Eur J Clin Nutr* 57:904–908.

Ibarra M, Perez-Vizcaino F, Cogolludo A, Duarte J, Zaragoza-Arnaez F, Lopez-Lopez JG, Tamargo J. 2002. Cardiovascular effects of isorhamnetin and quercetin in isolated rat and porcine vascular smooth muscle and isolated rat atria. *Planta Med* 68:307–310.

Jalili T, Carlstrom J, Kim S, Freeman D, Jin H, Wu TC, Litwin SE, David Symons J. 2006. Quercetin-supplemented diets lower blood pressure and attenuate cardiac hypertrophy in rats with aortic constriction. *J Cardiovasc Pharmacol* 47:531–541.

Joint Committee on Nomenclature of the American Society of Biological Chemists—American Institute of Nutrition. 1950. Term "vitamin P" recommended to be discontinued. *Science* 112:268.

Kamada C, da Silva EL, Ohnishi-Kameyama M, Moon JH, Terao J. 2005. Attenuation of lipid peroxidation and hyperlipidemia by quercetin glucoside in the aorta of high cholesterol-fed rabbit. *Free Radic Res* 39:185–194.

Knekt P, Jarvinen R, Reunanen A, Maatela J. 1996. Flavonoid intake and coronary mortality in Finland: A cohort study. *BMJ* 312:478–481.

Kobuchi H, Roy S, Sen CK, Nguyen HG, Packer L. 1999. Quercetin inhibits inducible ICAM-1 expression in human endothelial cells through the JNK pathway. *Am J Physiol* 277:C403–C411.

Kroon PA, Clifford MN, Crozier A, Day AJ, Donovan JL, Manach C, Williamson G. 2004. How should we assess the effects of exposure to dietary polyphenols in vitro? *Am J Clin Nutr* 80:15–21.

Kuhlmann CR, Schaefer CA, Kosok C, Abdallah Y, Walther S, Ludders DW, Neumann T, Tillmanns H, Schafer C, Piper HM, Erdogan A. 2005. Quercetin-induced induction of the NO/cGMP pathway depends on Ca^{2+}-activated K + channel-induced hyperpolarization-mediated Ca2 + -entry into cultured human endothelial cells. *Planta Med* 71:520–524.

Lee-Hilz YY, Stolaki M, van Berkel WJ, Aarts JM, Rietjens IM. 2008. Activation of EpRE-mediated gene transcription by quercetin glucuronides depends on their deconjugation. *Food Chem Toxicol.*

Lodi F, Jimenez R, Menedez C, Needs PW, Duarte J, Perez-Vizcaino F. 2008. Glucuronidated metabolites of the flavonoids quercetin do not auto-oxidize, do not genertate free radicals and do not decrease NO bioavalability Planta Medica.

Lodi F, Cogolludo A, Duarte J, Moreno L, Coviello A, Peral De Bruno M, Vera R, Galisteo M, Jimenez R, Tamargo J, Perez-Vizcaino F. 2006. Increased NADPH oxidase activity mediates spontaneous aortic tone in genetically hypertensive rats. *Eur J Pharmacol* 544:97–103.

Lopez-Lopez G, Moreno L, Cogolludo A, Galisteo M, Ibarra M, Duarte J, Tamargo J. 2004. NO scavenging and NO protecting effects of quercetin and their biological significance in vascular smooth muscle. *Mol Pharmacol* 65:851–859.

Lyseng-Williamson KA, Perry CM. 2003. Micronised purified flavonoid fraction: A review of its use in chronic venous insufficiency, venous ulcers and haemorrhoids. *Drugs* 63:71–100.

Mattagajasingh I, Kim CS, Naqvi A, Yamamori T, Hoffman TA, Jung SB, DeRicco J, Kasuno K, Irani K. 2007. SIRT1 promotes endothelium-dependent vascular relaxation by activating endothelial nitric oxide synthase. *Proc Natl Acad Sci USA* 104:14855–14860.

McGovern SL, Soichet BK. 2003. Kinase inhibitors: Not just for kinases anymore. *J Med Chem* 46:1478–1483.

Middleton E Jr, Kandaswami C, Theoharides CT. 2000. The effects of plant flavonoids on mammalian cells: Implications for inflammation, heart disease, and cancer. *Pharmacol Rev* 52:673–751.

Miyazaki R, Ichiki T, Hashimoto T, Inanaga K, Imayama I, Sadoshima J, Sunagawa K. 2008. SIRT1, a longevity gene, downregulates angiotensin II type 1 receptor expression in vascular smooth muscle cells. *Arterioscler Thromb Vasc Biol.*

Moon SK, Cho GO, Jung SY, Gal SW, Kwon TK, Lee YC, Madamanchi NR, Kim CH. 2003. Quercetin exerts multiple inhibitory effects on vascular smooth muscle cells: Role of ERK1/2, cell-cycle regulation and matrix metalloproteinase-9. Biochem Biophys. *Res Commun* 301:1069–1078.

Moon JH, Tsushida T, Nakahara K, Terao J. 2001. Identification of quercetin 3-O-beta-D-glucuronide as an antioxidative metabolite in rat plasma after oral administration of quercetin. *Free Radic Biol Med* 30:1274–1285.

Mursu J, Voutilainen S, Nurmi T, Tuomainen TP, Kurl S, Salonen JT. 2008. Flavonoid intake and the risk of ischaemic stroke and CVD mortality in middle-aged Finnish men: The Kuopio Ischaemic Heart Disease Risk Factor Study. *Br J Nutr* 100:890–895.

Naderi GA, Asgary S, Sarraf-Zadegan N, Shirvany H. 2003. Anti-oxidant effect of flavonoids on the susceptibility of LDL oxidation. *Mol Cell Biochem* 246:193–196.

Nicholson SK, Tucker GA, Brameld JM. 2008. Effects of dietary polyphenols on gene expression in human vascular endothelial cells. *Proc Nutr Soc* 67:42–47.

Ozgova S, Hermanek J, Gut I. 2003. Different antioxidant effects of polyphenols on lipid peroxidation and hydroxyl radicals in the NADPH-, Fe-ascorbate- and Fe-microsomal systems. *Biochem Pharmacol* 66:1127–1237.

Palmieri L, Mameli M, Ronca G. 1999. Effect of resveratrol and some other natural compounds on tyrosine kinase activity and on cytolysis. *Drugs Exp Clin Res* 25:79–85.

Perez-Vizcaino F, Bishop-Bailley D, Lodi F, Duarte J, Cogolludo A, Moreno L, Bosca L, Mitchell JA, Warner TD. 2006a. The flavonoid quercetin induces apoptosis and inhibits JNK activation in intimal vascular smooth muscle cells. *Biochem Biophys Res Commun* 346:919–925.

Perez-Vizcaino F, Duarte J, Andrilantsitohaina R. 2006b. Endothelial function and cardiovascular disease: Effects of quercetin and wine polyphenols. *Free Radic Res* 40:1054–1065.

Perez-Vizcaino F, Ibarra M, Cogolludo AL, Duarte J, Zaragoza-Arnaez F, Moreno L, Tamargo J. 2002. Endothelium-independent vasodilator effects of the flavonoid quercetin and its methylated metabolites in rat conductance and resistance arteries. *J Pharmacol Exp Ther* 302:66–72.

Picq M, Dubois M, Prigent AF, Nemoz G, Pacheco H. 1989. Inhibition of the different cyclic nucleotide phosphodiesterase isoforms separated from rat brain by flavonoid compounds. *Biochem Int* 18:47–57.

Pignatelli P, Di Santo S, Buchetti B, Sanguigni V, Brunelli A, Violi F. 2006. Polyphenols enhance platelet nitric oxide by inhibiting protein kinase C-dependent NADPH oxidase activation: Effect on platelet recruitment. *FASEB J* 20:1082–1089.

Pollard SE, Kuhnle GG, Vauzour D, Vafeiadou K, Tzounis X, Whiteman M, Rice-Evans C, Spencer JP. 2006. The reaction of flavonoid metabolites with peroxynitrite. *Biochem Biophys Res Commun* 350:960–968.

Reckelhoff JF, Romero JC. 2003. Role of oxidative stress in angiotensin-induced hypertension. *Am J Physiol* 284:R893–912.

Reshef N, Hayari Y, Goren C, Boaz M, Madar Z, Knobler H. 2005. Antihypertensive effect of sweetie fruit in patients with stage I hypertension. *Am J Hypertens* 18:1360–1363.

Rimm EB, Katan MB, Ascherio A, Stampfer MJ, Willett WC. 1996. Relation between intake of flavonoids and risk for coronary heart disease in male health professionals. *Ann Intern Med* 125:384–389.

Romero M, Jiménez R, Sánchez M, López-Sepúlveda R, Zarzuelo MJ, O'Valle F, Zarzuelo A, Pérez-Vizcaíno F, Duarte J. 2008. Quercetin inhibits vascular superoxide production induced by endothelin-1: Role of NADPH oxidase, uncoupled eNOS and PKC. Atherosclerosis.

Rusznyak S, Szent-Györgyi A. 1936. Vitamin P: Flavonols as vitamins. *Nature* 138:27–29.

Sadik CD, Sies H, Schewe T. 2003. Inhibition of 15-lipoxygenases by flavonoids: Structure-activity relations and mode of action. *Biochem Pharmacol* 65:773–781.

Sanchez M, Lodi F, Vera R, Villar IC, Cogolludo A, Jimenez R, Moreno L, Romero M, Tamargo J, Perez-Vizcaino F, Duarte J. 2007. Quercetin and isorhamnetin prevent endothelial dysfunction, superoxide production, and overexpression of p47phox induced by angiotensin II in rat aorta. *J Nutr* 137:910–915.

Sanchez M, Galisteo M, Vera R, Villar IC, Zarzuelo A, Tamargo J, Perez-Vizcaino F, Duarte J. 2006. Quercetin downregulates NADPH oxidase, increases eNOS activity and prevents endothelial dysfunction in spontaneously hypertensive rats. *J Hypertens* 24:75–84.

Saponara S, Testai L, Iozzi D, Martinotti E, Martelli A, Chericoni S, Sgaragli G, Fusi F, Calderone V. 2006. (+ /-)-Naringenin as large conductance Ca^{2+}-activated K + (BKCa) channel opener in vascular smooth muscle cells. *Br J Pharmacol* 149:1013–1021.

Saponara S, Sgaragli G, Fusi F. 2002. Quercetin as a novel activator of L-type Ca^{2+} channels in rat tail artery smooth muscle cells. *Br J Pharmacol* 135:1819–1827.

Shutenko Z, Henry Y, Pinard E, Seylaz J, Potier P, Berthet F. 1999. Influence of the antioxidant quercetin in vivo on the level of nitric oxide determined by electron paramagnetic resonance in rat brain during global ischemia and reperfusion. *Biochem Pharmacol* 57:199–208.

Stampfer M, Henekens CH, Manson JE, Colditz GA, Rosner B, Willett WC. 1993. Vitamin E consumption and the risk of coronary heart disease in women. *N Engl J Med* 328:1450–1455.

Steffen Y, Gruber C, Schewe T, Sies H. 2008. Mono-O-methylated flavanols and other flavonoids as inhibitors of endothelial NADPH oxidase. *Arch Biochem Biophys* 469:209–219.

Steinberg D, Witztum JL. 2002. Is the oxidative modification hypothesis relevant to human atherosclerosis? Do the antioxidant trials conducted to date refute the hypothesis? *Circulation* 105:2107–2111.

Tauber AI, Fay JR, Marletta MA. 1984. Flavonoid inhibition of the human neutrophil NADPH-oxidase. *Biochem Pharmacol* 33:1367–1369.

Taubert D, Roesen R, Lehmann C, Jung N, Schömig E. 2007. Effects of low habitual cocoa intake on blood pressure and bioactive nitric oxide: a randomized controlled trial. *JAMA* 298:49–60.

Taubert D, Berkels R, Roesen R, Klaus W. 2003. Chocolate and blood pressure in elderly individuals with isolated systolic hypertension. *JAMA* 290:1029–1030.

Tribolo S, Lodi F, Connor C, Suri S, Wilson VG, Taylor MA, Needs PW, Kroon PA, Hughes DA. 2008. Comparative effects of quercetin and its predominant human metabolites on adhesion molecule expression in activated human vascular endothelial cells. *Atherosclerosis* 197:50–56.

Walker EH, Pacold ME, Perisic O, Stephens L, Hawkins PT, Wymann MP. 2000. Structural determinants of phosphoinositide 3-kinase inhibition by wortmannin, LY294002, quercetin, myricetin, and staurosporine. *Mol Cell* 6:909–919.

Widlansky ME, Gokce N, Keaney JF, Vita JA. 2003. The clinical implications of endothelial dysfunction. *J Am Coll Cardiol* 42:1149–1160.

Yamamoto Y, Oue E. 2006. Antihypertensive effect of quercetin in rats fed with a high-fat high-sucrose diet. *Biosci Biotechnol Biochem* 70:933–939.

Yochum L, Kushi LH, Meyer K, Folsom AR. 1999. Dietary flavonoid intake and risk of cardiovascular disease in postmenopausal women. *Am J Epidemiol* 149:943–949.

Yoshizumi M, Tsuchiya K, Suzaki Y, Kirima K, Kyaw M, Moon JH, Terao J, Tamaki T. 2002. Quercetin glucuronide prevents VSMC hypertrophy by angiotensin II via the inhibition of JNK and AP-1 signaling pathway. *Biochem Biophys Res Commun* 293:1458–1465.

Zhao X, Gu Z, Attele AS, Yuan CS. 1999. Effects of quercetin on the release of endothelin, prostacyclin and tissue plasminogen activator from human endothelial cells in culture. *J Ethnopharmacol* 67:279–285.

10 Metabolism, Bioavailability, and Analysis of Dietary Isoflavones

ADRIAN A. FRANKE[1], BRUNHILD M. HALM[1,2], KERRY KAKAZU[1] AND XINGNAN LI[1]

[1]Cancer Research Center of Hawaii, Natural Products & Cancer Biology Program

[2]Cancer Research Center of Hawaii, Cancer Prevention & Control Program, and Kapiolani Medical Center for Women and Children (BMH), Honolulu, Hawaii

CONTENTS

INTRODUCTION

Considerable evidence exists from epidemiological and experimental studies for preventive effects of soy or isoflavones against chronic diseases including cancer (breast, prostate, colorectal, lung), osteoporosis, cardiovascular disorders, and menopausal symptoms, but this is not always consistent [Adlercreutz

Plant Phenolics and Human Health: Biochemistry, Nutrition, and Pharmacology, Edited by Cesar G. Fraga.

and Mazur, 1997; Cassidy, 2005; Duncan et al., 2003; Kurzer and Xu, 1997; Magee and Rowland, 2004; Setchell and Lydeking-Olsen, 2003; Yan and Spitznagel, 2004, 2005]. In four recent epidemiologic studies strong preventive effects were observed against breast cancer later in life when soy was consumed at young age [Shu et al., 2001; Thanos et al., 2006; Wu et al., 2002; Korde et al., 2009], and animal studies had shown this before [Lamartiniere, 2002]. The structural similarity of isoflavones to steroidal estrogens (Fig. 10.1) and the potent binding of genistein, one of the most predominant soy isoflavones, to the estrogen receptor-beta, including its transactivation [Kuiper et al., 1997], are believed to be the basis for many effects of soy intake [Cotroneo et al., 2002]. Recent findings on higher isoflavone bioavailability in children versus adults [Halm et al., 2007] and on cardioprotective effects independent from lipid profiles [Walker et al., 2008] strengthen the hypothesis that isoflavones play an important role in the biological effects of soy intake [Wiseman, 2006].

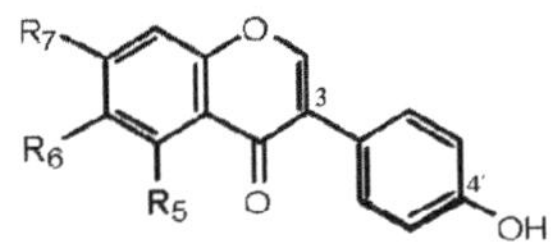

Aglucon	R_5	R_6	R_7
Genistein	OH	H	OH
Glycitein	H	OCH_3	OH
Daidzein	H	H	OH

Conjugates in soy foods	R_7
7-O-(6"-O-acetyl)-glucoside of daidzein of genistein of glycitein	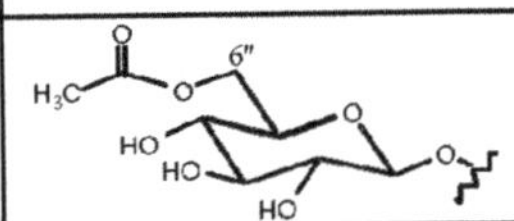
7-O-(6"-O-malonyl)-glucoside .... of daidzein .... of genistein .... of glycitein	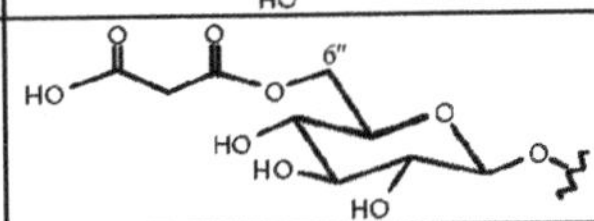
7-O-glucoside ... of daidzein = Daidzin .... of genistein = Genistin .... of glycitein = Glycitin	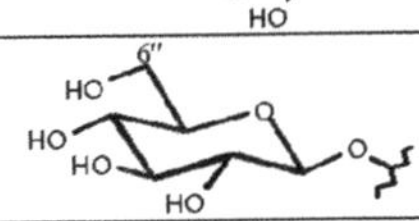

Metabolized isoflavonoids by intestinal bacteria

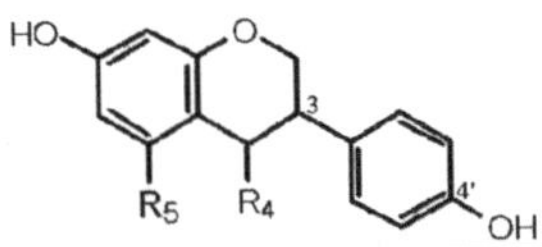

Aglucon	R_4	R_5
dihydrodaidzein	=O	H
dihydrogenistein	=O	OH
equol	H	H

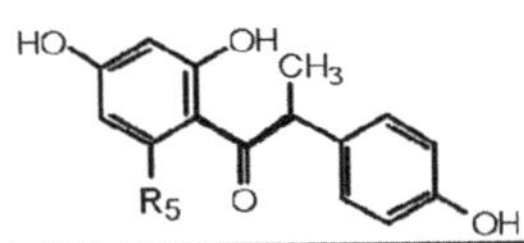

Aglucon	R_5
O-desmethylangolensin	H

Figure 10.1 Structures of Isoflavonoids Studied.

The basis for understanding the health benefits of isoflavones requires detailed knowledge on the absorption, distribution, metabolism, elimination, and bioavailability of these phytoestrogens that demands robust, precise, accurate, and affordable analyses from biological and other matrices.

OCCURRENCE

Isoflavone concentrations in foods are by far highest in soy and its products; therefore, but dependent on the population, dietary isoflavone exposure occurs mainly through intake of soy products. They typically contain a total of 0.01–0.3% isoflavones composed of glycosides of daidzein, genistein, and glycitein (Fig. 10.1) while fermented products contain predominantly aglycons [Adlercreutz and Mazur, 1997; Franke et al., 1995, 1999; Horn-Ross et al., 2000; Umphress et al., 2005]. Isoflavone concentrations in soy vary depending on a variety of factors such as environmental, genetic, harvesting, and processing conditions [Tsukamoto et al., 1995]. 7-*O*-malonylglucoside and 7-*O*-glucosides are the main native isoflavone conjugates in unprocessed soy, and 7-*O*-acetylglycosides are formed by dry heating [Barnes et al., 1994; Kudou et al., 1991]. Isoflavones occur in a variety of plants and are often used for chemotaxonomic purposes but are particularly concentrated in the Fabaceae [Andersen, 2006], for example, in kudzu (*Pueraria lobata*) or red clover (*Trifolium pratense*) or, of course, soy. Isoflavone-rich plants other than soy are often not consumed in large amounts or are not considered food plants at all. Updated databases for isoflavone content in commonly consumed foods became recently available [U.S. Department of Agriculture, 2007].

UPTAKE AND METABOLISM

As shown in Figure 10.2 isoflavonoid-*O*-glucosides require hydrolysis for absorption, which is followed by passive diffusion through the mucosa and reconjugation to glucuronides and sulfates on the basolateral side of the enterocyte and/or in the liver [Franke et al., 2004b; Setchell, 2000]. This hydrolytic step can happen either by lactate phloridzin hydrolase, an enzyme bound at the brush border of enterocytes, or by the gut bacteria. Hydrolysis could possibly occur in the oral cavity [Walle et al., 2005] or in the stomach but only to a small extent that is unlikely to contribute significantly to aglycon formation [Franke et al., 2004b]. Due to the absence of isoflavone-*O*-glucoside in the circulation [Setchell et al., 2002] or the bile [Prasain et al., 2006] transport of the intact glycoside through the apical membrane of the enterocyte via sodium-dependent glucose transporter 1 (SGLT1) does not take place. Whether the uptake can be interrupted by efflux of the aglycon from the enterocyte back into the lumen via multidrug resistance-associated protein 2 (MDRP2) is not known. To a minor extent further metabolism can take place in the liver to

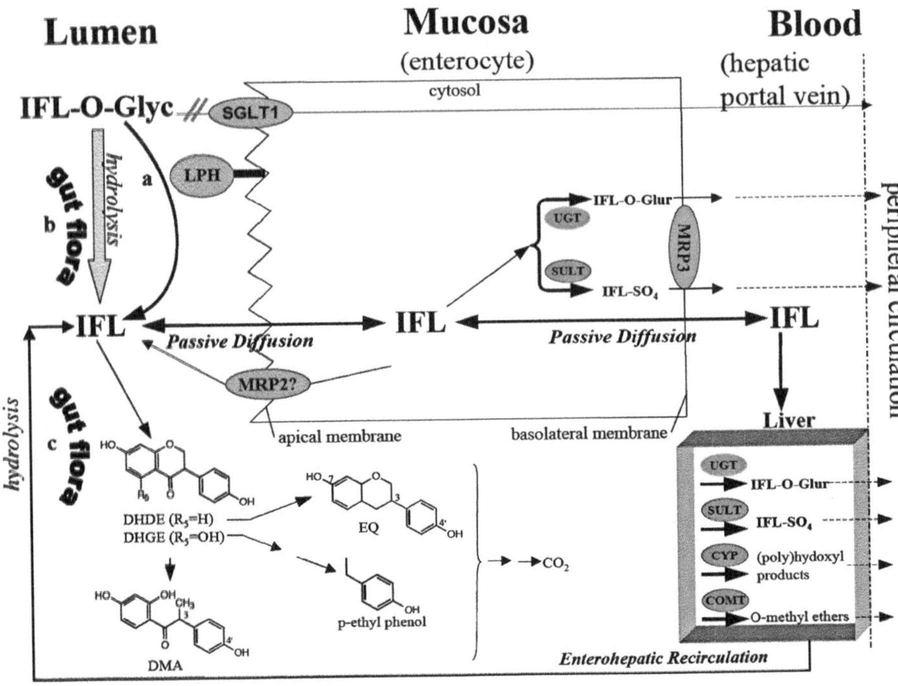

Figure 10.2 Absorption, transport, and metabolism of ingested isoflavone glycosides. See text for details. IFL-*O*-Glyc = isoflavone-*O*-glycoside; IFL = isoflavone aglycon; IFL-*O*-Glur = isoflavone-*O*-glucuronides; IFL-SO_4 = isoflavone sulfates; LPH = lactate phloridzin hydrolase; SDGLT1 = sodium-dependent glucose transporter 1; COMT = catecholamine ortho methyl transferase; MRP2/3 = multi-drug resistance-associated protein 2/3; UGT = UDP-glucuronyltransferase; SULT = sulfotransferases; DHDH = didhydrodaidzein; DHGE = didhydrogenistein; DMA = *O*-desmethylangolensin.

mono- or polyhydroxylated or *O*-methylated isoflavonoids [Heinonen et al., 2003; Kulling et al., 2001] while enterohepatic recirculation of isoflavone glucuronides/sulfates directly from the liver to the lumen can also occur [Sfakianos et al., 1997] followed by bacterial hydrolysis and reabsorption.

It is important to note that gut bacteria play a vital role because they not only hydrolyze but also form the major known human isoflavone metabolites dihydrodaidzein, dihydrogenistein, equol, *O*-desmethylangolensin, other minor compounds from isoflavones including *p*-ethyl phenol from genistein, and ultimately degrade isoflavones to CO_2 [Atkinson et al., 2005; Bannwart et al., 1984; Setchell et al., 2002]. This is highlighted by the extremely low levels of these metabolites in infants [Setchell et al., 1997] and germ-free animals [Bowey et al., 2003]. Urinary recovery of total isoflavones in humans including all major metabolites is usually in the range of 10–40% of ingested dose [Franke and Custer, 1994; Franke et al., 2004b; Setchell et al., 2003]. This, combined with the minimal fecal isoflavone excretion [Xu et al., 1995], suggests bioconversion of isoflavones to yet unknown metabolites (or CO_2). In vitro experiments identified bacterial species responsible for the known conversions, particularly for that to equol [Decroos et al., 2005; Ueno and Uchiyama, 2001; Wang et al., 2005], which, however, have not been confirmed in humans.

All metabolic studies after intake of soy or isoflavone preparations show great interindividual variations of all absorption, distribution, metabolism, and elimination parameters for isoflavones [Franke and Custer, 1994] [reviewed in Cassidy, 2006; Nielsen and Williamson, 2007]. Of the population 30–60% posses a gut flora that can produce equol [reviewed in Atkinson et al., 2005; Yuan et al., 2007] occurring naturally as the optical S-isomer, which has higher biological effects than its other isomers [Setchell et al., 2005]. Exact cut-offs for equol production have only recently been suggested based on a urinary equol/daidzein ratio of 0.018 [Setchell and Cole, 2006], but the assignment to an equol excretor in the literature remains still somewhat arbitrary. A carbohydrate rich diet may increase equol production while fat may reduce it [Rowland et al., 2000]. Fiber was reported to diminish isoflavone uptake [Tew et al., 1996], and oral antibiotics were shown to lead to a trend of decreased equol and variable *O*-desmethylangolensin production [Halm et al., 2008]. Vegetarians [Setchell and Cole, 2006] and Asians [Arai et al., 2000; Watanabe et al., 1998] who are usually habitual soy consumers, seem to have a greater ability to convert daidzein to equol (ca. 60%) than Western populations (ca. 30%) [Lampe et al., 1998]. Whether men are more likely to produce equol than women [Setchell and Cole, 2006] remains to be confirmed in larger studies. In contrast, 80–90% of populations are *O*-desmethylangolensin producers [Frankenfeld et al., 2005], whereas all animals except pigs [Gu et al., 2006] are the opposite, namely extensive equol and small *O*-desmethylangolensin producers. The health effects caused by the ability of the gut flora to produce equol, which is more estrogenic, more present as "free" aglycon (ca. 50%), and less protein bound than its precursor daidzein, or *O*-desmethylangolensin by intestinal bacteria are so far unclear and inconsistent [Atkinson et al., 2005; Frankenfeld et al., 2004, 2006; Horn-Ross et al., 2000; Setchell and Cole, 2006].

Metabolic studies by us [Fanti et al., 1999; Franke et al., 1999] and others [Anupongsanugool et al., 2005; de Pascual-Teresa et al., 2005; Kano et al., 2006; King and Bursill, 1998; Setchell et al., 2003; Vergne et al., 2008; Zubik and Meydani, 2003],) observed consistently a biphasic isoflavone appearance pattern in plasma and urine (Fig. 10.3) when humans consume soy or purified isoflavone preparations. Peak levels occur at 1–2 h and again at 4–7 h after intake. When this was investigated by simultaneous soy challenge and drastic reduction of the intestinal microbiota through oral antibiotic therapy combined with mechanical bowel preparation, it was concluded that only approximately 10% of overall uptake occurs in the proximal small intestine where probably lactate phloridzin hydrolase deconjugates isoflavones to the bioavailable aglycon [Franke et al., 2004b]. The majority, however, is absorbed after hydrolysis by the gut bacteria that start to occur in significant amounts in

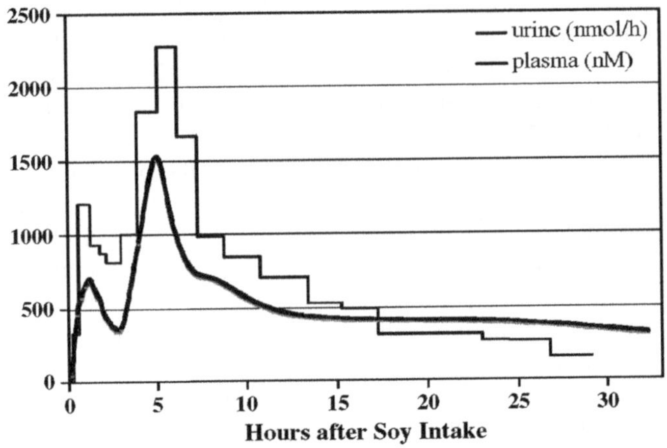

Figure 10.3 Typical pattern of isoflavone appearance in urine and plasma over time after soy intake. Plasma units are nanomole per liter and urine units are nanomole per hour. (From Franke et al. [2004b].)

the more distal intestine. Interestingly this biphasic isoflavone appearance pattern persisted with intake of aglycons indicating a saturation limit regarding uptake in the proximal small intestine and possibly preferred gut locations where absorption can take place.

BIOAVAILABILITY

Pharmacokinetic parameters from 16 studies summarized in Table 10.1 show greater bioavailability of genistein versus daidzein and of glycosides versus aglycons [reviewed in Nielsen and Williamson, 2007]. Relative to their *O*-glycosides, the aglycons daidzein and genistein show a C_{max} smaller by 43 and 37%, respectively, and about a. 17% shorter time to reach C_{max} (t_{max}) while the elimination half-lives ($t_{1/2}$) are similar. Relative to daidzein, genistein has a 28% higher C_{max} when dosed on a molar and 20% higher when dosed on a milligram basis while t_{max} is 8% shorter and $t_{1/2}$ 23% longer. The comparisons for genistin versus daidzin (the glucosides) show a similar trend (C_{max} 16 and 9% higher, t_{max} 9% shorter, $t_{1/2}$ 36% longer). According to these parameters doses of 25 mg of daidzein and genistein, or daidzin and genistin (these glucosides dosed in aglucon equivalents), applied to subjects weighing 50–75 kg would be expected to lead to peak plasma levels (C_{max}) for daidzein of 980-660 nM and genistein of 1190-790 nM, or 1730-1150 nM and 1890-1260 nM, respectively. Accordingly, relative to C_{max}, there will be residual isoflavone plasma levels of approximately 20–29%, 6–12%, and 2–5% 24, 36, and 48 h after isoflavone intake, respectively. As expected, isoflavone plasma accumulation over time

Table 10.1 Pharmacokinetic Parameters of Isoflavones

	C_{max}[a] (μM)	r[b]	t_{max}[c] (h)	$t_{1/2}$[d] (h)	C_{24}[e] Rel. to C_{max}	C_{36}[f] Rel. to C_{max}	C_{48}[g] Rel. to C_{max}
Daidzein	1.97±0.63	0.93	6.2±1.5	7.7±3.3	20%	7%	2%
Genistin	2.37±0.41	0.71	5.7±1.8	9.5±3.2	26%	11%	5%
Daidzin[h]	3.46±0.75	0.85	7.5±1.6	7.0±1.6	20%	6%	2%
Genistin[h]	3.78±1.00	0.85	6.8±1.4	9.5±1.7	29%	12%	5%

[a] Peak plasma levels in μM per mg isoflavone per kg body weight.
[b] Correlation between dose and plasma levels by linear regression.
[c] Time to reach peak plasma levels.
[d] Time to reach peak plasma levels.
[a-d] Elimination half life mean of 16 studies [Nielsen and Williamson 2007].
[e] Percent plasma level after 24 h relative to peak plasma levels.
[f] Percent plasma level after 36 h relative to peak plasma levels.
[g] Percent plasma level after 48 h relative to peak plasma levels.
[h] When dosed in aglycon units.

was observed by dosing soy 1–2 times per day [Gardner et al., 2008]. The correlation between dose and expected plasma levels in the circulation is almost linear with only slight saturation effects [Setchell et al., 2003]. However, it is still unclear whether fermented soy foods containing mostly isoflavone aglycons or unfermented soy foods containing mostly isoflavone *O*-glucosides result in better bioavailability. Aglycons as present in fermented soy products were found to result in higher isoflavone uptake by some [Cassidy et al., 2006; Hutchins et al., 1995; Izumi et al., 2000; Kano et al., 2006] but not others [Franke et al., 2004b; Setchell et al., 2003; Tsangalis et al., 2005] including nondistinguishable differences [Maskarinec et al., 2008b; Richelle et al., 2002]. In contrast, isoflavones from liquid relative to solid foods were consistently found to be absorbed more quickly and more extensively [Cassidy et al., 2006], but the overall isoflavone exposure as measured by urine excretion or area under the plasma curve is often not significantly different [de Pascual-Teresa et al., 2005; Franke et al., 2004b].

Intake Compared to Urine or Plasma Values

We found in several populations significant linear correlations of soy or isoflavone intake with urinary isoflavones [Franke et al., 1999; Jaceldo-Siegl et al., 2008]. This was confirmed by others who found these correlations with either urinary isoflavones [Atkinson et al., 2002; Slavin et al., 1998], plasma isoflavones [Frankenfeld et al., 2003; Heald et al., 2007; Wu et al., 2004], or both [Arai et al., 2000; Grace et al., 2004; Low et al., 2006; Ritchie et al., 2004; Yamamoto et al., 2001]. The correlations were highest (up to $r = 0.6$) for recent intake data (food diaries) and weaker (approximately $r = 0.2$) or disappeared for long-term intake measures (food frequency questionnaires). This is expected considering the fast elimination pattern of isoflavones. Very recent results indicate that fasting morning serum isoflavone estimates will provide a poor index of long-term soy intake, but that overnight urinary estimates perform much better [Fraser et al., 2008]. Therefore and due to high reproducibility within subjects over time [Frankenfeld et al., 2005; Maskarinec et al., 2008b], it is highly recommended to measure isoflavones in subjects, favorably urine (see below), as reliable biomarker for soy intake [Lampe, 2003].

Apparent Bioavailability

The same isoflavone appearance pattern (aglycon equivalents) was observed in plasma and urine [Fanti et al., 1999; Franke et al., 1998], and correlations between these matrices were very high [Arai et al., 2000; Grace et al., 2004; Low et al., 2006; Nettleton et al., 2004; Ritchie et al., 2004; Setchell and Cole, 2006; Yamamoto et al., 2001], particularly when the timing of sample collection was considered accurately [Franke et al., 2004b, 2006a; Halm et al., 2008]. The correlation between isoflavone values in urine and blood are much improved by using area under curve (AUC) units for both matrices instead of urinary isoflavone excretion rate (UIER) for urine, a time-based unit, and isoflavone level for plasma, a volume-based unit [Franke et al., 2004b; Halm et al., 2008]. This is largely due to the time domain being accurately considered by using the identical time intervals for the AUCs of both matrices. This is important to keep in mind because in living organisms isoflavone amounts in biological fluids change markedly over a given time period. It is vital to recognize that at similar doses more daidzein than genistein appears in urine while the opposite is true for plasma. Therefore, the AUC_{plasma} for daidzein is smaller than that for genistein (accordingly the bioavailability of daidzein is smaller than that of genistein) and the ratio AUC_{urine}/AUC_{plasma} much greater for daidzein than for genistein, but most importantly, this ratio remains constant over any given time interval for each isoflavone [Franke et al., 2004b].

Advantages of using urine over plasma include its noninvasiveness (particularly important for research in children), its superiority at defining an equol producer [Setchell and Cole, 2006], as well as the ability to collect a more concentrated matrix (particularly overnight urine) and large amounts, which leads to low quantification limits. Also, collections can be performed by participants themselves without medical supervision, in private, and, most importantly, can be accumulated over many hours (even days), reflecting exposures over much longer time periods compared to data from blood, which only reflects one given point in time per collection. This is confirmed as mentioned above by experimental findings of urinary estimates being a much better index of long-term soy intake than plasma values [Fraser et al., 2008]. We suggest to use UIERs as a reliable surrogate to determine circulating isoflavones and thereby assess isoflavone bioavailability. Since bioavailability is defined based on circulating levels, we propose to use the term apparent bioavailability when using urinary excretion data.

It has been recently suggested by our group [Halm et al., 2007, Halm et al., 2008] to examine the accuracy of timed urine collections by comparing the measured urinary creatinine excretion with established daily creatinine excretion data for each gender and age group (Table 10.2). Urinary excretion expressed relative to time (hour) is more accurate than expressed relative to creatinine, but timed urine is often not feasible to collect [Franke et al., 2006b]. Since creatinine excretion depends mostly on muscle mass, it varies largely depending on body weight (BW), gender, and age [1993; Remer et al., 2002]. This is particularly relevant in growing children, not only due to marked changes of muscle mass in absolute terms, but also after adjustment for body weight [Barr et al., 2005; Kampmann et al., 1974; Kesteloot and Joossens, 1996; Remer et al., 2002]. Urinary excretions adjusted solely for creatinine underestimate true excretion in heavier individuals—for example, in males versus females, or in older children versus younger children. The lack of conversion to time-based urinary excretion rates might explain the lower UIE in older than in younger soy-exposed children [Hoey et al., 2004] or the loss of differences in urinary sex steroid excretion in females of different reproductive ages [Hall Moran et al., 2001]. Therefore, as a general rule, it is best to have the collection time interval and the volume of the urine specimen available. The second choice is to convert a creatinine-based excretion value into a time-based excretion (in hours) by multiplying the former with the creatinine/hour ratio as determined from any timed urine collection for a given subject since the creatinine excretion remains quite constant within a subject. The third choice is to multiply the nmol/mg creatinine value with the established mg creatinine/kg BW/day ratio as shown in Table 10.2 [Kampmann et al., 1974; Remer et al., 2002] to arrive at a time-based unit. We suggest using 22 mg/kg/day and 24 mg/kg/day for 20- to 30-year-old women and men, respectively, to adjust the original data (19.7 mg/kg/day for women and 23.8 mg/kg/day for men) [Kampmann et al., 1974] to the more recent values (20.9 mg/kg/day and 23.3 mg/kg/day for 17-year-old girls and boys, respectively) [Remer et al., 2002].

Following this approach it was determined that the apparent isoflavone bioavailability in healthy children relative to healthy adults is ca. 30–40% higher [Halm et al., 2007]. This is in excellent agreement with previous reports on healthy minors, such as infants [Franke et al., 2006a], school-aged children [Franke et al., 2008], and (pre)pubertal girls [Maskarinec et al., 2005]. These findings suggest a higher systemic isoflavone exposure in children versus adults at the same relative soy dose. When considering that children eat generally much more per kilogram body weight, the exposure in children is probably up to twofold higher relative to adults. This could result in children experiencing more benefits from the health effects of soy [Shu et al., 2001; Wu et al., 2002]. We believe that the isoflavone exposure after soy intake will stay far below levels that would give rise to concern regarding adverse effects. Toxic activity is usually observed at much higher isoflavone levels, and adverse effects have not been reported in populations with high soy intake.

Table 10.2 Daily Creatinine Excretion Per kg of Body Weight as Function of Age and Gender[a]

Age (years)	Urinary Creatinine (mg/kg/d)		Age (years)	Urinary Creatinine (mg/kg/d)	
	Women	Men		Women	Men
3	14.5	15.2	25	22.0	24.0
4	15.7	17.1	30	20.4	21.9
5	15.7	17.1	35	20.4	21.9
6	17.9	19.3	40	17.6	19.7
7	17.9	19.3	45	17.6	19.7
8	17.9	19.3	50	14.9	19.3
9	18.9	20.7	55	14.9	19.3
10	18.9	20.7	60	12.9	16.9
11	18.9	20.7	65	12.9	16.9
12	18.9	20.7	70	11.8	14.2
13	18.9	20.7	75	11.8	14.2
14	20.9	23.3	80	10.7	11.7
15	20.9	23.3	85	10.7	11.7
16	20.9	23.3	90	8.4	9.4
17	20.9	23.3	95	8.4	9.4
20	22.0	24.0	30		

[a] Ages 3–17 from Remer et al. [2002]. Ages 20–95 Kampmann et al., [1974] except ages 20–25 where we suggest to use 22 mg/kg/d for women and 24 mg/kg/day for men to adjust the original Kampmann data (19.7 mg/kg/day for women and 23.8 mg/kg/day for men) to the more recently determined Remer values (20.9 mg/kg/day and 23.3 mg/kg/day for 17-year-old girls and boys, respectively).

Antibiotic Effects on Apparent Bioavailability

In soy intervention studies we found that children showed consistently a lower apparent isoflavone bioavailability when on oral antibiotics compared to when healthy; but the opposite was found in adults [Franke et al., 2004b; Halm et al., 2008]. We hypothesized that these observations can be explained by the changes in gut bacteria caused by oral antibiotics, or to unknown factors connected with the disease that led to oral antibiotic therapy. As pointed out above (Fig. 10.2), ingested isoflavone-*O*-glucosides are hydrolyzed to the bioavailable aglucons mainly by intestinal bacteria. However, gut bacteria also further metabolize the "free" bioavailable isoflavone aglucons to metabolic products such as equol, *O*-desmethylangolensin, *p*-ethylphenol, and other nonspecific phenolic agents [King, 2002]. Although definitive evidence is lacking, our seemingly paradoxical observation in adults versus children could be due to oral antibiotics in children reducing the isoflavone-hydrolyzing gut bacteria significantly (and possibly also the isoflavone aglycon degrading flora), which consequently leads to the isoflavones not becoming bioavailable. Vice versa, oral antibiotics in adults could possibly preferentially reduce the isoflavone aglycon degrading gut bacteria that are involved in isoflavone degradation to unknown and nonspecific metabolites, while the bacteria that hydrolyze isoflavone glycosides are comparably little affected; thereby the time during which bioavailable isoflavone aglucons can be absorbed is increased. The participants we investigated were treated with broad-spectrum oral antibiotics, but no correlation between specific types of antibiotics used, type of illness, gender, age, and ethnicity of the participant and UIER changes was observed. This might be due to the overall small sample size and/or inter-individual differences and the small numbers in each ethnic group. Similarly, our protocol did not allow for detailed evaluation of isoflavone metabolites because a 12-h urine sampling after soy exposure is insufficient to allow for efficient metabolite formation by the gut bacteria. With our relatively small studies we were unable to differentiate effects due to the bacterial infection, the antibiotics, or both. To further investigate the role of inflammation and infection on the isoflavone metabolism in humans, we recently initiated studies with participants suffering from viral or other acute and chronic infections not requiring antibiotics.

ANALYSIS

For accurate analyses serious attention needs to be paid to calibration of analytical instrumentation. For this purpose it is recommended to prepare stock solutions of authentic standards by determining concentrations through absorbance readings followed by checking purity via high-pressure liquid chromatography (HPLC). This avoids problems with weighing inaccuracies (hygroscopy), allows the repeated checking of existing stock solutions, and assures simple interlaboratory comparability. While the determination of the molar extinction coefficient (ε, the absorbance of a one-molar concentration) is critical, it remains a reliable reference point even if it needs to be changed because all final values can be adjusted linearly if ε value changes should become necessary. Table 10.3 shows that reported ε values for isoflavones as a function of wavelength and solvent vary greatly, which warrants an agreement on the true value to use for all analysts. For the sake of consistency we continue to use values reported by Ollis [1962] (daidzein, genistein) and Kelly et al. [1993] (glycitein) due to the lack of other published data when we started our isoflavone work in 1990.

Traditionally HPLC methods were used for isoflavone analysis from foods [Wang et al., 1990], and gas chromatography/mass spectrometry (GC/MS) to determine isoflavones and their metabolites in human biological fluids including urine [Adlercreutz et al., 1991; Kelly et al., 1993], plasma [Adlercreutz et al., 1993], and feces [Adlercreutz et al., 1995; Kurzer et al., 1995]. HPLC with photodiode array (PDA) detection was introduced in 1994 to measure these analytes in human urine [Franke and Custer, 1994; Xu et al., 1994]. Compared to GC/MS, HPLC methods require fewer steps for sample preparation and analysis and demand less technician time and less expensive instrumentation.

Table 10.3 Molar Extinction Coefficients (ε) of Isoflavones as a Function of Wavelength and Solvent[a]

	Solvent	λ (nm)	ε (L/mol/cm)	Reference
Daidzein	96% EtOH	250	20,893	Ollis, 1962
	100%EtOH	262	24,739	Wiseman et al., 2002
	NA	249	31,563	Murphy et al., 2002
	MeOH	248–249	27,100–27,200	Sigma-Aldrich 2006[b]
	80%MeOH	250	27,542	Walz, 1931
Genistein	96% EtOH	263	37,154	Ollis, 1962
	100%EtOH	262	35,842	Wiseman et al., 2002
	NA	263	35,323	Murphy et al., 2002
	EtOH	263	35,000–38,400	Sigma-Aldrich 2006[b]
	96% EtOH	262.5	37,291	Merck Index 10th ed.
	NA	261	33,113	Williams and Harborne, 1989
	85% MeOH	261	24,435	Walter, 1941
Glycitein	alcohol	256	22,387	Kelly et al., 1993
	NA	256	25,388	Murphy et al., 2002
Daidzin	NA	249	26,830	Murphy et al., 2002
	NA	NA	23,749	Univ.of St.Andrews[b]
	MeOH/Water	250	28,561	Purina/Nestle
Genistin	MeOH/Water	262.5	39,129	Nestle/Purina 2006[b]
	NA	262.5	35,323	Hendrich and Murphy, 2001
	85%aq EtOH	262	39,000–40,000	Sigma-Aldrich 2006[b]
Glycitin	NA	259	26,713	Murphy et al., 2002
Acetyldaidzin	NA	256	29,007	Murphy et al., 2002
Acetylgenistin	NA	261	38,946	Murphy et al., 2002
Acetylglycitin	NA	260	29,595	Murphy et al., 2002
Malonyldaidzin	NA	258	26,830	Murphy et al., 2002
Malonylgenistin	NA	260	29,895	Murphy et al., 2002
Malonylglycitin	NA	260	26,313	Murphy et al., 2002
Malonylglycitin	NA	260	26,313	Murphy et al., 2002
Dihydrodaidzein	NA	277	13,600	Wähälä et al., 1998[b]
Dihydrogenistein	NA	290	18,300	Wähälä et al., 1998[b]
O-desmethylangolensin	NA	280	12,023	Wähälä et al., 1998[b]
Equol	NA	281	6,761	Wähälä et al., 1994[b]
Formononetin	NA	256	29,512	Ollis, 1962
Biochanin-A	NA	263	27,542	Ollis, 1962

[a] λ(nm) = wavelength of absorbance maximum in nm; ε = molar extinction coefficient in L/mol/cm.

[b] Private communications; na = not available.

Higher sensitivity and selectivity for HPLC analyses can be achieved by detection with MS, which has gained extreme attention due to ease of use through electrospray ionization [Barnes et al., 1998; Franke et al., 2002; Twaddle et al., 2002]. For food isoflavone analyses we favor PDA monitoring over fluorescence or electrochemical detection (ECD) due to its much better diagnostic value, and over MS due its more affordable operation and extreme robustness except when levels become very low [Franke et al., 2004a]. Although ECD results in higher sensitivity than PDA, it is much less selective and can result in wrong positive results [Franke et al., 1998]. HPLC with PDA was applied for the analysis of isoflavones from foods [Franke et al., 1994, 1999; Umphress et al., 2005], urine [Franke and Custer, 1994] including epidemiologic studies [Zheng et al., 1999], plasma [Franke et al., 1998], and breast milk [Franke et al., 1998]. In all cases later projects that applied MS detection confirmed the earlier findings [Dai et al., 2002; Franke et al., 2006b] and were extended for analyses of nipple aspirate fluid [Maskarinec et al., 2008a].

Food analyses of isoflavones were improved very recently in our laboratory by the application of ultra-HPLC technology, which reduced run times of 45 min to less than 9 min, a more than five-fold gain (Fig. 10.4), while reducing solvent consumption and waste generation by a factor of more than 8. In brief, homogenized material was extracted with 80% aqueous methanol (v/v) containing 20 ppm flavone as internal standard by sonication (10 min) and stirring (2 h) at room temperature, yielding isoflavone conjugates and originally present aglycons. After centrifugation a clear aliquot was diluted 1:1 with 0.2 M acetate buffer (pH 5) and 5 μL were injected into a ultra-HPLC system (model Accela, ThermoFisher, San Jose, CA) using a Hypersil Gold C18 (100 × 2.1 mm; 1.9 μm) reversed-phase column (Waters; Milford, MA). Elution was performed at a flow rate of 0.5 mL/min (9600 psi) with the following linear gradient: A = 0.1% formic acid in 50/50 methanol/acetonitrile (v/v), B = 0.1% formic acid in water; A/B (v/v): from 22/78 to 35/65 in 6 min, then to 50/50 in 0.2 min, hold at the same ratio for 0.8 min, then change to 90/10 in 0.01 min, and hold at the same ratio for 1 min to eliminate hydrophobic components, then change back to the initial conditions in 0.01 min, and hold for 1 min for equilibration. Analytes were scanned between 200 and 400 nm by PDA detection for identification purposes and quantitated at 260 nm after adjustment of internal standard recovery. Further improvements of this method could include increasing the flow rate, applying a steeper gradient, or heating the column; but this was not warranted for our needs.

Isoflavone analyses from biological matrices including tissue are routinely analyzed in our laboratory by HPLC with electrospray ionization (negative mode) and has been improved by applying high-resolution tandem mass spectrometry (model TSQ Ultra, Thermo, San Jose, CA) (Fig. 10.5) [Blair et al., 2003; Franke et al., 2002a; Maskarinec et al., 2008a]. $^{13}C_3$-labeled daidzein, equol, and enterolactone (University of St. Andrews, UK) and $^{2}H_4$-labeled genistein (University of Helsinki, Finland) are added as internal standards to each homogenized specimen, and hydrolysis is performed with β-glucuronidase and arylsulfatase (Roche Applied Sciences, Indianapolis, IN) followed by phase separation with methyl tertiary butyl ether using a robotic extraction system (model Versa100, Aurora Biomed, Vancouver, Canada) [Franke et al., 1998]. The ether fractions are dried under nitrogen at room temperature and

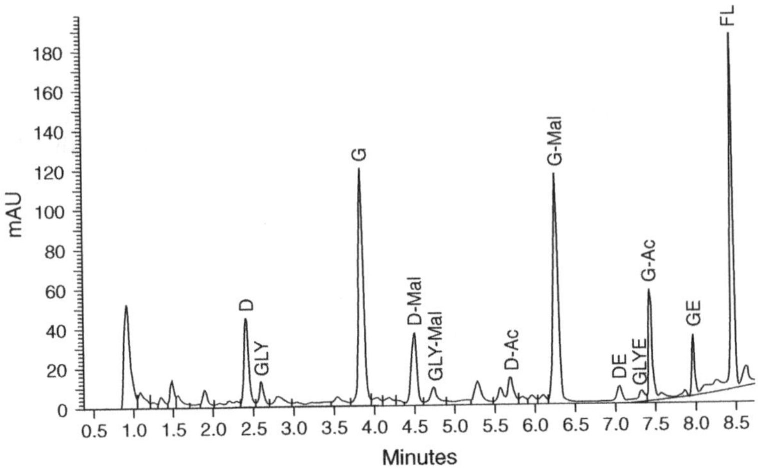

Figure 10.4 Typical UHPLC trace of isoflavones from a soy protein extract monitored at 260 nm using a Hypersil Gold C18 column (100 × 2.1 mm; 1.9 μm; Waters, Milford, MA) and a 0.5-mL/min flow rate (9600 psi) of a linear gradient of 0.1% aq. formic acid versus methanol/acetonitrile(1:1). FL = flavone (internal standard), other details in figure and in text. Concentrations of ISOFLAVONE peaks (μM): D = 17.66, GLY = 6.27, G = 26.49, D-Mal = 16.76, GLY-Mal = 3.62, D-Ac = 5.87, G-Mal = 26.70, DE = 3.37, GLYCITEIN = 1.74, G-Ac = 8.33, GE = 4.36.

redissolved in a 1:1 mixture of methanol:0.01% aq. formic acid. Of this extract 25 μL are separated on a Gemini C18 (150 × 2.0 mm; 5 μm) reversed-phase column coupled to a Gemini C18 (4.0 × 2.0 mm; 5 μm) direct-connect guard column (Phenomenex, Torrance, CA) using a linear gradient at 0.2 mL/min of MeOH/MeCN/Water = 15/15/70 to 30/30/40 in 2.5 min, to 40/40/20 in 5.5 min, and back to 15/15/70 in 0.1 min for equilibration before subsequent injections; 0.5% aq. ammonia at 40 μL/min is added postcolumn as dopant.

The general MS conditions we currently apply are as follows: source = electrospray ionization; ion polarity = negative; spray voltage = 3000 V; sheath and auxiliary and ion sweep gas = nitrogen; sheath gas pressure = 35 arbitrary units; auxiliary gas pressure = 10 arbitrary units; ion sweep gas pressure = 4 arbitrary units; ion transfer capillary temperature = 280°C; scan type = high-resolution selected reaction monitoring; collision gas = argon; collision gas pressure = 1.0 mTorr; source collision induced dissociation = 10V; scan width, 0.002u; scan time, 0.02–0.08 s; Q1 peak width = 0.1–0.3 u full width at half maximum (FWHM) and Q3 peak width at 0.70 u FWHM. Mass spectrometric monitoring is started 4.0 min after sample injection by multiple reaction monitoring as previously [Maskarinec et al., 2008a] using transitions (collision energies applied in brackets): daidzein from *m*/*z* 253.0 to *m*/*z* 223.0 (35 eV), 208.0 (33 eV), and 132.0 (40 eV); $^{13}C_3$-daidzein from *m*/*z* 256.0 to *m*/*z* 226.0 (31 eV), 211.0 (31 eV), and 182.0 (15 eV); genistein from *m*/*z* 269.1 to *m*/*z* 159.1 (30 eV), 133.0 (33 eV), and 132.0 (43 eV);

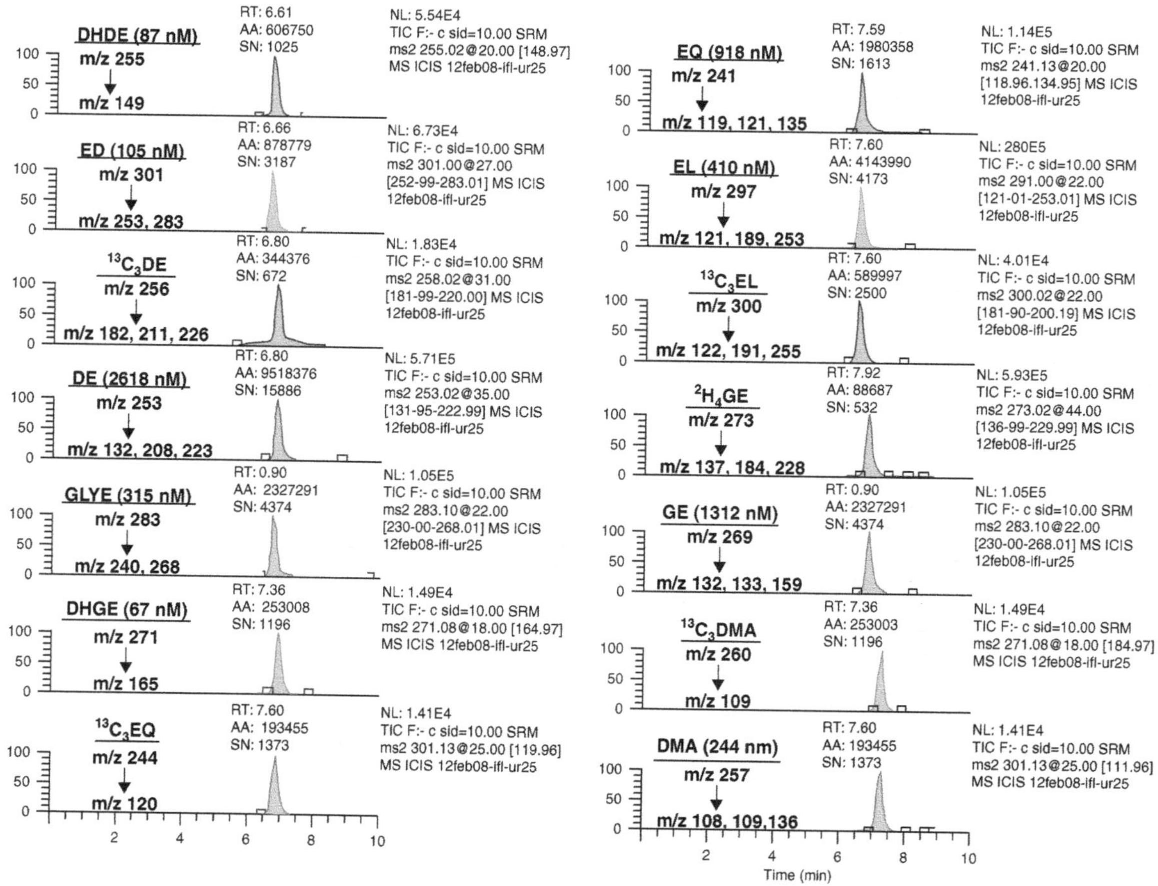

Figure 10.5 Typical LC–MS/MS trace by negative ESI of isoflavonoids from a urine extract. Measured concentrations in bracket. *Y*-axis units in relative intensity; EL = enterolactone, ED = enterodiol, other abbreviations see figure.

$^{2}H_4$-genistein from *m/z* 273.0 to *m/z* 228.0 (44 eV), 184.0 (44 eV), and 137.0 (42 eV); equol from *m/z* 241.1 to *m/z* 135.0 (20 eV), 121.0 (16 eV), and 119.0 (20 eV); $^{13}C_3$-equol from *m/z* 244.1 to *m/z* 120.0 (25 eV); *O*-desmethylangolensin from *m/z* 257.1 to *m/z* 135.9 (25 eV), *m/z* 109.0 (20 eV), and *m/z* 108.0 (28 eV); $^{13}C_3$-*O*-desmethylangolensin from *m/z* 260.1 to *m/z* 109.0 (35 eV); for glyciteinfrom m/z 283.1 to m/z 268.0 (22eV) and m/z 240.0 (25eV); for dihydrodaidzein from m/z 255.0 to m/z 149.0 (20eV); for dihydrogenistein from m/z 271.1 to m/z 165.0 (18eV); enterolactone from *m/z* 297.0 to *m/z* 253.0 (22 eV), 189.2 (39 eV), and 121.0 (39 eV); $^{13}C_3$-enterolactone from *m/z* 300.0 to *m/z* 255.2 (22 eV), 191.0 (39 eV), and 121.9 (39 eV); and enterodiol from *m/z* 301.0 to *m/z* 283.0 (27 eV) and 253.0 (27 eV). Limits of quantitation (LOQ) for all analytes were approximately 1 nM for daidzein and genistein, 2 nM for equol, and 5 nM for enterolactone; between-day coefficients of variation ranged 4–18% for all analytes, while intraday variation was half or less of that.

CONCLUSIONS

Reemerging interest in soy and isoflavones is due to recent findings on their potential cancer- and other disease-preventing effects, particularly when exposure happens at a young age and when intake persists throughout adulthood. Accurate, precise, and affordable analytical techniques will be the basis for future research aimed at defining the details of the biological activity of isoflavones.

ACKNOWLEDGMENTS

Supported in part by Physicians Pharmaceuticals, Inc., and NIH grants RR020890 and CA71789 (there are no conflicts of interest).

REFERENCES

Adlercreutz H, Mazur W. 1997. Phyto-estrogens and western diseases. *Ann Med* 29: 95–120.

Adlercreutz H, Fotsis T, Kurzer MS, Wähälä K, Mäkelä T, Hase T. 1995. Isotope dilution gas chromatographic–mass spectrometric method for the determination of unconjugated lignans and isoflavonoids in human feces, with preliminary results in omnivorous and vegetarian women. *Anal Biochem* 225:101–108.

Adlercreutz H, Fotsis T, Lampe J, Wahala T, Makela T, Brunow G, Hase T. 1993. Quantitative determination of lignans and isoflavonoids in plasma of omnivorous and vegetarian women by isotope dilution gas chromatography–mass spectrometry. *Scand J Clin Lab Invest* 53:5–18.

Adlercreutz H, Fotsis T, Bannwart C, Wahala K, Brunow G, Hase T. 1991. Isotope dilution gas chromatographic–mass spectrometric method for the determination of lignans and isoflavonoids in human urine, including identification of genistein. *Clin Chim Acta* 199:263–278.

Andersen O. 2006. Checklist for isoflavonoids. In: Andersen O, Markham K, Eds. *Flavonoids: Chemistry, Biochemistry, and Applications*. Boca Raton, FL: Taylor and Francis Group, PP. 1129–1198.

Anupongsanugool E, Teekachunhatean S, Rojanasthien N, Pongsatha S, Sangdee C. 2005. Pharmacokinetics of isoflavones, daidzein and genistein, after ingestion of soy beverage compared with soy extract capsules in postmenopausal thai women. *BMC Clin Pharmacol* 5:2.

Arai Y, Uehara M, Sato Y, Kimira M, Eboshida A, Adlercreutz H, Watanabe S. 2000. Comparison of isoflavones among dietary intake, plasma concentration and urinary excretion for accurate estimation of phytoestrogen intake. *J Epidemiol* 10:127–135.

Atkinson C, Frankenfeld CL, Lampe JW. 2005. Gut bacterial metabolism of the soy isoflavone daidzein: Exploring the relevance to human health. *Exp Biol Med (Maywood)* 230:155–170.

Atkinson C, Skor HE, Fitzgibbons ED, Scholes D, Chen C, Wahala K, Schwartz SM, Lampe JW. 2002. Overnight urinary isoflavone excretion in a population of women living in the United States, and its relationship to isoflavone intake. *Cancer Epidemiol Biomark Prev* 11:253–260.

Bannwart C, Adlercreutz H, Fotsis T, Wahala K, Hase T, Brunow G. 1984. Identification of o-desmethylangolensin, a metabolite of daidzein and of matairesinol, one likely plant precursor of the animal lignan enterolactone in human urine. *Finn Chem Lett* 4–5:120–125.

Barnes S, Coward L, Kirk M, Sfakianos J. 1998. HPLC-mass spectrometry analysis of isoflavones. *Proc Soc Exp Biol Med* 217:254–262.

Barnes S, Kirk M, Coward L. 1994. Isoflavones and their conjugates in soy foods: Extraction conditions and analysis by HPLC-mass spectrometry. *J Agric Food Chem* 42:2466–2474.

Barr DB, Wilder LC, Caudill SP, Gonzalez AJ, Needham LL, Pirkle JL. 2005. Urinary creatinine concentrations in the U.S. population: Implications for urinary biologic monitoring measurements. *Environ Health Perspect* 113:192–200.

Blair RM, Appt SE, Franke AA, Clarkson TB. 2003. Treatment with antibiotics reduces plasma equol concentration in cynomolgus monkeys (Macaca Fascicularis). *J Nutr* 133:2262–2267.

Bowey E, Adlercreutz H, Rowland I. 2003. Metabolism of isoflavones and lignans by the gut microflora: A study in germ-free and human flora associated rats. *Food Chem Toxicol* 41:631–636.

Cassidy A. 2006. Factors affecting the bioavailability of soy isoflavones in humans. *J AOAC Int* 89:1182–1188.

Cassidy A. 2005. Diet and menopausal health. *Nurs Stand* 19:44–52; quiz 54–45.

Cassidy A, Brown JE, Hawdon A, Faughnan MS, King LJ, Millward J, Zimmer-Nechemias L, Wolfe B, Setchell KD. 2006. Factors affecting the bioavailability of soy isoflavones in humans after ingestion of physiologically relevant levels from different soy foods. *J Nutr* 136:45–51.

Cotroneo MS, Wang J, Fritz WA, Eltoum IE, Lamartiniere CA. 2002. Genistein action in the prepubertal mammary gland in a chemoprevention model. *Carcinogenesis* 23:1467–1474.

Dai Q, Franke AA, Jin F, Shu X-O, Hebert JR, Custer LJ, Cheng J, Gao Y-T, Zheng W. 2002. Urinary excretion of phytoestrogens and risk of breast cancer among chinese women in shanghai. *Cancer Epidemiol Biomark Prev* 11:815–821.

Decroos K, Vanhemmens S, Cattoir S, Boon N, Verstraete W. 2005. Isolation and characterisation of an equol-producing mixed microbial culture from a human faecal sample and its activity under gastrointestinal conditions. *Arch Microbiol* 183:45–55.

de Pascual-Teresa S, Hallund J, Talbot D, Schroot J, Williams CM, Bugel S, Cassidy A. 2005. Absorption of isoflavones in humans: Effects of food matrix and processing. J Nutr Biochem

Duncan AM, Phipps WR, Kurzer MS. 2003. Phyto-oestrogens. *Best Pract Res Clin Endocrinol Metab* 17:253–271.

Fanti P, Sawaya PB, Custer LJ, Franke AA. 1999. Serum levels and metabolic clearance of the isoflavones genistein and daidzein in hemodialysis patients. *J Am Soc Nephrol* 40:382–389.

Fomon SJ. 1993. *Nutrition of Normal Infants*. St. Louis: Mosby.

Franke AA, Custer LJ. 1994. High-performance liquid chromatography assay of isoflavonoids and coumestrol from human urine. *J Chromatogr B* 662:47–60.

Franke AA, Halm BM, Ashburn LA. 2008. Urinary isoflavones are increased in adults, but decreased in children, consuming soy when on oral antibiotic therapy. Nutri Cancer.

Franke AA, Halm BM, Custer LJ, Tatsumura Y, Hebshi S. 2006a. Isoflavones in breastfed infants after mothers consume soy. *Am J Clin Nutr* 84:406–413.

Franke AA, Halm BM, Custer LJ, Tatsumura Y, Hebshi S. 2006b. Isoflavones in breastfed infants after mothers consume soy. *Am J Clin Nutr* 84:406–413.

Franke AA, Custer LJ, Arakaki C, Murphy S. 2004a. Vitamin C and flavonoid levels of fruits and vegetables consumed in Hawaii. *J Food Comp Anal* 17:1–35.

Franke AA, Custer LJ, Hundahl SA. 2004b. Determinants for urinary and plasma isoflavones in humans after soy intake. *Nutr Cancer* 50:141–154.

Franke AA, Custer LJ, Wilkens LR, Le Marchand L, Nomura AN, Goodman MT, Kolonel LN. 2002a. Liquid chromatographic analysis of dietary phytoestrogens from human urine and blood. *J Chromatogr B* 777:43–57.

Franke AA, Murphy SP, Le Marchand L, Zheng W, Custer L. 2002b Mar. Liquid chromatographic analysis of dietary phytoestrogens including isoflavonoids, flavonoids and lignans in foods and human body fluids. *J Nutr* 132:592S.

Franke AA, Hankin JH, Yu MC, Maskarinec G, Low SH, Custer LJ. 1999a. Isoflavone levels in soy foods consumed by multiethnic populations in Singapore and Hawaii. *J Agric Food Chem* 47:977–986.

Franke AA, Yu MC, Maskarinec G, Fanti P, Zheng W, Custer LJ. 1999b. Phytoestrogens in human biomatrices including breast milk. *Biochem Soc Trans* 27:308–318.

Franke AA, Custer LJ, Tanaka Y. 1998a. Isoflavones in human breast milk and other biological fluids. *Am J Clin Nutr* 68:1466S–1473S.

Franke AA, Custer LJ, Wang W, Shi SJ. 1998b. HPLC analysis of isoflavonoids and other phenolic agents from foods and from human fluids. *Proc Soc Exp Biol Med* 217:263–273.

Franke AA, Custer LJ, Cerna CM, Narala K. 1995. Rapid HPLC analysis of dietary phytoestrogens from legumes and from human urine. *Proc Soc Exp Biol Med* 208: 18–26.

Franke AA, Custer LJ, Cerna CM, Narala KK. 1994. Quantitation of phytoestrogens in legumes by HPLC. *J Agric Food Chem* 42:1905–1913.

Frankenfeld CL, McTiernan A, Thomas WK, LaCroix K, McVarish L, Holt VL, Schwartz SM, Lampe JW. 2006. Postmenopausal bone mineral density in relation to soy isoflavone-metabolizing phenotypes. *Maturitas* 53:315–324.

Frankenfeld CL, Atkinson C, Thomas WK, Gonzalez A, Jokela T, Wahala K, Schwartz SM, Li SS, Lampe JW. 2005. High concordance of daidzein-metabolizing phenotypes in individuals measured 1 to 3 years apart. *Br J Nutr* 94:873–876.

Frankenfeld CL, McTiernan A, Aiello EJ, Thomas WK, LaCroix K, Schramm J, Schwartz SM, Holt VL, Lampe JW. 2004. Mammographic density in relation to daidzein-metabolizing phenotypes in overweight, postmenopausal women. *Cancer Epidemiol Biomark Prev* 13:1156–1162.

Frankenfeld CL, Patterson RE, Horner NK, Neuhouser ML, Skor HE, Kalhorn TF, Howald WN, Lampe JW. 2003. Validation of a soy food-frequency questionnaire and evaluation of correlates of plasma isoflavone concentrations in postmenopausal women. *Am J Clin Nutr* 77:674–680.

Fraser GE, Franke AA, Jaceldo-Siegl K, Bennett H. 2008. Reliability of serum and urinary isoflavone estimates. J Nutr (in preparation).

Gardner C, Chatterjee L, Morris J, Franke AA. 2008. Effects of isoflavone supplements versus soy foods on blood concentrations of genistein and daidzein. J Nutr Biochem.

Grace PB, Taylor JI, Low YL, Luben RN, Mulligan AA, Botting NP, Dowsett M, Welch AA, Khaw KT, Wareham NJ, Day NE, Bingham SA. 2004. Phytoestrogen concentrations in serum and spot urine as biomarkers for dietary phytoestrogen intake and their relation to breast cancer risk in European prospective investigation of cancer and nutrition-Norfolk. *Cancer Epidemiol Biomark Prev* 13:698–708.

Gu L, House SE, Prior RL, Fang N, Ronis MJ, Clarkson TB, Wilson ME, Badger TM. 2006. Metabolic phenotype of isoflavones differ among female rats, pigs, monkeys, and women. *J Nutr* 136:1215–1221.

Hall Moran V, Leathard HL, Coley J. 2001. Urinary hormone levels during the natural menstrual cycle: The effect of age. *J Endocrinol* 170:157–164.

Halm BM, Franke AA, Ashburn LA, Hebshi SM, Wilkens LR. 2008. Oral antibiotics decrease urinary isoflavonoid excretion in children after soy consumption. *Nutrition and Cancer* 60:14–22.

Halm B, Ashburn L, Franke A. 2007. Isoflavones from soya foods are more bioavailable in children than adults. *Br J Nutr* 98:998–1005.

Heald CL, Ritchie MR, Bolton-Smith C, Morton MS, Alexander FE. 2007. Phyto-oestrogens and risk of prostate cancer in scottish men. *Br J Nutr* 98:388–396.

Hendrich S, Murphy SP. Isoflavones: Sources and Metabolism. In: Wildman REC, editor. Handbook of neutraceuticals and functional foods. *Boca Raton: CRC Press*; 2001. p. 55–72.

Heinonen SM, Hoikkala A, Wahala K, Adlercreutz H. 2003. Metabolism of the soy isoflavones daidzein, genistein and glycitein in human subjects. Identification of new metabolites having an intact isoflavonoid skeleton. *J Steroid Biochem Mol Biol* 87:285–299.

Hoey L, Rowland IR, Lloyd AS, Clarke DB, Wiseman H. 2004. Influence of soya-based infant formula consumption on isoflavone and gut microflora metabolite concentrations in urine and on faecal microflora composition and metabolic activity in infants and children. *Br J Nutr* 91:607–616.

Horn-Ross PL, Lee M, John EM, Koo J. 2000. Sources of phytoestrogen exposure among non-asian women in California, USA. *Cancer Causes Control* 11:299–302.

Hutchins AM, Slavin JL, Lampe JW. 1995. Urinary isoflavonoid phytoestrogen and lignan excretion after consumption of fermented and unfermented soy products. *J Am Diet Assoc* 95:545–551.

Izumi T, Piskula MK, Osawa S, Obata A, Tobe K, Saito M, Kataoka S, Kubota Y, Kikuchi M. 2000. Soy isoflavone aglycones are absorbed faster and in higher amounts than their glucosides in humans. *J Nutr* 130:1695–1699.

Jaceldo-Siegl K, Fraser GE, Chan J, Franke AA, Sabate J. 2008. Validation of soy protein estimates from a food-frequency questionnaire with repeated 24-h recalls and isoflavonoid excretion in overnight urine in a western population with a wide range of soy intakes. *Am J Clin Nutr* 87:1422–1427.

Kampmann J, Siersbaek-Nielsen K, Kristensen M, Hansen JM. 1974. Rapid evaluation of creatinine clearance. *Acta Med Scand* 196:517–520.

Kano M, Takayanagi T, Harada K, Sawada S, Ishikawa F. 2006. Bioavailability of isoflavones after ingestion of soy beverages in healthy adults. *J Nutr* 136:2291–2296.

Kelly GE, Nelson C, Waring MA, Joannou GE, Reeder AY. 1993. Metabolites of dietary (soya) isoflavones in human urine. *Clin Chim Acta* 223:9–22.

Kesteloot H, Joossens JV. 1996. On the determinants of the creatinine clearance: A population study. *J Hum Hypertens* 10:245–249.

King R. 2002. Digestion, absorption, and metabolism of isoflavones. In: Gilani G, Ed. *Phytoestrogens and Health.* Illinois: AOCS Press, PP. 209–234.

King RA, Bursill DB. 1998. Plasma and urinary kinetics of the isoflavones daidzein and genistein after a single soy meal in humans. *Am J Clin Nutr* 67:867–872.

Korde LA, Wu AH, Fears T, Nomura AM, West DW, Kolonel LN, Pike MC, Hoover RN, Ziegler RG. Childhood soy intake and breast cancer risk in Asian American women. Cancer Epidemiol Biomarkers Prev 2009; 18:1050–9.

Kudou S, Fluery Y, Welti D, Magnolato D, Uchida T, Kitamura K, Okubo M. 1991. Malonyl isoflavone glycosides in soybean seeds (Glycine Max Merrill). *Agric Biol Chem* 55:2227–2233.

Kuiper GGJM, Carlsson B, Grandien K, Enmark E, Haggblad J, Nilsson S, Gustafsson J-A. 1997. Comparison of the lignand binding specificity and transcript tissue distribution of estrogen receptors a and b. *Endocrinology* 138:863–870.

Kulling SE, Honig DM, Metzler M. 2001. Oxidative metabolism of the soy isoflavones daidzein and genistein in humans in vitro and in vivo. *J Agric Food Chem* 49:3024–3033.

Kurzer MS, Xu X. 1997. Dietary phytoestrogens. *Ann Rev Nutr* 17:353–381.

Kurzer MS, Lampe JW, Martini MC, Adlercreutz H. 1995. Fecal lignan and isoflavonoid excretion in premenopausal women consuming flaxseed powder. *Cancer Epidemiol Biomark Prevent* 4:353–358.

Lamartiniere CA. 2002. Timing of exposure and mammary cancer risk. *J Mammary Gland Biol Neoplasia* 7:67–76.

Lampe JW. 2003. Isoflavonoid and lignan phytoestrogens as dietary biomarkers. *J Nutr* 133:956S–964S.

Lampe JW, Karr SC, Hutchins AM, Slavin JL. 1998. Urinary equol excretion with a soy challenge: Influence of habitual diet. *Proc Soc Exp Biol Med* 217:335–339.

Low YL, Taylor JI, Grace PB, Mulligan AA, Welch AA, Scollen S, Dunning AM, Luben RN, Khaw KT, Day NE, Wareham NJ, Bingham SA. 2006. Phytoestrogen exposure, polymorphisms in comt, cyp19, esr1, and shbg genes, and their associations with prostate cancer risk. *Nutr Cancer* 56:31–39.

Magee PJ, Rowland IR. 2004. Phyto-oestrogens, their mechanism of action: Current evidence for a role in breast and prostate cancer. *Br J Nutr* 91:513–531.

Maskarinec G, Hebshi S, Custer L, Franke AA. 2008a. The relation of soy intake and isoflavone levels in nipple aspirate fluid. *Eur J Cancer Prev* 17:67–70.

Maskarinec G, Watts K, Kagihara J, Hebshi SM, Franke AA. 2008b. Urinary isoflavonoid excretion is similar after consuming soya milk and miso soup in Japanese American women. *Br J Nutr* 1–6.

Maskarinec G, Morimoto Y, Novotny R, Nordt FJ, Stanczyk FZ, Franke AA. 2005. Urinary sex steroid excretion levels during a soy intervention among young girls: A pilot study. *Nutr Cancer* 52:22–28.

Murphy PA, Barua K, Hauck CC. 2002. Solvent extraction selection in the determination of isoflavones in soy foods. *J Chromatogr B* 777:129–138.

Nettleton JA, Greany KA, Thomas W, Wangen KE, Adlercreutz H, Kurzer MS. 2004. Plasma phytoestrogens are not altered by probiotic consumption in postmenopausal women with and without a history of breast cancer. *J Nutr* 134:1998–2003.

Nielsen IL, Williamson G. 2007. Review of the factors affecting bioavailability of soy isoflavones in humans. *Nutr Cancer* 57:1–10.

Ollis WD. 1962. The isoflavonoids. In: Geissman TA, Ed. *The Chemistry of Flavonoid Compounds*. New York: MacMillan, PP. 353–405.

Prasain JK, Xu J, Kirk M, Smith Johnson M, Sfakianos J, Barnes S. 2006. Differential biliary excretion of genistein metabolites following intraduodenal and intravenous infusion of genistin in female rats. *J Nutr* 136:2975–2979.

Remer T, Neubert A, Maser-Gluth C. 2002. Anthropometry-based reference values for 24-h urinary creatinine excretion during growth and their use in endocrine and nutritional research. *Am J Clin Nutr* 75:561–569.

Richelle M, Pridmore-Merten S, Bodenstab S, Enslen M, Offord EA. 2002. Hydrolysis of isoflavone glycosides to aglycones by beta-glycosidase does not alter plasma and urine isoflavone pharmacokinetics in postmenopausal women. *J Nutr* 132:2587–2592.

Ritchie MR, Morton MS, Deighton N, Blake A, Cummings JH. 2004. Plasma and urinary phyto-oestrogens as biomarkers of intake: Validation by duplicate diet analysis. *Br J Nutr* 91:447–457.

Rowland IR, Wiseman H, Sanders TA, Adlercreutz H, Bowey EA. 2000. Interindividual variation in metabolism of soy isoflavones and lignans: Influence of habitual diet on equol production by the gut microflora. *Nutr Cancer* 36:27–32.

Setchell KD. 2000. Absorption and metabolism of soy isoflavones-from food to dietary supplements and adults to infants. *J Nutr* 130:654S–655S.

Setchell KD, Cole SJ. 2006. Method of defining equol-producer status and its frequency among vegetarians. *J Nutr* 136:2188–2193.

Setchell KD, Lydeking-Olsen E. 2003. Dietary phytoestrogens and their effect on bone: Evidence from in vitro and in vivo, human observational, and dietary intervention studies. *Am J Clin Nutr* 78:593S–609S.

Setchell KD, Clerici C, Lephart ED, Cole SJ, Heenan C, Castellani D, Wolfe BE, Nechemias-Zimmer L, Brown NM, Lund, TD, Handa RJ, Heubi JE. 2005. S-equol, a potent ligand for estrogen receptor beta, is the exclusive enantiomeric form of the soy isoflavone metabolite produced by human intestinal bacterial flora. *Am J Clin Nutr* 81:1072–1079.

Setchell KD, Brown NM, Desai PB, Zimmer-Nechimias L, Wolfe B, Jakate AS, Creutzinger V, Heubi JE. 2003a. Bioavailability, disposition, and dose-response effects of soy isoflavones when consumed by healthy women at physiologically typical dietary intakes. *J Nutr* 133:1027–1035.

Setchell KDR, Faughnan MS, Avades T, Zimmer-Nechemias L, Brown NM, Wolfe BE, Brashear WT, Desai P, Oldfield MF, Botting NP, Cassidy A. 2003b. Comparing the pharmacokinetics of daidzein and genistein with the use of 13C-labeled tracers in premenopausal women. *Am J Clin Nutr* 77:411–419.

Setchell KD, Brown NM, Zimmer-Nechemias L, Brashear WT, Wolfe, BE, Kirschner AS, Heubi JE. 2002. Evidence for lack of absorption of soy isoflavone glycosides in humans, supporting the crucial role of intestinal metabolism for bioavailability. *Am J Clin Nutr* 76:447–453.

Setchell KDR, Brown NM, Lydeking-Olsen E. 2002. The clinical importance of the metabolite equol-a clue to the effectiveness of soy and its isoflavones. *J Nutr* 132:3577–3584.

Setchell KDR, Zimmer-Nechemias L, Cai J, Heubi JE. 1997. Exposure of infants to phyto-oestrogens from soy-based infant formula. *Lancet* 350:23–27.

Sfakianos J, Coward L, Kirk M, Barnes S. 1997. Intestinal uptake and biliary excretion of the isoflavone genistein in the rat: Evidence for its enterohepatic circulation. *J Nutr* 127:1260–1268.

Shu XO, Jin F, Dai Q, Wen W, Potter JD, Kushi LH, Ruan Z, Gao YT, Zheng W. 2001. Soyfood intake during adolescence and subsequent risk of breast cancer among Chinese women. *Cancer Epidemiol Biomark Prev* 10:483–488.

Slavin JL, Karr SC, Hutchins AM, Lampe JW. 1998. Influence of soybean processing, habitual diet, and soy dose on urinary isoflavonoid excretion. *Am J Clin Nutr* 68:1492S–1495S.

Tew BY, Xu X, Wang HJ, Murphy PA, Hendrich S. 1996. A diet high in wheat fiber decreases the bioavailability of soybean isoflavones in a single meal fed to women. *J Nutr* 126:871–877.

Thanos J, Cotterchio M, Boucher BA, Kreiger N, Thompson LU. 2006. Adolescent dietary phytoestrogen intake and breast cancer risk (Canada). *Cancer Causes Control* 17:1253–1261.

Tsangalis D, Wilcox G, Shah NP, Stojanovska L. 2005. Bioavailability of isoflavone phytoestrogens in postmenopausal women consuming soya milk fermented with probiotic bifidobacteria. *Br J Nutr* 93:867–877.

Tsukamoto C, Shimada S, Igita K, Kudou S, Kokubun M, Okubo K, Kitamura K. 1995. Factors affecting isoflavone content in soybean seeds: Changes in isoflavones, saponins, and composition of fatty acids at different temperatures during seed development. *J Agric Food Chem* 43:1184–1192.

Twaddle NC, Churchwell MI, Doerge DR. 2002. High-throughput quantification of soy isoflavones in human and rodent blood using liquid chromatography with electrospray mass spectrometry and tandem mass spectrometry detection. *J Chromatogr B* 777:139–145.

Ueno T, Uchiyama S. 2001. Identification of the specific intestinal bacteria capable of metabolising soy isoflavone to equol. *Ann Nutr Metab* 45:113 (abstract).

Umphress ST, Murphy SP, Franke AA, Custer LJ, Blitz CL. 2005. Isoflavone content of foods with soy additives. *J Food Comp Anal* 18:533–550.

U.S. Department of Agriculture U. 2007. USDA-Iowa State University database on the isoflavone content of foods; release 1.4; http://www.Ars.Usda.Gov/sp2userfiles/place/12354500/data/isoflav/isoflav1–4.Pdf.

Vergne S, Bennetau-Pelissero C, Lamothe V, Chantre P, Potier M, Asselineau J, Perez P, Durand M, Moore N, Sauvant P. 2008. Higher bioavailability of isoflavones after a single ingestion of a soya-based supplement than a soya-based food in young healthy males. *Br J Nutr* 99:333–344.

Walker S, Adams M, Franke, AA, Register T. 2008. Effects of dietary soy protein on iliac and cartid artery atherosclerosis and gene expression in male monkeys. *Atherosclerosis* 196:106–113.

Walle T, Browning AM, Steed LL, Reed SG, Walle UK. 2005. Flavonoid glucosides are hydrolyzed and thus activated in the oral cavity in humans. *J Nutr* 135:48–52.

Walter ED. 1941. Genistin (an isoflavone glucoside) and its aglucone, genistein from soybean. *J Am Oil Chem Soc* 63:3273–3276.

Walz E. 1931. Isoflavone and saponin glucosides in soja hispida. *Justus Liebigs Ann Chem* 489:118–155.

Wang G, Kuan SS, Francis OJ, Ware GM, Carman AS. 1990. A simplified HPLC method for the determination of phytoestrogen in soybean and its processed products. *J Agric Food Chem* 38:185–190.

Wang XL, Hur HG, Lee JH, Kim KT, Kim SI. 2005. Enantioselective synthesis of s-equol from dihydrodaidzein by a newly isolated anaerobic human intestinal bacterium. *Appl Environ Microbiol* 71:214–219.

Watanabe S, Yamauchi M, Sobue T, Takahashi T, Miura T, Arai, Y, Mazur W, Wähälä K, Adlercreutz H. 1998. Pharmacokinetics of soybean isoflavones in plasma, urine and feces of men after ingestion of 60 g baked soybean powder (kinako). *J Nutr* 128:1710–1715.

Williams CA, Harborne JB. 1989. Isoflavonoids. In: Dey PM, Harborne JB, Eds. *Methods in Plant Biochemistry*. London: Academic Press, PP. 421–449.

Wiseman H. 2006. Isoflavonoids and human health. In: Andersen O, Markham K, Eds. *Flavonoids: Chemistry, Biochemistry, and Applications*. Boca Raton: Taylor and Francis Group, PP. 371–396.

Wiseman H, Casey K, Clarke DB, Barnes KA, Bowey E. 2002 Mar. Isoflavone aglycon and glucoconjugate content of high- and low-soy foods used in nutritional studies. *J Agric Food Chem* 50:1404–1410.

Wu AH, Yu MC, Tseng CC, Twaddle NC, Doerge DR. 2004. Plasma isoflavone levels versus self-reported soy isoflavone levels in Asian-American women in Los Angeles county. *Carcinogenesis* 25:77–81.

Wu AH, Wan P, Hankin J, Tseng CC, Yu MC, Pike MC. 2002. Adolescent and adult soy intake and risk of breast cancer in Asian-Americans. *Carcinogenesis* 23:1491–1496.

Xu X, Harris KS, Wang HJ, Murphy PA, Hendrich S. 1995. Bioavailability of soybean isoflavones depends upon gut microflora in women. *J Nutr* 125:2307–2315.

Xu X, Wang H-J, Murphy PA, Cook L, Hendrich S. 1994. Daidzein is a more bioavailable soymilk isoflavone than is genistein in adult women. *J Nutr* 124:825–832.

Yamamoto S, Sobue T, Sasaki S, Kobayashi M, Arai Y, Uehara M, Adlercreutz H, Watanabe S, Takahashi T, Iitoi Y, Iwase Y, Akabane M, Tsugane S. 2001. Validity and reproducibility of a self-administered food-frequency questionnaire to assess the isoflavone intake in a Japanese population in comparison with dietary records and blood and urine isoflavones. *J Nutr* 131:2741–2747.

Yan L, Spitznagel EL. 2005. Meta-analysis of soy food and risk of prostate cancer in men. Int J Cancer.

Yan L, Spitznagel E. 2004. A meta-analysis of soyfoods and risk of breast cancer in women. Internat. *J Cancer Prev* 1:281–293.

Yuan JP, Wang JH, Liu X. 2007. Metabolism of dietary soy isoflavones to equol by human intestinal microflora—implications for health. *Mol Nutr Food Res* 51:765–781.

Zheng W, Dai Q, Custer LJ, Shu X-O, Wen W-Q, Jin F, Franke AA. 1999. Urinary excretion of isoflavones and the risk of breast cancer. *Cancer Epidemiol Biomark Prev* 8:35–40.

Zubik L, Meydani M. 2003. Bioavailability of soybean isoflavones from aglycone and glucoside forms in American women. *Am J Clin Nutr* 77:1459–1465.

11 Phytoestrogens Up-regulate Antioxidant Genes

CONSUELO BORRÁS AND JOSE VIÑA
Department of Physiology, Faculty of Medicine, University of Valencia, Valencia, Spain

CONTENTS

SOURCES OF PHYTOESTROGENS IN FOOD

Phytoestrogens have been defined as any plant substance or metabolite that induces biological responses in vertebrates and can mimic or modulate the actions of endogenous estrogens, usually by binding to estrogen receptors [Madej et al., 2002]. The principal phytoestrogens in food are isoflavones, coumestrans, prenylated flavonoids, and lignans. Most of the available information on dietary phytoestrogen concentrations is related to isoflavones. Table 11.1 summarizes the principal dietary sources of each type of phytoestrogen and the concentrations found in them.

Plant Phenolics and Human Health: Biochemistry, Nutrition, and Pharmacology, Edited by Cesar G. Fraga.

The most prevalent dietary isoflavones include genistein, daidzein, glycitein, biochanin A, and formononetin [Klejdus et al., 2007]. These compounds are primarily found in legumes where they occur as glucosides. This fact may affect the absorption rate of isoflavones and will be discussed in the next section.

Soybeans and soy-based products are a particularly rich source of isoflavones, especially genistein and daidzein [Fukutake et al., 1996; Bingham et al., 1998; Mazur, 1998], reaching concentrations of up to 198 mg of total isoflavones per 100 g of soy-based product (Table 11.1). When referring to bioavailability of isoflavones, it is very important to consider the effect of food processing and cooking. In fact, commercial processing of soy into food products lowers isoflavone concentrations and may lead to chemical alterations, such as deacetylation or deglucosylation [Coward et al., 1998; Genovese et al., 2007]. An example for deglucosylation of isoflavones occurs in fermented soy products where the isoflavone content is halved when compared with nonfermented soy products [Slavin et al., 1998]. Nevertheless, it is very important to take into account that fermentation can alternatively increase bioavailability of isoflavones in soy: Fermented soy products are better absorbed than nonfermented ones [Cassidy et al., 2006]. Cooking can reduce phytoestrogen concentrations and alter their chemical characteristics. It has been shown that boiling and roasting affect phytoestrogen concentration [Coward et al., 1998; Liggins et al., 2000]. However, baking or frying does not appear to modify the total isoflavone content [Coward et al., 1998].

Table 11.1 Principal Dietary Sources of Each Class of Phytoestrogen and Concentrations Found in Them

Phytoestrogen class	Dietary Source	Concentration (mg/100 g)
Isoflavones	Fruits, cereals, vegetables	<0.1
	Legumes	0–0.58
	Soy milk	5–10
	Soy flour, soybeans	~150
Coumestrans	Bread	0
	Soybeans	0.05
	Beans	3.6
	Bean sprouts, alfalfa	4.7
Lignans	Peas	0.2
	Cereals	0.5
	Teas	2.04
	Linseeds	60–370

Source: Data from USDA–Iowa State University Isoflavone Database, 2002.

PHARMACOKINETICS OF PHYTOESTROGENS

Pharmacokinetic data of phytoestrogens given are basically centered in their excretion profiles. In general, the conclusions are that genistein and daidzein begin to rise within 2 h of an ingested dose [King and Bursill, 1998] and reach peak concentrations within 5–8 h after ingestion [King and Bursill, 1998; Setchell et al., 2001].

It is not clear whether absorption of aglucones is better than glucosides or not. Piskula et al. [1999] reported that aglucones can be absorbed directly from the stomach unlike their glucoside forms. It has been showed that glucosides need gut microflora to be absorbed from the gut [Xu et al., 1995; Rowland et al., 2003; Turner et al., 2003]. However, germ-free rats fed a soy-isoflavone-containing diet excrete large quantities of daidzein and genistein in urine, indicating that gut microflora is not absolutely required for the absorption of isoflavones. Isoflavone metabolites equol, *O*-desmethylangolensin, and lignan enterolactone were not detectable in urine from the germ-free rats but were present in human flora-associated rat urine, indicating that they were products of gut microflora activity [Bowey et al., 2003]. Interindividual variation of gut microflora population may be one of the reasons for the different bioavailability found in different individuals [Rowland et al., 2003]. In this regard, when studying the pharmacokinetics of isoflavones, it is important to consider gender, age, and dietary habits.

EFFECTS OF PHYTOESTROGENS ON HEALTH AND DISEASE

The significance of the structural similarity of lignans and isoflavones to mammalian estrogens and possible effects on cancer prevention were first postulated in the early 1980s [Setchell et al., 1984; Adlercreutz et al., 1987]. Heart disease, osteoporosis, breast cancer, prostate cancer, and menopausal symptoms share a common epidemiology: They are rare in Far Eastern populations eating traditional diets containing soybean products when compared with Western populations [Bingham et al., 1998]. Cross-sectional studies have shown higher phytoestrogen levels in urine and plasma of populations at lower risk of these diseases [Adlercreutz et al., 1982, 1991]. It was concluded that beneficial effects of phytoestrogens on climacteric women are either very weak or barely noticeable [Wuttke et al., 2007]. Only a few double-blind placebo-controlled trials document any significant relief of the symptoms. The situation concerning the risk to the cardiovascular system is similar. Isoflavones share the beneficial effects on high-density (HDL) and low-density lipoproteins (LDL) with estrogens. Clearly, there is a current necessity of clinical endpoint studies investigating the influence of preparations containing isoflavones on cardiovascular events. Isoflavones are likely to have a mild bone protective (antiosteoporotic) effect [Atkinson et al., 2004].

CELLULAR AND MOLECULAR MECHANISMS OF PHYTOESTROGEN ACTION

Phytoestrogens have similar chemical structures to estrogens and have been found to bind to estrogen receptors [Cos et al., 2003]. Therefore, there is reason to believe they will elicit estrogenic effects by similar, if not identical, mechanisms.

Many, but not all, the classical physiological effects of estrogens are mediated by ligand-activated nuclear transcription factors, the estrogen receptors (ERs). ERs are predominantly located in the nucleus under normal conditions, but they can move between the nucleus and the cytoplasm bound to heat shock proteins (HSPs). Upon entering the cell, estrogens bind to ERs, causing the dissociation from the heat shock proteins. Subsequently, ERs form a dimer and bind to DNA containing estrogen responsive elements (ERE). Receptor, estrogen-responsive elements, and the corresponding coactivators and corepressors promote gene transcription and synthesis of proteins, which produces an estrogenic response [Parker, 1995].

Three estrogen receptor subtypes have been described: ERα, ERβ, and recently ERγ [Hawkins et al., 2005], but the significance of this latter receptor is still unclear. Estrogen receptor subtypes can mediate different biological effects and display different intracellular and tissue distribution patterns [Parker, 1995]. The ERα and ERβ subtypes can interact with a wide variety of different compounds, although some ligands have different affinities for the subtypes. Indeed, phytoestrogens bind preferentially to ERβ, although they can also bind weakly to ERα [Kuiper et al., 1997; Pike et al., 1999; Branham et al., 2002]. Cell-surface forms of ERα and ERβ coupled to cytosolic signal transduction pathways have been described [Collins and Webb, 1999] and are discussed in the next section.

PHYTOESTROGENS UP-REGULATE ANTIOXIDANT GENES

Estrogens were long ago recognized as in vitro antioxidants [Ruiz-Larrea et al., 1997]. However, at physiological concentrations it is very unlikely that they may act as such, especially due to their low concentration in plasma. A simple calculation indicates that if the recommended dose of estradiol in estrogen replacement therapy is 50 micrograms/day and the recommended dose of vitamin E as supplement is 500 mg/day, estrogen ought to be approximately 10,000 times more potent than vitamin E to have a similar antioxidant effect; and this is obviously not the case when chemical structures, that is, capacity to trap radicals, are compared.

Yet, in vivo experiments show that estrogens have a powerful antioxidant effect: Mitochondrial H_2O_2 production is significantly increased (by more than 50%) after ovariectomy, and this is completely prevented when ovariectomized rats are treated with estradiol at doses similar to those used in estrogen replacement therapy [Borras et al., 2003]. Therefore, estrogens clearly do not act as chemical antioxidants in vivo, but rather they exert their antioxidant effect by up-regulating the expression of antioxidant genes and enzymes. In vitro studies using MCF-7 cells confirmed the results in vivo and demonstrated that estradiol up-regulates the expression of antioxidant genes by binding to estrogen receptors, which in turn activates mitogen-activated protein (MAP) kinases and nuclear factor kappa B (NF-κB) signaling pathways [Borras et al., 2005].

The effect of estradiol as an up-regulator of antioxidant and longevity-related genes indicates that its administration might be beneficial to increase life span, particularly for males who should then have a life span similar to females. However, considerable evidence has shown that estrogen replacement therapy after menopause may have setbacks (on top of their obvious feminizing effects in males) [Kasai et al., 1986]. Phytoestrogens constitute an interesting alternative. Their beneficial health effects have been reported repeatedly [Munro et al., 2003; Mahn et al., 2005] and, to our knowledge, very few, if any, major reports have shown detrimental effects. Thus, to find the molecular mechanisms by which phytoestrogens exert their antioxidant action, we tested the effect of 0.5 μM genistein on the H_2O_2 levels in MCF-7 cells [Vina et al., 2005; Borras et al., 2006]. This level of genistein can be considered as nutritionally relevant as it is the concentration normally found in plasma of people at the Far East, who eat relatively large quantities of soy in their normal diet. This concentration is, however, significantly higher than that found in people living in the Western world. When these cells were incubated with genistein, the levels of H_2O_2 were significantly decreased. However, when the cells were incubated with both genistein and tamoxifen (an estrogen receptor modulator) levels of H_2O_2 were similar to controls, thus confirming that the antioxidant effect of genistein is mediated by the interaction of genistein with the estrogen receptor. This observation posed the question about the intracellular mechanism by which genistein acts to increase the expression of mitochondrial antioxidant enzymes. A direct genomic effect of genistein was unlikely because neither Mn-SOD nor GPx has estrogen-responsive elements in their promoter region. Thus, it is likely that the action of genistein is mediated via intracellular signaling cascades. We tested the effect of mitogen-activated protein kinases (MAPK) by using UO126, an inhibitor of the phosphorylation of these kinases. Our experiments showed that UO126 completely inhibited the lowering effect of genistein on the level of H_2O_2 in cells. MAPK are known to activate the NF-κB. Therefore, we tested whether genistein acts by activating NF-κB in its up-regulation of the expression of both Mn-SOD and GPx genes, whose promoters contain putative NF-κB-binding motifs. This was indeed the case: When cells were incubated with pyrrolidine dithiocarbamate, an inhibitor of the I-κB degradation, and therefore an inhibitor of the NF-κB translocation to the nucleus, the effect of genistein on the up-regulation of antioxidant enzyme expression was prevented. Integrating the observed effects of these pharmacological inhibitors of the signaling pathways, we concluded that genistein up-regulates the expression of

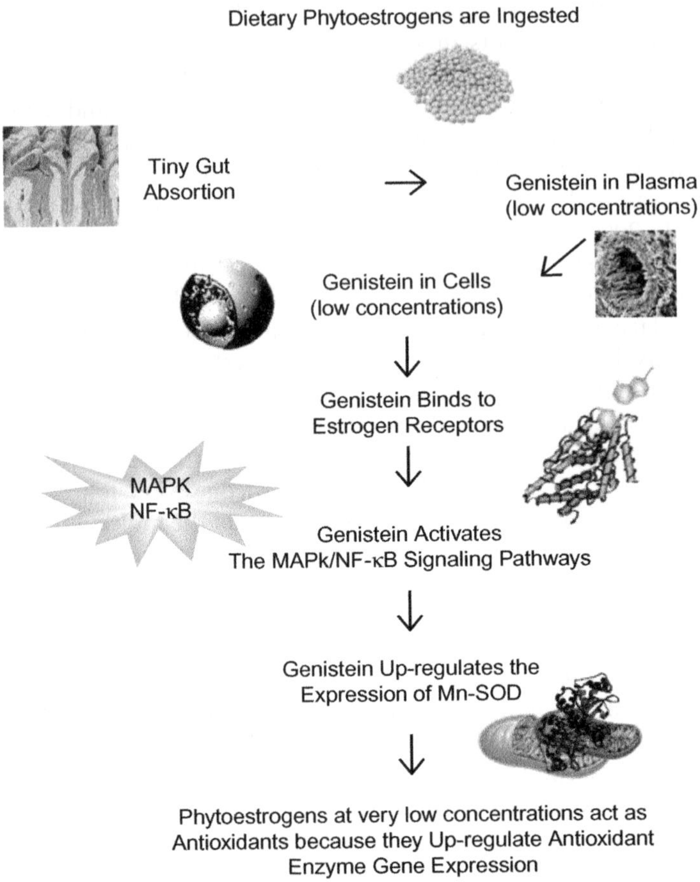

Figure 11.1 Mechanism of the up-regulation of antioxidant genes by phytoestrogens.

Mn-SOD mediated by the following pathway: interaction with membrane estrogen receptor, activation of MAPK, activation of NF-κB, and up-regulation of gene expression [Borras et al., 2006; Vina et al., 2008].

To sum up, we found that genistein significantly decreased H_2O_2 levels in cells and that effect, just as with estradiol, is mediated by estrogen receptors signaling pathways involving MAPK and NF-κB activations and up-regulation of the antioxidant gene for Mn-SOD. A schematic rendition of these pathways is shown in Figure 11.1.

CONCLUDING REMARKS

Work from our laboratory has shown that the difference in longevity between males and females may be traced to the beneficial effects of estrogens, which up-regulate the expression of longevity-associated genes. However, estrogens present the obvious problems of their feminizing action in males, and, when given to postmenopausal women, they may increase the incidence of cardiovascular events. Phytoestrogens, especially isoflavones, offer an interesting alternative: They are not feminizing and their use has proved very safe. We have shown that phytoestrogens exert their antioxidant action by up-regulating the expression of antioxidant genes. A conclusion that can be drawn is that even if phytoestrogens are absorbed from the gut in tiny concentrations, since their effect is catalytic and mediated by interaction with specific receptors, it may be significant. The effect of the oral administration of phytoestrogens and, in general, of flavonoids has to be tested in specific experimental or clinical model. Low rates of absorption and/or metabolism should not deter from studying biological effects that, as stated above, may occur significantly at catalytic concentrations.

ACKNOWLEDGMENTS

This work was supported by grants SAF 2004-03755, BFU2007-65803/BFI, and ISCIII2006-RED13-027 from the "Red Temática de investigación cooperativa en envejecimiento y fragilidad (RETICEF)" to J.V. and by grant GV/2007/263 to C.B. C.B. is recipient of a Bancaja postdoctoral fellowship.

REFERENCES

Adlercreutz H, Honjo H, Higashi A, Fotsis T, Hamalainen E, Hasegawa T, Okada H. 1991. Urinary excretion of lignans and isoflavonoid phytoestrogens in Japanese men and women consuming a traditional Japanese diet. *Am J Clin Nutr* 54:1093–1100.

Adlercreutz H, Hockerstedt K, Bannwart C, Bloigu S, Hamalainen E, Fotsis T, Ollus, A. 1987. Effect of dietary components, including lignans and phytoestrogens, on enterohepatic circulation and liver metabolism of estrogens and on sex hormone binding globulin (SHBG). *J Steroid Biochem* 27:1135–1144.

Adlercreutz H, Fotsis T, Heikkinen R, Dwyer JT, Woods M, Goldin BR, Gorbach SL. 1982. Excretion of the lignans enterolactone and enterodiol and of equol in omnivorous and vegetarian postmenopausal women and in women with breast cancer. *Lancet* 2:1295–1299.

Atkinson C, Compston JE, Day NE, Dowsett M, Bingham SA. 2004. The effects of phytoestrogen isoflavones on bone density in women: A double-blind, randomized, placebo-controlled trial. *Am J Clin Nutr* 79:326–333.

Bingham SA, Atkinson C, Liggins J, Bluck L, Coward A. 1998. Phyto-oestrogens: Where are we now? *Br J Nutr* 79:393–406.

Borras C, Gambini J, Gomez-Cabrera MC, Sastre J, Pallardo FV, Mann GE, Vina J. 2006. Genistein, a soy isoflavone, up-regulates expression of antioxidant genes: Involvement of estrogen receptors, ERK1/2, and NF B. *Faseb J* 20:2136–2138.

Borras C, Gambin J, Gomez-Cabrera MC, Sastre J, Pallardo FV, Mann GE, Vina J. 2005. 17beta-oestradiol up-regulates longevity-related, antioxidant enzyme expression via the ERK1 and ERK2[MAPK]/NfkappaB cascade. *Aging Cell* 4:113–118.

Borras C, Sastre J, Garcia-Sala D, Lloret Z, Pallardo FV, Vina J. 2003. Mitochondria from females exhibit higher antioxidant gene expression and lower oxidative damage than males. *Free Radic Biol Med* 34:546–552.

Bowey E, Adlercreutz H, Rowland I. 2003. Metabolism of isoflavones and lignans by the gut microflora: A study in germ-free and human flora associated rats. *Food Chem Toxicol* 41:631–636.

Branham WS, Dial SL, Moland CL, Hass BS, Blair RM, Fang H, Shi L, Tong W, Perkins RG, Sheehan DM. 2002. Phytoestrogens and mycoestrogens bind to the rat uterine estrogen receptor. *J Nutr* 132:658–664.

Cassidy A, Brown JE, Hawdon A, Faughnan MS, King LJ, Millward J, Zimmer-Nechemias L, Wolfe B, Setchell KD. 2006. Factors affecting the bioavailability of soy isoflavones in humans after ingestion of physiologically relevant levels from different soy foods. *J Nutr* 136:45–51.

Collins P, Webb C. 1999. Estrogen hits the surface. *Nat Med* 5:1130–1131.

Cos P, De Bruyne T, Apers S, Vanden Berghe D, Pieters L, Vlietinck AJ. 2003. Phytoestrogens: Recent developments. *Planta Med* 69:589–599.

Coward L, Smith M, Kirk M, Barnes S. 1998. Chemical modification of isoflavones in soyfoods during cooking and processing. *Am J Clin Nutr* 68:1486S–1491S.

Fukutake M, Takahashi M, Ishida K, Kawamura H, Sugimura T, Wakabayashi K. 1996. Quantification of genistein and genistin in soybeans and soybean products. *Food Chem Toxicol* 34:457–461.

Genovese MI, Barbosa AC, Pinto Mda S, Lajolo FM. 2007. Commercial soy protein ingredients as isoflavone sources for functional foods. *Plant Foods Hum Nutr* 62: 53–58.

Hawkins MB, Godwin J, Crews D, Thomas P. 2005. The distributions of the duplicate oestrogen receptors ER-beta a and ER-beta b in the forebrain of the Atlantic croaker (Micropogonias undulatus): Evidence for subfunctionalization after gene duplication. *Proc Biol Sci* 272:633–641.

Kasai H, Crain PF, Kuchino Y, Nishimura S, Ootsuyama A, Tanooka H. 1986. Formation of 8-hydroxyguanine moiety in cellular DNA by agents producing oxygen radicals and evidence for its repair. *Carcinogenesis* 7:1849–1851.

King RA, Bursill DB. 1998. Plasma and urinary kinetics of the isoflavones daidzein and genistein after a single soy meal in humans. *Am J Clin Nutr* 67:867–872.

Klejdus B, Vacek J, Benesova L, Kopecky J, Lapcik O, Kuban V. 2007. Rapid-resolution HPLC with spectrometric detection for the determination and identification of isoflavones in soy preparations and plant extracts. *Anal Bioanal Chem* 389:2277–2285.

Kuiper GG, Carlsson B, Grandien K, Enmark E, Haggblad J, Nilsson S, Gustafsson JA. 1997. Comparison of the ligand binding specificity and transcript tissue distribution of estrogen receptors alpha and beta. *Endocrinology* 138:863–870.

Liggins J, Bluck LJ, Runswick S, Atkinson C, Coward WA, Bingham SA. 2000. Daidzein and genistein contents of vegetables. *Br J Nutr* 84:717–725.

Madej A, Persson E, Lundh T, Ridderstrale Y. 2002. Thyroid gland function in ovariectomized ewes exposed to phytoestrogens. *J Chromatogr B Analyt Technol Biomed Life Sci* 777:281–287.

Mahn K, Borras C, Knock GA, Taylor P, Khan IY, Sugden D, Poston L, Ward JP, Sharpe RM, Vina J, Aaronson PI, Mann GE. 2005. Dietary soy isoflavone induced increases in antioxidant and eNOS gene expression lead to improved endothelial function and reduced blood pressure in vivo. *Faseb J* 19:1755–1757.

Mazur W. 1998. Phytoestrogen content in foods. *Baillieres Clin Endocrinol Metab* 12:729–742.

Munro IC, Harwood M, Hlywka JJ, Stephen AM, Doull J, Flamm WG, Adlercreutz H. 2003. Soy isoflavones: A safety review. Nutr Rev 61:1–33.

Park D, Huang T, Frishman WH. 2005. Phytoestrogens as cardioprotective agents. *Cardiol Rev* 13:13–17.

Parker MG. 1995. Structure and function of estrogen receptors. *Vitam Horm* 51:267–287.

Pike AC, Brzozowski AM, Hubbard RE, Bonn T, Thorsell AG, Engstrom O, Ljunggren J, Gustafsson JA, Carlquist M. 1999. Structure of the ligand-binding domain of oestrogen receptor beta in the presence of a partial agonist and a full antagonist. *Embo J* 18:4608–4618.

Piskula MK, Yamakoshi J, Iwai Y. 1999. Daidzein and genistein but not their glucosides are absorbed from the rat stomach. *FEBS Lett* 447:287–291.

Reinli K, Block G. 1996. Phytoestrogen content of foods—a compendium of literature values. *Nutr Cancer* 26:123–148.

Rowland I, Faughnan M, Hoey L, Wahala K, Williamson G, Cassidy A. 2003. Bioavailability of phyto-oestrogens. *Br J Nutr* 89:S45–S58.

Ruiz-Larrea MB, Leal AM, Martin C, Martinez R, Lacort M. 1997. Antioxidant action of estrogens in rat hepatocytes. *Rev Esp Fisiol* 53:225–229.

Setchell KD, Brown NM, Desai P, Zimmer-Nechemias L, Wolfe BE, Brashear WT, Kirschner AS, Cassidy A, Heubi JE. 2001. Bioavailability of pure isoflavones in healthy humans and analysis of commercial soy isoflavone supplements. *J Nutr* 131:1362S–1375S.

Setchell KD, Borriello SP, Hulme P, Kirk DN, Axelson M. 1984. Nonsteroidal estrogens of dietary origin: Possible roles in hormone-dependent disease. *Am J Clin Nutr* 40:569–578.

Slavin JL, Karr SC, Hutchins AM, Lampe JW. 1998. Influence of soybean processing, habitual diet, and soy dose on urinary isoflavonoid excretion. *Am J Clin Nutr* 68:1492S–1495S.

Turner NJ, Thomson BM, Shaw IC. 2003. Bioactive isoflavones in functional foods: The importance of gut microflora on bioavailability. *Nutr Rev* 61:204–213.

Vina J, Sastre J, Pallardo FV, Gambini J, Borras C. 2008. Modulation of longevity-associated genes by estrogens or phytoestrogens. *Biol Chem* 389:273–277.

Vina J, Borras C, Gambini J, Sastre J, Pallardo FV. 2005. Why females live longer than males? Importance of the upregulation of longevity-associated genes by oestrogenic compounds. *FEBS Lett* 579:2541–2545.

Wuttke W, Jarry H, Seidlova-Wuttke D. 2007. Isoflavones—safe food additives or dangerous drugs? *Ageing Res Rev* 6:150–188.

Xu X, Harris KS, Wang HJ, Murphy PA, Hendrich S. 1995. Bioavailability of soybean isoflavones depends upon gut microflora in women. *J Nutr* 125:2307–2315.

12 Dietary Isoflavones: Cardiovascular Actions and Activation of Cellular Signaling Pathways

RICHARD C. M. SIOW AND GIOVANNI E. MANN
Cardiovascular Division, School of Medicine, King's College London, London, United Kingdom

CONTENTS

INTRODUCTION

Cardiovascular diseases, such as hypertension, coronary heart disease (CHD), and atherosclerosis, are associated with increased oxidative stress and are more common in men than in premenopausal women of similar age, with the incidence of CHD increasing significantly after menopause, with loss of cardiovascular protection attributed to estrogen deficiency [Forte et al., 1998]. Differences in the regulation of blood pressure and vascular function between males and females have been investigated extensively over the past

Plant Phenolics and Human Health: Biochemistry, Nutrition, and Pharmacology,
Edited by Cesar G. Fraga.

decade, and female hormones are implicated in the protective effects of gender on the vasculature [Sader et al., 2002; Orshal et al., 2004; Mendelsohn et al., 2005] and expression of longevity associated genes [Viña et al., 2008]. Although estrogens and dietary phytoestrogen isoflavones have antioxidant properties due to the presence of a phenolic group in their steroid structure, these compounds can also activate intracellular kinase cascades and transcription of genes [Mann et al., 2007; Siow et al., 2007], potentially accounting for their beneficial effects on hypertension, thrombosis, and cardiovascular diseases [Chahoud et al., 2000; Visioli et al., 2000]. Nutrient–gene interactions have gained wide acceptance in the "postgenomic" era, and nutrigenomic approaches are yielding insights at genomic, proteomic, and metabolomic levels to evaluate the therapeutic potential of diets in the prevention and/or treatment of cardiovascular diseases [Muller and Kersten, 2003].

The cardiovascular benefits of naturally occurring diphenolic plant compounds, such as the isoflavones genistein, daidzein, and its metabolite equol, have recently been extensively examined in view of the lower incidence of CHD in populations consuming high amounts of soy-based foods. The U.S. Food and Drug Administration initially approved health claims for the addition of soy protein to foods for cardiovascular protection [Erdman, 2000], yet re-evaluation of the benefits of soy protein and isoflavones by an American Heart Association study group in 2006 concluded that "direct cardiovascular health benefits of soy protein or isoflavone supplements is minimal at best", suggesting that the high content of polyunsaturated fats, fiber, vitamins, minerals, and low content of saturated fat in soy products may account for their observed benefits [Sacks et al., 2006]. Although numerous studies have evaluated the benefits of dietary antioxidants against CHD, there is limited direct evidence linking soy isoflavones with the induction of antioxidant defense genes in the vasculature. In this chapter, we provide an up-to-date review of the molecular mechanisms by which estrogens and dietary isoflavones activate intracellular signal transduction pathways involved in the transcriptional activation of endothelial nitric oxide synthase (eNOS) and antioxidant defense genes in the vasculature. We evaluate the evidence that estrogens and phytoestrogens modulate vascular reactivity in vitro and in vivo and suggest that these compounds protect against cardiovascular disease by virtue of their ability to activate intracellular signaling pathways, leading to increased bioavailability of the vasodilator nitric oxide (NO) and up-regulation of the activity and/or expression of antioxidant defense genes mediated by the activation of kinase cascades and the key antioxidant transcription factor Nrf2. Dietary polyphenols contained in soy, red wine, green tea, and chocolate influence vascular reactivity by targeting eNOS and redox-sensitive gene expression [Mann et al., 2007]. Increased endothelium-dependent NO generation in response to dietary isoflavones may modulate cellular sensors for oxidative stress and thereby increase NO bioavailability [Liu et al., 2005] in cardiovascular diseases associated with enhanced generation of reactive oxygen species and diminished antioxidant defenses [Stocker and Keaney, 2004].

VASCULAR ACTIONS OF DIETARY ISOFLAVONES IN VIVO

Plasma concentrations of the isoflavones genistein and daidzein range between 50 and 800 ng mL^{-1} in adults consuming soy-rich foods and can achieve levels found in Japanese consuming their traditional soy-based diet [Adlercreutz et al., 1993; Setchell and Cassidy, 1999; Clarkson, 2002]. The protection against CHD in Asian populations consuming a soy-enriched diet may be due to increased eNOS expression and enhanced bioavailability of NO [Lissin and Cooke, 2000; Hwang et al., 2003; Joy et al., 2006]. Clinical trials evaluating the benefits of soy isoflavones report improvements in systemic arterial compliance in men and postmenopausal women [Teede et al., 2003], increased plasma nitrite/nitrate and decreased endothelin-1 levels, and increased brachial artery flow-mediated dilatation [Squadrito et al., 2003]. Dietary isoflavones have been shown to acutely modulate vascular reactivity in healthy postmenopausal women and male subjects in vivo. Infusion of genistein or dihydroequol directly into brachial arteries, at concentrations achieved in populations consuming an isoflavone-rich diet [Adlercreutz et al., 1993], evoked concentration- and endothelium-dependent increases in forearm blood flow [Walker et al., 2001; Chin-Dusting et al., 2004], while consuming test meals supplemented with 80 mg of soybean isoflavones resulted in significantly greater brachial artery flow-mediated dilatation and plasma levels of NO metabolites after 5–7 h compared to values in subjects consuming control meals [Hall et al., 2008].

As a significant component of the genistein-mediated dilation was shown to be insensitive to the inhibition of eNOS by N^G-monomethyl-L-arginine [Walker et al., 2001], however, it is possible that isoflavones can also modulate vascular tone by inhibiting Ca^{2+} influx via L-type Ca^{2+} channels in vascular smooth muscle cells [Ruehlmann et al., 1998; Figtree et al., 2000; Mishra et al., 2000]. Several randomized, double-blind, placebo-controlled trials report cardiovascular benefits of longer-term (2–12 months) supplementation with genistein, tetrahydrodaidzein, or soy isoflavones [Nestel et al., 1997, 2006; Squadrito et al., 2002, 2003; Hallund et al., 2006; Teede et al., 2001]. The consensus of such studies in healthy postmenopausal women is that isoflavone supplementation increases brachial artery flow-mediated dilation and improves systemic arterial compliance. Despite numerous clinical studies, there is still controversy whether isoflavone supplementation lowers arterial blood pressure in vivo [Nestel et al., 1997, 2006; Teede et al., 2001; Kreijkamp-Kaspers et al., 2004, 2005]. It is possible that vascular responses to isoflavone metabolites may be limited to individuals capable of metabolizing ingested daidzein to equol [Cassidy et al. 2006; Setchell et al., 2005; Vergne et al., 2006]. In summary, the clear benefits of soy isoflavone supplementation on endothelial function have largely been observed in humans and animal models with mild to moderate hypertension [Teede et al., 2003; Rivas et al., 2002; Vera et al., 2007; Xu et al., 2004].

Randomized clinical studies in healthy human volunteers, smokers, and patients with coronary artery disease have shown that consumption of black and green tea polyphenols or flavanol-rich chocolate/cocoa enhances flow-mediated brachial artery dilation within 1–24 h [Duffy et al., 2001; Heiss et al., 2005, 2006; Schroeter et al. 2006; Vlachopoulos et al., 2005]. Clinical studies with healthy volunteers report that ingestion of flavanol-rich chocolate/cocoa results in NO-mediated peripheral vasodilation [Fisher et al., 2003], increased insulin sensitivity, and reduced blood pressure [Grassi et al., 2005a]. Moreover, patients with essential hypertension consuming flavanol-rich dark chocolate exhibit increased brachial artery flow-mediated dilation and a significant reduction in systolic blood pressure [Grassi et al., 2005b]. Although flavanol-rich chocolate/cocoa increases plasma epicatechin concentrations and endothelium-dependent dilation and pure epicatechin mimics these effects [Schroeter et al, 2006; Engler et al., 2004], other bioactive components in chocolate could contribute to vascular relaxation in vivo. Apart from having direct antioxidant effects, flavanols and green tea polyphenols can modulate redox signaling in vascular cells providing additional beneficial mechanisms for increasing NO bioavailability and reducing smooth muscle proliferation [Locher et al., 2002; Fraga, 2005; Mackenzie et al., 2004].

NITRIC OXIDE SYNTHESIS AS A VASCULAR TARGET FOR DIETARY ISOFLAVONES

The vasodilator NO is synthesized by eNOS from the cationic amino acid L-arginine in the presence of molecular oxygen and other cofactors [Palmer et al., 1987]. As a Ca^{2+}/calmodulin (CaM)-dependent enzyme, eNOS is regulated by its association with the plasmalemmal scaffolding protein caveolin-1 [Shaul et al., 1996], posttranslational modification via phosphorylation by protein kinases, and an interaction with the molecular chaperone heat shock protein 90 (Hsp90) [Garcia-Cardena et al., 1998; Boo & Jo 2003; Takahashi and Mendelsohn, 2003]. Tethering of eNOS to caveolin-1 maintains the enzyme in an inactive state, and an elevation in intracellular Ca^{2+} leads to vasoactive agonists promoting association of CaM with eNOS and dissociation of the enzyme from caveolin-1 [Ghosh et al., 1998]. In addition to Ca^{2+}-dependent activation of eNOS, laminar shear stress [Boo et al., 2003], 17β-estradiol [Russell et al., 2000], and isoflavones [Joy et al., 2006] can stimulate NO production via phosphorylation of eNOS at cytosolic Ca^{2+} levels, with some studies reporting rapid dissociation of eNOS from caveolin-1 and association with heat shock protein Hsp90 (Fig. 12.1). The major eNOS phosphorylation sites include Ser^{1177} in the reductase domain and Thr^{495} and Thr^{497} in the CaM binding domain. Protein kinase (Akt) [Dimmeler et al., 1999], protein kinase A [Boo et al., 2003], mitogen-activated protein kinases ERK1/2 [Cai et al., 2003], and $p38^{MAPK}$ [Anter et al., 2005] have been reported to phosphorylate Ser^{1177}. Phosphorylation at Thr^{495} antagonizes CaM binding to eNOS [Fleming et al., 2001], and thus concerted phosphorylation at Ser^{1177} and dephosphorylation of Thr^{495} occurs in response to different stimuli to modulate eNOS activity.

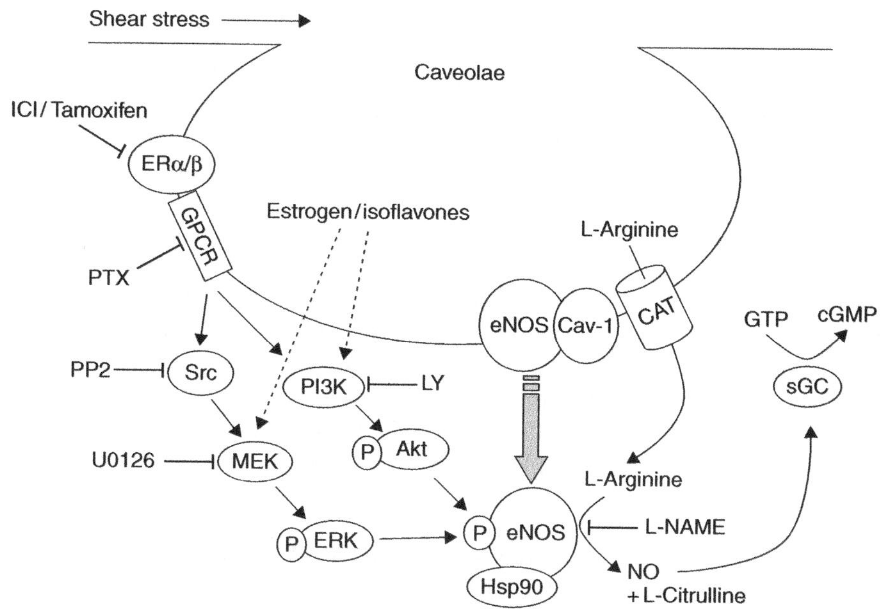

Figure 12.1 Ca^{2+}-independent stimuli such as fluid shear stress, adenosine, isoprenaline, estrogen, and soy isoflavones acutely stimulate eNOS phosphorylation at resting cytosolic Ca^{2+} levels, involving activation of ERK1/2, PI(3)-kinase/Akt and/or PKA (not shown) signaling pathways. Activation of eNOS involves rapid dissociation of the enzyme from membrane-bound caveolin-1 (Cav-1) and association with Hsp90 and metabolism of the cationic amino acid L-arginine to NO and L-citrulline. NO binds to the heme moiety of soluble guanylyl cyclase in target smooth muscle cells to increase intracellular cGMP levels, leading to vascular relaxation. Although ERα and ERβ (and truncated ER receptor) are expressed in HUVEC, rapid actions (2 min) of isoflavones appear not to involve coupling of ER receptors to Gαi in membrane caveolae in endothelial cells. Inhibitors for specific kinase pathways are indicated, e.g., PP2, U0126, LY294,002, and L-NAME. (Adapted from Siow et al. [2007] and Mann et al. [2007].)

The signaling pathways underlying eNOS activation by estrogens have been reviewed previously [Chambliss and Shaul, 2002; Mendelsohn, 2002], and co-localization of estrogen receptors (ERα and ERβ) with eNOS in membrane caveolae in endothelial cells provides a functional signaling unit for enzyme activation [Haynes et al., 2000]. Moreover, Hsp90 can also interact directly with ERs, enabling 17β-estradiol to rapidly activate eNOS [Russell et al., 2000]. Although a subpopulation of estrogen-like receptors in the plasma membrane of endothelial cells may also mediate rapid activation of intracellular signaling pathways [Joy et al., 2005; Figtree et al., 2003], there is no direct evidence that isoflavones interact with plasma membrane "estrogen receptors". Isoflavones elicit both endothelium-dependent and -independent relaxation of preconstricted arterial rings in vitro, mimicking their dilator actions in the human forearm vasculature [Walker et al., 2001; Ching-Dusting et al., 2004; Hall et al.,

2008]. We previously reported that the daidzein metabolite equol causes acute relaxation of preconstricted rat aortic rings [Joy et al., 2006]. Equol-mediated ring relaxation was inhibited by the eNOS inhibitor L-NAME, whereas the environmental estrogenic pollutant 4-octylphenol and ICI 182,780 reduced coronary perfusion pressure in isolated hearts via an inhibition of L-type Ca^{2+} channels in vascular smooth muscle [Ruehlmann et al., 1998]. Low nanomolar concentrations of isoflavones rapidly stimulate NO release from cultured endothelial cells [Liu et al., 2004; Joy et al., 2006]. As summarized in Figure 12.1, studies in human endothelial cells established that genistein, daidzein, and equol (100 nM) rapidly (30 s–2 min) stimulate ERK1/2 and Akt-dependent eNOS phosphorylation at basal cytosolic Ca^{2+} levels. Isoflavone-stimulated NO release was unaffected by the ER antagonists ICI 182,780 (Faslodex) and tamoxifen or uncoupling of G-protein receptors. It was further established that equol induces a rapid dissociation of eNOS from caveolin-1 and association with the chaperone protein Hsp90 [Joy et al., 2006].

ACTIVATION OF CELLULAR SIGNALING PATHWAYS AND ANTIOXIDANT DEFENSE GENES BY ISOFLAVONES

Extensive clinical and experimental evidence links hypertension and atherosclerotic vascular disease with the accumulation of oxidized low-density lipoproteins and enhanced generation of reactive oxygen species within the vessel wall [Stocker and Keaney, 2004]. ROS are short-lived molecules generated as by-products of metabolism and rapidly react with cellular components, causing damage to membrane lipids, proteins, and DNA [Dröge, 2002]. It is now apparent that numerous cellular targets for isoflavones exist that lead to modulation of signaling pathways, including those regulating protective antioxidant and anti-inflammatory transcription factor activity and gene expression [Siow et al., 2007; Rimbach et al., 2008]. Although estrogens and phytoestrogens can act as direct free-radical scavengers [Williams et al., 2004], recent evidence suggests that soy isoflavones [Mahn et al., 2005], red wine, tea and dark chocolate polyphenols [Wallerath et al., 2002, 2005; Leikert et al., 2002] can modulate pathways leading to increased messenger ribonucleic acid (mRNA) and protein levels for eNOS in endothelial cells. Moreover, polyphenols have been implicated in antioxidant response element (ARE)-mediated gene transcription [Ansell et al., 2004; Bianco et al., 2005] and up-regulation of manganese superoxide dismutase (MnSOD) expression in vascular cells [Strehlow et al., 2003].

Induction of antioxidant genes is regulated through a cis-acting ARE element within the regulatory region of target genes [Nguyen et al., 2003]. The redox-sensitive basic leucine zipper protein transcription factor Nrf2 is involved in the regulation of many detoxification and antioxidant genes. Nrf2 belongs to the Cap-N-Collar family of transcription factors, forming heterodimers with small Maf proteins, with subsequent binding to ARE leading to gene transcription

[Ishii et al., 2000]. Since the expression of a wide array of protective antioxidant genes are positively regulated by the ARE sequence, Nrf2 may serve as a regulator of ARE-driven cellular defense systems against oxidative stress induced by dietary isoflavones [Mann et al., 2007; Siow et al., 2007].

Exposure of endothelial cells to donors of NO and/or peroxynitrite leads to adaptive increases in synthesis of the major intracellular antioxidant glutathione (GSH) and Nrf2/ARE-regulated expression of antioxidant genes including heme oxygenase-1 (HO-1) [Foresti et al., 2003]. The microsomal enzyme HO-1 metabolizes heme to generate biliverdin/bilirubin and carbon monoxide, which like NO can inhibit platelet aggregation and act as a vasodilator when the bioavailability of NO is limited [Siow et al., 1999; Ryter et al., 2006]. Biliverdin is subsequently converted by biliverdin reductase to bilirubin, a chain-breaking antioxidant that scavenges lipid peroxyl radicals [Baranno et al., 2002]. Accumulating evidence suggests that Nrf2 constitutes a sensor of oxidative stress involved in triggering ARE-mediated gene expression to restore cellular redox status [Siow et al., 2007].

Although protective actions of dietary antioxidants and isoflavones against CHD have been described, few reports have adequately addressed the molecular and cellular targets involved in the induction of antioxidant defense genes in vascular cells. There is limited evidence in vascular endothelial and smooth muscle cells that estrogens, isoflavones (genistein, daidzein, equol), and polyphenols modulate antioxidant gene expression via the transcription factors Nrf2 and nuclear factor-κB (NF-κB), however several recent studies have used nonphysiological concentrations of isoflavones (>20 μM) that are well recognized to inhibit tyrosine kinase activity [Akiyama et al., 1987]. Thus, it is important to critically evaluate both concentration- and time-dependent actions of estrogenic-like compounds on intracellular signaling cascades and gene expression. It is conceivable that acute activation of NO generation elicited by isoflavones in endothelial cells may lead to nuclear translocation of Nrf2 [Buckley et al., 2003]. As summarized in Figure 12.2, modulation of the Nrf2/Keap1 complex by NO, peroxynitrite, or other reactive oxygen and nitrogen species will lead to induction of ARE-gene transcription [Kang et al., 2002]. In a similar manner, CO has been shown to enhance nuclear translocation of Nrf2 and HO-1 induction in primary human hepatocytes through activation of MAPK pathways [Lee et al., 2006]. It is therefore likely that isoflavone-stimulated NO and CO production in endothelial and smooth muscle cells may sensitize/modulate antioxidant gene expression through Nrf2-mediated activation of the ARE in target defense genes.

The promoter sequences of MnSOD and GPx contain putative NF-κB-binding motifs, and in MCF-7 cells inhibition of ERK1/2 and NF-κB signaling pathways prevents up-regulation of antioxidant and longevity-related gene expression induced by both estrogen and genistein [Borrás et al., 2005, 2006]. Although it is well known that NF-κB coordinates the induction of numerous pro-inflammatory genes, recent evidence establishes that activation of NF-κB modulates the expression of antioxidant genes, such as MnSOD and HO-1, and

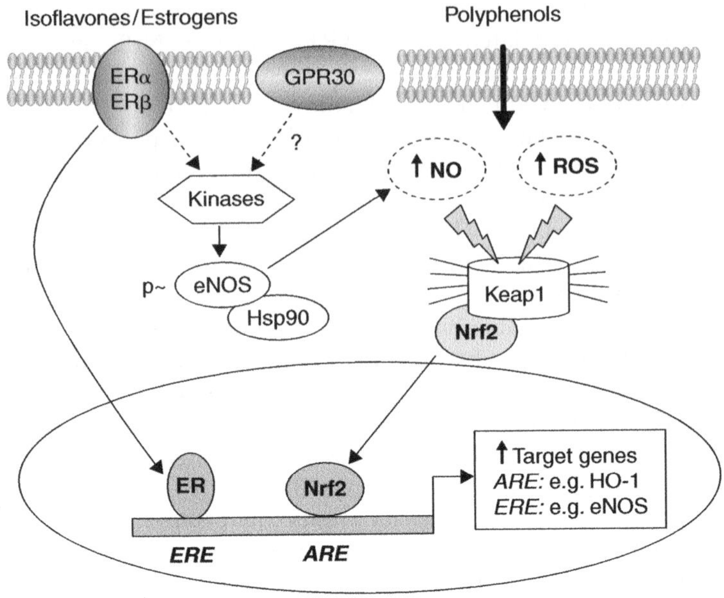

Figure 12.2 Nitric oxide as a mediator of polyphenol-induced transcriptional activation of antioxidant genes. Isoflavones, estrogens, and other polyphenols activate intracellular kinase cascades, leading to acute activation of eNOS and NO and/or ROS, generation. Increased NO, ROS, or peroxynitrite levels will modify cysteine residues on Keap-1 leading to nuclear accumulation of the redox-sensitive transcription factor Nrf2. Nrf2 binds to the antioxidant response element (ARE) or electrophile response element (EpRE) in the promoter region of target genes (e.g., phase II and antioxidant enzymes NQO1, HO-1) while estrogen receptors bind to estrogen response element (ERE) to enhance eNOS expression. Induction of other antioxidant genes such as MnSOD may involve rapid phosphorylation of ERK1/2 and I-κB and translocation of the p50 subunit of NF-κB to the nucleus and transactivation of MnSOD expression. (Adapted from Mann et al. [2007].)

confers protection against apoptosis by up-regulating antiapoptotic proteins [Collins and Cybulsky, 2001]. In this context, induction of HO-1 by the dietary polyphenol resveratrol is abrogated by inhibitors of NF-κB activation or IκB phosphorylation [Baur et al., 2006] and deletion of NF-κB binding sites in the HO-1 promoter reduces transcriptional activity [Juan et al., 2005]. Modulation of NF-κB and Nrf2 signaling pathways by polyphenols via mitogen-activated protein kinases can lead to transcriptional activation of antioxidant defense genes [Chen et al., 2005; Andreadi et al., 2006].

CONCLUSIONS AND FUTURE PERSPECTIVES

The molecular mechanisms by which estrogens and the isoflavones genistein, daidzein, and equol up-regulate antioxidant gene expression in different vascular cell types merit further investigation. Ongoing studies of the non-genomic and genomic actions of soy-derived isoflavones on vascular reactivity and antioxidant gene expression in human endothelial cells and rodent blood vessels [Mahn et al., 2005; Knock et al., 2006; Joy et al., 2006] have provided an insight into the molecular mechanisms underlying the increased risk of coronary heart disease in postmenopausal women and men consuming a Western diet. These findings have important implications for the design of alternative estrogen receptor modulators in the treatment of postmenopausal women, as studies in rodents have established that improved antioxidant and eNOS gene expression in response to soy isoflavones is associated with decreased ROS production, improved endothelial function, and lower blood pressure in vivo. These findings suggest that patients at risk of coronary heart disease may potentially benefit from an increased intake of dietary soy isoflavones leading to an increased expression and/or activity antioxidant enzymes.

In summary, dietary polyphenols may counteract oxidative stress in vascular and inflammatory diseases [Rahman et al., 2006] by modulating key redox-sensitive gene transcription via NF-κB and Nrf2/ARE [Siow et al., 2007] signaling pathways. In view of the inherent genetic variability and the potential for noncompliance in human supplementation trials, further studies in genetically modified mice should provide important insights into the mechanisms by which dietary soy isoflavones and other polyphenols regulate eNOS and antioxidant gene expression in endothelial and other vascular cell types, including smooth muscle cells, monocytes, and platelets.

ACKNOWLEDGMENTS

We gratefully acknowledge research funding from the British Heart Foundation (FS/9075, FS/05/005), Biotechnology and Biological Sciences Research Council (BBS/S/K/204/11207), Medical Research Council (G9820929), Heart Research UK (RG2489/04/08), EU COST Action B35, and Wing Yip & Brothers Charitable Foundation. We are indebted to our collaborators whose names appear in the cited publications.

REFERENCES

Adlercreutz H, Markkanen H, Watanabe S. 1993. Plasma concentrations of phyto-oestrogens in Japanese men. *Lancet* 342:1209–1210.

Akiyama T, Ishida J, Nakagawa S, Ogawara H, Watanabe S, Itoh N, Shibuya M, Fukami Y. 1987. Genistein, a specific inhibitor of tyrosine specific protein kinases. *J Biol Chem* 262:5592–5595.

Andreadi CK, Howells LM, Atherfold PA, Manson MM. 2006. Involvement of Nrf2, p38, B-Raf, and nuclear factor-κB, but not phosphatidylinositol 3-kinase, in induction of heme oxygenase-1 by dietary polyphenols. *Mol Pharmacol* 69:1033–1040.

Ansell PJ, Espinosa-Nicholas C, Curran EM, Judy BM, Philips BJ, Hannink M, Lubahn DB. 2004. In vitro and in vivo regulation of antioxidant response element-dependent gene expression by estrogens. *Endocrinology* 145:311–317.

Anter E, Chen K, Shapira OM, Karas RH, Keaney JF Jr. 2005. p38 mitogen activated protein kinase activates eNOS in endothelial cells by an estrogen receptor alpha-dependent pathway in response to black tea polyphenols. *Circ Res* 96:1072–1078.

Baranano DE, Rao M, Ferris CD, Snyder SH. 2002. Biliverdin reductase: A major physiologic cytoprotectant. *Proc Natl Acad Sci* 99:16093–16098.

Baur JA, Pearson KJ, Price NL, Jamieson HA, Lerin C, Kalra A, Prabhu VV, Allard JS, Lopez-Lluch G, Lewis K, Pistell PJ, Poosala S, Becker KG, Boss O, Gwinn D, Wang M, Ramaswamy S, Fishbein KW, Spencer RG, Lakatta EG, Le Couteur D, Shaw RJ, Navas P, Puigserver P, Ingram DK, de Cabo R, Sinclair DA. 2006. Resveratrol improves health and survival of mice on a high-calorie diet. *Nature* 444:337–342.

Bianco NR, Chaplin LJ, Montano MM. 2005. Differential induction of quinone reductase by phytoestrogens and protection against oestrogen-induced DNA damage. *Biochem J* 385:279–287.

Boo YC, Jo H. 2003. Flow-dependent regulation of endothelial nitric oxide synthase: Role of protein kinases. *Am J Physiol* 285:C499–C508.

Borrás, C, Gambini J, Gómez-Cabrera MC, Sastre J, Pallardó FV, Mann GE, Vina J. 2006. Genistein, a soy isoflavone, upregulates expression of antioxidant genes: involvement of estrogen receptors, ERK1/2 and NF-κB. *FASEB J* 20:E1–E6.

Borrás C, Gambini J, Gomez-Cabrera M, Sastre J, Pallardo FV, Mann GE, Viña J. 2005. 17β-estradiol upregulates longevity-related, antioxidant enzyme expression via the ERK1 and ERK2[MAPK]/ NF-κB cascade. *Aging Cell* 4:113–118.

Buckley BJ, Marshall ZM, Whorton AR. 2003. Nitric oxide stimulates Nrf2 nuclear translocation in vascular endothelium. *Biochem Biophys Res Commun* 307:973–979.

Cai H, Li Z, Davis ME, Kanner W, Harrison DG, Dudley SC. Jr 2003. Akt dependent phosphorylation of serine 1179 and mitogen-activated protein kinase kinase/extracellular signal-regulated kinase 1/2 cooperatively mediate activation of the endothelial nitric-oxide synthase by hydrogen peroxide. *Mol Pharmacol* 63:325–331.

Cassidy A, Brown JE, Hawdon A, Faughnan MS, King LJ, Millward J, Zimmer-Nechemias L, Wolfe B, Setchell KD. 2006. Factors affecting the bioavailability of soy isoflavones in humans after ingestion of physiologically relevant levels from different soy foods. *J Nutr* 136:45–51.

Chahoud G, Aude YW, Mehta JL. 2004. Dietary recommendations in the prevention and treatment of coronary heart disease: Do we have the ideal diet yet? *Am J Cardiol* 94:1260–1267.

Chambliss KL, Shaul PW. 2002. Estrogen modulation of endothelial nitric oxide synthase. *Endocrine Rev* 23:665–686.

Chen CY, Jang JH, Li MH, Surh YJ. 2005. Resveratrol upregulates heme oxygenase-1 expression via activation of NF-E2-related factor 2 in PC12 cells. *Biochem Biophys Res Commun* 331:993–1000.

Chin-Dusting JP, Boak L, Husband A, Nestel PJ. 2004. The isoflavone metabolite dehydroequol produces vasodilatation in human resistance arteries via a nitric oxide dependent mechanism. *Atherosclerosis* 176:45–48.

Clarkson TB. 2002. Soy, soy phytoestrogens and cardiovascular disease. *J Nutr* 132:566S–569S.

Collins T, Cybulsky MI. 2001. NF-κB: A pivotal mediator or innocent bystander in atherogenesis? *J Clin Invest* 107:255–264.

Dröge W. 2002. Free radicals in the physiological control of cell function. *Physiol Rev* 82:47–95.

Duffy SJ, Keaney JF Jr, Holbrook M, Gokce N, Swerdloff PL, Frei B, Vita JA. 2001. Short- and long-term black tea consumption reverses endothelial dysfunction in patients with coronary artery disease. *Circulation* 104:151–156.

Engler MB, Engler MM, Chen CY, Malloy MJ, Browne A, Chiu EY, Kwak HK, Milbury P, Paul SM, Blumberg J, Mietus-Snyder ML. 2004. Flavonoid-rich dark chocolate improves endothelial function and increases plasma epicatechin concentrations in healthy adults. *J Am Coll Nutr* 23:197–204.

Erdman JW Jr. 2000. AHA Science Advisory: Soy protein and cardiovascular disease: A statement for healthcare professionals from the Nutrition Committee of the AHA. *Circulation* 102:2555–2559.

Figtree GA, McDonald D, Watkins H, Channon KM. 2003. Truncated estrogen receptor alpha 46-kDa isoform in human endothelial cells: Relationship to acute activation of nitric oxide synthase. *Circulation* 107:120–126.

Figtree GA, Griffiths H, Lu YQ, Webb CM, MacLeod K, Collins P. 2000. Plant-derived estrogens relax coronary arteries in vitro by a calcium antagonistic mechanism. *J Am Coll Cardiol* 35:1977–1985.

Fisher ND, Hughes M, Gerhard-Herman M, Hollenberg NK. 2003. Flavanol-rich cocoa induces nitric-oxide-dependent vasodilation in healthy humans. *J Hypertens* 21:2281–2286.

Fleming I, Fisslthaler B, Dimmeler S, Kemp BE, Busse R. 2001. Phosphorylation of Thr(495) regulates Ca^{2+}/calmodulin-dependent endothelial nitric oxide synthase activity. *Circ Res* 88:E68–E75.

Foresti R, Hoque M, Bains S, Green CJ, Motterlini R. 2003. Haem and nitric oxide: Synergism in the modulation of the endothelial haem oxygenase-1 pathway. *Biochem J* 372:381–390.

Forte P, Kneale BJ, Milne E, Chowienczyk PJ, Johnston A, Benjamin N, Ritter JM. 1998. Evidence for a difference in nitric oxide biosynthesis between healthy women and men. *Hypertension* 32:730–734.

Fraga CG. 2005. Cocoa, diabetes, and hypertension: Should we eat more chocolate? *Am J Clin Nutr* 81:541–542.

Garcia-Cardena G, Fan R, Shah V, Sorrentino R, Cirino G, Papapetropoulos A, Sessa W. 1998. Dynamic activation of endothelial nitric oxide synthase by Hsp90. *Nature* 392:821–824.

Ghosh S, Gachhui R, Crooks C, Wu C, Lisanti MP, Stuehr DJ. 1998. Interaction between caveolin-1 and the reductase domain of endothelial nitric-oxide synthase. Consequences for catalysis. *J Biol Chem* 273:22267–22271.

Grassi D, Lippi C, Necozione S, Desideri G, Ferri C. 2005a. Short-term administration of dark chocolate is followed by a significant increase in insulin sensitivity and a decrease in blood pressure in healthy persons. *Am J Clin Nutr* 81:611–614.

Grassi D, Necozione S, Lippi C, Croce G, Valeri L, Pasqualetti P, Desideri G, Blumberg JB, Ferri C. 2005b. Cocoa reduces blood pressure and insulin resistance and improves endothelium-dependent vasodilation in hypertensives. *Hypertension* 46:398–405.

Hall WL, Formanuik NL, Harnpanich D, Cheung M, Talbot D, Chowienczyk PJ, Sanders TA. 2008. A meal enriched with soy isoflavones increases nitric oxide-mediated vasodilation in healthy postmenopausal women. *J Nutr* 138:1288–1292.

Hallund J, Bugel S, Tholstrup T, Ferrari M, Talbot D, Hall WL, Reimann M, Williams CM, Wiinberg N. 2006. Soya isoflavone-enriched cereal bars affect markers of endothelial function in postmenopausal women. *Br J Nutr* 95:1120–1126.

Haynes MP, Sinha D, Russell KS, Collinge M, Fulton D, Morales-Ruiz M, Sessa WC, Bender JR. 2000. Membrane estrogen receptor engagement activates endothelial nitric oxide synthase via the PI3-kinase-Akt pathway in human endothelial cells. *Circ Res* 87:677–682.

Heiss C, Schroeter H, Balzer J, Kleinbongard P, Matern S, Sies H, Kelm M. 2006. Endothelial function, nitric oxide, and cocoa flavanols. *J Cardiovasc Pharmacol* 47:S128–S135.

Heiss C, Kleinbongard P, Dejam A, Perre S, Schroeter H, Sies H, Kelm M. 2005. Acute consumption of flavanol-rich cocoa and the reversal of endothelial dysfunction in smokers. *J Am Coll Cardiol* 46:1276–1283.

Hwang J, Wang J, Morazzoni P, Hodis HN, Sevanian A. 2003. The phytoestrogen equol increases nitric oxide availability by inhibiting superoxide production: An antioxidant mechanism for cell-mediated LDL modification. *Free Radic Biol Med* 34:1271–1282.

Ishii T, Itoh K, Takahashi S, Sato H, Yanagawa T, Katoh Y, Bannai S, Yamamoto M. 2000. Transcription factor Nrf2 coordinately regulates a group of oxidative stress-inducible genes in macrophages. *J Biol Chem* 275:16023–16029.

Joy S, Siow RC, Rowlands DJ, Becker M, Wyatt AW, Aaronson PI, Coen CW, Kallo I, Jacob R, Mann GE. 2006. The isoflavone Equol mediates rapid vascular relaxation: Ca^{2+}-independent activation of endothelial nitric-oxide synthase/Hsp90 involving ERK1/2 and Akt phosphorylation in human endothelial cells. *J Biol Chem* 281:27335–27345.

Juan SH, Cheng TH, Lin HC, Chu YL, Lee WS. 2005. Mechanism of concentration-dependent induction of heme oxygenase-1 by resveratrol in human aortic smooth muscle cells. *Biochem Pharmacol* 69:41–48.

Kang KW, Choi SH, Kim SG. 2002. Peroxynitrite activates NF-E2-related factor 2/antioxidant response element through the pathway of phosphatidylinositol 3-kinase: The role of nitric oxide synthase in rat glutathione S-transferase A2 induction. *Nitric Oxide* 7:244–253.

Knock GA, Mahn K, Mann GE, Ward JP, Aaronson PI. 2006. Dietary soy modulates endothelium-dependent relaxation in aged male rats: Increased agonist-induced endothelium-derived hyperpolarising factor and basal nitric oxide activity. *Free Radic Biol Med* 41:731–739.

Kreijkamp-Kaspers S, Kok L, Bots ML, Grobbee DE, Lampe JW, van der Schouw YT. 2005. Randomized controlled trial of the effects of soy protein containing isoflavones on vascular function in postmenopausal women. *Am J Clin Nutr* 81:189–195.

Kreijkamp-Kaspers S, Kok L, Bots ML, Grobbee DE, van der Schouw YT. 2004. Dietary phytoestrogens and vascular function in postmenopausal women: A cross-sectional study. *J Hypertens* 22:1381–3188.

Lee BS, Heo J, Kim YM, Shim SM, Pae HO, Kim YM, Chung HT. 2006. Carbon monoxide mediates heme oxygenase 1 induction via Nrf2 activation in hepatoma cells. *Biochem Biophys Res Commun* 343:965–972.

Leikert JF, Rathel TR, Wohlfart P, Cheynier V, Vollmar AM, Dirsch VM. 2002. Red wine polyphenols enhance endothelial nitric oxide synthase expression and subsequent nitric oxide release from endothelial cells. *Circulation* 106:1614–1617.

Lissin LW, Cooke JP. 2000. Phytoestrogens and cardiovascular health. *J Am Coll Cardiol* 35:1403–1410.

Liu H, Colavitti R, Rovira II, Finkel T. 2005. Redox-dependent transcriptional regulation. *Circ Res* 97:967–974.

Liu D, Homan LL, Dillon JS. 2004. Genistein acutely stimulates nitric oxide synthesis in vascular endothelial cells by a cyclic adenosine 5′-monophosphate-dependent mechanism. *Endocrinology* 145:5532–5539.

Locher R, Emmanuele L, Suter PM, Vetter W, Barton M. 2002. Green tea polyphenols inhibit human vascular smooth muscle cell proliferation stimulated by native low-density lipoprotein. *Eur J Pharmacol* 434:1–7.

Mackenzie GG, Carrasquedo F, Delfino JM, Keen CL, Fraga CG, Oteiza PI. 2004. Epicatechin, catechin, and dimeric procyanidins inhibit PMA-induced NF-kappaB activation at multiple steps in Jurkat T cells. *FASEB J* 18:167–169.

Mahn K, Borrás C, Knock GA, Taylor P, Khan IY, Sugden D, Poston L, Ward JP, Sharpe RM, Viña J, Aaronson PI, Mann GE. 2005. Dietary soy isoflavone induced increases in antioxidant and eNOS gene expression lead to improved endothelial function and reduced blood pressure in vivo. *FASEB J* 19:1755–1777.

Mann GE, Rowlands DJ, Li FY, de Winter P, Siow RC. 2007. Activation of endothelial nitric oxide synthase by dietary isoflavones: Role of NO in Nrf2-mediated antioxidant gene expression. *Cardiovasc Res* 75:261–274.

Mendelsohn ME. 2002. Genomic and nongenomic effects of estrogen in the vasculature. *Am J Cardiol* 90:3F–6F.

Mendelsohn ME, Karas RH. 2005. Molecular and cellular basis of cardiovascular gender differences. *Science* 308:1583–1587.

Mishra SK, Abbot SE, Choudhury Z, Cheng M, Khatab N, Maycock NJ, Zavery A, Aaronson PI. 2000. Endothelium-dependent relaxation of rat aorta and main pulmonary artery by the phytoestrogens genistein and daidzein. *Cardiovasc Res* 46:539–546.

Muller M, Kersten S. 2003. Nutrigenomics: Goals and strategies. *Nat Rev Genet* 4:315–322.

Nestel P, Fujii A, Zhang L. 2006. An isoflavone metabolite reduces arterial stiffness and blood pressure in overweight men and postmenopausal women. *Atherosclerosis* 192:184–189.

Nestel PJ, Yamashita T, Sasahara T, Pomeroy S, Dart A, Komesaroff P, Owen A, Abbey M. 1997. Soy isoflavones improve systemic arterial compliance but not plasma lipids in menopausal and perimenopausal women. *Arterioscler Thromb Vasc Biol* 17:3392–3398.

Nguyen T, Sherratt PJ, Pickett CB. 2003. Regulatory mechanisms controlling gene expression mediated by the antioxidant response element. *Annu Rev Pharmacol Toxicol* 43:233–260.

Orshal JM, Khalil RA. 2004. Gender, sex hormones, and vascular tone. *Am J Physiol* 286:R233–R249.

Palmer RM, Ferrige AG, Moncada S. 1987. Nitric oxide release accounts for the biological activity of endothelium-derived relaxing factor. *Nature* 327:524–526.

Rahman I, Biswas SK, Kirkham PA. 2006. Regulation of inflammation and redox signaling by dietary polyphenols. *Biochem Pharmacol* 72:1439–1452.

Rimbach G, Boesch-Saadatmandi C, Frank J, Fuchs D, Wenzel U, Daniel H, Hall W, Weinberg P. 2008. Dietary isoflavones in the prevention of cardiovascular disease—A molecular perspective. *Food Chem Toxicol* 46:1308–1319.

Rivas M, Garay RP, Escanero JF, Cia P Jr, Cia P, Alda JO. 2002. Soy milk lowers blood pressure in men and women with mild to moderate essential hypertension. *J Nutr* 132:1900–1902.

Ruehlmann DO, Steinert JR, Valverde MA, Jacob R, Mann GE. 1998. Environmental estrogenic pollutants induce acute vascular relaxation by inhibiting L-type Ca^{2+} channels in smooth muscle cells. *FASEB J* 12:613–619.

Russell KS, Haynes MP, Caulin-Glaser T, Rosneck J, Sessa WC, Bender JR. 2000. Estrogen stimulates heat shock protein 90 binding to endothelial nitric oxide synthase in human vascular endothelial cells. Effects on calcium sensitivity and NO release. *J Biol Chem* 275:5026–5030.

Ryter SW, Alam J, Choi AM. 2006. Heme oxygenase-1/carbon monoxide: From basic science to therapeutic applications. *Physiol Rev* 86:583–650.

Sacks FM, Lichtenstein A, Van HL, Harris W, Kris-Etherton P, Winston M. 2006. Soy protein, isoflavones, and cardiovascular health: An American Heart Association Science Advisory for professionals from the Nutrition Committee. *Circulation* 113:1034–1044.

Sader MA, Celermajer DS. 2002. Endothelial function, vascular reactivity and gender differences in the cardiovascular system. *Cardiovasc Res* 53:597–604.

Schroeter H, Heiss C, Balzer J, Kleinbongard P, Keen CL, Hollenberg NK, Sies H, Kwik-Uribe C, Schmitz HH, Kelm M. 2006. (−)-Epicatechin mediates beneficial effects of flavanol-rich cocoa on vascular function in humans. *Proc Natl Acad Sci* 103:1024–1029.

Setchell KD, Cassidy A. 1999. Dietary isoflavones: Biological effects and relevance to human health. *J Nutr* 129:758S–767S.

Setchell KD, Clerici C, Lephart ED, Cole SJ, Heenan C, Castellani D, Wolfe BE, Nechemias-Zimmer L, Brown NM, Lund TD, Handa RJ, Heubi JE. 2005. S-equol, a potent ligand for estrogen receptor beta, is the exclusive enantiomeric form of the soy isoflavone metabolite produced by human intestinal bacterial flora. *Am J Clin Nutr* 81:1072–1079.

Shaul PW, Smart EJ, Robinson LJ,German Z,Yuhanna IS,Ying Y, Anderson RG, Michel T. 1996. Acylation targets endothelial nitric-oxide synthase to plasmalemmal caveolae. *J Biol Chem* 271:6518–6522.

Siow RC, Li FY, Rowlands DJ, de Winter P, Mann GE. 2007. Cardiovascular targets for estrogens and phytoestrogens: Transcriptional regulation of nitric oxide synthase and antioxidant defense genes. *Free Radic Biol Med* 42:909–925.

Squadrito F, Altavilla D, Crisafulli A, Saitta A, Cucinotta D, Morabito N, D'Anna R, Corrado F, Ruggeri P, Frisina N, Squadrito G. 2003. Effect of genistein on endothelial function in postmenopausal women: A randomized, double-blind, controlled study. *Am J Med* 114:470–476.

Squadrito F, Altavilla D, Morabito N, Crisafulli A, D'Anna R, Corrado F, Ruggeri P, Campo GM, Calapai G, Caputi AP, Squadrito G. 2002. The effect of the phytoestrogen genistein on plasma nitric oxide concentrations, endothelin-1 levels and endothelium dependent vasodilation in postmenopausal women. *Atherosclerosis* 163:339–347.

Stocker R, Keaney JF Jr. 2004. Role of oxidative modifications in atherosclerosis. *Physiol Rev* 84:1381–1478.

Strehlow K, Rotter S, Wassmann S, Adam O, Grohe C, Laufs K, Böhm M, Nickenig G. 2003. Modulation of antioxidant enzyme expression and function by estrogen. *Circ Res* 93:170–177.

Takahashi S, Mendelsohn ME. 2003. Synergistic activation of endothelial nitric-oxide synthase (eNOS) by HSP90 and Akt: calcium-independent eNOS activation involves formation of an HSP90-Akt-CaM-bound eNOS complex. *J Biol Chem* 278:30821–30827.

Teede HJ, McGrath BP, DeSilva L, Cehun M, Fassoulakis A, Nestel PJ. 2003. Isoflavones reduce arterial stiffness: A placebo-controlled study in men and postmenopausal women. *Arterioscler Thromb Vasc Biol* 23:1066–1071.

Teede HJ, Dalais FS, Kotsopoulos D, Liang YL, Davis S, McGrath BP. 2001. Dietary soy has both beneficial and potentially adverse cardiovascular effects: A placebo-controlled study in men and postmenopausal women. *J Clin Endocrinol Metab* 86: 3053–3060.

Vera R, Sanchez M, Galisteo M, Villar IC, Jimenez R, Zarzuelo A, Pérez-Vizcaíno F, Duarte J. 2007. Chronic administration of genistein improves endothelial dysfunction in spontaneously hypertensive rats: Involvement of eNOS, caveolin and calmodulin expression and NADPH oxidase activity. *Clin Sci (Lond)* 112:183–191.

Vergne S, Titier K, Bernard V, Asselineau J, Durand M, Lamothe V, Potier M, Perez P, Demotes-Mainard J, Chantre P, Moore N, Bennetau-Pelissero C, Sauvant P. 2006. Bioavailability and urinary excretion of isoflavones in humans: Effects of soy-based supplements formulation and equol production. *J Pharm Biomed Anal* 43:1488–1494.

Viña J, Sastre J, Pallardó FV, Gambini J, Borrás C. 2008. Modulation of longevity-associated genes by estrogens or phytoestrogens. *Biol Chem* 389:273–277.

Visioli F, Keaney JF, Halliwell B. 2000. Antioxidants and cardiovascular disease; panaceas or tonics for tired sheep? *Cardiovasc Res* 47:409.

Vlachopoulos C, Aznaouridis K, Alexopoulos N, Economou E, Andreadou I, Stefanadis C. 2005. Effect of dark chocolate on arterial function in healthy individuals. *Am J Hypertens* 18:785–791.

Walker HA, Dean TS, Sanders TA, Jackson G, Ritter JM, Chowienczyk PJ. 2001. The phytoestrogen genistein produces acute nitric oxide dependent dilation of human forearm vasculature with similar potency to 17beta-estradiol. *Circulation* 103:258–262.

Wallerath T, Li H, Godtel-Ambrust U, Schwarz PM, Forstermann U. 2005. A blend of polyphenolic compounds explains the stimulatory effect of red wine on human endothelial NO synthase. *Nitric Oxide* 12:97–104.

Wallerath T, Deckert G, Ternes T, Anderson H, Li H, Witte K, Förstermann U. 2002. Resveratrol, a polyphenolic phytoalexin present in red wine, enhances expression and activity of endothelial nitric oxide synthase. *Circulation* 106:1652–1658.

Williams RJ, Spencer JP, Rice-Evans C. 2004. Flavonoids: Antioxidants or signalling molecules? *Free Radic Biol Med* 36:838–849.

Xu JW, Ikeda K, Yamori Y. 2004. Genistein inhibits expressions of NADPH oxidase p22phox and angiotensin II type 1 receptor in aortic endothelial cells from stroke-prone spontaneously hypertensive rats. *Hypertens Res* 27:675–83.

13 Bioavailability and Metabolism of Resveratrol

CRISTINA ANDRES-LACUEVA[1], MIREIA URPI-SARDA[1], RAUL ZAMORA-ROS[1], AND ROSA M. LAMUELA-RAVENTOS[1]

[1]Department of Nutrition and Food Science, XaRTA, INSA, Pharmacy Faculty, University of Barcelona, Barcelona, Spain

CONTENTS

INTRODUCTION

The essential chemical structure of stilbenes is the *trans*-1,2-diphenylethylene. The parent molecule of this group is the resveratrol (3,4′,5- trihydroxystilbene) that exists as two geometric isomers: *cis*- (*Z*) and *trans*- (*E*). The *trans*-form can undergo isomerization to the *cis*-form when exposed to ultraviolet irradiation. Piceid (resveratrol glucoside) is the major resveratrol derivative in plants. Other stilbenes in the vegetal kingdom are pterostilbene, piceatannol, astringin, and viniferins (Fig. 13.1).

Resveratrol and piceid are mainly present in grape and grape products, and its composition is affected by grape variety, maturity degree at harvest, fungal

Plant Phenolics and Human Health: Biochemistry, Nutrition, and Pharmacology, Edited by Cesar G. Fraga.

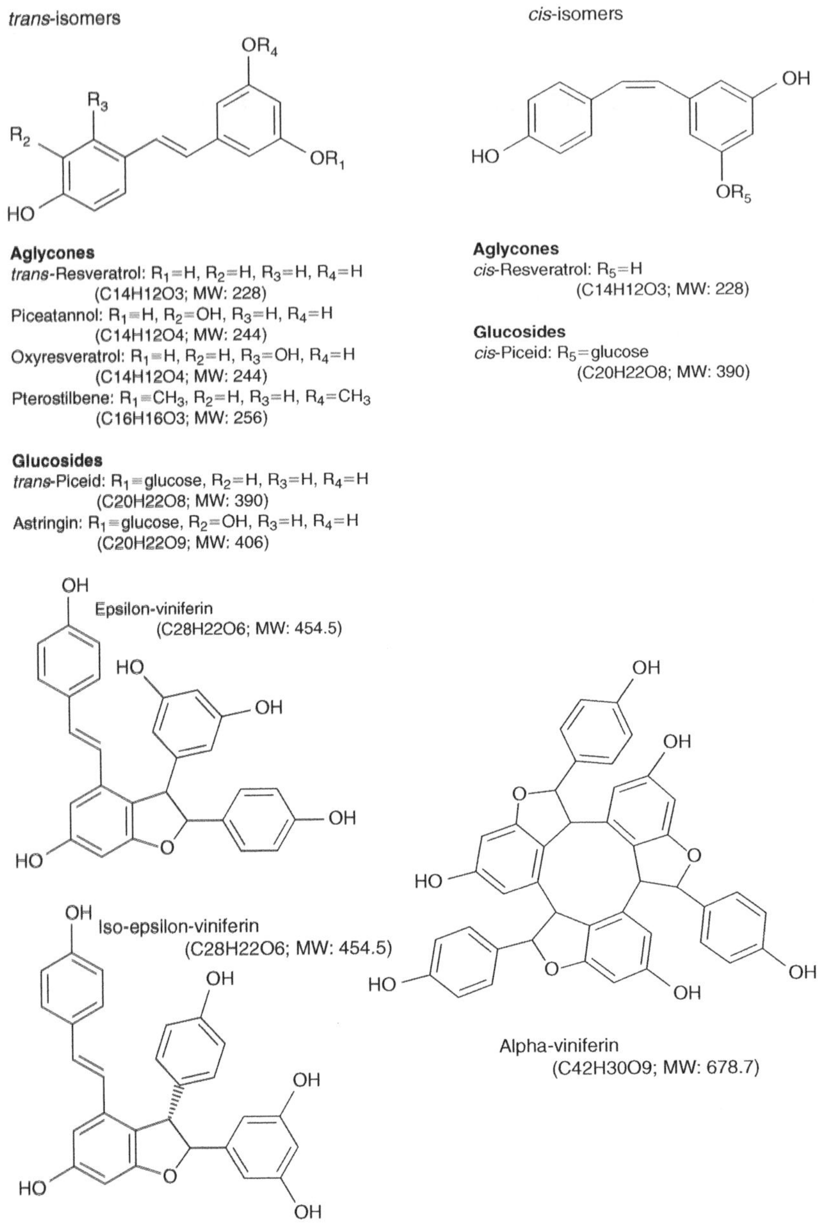

Figure 13.1 Chemical structure of stilbenes present in foods.

stress, climate, soil characteristics (*terroir*), and wine-making process as well as technology [de Andrés-de Prado et al., 2007; González-Barrio et al., 2006; Romero-Perez et al., 2001]. Other slight food sources of stilbenes are peanuts, pistachios, and berries such as bilberry, blueberry, and cranberry [Burns et al.,

2002; Rimando et al., 2004; Sobolev and Cole, 1999; Tokusoglu et al., 2005]. In an adult Spanish cohort resveratrol and piceid, *trans* and *cis*-forms, were evaluated. Estimated median and mean of resveratrol and piceid were 100 and 933 μg/day, respectively, of which 98.4, 1.6, and less than 0.1% come from wines, grape and grape juice, and peanuts, pistachios, and berries, respectively [Zamora-Ros et al., 2008].

IN VITRO AND EX VIVO STUDIES

Several studies have investigated the absorption, transport, and metabolism of resveratrol in vitro and ex vivo. They are summarized in Table 13.1.

Caco-2 cells and isolated small intestine are models of basic nutrition that contributed to the understanding of resveratrol absorption and bioavailability. While the Caco-2 absorption model is a well-defined cellular in vitro system based on a human colonic adenocarcinoma cell line, the isolated small intestine model is nearer to in vivo conditions and is also simpler to handle. It also avoids the methodological problems of in vivo perfusion models [Barthe et al., 1998, 1999].

The Caco-2 cells incubated in vitro with *trans*-resveratrol (5–40 μM) showed a concentration-dependent transcellular absorption up to 3 h, with a linear rate for the first hour [Kaldas et al., 2003]. At 3 h of incubation, the concentration of resveratrol reached a plateau [Kaldas et al., 2003; Maier-Salamon et al., 2006]. However, incubations with higher amounts of this polyphenol (40 μM) increased their concentration in the Caco-2 cells, so there was no saturation of the transport systems [Kaldas et al., 2003]. The cellular uptake of *trans*-piceid was also investigated in Caco-2 cells and its transport was slower and about 20% of the resveratrol aglycone [Henry et al., 2005]. *trans*-Resveratrol crosses the apical membrane of the Caco-2 cells using a passive transport, whereas *trans*-piceid seems to use both the active transporter sodium-dependent glucose transporter 1 (SGLT1) and the multidrug resistance protein 2 (MRP2) [Henry et al., 2005]. After piceid absorption, it can be hydrolized to resveratrol by the cytosolic-β-glucosidase [Henry-Vitrac et al., 2006]. Another possible pathway to absorb piceid is through its deglycosilation by membrane-bound lactase phlorizin hydrolase, since it then goes across the apical membrane as resveratrol [Henry-Vitrac et al., 2006]. The basolateral to apical efflux also occurs at similar concentrations to apical and basolateral efflux [Henry-Vitrac et al., 2006; Kaldas et al., 2003]. After resveratrol absorption, this is conjugated rapidly in intestinal cells. At 10 μM concentration, *trans*-3-*O*-sulfate was the main metabolite of resveratrol, however, its formation drastically decreased at higher resveratrol concentrations (200 μM), possibly due to saturation or inhibition of metabolism at higher stilbene concentrations [Maier-Salamon et al., 2006]. Glucuronidate forms such as *trans*-resveratrol-4′-*O*-glucuronide and *trans*-resveratrol-3-*O*-glucuronide were also released at less levels than sulfate forms (Fig.13.2) [Maier-Salamon et al., 2006].

Table 13.1 Metabolism of Resveratrol in Vitro and Ex Vivo Models[a]

Tissue	Dose, Source (time)	Metabolites (C found)	Reference
Human erythrocytes (1.6×10^9)	21.9 nmol *t*-Resv (0.25 h)	**t*-Resv 10.0 ± 1.7 nmol/10^9	Blache et al., 1997
Rat erythrocytes (1.6×10^9)		**t*-Resv 10.8 ± 2.2 nmol/10^9	
Rat platelets (10^9)		**t*-Resv 2.2 ± 1.2 nmol/10^9	
Human LDL (0.5 mg/mL)	17.5 μM *t*-Resv (0.5 h)	**t*-Resv 3.8 ± 0.9 nmol/mg protein	
Jejunum and ileum of Sprague–Dawley male rats	200 μM Resv (1.5 h)	Resv (0.03 nmol/cm jejunum) Gluc (1.19 nmol/cm jejunum)	Kuhnle et al., 2000
Small intestine of male Sprague–Dawley rats	28, 34, 57 μM Resv (1 h)	Vascular effluent: Gluc (16.8%), Resv (3.4%), Sulf (0.3%) Luminal effluent: Gluc (11.2%), Resv (39.7%), Sulf (3.0%) Intestinal tissue: Gluc (0.1%), Resv (1.5%), Sulf (0.3%)	Andlauer et al., 2000
Human liver microsomes	1 mM *cis* and *t*-Resv	*c*-3-Gluc (+), *t*-3-Gluc (+), *c*-4′-Gluc (-), *t*-4′-Gluc (-)	Aumont et al., 2001
Rat hepatocytes	20 μM *t*- Resv (1 h)		Asensi et al., 2002
Human liver microsomes	5 mM Resv (1 h)	Free Resv	Yu et al., 2002
Human hepatocytes	0.1 mL of 0.1 mM *t*-resveratrol (4 h)	*t*-3-Gluc, *t*-4′-Gluc, *c*-3-Gluc, *t*-3-Sulf	
Caco-2 cells	5–40 μM Resv (6 h)	Resv (200–4000 pmol)	Kaldas et al., 2003
Caco-2 cells	150–300 μM *t*-Resv and *t*-piceid (0.03–0.5 h)	*t*-Resv > *t*-piceid	Henry et al., 2005
Caco-2 cells	10–200 μM Resv	*t*-4′-Gluc, *t*-3-Gluc, *t*-3-Sulf	Maier-Salamon et al., 2006
Human liver microsomes	500 μM *t*-Resv (5 h)	3-Gluc > 4′-Gluc	Brill et al., 2006
Human intestional microsomes		3-Gluc < 4′-Gluc	

[a] Resv: resveratrol; Gluc: glucuronide; Sulf: sulfate; *t*-3-Gluc: *trans*-resveratrol-3-*O*-glucuronide; *t*-4′-Gluc: *trans*-resveratrol-4′-*O*-glucuronide; *c*-4′-Gluc: *cis*-resveratrol-4′-*O*-glucuronide; *c*-3-Gluc: *cis*-resveratrol-3-*O*-glucuronide, *t*-3-Sulf: *trans*-resveratrol-3-sulfate.

*Normalized values.

INTESTINAL METABOLISM

trans-Isomers

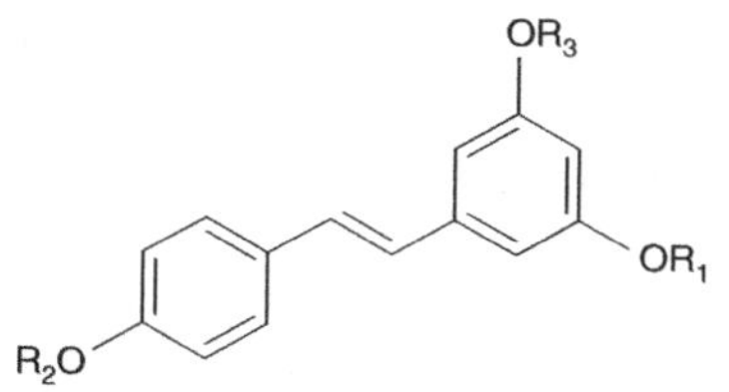

cis-Isomers

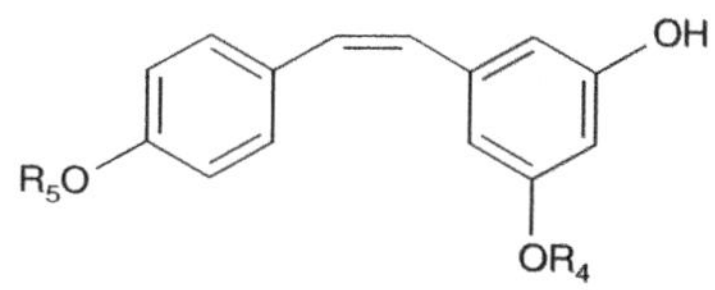

***trans*-Isomers**

trans-Resveratrol-3-O-glucuronide: R_1=glucuronic acid, R_2=H, R_3=H ($C_{20}H_{20}O_9$; MW: 404)
trans-Resveratrol-4′-O-glucuronide: R_1=H, R_2=glucuronic acid, R_3=H, ($C_{20}H_{20}O_9$; MW: 404)
trans-Resveratrol-3,4′-diglucuronide: R_1=glucuronic acid, R_2=glucuronic acid, R_3=H ($C_{26}H_{28}O_{15}$; MW: 580)
trans-Resveratrol-3-sulfate: R_1=SO_3H, R_2=H ($C_{14}H_{12}O_6S$; MW: 308)
trans-Resveratrol-4′-sulfate: R_1=H, R_2=SO_3H, R_3=H ($C_{14}H_{12}O_6S$; MW: 308)
trans-Resveratrol-3,4′-sulfate: R_1=SO_3H, R_2=SO_3H, R_3=H ($C_{14}H_{12}O_9S_2$; MW: 388)
trans-Resveratrol-3,5′-disulfate: R_1=SO_3H, R_2=H, R_3=SO_3H ($C_{14}H_{12}O_9S_2$; MW: 388)
trans-Resveratrol-3,5,4′-trisulfate: R_1=SO_3H, R_2=SO_3H, R_3=SO_3H ($C_{14}H_{12}O_{12}S_3$; MW: 468)

***cis*-Isomers**

cis-Resveratrol-3-O-glucuronide: R_4=glucuronic acid, R_5=H ($C_{20}H_{20}O_9$; MW: 404)
cis-Resveratrol-4′-O-glucuronide: R_4=H, R_5=glucuronic acid ($C_{20}H_{20}O_9$; MW: 404)
cis-Resveratrol-3-4′-diglucuronide: R_4=glucuronic acid, R_5=glucuronic acid ($C_{26}H_{28}O_{15}$; MW: 580)
cis-Resveratrol-3-sulfate: R_4=SO_3H, R_5=H ($C_{14}H_{12}O_6S$; MW: 308)
cis-Resveratrol-4′-sulfate: R_4=H, R_5=SO_3H ($C_{14}H_{12}O_6S$; MW: 308)
cis-Resveratrol-3,4′-disulfate: R_4=SO_3H, R_5=SO_3H ($C_{14}H_{12}O_6S_2$; MW: 388)

MICROBIAL METABOLISM

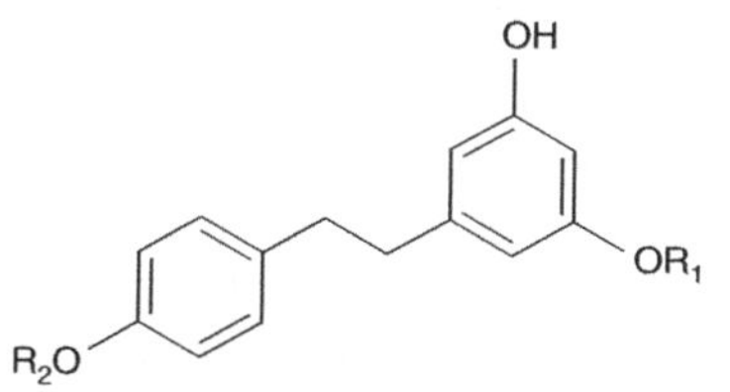

Dihydroresveratrol: R_1=H, R_2=H, ($C_{14}H_{14}O_3$; MW: 230)
Dihydroresveratrol-glucuronide: R_1=H or glucuronic acid, R_2=glucuronic acid or H ($C_{20}H_{22}O_9$; MW: 406)
Dihydroresveratrol-sulfate: R_1=H or SO_3H, R_2=SO_3H or H ($C_{14}H_{14}O_6S$; MW: 310)

Figure 13.2 Chemical structures of resveratrol metabolites.

Maier-Salomon et al. [2006] showed the high influence of metabolized resveratrol on the transepithelial transport of resveratrol and its intracellular accumulation. At 10 μM concentration in the Caco-2 cells, resveratrol was 84, 8, and 12% conjugated, transported, and accumulated, respectively. Whereas at higher doses (200 μM) resveratrol was 8, 26, and 61% conjugated, transported, and accumulated, respectively. Moreover, Kaldas et al. [2003] also found a resveratrol accumulation higher than 35-fold when transported through the Caco-2 cells, which suggests enterocytes as a major target site for this polyphenol.

Perfusion of 200 μM *trans*-resveratrol in isolated jejunum and ileum only transferred 6 and 2% of the total amount of resveratrol available, respectively. Resveratrol passed across jejunum as a glucuronide form (1.1 nmol/cm) and in minor amounts as free resveratrol (0.03 nmol/cm) [Kuhnle et al., 2000]. In a single-pass perfusion at low concentrations (28, 34, or 57 μM of *trans*-resveratrol) in the vascular and the luminar side, tissue and blood vessels of rat small intestine, 54, 21, 2, and 0.1% of the doses were recovered, respectively. In isolated luminal small intestine, the stilbene compounds found were free resveratrol (74%), 2-glucuronides (21%), and a sulfate (6%). Whereas in isolated vascular small intestine, the main metabolites were the glucuronidate forms (82%), free resveratrol (17%), and the sulfate form (1.5%). No differences were found in the recoveries using several different doses of resveratrol [Andlauer et al., 2000]. *trans*-Resveratrol was able to bind to human and rat erythrocytes, rat platelets, and low-density lipoprotein (LDL) as free resveratrol after in vitro incubations, confirming that resveratrol can diffuse throughout the body by means of its incorporation into blood cells and lipoproteins [Blache et al., 1997].

Conjugation forms of resveratrol can occur in both enterocytes and hepatocytes. After intestinal microsome incubation with 500 μM *trans*-resveratrol, only glucuronides were formed, in which *trans*-resveratrol-4′-*O*-glucuronide was more abundant than *trans*-resveratrol-3-*O*-glucuronide. In comparison with liver ability to metabolize, glucuronides were formed at higher levels (up to 10-fold) by the intestinal microsomes than the liver microsomes [Brill et al., 2006]. Likewise in liver microsomes, only glucuronidate forms and aglycones were observed after incubation with *trans*- and *cis*-resveratrol. Glucuronidation in liver was stereoselective (*cis*-isomer was 5 to 10-fold faster than *trans*-isomer) at both positions and regioselective (3-position was greater than 4′-position) for both isomers [Aumont et al., 2001]. Moreover Aumont et al. [2001] and Brill et al. [2006] found that resveratrol was only glucuronidized by the UDP-glucuronosyltransferase family 1A (UGT1A). In particular, UGT1A1 and UGT1A9 were mainly involved in the formation of *trans*-resveratrol-4′-*O*-glucuronide and *trans*-resveratrol-3-*O*-glucuronide, respectively. Likewise, in biopsies of human liver it was found that resveratrol is a better substrate for glucuronosyl transferase (K_m 0.15 mM) [de Santi et al., 2000a] than sulfotransferases (K_m 0.60 microM)) [de Santi et al., 2000b].

trans-Resveratrol was rapidly metabolized in a dose-dependent manner in rat hepatocytes, approximately 80 and 100% of the resveratrol (20 μM) incubated were conjugated at 20 min and at 1 h, respectively [Asensi et al., 2002]. In rat hepatocytes the main metabolite formed was *trans*-resveratrol-3-*O*-sulfate followed by *trans*-resveratrol-3-*O*-glucuronide. In contrast, in human hepatocytes glucuronides, that is, *trans*-resveratrol-4′-*O*-glucuronide, *trans*-resveratrol-3-*O*-glucuronide, and *cis*-resveratrol-3-*O*-glucuronide, were more abundant than sulfates (*trans*-resveratrol-3-*O*-sulfate) [Yu et al., 2002]. Based on these results, sulfate forms seem to be a minor human hepatic metabolite.

In summary, resveratrol absorption is higher as aglycon than piceid, although piceid can be hydrolyzed by β-glucosidases in the enterocytes [Henry-Vitrac et al., 2006]. Enterocytes metabolize resveratrol up to a certain extent. At low concentrations it seems that there is mainly a sulfate pathway, but at higher amounts glucuronidate forms become more abundant [Maier-Salamon et al., 2006]. Finally, the accumulation of resveratrol in Caco-2 cells suggests that enterocytes is an important target site [Kaldas et al., 2003]. Resveratrol is transported into the bloodstream bounded to blood cells and lipoproteins [Blache et al., 1997]. Hepatocytes also metabolize resveratrol to facilitate its excretion, mainly to glucuronides and, in fewer amounts, to sulfates [Asensi et al., 2002; Yu et al., 2002].

METABOLISM OF RESVERATROL IN VIVO IN ANIMAL MODELS

There is an extensive amount of literature on the in vitro activities of resveratrol. Bertelli et al. [1996] were the first to study the bioavailability of resveratrol in rats. They assessed analysis of plasma and tissues from rats that were administered red wine in a single dose or a regular dose for 15 day. Results indicated that resveratrol is quickly absorbed with a maximum peak at 1 h in plasma as well as in liver and kidney, although in the heart it peak at 2 h. Values obtained after regular consumption of red wine were higher than without the regular consumption in the different organs studies, in particular in the liver. The kidney seemed to be the main route of excretion (Table 13.2) [Bertelli et al., 1996].

Soleas et al. [2001a] performed experiments using *trans*-resveratrol radiolabeled with [^{3}H] in a fixed position in the first benzene ring. Male Wistar rats were administered by gavage 120 nCi as part of three different matrices: 10% (v/v) ethanol, a homogenized vegetable cocktail, or white grape juice. Short-term experiments at 2 h and longer-term experiments at 24 h were carried out. Over a 24-h period only trace amounts of resveratrol (<1%) were detected in tissues such as liver, kidney, heart, or spleen. However, it appeared that 77–80% of *trans*-resveratrol could be absorbed in the rat intestine, with no differences among the three food matrices when absorption was measured as the difference between the dose of radioactivity given and the recovered radioactivity in the stool (24-h feces plus colonic contents and colon). Nevertheless, values ranging from 49 to 61% were present in the urine, regardless the matrix. This 25% difference could be accounted for by excretion via sweat and the respiratory system, metabolism to CO_2, and accumulation in fat-rich tissues such as the brain and nervous system [Soleas et al., 2001a]. Short-term experiments (2 h) showed that over the first 30 min after administration, a significant amount of radioactivity accumulated blood and serum. The radioactivity remained at similar levels during the following 90 min. These observations included the parent compound as well as the metabolites (Table 13.3).

Experiments with unlabeled resveratrol were also conducted in rats after administration of 5 mg of resveratrol. *trans*-Resveratrol concentrations appeared in serum at 15 min, peaked at 30 min, and declined abruptly over the next 30 min. Soleas et al. [2001a] also measured the competition of several polyphenols

Table 13.2 Tissue Distribution in Animal Models[a]

Species	Source and Dose	Administration	Mean (SD) nmol/g												Plasma (μmol/l)	Feces (%)	Urine (%)	Time	Reference
			Brain	Lung	Liver	Kidney	Heart	Spleen	Gastrointestinal tract	Small Intestine	Stomach	Testis	Colonic Mucosae	Bile					
6 Wistar rats	4 ml red wine (0.09 mg/kg)	Intragastric	–	–	0.09	–	–	–	–	–	–	–	–	–	–	–	–	0.5h	Bertelli et al. 1996
6 Wistar rats			–	–	–	0.09	–	–	–	–	–	–	–	–	0.09	–	–	1h	
6 Wistar rats			–	–	–	–	0.009	–	–	–	–	–	–	–	–	–	–	2h	
36 Wistar rats	2 ml red wine/day (0.04 mg/kg/day) during 15 days	Intragastric			0.23 (0.006)	0.19 (0.005)	0.01 (0.001)	–	–	–	–	–	–	–	0.03 (0.002)	–	0.29 (0.007)	24h	
5 C57BL/6J male mice	20 mg/kg *t*-Resv	Intragastric	0.50 (0.35)	0.40 (0.23)	1.03 (0.80)	0.17 (0.06)		–	–	–	–	–	–	–	1.7 (0.9)	–	–	5 min	Asensi et al. 2002
								–	–	–	–	–	–	–		–	2% Gluc	24 h	
5 Wistar male rats		Intragastric	0.11 (0.05)	0.10 (0.05)	0.45 (0.11)	0.06 (0.03)		–	–	–	–	–	–	–	1.2 (0.4)	–	–	5 min	
								–	–	–	–	–	–	–		–	Gluc	24 h	
5 ESD NZW male rabbits		Intragastric	0.08 (0.03)	0.62 (0.17)	0.35 (0.14)	0.05 (0.02)		–	–	–	–	–	–	–	0.7 (0.5)	–	–	5 min	

Table 13.2 *(Continued)*

Species	Source and Dose	Administration	Mean (SD) nmol/g												Plasma (µmol/l)	Feces (%)	Urine (%)	Time	Reference
			Brain	Lung	Liver	Kidney	Heart	Spleen	Gastrointestinal tract	Small intestine	Stomach	Testis	Colonic Mucosae	Bile					
3 Mice per time point	240 mg/kg *t*-Resv	Intragastric	1.2 (0.1)	50 (25)	51 (15)	16 (10)	75 (27)	–	–	960 (100)	–	–	30 (5)	–	32 (14)	–	–	10 min	Sale et al. 2004
	240 mg/kg *t*-3,4,5,4′-tetramethoxystilbene	Intragastric	5 (1)	11 (0)	8 (3)	11 (4)	10 (5)	–	–	7600 (6500)	–	–	330 (300) 30 min	–	5 (2)	–	–	10 min	
10 Male Wistar rats	300 mg/kg/day *t*-Resv during 8 weeks	Oral	–	–	triSulf; n.d. diSulf: n.d. 3-Sulf: 1.92 (0.78) 4′-Sulf: 5.75 (2.08) 3-Gluc: 1.58 (0.59) Resv: 3.20 (1.01)	triSulf: n.d. diSulf: 5.00 (1.31) 3-Sulf: 1.10 (0.36) 4′-Sulf: n.d. 3-Gluc: 6.71 (1.49) Resv: n.d.		–	–	–	–	–	–	–	TriSulf: 7.0 (2.6) 3,4′-diSulf: 19.23 (6.0) 3,5-diSulf: 3.30 (0.8) 3-Sulf: 1.20 (0.3) 3-Gluc: 7.75 (2.2) Resv: n.d.	TriSulf: n.d. 3,4′-diSulf: 3.42% 3,5-diSulf: 0.80% 3-Sulf: 5.68% 4′-Sulf: 3.83% 3-Gluc: n.d. Resv: 17.2%	TriSulf: 1.42% 3,4′-diSulf: 3.5% 3,5-diSulf: 10% 3-Sulf: 5.93% 4′-Sulf: n.d. 3-Gluc: 28.6% Resv: 3.99%	24 h	Wenzel et al. 2005
10 Male Wistar rats	50 mg/kg/day *t*-Resv during 8 weeks	Oral	–	–	n.d.	n.d.		–	–	–	–	–	–	–	n.d.	TriSulf: n.d. 3,4′-diSulf: n.d. 3,5-diSulf: n.d. 3-Sulf: 1.73% 4′-Sulf: n.d. 3-Gluc: n.d. Resv: 13.1%	TriSulf: n.d. 3,4′-diSulf: 0.56% 3,5-diSulf: 0.26% 3-Sulf: 0.60% 4′-Sulf: n.d. 3-Gluc: 8.90% Resv: 5.03%		
18 Male Wistar rats	50 mg/kg Piceid	Oral	15.56 (7.3)	26.72 (9.9)	11.46 (6.4)	6.62 (3.1)	1.28 (0.7)	71.87 (35.4)	-	278.60 (76.4)	432.78 (198.6)	13.60 (6.2)	–	–	–	–	–	10 min	Lv et al. 2006

Table 13.2 (Continued)

Species	Source and Dose	Administration	Mean (SD) nmol/g												Plasma (µmol/l)	Feces (%)	Urine (%)	Time	Reference
			Brain	Lung	Liver	Kidney	Heart	Spleen	Gastrointestinal tract	Small intestine	Stomach	Testis	Colonic Mucosae	Bile					
6 Male Sprague-Dawley rats	50 mg/kg resveratrol + 1.85 MBq [3H]resveratrol	Gavage	< 0.1% dpm	< 0.1% dpm	0.98% dpm	0.59% dpm	< 0.1% dpm	< 0.1% dpm	76.2% dpm	–	–	–	–	–	1.7% dpm	–	–	2 h	El-Mohsen et al. 2006
			Gluc: 0.2 (0.0) Resv: 0.1 (0.0)	Gluc: 0.5 (0.2) Rev: 0.2 (0.2)	Gluc: 1.2 (0.3)	Gluc: 4.0 (0.6)	Gluc: 0.4 (0.0)	n.d.	–	–	–	–	–	–	Gluc: 7.0 (1.0) Resv	2 Gluc Resv	Resv 2 Gluc		
6 Male Sprague-Dawley rats	50 mg/kg resveratrol + 1.85MBq [3H]resveratrol	Gavage	0.35% dpm						5.1% dpm	–	–	–	1.55% dpm	–	0.48% dpm	–	3.3% dpm	18 h	El-Mohsen et al. 2006
			Gluc: 0.13 (0.05) Resv: 0.07(0.05)	Gluc: 0.1 (0.03) Resv: 0.2 (0.03)	Gluc: 0.1 (0.03) Resv: 0.15 (0.03)	n.d.	Resv: 0.09 (0.02)	n.d.	–	–	–	–	–	–	n.d.	–	–		

Table 13.2 *(Continued)*

Species	Source and Dose	Administration	Mean (SD) nmol/g												Plasma (μmol/l)	Feces (%)	Urine (%)	Time	Reference
			Brain	Lung	Liver	Kidney	Heart	Spleen	Gastrointestinal tract	Small intestine	Stomach	Testis	Colonic Mucosae	Bile					
3 Balb/c male mice			n.c.	n.c.	441 (90) dpm/100 mg	342 (165)	n.c.	n.c.	–	2208 (1436) dpm/100 mg	–	n.c.	106 (80) dpm/100 mg	460 (70) $\cdot 10^3$dpm/ml	0.5 (0.2) $\cdot 10^3$dpm/ml	–	100 (26.7) $\cdot 10^3$dpm/ml	1.5h	
3 Balb/c male mice	7.4 kBq ^{14}C- *t*-Resv (5 mg/kg)	Intragastric	196 (47) dpm/100 mg	380 (148) dpm/100 mg	374 (48) dpm/100 mg	552 (51) dpm/100 mg	210 (132) dpm/100 mg	312 (62) dpm/100 mg	–	1841 (183) dpm/100 mg	–	**193 (9)**	300 (83) dpm/100 mg	1170 (500) $\cdot 10^3$dpm/ml	0.6 (0.8) $\cdot 10^3$dpm/ml	–	286 (16.7) $\cdot 10^3$dpm/ml	3h	
3 Balb/c male mice			n.c.	n.c.	189 (28) dpm/100 mg	263 (123) dpm/100 mg	n.c.	n.c.	–	933 (201) dpm/100 mg	–	n.c.	88 (26) dpm/100 mg	340 (330) $\cdot 10^3$dpm/ml	1.25 (0.8) $\cdot 10^3$dpm/ml	–	235 (23.3) $\cdot 10^3$dpm/ml	6h	
3 Balb/c male mice			-	-	0.09 (0.03) cpm/mm^2	0.05 (0.02) cpm/mm^2	–	–	–	**n.d.**	1.09 (0.72) cpm/mm^2	–	–	–	–	–	–	1.5h	Vitrac et al. 2003
3 Balb/c male mice	74 kBq ^{14}C- *t*-Resv (50 mg/kg)	Intragastric	-	-	0.1 (0.01) cpm/mm^2	0.07 (0.02) cpm/mm^2	–	–	–	1.7 (0.16) cpm/mm^2	0.93 (0.55) cpm/mm^2	–	–	–	–	–	–	3h	
3 Balb/c male mice			-	-	0.05 (0.01) cpm/mm^2	0.03 (0.01) cpm/mm^2	–	–	–	0.2 (0.04) cpm/mm^2	0.4 (0.04) cpm/mm^2	–	–	–	–	–	–	6h	
3 Balb/c male mice	92.5 kBq ^{14}C- *t*-Resv (66 mg/kg)	Intragastric			^{14}C- *t*-Resv: 25 μmol/l	^{14}C- *t*-Resv: 30 μmol/l	–	–	–	-	-	–	–	–	–	–	–	3h	

[a]TriSulf: resveratrol-3,4′,5-trisulfate; diSulf: resveratrol-disulfate; Sulf: resveratrol-sulfate; Gluc: resveratrol-glucuronide; *t*-Resv: *trans*-resveratrol; Radioactivity was measured by dpm or cpm: dpm: disintegrations per minute and cpm: counts per minute; MBq: megabecquerel; kBq: kilobecquerel; n.d.: not detected; n.c.: not collected.

Table 13.3 Metabolism of Resveratrol in Vivo in Animal Models[a]

Animals	Source	Administration	Dose mg/kg	Blood or Serum or Plasma	Urinary excretion		Time (h)	Reference
				Metabolites (Concentration)	%	Metabolites		
8 male Wistar rats	*trans*-Resveratrol tritiated in 4-position with 10%EtOH or V-8 homogenized vegetable cocktail or white grape juice	Intragastric	2 μL of 120 nCi/mL	—	49–61%	—	24	Soleas et al., 2001
2 male Wistar rats				Serum: 690 dpm/mL Blood: 600 dpm/ml	—	—	0.5	
Male Wistar rats	*t*-Resv	Intragastric	1.43 mg/kg	*t*-Resv: Serum: 0.01 μM Blood: 0.01 μM	—	—	1	
			4.29 mg/kg	*t*-Resv: Serum: 0.02 μM Blood: 0.01 μM	—	—	1	
			7.14 mg/kg	*t*-Resv: Serum: 0.03 μM Blood: 0.02 μM	—	—	1	
2 male Wistar rats	*t*-Resv	Intragastric	14.28 mg/kg	*t*-Resv Serum: 1.44 μM Blood: 1.00 μM	—	—	0.5	
3 Sprague–Dawley female rats	Resv	Intraperitoneal	20 mg/kg	—	—	*t*-3-Gluc	2 h	Yu et al., 2002
12 Balb/c female mice	Resv	Intraperitoneal	20 mg/kg	*t*-3-Gluc (3 μM), *t*-3-Sulf (13 μM)	—	—	0.25	
12 Balb/c female mice		Gavage	20 mg/kg	*t*-3-Gluc (1 μM), *t*-3-Sulf (5 μM)	—	—	0.25	
12 Balb/c female mice		Gavage	60 mg/kg	*t*-3-Gluc (175 μM), *t*-3-Sulf (300 μM)	—	—	0.25, 0.5	

(*Continued*)

Table 13.3 *(Continued)*

Animals	Source	Administration	Dose mg/kg	Blood or Serum or Plasma	Urinary excretion		Time (h)	Reference
				Metabolites (Concentration)	%	Metabolites		
5 male Sprague–Dawley rats	*t*-Resv	Oral	20 mg/kg	—	—	Gluc, Sulf, DHR, DHR-Sulf	—	Wang et al., 2005
Female CF-1 mice	Grape juice	Oral	~0.7 mg/kg Resv	n.d.	n.d.	—	—	Meng et al., 2004
			~1.4 mg/kg Resv	n.d.	1–2%	*t*-Resv *	24	
			~2.7 mg/kg Resv	n.d.	0.9–2.3%	*t*-Resv *	24	
2 female Wistar rats	*t*-Resv	Intragastric	2 mg/kg	Resv: 0.09 μM	—	—	4	
				Total: 1.2 μM	—	—	4	
			5 mg/kg	Resv: 0.11 μM	—	—	1.5	
				Total:1.5μM	—	—	4	
5 Sprague–Dawley rats	*t*-Resv	Intravenous	5.13 mg/kg	*t*-Resv (~21.9 μM)	—	—	0.08	Chen et al., 2007
	c-Resv		4.87 mg/kg	*c*-Resv (~17.5 μM)	—	—	0.08	
Male Sprague–Dawley rats		Oral	2 mg/kg	*t*-Resv (0.77 μM)	—	—	0.25	Juan et al., 1999

[a] t-Resv: *trans*-resveratrol; *t*-3-Gluc: *trans*-resveratrol-3-*O*-glucuronide; *t*-3-Sulf: *trans*-resveratrol-3-sulfate; Gluc: resveratrol glucuronide; Sulf: resveratrol-sulfate; DHR: 7,8-dihydroresveratrol; DHR-Sulf: 7,8-dihydroresveratrol sulfate; radioactivity was measured by dpm: desintegrations per minute.
* Quantified after hydrolysis.

when catechin, quercetin, and resveratrol (1–10 nM) were co-administered in rats. They found no competition between the three polyphenols and the absorption of *trans*-resveratrol was not saturable in the used concentrations [Soleas et al., 2001a].

In 2006, El-Mohsen et al. investigated the time-dependent appearance and disappearance in various organs of metabolic products of [^{3}H] *trans*-resveratrol (2 and 18 h following gastric administration of 50 mg/kg + 1.85 MBq) to Sprague–Dawley rats. At 2 h postgavage, most of the recovered radioactivity was still present in the gastrointestinal tract. The total dose administered reached only 1.7% in plasma. The only tissues with high concentrations were the liver and the kidney while the amount detected in other tissues was <0.1%. In contrast, at 2 and 18 h postadministration, approximately 11% of total dose was accounted for in all of the studied tissues (Table 13.2). They found that around 90% of the administered dose was absorbed, however, only 3.3% of absorbed resveratrol was detected in urine. The undetected radioactivity is expected to be lost via respiration and/or accumulation in other tissues, such as skeletal muscle and adipose tissue [El-Mohsen et al., 2006; Soleas et al., 2001a].

The metabolites of [^{3}H] *trans*-resveratrol detected in tissues and plasma were also investigated. In kidney, liver, heart, lungs, brain, and plasma (2-h samples), the only metabolite found was resveratrol–glucuronide. Glucuronides in plasma and kidney disappeared completely at 18 h. In lungs, liver, heart, and brain, the main detected metabolite at 18 h was the *trans*-resveratrol (Table 13.2). This study provided data on the metabolic fate of resveratrol. While glucuronides were predominant in plasma and tissues at the earlier times, the aglycone represented the main form at later times [El-Mohsen et al., 2006].

The kinetics of absorption, tissue distribution, and excretion was assessed after a single oral dose of ^{14}C-*trans*-resveratrol to male Balb/c mice. Blood and tissue samples were collected at 1.5, 3, and 6 h postadministration [Vitrac et al., 2003]. The autoradiographic examination of mice tissue sections and the radioactivity quantification revealed a higher fixation of ^{14}C-*trans*-resveratrol in the stomach, liver, kidney, intestine, bile, and urine and other organs of absorption and elimination. The concentration of radioactivity in blood was low and regular during the experimental period. During the entire experimental period, nearly complete absorption occurred in the small intestine as suggested by the higher concentration found in the proximal section compared to the distal section. After 6 h of oral administration the high concentration in stomach and duodenum was perhaps located in the mucosa, although this was also probably due to residual stomach content. The low concentrations found in the colon suggests that it was a minor way of elimination. The major concentrations found in urine and the decreasing concentrations in kidney showed that renal excretion was a major way of elimination. The kidney and the liver were the organs with highest deposition of ^{14}C-*trans*-resveratrol. In contrast to the kidney in which the parent drug was the major radioactive product, liver extracts 3 h after administration showed the presence of ^{14}C-*trans*-resveratrol with a high concentration of radioactive glucuronide or sulfated conjugated [Vitrac et al., 2003].

Contrasting with other studies that showed the accumulation of labeled resveratrol in tissues, the following studies addressed the search for the formed metabolites in biological fluids and tissues. Yu et al. [2002] synthesized and identified for the first time, the monosulfate isomers of resveratrol, resveratrol-3 and 4′-sulfate. These studies were performed in female Sprage–Dawley rats after intraperitoneal administration of 20 mg/kg of resveratrol with urine collection at 2 h and in female Balb/c mice after intraperitoneal (20 mg/kg) and oral administration (20 and 60 mg/kg) with collection of serum samples up to 4 h. The mass spectrometry analysis of rat urine samples only detected *trans*-resveratrol-3-glucuronide. In the mouse serum samples, after administration of 20 mg/kg via intraperitoneal or oral, maximum concentrations of resveratrol-3-glucuronide and resveratrol-3-sulfate were observed at 15 min. The sulfate was almost three-fold greater than the glucuronide (Table 13.3). Only traces of free resveratrol were observed. Furthermore, no resveratrol or metabolites were detected after 1 h. When higher doses were orally administered to mouse (60 mg/kg), the same metabolites as before were observed, but the maximum value for sulfate metabolite were reached after 30 min instead of 15 min, probably because more time was required to absorb the large volume that was administered. This was not the case for the minor dosage both glucuronide and sulfate metabolite were still detected after 3 h, suggesting that resveratrol was distributed to tissues and cleared slowly.

Wenzel et al. [2005] synthesized the same resveratrol-3 and 4′-sulfates as Yu et al [2002] but also the 3 and 4′-glucuronide, 3,4′-disulfate, 3,5-disulfate, and 3,4′,5-trisulfate metabolites of resveratrol. To search for these compounds in in vivo conditions, two experiments were carried out. Resveratrol aglycone was administered to male Wistar rats in a dosage of 50 or 300 mg/kg/day during 8 weeks, and urine, feces, and tissue samples were collected. As shown in Table 13.2, the administration of 50 mg/kg of resveratrol resulted in the formation of *trans*-resveratrol-3-sulfate, *trans*-resveratrol-disulfate, *trans*-resveratrol-3-glucuronide, and resveratrol in urine. Furthermore, an increase of the dosage (sixfold) showed the additional formation of *trans*-resveratrol-trisulfate. In both experiments, the 3-glucuronide was the main metabolite, and sulfate (3-sulfate and 3,5-disulfate) derivatives were 100- and 50-fold less, in relation to the dosage, respectively. The total recovery in urine of rats on 50 and 300 mg/kg was 15 and 54%, respectively. In feces samples of rats of the 50-mg/kg group, only 3-sulfate and resveratrol were determined, and 300 mg/kg administration resulted in the formation of all possible sulfates except trisulfate metabolite. The total recovery in feces samples was 15 and 31%, respectively, for both dosages [Wenzel et al., 2005]. They also studied the different distribution of these metabolites in plasma, kidney, and liver tissues and only after feeding 300 mg/kg were metabolites observed in 50% of the animals. The main metabolite in plasma samples was 3,4′-disulfate followed by 3,4′,5-trisulfate and 3-glucuronide metabolites (Fig. 13.2). In liver samples only 3 and 4′-monosulfates and 3-monoglucuronide were identified. The main metabolite in kidneys was the 3-glucuronide metabolite, followed by disulfates and minor 3-sulfate. In contrast to plasma and kidney, free resveratrol was only observed in liver samples.

Similarly in 2005, Wang et al. identified the microbial metabolites of resveratrol in rats (Table 13.3). Urine samples were obtained after oral administration of 20 mg/kg to Sprague–Dawley rats. They identified resveratrol–glucuronide, resveratrol–sulfate, 7,8-dihydroresveratrol, and 7,8-dihydroresveratrol–sulfate as the main 12-h urinary metabolite in rats by mass spectrometry after are SPE treatment [Wang et al., 2005].

In 2004, Meng et al. investigated the urinary and plasma levels of resveratrol and quercetin after their administration as constituents of grape juice or as pure aglycones. The first study was carried out during 4 days with female CF-1 mice receiving solutions containing 18.4 and 36.8% of grape preparation. The urinary excretion of resveratrol increased gradually during the study period, excreting a cumulative amount of approximately 1–2% of the ingested dose when receiving 18.4% grape juice and 0.9–2.3% when the dose was 36.8%. The second study was done after the oral dose of 2 and 5 mg/kg resveratrol to female Wistar rats. In plasma for both doses, resveratrol was mainly present as conjugates and the resveratrol aglycone constituted around 10–11% of total resveratrol at the beginning and declined to 5–7% at 4 h (Table 13.3) [Meng et al., 2004].

PHARMACOKINETICS OF RESVERATROL AND DERIVATIVES

The evaluation of pharmacokinetic analysis of resveratrol was carried out in different animal models such as rats, mice, and rabbits [Asensi et al., 2002; Chen et al., 2007; Juan et al., 2002; Marier et al., 2002; Sale et al., 2004]; furthermore, pharmacokinetic analysis was also evaluated with other resveratrol derivatives such as piceid [Lv et al., 2006; Zhou et al., 2007], a *Smilax china* root extract [Huang et al., 2008], 3,4,5,4′-tetramethoxystilbene [Sale et al., 2004], piceatannol, pinosylvin, and rhapontigenin [Roupe et al., 2006] (Table 13.4).

In the first kinetic study, Juan et al. [2002] focused on the determination of the time course of *trans*-resveratrol level in plasma after the 2-mg/kg orally administrated dose to rats. Resveratrol had already reached the bloodstream at 5 min, presented its maximum level at around 10 min, and was still detected after 60 min [Juan et al., 2002].

Another pharmacokinetic study was carried out after intravenous administration of 4.87 mg/kg of *cis*-resveratrol and 5.13 mg/kg of *trans*-resveratrol to Sprague–Dawley rats. The study showed that both isomers showed a rapid eliminate disposition in 90 min [Chen et al., 2007].

There is a further pharmacokinetic study in which *trans*-resveratrol in the aglycone form and the glucuronide forms were examined following intravenous (15 mg/kg) and oral (50 mg/kg) administration of *trans*-resveratrol to rats [Marier et al., 2002]. After intravenous administration, the plasmatic resveratrol concentrations declined rapidly over the first 2 h. Then, concentration profiles of resveratrol and resveratrol–glucuronide from intravenous or oral administration increased abruptly due to enterohepatic recirculation over the 4- to 8-h time period that resulted in a significant maintenance in the terminal elimination

Table 13.4 Pharmacokinetic Studies of Resveratrol in Animal Models and Humans[a]

Species	Source (dose)	Administration	C_{max} (μmol/L)	T_{max} (h)	$T_{1/2}$ (h)	AUC $_{0-\infty}$ (μmol · h/L)	Urinary Excretion (%)	References
6 male Wistar rats	Piceid (50 mg/kg)	Oral	0.93 (0.39)	0.35 (0.14)	1.68 (0.3)	2.23 (0.69)	—	Lv et al., 2006
C57BL/6J male mice	*t*-Resv (20 mg/kg)	Intragastric	2.6 (1.0)	2.5	—	—	—	Asensi et al., 2002
Wistar male rats		Intragastric	1.2 (0.4)	5	—	—	2	
6 ESD NZW male rabbits		Intragastric	1.1 (0.8)	2.5	—	—	—	
		IV	—	—	0.24	—	—	
3 mice	*t*-Resv (240 mg/kg)	Intragastric	32	0.2	—	—	—	Sale et al., 2004
6 male Sprague–Dawley rats	*t*-Resv (15 mg/kg)	Intravenous	—	—	Resv: 0.13 (0.02) Gluc: n.a.	Resv: 5.64 (0.5) Gluc: 38.7 (5.5)	—	Marier et al., 2002
6 male Sprague–Dawley rats	*t*-Resv (50 mg/kg)	Oral	Resv: 6.57 (1.55) Gluc:105.2 (32.4)	Resv: 0.29 (0.1) Gluc: 0.42 (0.3)	Resv: 1.48 (0.4) Gluc:1.55 (0.4)	Resv: 7.1 (2.0) Gluc: 324.7 (57.6)	—	
5 female Sprague–Dawley rats	1 g/kg *Smilax china* root extract equivalent to 180 mg/kg *O*-Resv and 80 mg/kg Resv	Oral	*O*-Resv: 21.93 (3.1) Resv: 9.61 (1.5)	0.25 0.25	— —	91.78 (13.7) 17.81 (0.6)	— —	Huang et al., 2008
Male Sprague–Dawley rats	*t*-Resv (2 mg/kg)	Intragastric	Resv: 2.57	*t*-Resv 0.16	0.25	—	—	Juan et al., 2002
8 Wistar rats	*t*-Piceid (150 mg/kg)	Oral	*t*-Gluc: 64.85 (18.5) *t*-Resv: 3.55 (0.7) *t*-Piceid: 4.35 (1.3)	2 1 0.5	4 2 1	— — —	— — —	Zhou et al., 2007
5 male Sprague–Dawley rats	Piceatannol: (10 mg/kg)	Intravenous	—	—	4.23 (1.25)	34.75 (10.2)	32.8	Roupe et al., 2006
	Pinosylvin: (10 mg/kg)		—	—	0.82 (0.05)	24.67 (5.7)	9.46	

(*Continued*)

Table 13.4 *(Continued)*

Species	Source (dose)	Administration	C_{max} (μmol/L)	T_{max} (h)	$T_{1/2}$ (h)	AUC $_{0-\infty}$ (μmol · h/L)	Urinary Excretion (%)	References
5 male Sprague–Dawley rats								
5 male Sprague–Dawley rats	Rhapontigenin: (10 mg/kg)		—	—	3.0 (1.35)	32.52 (0.4)	1.25	
10 humans (45% males)	*t*-Resv (7.7 mg/kg)	Oral	*t*-Resv: 0.32 (0.16) Gluc 1: 1.00 (0.35) Gluc 2: 0.91 (0.36) 3-Sulf: 3.69 (0.95)	*t*-Resv: 0.83 (0.5-1.5) Gluc 1: 2.00 (1.0-6.0) Gluc 2: 1.50 (1.0-5.0) 3-Sulf: 1.50 (1.0-5.0)	*t*-Resv: 2.85 Gluc 1: 2.85 Gluc 2: 3.09 3-Sulf: 3.21	*t*-Resv: 0.98 Gluc 1: 8.4 (0.1) Gluc 2: 5.6 (0.1) 3-Sulf: 17.8(0.1)	*t*-Resv: 0.04 (0.05) Gluc 1: 2.0 (0.4) Gluc 2: 8.9 (2.6) 3-Sulf: 11.4 (2.3)	Boocock et al 2007
10 humans (45% males)	*t*-Resv (15.4 mg/kg)		*t*-Resv: 0.51 (0.38) Gluc 1: 1.17 (0.90) Gluc 2: 1.66 (1.35) 3-Sulf: 6.82(21.39)	*t*-Resv: 0.759 (0.5-4.0) Gluc 1: 2.25 (1.0-6.0) Gluc 2: 1.75 (1.0-5.1) 3-Sulf: 2.00 (1.0-5.0)	*t*-Resv: 8.87 Gluc: 1 7.27 Gluc: 2 6.64 3-Sulf: 4.51	*t*-Resv: 2.4 (0.3) Gluc 1: 13.4(0.3) Gluc 2: 11.4(0.3) 3-Sulf: 44.1(0.3)	*t*-Resv: 0.1 (0.1) Gluc 1: 2.1 (1.1) Gluc 2: 3.2 (1.7) 3-Sulf: 7.3 (3.1)	
10 humans (45% males)	*t*-Resv (38.5 mg/kg)		*t*-Resv: 1.18 (0.65) Gluc 1: 2.16 (0.81) Gluc 2: 4.02 (2.88) 3-Sulf: 9.05 (2.46)	*t*-Resv: 1.38 (0.5-4.0) Gluc 1: 2.375 (1.0-8.0) Gluc 2: 2.00	*t*-Resv: 4.22 Gluc 1: 10.6 Gluc 2: 8.42 3-Sulf: 11.5	*t*-Resv: 3.4 (0.2) Gluc 1: 24.8(1.2) Gluc 2: 18.9(0.1) 3-Sulf: 74.5(0.2)	*t*-Resv: 0.1 (0.1) Gluc 1: 1.7 (1.7) Gluc 2: 3.1 (1.4)	

(*Continued*)

Table 13.4 *(Continued)*

Species	Source (dose)	Administration	C_{max} (μmol/L)	T_{max} (h)	$T_{1/2}$ (h)	AUC $_{0-\infty}$ (μmol · h/L)	Urinary Excretion (%)	References
				(1.0-6.0) 3-Sulf: 2.00 (1.0-5.2)			3-Sulf: 5.2 (2.6)	
10 Humans (45% males)	*t*-Resv (76.9 mg/kg)		*t*-Resv: 2.36 (1.71) Gluc 1: 3.18 (1.47) Gluc 2: 4.29 (2.85) 3-Sulf: 13.94(6.69)	*t*-Resv: 0.83 (0.5-1.5) Gluc 1: 2.00 (1.0-6.0) Gluc 2: 1.50 (1.0-5.0) 3-Sulf: 1.50 (1.0-5.0)	*t*-Resv: 8.52 Gluc 1: 7.90 Gluc 2: 5.83 3-Sulf: 7.71	*t*-Resv: 5.8(0.3) Gluc 1: 43.5(0.2) Gluc 2: 37.5(0.3) 3-Sulf: 135.5(0.2)	*t*-Resv: 0.1 (0.1) Gluc 1: 0.5 (0.3) Gluc 2: 3.0 (1.4) 3-Sulf: 5.0 (1.6)	

[a] *t*-Resv: *trans*-resveratrol; *O*-Resv: oxyresveratrol; Gluc: resveratrol-glucuronide; 3-Sulf: resveratrol-3-sulfate.

half-life of resveratrol aglycone. The clearance of resveratrol after oral or intravenous administration was higher than that of resvertarol–glucuronide, which resulted in a systemic exposure of approximately 46- or 7-fold, respectively, lower than that of glucuronide.

Asensi et al. [2002] studied tissue levels and pharmacokinetics of resveratrol after intravenous (20 mg/kg) and oral (20 mg/kg) administration to rabbits, rats, and mice. The highest concentration levels in plasma of *trans*-resveratrol were reached within the first 5 min in all animals studied but showed a short half-life and a rapid clearance. They found extravascular levels of resveratrol after its oral administration to rabbits, rats, and mice with the highest levels occurring within the first 10 min in liver, lung, brain, and kidney; therefore, it appears that resveratrol does not accumulate extravascularly and its presence in the tissues is parallel in time to its bioavailability in blood [Asensi et al., 2002]. They suggested that most or all circulating resveratrol may be removed by liver metabolism and if high doses are administered, high rates of hepatic metabolism could be occurring [Asensi et al., 2002].

Sale et al. [2004] studied the pharmacokinetic properties of 3,4,5,4′-tetramethoxystilbene compared with those of resveratrol in the plasma and mice tissues. This analog compound was a novel congener of pharmacological interest, and it was under preclinical evaluation as a potential antitumor prodrug. This tetramethoxystilbene was capable of preferentially interfering with the proliferation and survival of transformed human lung-derived cells, with much lower growth inhibitory and apoptotic properties in its untransformed counterparts than resveratrol, which does not possess this discriminatory potential. Therefore, a kinetic evaluation and a tissue distribution for both were applied after a single dose (240 mg/kg) oral administration to mice. The results suggested that the introduction of four methoxy groups into the stilbene structure, three of which replaced the hydroxyl moieties in resveratrol, did not increase the systemic availability of the molecule in comparison to resveratrol [Sale et al., 2004].

They also evaluated the pharmacokinetic properties of resveratrol compared to the synthetic analog in different mice tissues in which resveratrol might prevent malignancy or delay its onset. The availability of the synthetic analog was inferior in plasma, liver, kidney, lung, and heart than resveratrol; meanwhile, it was more available in intestinal and colonic mucosa and in brain [Sale et al., 2004]. These results provided a good argument to assess 3,4,5,4′-tetramethoxystilbene as a colorectal cancer chemo preventive agent. Furthermore, in the search for the main conjugated forms, resveratrol showed its glucuronided and sulfated conjugates, while, the 3,4,5,4′-tetramethoxystilbene underwent metabolic hydroxylation or single and double *O*-demethylation [Sale et al., 2004].

The pharmacokinetic dispositions of other stilbenes that are structurally similar to resveratrol and have pharmacological activity across many anticancer, anti-inflammatory, and antioxidant assays have been studied. The pharmacokinetics was characterized in male Sprague–Dawley rats after single intravenous doses of 10 mg/kg of piceatannol, pinosylvin, or rhanpontigenin. The detectable plasma half-lives of these compounds appear to be relatively short. The estimates of oral bioavailability characterize these stilbenes as poorly bioavailable compounds. All three stilbenes undergo extensive glucuronidation upon intravenous administration, as was determined by plasma and urine concentrations. The total amount excreted shows that the three stilbenes excreted in urine—32.8, 9.5, and 1.3%, respectively—are very small compared with the overall dose given of each one (3.5 mg). This indicates that the three stilbenes in contrast to resveratrol are eliminated predominantly via nonrenal excretion [Roupe et al., 2006].

Recently, the pharmacokinetics of resveratrol from *Smilax china,* a rhizome extensively used in traditional Chinese medicine was evaluated [Huang et al., 2008]. Forty-five female rats were orally administered with 1 g/kg *S. china* extract equivalent to 180 mg/kg of oxyresveratrol and 80 mg/kg of resveratrol. The pharmacokinetic parameters showed that the two stilbenes were rapidly absorbed into the body fluid from the gastrointestinal tract and could still be detected in the plasma at least 6 h after the administration [Huang et al., 2008], probably due to the high dose administered.

PICEID, THE GLUCOSIDE OF RESVERATROL

Various studies have focused on the metabolism of piceid, the glucoside of resveratrol. These studies are of great interest due to the higher amount of piceid compared to resveratrol in food. Therefore, bioavailability studies of this compound are required. In vitro studies have already observed the absorption of piceid in the enterocytes, but in vivo studies in animal models are still scarce.

The bioavailability and tissue distribution of *trans*-piceid was studied in Wistar rats after its oral administration [Lv et al., 2006; Zhou et al., 2007]. At present, the studies of the pharmacokinetics and distribution of piceid have been poorly documented. The first study with piceid administration to rats, measured its pharmacokinetics and its tissue distribution after a single oral administration of 50 mg/kg to 6 male Wistar rats. This was the first in vivo study that demonstrated the absorption of piceid with maximum plasma concentrations of 0.93 (0.4) μM at 21 min (Table 13.2) [Lv et al., 2006]. Another relevant result of this study was the diffusion of piceid to tissues. The highest concentrations were found in the stomach, then, in the small intestine, followed by spleen, lung, brain, testis, liver, kidney, and heart at 10 min. At 30 min, high concentrations were still detected in stomach and relatively high concentrations were present at 120 min. The absorption, distribution, and elimination of piceid were quick after the oral administration. The major distribution organs in rats were stomach, small intestine, and spleen; furthermore, it was able to cross the blood–brain and blood–testis barriers. Nevertheless, no long-term accumulation of piceid in tissues took place [Lv et al., 2006].

A recent study showed the bioavailability of piceid after oral administration of 150 mg/kg of piceid to rats. *trans*-Piceid was absorbed, with maximum plasma levels at 30 min, and metabolized to *trans*-resveratrol, with a maximum plasma concentration at 60 min, and this to *trans*-resveratrol–glucuronide, with maximum concentration at 120 min [Zhou et al., 2007]. The resveratrol–glucoside was absorbed by transepithelial transport across the intestine with maximum concentration occurring at 30 min after administration. This glycosytated derivative is deglycosylated in *trans*-resveratrol in the intestine with a cleavage by the CBG after passing the brush–border membrane by SGLT1 or by the membrane-bound enzyme LPH followed by passive diffusion of the released *trans*-resveratrol, which is further metabolized inside the cells into glucuronoconjugates [Henry-Vitrac et al., 2006]. The constant absorption of piceid from the first minutes of ingestion is reflected in the *trans*-resveratrol–glucuronide with the highest concentrations (30-fold higher) taking place in plasma and having a relatively long elimination half-life. Furthermore, the parent drug and the metabolites, *trans*-resveratrol and *trans*-resveratrol–glucuronide were detected at 8, 12, and 24 h after the oral dose [Zhou et al., 2007].

In conclusion, resveratrol is absorbed and already shows plasmatic and serum levels between 5 min and 4 h, depending on the dose and the animal species. Furthermore, plasmatic levels increase between 4 and 8 h due to the enterohepatic recirculation [Marier et al., 2002]. When absorbed, resveratrol is metabolized and the major conjugated forms in plasma consisted of sulfate conjugates, in which minor concentrations of the 3-glucuronide were observed. Further studies are required to obtain more standards of metabolites and more data about the major sulfate metabolite since Wenzel et al. [2005] and Yu et al. [2002] found major amounts for 3,4′-disulfate and 3-sulfate conjugates, respectively. Resveratrol is also distributed to different tissues such as liver and kidney, the major organs of deposition, and also in the lung, spleen, and heart. It crosses the blood–brain and blood–testis barriers showing major concentration levels in the intestinal tract. Maximum concentrations were found at early hours and trace amounts at later hours, showing no accumulation extravascularly [Asensi et al., 2002]. The main metabolite found in tissues at early hours was the glucuronide form, and the free form of resveratrol predominated at later hours [El-Mohsen et al., 2006].

Renal excretion was the major way of elimination compared to the colonic one. Urinary excretion varied between 3 and 61%, depending on the study, animal species, and dose. Some studies had also shown a possible excretion via sweat and respiratory system and metabolism to CO_2 [El-Mohsen et al., 2006; Soleas et al., 2001a]. The major conjugated form present in urine is the 3-glucuronide metabolite [Wenzel et al., 2005; Yu et al., 2002] although mono-, di-, and tri-sulfated metabolites and free resveratrol were also determined [Wang et al., 2005; Wenzel et al., 2005]. Furthermore, microbial metabolites such as dihydroresveratrol and its sulfate conjugate were also identified in 12 h rat urine [Wang et al., 2005].

HUMAN STUDIES

Studies that investigate the bioavailability of resveratrol in humans are scarce. Moreover, the research in this area is quite recent. It has been summarized in Table 13.5. The experimental approaches have been improved with the use of new analytical techniques such as mass spectrometry to identify and quantify metabolites present in very low concentrations. Resveratrol and its metabolites have been measured in several biofluids: plasma or serum, urine, LDL, and feces.

In 2001, the group of Soleas, Yang, and Goldberg [Soleas et al., 2001b] was the first to investigate the oral administration of resveratrol: 25 mg of *trans*-resveratrol standard dissolved in 120 mL of white wine were consumed by 10 healthy volunteers. The analyses were performed by gas chromatography–mass spectrometry (GC-MS) after treatment with β-glucuronidases and sulfatases. After stilbene intake, free resveratrol and its conjugates were found in all subjects and at all times, even after an abstinence of at least 24 h from food sources of this polyphenol. In plasma samples, they found that the highest resveratrol concentration (345.1 μg/L) occurred at 30 min. Moreover, resveratrol conjugates were 20- to 50-fold more abundant than free resveratrol. After resveratrol intake, the recovery in urine of 24 h was 24.6% as free and conjugated forms. Likewise, the urine concentrations of resveratrol conjugates were 30- to 50-fold higher than aglycone. At 2 h the highest concentration of resveratrol was observed, it was nearly 8 mg/L.

Three years later the same group [Goldberg et al., 2003] tested the absorptive efficiency of *trans*-resveratrol standard (25 mg) dissolved in three different matrices: white wine, grape juice, and vegetable juice. The conditions of study were the same as in the previous work. The results were also similar and serum showed the highest level at 30 min. Furthermore, the total absorption curves were similar regardless the matrices. In this study slight amounts of resveratrol at basal time were detected too. Urinary 24-h resveratrol excretions were 17.0, 16.8, and 16.0% after oral administration of vegetable juice, wine, and grape juice, respectively. The results in plasma and urine supported that there were no differences in resveratrol absorption by using the different matrices.

In 2004, Meng et al. published the first study that investigated the bioavailability of grape juice (ranged from 200 to 1200 mL) after oral ingestion, whose composition was 1.6 mg/L of stilbene, mainly as piceid [Meng et al., 2004]. Oral consumption of *trans*-resveratrol standard at several concentrations (0.03, 0.5, and 1 mg/kg) were also studied. In this case, the analyses were performed by liquid chromatography–tandem mass spectrometry (LC-MS/MS) after enzymatic hydrolysis. Resveratrol was only found in plasma when high doses (1 mg/kg) were ingested. At lower concentrations of standard or grape juice were ingested , no peaks were reached. However, urinary recoveries were 52, 34, and 26% after 0.03, 0.5, and 1 mg/kg, respectively. These results could suggest an inversely dose-dependent manner. Furthermore with a dose of 0.03 mg/kg, resveratrol was mainly excreted in the first 2–3 h; however, with 1 mg/kg, 7–10 h were necessary to excrete most of the resveratrol. Resveratrol was mainly found in the conjugated form as

Table 13.5 Human Bioavailability Studies[a]

Participants	Gender (%)	Age (y[c])	Resveratrol Source	Dose (mg/kg body weight)	Concentration (μmol/L)	Time (h)	Urinary Excretion (%) (Time, h)	Reference
10	Male (45%)	19–61	RESV Standard in 120 mL of white wine	0.357 mg/kg (25 mg total)	Resv 0.031 Conjugated[b] 1.48	0.5–1	Total 24.6% (24 h)	Soleas et al., 2001
4	Males (100%)	25–45	RESV Standard dissolved in 100 mL vegetable juice	0.357 mg/kg (25 mg total)	Resv 0.037 Conjugated[b] 2.03	0.5	17.0 (24 h)	Goldberg et al., 2003
4			RESV Standard dissolved in 100 mL white wine	0.357 mg/kg (25 mg total)	Resv 0.031 Conjugated[b] 1.80	0.5	16.8 (24 h)	
4			RESV Standard dissolved in 100 mL grape juice	0.357 mg/kg (25 mg total)	Resv 0.035 Conjugated[b] 1.82	0.5	16.0 (24 h)	
1	Male	30–50	RESV Standard	0.03 mg/kg	n.d.		Gluc: 52 (24 h)	Meng et al., 2004
1				0.5 mg/kg	n.d.		Gluc: 34 (24 h)	
1				1 mg/kg	Gluc 1.86	1.5	Gluc: 26 (24 h)	
1			200 mL grape juice	0.005 mg/kg[c] (0.32 mg total)	n.d.		n.d. (24 h)	
1			400 mL grape juice	0.009 mg/kg[c] (0.64 mg total)	n.d.		n.d. (24 h)	
1			600 mL grape juice	0.014 mg/kg[c] (0.96 mg total)	n.d.		Gluc: (24 h)	
1			1200 mL grape juice	0.027 mg/kg[c] (1.92 mg total)	n.d.		Gluc: 5.0 (24 h)	
6	3 Males (50%)	23–34	^{14}C- RESV standard Oral	0.385 mg/kg[cc] (25 mg total)			53.4–84.9% (72 h)	Walle et al., 2004
5			^{14}C- RESV standard Intravenous	0.023 mg/kg[c] (1.5 mg total)			42.3–83.2% (72 h)	
1			RESV Standard ORAL	1.538 mg/kg[c] (100 mg total)	Tr.		Gluc: 13 (1) Sulf: 24 (3) Total 37 (12 h)	
10	Males (100%)	30 (25–40)	300 mL red wine Lambrusco + meal	0.0034 g/kg	Gluc 0.096	1		Vitaglione et al., 2005
5	1 Male (20%)	29 (24–38)	600 mL red wine Cabernet Franc	0.0329 μg/kg	Gluc 0.687	0.5–2		
10	3 Males (30%)	31 (24–54)	600 mL red wine Aglianico + meal	0.0075 μg/kg	Resv 0.004 Gluc 0.150	0.5 1–2		
10	Male (100%)	28.2 (25–41)	300 mL/d sparkling wine (28 d)	0.005 mg/kg/d (0.357 mg/d)[c]	n.d.		*t*-3-Gluc: 4.8 (2.5) *c*-3-Gluc: 2.4 (1.3) Total 7.2 (2.7) (morning urine)	Zamora-Ros et al., 2006

(Continued)

Table 13.5 *(Continued)*

Participants	Gender (%)	Age (y[c])	Resveratrol Source	Dose (mg/kg body weight)	Concentration (μmol/L)	Time (h)	Urinary Excretion (%) (Time, h)	Reference
10	Women (100%)	38.1 (25–50)	200 mL/d white wine (28 d)	0.007 mg/kg/d (0.398 mg/d)[c]	n.d.		*t*-3-Gluc: 11.7 (2.8) *c*-3-Gluc: 3.3 (3.6) Total 15.0 (4.3) (morning urine)	
10	Women (100%)	38.1 (25–50)	200 mL/d red wine (28 d)	0.043 mg/kg/d (2.56 mg/d)[c]	n.d.		*t*-3-Gluc: 4.2 (3.2) *c*-3-Gluc: 1.2 (1.2) Total 5.4 (3.9) (morning urine)	
5	Male (100%)	25–28	250 mL red wine	0.077 mg/kg[c] (5.4 mg total)			*t*-4' Gluc: 0.13 (0.19) *t*-3 Gluc: 0.38 (0.59) *c*-4' Gluc: 0.75 (1.2) *c*-3 Gluc: 1.9 (1.9) *t*-4' Sulf: 0.01 (0.03) *t*-3 Sulf: 0.16 (0.67) *c*-4' Sulf: 19.6 (17.4) *c*-3 Sulf: 0.47 (2.2) Total 23.4 (4 h)	Urpi-Sarda et al., 2007
9	Males (100%)	23–41	PICEID Standard dissolved in 100 mL ethanol (15%) + 400 mL milk (1.5% fat)	1.22 mg/kg (85.5 mg/70 kg)	*t*-3-Sulf: 0.95 (0.16) *t*-3,4′-Disulf: 0.33 (0.07) *t*-3,5-Disulf: 0.94 (0.17) 3-Gluc: 0.16 (0.04) 4′-Gluc: 0.19 (0.05) (2) *t*-Digluc: 0.35 (0.09)	*t*-3-Sulf: 1 *t*-3,4′-Disulf: 6- *t*-3,5-Disulf: 8 3-Gluc: 6-8 4′-Gluc: 6 (2) *t*-Digluc: 6	*t*-3-Sulf: 4.53 *t*-3,4′-Disulf: 1.71 *t*-3,5-Disulf: 7.18 3-Gluc: 2.99 4′-Gluc: 0.69 (2) *t*-Digluc: 2.65 Total 13.6-35.7 (48 h)	Burkon and Somoza 2008

[a] Resv: resveratrol; Gluc: glucuronide; Sulf: sulfate; *t*-4 Gluc: *trans*-resveratrol-4′-*O*-glucuronide; *t*-3 Gluc: *trans*-resveratrol-3-*O*-glucuronide; *c*-4 Gluc: *cis*-resveratrol-4′-*O*-glucuronide; *c*-3 Gluc: *cis*-resveratrol-3-*O*-glucuronide; *t*-4 Sulf: *trans*-resveratrol-4′-sulfate; *t*-3 Sulf: *trans*-resveratrol-3-sulfate; *c*-4 Sulf: *cis*-resveratrol-4′-sulfate; *c*-3 Sulf: *cis*-resveratrol-3-sulfate; *t*-3,4′-Disulf: *trans*-resveratrol-3,4′-disulfates; *t*-3,5-Disulf: *trans*-resveratrol-3,5-disulfates; *t*-digluc:*trans*-resveratrol-*C*/*O*-diglucuronides.

[b] Quantified after hydrolysis of resveratrol conjugates.

[c] Calculated as weight estimation 70 and 60 kg for males and females, respectively. n.d. non detected.

glucuronide. After grape juice consumption, at low doses (200 and 400 mL) peaks of resveratrol in urine were not detected. Although at high doses (600 and 1200 mL) only conjugated forms were found. Moreover, after 1200 mL of grape juice, the recovery was only about 5% of the dose administered. This study showed that the glycoside forms are absorbed less than aglycones.

In the same year, Walle et al. [2004] were the first to administer intravenous and oral labeled resveratrol in humans. After 25 mg of an oral ^{14}C-resveratrol dose (6 healthy subjects), total radioactivity in plasma was maximum (491 ng/mL) at approximately 1 h after the intake, and then it kept falling during the following 72 h over the study. After 1.5 mg intravenous ^{14}C-resveratrol (5 healthy subjects) total radioactivity fell rapidly, but plasma radioactivity remained for the following 72 h. Moreover, both half-lives ranged from 7 to 14 h after any dose. This data is important because a single high dose of resveratrol can be active in plasma at least half a day. After oral dosage, 53–85 and 0.3–38 radioactivity were recovered in urine and feces, respectively. Similar results were observed after intravenous doses: 42–83 and 0.6–23 of total radioactivity were found in urine and feces, respectively. High variability was observed in the urinary and fecal recoveries. Elimination half-lives in urine ranged from 6.5 to 18.8 h after oral or intravenous doses. The authors also tested the metabolites formed after a large unlabeled oral dose of 100 mg of resveratrol. The analyses were performed by LC-MS-UV. This was the first study of human urine that analyzed resveratrol metabolite profile, showing the presence of two monoglucuronides, a monosulfate, a dihydroresveratrol monoglucuronide, and a dihydroresveratrol monosulfate. Dihydroresveratrol metabolites could be formed by the intestinal microbiota as occurs with other polyphenols [Gonthier et al., 2003]. The sulfate and glucuronide conjugates excreted in the urine accounted for 24 and 13% of the dose, respectively. However, in plasma resveratrol or its metabolites were not detected at any time. Only trace amounts (less than 5 ng/mL) could be found in plasma after an oral dose of the 100 mg.

Vitaglione et al. [2005] evaluated the bioavailability of red wine resveratrol consumed with several meals: standard, fat, or lean meal. Identification and quantification of resveratrol and its metabolites in serums were done by LC-MS/MS. In the first experiment, 10 healthy males were involved in the assessment of the bioavailability of Lambrusco red wine (0.82 mg *trans*-resveratrol/L) consumed with a standard meal (Milanese beef cutlet and chips). Only in 4 of the volunteers at 1 h were some amount of resveratrol glucuronides that ranged from 15 to 168 ng/mL found. In the second experiment, 5 healthy volunteers were recruited to intake 600 mL of Cabernet Franc red wine (3.2 mg *trans*-resveratrol/L) over night while fasting. Only 3 of the 5 subjects showed resveratrol free or metabolites in serum. In 2 volunteers resveratrol aglycone was detected but not in quantifiable amounts. Resveratrol glucuronides (isomers 3 and 4′) were reached at different times (0.5–2 h) and different concentrations (77–900 ng/mL). In the third experiment, 10 healthy subjects consumed 600 mL of aglianico red wine (0.8 mg *trans*-resveratrol/L) with either a lean meal or with a fat meal. Free resveratrol was detected in 2 of the subjects at concentrations ranging from 1 to 6 ng/mL at 30 min after wine consumption. Resveratrol–glucuronides were only detected in one subject of each intervention at 1–2 h after intake. This study clearly showed a high interindividual variation in the absorption and bioavailability of resveratrol.

Zamora-Ros et al [2006] carried out the first work that assessed the bioavailability of resveratrol (provided by different wines) in a regular intervention during 28 day. The analyses were performed by LC-MS/MS. In the first study, 10 healthy males were recruited to consume 300 mL/day of sparkling wine (1.19 mg resveratrol/l). After 28 day of supplementation, urinary *trans*- and *cis*-resveratrol-3-*O*-glucuronides were 75 and 38 nmol/g creatinine, respectively. In the second study, 10 healthy females were selected to consume 200 mL of white wine (1.99 mg resveratrol/L) or 200 mL of red wine (12.8 mg resveratrol/L) in a crossover clinical trial. Likewise after 28 days only resveratrol metabolites were detected in morning urine. *trans*- (205 and 473 nmol/g creatinine) and *cis*-resveratrol-3-*O*-glucuronides (58 and 140 nmol/g creatinine) were found after white and red wine intake, respectively. Those studies showed that urinary excretion was dose dependent. Furthermore, slight amounts of resveratrol metabolites were also detected at baseline periods. No free resveratrol or piceid were detected in any of the studies.

Urpi-Sarda et al. [2005, 2007] published the first works that investigated the presence of resveratrol in LDL. The analyses were performed by LC-MS/MS. Eleven healthy males were recruited to consume 250 mL of merlot wine (10.2 mg resveratrol/L). Free resveratrol, glycoside, glucuronidate, and sulfate forms were found in 24-h LDL. The detected metabolites were *cis*- and *trans*-, 3- and 4′-position, glucuronides, and sulfates. The more abundant metabolites were 88% glucuronides (*trans*-resveratrol-3-*O*-glucuronide, 112 pmol/mg LDL protein), 10.4% sulfates, and 2.0% *trans*-aglycone. Piceid was also found in LDL 24 h in lower concentrations (1.1–28.5 pmol/mg LDL protein). In the second experiment the metabolic profile was assessed at low resveratrol doses. Five healthy males were recruited to consume 250 mL of merlot wine (10.2 mg/L). Only conjugated forms were detected in urine 4 h after wine consumption. The more abundant metabolites were *cis*-resveratrol-4′-*O*-sulfate (9.3 μmol/g creatinine) and *cis*-resveratrol-3-*O*-glucuronide (0.9 μmol/g creatinine). Sulfation and glucuronidation represented 86.6 and 13.4% of total urinary resveratrol excretion, respectively.

In 2007, Boocock et al. [2007] were the first to publish a complete phase I dose pharmacokinetic study in humans. Ten healthy volunteers were recruited to consume single doses of oral resveratrol (0.5, 1, 2.5, or 5 g). Consumption of resveratrol did not cause serious adverse events. Analyses of resveratrol and its metabolites were performed by LC-MS/MS. In plasma in all intake doses resveratrol-3-sulfate (56%) was the highest metabolite, the second and third metabolites were monoglucuronides (17 and 23%, respectively), and, finally, the lowest was free resveratrol (5%). Resveratrol was rapidly absorbed, the $T_{\max}$ for all metabolites ranged between 0.8 and 2.4 h, although the half-lives of free resveratrol and the conjugated forms remained for a long time in plasma, between 2.9 and 11.5 h.

Urinary excretion mainly took place in the first 4 h after consumption (77% of total excretion), although resveratrol metabolites remained in urine between 12and24 h after intake. Free resveratrol, 2 glucuronides, and the 3-sulfate excreted in the urine 24 h after intake were below 0.04, 2, 9, and 11% of the 0.5 mg provided, respectively. At higher dose (5 mg) resveratrol, glucuronides and sulfate recoveries in the urine at 24 h were 0.1, 0.5, 3, and 5% of the dose, respectively. In urinary excretion, the sulfate forms were also higher than the glucuronide and free forms.

The piceid absorption was recently investigated for the first time. Nine healthy males participated in this controlled trial, which consisted of the administration of a single oral dose of 85.5 mg piceid standard per 70 kg. Resveratrol metabolites in plasma and urine were identified by LC-MS/MS, although these were quantified by HPLC-DAD. The same number of resveratrol metabolites were detected in both urine and plasma: *trans*-resveratrol-3-sulfate, *trans*-resveratrol-3,4′-disulfate, *trans*-resveratrol-3,5-disulfate, *trans*-3-*O*-glucuronide, *trans*-4′-*O*-glucuronide, and two resveratrol diglucuronides (*trans*-resveratrol-2-C-β-/4′-β-*O*-diglucuronide and *trans*-resveratrol-2-C-β-/5-β-*O*-diglucuronide). The two disulfates, previously identified in animals, and the two diglucuronides have been found in humans for the first time, thereby increasing the classical metabolic profile (monosulfates and monoglucoronides). The authors did not detect piceid nor resveratrol aglycone in any sample. Piceid was absorbed rapidly (1 h) in the form of *trans*-resveratrol-3-sulfate, whereas the other resveratrol metabolites reached their maximum concentration between 6 and 8 h after piceid administration. Sulfation pathway was more efficient than glucuronide. These authors also observed that 34, 44, and 46% of sulfates, disulfates, and diglucuronides, respectively, were non-covalently bound to plasma proteins; the rest of the percentage of conjugates were transported freely in plasma. After 24 h of piceid intake, no resveratrol metabolites were detected. The total urinary recovery ranged between 14 and 36%. The metabolic profile was approximately 15 and 8% as sulfate and glucuronide conjugates, respectively. Urinary excretion was completed within 48 h of oral piceid administration (Fig. 13.3).

In conclusion, resveratrol seems to have a greater absorption than piceid at nutritional doses [Meng et al., 2004], although pharmacological doses of piceid standard is also recovered at similar percentages than resveratrol standard [Burkon and Somoza, 2008]. Moreover, resveratrol and piceid are absorbed and metabolized quickly. In blood samples, the highest resveratrol peak is detected at around 30–60 min [Boocock et al., 2007; Goldberg et al., 2003; Soleas et al., 2001b; Vitaglione et al., 2005] and 6 h [Burkon and Somoza, 2008] after consumption of resveratrol or piceid, respectively. Nevertheless traces of resveratrol could remain in plasma for at least 72 h after ingestion [Walle et al., 2004]. Resveratrol conjugates are more abundant than the free form. It seems that sulfation is a more efficient metabolic pathway than glucuronidation (56 vs. 39%) [Boocock et al., 2007; Burkon and Somoza, 2008]. Part of resveratrol is transported through the body bound to LDL mainly as glucuronides (88%)

INTESTINAL METABOLISM

trans-isomers

cis-isomers

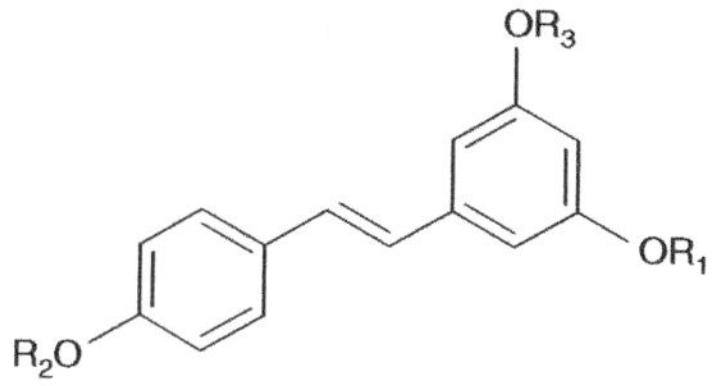

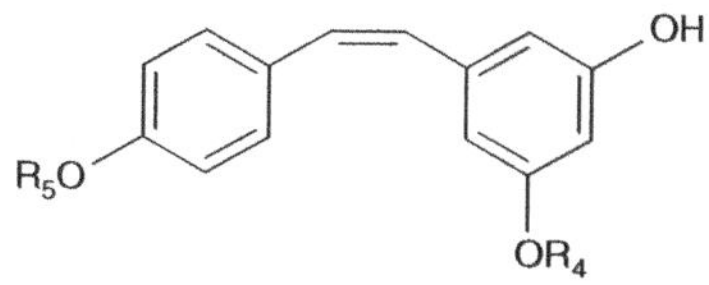

trans-Isomers

trans-Resveratrol-3-O-glucuronide: R_1=glucuronid acid, R_2=H, R_3=H (C20H20O9; MW: 404)
trans-Resveratrol-4′-O-glucuronide: R_1=H, R_2=glucuronid acid, R_3=H, (C20H20O9; MW: 404)
trans-Resveratrol-3,4′-diglucuronide: R_1=glucuronid acid, R_2=glucuronid acid, R_3=H (C26H28O15; MW: 580)
trans-Resveratrol-3-sulfate: R_1=SO_3H, R_2=H (C14H12O6S; MW: 308)
trans-Resveratrol-4′-sulfate: R_1=H, R_2=SO_3H, R_3=H (C14H12O6S; MW: 308)
trans-Resveratrol-3,4′-sulfate: R_1=SO_3H, R_2=SO_3H, R_3=H (C14H12O9S2; MW: 388)
trans-Resveratrol-3,5′-disulfate: R_1=SO_3H, R_2=H, R_3=SO_3H (C14H12O9S2; MW: 388)
trans-Resveratrol-3,5,4′-trisulfate: R_1=SO_3H, R_2=SO_3H, R_3=SO_3H (C14H12O12S3; MW: 468)

cis-Isomers

cis-Resveratrol-3-O-glucuronide: R_4=glucuronid acid, R_5=H (C20H20O9; MW: 404)
cis-Resveratrol-4′-O-glucuronide: R_4=H, R_5=glucuronid acid (C20H20O9; MW: 404)
cis-Resveratrol-3-4′-diglucuronide: R_4=glucuronid acid, R_5=glucuronid acid (C26H28O15; MW: 580)
cis-Resveratrol-3-sulfate: R_4=SO_3H, R_5=H (C14H21O6S; MW: 308)
cis-Resveratrol-4′-sulfate: R_4=H, R_5=SO_3H (C14H12O6S; MW: 308)
cis-Resveratrol-3,4′-disulfate: R_4=SO_3H, R_5=SO_3H (C14H12O9S2; MW: 388)

MICROBIAL METABOLISM

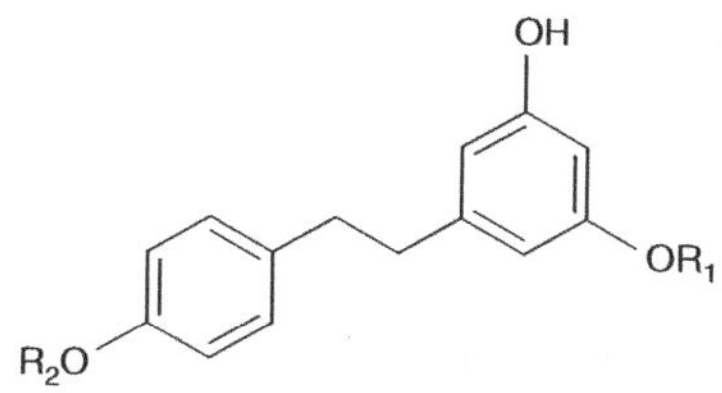

Dihydroresveratrol: R_1=H, R_2=H, (C14H14O3; MW: 230)
Dihydroresveratrol-glucuronide: R_1=H or glucuronide acid, R_2=glucuronide acid or H (C20H22O9; MW: 406)
Dihydroresveratrol-sulfate: R_1=H or SO_3H, R_2=SO_3H or H (C14H14O6S; MW: 310)

Figure 13.3 Pathways of resveratrol absorption, distribution, metabolism, and excretion.

and sulfates (11%) [Urpi-Sarda et al., 2005, 2007]. In a recent study, it was shown that more than 50% of resveratrol conjugates (sulfates, disulfates, and *C*/*O*-diglucuronides) were bound to proteins in plasma [Burkon and Somoza, 2008]. Urine (53–85%) and fecal (0.3–38%) were the most important ways of excretion of resveratrol measured by total radioactivity [Walle et al., 2004]. However, urinary recoveries by mass spectrometry ranged from 5 to 37%, depending on dose and the kind of resveratrol source [Boocock et al., 2007; Burkon and Somoza, 2008; Goldberg et al., 2003; Meng et al., 2004; Soleas et al., 2001b; Urpi-Sarda et al., 2007; Walle et al., 2004; Zamora-Ros et al., 2006]. Metabolites identified in urine are four monoglucuronides, four monosulfates, two disulfates, two C/*O*-diglucuronides, free aglycone, dihydroresveratrol monogucuronide, and monosulfate. Likewise in urinary excretion the sulfate forms seem to be higher than the glucuronide and free forms [Boocock et al., 2007; Burkon and Somoza, 2008; Urpi-Sarda et al., 2007; Walle et al., 2004]. The most important limitation of the bioavailability is the great individual variability; for this reason, further investigation with a higher number of volunteers is necessary in order to assess the percentage of absorption and excretion, and the metabolite profile of this polyphenol.

REFERENCES

Andlauer W, Kolb J, Siebert K, Furst P. 2000. Assessment of resveratrol bioavailability in the perfused small intestine of the rat. *Drugs Exp Clin Res* 26:47–55.

Asensi M, Medina I, Ortega A, Carretero J, Bano MC, Obrador E, Estrela J. 2002. Inhibition of cancer growth by resveratrol is related to its low bioavailability. *Free Radic Biol Med* 33:387–398.

Aumont V, Krisa S, Battaglia E, Netter P, Richard T, Merillon JM, Magdalou J, Sabolovic N. 2001. Regioselective and stereospecific glucuronidation of *trans*- and *cis*-resveratrol in human. *Arch Biochem Biophys* 393:281–289.

Barthe L, Woodley J, Houin G. 1999. Gastrointestinal absorption of drugs: Methods and studies. *Fundam Clin Pharmacol* 13:154–168.

Barthe L, Woodley JF, Kenworthy S Houin G. 1998. An improved everted gut sac as a simple and accurate technique to measure paracellular transport across the small intestine. *Eur J Drug Metab Pharmacokinet* 23:313–323.

Bertelli AA, Giovannini L, Stradi R, Bertelli A, Tillement JP. 1996. Plasma, urine and tissue levels of *trans*- and *cis*-resveratrol (3,4′,5-trihydroxystilbene) after short-term or prolonged administration of red wine to rats. *Int J Tissue React* 18:67–71.

Blache D, Rustan I, Durand P, Lesgards G, Loreau N. 1997. Gas chromatographic analysis of resveratrol in plasma, lipoproteins and cells after in vitro incubations. *J Chromatogr B Biomed Sci Appl* 702:103–110.

Boocock DJ, Faust GE, Patel KR, Schinas AM, Brown VA, Ducharme MP, Booth TD, Crowell JA, Perloff M, Gescher AJ, Steward WP, Brenner DE. 2007. Phase I dose escalation pharmacokinetic study in healthy volunteers of resveratrol, a potential cancer chemopreventive agent. *Cancer Epidemiol Biomarkers Prev* 16:1246–1252.

Brill SS, Furimsky AM, Ho MN, Furniss MJ, Li Y, Green AG, Bradford WW, Green CE, Kapetanovic IM, Iyer LV. 2006. Glucuronidation of *trans*-resveratrol by human liver and intestinal microsomes and UGT isoforms. *J Pharm Pharmacol* 58:469–479.

Burkon A, Somoza V. 2008. Quantification of free and protein-bound *trans*-resveratrol metabolites and identification of *trans*-resveratrol-C/O-conjugated diglucuronides—Two novel resveratrol metabolites in human plasma. *Mol Nutr Food Res* 52:549–557.

Burns J, Yokota T, Ashihara H, Lean ME, Crozier A. 2002. Plant foods and herbal sources of resveratrol. *J Agric Food Chem* 50:3337–3340.

Chen X, He H, Wang G, Yang B, Ren W, Ma L, Yu Q. 2007. Stereospecific determination of *cis*- and *trans*-resveratrol in rat plasma by HPLC: Application to pharmacokinetic studies. *Biomed Chromatogr* 21:257–265.

de Andrés-de Prado R, Yuste-Rojas M, Sort X, Andres-Lacueva C, Torres M, Lamuela-Raventos RM. 2007. Effect of soil type on wines produced from Vitis vinifera L. cv. Grenache in commercial vineyards. *J Agric Food Chem* 55:779–786.

de Santi C, Pietrabissa A, Mosca F, Pacifici GM. 2000a. Glucuronidation of resveratrol, a natural product present in grape and wine, in the human liver. *Xenobiotica* 30:1047–1054.

de Santi C, Pietrabissa A, Spisni R, Mosca F, Pacifici GM. 2000b. Sulphation of resveratrol, a natural product present in grapes and wine, in the human liver and duodenum. *Xenobiotica* 30:609–617.

El-Mohsen M, Bayele H, Kuhnle G, Gibson G, Debnam E, Kaila SS, Rice-Evans C, Spencer JP. 2006. Distribution of [3H]*trans*-resveratrol in rat tissues following oral administration. *Br J Nutr* 96:62–70.

Goldberg DM, Yan J, Soleas GJ. 2003. Absorption of three wine-related polyphenols in three different matrices by healthy subjects. *Clin Biochem* 36:79–87.

Gonthier MP, Cheynier V, Donovan JL, Manach C, Morand C, Mila I, Lapierre C, Remesy C, Scalbert A. 2003. Microbial aromatic acid metabolites formed in the gut account for a major fraction of the polyphenols excreted in urine of rats fed red wine polyphenols. *J Nutr* 133:461–467.

González-Barrio R, Beltran D, Cantos E, Gil MI, Espín JC, Tomás-Barberán FA. 2006. Comparison of ozone and UV-C treatments on the postharvest stilbenoid monomer, dimer, and trimer induction in var. "Superior" white table grapes. *J Agric Food Chem* 54:4222–4228.

Henry C, Vitrac X, Decendit A, Ennamany R, Krisa S, Merillon JM. 2005. Cellular uptake and efflux of trans-piceid and its aglycone *trans*-resveratrol on the apical membrane of human intestinal Caco-2 cells. *J Agric Food Chem* 53:798–803.

Henry-Vitrac C, Desmouliere A, Girard D, Merillon JM, Krisa S. 2006. Transport, deglycosylation, and metabolism of trans-piceid by small intestinal epithelial cells. *Eur J Nutr* 45:376–382.

Huang H, Zhang J, Chen G, Lu Z, Wang X, Sha N, Shao B, Li P, Guo DA. 2008. High performance liquid chromatographic method for the determination and pharmacokinetic studies of oxyresveratrol and resveratrol in rat plasma after oral administration of Smilax china extract. *Biomed Chromatogr* 22:421–427.

Juan ME, Buenafuente J, Casals I, Planas JM. 2002. Plasmatic levels of trans-resveratrol in rats. *Food Res Int* 35:195–199.

Kaldas MI, Walle UK, Walle T. 2003. Resveratrol transport and metabolism by human intestinal Caco-2 cells. *J Pharm Pharmacol* 55:307–312.

Kuhnle G, Spencer JPE, Chowrimootoo G, Schroeter H, Debnam ES, Srai SKS, Rice-Evans C, Hahn U. 2000. Resveratrol is absorbed in the small intestine as resveratrol glucuronide. *Biochem Biophys Res Commun* 272:212–217.

Lv C, Zhang L, Wang Q, Liu W, Wang C, Jing X, Liu Y. 2006. Determination of piceid in rat plasma and tissues by high-performance liquid chromatographic method with UV detection. *Biomed Chromatogr* 20:1260–1266.

Maier-Salamon A, Hagenauer B, Wirth M, Gabor F, Szekeres T, Jager W. 2006. Increased transport of resveratrol across monolayers of the human intestinal Caco-2 cells is mediated by inhibition and saturation of metabolites. *Pharm Res* 23:2107–2115.

Marier JF, Vachon P, Gritsas A, Zhang J, Moreau JP, Ducharme MP. 2002. Metabolism and disposition of resveratrol in rats: Extent of absorption, glucuronidation, and enterohepatic recirculation evidenced by a linked-rat model. *J Pharmacol Exp Ther* 302:369–373.

Meng X, Maliakal P, Lu H, Lee MJ, Yang CS. 2004. Urinary and plasma levels of resveratrol and quercetin in humans, mice, and rats after ingestion of pure compounds and grape juice. *J Agric Food Chem* 52:35–942.

Rimando AM, Kalt W, Magee JB, Dewey J, Ballington JR. 2004. Resveratrol, pterostilbene, and piceatannol in vaccinium berries. *J Agric Food Chem* 52:4713–4719.

Romero-Perez AI, Lamuela-Raventos RM, Andres-Lacueva C, Torre-Boronat MC. 2001. Method for the quantitative extraction of resveratrol and piceid isomers in grape berry skins. Effect of powdery mildew on the stilbene content. *J Agric Food Chem* 49:210–215.

Roupe KA, Yáñez JA, Teng XW, Davies NM. 2006. Pharmacokinetics of selected stilbenes: rhapontigenin, piceatannol and pinosylvin in rats. *J Pharm Pharmacol* 58:1443–1450.

Sale S, Verschoyle RD, Boocock D, Jones DJ, Wilsher N, Ruparelia KC, Potter GA, Farmer PB, Steward WP, Gescher AJ. 2004. Pharmacokinetics in mice and growth-inhibitory properties of the putative cancer chemopreventive agent resveratrol and the synthetic analogue trans 3,4,5,4′-tetramethoxystilbene. *Br J Cancer* 90:736–744.

Sobolev VS, Cole RJ. 1999. *trans*-Resveratrol content in commercial peanuts and peanut products. *J Agric Food Chem* 47:1435–1439.

Soleas GJ, Angelini M, Grass L, Diamandis EP, Goldberg DM. 2001a. Absorption of *trans*-resveratrol in rats. *Methods Enzymol* 335:145–154.

Soleas GJ, Yan J, Goldberg DM. 2001b. Ultrasensitive assay for three polyphenols (catechin, quercetin and resveratrol) and their conjugates in biological fluids utilizing gas chromatography with mass selective detection. *J Chromatogr B Biomed Sci Appl* 757:161–172.

Tokusoglu O, Unal MK, Yemis F. 2005. Determination of the phytoalexin resveratrol (3,5,4′-trihydroxystilbene) in peanuts and pistachios by high-performance liquid chromatographic diode array (HPLC-DAD) and gas chromatography-mass spectrometry (GC-MS). *J Agric Food Chem* 53:5003–5009.

Urpi-Sarda M, Zamora-Ros R, Lamuela-Raventós RM, Cherubini A, Jauregui O, de la Torre R, Covas M, Estruch R, Jaeger W, Andres-Lacueva C. 2007. HPLC-tandem mass spectrometric method to characterize resveratrol metabolism in humans. *Clin Chem* 53:292–299.

Urpi-Sarda M, Jauregui O, Lamuela-Raventós RM, Jaeger W, Miksits M, Covas M, Andres-Lacueva C. 2005. Uptake of diet resveratrol into the human low density lipoprotein. Identification and quantification of resveratrol metabolites by liquid chromatography coupled with tandem mass spectrometry. *Anal Chem* 77:3149–3155.

Vitaglione P, Sforza S, Galaverna G, Ghidini C, Caporaso N, Vescovi PP, Fogliano V, Marchelli R. 2005. Bioavailability of *trans*-resveratrol from red wine in humans. *Mol Nutr Food Res* 49:495–504.

Vitrac X, Desmouliere A, Brouillaud B, Krisa S, Deffieux G, Barthe N, Rosenbaum J, Merillon JM. 2003. Distribution of [^{14}C]-trans-resveratrol, a cancer chemopreventive polyphenol, in mouse tissues after oral administration. *Life Sci* 72:2219–2233.

Walle T, Hsieh F, DeLegge MH, Oatis JE, Walle UK. 2004. High absorption but very low bioavailability of oral resveratrol in humans. *Drug Metab Dispos* 32:1377–1382.

Wang D, Hang T, Wu C, Liu W. 2005. Identification of the major metabolites of resveratrol in rat urine by HPLC-MS/MS. *J Chromatogr B Analyt Technol Biomed Life Sci* 829:97–106.

Wenzel E, Soldo T, Erbersdobler H, Somoza V. 2005. Bioactivity and metabolism of trans-resveratrol orally administered to Wistar rats1847. *Mol Nutr Food Res* 49:482–494.

Yu C, Shin YG, Chow A, Li Y, Kosmeder, JW, Lee YS, Hirschelman WH, Pezzuto JM, Mehta RG, van Breemen RB. 2002. Human, rat, and mouse metabolism of resveratrol. *Pharm Res* 19:1907–1914.

Zamora-Ros R, Andres-Lacueva C, Lamuela-Raventós RM, Berenguer T, Jakszyn P, Martínez C, Sánchez MJ, Navarro C, Chirlaque MD, Tormo MJ, Quirós JR, Amiano P, Dorronsoro M, Larrañaga N, Barricarte A, Ardanaz E, González CA. 2008. Concentrations of resveratrol and derivatives in foods and estimation of dietary intake in a Spanish population: European Prospective Investigation into Cancer and Nutrition (EPIC)—Spain cohort. *Br J Nutr* 100:188–196.

Zamora-Ros R, Urpi-Sarda M, Lamuela-Raventós RM, Estruch R, Vázquez-Agell M, Serrano-Martínez M, Jaeger W, Andres-Lacueva C. 2006. Diagnostic performance of urinary resveratrol metabolites as a biomarker of moderate wine consumption. *Clin Chem* 52:1373–1380.

Zhou M, Chen X, Zhong D. 2007. Simultaneous determination of *trans*-resveratrol-3-*O*-glucoside and its two metabolites in rat plasma using liquid chromatography with ultraviolet detection. *J Chromatogr B Analyt Technol Biomed Life Sci* 854:219–223.

14 Resveratrol: Biochemistry and Functions

SAMARJIT DAS[1], HANNAH R. VASANTHI[2], AND DIPAK K. DAS[1]

[1]Cardiovascular Research Center, University of Connecticut School of Medicine, Farmington, Connecticut

[2]Department of Biochemistry, Sri Ramachandra Medical College & Research Institute, Sri Ramachandra University, Chennai, India

CONTENTS

Plant Phenolics and Human Health: Biochemistry, Nutrition, and Pharmacology,
Edited by Cesar G. Fraga.

INTRODUCTION

Interest in the potential health benefits associated with dietary consumption of plant polyphenols has increased significantly in the past decade. Moreover, the role of these phytochemicals to combat oxidative stress, which plays a major role in the pathophysiology associated with various diseases, is enormous. Resveratrol (3,4′,5-trihydroxy stilbene) is one of the numerous polyphenolic compounds which is found in nearly 72 vegetal sources [Dercks and Creasy, 1989]. After the French paradox study in 1992 [Renaud and de Lorgeril, 1992], resveratrol got recognized proving best among all the natural polyphenolic compounds. Resveratrol is a member of a family of polyphenols called viniferins. A mild-to-moderate wine-drinking habit attenuates cardiovascular, cerebrovascular, and peripheral vascular risk due to reduced platelet and monocyte adhesion and attenuates the risk of prostate as well as a variety of cancers including pancreatic, gastric, and thyroid cancer. Resveratrol, a polyphenol phytoalexin, possesses diverse biochemical and physiological properties including estrogenic, antiplatelet, and anti-inflammatory actions. Several recent studies have revealed that resveratrol mediates protection from a wide variety of degenerative diseases and hence increases the life span. Resveratrol inhibits apoptotic cell death at a lower concentration, thereby providing protection from various diseases including myocardial ischemic reperfusion injury, atherosclerosis, and ventricular arrhythmias. Both in acute and in chronic models, resveratrol-mediated cardioprotection is achieved through the preconditioning effect, rather than direct effect as found in conventional medicine. Used in higher doses resveratrol facilitates apoptotic cell death, behaving in contrast, as a chemopreventive agent. Resveratrol likely fulfills the definition of a pharmacological preconditioning compound and gives hope for its therapeutic promise from alternative medicine. The purpose of this review is to provide evidence on how the biochemical properties of resveratrol is helping it to use as a preventive medicine for the maintenance of health.

CHEMICAL OVERVIEW OF RESVERATROL

Resveratrol is a member of the stilbene family, a group of compounds which consist of two aromatic rings joined by a methylene bridge. 3,4′,5-trihydroxystilbene is the IUPAC (International Union of Pure and Applied Chemistry) nomenclature for resveratrol but is also known as 3,4′,5-stilbenetriol, and it has

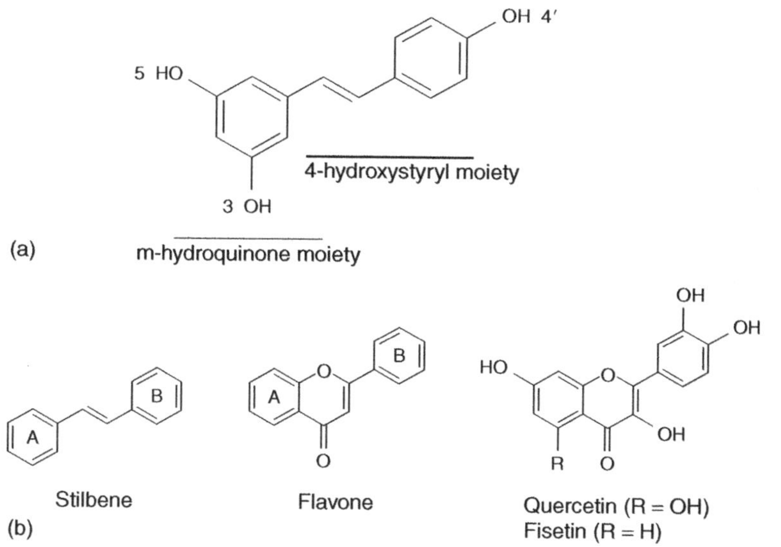

Figure 14.1 (*a*) Chemical structure of resveratrol. (*b*) Chemical structure of stilbene and the structurally similar polyphenols (quercetin and fisetin). The common rings are lettered.

the molecular formula $C_{14}H_{12}O_3$ with a molecular weight of 228.25 Da (Fig. 14.1).

Resveratrol exists in two structural isomeric forms, *cis* and *trans* (Fig. 14.2). Although both *cis*- and *trans*-resveratrol (with their glucosides) occur naturally and seem to exert similar biological effects, but the actions of the *trans*-isoform are more widely investigated and are better known [Orallo, 2006]. It has been identified that the greater biological activity of the *trans*-isomer is due to the 4′-hydroxystyryl group [Stivala et al., 2001]. *Trans*-resveratrol in powder form is found to be stable under "accelerated stability" conditions of 75% humidity and 40°C in the presence of air [Baur et al., 2006]. The nuclear magnetic resonance (NMR)-derived solution structure of 3,4′,5-trihydrostilbene is an accurate representation of the line segment 1–α–α′–1′, the angles 1–α–α′ and 1′– α′–α, and the dihedral angle 2–1–1′–2′, all of which are related to the *trans*-stereochemistry of the phenolic rings, which are free to rotate about the C1–Cα and C1′–Cα′ bonds [Commodari et al., 2005].

Resveratrol-3-*O*-β-D-glucoside is called piceid. Resveratrol is a naturally occurring phytoalexin ("defender of the plant") which is produced in response to an injury, such as mechanical trauma, ultraviolet light, and infection by pathogenic microorganisms, especially fungi, providing means for defense [Bertelli et al., 1995; Baolin et al., 2004; Bak et al., 2006; Baur et al., 2006]. It is formed via a condensation reaction between three molecules of malonyl CoA and one molecule of 4-coumaroyl CoA [Soleas et al., 1997]. Resveratrol synthase facilitates this condensation reaction, which also produces four

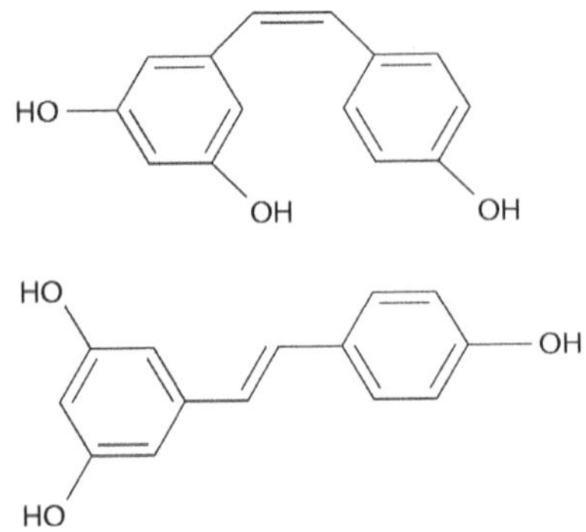

Figure 14.2 *Cis* (top) and *trans* (bottom) isomers of resveratrol.

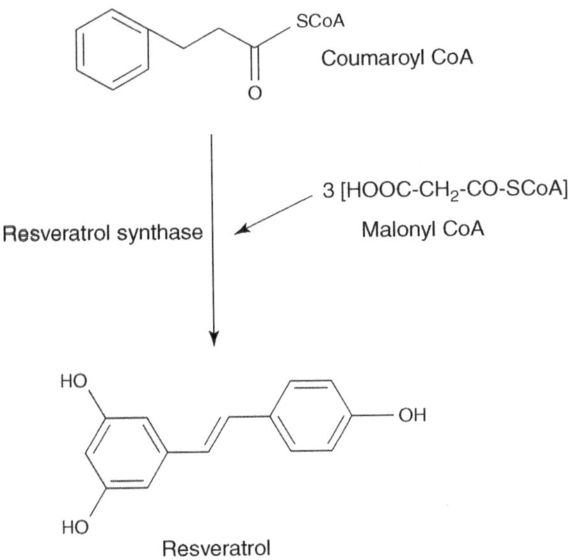

Figure 14.3 Synthesis of resveratrol.

molecules of CO_2 (Fig. 14.3). The biosynthesis of resveratrol is similar to that of the chalcone, which is the precursor of ubiquitous flavonoids and anthocyanins, except for the formation of the final product, which is due to the ring folding after the formation of a tetraketide.

SOURCES OF RESVERATROL

Polygonum cuspidatum, a weed used in Chinese and Japanese medicine, is one of the first plants identified to extract resveratrol and is one of the richest sources of resveratrol. In certain higher order plants, such as eucalyptus, spruce, lily, mulberry and peanut, stilbene synthase (STS) genes produce resveratrol from the reaction between one molecule of *p*-coumaroyl CoA and three molecules of malonyl CoA [Cellotti et al., 1996]. External stimuli such as fungal attack and ultraviolet (UV) radiation activate the stilbene synthase genes in grapes to produce resveratrol in order to adequately protect the grapes from those stimuli [Soleas et al., 1997; Pervaiz, 2003]. This STS gene is also found in peanut root, strawberry, blueberry, mulberry, grapes, and other plants such as eucalyptus, spruce, and lily. *Vitis vinifera*, *V. labrusca*, and *V. muscadine* are three different types of grapes, that contain a very high concentration of resveratrol. The skin and seeds of these types of grapes contain 50–100 μg/g of resveratrol [Koop, 1998]. These varieties of grapes are particularly suitable to make red wine. Thus, grapes and red wines are considered as the major sources of resveratrol. In addition to grapes, a large variety of fruits including mulberry, bilberry, lingo berry, sparkle-berry, deer berry, partridgeberry, cranberry, blueberry, jackfruit, peanut a wide variety of flowers and leaves including gnetum, white hellebore, corn lily, butterfly orchid tree, eucalyptus, spruce, poaceae, and rheum also contain resveratrol [Burns et al., 2002]. In contrast to the constitutive isoform of stilbene synthase occurring in the *Rheum rhaponticum* (rhubarb), an inducible enzyme is expressed in the Vitaceae family [Hain et al., 1990]. Since fungal infections are more common in cooler climates, grapes grown in cooler climates have a higher concentration of resveratrol [Koop, 1998]. However, grapes cultured in the zone of the equator also contain resveratrol in high concentration because of higher ultraviolet irradiation [Schubert et al., 1997].

EPIDEMIOLOGIC OBSERVATIONS OF RESVERATROL

There are several lines of epidemiologic data suggesting the beneficial effects of resveratrol. There are some circumscribed areas around the globe where average age of people is remarkably high and there is an unusual presence of centenarians.

The well-known French paradox was first noted by Irish physician Samuel Black in 1819. Back in the nineteenth century, he was the first to observe the fact that people in France suffer relatively low incidence of coronary heart disease, despite their diet being rich in saturated fats. It was proposed that France's profound red wine consumption is a primary factor contributing to the protective effect. The first scientific evidence, however, for the cardiovascular benefits of red wine was put forward by Renaud and his associates in 1992 [Renaud and Lorgeril, 1992]. In this study, popularly known as the French paradox, the researchers found that there had been a low mortality rate from, and incidence of, coronary heart disease among French men above the age of 40 years compared to men in the United Kingdom and the United States, despite their high consumption of saturated fats and the prevalence of other risk factors such as smoking. This was attributed to their so-called Mediterranean diet, which includes a moderate intake of wine. It has been proposed that one of the active ingredients potentially related to this effect in red wine is resveratrol [Corder et al., 2001].

METABOLISM OF RESVERATROL

The potential health benefits of resveratrol depend, in part, upon its absorption, bioavailability, and metabolism. Several in vitro and in vivo models have been utilized to characterize the absorption and bioavailability of resveratrol. Using the Caco-2 human intestinal cell model, it was demonstrated that resveratrol uptake remained linear, for an hour, and transportation was nondirectional [Kaldas et al., 2003]. Metabolites identified in the Caco-2 cells were resveratrol sulfate and resveratrol glucuronide, with resveratrol sulfate being predominant. The ease of absorption and accumulation of resveratrol suggests gastrointestinal cells as a possible biological target in vivo. Henry and co-workers [2005] also observed passive diffusion of resveratrol in Caco-2 cells; however, the transport of trans-piceid required the sodium-dependent glucose co-transporter SGLT1. Similar to Caco-2 cell uptake, normal and tumor human hepatic cell lines utilize both passive diffusion and active transport for resveratrol uptake [Lancon et al., 2004]. In the rat small intestine model, resveratrol was absorbed on the serosal side of the jejunum, with the majority being metabolized to resveratrol glucuronide [Kuhnle et al., 2000]. Recently, several studies have examined resveratrol absorption in vivo. A detailed study on the metabolism of resveratrol in humans have demonstrated increased absorption after oral administration. Extensive metabolism of resveratrol in the gut and liver leads to it's trace amount in the plasma (5 ng/mL). Rapid metabolism is one of the main reasons for short initial half-life of the primary molecule (8–14 min) [Asensi et al., 2002; Marier et al., 2002]. The bulk of an intravenous dosing of resveratrol was converted to sulfate conjugates within ~30 min in humans. However, glucuronide conjugates were also detected in the plasma [Walle et al., 2004]. The food matrix in which resveratrol is consumed may have an effect on absorption and bioavailability. Following oral administration of pure resveratrol to human subjects, resveratrol glucuronide was the major metabolite detected in the plasma and urine. High oral doses of grape juice resulted in the detection of the glucuronide and sulfate conjugates in the subjects plasma and urine [Meng et al., 2004]. Grape juice consists of mostly resveratrol glucosides, *cis*- and *trans*-piceid, with low amounts of the free resveratrol, suggesting a lower bioavailability of the glucosides compared to the pure compound.

More recently, the sulfation of resveratrol in human liver cytosol has been examined [Miksits et al., 2005]. In the presence of 30-phosphoadenosine-50-phosphosulfate, three metabolites occurred whose structures were identified by mass spectrometry and NMR as *trans*-resveratrol-3-*O*-4′-*O*-disulfate (S1), *trans*-resveratrol-4′-*O*-sulfate (S2), and *trans*-resveratrol-3-*O*-sulfate (S3), respectively. Incubation in the presence of human recombinant sulfotransferases (SULTs) demonstrated that S1 is almost exclusively catalyzed by SULT1A1 and only to a minor extent by SULT 1A2, 1A3, and 1E1, whereas S2 is selectively formed by SULT1E1. S3 is catalyzed by SULT1A1, SULT1E1, SULT1A2, and 1A3 (depending on the resveratrol concentration) [Miksits et al., 2005]. The metabolic pathway of resveratrol in human liver cytosol in vitro is shown in Figure 14.4.

In addition, the dose-dependent effect of sulfation and glucuronidation on intestinal absorption of resveratrol was investigated [Maier-Salamon et al., 2006]. The intestinal epithelial membrane transport kinetics and metabolism of resveratrol (10–200 mM) was studied using Caco-2 monolayers cultured in transwells.

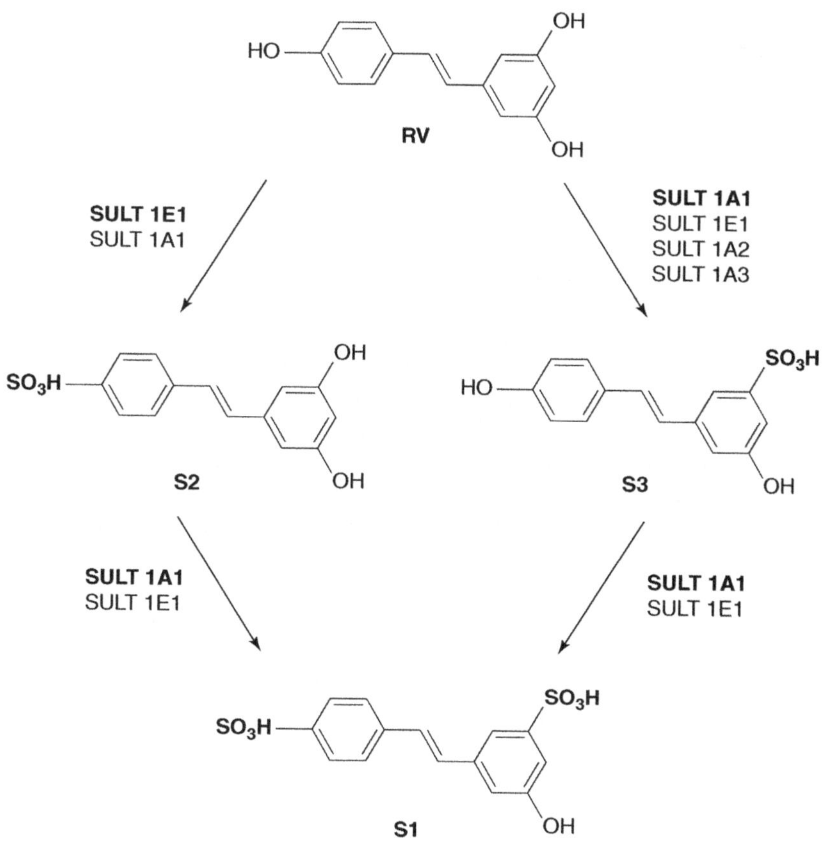

Figure 14.4 Resveratrol (RV) metabolic pathway proposed in human liver cytosol. SULT enzymes shown in bold represent the predominant contribution to sulfate formation.

Along with resveratrol, it was possible to identify three metabolites, namely *trans*-resveratrol-4′-*O*-glucuronide (G1), *trans*-resveratrol-3-*O*-glucuronide (G2), and *trans*-resveratrol-3-*O*-sulfate (S3) by liquid chromatography/mass spectrometry (LC/MS) and NMR. Efflux of the glucuronides G1 and G2 significantly followed Michaelis–Menten kinetics favoring basolateral efflux. The predominant metabolite was the monosulfate S3, however, its formation was strongly inhibited at higher resveratrol concentrations. As metabolism was either inhibited or saturated, the total amount of resveratrol transported across the Caco-2 monolayers increased as much as 3.5-fold at 200 mM resveratrol, demonstrating a concentration-dependent biotransformation of resveratrol in Caco-2 cells, which may also apply to human enterocytes, thereby affecting oral bioavailability [Maier-Salamon et al., 2006]. The intestinal metabolic pathway of resveratrol in humans is depicted in Figure 14.5. Given that in vivo concentrations of individual metabolites can be much higher than those of the native compound, further studies are needed to determine whether the metabolites represent inactivated forms of the drug, act as a pool from which free resveratrol can be released into various tissues, or are themselves active in promoting many of the health benefits attributed to resveratrol. [Saiko et al. 2008].

INTERACTION OF RESVERATROL WITH DIETARY CONSTITUENTS

The diet consists of varied ingredients that impart their potential benefits to human health. Potential interaction of resveratrol with other constituents of the diet is worth considering. Resveratrol has been shown to synergize with both quercetin and ellagic acid present in most of our natural products, in the induction of apoptosis in human leukemia cells [Mertens-Talcott and Percival 2005], with ethanol in the inhibition of inducible nitric oxide synthase (iNOS) expression [Chan et al., 2000], with vitamin E in the prevention of lipid peroxidation (Fang et al., 2002], with catechin in the protection of PC12 cells from β-amyloid toxicity [Conte et al., 2003], and with nucleoside analogs in the inhibition of HIV1 replication in cultured T lymphocytes [Heredia et al., 2000]. These effects could help to explain how a relatively low dose of resveratrol obtained from red wine or other dietary sources could produce a measurable health benefit. However, no adverse action of resveratrol with other chemical constituents in the body has been reported so far, but instead it detoxifies the chemical carcinogens and proves to be chemopreventive [Baur and Sinclair, 2006].

TARGET ENZYMES OF RESVERATROL ACTION

Resveratrol is commonly referred to as a “dirty” molecule in the pharmaceutical industry, meaning that it seems to interact with many different proteins,

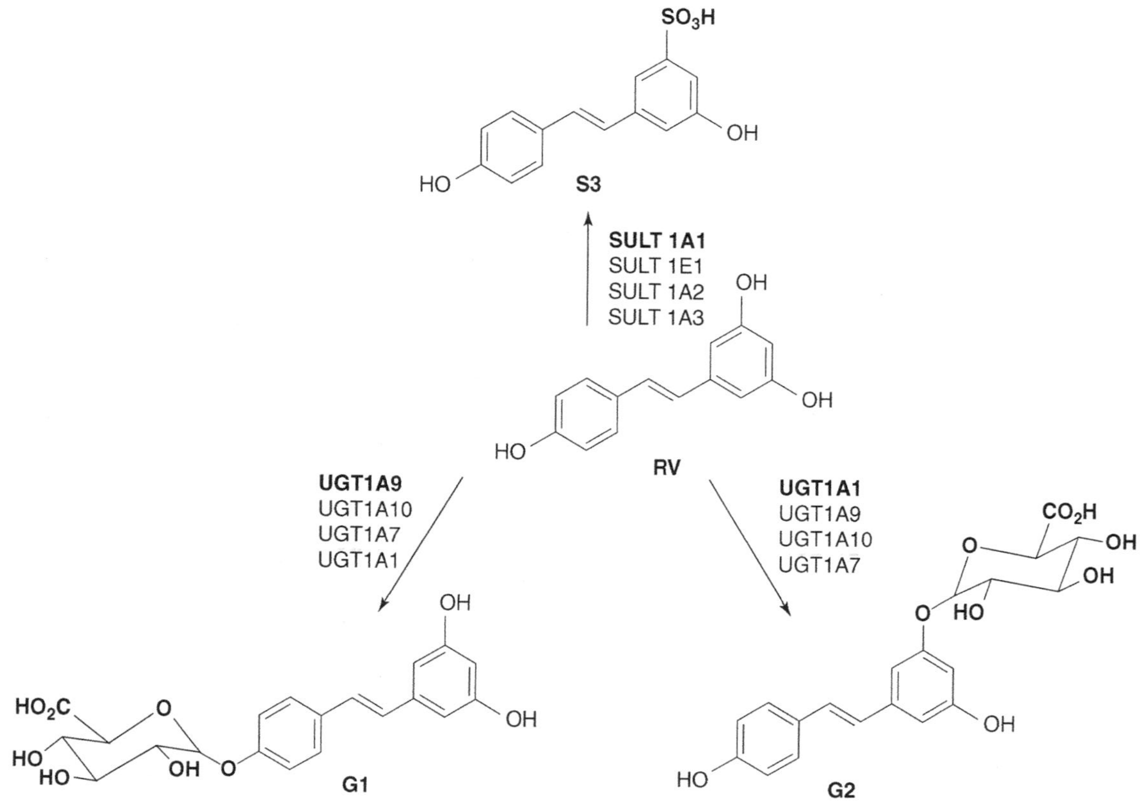

Figure 14.5 Intestinal resveratrol metabolic pathway proposed after oral administration using Caco-2 monolayers. SULT and UGT enzymes shown in bold represent the predominant contribution to sulfate and glucuronide formation.

including cyclooxygenases, ribonucleotide reductase, and DNA polymerases [Saiko et al., 2008]. Thus, its activity cannot be resumed in a unique mode of action but likely results from many complementary actions of different biochemical pathways. Resveratrol has been shown to inhibit a plethora of enzymes belonging to different classes, including (but not limited to) kinases, lipo- and cyclooxygenases, sirtuins, and other proteins [Pirola and Frojdo, 2008]. From a mechanistic point of view, both the *m*-hydroxyquinone and 4-hydroxystyryl moieties (Fig. 14.1) of the molecule have been shown to be important for the determination of resveratrol's inhibitory properties toward various enzymes such as STAT3, JNK1, PKCs, ERK1, Src, Cyclo and lipooxygenases, aromatases, and DNA polymerases [Pirola and Frojdo, 2008]. The current evidence about the various resveratrol targets, which provide a framework of possible pathways mediating some of the endpoint responses elicited by resveratrol treatment, is overwhelming, which gives an idea that the effects of resveratrol may reflect its simultaneous action on several targets by both activating and inhibiting.

STRUCTURE–ACTIVITY RELATIONS OF RESVERATROL

Quantitative structure–activity relationships (QSAR) are powerful tools in predicting the action of chemicals in biological systems [Karcher et al., 1990; Bradbury, 1994]. The chemicals may be synthetic or a derivative of a natural product, which exhibit a wide array of biological activity exhibiting potential health benefits. As evidenced by many reports, resveratrol and its analogs are a therapeutic promise of alternative medicine [Das and Maulik, 2006]. Activity of resveratrol cannot be resumed in a unique mechanism of action, but it likely results but it results from various complementary actions of different biochemical pathways. PAGE analysis (parametric analysis of gene set enrichment) indicated that resveratrol caused a significant alteration in 127 pathways, including the TCA cycle, glycolysis, the classic and alternative complement pathways, butanoate and propanoate metabolism, sterol biosynthesis, Stat3 signaling, insulin signaling, IGF-1 and mTOR signaling, oxidative phosphorylation, and electron transport [Baur et al., 2006]. Such a diverse range of effects may depend on the interplay between resveratrol and a very specific cellular target that must be an upstream controller of many cellular events.

Resveratrol is one of the many naturally occurring hydroxylstilbenes consisting of two benzene rings which are connected with an olefin. The structural features of hydroxystilbenes (C6–C2–C6) are similar to those of flavonoids (C6–C3–C6), which have various biological activities including the inhibition of mast cell degranulation [Cheong et al., 1998]. Resveratrol exerts an inhibitory effect on the nitric oxide production in macrophage cells not through estrogen-mediated pathway but through direct non-receptor-mediated pathway, and the hydroxystilbene moiety is the main constituent involved in its activity [Cho et al., 2002].

PHARMACOLOGICAL PRECONDITIONING EFFECTS OF RESVERATROL

A state-of-the-art technique, preconditioning, was initially reported by Murry et al. in 1986, where several brief periods of ischemia followed by short interval of reperfusion for a few cycles caused cardioprotection against ischemic heart injury [Das et al., 2005]. There are numerous attempts to mimic the protective effects of preconditioning by exogenous agents, and thereby to produce pharmacological preconditioning. Resveratrol has also, cardioprotective effect which can be related to its antioxidant activity. Therefore, Bradamante et al. [2000] and De Jonge et al. [2002] have investigated how resveratrol affects high-energy phosphate metabolism, using ^{31}P-NHR, and contractility of isolated Langendorff perfused rat hearts subjected to 20 min no-flow ischemia and 30 min reperfusion. It is well established from various studies that adenosine (A) receptors play a crucial role in ischemic preconditioning [Das et al., 2005]. In this study beside the resveratrol pretreatment, all the adenosine receptor blockers were used with the resveratrol, and then the heart was subjected to 30 min global ischemia followed by 120 min reperfusion. In addition to this, Akt, p-Akt, Bcl-2, p-Bcl-2, Bad, p-Bad, CREB, p-CREB were determined. The results of this study are parallel with many others reports indicating the role of A_1 and A_3, but not A_{2a} receptors, in the resveratrol-mediated preconditioning [De Jonge et al., 2002]. The main finding of this study and another follow-up study by the same group [Das et al., 2005b] is that resveratrol-mediated preconditioning is related to the activation of A_1 and A_3receptors, which triggers a survival pathway through the phosphatidylinositol (PI) 3-kinase-Akt-Bad/Bcl-2 and by activating adenosine A_3-CREB-Bcl-2 (antiapoptotic) signaling.

Hattori and co-workers [2002] also proved the preconditioning effects of resveratrol which is to be mediated by nitric oxide (NO) or NOS. To confirm this hypothesis, a NO inhibitor, N^G-nitro-L-arginine methyl ester (L-NAME), as well as an iNOS blocker, aminoguanine (AG), were used in their studies. The L-NAME and AG pretreatment abolished the resveratrol-mediated cardioprotection, suggesting the role of NO in resveratrol-induced cardioprotection. To gather further evidence, Imamura and collaborators [2002] used isolated iNOS knockout mouse hearts. Both resveratrol treated and nontreated iNOS knockout mice hearts displayed relatively poor recovery in cardiac function as compared to the wild-type mice. Obviously, resveratrol induced an increased expression of iNOS in the wild-type mice but not in the iNOS knockout hearts, indicating an essential role of iNOS in resveratrol-mediated preconditioning. In an additional study, the expression of the molecular targets of NO [e.g., vascular endothelial growth factor (VEGF), KDR, iNOS, eNOS] were examined by western blot and immunohistochemistry [Das et al., 2005c]. Recent studies have also demonstrated that NO can induce the expression of heme-oxygenase (HO-1). Das and co-workers showed that protoporphyrin (SnPP), an HO-1 inhibitor treatment, was able to abolish the cardioprotective effects of resveratrol (i.e., increased cardiac function parameters, reduced myocardial infarct size, decreased cardiomyocytes apoptosis) [Das et al., 2006]. The HO-1, mediated mechanisms were related to the p38MAP kinase and Akt survival signaling, but independent of NF-κB activation using p38MAPK and Akt inhibitor. The cardioprotective mechanisms of resveratrol using sour cherry seed kernel extract, against ischemia-reperfusion-induced damage through the reduced caspase-3 activity was confirmed by Bak et al. [2006]. Similarly HO-1 up-regulation via the antioxidant response element (ARE)-mediated transcriptional activation of NF-E2-related factor-2 (Nrf2) was demonstrated [Chen et al., 2005].

ANTIOXIDANT POWER OF RESVERATROL

Many compounds with aromatic groups are able to function as antioxidants by forming stable radicals via resonance structures, thereby preventing continued oxidation. Resveratrol contains two aromatic groups and has been shown to have a higher 2,2′-azinobis (3-ethylbenzthiazoline-6-sulfonic acid) (ABTS), 1,1-diphenyl-2-picrylhydrazyl (DPPH), and hydroxyl radical scavenging capacity than propyl gallate, vitamin C, and vitamin E [Soares et al., 2003]. A positive correlation has been established between the antioxidant power of wine and resveratrol content [Alonso et al., 2002]. Molecules bearing ortho-dihydroxyl or 4-hydroxy-3-methoxyl functionality are appreciably more active than those bearing no such functionalities. It has been proven that the ortho-methoxyl group can form an intramolecular hydrogen bond with the phenolic hydrogen, making the H-atom abstraction from the ortho-methoxyphenols surprisingly easy [de Heer et al., 2000]. It is also known that the ortho-hydroxyl substitution on phenol would make the oxidation intermediate, ortho-hydroxyphenoxyl radical, more stable due to the intramolecular hydrogen-bonding interaction as reported recently from both experiments by spectrophotometric measurements [Foti and Ruberto; 2001] and theoretical calculations. The theoretical calculations show that the intramolecular H bond in ortho-hydroxyphenoxyl radical is ca. 4 kcal/mol stronger than that in the parent molecule catechol, and the bond dissociation energy (BDE) of catechol is 9.1 kcal/mol lower than that of phenol [Wright et al., 2001]. In addition, ortho-hydroxyphenoxyl radical and/or ortho-semiquinone radical anion will be more easily further oxidized to form the final product ortho-quinone as exemplified in Figure 14.6. As an antioxidant, resveratrol may delay and/or prevent oxidative stress-induced cellular damage and disease. Resveratrol is thus a promising source to combat most of the human ailments and to increase longevity of life.

ANTI-INFLAMMATION WITH RESVERATROL

Recently, numerous possible ways have been identified by which resveratrol may attenuate inflammation in various organs. The anti-inflammatory response of resveratrol was realized from its ability to down-regulate pro-inflammatory

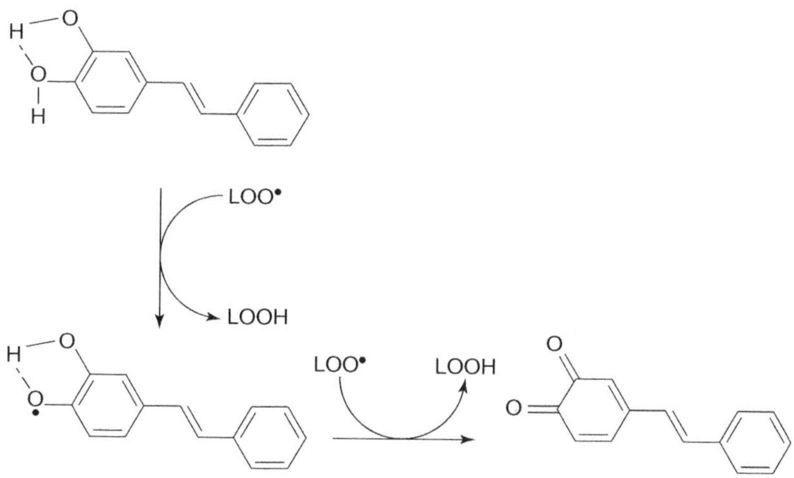

Figure 14.6 Mechanism of 3,4-DHS-inhibited peroxidation.

cytokines [Pervaiz, S 2003]. Tissue injury is a multistep process and such processes last till the time the region is healed. The first step of inflammation includes the activation and migration of leukocytes, including neutrophils, monocytes, and eosinophils, to the injured tissue site from the various parts of the body. Resveratrol is found to reduce inflammatory responses induced by formyl methionyl leucyl phenylalanine (fMLP), component fragment C5a, or calcium ionophore A23187 [Rotondo et al., 1998]. The next step, the macrophages, takes over the process of inflammation. NO production can have some beneficiary effects on the activation of macrophages. Inducible NO synthase (iNOS) plays an important role in this process. As mentioned earlier, resveratrol has been found to induce iNOS [Sharma et al., 2007; and Chander et al., 2005]. Resveratrol also interferes with the pro-inflammatory signaling of thrombin, resulting in the inhibition of adenosine nucleotide secretion from activated platelets and decreased neutrophils functions via inhibition of PAP and P2-receptor signaling through MAPK and cJun and JNK [Kaneider et al., 2004]. A recent study showed that in mouse epidermal cells, resveratrol activated extracellular signal-regulated kinases (ERK), c-Jun NH_2-terminal kinases (JNK), and p38 MAPK, leading to the serine 15 phosphorylation of p53 [She et al., 2001].

Recent studies revealed that resveratrol protects from inflammation by acting at different phases of inflammation. Protection at the pro-inflammatory phase appears to be very important for reducing inflammation effectively and promptly. A recent study [Ge et al., 2006] showed that resveratrol inhibits macrophage expression of EMMPRIN by activating PPAR-γ. In another similar study, Ma et al., [2006] showed a similar observation, but additionally found a role of nuclear transcription factor κB (NF-κB) in macrophage inhibition. Numerous studies confirmed that resveratrol suppresses the TNF-α activation [Marier et al., 2005; Moon et al., 2006]. Some preclinical as well as clinical studies have shown significant suppression of interleukin-1 (IL-1) and IL-6 with resveratrol [Baggiolini et al., 1995; Marier et al., 2005]. Similarly, IL-8 plays an important role in inflammation as it recruits leucocytes [Olas et al., 2001]. Donnelly et al. [2004] showed that resveratrol could inhibit IL-8. In another study, resveratrol also inhibited the TNF-α and histamine release by immunoglobulin E (IgE) in mouse mast cells [Baolin et al., 2004]. There are three different soluble adhesive molecules present in the leukocytes and endothelium; integrins (α- and β-integrins), selectins (L-, P-, and E-selectin), and immunoglobulin subfamily of cell surface proteins (vascular cell adhesion molecule-1, VCAM-1 and intracellular adhesion molecule-1, ICAM-1). In an acute in vivo study, Das et al. [2006a] successfully showed that resveratrol significantly inhibited the generation of adhesion molecules.

There are two different enzyme systems that are involved in the synthesis of pro-inflammatory mediators, such as cyclooxygenase (COX) and lipoxygenase, which finally release inflammatory substances, prostanoids and leukotrienes, respectively. Resveratrol inhibited the induction of LTB_4 and attenuated PAF in rats [Shigematsu et al., 2003]. Similarly, it has been found that resveratrol has significant effect on the production of the peptide leukotriene LTC_4 [Kimura et al., 1995]. The anti-inflammatory response of resveratrol was showed by Jang et al. [1997] who found inhibition of prostaglandin synthesis by the inhibitory effect of COX-1 with resveratrol. Subsequently, the same group found that resveratrol could selectively inhibit COX activity by inhibiting COX-1 pathway, but not through COX-2 pathway. Szewczuk et al. [2004] confirmed this finding. In contrast, another study found that resveratrol suppressed the synthesis PGE_2 by the inhibition of COX-2 but not altering the COX-1 [Subbaramaiah et al., 1998].

In the process of evaluating certain derivatives of resveratrol, inhibition assays for COX-1 and COX-2 were done on hydroxylated and methoxylated resveratrol analogs and compared to celecoxib, a known COX inhibitor. Quantitative structure–activity relationship (QSAR) analysis exhibited a high correlation with the topological surface area TPSA ($r = 0.93$) and the percentage of inhibition. Docking studies on both COX-1 and COX-2 protein structures also revealed that hydroxylated but not methoxylated resveratrol analogs are able to bind to the previously identified binding sites of the enzymes [Murias et al., 2004]. Hydroxylated resveratrol analogs, therefore, represent a novel class of highly selective COX-2 inhibitors and promising candidates for in vivo studies. In another study, resveratrol was found to be a potent inhibitor of both the cyclooxygenase and peroxidase activities of COX-1, but the drug acted only as a reducing co-substrate for COX-2 [Szewczuk et al., 2004]. The observation that resveratrol acted as a noncompetitive inhibitor of the cyclooxygenase activity of COX-1 suggests that arachidonic acid and resveratrol bind at different sites that correspond to the cyclooxygenase and peroxidase active sites, respectively. The observation that resveratrol acted as an uncompetitive inhibitor of the peroxidase activity of COX-1 suggests that resveratrol requires a peroxide substrate to exert its inhibitory effects via the formation of an E-S-I complex and confirms a novel mode of inhibition for resveratrol. It has been confirmed that the *m*-hydroquinone moiety (3,5-di-OH group) of resveratrol is required for mechanism-based inactivation of COX-1. The *m*-hydroquinone is unique because oxidation of one hydroxy group results in a semiquinone radical that cannot be stabilized through the ring structure to the remaining hydroxy group, as is the case for *o*- and *p*-hydroquinones. However, with COX-2, all of the hydroxy groups on resveratrol can serve as reducing co-substrates.

In a related study, resveratrol significantly reduced colonic injury, neutrophil infiltration, and drastically reduced the PGD_2 concentration by inhibiting COX-2, but not affecting COX-1 [Martin et al., 2004]. It appears that the anti-inflammatory activity of resveratrol may be realized through the inhibition of both COX-1 and COX-2-mediated pro-inflammatory signaling, suppression of pro-inflammatory mediator production, as well as from its potent antioxidative effects.

DIABETES CURE WITH RESVERATROL

Diabetes is a disease when islets of Langerhans cells in the pancreas do not produce, or poorly produce, insulin or the body does not properly use the insulin present in the system. Insulin is a hormone which mobilizes sugar and starches to get the bloodstream to get converted into energy which is needed for the daily activity of the body. The cause of diabetes is still a mystery to the research world. Although both genetics and environmental factors such as obesity, food habit, or lack of exercise appears to play an important role, still the underling mechanism remains unclear.

According to the latest report of the American Diabetes Association in 2006, there are at least 20.8 million children and adults in the United States, or 7% of the population, who have diabetes. While an estimated 14.6 million have been diagnosed with diabetes, unfortunately, 6.2 million people (or nearly one-third) are unaware that they have the disease. That is why diabetes is also known as the "silent killer." The rapid increase in the number of silent killers has caught the attention of the research world and recently almost double the number of scientists are working in this area compared to the last decade.

Recently, resveratrol was found to reverse fat-induced insulin resistance [McCarty, 2005]. This observation provides more enthusiasm for researchers to use resveratrol as an antidiabetic agent. Su and associates [2006] showed that resveratrol significantly reduced the plasma glucose concentration as well as the dramatic reduction of triglyceride concentration in streptozotocin (STZ)-induced diabetic mellitus rats in 14 days treatment. They concluded from this observation that resveratrol possesses hypoglycemic and hypolipidemic properties [Su et al., 2006]. Baur et al., [2006] added more value to this conclusion by showing that resveratrol increases insulin sensitivity by lowering the blood glucose level in a group of high-calorie diet mice. They have pointed out that reduced insulin-like growth factor-1 (IGF-I) levels and increased AMP-activated protein kinase (AMPK) probably play a major role.

Some other related studies established that resveratrol can attenuate diabetic nephropathy in rats [Sharma et al., 2006a] and reduce thermal hyperalgesia and cold allodynia in streptozotozin-induced diabetic rats [Sharma et al., 2006b].

CANCER CURE WITH RESVERATROL

Resveratrol plays critical roles in signaling mechanisms of different tumors. Currently, it is well documented that many factors play a role in the pathomechanisms of cancers. However, the exact mechanisms are not precisely known. If resveratrol could provoke an influence in only one pathological factor, then the effects of anticancer drugs (i.e., in combination therapy) might beneficially modulate the signal transduction pathways. This is the basic reason to investigate the precise mechanisms and effects of resveratrol and its derivatives on various cancers.

The inhibition of COX-2 (inducible form of COX) could be an important strategy for preventing the development of cancers. Subbaramaiah and his team treated human mammary and oral epithelial cells with a phorbol ester, PMA, which induced the COX-2 mRNA, COX-2 protein, protein-kinase C, c-Jun (promoter of COX-2), and prostaglandin synthesis. Phorbol ester-mediated inductions of these agents were inhibited by resveratrol [Subbaramaiah et al., 1999]. Nakagawa et al. [2001] examined the effects of resveratrol on the growth of estrogen receptor (ER)-positive (KPL-1 and MCF-7) and negative (MKL-F) human breast cancer cell lines. Synthetic resveratrol at high concentration ($\geq$44 μM) suppressed the cell growth in all cell lines. The chemical structure of resveratrol is similar to the 4,4′-dihydroxy-*trans*-diethylstilbene that belongs to the type I class of estrogens and acts as an ER agonist [Levenson et al., 2003]. The effect of resveratrol was related to the apoptosis cascade (up-regulated Bax and Bak proteins, down-regulated Bcl-xL proteins, and activated caspase-3). Jazirehi and Bonavida, [2004] have also shown that resveratrol selectively down-regulated the expression of antiapoptotic proteins Bcl-xL and myeloid cell differentiation factor-1 (Mcl-1) and up-regulated the expression of proapoptotic proteins (Bax) and apoptosis protease activating factor-1 (Apaf-1). Scarlatti et al. [2003] found that resveratrol could provoke growth inhibition and apoptosis in metastatic breast cancer cells (MDA-MB-231) by activating the ceramide synthesis pathway. Furthermore, Dorrie et al. [2001] showed that resveratrol induced extensive apoptotic cell death by depolarizing mitochondrial membranes in CD95-sensitive leukemia cell lines as well as B-lineage leukemic cells, which are resistant to CD95-signaling. In addition, resveratrol induced the activation of caspase-9 in all lines. However, the benzyloxycarbonyl-Val-Ala-Asp-fluoromethylketone (wide-spectrum caspase-inhibitor) failed to inhibit the depolarization of mitochondrial membranes elicited by resveratrol, suggesting that resveratrol activation is independent of upstream caspase-8 activation via receptor ligation.

Resveratrol is a sensitizer of tumor necrosis factor-related apoptosis through p53-independent induction of p21 and p21-mediated cell cycle (G_1) arrest associated with down-regulation of surviving cells [Fulda et al., 2004]. Kuwajerwala et al. [2002] reported that resveratrol-induced increase in DNA synthesis was associated with enrichment of LNCaP cells in S phase, and a concurrent decrease in nuclear p21Cipl and p27Kip1 levels. In addition, Cao et al. [2004] have shown that resveratrol inhibited hypoxia-inducible-factor 1α (HIF-1α) and VEGF expression in human ovarian cancer through multiple mechanisms: (i) inhibition of Akt and mitogen-activated protein kinase activation, (ii) inhibition of protein translational regulators, and (iii) proteosome pathway [Kim et al., 1993]. Resveratrol also reduced the antiapoptotic kinase Akt in various uterine cancer cell lines [Sexton et al., 2006]. Similar effects occurred in human tongue squamous cell carcinoma and hepatoma cells, [Zhang et al., 2005]. It was reported that resveratrol down-regulated the expression of NF-κB, TNF-α, interleukin 1β, interleukin-6, Bcl-2, Bcl-xL, XIAP, c-IAP, VEGF, and matrix metalloproteinase-9 (MMP-9) in human multiple myeloma (MM) cells or ex vivo studies [Millauer et al., 1994]. In a further study, it was investigated whether resveratrol-Cu (II) system is capable of causing DNA degradation in lymphocytes isolated from human peripheral blood [Azmi et al., 2005]. This study partially supports the hypothesis that anticancer properties of resveratrol may involve mobilization of endogenous copper and the consequent pro-oxidant action. Kotha et al. [2006] was among the initial group to identify Src-Stat3 signaling as a target of resveratrol in tumors. Beside the effects of resveratrol and its derivates alone, it is important to emphasize the combined effects of resveratrol with other anticancer drugs. Therefore, resveratrol combined with cisplatin or doxorubicin produced an additive growth-inhibitory and anticancer effects and protection against doxorubicin-induced cardiac toxicity both in human gynecologic cancer cell lines and *in* vivo [Rezk et al., 2006]. Resveratrol analogs with hydroxyl groups in positions 3′,4, and 5′ showed an even higher cytotoxicity in HL-60 cells than resveratrol, indicating an increased antitumor activity [Murias et al., 2004]. Their data were consistent with another group who also found that 3,4-dihydroxyl groups are important for enhanced antioxidant and anticancer activities of *trans*-resveratrol analogs [Cai et al., 2004].

ANTIAGING EFFECTS OF RESVERATROL

In many eukaryotic cells such as rodents, flies, and nematode worms, and even single-celled organisms such as baker's yeast, only SIR2 among many longevity regulatory genes has received significant attention from researchers [Hall, 2003]. In human cells, the analog gene of SIR2 is SIRT1, which can extend life span [Howitz et al., 2003].

In a recent study, Howitz et al. [2003] showed red-wine-derived resveratrol could increase the SIRT1 activity 13-fold by inhibiting apoptotic cell death through the deacetylation of p53. In the same study they also showed that resveratrol extended the life span by 70% by mimicking calorie restriction by stimulating SIR2 in the yeast cells and, thus, increased DNA stability. In another very recent study on calorie restriction experiments on monkey and other mammalian species, the concept of wine as "fountain of youth" was reconfirmed [Kujoth et al., 2005].

ANTIULCER EFFECTS OF RESVERATROL

In 1982, the bacterium *Helicobacter pylori* was discovered by Marshall and Warren which was identified as the major cause of peptic ulcer rather than only stress and lifestyle factors which was thought before. This bacterium ruptures the mucus membrane of the stomach. In an in vitro study, it was shown that wine could effectively reduce the growth of *H. pylori* propagation [Marimon et. al., 1998]. A few other studies showed that red wine is more effective over white wine to inhibit the growth of *H. pylori* [Mahady and Pendland, 2000; Daroch et al., 2001]. Mahady and associates [2003] showed in their study that resveratrol is the active component that causes the inhibitory effect of 15 different stained *H. pylori* propagations [Mahady et al., 2003]. This observation was also confirmed by another in vitro study [Mahady et. al., 2003] by the same group of researchers.

ANTIOBESE EFFECTS OF RESVERATROL

Adult-onset obesity is a serious health problem because it is implicated in various diseases including type II diabetes, hypertension, coronary heart disease, and cancer [Kopelman, 2000]. Obesity is characterized by an increase in adipose tissue mass which results both from increase fat cell numbers and increased fat cell size [Couillard et al., 2000]. Picard et al. [2004] showed that resveratrol activated the expression of Sirt1, and this, in turn, resulted in decreased lipid accumulation by repression of PPAR-γ in differentiated adipocytes. Chronic low-grade inflammation characterized by adipose tissue macrophage accumulation and abnormal cytokine production is a key feature of obesity and type II diabetes. Adipose-tissue-derived monocyte chemoattractant protein (MCP)-1, induced by cytokines, has been shown to play an essential role in the early events during macrophage infiltration into adipose tissue. In a recent study, resveratrol was found to inhibit TNF-α-induced MCP-1 secretion and gene transcription, as well as promoter activity, which is based on down-regulation of TNF-α induced MCP-1 transcription [Zhu et al., 2008]. One of the latest studies from our laboratory reveals that resveratrol induced cardiac protection in Zucker obese rats. It was confirmed by the increased GLUT-4 expression and reduced endothelin expression and cardiac apoptosis in ischemic-reperfused hearts in the presence or absence of glucose intake [Lekli et al., 2008]. Parametric analysis of gene set enrichment revealed that resveratrol opposed the effects of the high-calorie diet in 144 out of 153 significantly altered pathways such as increased insulin sensitivity, reduced insulin-like growth factor-1 (IGF-I) levels, increased AMP-activated protein kinase (AMPK), and peroxisome proliferator-activated receptor-c coactivator 1α (PGC-1α) activity, increased mitochondrial number, and improved motor function [Baur et al., 2006]. These data show that improving general health in mammals using small molecules such as resveratrol is an attainable goal, and point to new approaches for treating obesity-related disorders.

OTHER HEALTH BENEFITS OF RESVERATROL

Resveratrol has been recognized as a phytoestrogen since it possesses structural similarities with estrogenic compounds and may exert some biological activities through estrogen receptors [Cheng et al., 1996; Mgbonyebi et al., 1998; Gehm et al., 1997]. In some studies, resveratrol has been shown to bind to estrogen receptors as an agonist [Lu and Serrero, 1999; Henry and Witt, 2002; Klinge et al., 2003].

The resveratrol-induced protection against ischemia/reperfusion has been documented in various organs (e.g., kidney, brain, heart, liver, etc.). In a study of spinal cord ischemia model (SCI), resveratrol-induced neuroprotection was found to be mediated by decreased oxidative stress, and increased NO release. Organotypic hippocampal slice cultures were exposed to oxygen–glucose deprivation (OGD) using LY294002 and PD98059 treatment to confirm the exact mechanism. Neuroprotection was arrested by LY294002 but not by PD98059. Thus, these findings suggest that PI3-k/Akt pathway but not mitogen-activated protein kinase (MAPK) are involved in neuroprotection [Zamin et al., 2006]. Ischemic preconditioning was mimicked by resveratrol in brain via the SIRT1 pathway, which was confirmed by sirtinol [Raval et al., 2006]. Recent studies have suggested that resveratrol may act as not only an antioxidant agent but as a signaling molecule also [Dore, 2005] within tissues and cells resveratol acts not only as an antioxidant agent but also as a signaling molecule within cells and tissues to modulate the expression of genes and proteins (i.e., HO-1, COX-2, eNOS, iNOS, endotelin-1,TNF-, insulin-like growth factor binding protein). HO-1 protein has been documented with different functions in the brain. Since HO-1 is a heat-shock protein, which is induced in several cells and following different stimuli (e.g., hypothermia, global ischemia, subarachnoid hemorrhage, Parkinson disease, Alzheimer disease, etc.), the modulation of the different genes may explain the cytoprotective actions afforded by resveratrol, as well as the influence of resveratrol on blood flow, cell death, and inflammatory cascades [Sener et al., 2006].

Protection against ischemia/reperfusion-induced oxidative injury occurred in rat kidney as well. Resveratrol caused a significant increase in tissue glutathione (GSH), and decreased the renal luminal, lucigenin CL, MDA levels, MPO activity, and collagen content [Sener et al., 2006]. Acute renal failure (ARF) can be induced by glycerol, and glycerol provokes a marked decrease in tissue and urine nitric oxide levels, causes renal oxidative stress, deterioration of renal morphology, and renal function. Resveratrol could reverse the aforementioned effects. However, the renoprotective effect was abolished using N^G-nitro-L-arginine methyl ester, a nonspecific NO-synthase-inhibitor (L-NAME) co-treatment, suggesting that resveratrol exerts the protective effects through the NO release [Chander et al., 2005, 2006]. Leikert et al., [2002] also reported that renoprotective effects of resveratrol were attributed to the release of endothelial nitric oxide. It is well known that reactive oxygen species (ROS) are harmful for glomeruli and tubular epithelial cells. The up-regulation of NF-κB activates the cytokines, chemokines, and adhesion molecules. These are very harmful mediators of inflammation, leading to fibrosis, glomerular damage, and tubulointerstitial damage. Resveratrol reinforces the endogenous antioxidant systems that depurates ROS and thereby contributes to cytoprotective effects [Rodrigo et al., 2006]. The deterioration of lipid peroxidation caused by resveratrol is also related to lung, liver, and kidney protection [Bertelli et al., 2002]. Angiotensin II (Ang II) also plays a crucial role in the progression of renal diseases by activating membrane-bound nicotinamide adenine dinucleotide phosphate (NADPH) oxidase and superoxide anion leading to hypertrophy of renal tubular cells. Since Ang II regulates the overexpression of TGF-β and TNF-α, this "side" mechanism also contributes to the obstructive nephropathy. This pathomechanism pathway (Ang II) is inhibited by resveratrol and plays a part in nephroprotection [Klahr, 1998].

The oxidative stress also plays a key role in the pathogenesis of various liver diseases and their progression. The use of resveratrol has been proposed as a therapeutic agent as well as coadjuvant, to prevent liver damage [Vitaglione et al., 2004]. The antioxidant activity was evaluated by measuring the inhibition of citronellal thermo oxidation or the reduction of 2,2-diphenil-1-picrylhydrazy radical [Vitaglione et al., 2004]. Furthermore, resveratrol can directly interact with DNA polymerases alpha and delta beyond the antioxidant activity [Stivala et al., 2001]. In a further study, hepatotoxicity was induced by acetaminophen. Alanine aminotransferase (ALT), aspartate aminotransferase (AST), MDA, TNF-α levels, and MPO activity were significantly increased, and GSH level was significantly decreased after the acetaminophen treatment [Sener et al., 2006]. Co-treatment with resveratrol reversed these effects. Acetaminophen is not the only medicine that can provoke liver damage. Therefore, these results suggest that resveratrol and natural antioxidant nutrients can be used as therapeutic agents in order to prevent hepatic oxidative injury due to hepatotoxic drugs.

It is also well documented that an antioxidant-rich diet protects against age-related macular degeneration (AMD), the leading cause of vision loss among the elderly. Development of AMD in the retinal pigment epithelium (RPE) is associated with oxidative stress. King and co-workers [2005] reported that resveratrol treatment (100 μmol/L) inhibited basal and H_2O_2-induced intracellular oxidation, and protected RPE cells from cell death via the MAPK/ERK and extracellular signal-regulated kinase (ERK 1/2) pathways. Moreover, Sparrow and colleagues showed that resveratrol reduced A2E-epoxide formation in RPE [Sparrow et al., 2003]. Doganay et al. [2006] demonstrated that resveratrol suppressed selenite-induced oxidative stress and cataract formation in the rat eye. The protective effect was associated with higher GSH and lower MDA in lens and erythrocytes. The main point of this study is that resveratrol-rich diet could prevent human senile cataract. The protection of resveratrol against the ischemia/reperfusion injury also occurs in the eye (e.g., in retina). Retinal ischemia/reperfusion resulted in a significant reduction in HO-1-protein expression, and HO-1-related endogenous CO production which affects the ion levels in retina. Cellular Na^+ and Ca^{2+} levels were increased, and K^+ content was decreased. In rats treated with resveratrol-rich extract, tissue Na^+ and Ca^{2+} accumulation and K^+ loss were prevented through the HO-1-related endogenous CO production [Szabo 2004].

Huang et al. [2005] reported that resveratrol induced HO-1 protein up-regulation in rat aortic smooth muscle cells via the NF-κB pathway. Pendurthi and co-workers [1999] showed that resveratrol inhibited the tissue factor (TF) expression, LPS-induced expression of TNF-α, and IL-1β in vascular cells. In contrast, Di Santo and associates [2003] demonstrated that resveratrol significantly reduced TF activity, which originated from a reduction in nuclear binding activity of the transacting factor c-Rel/p65 in human endothelial and mononuclear cells. Antiplatelet activity of resveratrol is beneficial in many diseases. Olas et al. [2001] showed that resveratrol inhibited chemiluminescence and generation of $O_2^{\cdot}$in blood platelets caused by LPS or thrombin. This antiplatelet effect is in agreement with the results obtained by another study [Bertelli et al., 1995]. Moreover, resveratrol significantly inhibited ICAM-1 and VCAM-1 expression on TNF-α-stimulated human umbilical vein endothelial cells and LPS-stimulated human saphenous vein endothelial cells [Ferrero et al., 1998]. This group showed that resveratrol induced a significant inhibition in the adhesion of U937 monocytoid cells and neutrophils to TNF-α-stimulated NIH/3T3 ICAM-1-transfected cells. These results suggest an independent pathway of its antioxidant function [Ferrero et al., 1998].

ACKNOWLEDGMENT

The study was supported by the following grants: NIH HL 22559, HL33889, and HL69910 to DKD.

REFERENCES

Adamis AP, Miller JW, Bernal MT, Damico DJ, Folkman J, Yeo TK, Yeo KT. 1994. Increased vascular endothelial growth factor levels in the vitreous of eyes with proliferative diabetic retinopathy. *Am J Ophthalmol* 118:445–450.

Aggarwal BB, Bhardwaj A, Aggarwal RS, Seeram NP, Shishodia S, Takada Y. 2004. Role of resveratrol in prevention and therapy of cancer: Preclinical and clinical studies. *Anticancer Res* 24:2783–2840.

Aiello LP, Avery RL, Arrigg PG, Keyt BA, Jampel HD, Shah ST, Pasquale LR, Thieme H, Iwamoto MA, Park JE, Nguyen HV, Aiello LM, Ferrara N, King GL. 1994. Vascular endothelial growth factor in ocular fluid of patients with diabetic retinopathy and other retinal disorders. *N Engl J Med* 331:1480–1487.

Alonso AM, Domínguez C, Guillén DA, Barroso CG. 2002. Determination of antioxidant power of red and white wines by a new electrochemical method and its correlation with polyphenolic content. *J Agric Food Chem* 50:3112–3115.

Asensi M, Medina I, Ortega A, Carretero J, Bano MC, Obrador E, Estrela JM. 2002. Inhibition of cancer growth by resveratrol is related to its low bioavailability, *Free Radic Biol Med* 33:387–398.

Azmi AS, Bhat SH, Hadi SM. 2005. Resveratrol-Cu (II) induced DNA breakage in human peripheral lymphocytes: Implications for anticancer properties. *FEBS Lett* 579:3131–3135.

Baggiolini M, Loetscher P, Moser B. 1995. Interleukin-8 and the chemokine family. *Int J Immunopharmacol* 17:103–108.

Bak I, Lekli I, Juhasz B, Nagy N, Varga E, Varadi J, Gesztelyi R, Szabo G, Szendrei L, Bacskay I, Vecsernyes M, Antal M, Fesus L, Boucher F, de Leiris J, Tosaki A. 2006. Cardioprotective mechanisms of Prunus cerasus (sour cherry) seed extract against ischemia-reperfusion-induced damage in isolated rat hearts. *Am J Physiol Heart Circ Physiol* 291:H1329–H1336.

Baolin L, Inami Y, Tanaka H, Inagaki N, Inagaki N, Iinuma M, Nagai H. 2004. Resveratrol inhibits the release of mediators from bone marrow-derived mouse mast cells in vitro. *Planta Med* 70:305–309.

Baur JA, Sinclair DA. 2006. Therapeutic potential of resveratrol: The in vivo evidence. *Nat Rev Drug Discov* 5:493–506.

Baur JA, Pearson KJ, Price NL, Jamieson HA, Lerin C, Kalra A, Prabhu VV, Allard JS, Lopez-Lluch G, Lewis K, Pistell PJ, Poosala S, Becker KG, Boss O, Gwinn D, Wang M, Ramaswamy S, Fishbein KW, Spencer RG, Lakatta EG, Le Couteur D, Shaw RJ, Navas P, Puigserver P, Ingram DK, de Cabo R, Sinclair DA. 2006. Resveratrol improves health and survival of mice on a high-calorie diet. *Nature* 444:337–342.

Bertelli AA., Migliori M, Panichi V, Origlia N, Filippi C, Das DK, Giovannini L. 2002. Resveratrol, a component of wine and grapes, in the prevention of kidney disease. *Ann N Y Acad Sci* 957:230–238.

Bertelli AA, Giovannini L, Giannessi D, Migliori M, Bernini W, Fregoni M, Bertelli A. 1995. Antiplatelet activity of synthetic and natural resveratrol in red wine. *Int J Tissue React* 17:1–3.

Bradamante S, Piccinini F, Barenghi L, Bertelli AA, De Jonge R, Beemster P, De Jong JW. 2000. Does resveratrol induce pharmacological preconditioning? *Int J Tissue React* 22:1–4.

Bradbury SP. 1994. Predicting modes of toxic action from chemical structure: An overview, SAR QSAR. *Environ Res* 2:89–104.

Brown LF, Yeo KT, Berse B, Yeo TK, Senger DR, Dvorak HF, Van de Water L. 1992. Expression of vascular permeability factor (vascular endothelial growth factor) by epidermal keratinocytes during wound healing. *J Exp Med* 176:1375–1379.

Burns J, Yokota T, Ashihara H, Lean ME, Crozier A. 2002. Plant foods and herbal sources of resveratrol. *J Agric Food Chem* 50:3337–3340.

Cabot RC. 1904. The relation of alcohol to atherosclerosis. *J Am Med Assoc* 43:774–775.

Cai YJ, Wei QY, Fang JG, Yang L, Liu ZL, Wyche JH. 2004. The 3,4-dihydroxyl groups are important for trans-resveratrol analogs to exhibit enhanced antioxidant and apoptotic activities. *Anticancer Res* 24:999–1002.

Cao Z, Fang J, Xia C, Shi X, Jiang BH. 2004. *Trans*-3,4,5′-trihydroxystilbene inhibits hypoxia-inducible factor 1alpha and vascular endothelial growth factor expression in human ovarian cancer cells. *Clin Cancer Res* 10:5253–5263.

Cellotti E, Ferrarini R, Zironi R, Conte LS. 1996 Resveratrol content of some wines obtained from dried Valpolicella grapes: Recioto and Amarone. *J Chromatogr A* 730:47–52.

Chan MM, Mattiacci JA, Hwang HS, Shah A, Fong D. 2000. Synergy between ethanol and grape polyphenols, quercetin, and resveratrol, in the inhibition of the inducible nitric oxide synthase pathway. *Biochem Pharmacol* 60:1539–1548.

Chander V, Chopra K. 2006. Protective effect of resveratrol, a polyphenolic phytoalexin on glycerol-induced acute renal failure in rat kidney. *Renal Failure* 28:161–169.

Chander V, Chopra K. 2005. Role of nitric oxide in resveratrol-induced renal protective effects of ischemic preconditioning. *J Vasc Surg* 42:1198–205.

Chen CK, Pace-Asciak CR. 1996. Vasorelaxing activity of resveratrol and quercetin in isolated rat aorta. *Gen Pharmacol* 27:363–366.

Chen CY, Jang JH, Li MH, Surh YJ. 2005. Resveratrol upregulates heme oxygenase-1 expression via activation of NF-E2-related factor 2 in PC12 cells. *Biochem Biophys Res Commun* 331:993–1000.

Cheong H, Ryu SY, Oak MH, Cheon SH, Yoo GS, Kim KM. 1998. Studies of structure activity relationship of flavonoids for the anti-allergic actions. *Arch Pharmacol Res* 21:478–480.

Cho DI, Koo NY, Chung WJ, Kim TS, Ryu SY, Im SY, Kim KM. 2002. Effects of resveratrol-related hydroxystilbenes on the nitric oxide production in macrophage cells: Structural requirements and mechanism of action. *Life Sci* 71:2071–2082.

Commodari F, Khiat A, Ibrahimi S, Brizius AR, Kalkstein N. 2005. Comparison of the phytoestrogen, trans-resveratrol (3,4′,5 trihydroxystilbene) structures from X-Ray diffraction and solution NMR. *Magn Reson Chem* 43:567–572.

Conte A, Pellegrini S, Tagliazucchi D. 2003. Synergistic protection of PC12 cells from β-amyloid toxicity by resveratrol and catechin. *Brain Res Bull* 62:29–38.

Corder R, Douthwaite JA, Lees DM. 2001. Endothelin-1 synthesis reduced by red wine. *Nature* 414:863–864.

Couillard C, Mauriege P, Imbeault P, Prud'homme D, Nadeau A, Tremblay A, Bouchard C, Despres JP, 2000. Hyperleptinemia is more closely associated with adipose cell hypertrophy than with adipose tissue hyperplasia. *Int J Obes Relat Metab Disord* 24:782–788.

Daroch F, Hoeneisen M, Gonzalez CL, Kawaguchi F, Salgado F, Solar H, Garcia A. 2001. In vitro antibacterial activity of Chilean red wines against Helicobacter pylori. *Microbios* 104:79–85.

Das S, Das DK. 2006a. Wine and the heart: A journey from grapes to resveratrol. *S Afr J Enol Vitic* 27:127–132.

Das DK, Maulik N. 2006. Resveratrol in cardioprotection: A therapeutic promise of alternative medicine. *Mol Interv* 6:36–47.

Das S, Falchi M, Bertelli A, Maulik N, Das DK. 2006a. Attenuation of ischemia/reperfusion injury in rats by the anti-inflammatory action of resveratrol. *Arzneimittelforschung* 56:700–706.

Das S, Fraga CG, Das DK. 2006b. Cardioprotective effect of resveratrol via HO-1 expression involves p38 map kinase and PI-3-kinase signaling, but does not involve NFkappaB. *Free Radic Res* 40:1066–1075.

Das S, Cordis GA, Maulik N, Das DK. 2005a. Pharmacological preconditioning with resveratrol: Role of CREB-dependent Bcl-2 signaling via adenosine A3 receptor activation. *Am J Physiol Heart Circ Physiol* 288:H328–335.

Das S, Tosaki A, Bagchi D, Maulik N, Das DK. 2005b. Resveratrol-mediated activation of cAMP response element-binding protein through adenosine A3 receptor by Akt-dependent and -independent pathways. *J Pharmacol Exp Ther* 314:762–769.

Das S, Alagappan VK, Bagchi D, Sharma HS, Maulik N, Das DK. 2005c. Coordinated induction of iNOS-VEGF-KDR-eNOS after resveratrol consumption: A potential mechanism for resveratrol preconditioning of the heart. *Vascul Pharmacol* 42:281–289.

Das DK, Sato M, Ray PS, Maulik G, Engelman RM, Bertelli AA, Bertelli A. 1999. Cardioprotection of red wine: Role of polyphenolic antioxidants. *Drugs Exp Clin Res* 25:115–120.

de Heer MI, Mulder P, Korth HG, Ingold KU, Lusztyk J. 2000. Hydrogen atom abstraction kinetics from intramolecularly hydrogen bonded ubiquinol-0 and other (poly)methoxy phenols. *J Am Chem Soc* 122:2355–2360.

De Jonge R, Out M, Maas WJ, De Jong JW. 2002. Preconditioning of rat hearts by adenosine A1 or A3 receptor activation. *Eur J Pharmacol* 441:165–172.

Dercks W, Creasy LL. 1989. Influence of fosetyl-A1 on phytoalexin accumulation in the *Plasmopara viticola*-grapevine interaction. *Physiol Mol Plant Pathol* 34:203–13.

Devries C, Escobedo JA, Ueno H, Houck K, Ferrara N, Williams LT. 1992. The fms-like tyrosine kinase, a receptor for vascular endothelial growth factor. *Science* 255:989–991.

Di Santo A, Mezzetti A, Napoleone E, Di Tommaso R, Donati MB, De Gaetano G, Lorenzet R. 2003. Resveratrol and quercetin down-regulate tissue factor expression by human stimulated vascular cells. *J Thromb Haemost* 1:1089–1095.

Doganay S, Borazan M, Iraz M, Cigremis Y. 2006. The effect of resveratrol in experimental cataract model formed by sodium selenite. *Curr Eye Res* 31:147–153.

Donnelly LE, Newton R, Kennedy GE, Fenwick PS, Fenwick PS, Leung RH, Ito K, Russell RE, Barnes PJ. 2004. Anti-inflammatory effects of resveratrol in lung epithelial cells: Molecular mechanisms. *Am J Physiol Lung Cell Mol Physiol* 287:L774–L783.

Dore S. 2005. Unique properties of polyphenol stilbenes in the brain: More than direct antioxidant actions; gene/protein regulatory activity. *Neurosignals* 14:61–70.

Dorrie J, Gerauer H, Wachter Y, Zunino SJ. 2001. Resveratrol induces extensive apoptosis by depolarizing mitochondrial membranes and activating caspase-9 in acute lymphoblastic leukemia cells. *Cancer Res* 61:4731–4739.

Fang JG, Lu M, Chen ZH, Zhu HH, Li Y, Yang L, Wu LM, Liu ZL. 2002. Antioxidant effects of resveratrol and its analogues against the free-radical-induced peroxidation of linoleic acid in micelles. *Chemistry* 8:4191–4198.

Ferrara N, Henzel WJ. 1989. Pituitary follicular cells secrete a novel heparin-binding growth factor specific for vascular endothelial cells. *Biochem Biophys Res Commun* 161:851–858.

Ferrero ME, Bertelli AE, Fulgenzi A, Pellegatta F, Corsi MM, Bonfrate M, Ferrara F, De Caterina R, Giovannini L, Bertelli A. 1998. Activity in vitro of resveratrol on granulocyte and monocyte adhesion to endothelium. *Am J Clin Nutr* 68:1208–1214.

Foti M, Ruberto G, 2001. Kinetic solvent effects on phenolic antioxidants determined by spectrophotometric measurements. *J Agric Food Chem* 49:342–348.

Fulda S, Debatin KM. 2004. Sensitization for tumor necrosis factor-related apoptosis-inducing ligand-induced apoptosis by the chemopreventive agent resveratrol. *Cancer Res.* 64:337–346.

Ge H, Zhang JF, Guo BS, He Q, Wang BY, He B, Wang CQ. 2006. Resveratrol inhibits macrophage expression of EMMPRIN by activating PPARgamma. *Vasc Pharmacol* 46:114–121.

Gehm BD, McAndrews JM, Chien PY, Jameson JL. 1997. Resveratrol, a polyphenolic compound found in grapes and wine, is an agonist for the estrogen receptor. *Proc Natl Acad Sci USA* 94:14138–14143.

Gospodarowicz D, Abraham JA, Schilling J. 1989. Isolation and characterization of a vascular endothelial cell mitogen produced by pituitary-derived folliculo stellate cells. *Proc Natl Acad Sci USA.* 86:7311–7315.

Gronbaek M, Becker U, Johansen D, Gottschau A, Schnohr P, Hein HO, Jensen G, Sorensen TI. 2000. Type of alcohol consumed and mortality from all causes, coronary heart disease, and cancer. *Ann Intern Med* 133:411–419.

Hain R, Bieseler B, Kindl H, Schroder G, Stocker R. 1990 Expression of a stilbene synthase gene in Nicotiana tabacum results in synthesis of the phytoalexin resveratrol. *Plant Mol Biol* 15:325–335.

Hall SS. 2003. Longevity research. In vino vitalis? Compounds activate life-extending genes. *Science* 301:1165.

Hattori R, Otani H, Maulik N, Das DK. 2002. Pharmacological preconditioning with resveratrol: Role of nitric oxide. *Am J Physiol Heart Circ Physiol.* 282:H1988–1995.

Henry LA, Witt DM. 2002. Resveratrol: Phytoestrogen effects on reproductive physiology and behavior in female rats. *Horm Behav* 41:220–228.

Henry C, Vitrac X, Decendit A, Ennamany R, Krisa S, Merillon JM. 2005. Cellular uptake and efflux of trans-piceid and its aglycone trans-resveratrol on the apical membrane of human intestinal Caco-2 cells. *J Agric Food Chem* 53:798–803.

Heredia A, Davis C, Redfield R. 2000. Synergistic inhibition of HIV-1 in activated and resting peripheral blood mononuclear cells, monocyte-derived macrophages, and selected drug-resistant isolates with nucleoside analogues combined with a natural product, resveratrol. *J Acquir Immune Defic Syndr* 25:246–255.

Howitz KT, Bitterman KJ, Cohen HY, Lamming DW. Lavu S, Wood JG, Zipkin RE, Chung P, Kisielewski A, Zhang LL, Scherer B, Sinclair DA. 2003. Small molecule activators of sirtuins extend Saccharomyces cerevisiae lifespan. *Nature* 425:191–196.

Huang HM, Liang YC, Cheng TH, Chen CH, Juan SH. 2005. Potential mechanism of blood vessel protection by resveratrol, a component of red wine. *Ann N Y Acad Sci* 1042:349–356.

Imamura G, Bertelli AA, Bertelli A, Otani H, Maulik N, Das DK. 2002. Pharmacological preconditioning with resveratrol: An insight with iNOS knockout mice. *Am J Physiol Heart Circ Physiol* 282:H1996–H2003.

Jang M, Cai L, Udeani GO, Slowing KV, Thomas CF, Beecher CW, Fong HH, Farnsworth NR, Kinghorn AD, Mehta RG, Moon RC, Pezzuto JM. 1997. Cancer chemopreventive activity of resveratrol, a natural product derived from grapes. *Science* 275:218–220.

Jazirehi AR, Bonavida B. 2004. Resveratrol modifies the expression of apoptotic regulatory proteins and sensitizes non-Hodgkin's lymphoma and multiple myeloma cell lines to paclitaxel-induced apoptosis. *Mol Cancer Ther* 3:71–84.

Kaga S, Zhan L, Matsumoto M, Maulik N. 2005. Resveratrol enhances neovascularization in the infarcted rat myocardium through the induction of thioredoxin-1, heme oxygenase-1 and vascular endothelial growth factor. *J Mol Cell Cardiol* 39:813–822.

Kaldas MI, Walle UK, Walle T. 2003. Resveratrol transport and metabolism by human intestinal Caco-2 cells. *J Pharm Pharmacol* 55:307–312.

Kaneider NC, Mosheimer B, Reinisch N, Patsch JR, Wiedermann CJ. 2004. Inhibition of thrombin-induced signaling by resveratrol and quercetin: Effects on adenosine nucleotide metabolism in endothelial cells and platelet-neutrophil interactions. *Thromb Res* 114:185–194.

Karcher W, Devillers J, 1990. In: Karcher W, Devillers J, Eds., *Practical Applications of Quantitative Structure-Activity Relationships (QSAR) in Environmental Chemistry and Toxicology*. Dordrecht, Netherlands: Kluwer Academic, pp. 1–12.

Kim KJ, Li B, Winer J, Armanini M, Gillett N, Phillips HS, Ferrara N. 1993. Inhibition of vascular endothelial growth factor induced angiogenesis suppresses tumor growth in vivo. *Nature (London)* 362:841–844.

Kimura Y, Okuda H, Kubo M. 1995 Effects of stilbenes isolated from medicinal plants on arachidonate metabolism and degranulation in human polymorphonuclear leukocytes. *J Ethnopharmacol* 45:131–139.

King RE, Kent KD, Bomser JA. 2005 Resveratrol reduces oxidation and proliferation of human retinal pigment epithelial cells via extracellular signal-regulated kinase inhibition. *Chem Biol Interact* 151:143–149.

Kiziltepe U, Turan NN, Han U, Ulus AT, Akar F. 2004. Resveratrol, a red wine polyphenol, protects spinal cord from ischemia-reperfusion injury. *J Vasc Surg* 40:138–145.

Klahr S. 1998. Obstructive nephropathy. *Kidney Int* 54:286–300.

Klinge CM, Risinger KE, Watts MB, Beck V, Eder R, Jungbauer A. 2003. Estrogenic activity in white and red wine extracts. *J Agri Food Chem* 51:1850–1857.

Kopelman PG, 2000. Obesity as a medical problem. *Nature* 404:635–643.

Koop P, 1998. Resveratrol, a phytoestrogen found in red wine. A possible explanation for the conundrum of the "French paradox"? *Eur J Endocrinol* 138:619–620.

Kotha A, Sekharam M, Cilenti L, Siddiquee K, Khaled A, Zervos AS, Carter B, Turkson J, Jove R. 2006. Resveratrol inhibits Src and Stat3 signaling and induces the apoptosis of malignant cells containing activated Stat3 protein. *Mol Cancer Ther* 5:621–629.

Kuhnle G, Spencer JPE, Chowrimootoo G, Schroeter H, Debnam ES, Srai SKS, Rice-Evans C, Hahn U. 2000. Resveratrol is absorbed in the small intestine as resveratrol glucuronide. *Biochem Biophys Res Commun* 272:212–217.

Kujoth GC, Hiona A, Pugh TD, Someya S, Panzer K, Wohlgemuth SE, Hofer T, Seo AY, Sullivan R, Jobling WA, Morrow JD, Van Remmen H, Sedivy JM, Yamasoba T, Tanokura M, Weindruch R, Leeuwenburgh C, Prolla TA. 2005. Mitochondrial DNA mutations, oxidative stress, and apoptosis in mammalian aging. *Science* 309:481–484.

Kuwajerwala N, Cifuentes E, Gautam S, Menon M, Barrack ER, Reddy GP. 2002. Resveratrol induces prostate cancer cell entry into S phase and inhibits DNA synthesis. *Cancer Res* 62:2488–2492.

Lancon A, Delmas D, Osman H, Thenot JP, Jannin B, Latruffe N. 2004. Human hepatic cell uptake of resveratrol: Involvement of both passive diffusion and carrier-mediated process. *Biochem Biophys Res Commun* 316:1132–1137.

Langcake P, Pryce RJ. 1977. A new class of phytoalexins from grapevines. *Experientia* 33:151–152.

Lee B, Moon SK. 2005. Resveratrol inhibits TNF-alpha-induced proliferation and matrix metalloproteinase expression in human vascular smooth muscle cells. *J Nutr* 135:2767–2773.

Leikert JF, Rathel TR, Wohlfart P, Cheynier V, Vollmar AM, Dirsch VM. 2002. Red wine polyphenols enhance endothelial nitric oxide synthase expression and subsequent nitric oxide release from endothelial cells. *Circulation* 106:1614–1617.

Lekli I, Szabo G, Juhasz B, Das S, Das M, Varga E, Szendrei L, Gesztelyi R, Varadi J, Bak I, Das DK, Tosaki A. 2008. Protective mechanisms of resveratrol against ischemia-reperfusion-induced damage in hearts obtained from Zucker obese rats: The role of GLUT-4 and endothelin. *Am J Physiol Heart Circ Physiol* 294:H859–H866.

Levenson AS, Gehm BD, Pearce ST, Horiguchi J, Simons LA, Ward JE 3rd, Jameson JL, Jordan VC. 2003. Resveratrol acts as an estrogen receptor (ER) agonist in breast cancer cells stably transfected with ER alpha. *Int J Cancer* 104:587–596.

Li JM, Mukamal KJ. 2004. An update on alcohol and atherosclerosis. *Curr Opin Lipidol.* 15:673–680.

Lu R, Serrero G. 1999. Resveratrol, a natural product derived from grape, exhibits antiestrogenic activity and inhibits the growth of human breast cancer cells. *J. Cell Physiol* 179:297–304.

Ma ZH, Ma QY, Wang LC, Sha HC, Wu SL, Zhang M. 2006. Effect of resveratrol on NF-kappaB activity in rat peritoneal macrophages. *Am J Chin Med.* 34:623–630.

Mahady GB, Pendland SL. 2000. Resveratrol inhibits the growth of Helicobacter pylori in vitro. *Am J Gastroenterol* 95:1849.

Mahady GB, Pendland SL, Chadwick LR. 2003. Resveratrol and red wine extracts inhibit the growth of CagA+ strains of Helicobacter pylori in vitro. *Am J Gastroenterol* 98:1440–1441.

Maier-Salamon A, Hagenauer B, Wirth M, Gabor F, Szekeres T, Jäger W. 2006. Increased transport of resveratrol across monolayers of the human intestinal Caco-2 cells is mediated by inhibition and saturation of metabolites. *Pharm Res* 23:2107–2115.

Marier JF, Chen K, Prince P, Scott G, del Castillo JR, Vachon P. 2005. Production of ex vivo lipopolysaccharide-induced tumor necrosis factor-alpha, interleukin-1beta, and interleukin-6 is suppressed by *trans*-resveratrol in a concentration-dependent manner. *Can J Vet Res* 69:151–154.

Marier JF, Vachon P, Gritsas A, Zhang J, Moreau JP, Ducharme MP. 2002. Metabolism and disposition of resveratrol in rats: Extent of absorption, glucuronidation, and enterohepatic recirculation evidenced by a linked-rat model, *J Pharmacol Exp Ther* 302:369–373.

Marimon JM, Bujanda L, Gutierrez-Stampa MA, Cosme A, Arenas JI. 1998. In vitro bactericidal effect of wine against Helicobacter pylori. *Am J Gastroenterol* 93:1392.

Martín AR, Villegas I, La Casa C, de la Lastra CA. 2004. Resveratrol, a polyphenol found in grapes, suppresses oxidative damage and stimulates apoptosis during early colonic inflammation in rats. *Biochem Pharmacol* 67:1399–1410.

McCarty MF. 2005. Potential utility of natural polyphenols for reversing fat-induced insulin resistance. *Med Hypotheses* 64:628–635.

Meng X, Maliakal P, Lu H, Lee MJ, Yang CS. 2004. Urinary and plasma levels of resveratrol and quercetin in humans, mice, and rats after ingestion of pure compounds and grape juice, *J Agric Food Chem* 52:935–942.

Mertens-Talcott SU, Percival SS. 2005. Ellagic acid and quercetin interact synergistically with resveratrol in the induction of apoptosis and cause transient cell cycle arrest in human leukemia cells. *Cancer Lett* 218:141–151.

Mgbonyebi OP, Russo J, Russo IH. 1998. Antiproliferative effect of synthetic resveratrol on human breast epithelial cells. *Int J Oncol* 12:865–869.

Miksits M, Maier-Salamon A, Aust S, Thalhammer T, Reznicek G, Kunert O, Haslinger E, Szekeres T, Jaeger W. 2005. Sulfation of resveratrol in human liver: Evidence of a major role for the sulfotransferases SULT1A1 and SULT1E1. *Xenobiotica* 35:1101–1119.

Millauer B, Shawver LK, Plate KH, Risau W, Ullrich A. 1994. Glioblastoma growth inhibited in vivo by a dominant-negative Flk-1 mutant. *Nature* 367:576–579.

Moon SO, Kim W, Sung MJ, Lee S, Kang KP, Kim DH, Lee SY, So JN, Park SK. 2006. Resveratrol suppresses tumor necrosis factor-alpha-induced fractalkine expression in endothelial cells. *Mol Pharmacol* 70:112–119.

Morelli R, Das S, Bertelli A, Bollini R, Scalzo RL, Das DK, Falchi M. 2006. The introduction of the stilbene synthase gene enhances the natural antiradical activity of Lycopersicon esculentum mill. *Mol Cell Biochem* 282:65–73.

Murakami A, Ishida H, Kobo K, Furukawa I, Ikeda Y, Yonaha M, Aniya Y, Ohigashi H. 2005. Related suppressive effects of Okinawan food items on free radical generation from stimulated leukocytes and identification of some active constituents:

Implications for the prevention of inflammation-associated carcinogenesis. *Asian Pac J Cancer Prev* 6:437–448.

Murias M, Jager W, Handler N, Erker T, Horvath Z, Szekeres T, Nohl H, Gille L. 2005. Antioxidant, prooxidant and cytotoxic activity of hydroxylated resveratrol analogues: Structure–activity relationship, *Biochem Pharmacol* 69:903–912.

Murry CE, Jennings RB, Reimer KA. 1986. Preconditioning with ischemia: A delay of lethal cell injury in ischemic myocardium. *Circulation* 74:1124–1136.

Nakagawa H, Kiyozuka Y, Uemura Y, Senzaki H, Shikata N, Hioki K, Tsubura A. 2001. Resveratrol inhibits human breast cancer cell growth and may mitigate the effect of linoleic acid, a potent breast cancer cell stimulator. *J Cancer Res Clin Oncol* 127:258–264.

Olas B, Wachowicz B, Saluk-Juszczak J, Zielinski T, Kaca W, Buczynski A. 2001. Antioxidant activity of resveratrol in endotoxin-stimulated blood platelets. *Cell Biol Toxicol* 17:117–125.

Orallo F. 2006. Comparative studies of the antioxidant effects of cis- and trans-resveratrol. *Curr Med Chem* 13:87–98.

Paul B, Masih I, Deopujari J, Charpentier C. 1999. Occurrence of resveratrol and pterostilbene in age-old darakchasava, an ayurvedic medicine from India. *J Ethnopharmacol* 68:71–76.

Pearl R. 1926. *Alcohol and Longevity*. New York: Knopf.

Pedersen A, Johansen C, Gronbaek M. 2003. Relations between amount and type of alcohol and colon and rectal cancer in a Danish population based cohort study. *Gut* 52:861–867.

Pendurthi UR, Williams JT, Rao LV. 1999. Resveratrol, a polyphenolic compound found in wine, inhibits tissue factor expression in vascular cells: A possible mechanism for the cardiovascular benefits associated with moderate consumption of wine. *Arterioscler Thromb Vasc Biol* 19:419–26.

Pervaiz S. 2003. Resveratrol: From grapevines to mammalian biology. *FASEB J.* 17:1975–1985.

Picard F, Kurtev M, Chung N, Topark-Ngarm A, Senawong T, Machado De Oliveira R, Leid M, McBurney MW, Guarente L. 2004. Sirt1 promotes fat mobilization in white adipocytes by repressing PPAR-gamma. *Nature* 429:771–776.

Pirola L, Frojdo S. 2008. Resveratrol: One molecule, many targets. *IUBMB Life* 60: 323–332.

Raval AP, Dave KR, Perez-Pinzon MA. 2006. Resveratrol mimics ischemic preconditioning in the brain. *J Cereb Blood Flow Metab* 26:1141–1147.

Renaud S, de Lorgeril M. 1992. Wine, alcohol, platelets, and the French paradox for coronary heart disease. *Lancet* 339:1523–1526.

Rezk YA, Balulad SS, Keller RS, Bennett JA. 2006. Use of resveratrol to improve the effectiveness of cisplatin and doxorubicin: Study in human gynecologic cancer cell lines and in rodent heart. *Am J Obstet Gynecol* 194:e23–e26.

Rodrigo R, Bosco C. 2006. Oxidative stress and protective effects of polyphenols: Comparative studies in human and rodent kidney. *A review. Comp Biochem Physiol C Toxicol Pharmacol* 142:317–327.

Rotondo S, Rajtar G, Manarini S, Celardo A, Rotillo D, de Gaetano G, Evangelista V, Cerletti C. 1998. Effect of trans-resveratrol, a natural polyphenolic compound, on human polymorphonuclear leukocyte function. *Br J Pharmacol* 123:1691–1699.

Rupprich N, Hildebrand H, Kindl H. 1980. Substrate specificity in vivo and in vitro in the formation of stilbenes. Biosynthesis of rhaponticin. *Arch Biochem Biophys* 200:72–78.

Scarlatti F, Sala G, Somenzi G, Signorelli P, Sacchi N, Ghidoni R. 2003. Resveratrol induces growth inhibition and apoptosis in metastatic breast cancer cells via de novo ceramide signaling. *FASEB J* 17:2339–2341.

Schoeppner A, Kindl H. 1979. Stilbene synthase (pinosylvine synthase) and its induction by ultraviolet light. *FEBS Lett* 108:349–352.

Schubert R, Fischer R, Hain R, Schreier PH, Bahnweg G, Ernst D, Sandermann HJr. 1997. An ozone-responsive region of the grapevine resveratrol synthase promoter differs from the basal pathogen-responsive sequence. *Plant Mol Biol* 34:417–426.

Sener G, Toklu HZ, Sehirli AO, Velioglu-Ogunc A, Cetinel S, Gedik N. 2006a. Protective effects of resveratrol against acetaminophen-induced toxicity in mice. *Hepatol Res.* 35:62–68.

Sener G, Tugtepe H, Yuksel M, Cetinel S, Gedik N, Yegen BC. 2006b. Resveratrol improves ischemia/reperfusion-induced oxidative renal injury in rats. *Arch Med Res* 37:822–829.

Sexton E, Van Themsche C, LeBlanc K, Parent S, Lemoine P, Asselin E. 2006. Resveratrol interferes with AKT activity and triggers apoptosis in human uterine cancer cells. *Mol Cancer* 5:45–58.

Sharma S, Anjaneyulu M, Kulkarni SK, Chopra K. 2006a. Resveratrol, a polyphenolic phytoalexin, attenuates diabetic nephropathy in rats. *Pharmacology* 76:69–75.

Sharma S, Chopra K, Kulkarni SK. 2007. Effect of insulin and its combination with resveratrol or curcumin in attenuation of diabetic neuropathic pain: Participation of nitric oxide and TNF-alpha. *Phytother Res* 21:278–283.

Sharma S, Kulkarni SK, Chopra K. 2006b. Resveratrol, a polyphenolic phytoalexin attenuates thermal hyperalgesia and cold allodynia in STZ-induced diabetic rats. *Indian J Exp Biol* 44:566–569.

She QB, Bode AM, Ma WY, Chen NY, Dong Z. 2001. Resveratrol-induced activation of p53 and apoptosis is mediated by extracellular-signal-regulated protein kinases and p38 kinase. *Cancer Res* 61:1604–1610.

Shigematsu S, Ishida S, Hara M, Takahashi N, Yoshimatsu H, Sakata T, Korthuis RJ. 2003. Resveratrol, a red wine constituent polyphenol, prevents superoxide-dependent inflammatory responses induced by ischemia/reperfusion, platelet-activating factor, or oxidants. *Free Radic Biol Med* 34:810–817.

Sipos P, Gyory H, Hagymasi K, Ondrejka P, Blazovics A. 2004. Special wound healing methods used in ancient Egypt and the mythological background. *World J Surg* 28:211–221.

Soares DG, Andreazza AC, Salvador M. 2003. Sequestering ability of butylated hydrooxytoluene, propyl gallate, resveratrol, and vitamin C and E against ABTS, DPPH, and hydroxyl free radicals in chemical and biological systems. *J Agric Food Chem* 51:1077–1080.

Soleas JS, Diamandis EP, Goldberg DM. 1997. Resveratrol: A molecule whose time has come? *And gone? Clin Biochem* 30:91–113.

Sparrow JR, Vollmer-Snarr HR, Zhou J, Jang YP, Jockusch S, Itagaki Y, Nakanishi K. 2003. A2E-epoxides damage DNA in retinal pigment epithelial cells. Vitamin E and other antioxidants inhibit A2E-epoxide formation. *J Biol Chem* 278:18207–18213.

Stervbo U, Vang O, Bonnesen C. 2007. A review of the content of the putative chemopreventive phytoalexin resveratrol in red wine. *Food Chem* 101:449–457.

Stivala LA, Savio M, Carafoli F, Perucca P, Bianchi L, Maga G, Forti L, Pagnoni UM, Albini A, Prosperi E, Vannini V. 2001. Specific structural determinants are responsible for the antioxidant activity and the cell cycle effects of resveratrol. *J Biol Chem* 276:22586–22594.

Su HC, Hung LM, Chen JK. 2006. Resveratrol, a red wine antioxidant, possesses an insulin-like effect in streptozotocin-induced diabetic rats. *Am J Physiol Endocrinol Metab* 290:E1339–E1346.

Subbaramaiah K, Michaluart P, Chung WJ, Tanabe T, Telang N, Dannenberg AJ. 1999. Resveratrol inhibits cyclooxygenase-2 transcription in human mammary epithelial cells. *Ann N Y Acad Sci* 889:214–223.

Subbaramaiah K, Chung WJ, Michaluart P, Telang N, Tanabe T, Inoue H, Jang M, Pezzuto JM, Dannenberg AJ. 1998. Resveratrol inhibits cyclooxygenase-2 transcription and activity in phorbol ester-treated human mammary epithelial cells. *J Biol Chem* 273:21875–21882.

Sun C, Hu Y, Liu X, Wu T, Wang Y, He W, Wei W. 2006. Resveratrol downregulates the constitutional activation of nuclear factor-kappaB in multiple myeloma cells, leading to suppression of proliferation and invasion, arrest of cell cycle, and induction of apoptosis. *Cancer Genet Cytogenet* 165:9–19.

Sun AY, Simonyi A, Sun GY. 2002. The "French Paradox" and beyond: Neuroprotective effects of polyphenols. *Free Radic Biol Med* 32:314–318.

Szabo ME, Gallyas E, Bak I, Rakotovao A, Boucher F, de Leiris J, Nagy N, Varga E, Tosaki A. 2004 Heme oxygenase-1-related carbon monoxide and flavonoids in ischemic/reperfused rat retina. *Invest Ophthalmol Vis Sci* 45:3727–3232.

Szewczuk LM, Forti L, Stivala LA, Penning TM. 2004. Resveratrol is a peroxidase mediated inactivator of COX-1 but not COX-2. *J Biol Chem* 279:22727–22737.

Takada Y, Bhardwaj A, Potdar P, Aggarwal BB. 2004. Nonsteroidal anti-inflammatory agents differ in their ability to suppress NF-kappaB activation, inhibition of expression of cyclooxygenase-2 and cyclin D1, and abrogation of tumor cell proliferation. *Oncogene* 23:9247–9258.

Terman BI, Dougher-Vermazen M, Carrion ME, Dimitrov D, Armellino DC, Gospodarowicz D, Bohlen P. 1992. Identification of the KDR tyrosine kinase as a receptor for vascular endothelial cell growth factor. *Biochem Biophys Res Commun* 187: 1579–1586.

Vitaglione P, Morisco F, Caporaso N, Fogliano V. 2004. Dietary antioxidant compounds and liver health. *Crit Rev Food Sci Nutr* 44:575–586.

Walle T, Hsieh F, DeLegge MH, Oatis JEJr, Walle UK. 2004. High absorption but very low bioavailability of oral resveratrol in humans. *Drug Metab Dispos* 32:1377–1382.

Wright JS, Johnson ER, DiLabio GA. 2001. Predicting the activity of phenolic antioxidants: Theoretical methods, analysis of substituent effects, and application to major families of antioxidants, *J Am Chem Soc* 123:1173–1183.

Zamin LL, Dillenburg-Pilla P, Argenta-Comiran R, Horn AP, Simao F, Nassif M, Gerhardt D, Frozza RL, Salbego C. 2006. Protective effect of resveratrol against oxygen-glucose deprivation in organotypic hippocampal slice cultures: Involvement of PI3-K pathway. *Neurobiol Dis* 24:170–182.

Zhang Q, Tang X, Lu QY, Zhang ZF, Brown J, Le AD. 2005. Resveratrol inhibits hypoxia-induced accumulation of hypoxia-inducible factor-1alpha and VEGF expression in human tongue squamous cell carcinoma and hepatoma cells. *Mol Cancer Ther* 4:1465–1474.

Zhu J, Yong W, Wu X, Yu Y, Lv J, Liu C, Mao X, Zhu Y, Xu K, Han X, Liu C. 2008. Anti-inflammatory effect of resveratrol on TNF-a-induced MCP-1 expression in adipocytes. *Biochem Biophys Res Commun* 369:471–477.

15 Resveratrol: The Biochemistry Behind Its Anticancer Effects

JOYDEB K. KUNDU[1] AND YOUNG-JOON SURH[1,2]

[1]National Research Laboratory of Molecular Carcinogenesis and Chemoprevention, College of Pharmacy, Seoul National University, Seoul, South Korea

[2]Cancer Research Institute, Seoul National University, Seoul, South Korea

CONTENTS

Plant Phenolics and Human Health: Biochemistry, Nutrition, and Pharmacology,
Edited by Cesar G. Fraga.

INTRODUCTION

Cancer is a group of heterogeneous diseases characterized by uncontrolled growth and spread of transformed cells. Although numerous anticancer therapies are currently under clinical use, the mortality from cancer as well as new cancer cases are still alarmingly high. According to a report published by American Cancer Society, 7.6 million people around the world have died of cancer in 2007, and this figure is expected to rise to 17.5 million by the year 2050. In this context, the fight against cancer has now been shifted to a new front, that is, the prevention of cancer. With the progress in our understanding of the molecular biology of cancer, it is now evident that the multistage carcinogenesis involves alterations in biochemical processes, which disrupt cellular homeostasis. Physiologically, the cellular biochemistry is under strict control of a divergent network of intracellular signaling pathways, which often functions inappropriately during tumorigenesis. Many of these altered cell signaling pathways can reversibly be restored to a normal state by chemopreventive agents, which can block the onset and the progression of cancer [Bode and Dong, 2004; Chen and Kong, 2005; Kundu and Surh, 2005]. Therefore, the prevention of cancer at early stages through restoration of altered biochemical processes might be the most practical strategy to fight against cancer. Population-based and laboratory studies suggest that regular consumption of fruits and vegetables can reduce the risk of certain malignancies [Chen and Kong, 2005; Surh, 2003]. In fact, the search for plant-originated anticancer principles has led to the identification of diverse classes of chemopreventive agents.

Resveratrol (trans-3,5,4′-trihydroxystilbene), a polyphenolic phytoalexin present in grapes, red wine, peanuts, mulberries, and in many other medicinal plants is an extensively investigated phytochemical with chemopreventive and chemotherapeutic potential [Kundu and Surh, 2004; Kundu and Surh, 2008a]. Since its first isolation in 1940 from the roots of white hellebore (*Veratrum glandiflorum* O. Loes), resveratrol has received the attention of the scientific community in 1990s, when the compound has been shown to be responsible for the cardioprotective effects of red wine [Wu et al., 2001]. Resveratrol emerged as nature's gift with a wide spectrum of health beneficial effects including antibacterial, antifungal, antioxidant, anti-inflammatory, and cardioprotective functions [Baur and Sinclair, 2006]. Momentum in the research on the anticancer activity of resveratrol has flared up by John M. Pezzuto and colleagues [Jang et al., 1997], who have reported that the compound interferes with multiple biochemical events associated with the initiation, promotion, and progression stages of carcinogenesis. Since this pioneering study, there has been rapid progress in understanding how resveratrol modulates the altered cellular biochemistry associated with tumorigenesis [Aggarwal et al., 2004]. The transformation of cells and their subsequent journey to progressive cancer involve improper cell-signaling pathways. Resveratrol modulates various components of abnormal signal transduction cascades implicated in the tumorigenesis [Aggarwal et al., 2004; Kundu and Surh, 2004].

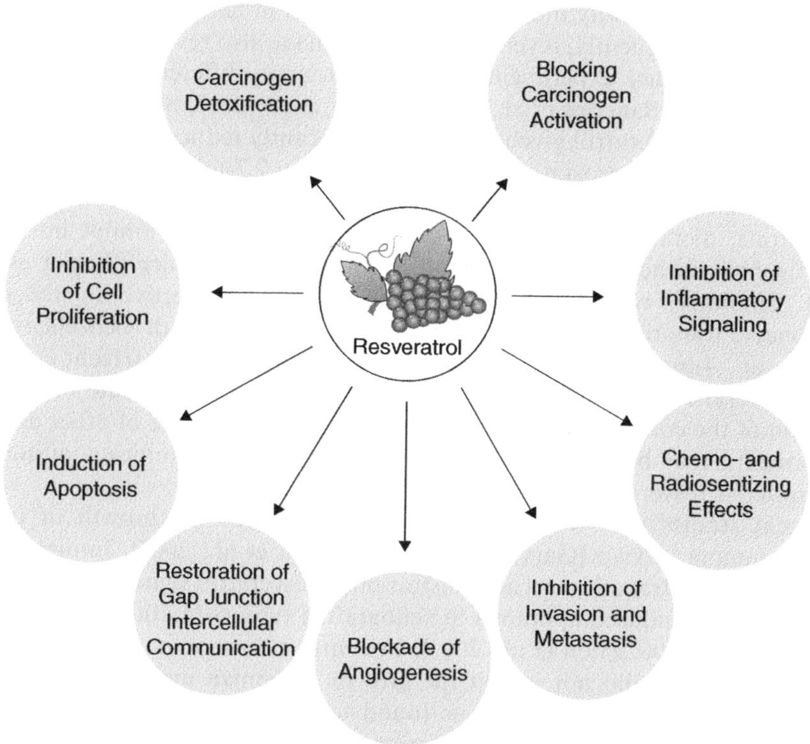

Figure 15.1 Biochemical mechanisms of anticancer effects of resveratrol.

Over the last 10 years, multidirectional studies with resveratrol have resulted in several key findings, which can explain the biochemistry behind its anticancer effects (Fig. 15.1). Resveratrol prevents oxidative damage of cellular macromolecules, blocks the activation of carcinogens, stimulates carcinogen detoxification, reduces inflammatory responses, diminishes proliferation, induces apoptosis in cancer cells, inhibits angiogenesis and metastasis, and sensitizes cancer cells to chemotherapy and radiotherapy [Aggarwal et al., 2004; Kundu and Surh, 2004; Shankar et al., 2007]. This chapter focuses on the biochemical basis of anticancer effects of resveratrol with particular emphasis on the components of cell-signaling pathways as major targets.

RESVERATROL IN CANCER PREVENTION AND THERAPY

The chemopreventive and chemotherapeutic potential of resveratrol has been extensively investigated in different in vivo models of tumorigenesis and in vitro cell culture systems [Aggarwal et al., 2004; Kundu and Surh, 2004]. Repeated topical application of resveratrol significantly reduced the incidence and the multiplicity of chemically induced skin papillomas in female CD-1 mice [Jang et al., 1997]. Subsequently, several other studies have also revealed the ability of resveratrol to reduce experimental carcinogenesis in different rodent models [Aziz et al., 2005; Banerjee et al., 2002; Harper et al., 2007; Sengottuvelan and Nalini, 2006]. Resveratrol given with diet significantly reduced the incidence of poorly differentiated prostatic adenocarcinomas by 7.7-fold in the TRansgenic Adenocarcinoma Mouse Prostate (TRAMP) model [Harper et al., 2007]. In contrast, dietary administration of resveratrol failed to inhibit intestinal tumorigenesis in adenomatous polyposis coli $(Apc)^{Min/+}$ mice [Ziegler et al., 2004]. Similarly, resveratrol given with diet was ineffective in reducing pulmonary tumorigenesis in female A/J mice treated with benzo[*a*]pyrene (B[*a*]P) or 4-(methyl-nitrosamino)-1-(3-pyridyl)-1-butanone [Hecht et al., 1999]. Several factors, such as the dosage of resveratrol, the route of administration of the compound, the tumor origin, and the presence of other dietary components, may be attributed to such discrepancies in the efficacy of dietary administered resveratrol.

Recent studies reported that resveratrol suppressed the growth of transplanted tumors in vivo [Garvin et al., 2006; Zhou et al., 2005]. Intratumoral, peritumoral or intraperitoneal administration of resveratrol significantly arrested the growth and induced apoptosis in xenografted tumors in athymic nude mice [Garvin et al., 2006; Zhou et al., 2005]. In contrast, intraperitoneal administration of resveratrol did not inhibit the growth of human melanoma xenograft [Niles et al., 2006]. Resveratrol was found to inhibit proliferation of various cancer cells in culture. The transformation of mouse epidermal JB6 C141 cells stimulated with epidermal growth factor (EGF) or a tumor promoter 12-*O*-tatradecanoylphorbol-13-acetate (TPA) was blunted by resveratrol [Huang et al., 1999; She et al., 2003]. Although, resveratrol attenuated the growth of mouse breast cancer (4T1) cells in culture [Bove et al., 2002], intraperitoneal administration of the compound failed to affect the growth and metastasis of these cells xenografted in Balb/c mice [Bove et al., 2002]. Resveratrol also induced apoptosis in various cancerous or transformed cells, sensitized cancer cells to chemotherapy- or radiotherapy-induced growth arrest and apoptosis, and suppressed tumor-associated angiogenic and metastatic processes [Aggarwal et al., 2004; Garg et al., 2005; Kundu and Surh, 2004]. Based on the promising anticancer effects of resveratrol in numerous preclinical models, the compound has been subjected to the first phase of clinical trials [Boocock et al., 2007].

BIOCHEMICAL BASIS OF CHEMOPREVENTION AND CHEMOTHERAPY WITH RESVERATROL

Inhibition of Metabolic Activation of Carcinogens and DNA Adduct Formation

Many exogenous and endogenous carcinogens undergo oxidative metabolism by phase I enzymes, especially by the members of the cytochrome P450 (CYP) superfamily. Some chemical carcinogens are naturally inactive but are converted to reactive species through phase I metabolism. The phase I metabolic end products are usually eliminated through a second phase of metabolic reactions, commonly known as detoxification. Inadequate detoxification leads to the accumulation of metabolically activated carcinogens, which can directly attack the target cell DNA, thereby contributing to tumor initiation [Perwez Hussain and Harris, 2007].

Genes encoding CYP enzymes are transcriptionally regulated by arylhydrocarbon receptor (AhR), a member of the nuclear receptor transcription factor superfamily. For example, the transactivation of *CYP1A1*, which encodes an enzyme frequently involved in metabolic activation of a wide spectrum of polycyclic aromatic hydrocarbons (PAHs), requires the binding of activated AhR to the promoter segment of the gene. Resveratrol, by acting as an AhR antagonist, inhibits expression and/or activities of different isoforms of CYP enzymes (Fig. 15.2). Resveratrol inhibits the activation of AhR by blocking the conversion of ligand-bound cytosolic AhR into its nuclear DNA-binding form [Ciolino et al., 1998] and by suppressing the interaction between the AhR and the transcription initiation complex at the *CYP1A1* gene promoter [Ciolino and Yeh, 1999]. Resveratrol inhibited the induction of CYP1A1 expression in rat primary hepatocytes, suggesting that the compound acts as an AhR antagonist [Andrieux et al., 2004]. A study from our laboratory also revealed that resveratrol strongly inhibited 2,3,7,8-tetrachlorodibenzo-*p*-dioxin-(TCDD)-induced AhR DNA-binding activity in human mammary epithelial (MCF-10A) cells [Chen et al., 2004]. In contrast, Lee and Safe [2001] demonstrated that resveratrol enhanced the degradation of CYP1A1 (messenger ribonucleic acid), and diminished the expression and the activity of CYP1A1 in TCDD-stimulated human breast cancer (T47D and MCF-7) cells in an AhR-independent manner. Resveratrol suppressed the expression and/or the activity of CYP1A1, CYP1B1, and CYP1A2 in murine hepatoma (Hepa1c1c7) cells [Gerhauser et al., 2003], TCDD-stimulated mammary epithelial (MCF-10A) cells [Chen et al., 2004], 7,12-dimethylbenz[*a*]anthracene (DMBA)-treated MCF-7 cells [Ciolino and Yeh, 1999], and B[*a*]P-treated human hepatoma (HepG2) cells [Ciolino and Yeh, 1999]. Chang et al. [2001] demonstrated that resveratrol exerted direct inhibitory effects on the activities of CYP1A1 and CYP1B1 but it inactivated CYP1A2 indirectly. In contrast, treatment of normal human esophageal (HET-1A) cells with resveratrol failed to inhibit B[*a*]P-induced expression of CYP1B1 mRNA and protein [Wen and Walle, 2007].

Resveratrol also inhibits the activity of CYP19, alternatively known as aromatase, which is a rate-limiting enzyme in the biosynthesis of estrogens. A recent molecular modeling and docking study revealed that resveratrol binds to the active site of aromatase [Neves et al., 2007]. Since estrogens function as a risk factor for breast cancer, the inhibitory effect of resveratrol on the aromatase/CYP19 activity in MCF-7 cells [Wang et al., 2006] suggests that the compound exerts chemopreventive effects on mammary carcinogenesis partly due to its antiestrogenic property. Resveratrol protects MCF-10A cells from carcinogenic estrogen metabolites as evidenced by the reduced formation of DNA adducts from 4-hydroxyestradiol or estradiol-3,4-quinone [Zahid et al., 2008]. However, a weak estrogenic effect of resveratrol has also been reported [Bhat et al., 2001].

As a consequence of its inhibitory effects on the expression and/or activity of CYP enzymes, resveratrol suppressed carcinogen-DNA adduct formation. Resveratrol abrogated the B[*a*]P-diol epoxide-DNA adduct formation in mouse lung tissue by down-regulating the expression of CYP1A1 [Revel et al., 2003]. Although, resveratrol diminished DMBA-DNA adduct formation and B[*a*]P-induced arylhydrocarbon hydrolase activity, it did not inhibit B[*a*]P-diol epoxide-DNA adduct formation in mouse epidermis [Szaefer et al., 2004]. The compound attenuated the genotoxicity induced by *N*-hydroxy-2-amino-1-methyl-6-phenylimidazo-[4,5-*b*]pyridine (*N*-hydroxy-PhIP) in human CYP1A2 overexpressing Chinese hamster lung fibroblast V79 cells, possibly by blocking the activity of CYP1A2 [Boyce et al., 2004]. Since the AhR and various CYP enzymes are involved in the detoxification of many xenobiotics, the inactivation of AhR and the inhibition of CYP enzymes by resveratrol should be carefully considered in achieving chemoprevention.

Induction of Phase II Detoxification or Antioxidant Enzymes

The biochemical basis of anticancer effects of resveratrol is partly ascribed to its antioxidant potential. It is generally accepted that the ROS-induced DNA damage plays a critical role in oncogenesis. Resveratrol functions as a natural antioxidant by inducing antioxidant enzymes, restoring cellular reduced glutathione (GSH) levels and scavenging ROS [de la Lastra and Villegas, 2007; Kode et al., 2007; Kundu and Surh, 2004]. The activation of signaling pathways involved in the induction of phase II detoxification or antioxidant enzymes is one of the well-defined mechanisms underlying chemoprevention with resveratrol (Fig. 15.2).

Several studies have demonstrated that resveratrol enhances the expression and/or the activity of phase II antioxidant or detoxification enzymes, such as glutathione-*S*-transferase (GST) [Hebbar et al., 2005], glutathione peroxidase (GPx) [Kluth et al., 2007], UDP glucuronosyl transferase (UGT)-1A [Hebbar et al., 2005], NAD(P)H:quinone oxidoreductase (NQO) [Gerhauser et al., 2003; Hebbar et al., 2005; Hsieh et al., 2006; Jang et al., 1997], heme oxygenase-1 (HO-1) [Chen et al., 2005; Juan et al., 2005], and glutamate cysteine ligase (GCL) [Kode et al., 2007]. Resveratrol induces aforementioned cytoprotective enzymes largely by activating nuclear factor erythroid E2-related factor-2 (Nrf2) [Chen et al., 2005; Hsieh et al., 2006; Kode et al., 2007], which is a basic-region leucine zipper (bZIP) transcription factor.

Resveratrol induced the activity as well as the expression of NQO-1 at both protein and mRNA levels in human K562 cells by promoting nuclear translocation and subsequent binding of Nrf2 to antioxidant response element (ARE) [Hsieh et al., 2006]. The compound induced the protein and mRNA expression

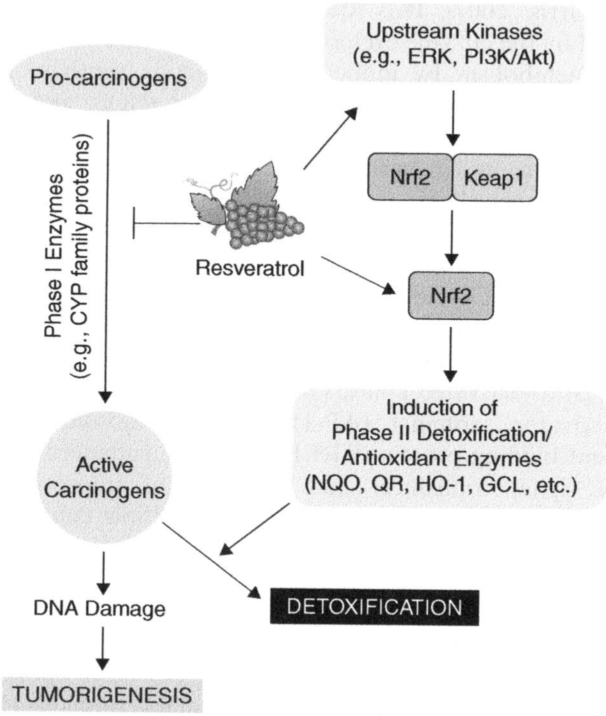

Figure 15.2 Inhibition of carcinogen activation and induction of phase II detoxification enzymes by resveratrol.

of HO-1 in human aortic smooth muscle [Juan et al., 2005] and rat pheochromocytoma (PC12) cells [Chen et al., 2005] by activating nuclear factor-kappaB (NF-κB) and Nrf2, respectively. According to the former study, resveratrol (1–10 μM) activated NF-κB and increased the expression and the promoter activity of HO-1, while the compound at a concentration higher than 20 μM suppressed NF-κB activation, accounting for its anti-inflammatory effect [Juan et al., 2005]. Chen et al. [2005] demonstrated that resveratrol induced HO-1 expression in PC12 cells via Akt-mediated activation of Nrf2. Resveratrol restored cigarette smoke extract (CSE)-induced depletion of cellular GSH level by inducing Nrf2-driven GCL expression and subsequent GSH biosynthesis in human primary small airway epithelial cells (SAEC) and human alveolar epithelial cells (A549) [Kode et al., 2007]. Since Nrf2 inhibitory protein Keap1 is a zinc metalloprotein [Dinkova-Kostova et al., 2005], the resorcin moiety of resveratrol may form a zinc ion chelate, thereby leading to inactivation of Keap1 and subsequent activation of Nrf2.

Suppression of Pro-inflammatory Signaling Pathways

Pro-inflammatory mediators, such as prostaglandins (PGs), nitric oxide (NO), and cytokines are involved in carcinogenesis [Kundu and Surh, 2005; Perwez

Hussain and Harris, 2007]. PGs are produced through the breakdown of arachidonic acid by the enzyme cyclooxygenase (COX), while NO is generated from arginine metabolism by inducible nitric oxide synthase (iNOS). The elevated expression and/or activity of COX-2 and iNOS, and the enhanced production of various pro-inflammatory mediators have been implicated in the initiation, promotion, and progression stages of tumorigenesis [Kundu and Surh, 2008b]. The intracellular signaling pathways involved in regulating the expression of COX-2, iNOS and various pro-inflammatory mediators are composed of cell surface receptors (e.g., epidermal growth factor receptor, cytokine receptors), a distinct set of serine/threonine kinases and tyrosine kinases, and transcription factors. These include members of the mitogen-activated protein (MAP) kinase family, PI3K/Akt, PKC, glycogen synthase kinase (GSK), janus-activated kinase (JAK), and DNA-binding proteins, such as NF-κB, activator protein-1 (AP-1), cyclic ade;nosine monophosphate response element binding protein (CREB), CCAAT/enhancer binding protein (C/EBP), and signal transducer and activator of transcription (STAT). Resveratrol has been reported to interfere with multiple components of pro-inflammatory signaling pathways (Fig. 15.3).

Pro-inflammatory Mediators Resveratrol attenuated the expression and/or the activity of various components of pro-inflammatory signaling pathways. The compound significantly inhibited the expression of COX-2 in lipopolysachaaride (LPS)-, TPA-, or H_2O_2-stimulated mouse peritoneal macrophages (Martinez and Moreno, 2000), LPS plus interferon (IFN)-γ-treated RAW 264.7 macrophages [Murakami et al., 2003], and TPA-stimulated mouse skin [Kundu et al., 2006a] and down-regulated the expression of COX-2 mRNA transcript in *N*-nitrosomethylbenzylamine (NMBA)-induced esophageal tumors in F344 rats [Li et al., 2002]. The inhibition of COX-2 promoter activity in both unstimulated and transforming growth factor-α-stimulated colon cancer (DLD-1) cells by resveratrol was partly attributed to the resorcin moiety present in the molecule [Mutoh et al., 2000]. Resveratrol also diminished the COX-2 activity and reduced the production of PGE_2 in peripheral blood leukocytes stimulated with LPS plus IFN-γ [Richard et al., 2005] and human mammary epithelial cells treated with TPA [Subbaramaiah et al., 1998]. A recent study by Zykova et al. [2008] demonstrated that resveratrol inhibited anchorage-independent growth of human colon cancer (HT-29) cells by directly binding to COX-2 and decreasing the production of PGE_2. The study also revealed that resveratrol reduced the growth of COX-$2^{+/+}$ cells in culture, but did not affect the growth of cells deficient in COX-2.

The expression of iNOS and the generation of NO in LPS-activated RAW 264.7 cells were inhibited by resveratrol [Murakami et al., 2003; Tsai et al., 1999]. The compound significantly diminished the expression of tumor necrosis factor (TNF)-α mRNA in LPS-activated J774.2 macrophage cells [Kowalski et al., 2005] and peripheral blood leukocytes [Richard et al., 2005]. The expression of interleukin (IL)-8 protein and/or its mRNA in human monocytic leukemia (U937) cells [Shen et al., 2003] and human peripheral blood

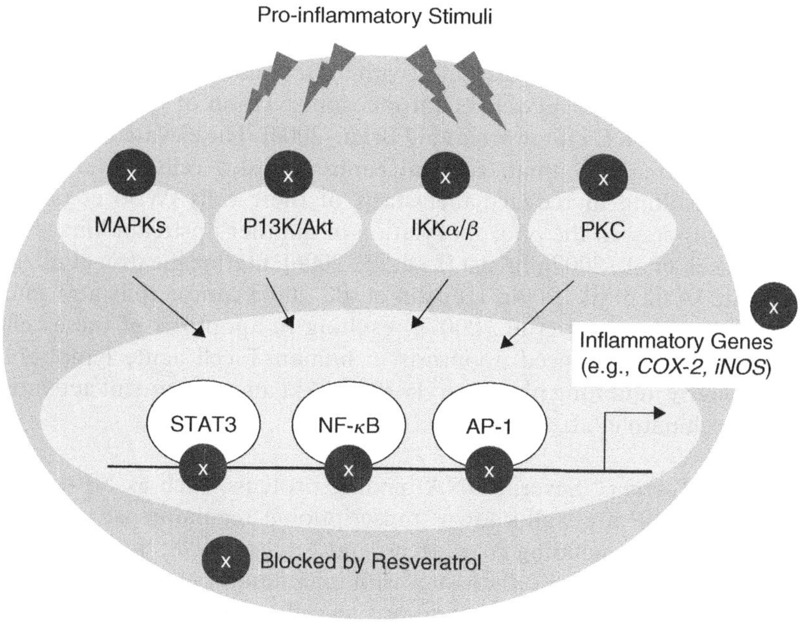

Figure 15.3 Resveratrol modulation of pro-inflammatory signaling pathways. Resveratrol represses transcriptional activation of various pro-inflammatory genes by inhibiting pro-inflammatory stimuli-induced activation of upstream kinases, such as MAP kinases, PI3K/Akt, IKK, PKC, etc., and blocking the DNA binding of eukaryotic transcription factors, such as NF-κB, AP-1, and STAT3.

leukocytes [Richard et al., 2005] was also attenuated by resveratrol. Li et al. [2007] demonstrated that resveratrol reduced the serum IL-6 level in mice transplanted with lymphocytic leukemia L1210 cells.

Upstream Kinases Resveratrol exerts antiinflammatory and antitumor promoting effects partly by suppressing the phosphorylation of MAP kinases [Kundu et al., 2006a, 2006b; Reagan-Shaw et al., 2004; Tang et al., 2008; Yu et al., 2001]. The TPA-induced phosphorylation of all three representative MAP kinases such as ERK1/2, p38 MAP kinase, and JNK was diminished by resveratrol in HeLa cells [Yu et al., 2001] and mouse skin in vivo [Kundu et al., 2006a, 2006b]. Likewise, resveratrol inhibited the phosphorylation of ERK1/2 in human epidermoid carcinoma (A431) cells [Kim et al., 2006] and in heregulin-β1-stimulated human breast cancer cells [Tang et al., 2008]. Resveratrol attenuated ultraviolet B (UVB)-induced expression of MAP kinase kinase (MEK) in SKH-1 hairless mouse skin [Reagan-Shaw et al., 2004]. Subbaramaiah et al. [1998] demonstrated that resveratrol down-regulated the expression of PKC in human mammary and oral epithelial cells stimulated with TPA. The compound attenuated the proliferation of human gastric adenoma cells by blocking the PKC activity, without influencing the phosphorylation of ERK1/2 [Atten et al., 2001]. Moreover, the inhibition of TPA-induced growth of human prostate cancer (PC3) cells by resveratrol was partly associated with the translocation of cytosolic PKCα to the plasma membrane as well as autophosphorylation of both cytosolic and membrane-bound PKCα [Stewart and O'Brian, 2004]. The elevated expression of PKCδ in TPA-treated human cervical cancer (Caski) cells was blunted by resveratrol, resulting in reduced metastasis of these cells [Woo et al., 2004]. Resveratrol suppressed the phosphorylation of another upstream kinase Akt in ovarian [Kueck et al., 2007], breast [Li et al., 2006], uterine [Sexton et al., 2006], and prostate [Aziz et al., 2006; Hudson et al., 2007] cancer cells and multiple myeloma cells [Bhardwaj et al., 2007], resulting in apoptosis of these cells. In addition, resveratrol induced apoptosis in human T cell acute lymphoblastic leukemia cells by inhibiting phosphorylation of Akt and subsequent activation of GSK3β [Cecchinato et al., 2007].

Transcription Factors Several DNA-binding proteins, such as NF-κB, AP-1, CREB, and C/EBP are well-known transcriptional regulators of various pro-inflammatory genes including *cox-2* [Kim and Fischer, 1998; Surh and Kundu, 2005]. Upon stimulation of cells/tissues with inflammatory stimuli, a distinct set of intracellular signaling pathways comprising cell surface receptors and protein kinases are turned on and activate these transcription factors to bind to target gene promoter. Detailed molecular mechanisms of activation of these transcription factors have been discussed elsewhere [Chun and Surh, 2004]. Resveratrol exerted its anti-inflammatory effects by targeting some of these transcription factors. Manna et al. [2000] demonstrated that resveratrol inhibited the activation of NF-κB in Jurkat-T, HeLa, and glioma cells treated with different stimuli such as TPA, LPS, H_2O_2, okadaic acid, and ceramide. Resveratrol also suppressed the activation of NF-κB in IL-1β- and Cr (VI)-stimulated acute myeloid leukemia (OCIM2) cells [Estrov et al., 2003] and mouse epidermal (JB6) cells [Leonard et al., 2003], respectively. We have reported that resveratrol diminished TPA-induced NF-κB activation in mouse skin in vivo by blocking the activation of IKK, phosphorylation of IκBα and p65, nuclear translocation of p65 and interaction of p65 with a transcriptional co-activator CREB-binding protein (CBP) [Kundu et al., 2006a]. Likewise, the compound inhibited UVB-induced activation of NF-κB in normal human epidermal keratinocytes by blocking the activation of upstream IKKα as well as phosphorylation and degradation of IκBα [Adhami et al., 2003]. Resveratrol also suppressed proliferation and induced apoptosis in human multiple myeloma cells by inhibiting the constitutive activation of NF-κB via blockade of IKK activity and subsequent phosphorylation of IκBα and p65 [Bhardwaj et al., 2007]. The compound also attenuated the nuclear translocation and the DNA binding of NF-κB in LPS-stimulated RAW 264.7 cells by blocking phosphorylation and degradation of IκBα [Cho et al., 2002; Tsai et al., 1999]. Although resveratrol negated TNF-α-induced activation of NF-κB in U937 cells by suppressing phosphorylation and nuclear translocation of p65, the compound did not affect the phosphorylation and degradation of IκBα [Manna et al., 2000]. Pendurthi et al. [2002] reported that resveratrol suppressed LPS-induced activation of NF-κB by inhibiting phosphorylation and transactivation potential of p65, but it failed to inhibit nuclear translocation of NF-κB/Rel proteins.

Alternatively, resveratrol exerted epigenetic control on NF-κB activation via SIRT1 activation [Yang et al., 2007; Yeung et al., 2004]. It was reported that SIRT1, a nicotinamide adenosine dinucleotide-dependent histone deacetylase, interacted physically with the RelA/p65 and negated NF-κB-driven gene transcription by deacetylating RelA/p65 at lysine 310 [Yeung et al., 2004]. Resveratrol inhibited CSE-induced NF-κB activation and NF-κB-regulated pro-inflammatory gene expression by activating SIRT1 in monocyte-macrophage (MonoMac6) cells, bronchoalveolar lavage fluid, and rat lungs [Yang et al., 2007].

AP-1 is another molecular target of resveratrol in its suppression of inflammatory responses. Resveratrol abrogated TPA-induced activation of AP-1 in human mammary epithelial cells [Subbaramaiah et al., 1998, 1999], U937 human leukemia cells (Li et al., 2003), and mouse skin in vivo [Kundu et al., 2006b] and inhibited TNF-α-induced AP-1 DNA binding in U937 cells [Manna et al., 2000]. Resveratrol diminished UVC- or TPA-induced AP-1 reporter gene activity in HeLa cells by blocking the activation of MAP kinases and PKC (Yu et al., 2001). In contrast, resveratrol failed to suppress AP-1-driven transcriptional activity in LPS-stimulated human monocytic (THP-1) cells [Pendurthi et al., 2002].

Restoration of Gap–Junction Intercellular Communication (GJIC)

The dysfunction of GJIC is a characteristic feature of premalignant and malignant cells [Upham et al., 2007]. Many tumor promoters and growth factors have been shown to inhibit GJIC [Upham et al., 2007]. Several studies suggest that the anticancer activity of resveratrol is, partly, ascribed to its ability to restore GJIC [Kim et al., 2008; Nielsen et al., 2000; Upham et al., 2007]. Resveratrol has been reported to restore the decreased level of GJIC in WB-F344 rat liver epithelial cells treated with TPA (Nielsen et al., 2000), dicumyl peroxide [Upham et al., 2007], or gallic acid [Kim et al., 2008].

Inhibition of Cell Proliferation

Resveratrol exerts antitumor effects partly by arresting the growth of various cancer cells in culture [Kundu and Surh, 2004]. The inhibition of ornithine decarboxylase (ODC), a biochemical hallmark of tumor promotion, has been shown to account for the antiproliferative and antitumor effects of resveratrol [Schneider et al., 2000; Ulrich et al., 2007]. Aberrant changes in cell-cycle machinery are considered as the biochemical basis of abnormal proliferation of transformed cells. Major cell-cycle regulatory proteins include various cyclins, cyclin-dependent kinases (Cdk), Cdk inhibitors, and check point kinases (Chk1 and Chk2), which maintain the homeostatic control of cell growth and differentiation [Collins and Garrett, 2005; Gali-Muhtasib and Bakkar, 2002]. Resveratrol blocks inappropriate signaling through modulation of these cell-cycle regulatory proteins and suppresses the growth of various cancer cells in culture [Kundu and Surh, 2004] (Table 15.1). Moreover, the antiproliferative and growth inhibitory effects of resveratrol have been attributed to its ability to inhibit DNA synthesis [Fontecave et al., 1998] and telomerase activity [Fuggetta et al., 2006].

Resveratrol inhibited the expression and/or activities of various cyclins (D1, D2, and E) and Cdks (-2, -4, and -6), and induced the expression of the Cdk

Table 15.1 Molecular Targets of Resveratrol as an Antiproliferative Agent

Cells/Tissues/ Stimuli	Effects	References
A431 cells	G1 phase arrest, ↑$p21^{WAF1/CIP1}$ expression, ↓cyclin D1/D2-cdk6, ↓cyclin D1/D2-cdk4 and cyclin E-cdk2 complex formation	Ahmad et al., 2001
	Block G1→S phase transition, ↓cyclin E-Cdk2 complex formation↓hyperphosphorylated Rb, ↑hypophosphorylated Rb, ↓expression of E2Fand DP, ↓free E2F level	Adhami et al., 2001
	G1 phase arrest, ↓expression of cyclin A, D1, and cdk-6, ↑accumulation of hypophosphorylated Rb, ↑expression of $p21^{WAF1/CIP1}$ (independent of p53), ↑$p27^{KIP1}$↓c-Jun expression, ↓AP-1 DNA binding, ↓MEK1, ↓ERK1/2	Kim et al., 2006
A549 cells	S phase arrest, ↓phosphorylation of Rb, ↑$p21^{WAF1/CIP1}$ (p53-dependent)	Kim et al., 2003a
DU-145 cells	↓Expression of cyclin D1, and cdk-4, ↓activity of cyclin E and cdk-2,↑expression of $p21^{WAF1/CIP1}$ and p53	Kim et al., 2003b
SW480 cells	S phase arrest, ↓expression of cyclin B1, D1, A1, ↓β-catenin expression, No inhibitory effect on cyclin D1 promoter activity,↓cyclin D1 mRNA expression only at 300 μM	Joe et al., 2002
OVCAR-3 cells	S phase arrest, ↑phosphorylation of Cdc25c at tyrosine 15 residue,↑phosphorylation of H2A.X at serine 139 residue, ↑chk1/2 expression,↑ATM/ATR kinase activity	Tyagi et al., 2005
Caco2 and HCT-116 cells	Blocks S→G2 transition, ↓expression of cyclin D1 and Cdk4,↑cyclin E and A expression, ↓hyperphosphorylated Rb,↑ hypophosphorylated Rb	Wolter et al., 2001
UVB-irradiated SKH-1 hairless mouse skin	↓Phosphorylation of ERK, ↓expression of cylcin D1/D2 and cdk-2,-4,-6, ↑expression of p21 and p53	Reagan-Shaw et al., 2004

inhibitor $p21^{WAF1/CIP1}$ in human epidermoid carcinoma (A431) cells, thereby blocking growth of these cells [Ahmad et al., 2001]. The antiproliferative effect of resveratrol in A431 cells was also associated with a decrease in the expression of E2F transcription factor as well as the reduced level of the hyperphosphorylated form of retinoblastoma (Rb) protein [Adhami et al., 2001]. Likewise, resveratrol suppressed UV-induced expression of Cdks (-2, -4, -6) and cyclins (D1 and D2), and up-regulated the expression of $p21^{WAF1/CIP1}$ and p53 in SKH-1 hairless mouse skin [Reagan-Shaw et al., 2004]. Moreover, the compound diminished the expression and activities of cyclin D1 and Cdk4 in association with the induction of p53 and $p21^{WAF1/CIP1}$ in human prostate (DU-145) [Kim et al., 2003b] and breast (MCF-7) [Kim et al., 2004] cancer cells. Although resveratrol attenuated the expression of cyclin D1 in SW480 cells, it failed to inhibit the cyclin D1 promoter activity [Joe et al., 2002]. Resveratrol inhibited the formation of a complex between cyclin E-Cdk-2 in human prostate cancer (DU-145) cells without altering the levels of interacting proteins [Kim et al., 2003b]. By inactivating AP-1, resveratrol suppressed proliferation of human epidermoid carcinoma (A431) cells [Kim et al., 2006]. The inhibition of Rb phosphorylation and induction of $p21^{WAF1/CIP1}$ and p53 proteins by resveratrol led to S-phase arrest in human lung cancer (A549) cells in culture [Kim et al., 2003a]. Treatment of a series of human cancer cells (MCF-7, SW480, HCE7, Seg-1, Bic-1, and HL60) with resveratrol reduced cell proliferation via down-regulation of cyclin B1 [Joe et al., 2002]. Resveratrol exerted antiproliferative effects in human multiple myeloma cells by suppressing NF-κB activation by inhibiting upstream kinases IKK and Akt, and blocking the phosphorylation of IκBα and p65 [Bhardwaj et al., 2007]. This study also demonstrated that resveratrol repressed phosphorylation and DNA binding of STAT3 and down-regulated STAT3-regulated cyclin D1, Bcl-xl, and Mcl-1 genes in various human multiple myeloma cells [Bhardwaj et al., 2007].

Resveratrol activated ataxia telangiectasia mutated (ATM)/ataxia telangiectasia-Rad3-related (ATR)-Chk1/2, phosphorylated cell division cycle (Cdc)-25C, Cdc-2 and H2A.X, and induced S phase arrest in human ovarian cancer (OVCAR-3) cells, while it caused only marginal S phase arrest in normal human foreskin fibroblasts [Tyagi et al., 2005], suggesting that the compound selectively inhibited proliferation of cancer cells. Besides modulating the expression and/or activities of various cell-cycle regulatory proteins, resveratrol attenuated tumor growth by targeting several components of protein translation machinery. For example, resveratrol inhibited insulin-like growth factor-1-induced phosphorylation of protein translation regulators, such as 4E-BP1, eIF4E, and 70S6K1, and expression of S6 ribosomal kinase in ovarian cancer (A2780/CP70 and OVCAR-3) cells (Cao et al., 2004).

Induction of Apoptosis

The biochemistry behind the anticancer effects of resveratrol has been corroborated by its ability to induce apoptotic death in various cancer cells in culture, xenografted tumors, and chemically induced mouse skin tumors in vivo [Garvin et al., 2006; Kalra et al., 2007; Kundu and Surh, 2004]. Resveratrol activates both extrinsic (death receptor-mediated) and intrinsic (mitochondria-dependent) pathways of cell death machinery (Fig. 15.4).

Extrinsic Pathway The induction of apoptosis in human promyelocytic leukemia (HL-60) and human breast cancer (T47D) cells by resveratrol was mediated via activation of the CD95-CD95L signaling, while it did not affect the survival of normal peripheral blood lymphocytes [Clement et al., 1998]. Delmas and colleagues [2003, 2004] demonstrated that resveratrol did not cause any change in the expression level of FAS or FAS-L in human colon cancer (SW480) cells, but rather it redistributed cell surface receptors such as CD95, and death receptor-4 and -5 in membrane lipid rafts and induced apoptosis in a

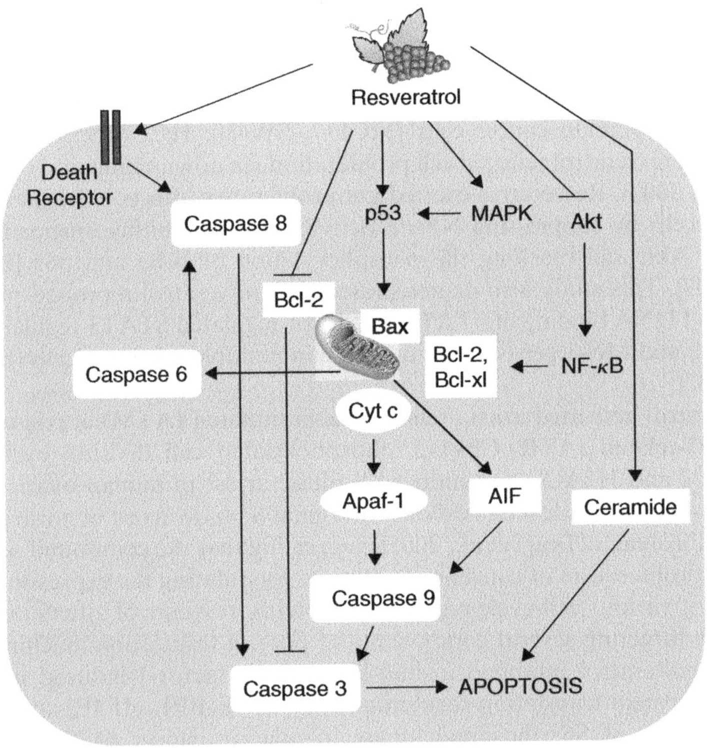

Figure 15.4 Mechanisms of apoptosis induction by resveratrol. Resveratrol selectively induces apoptosis in various cancer cells in culture by several mechanisms including activation of death receptor-mediated signaling, mitochondria-dependent cytochrome *c* release and caspase activation, induction of *p53* and *p53*-regulated proapototic genes, activation of MAP kinase-mediated *p53* phosphorylation, blockade of Akt-mediated cell survival pathways, and accumulation of intracellular ceramide level.

caspase-dependent and Bcl-2-independent manner. Co-treatment of these cancer cells with nystatin, a cholesterol sequestering agent, abrogated resveratrol-induced redistribution of death receptors into membrane lipid rafts and death-receptor-mediated cell death [Delmas et al., 2004]. Pretreatment of human prostate cancer (PC-3 and DU-145) cells with resveratrol resulted in TNF-related apoptosis inducing ligand (TRAIL)-, Fas-, or TNF-α-mediated cell death by multiple mechanisms involving down-regulation of inhibitor of apoptotic protein (IAPs), suppression of Akt phosphorylation, and subsequent activation of caspases (Gill et al., 2007).

Intrinsic Pathway Resveratrol induced apoptosis by targeting the mitochondria-dependent pathway. The activation of intrinsic pathway of cell death by resveratrol is supported by its inhibitory effect on the mitochondrial F1F0-ATPase, which is an enzyme involved in mitochondrial ATP synthesis [Gledhill et al., 2007]. Resveratrol activated p53, up-regulated proapoptotic Bax, down-regulated antiapoptotic Bcl-2, reduced the mitochondrial membrane potential, stimulated cytochrome *c* release and induced various caspases, thereby inducing apoptosis [Gusman et al., 2001; Kim et al., 2004; Zhou et al., 2005]. The induction of apoptosis by resveratrol has been shown to be cell type specific, being p53- dependent in certain cancer cells [Heiss et al., 2007; Huang et al., 1999], while p53- independent in others [Mahyar-Roemer et al., 2001, 2002]. Resveratrol increased the MAP kinase-mediated phosphorylation of p53 at serine 15 residue, thereby inducing apoptosis in JB6 Cl41 cells [She et al., 2001, 2002]. Moreover, resveratrol induced apoptosis in HepG2 cells by a mechanism involving p53-dependent increases in Bax and p21 [Kuo et al., 2002]. Likewise, the induction of p53-responsive genes, such as $p21^{WAF1/CIP1}$, *p300/CBP*, and *Apaf1*, by resveratrol was attributed to apoptosis of prostate cancer (LNCaP) cells [Narayanan et al., 2003]. Besides inducing apoptosis in cancer cells, resveratrol also induced apoptosis in chemically induced mouse skin papillomas in vivo via the induction of p53, release of cytochrome *c*, activation of Bax, inhibition of Bcl-2, and proteolytic cleavage of various caspases, such as caspase 9 and caspase 3 as well as poly-(ADP)ribosylpolymerase (PARP) [Kalra et al., 2007].

Other Mechanisms Resveratrol activated caspase-2 and -8, resulting in the activation of downstream caspases, and induced cell death in a death receptor- or mitochondria-independent manner [Mohan et al., 2006]. The induction of apoptosis in HCT-116 ($Bax^{+/-}$) cells by resveratrol was mediated via activation of caspase 6 and subsequent degradation of nuclear coat protein lamin A [Lee et al., 2006]. Resveratrol induced a senescence-like growth arrest in HCT-116 cells via ROS-dependent activation of p38 MAP kinase and ATM kinase, and subsequent phosphorylation of p53 and induction of $p21^{WAF1/CIP1}$ [Heiss et al., 2007]. In contrast, a p53-independent mechanism for resveratrol-induced apoptosis of HCT-116 cells was reported [Mahyar-Roemer et al., 2001, 2002].

Lin et al. [2006] demonstrated that resveratrol induced apoptosis in MCF-7 cells by interacting with the β3 domain of a cell surface receptor integrin-αVβ3, thereby transducing the activating signals to ERK and p53. Another interesting mechanism involved in p53-dependent apoptosis was observed in MCF-7 and MDA-MB-231 cells treated with resveratrol [Tang et al., 2006]. According to this study, resveratrol induced apoptosis through enhanced co-localization of COX-2, phosphorylated p53 (at serine 15 residue), and transcriptional co-activator p300. Blocking the COX-2 function with a pharmacologic inhibitor or small interfering RNA (COX-2 siRNA) reduced resveratrol-induced phosphorylation of p53 and induction of apoptosis in MCF-7 cells [Tang et al., 2006]. The inhibition of Akt, activation of GSK-3β, and attenuation of Notch-mediated survival signals partly accounted for resveratrol-induced apoptosis in T-cell acute lymphoblastic leukemia cells [Cecchinato et al., 2007].

Resveratrol inhibited proliferation and/or induced apoptosis of different cancer cells by accumulating intracellular ceramide, which is a key component of sphingolipids and a mediator of many cellular processes including proliferation, differentiation, and senescence [Scarlatti et al., 2003, 2007]. Resveratrol induced apoptosis in MDA-MB-231 cells by inducing de novo synthesis of ceramide and hydrolysis of sphingomyelin, thereby increasing ceramide level [Scarlatti et al., 2003]. In another study, resveratrol-induced accumulation of ceramide sensitized DU-145 cells to apoptosis upon exposure to ionizing radiation [Scarlatti et al., 2007].

Inhibition of Tumor Angiogenesis

Angiogenesis, the formation of new blood vessels, is an important feature of rapidly proliferating tumors to meet the requirement of adequate oxygen and nutrient supply for their survival. Increased metabolic activities and oxygen consumption by rapidly proliferating cells in a solid tumor create an intratumoral hypoxic environment [Liao and Johnson, 2007; Lopez-Lazaro, 2006], which leads tumor cells to induce hypoxia-responsive genes [Hickey and Simon, 2006; North et al., 2005]. One of the master regulators of cellular oxygen homeostasis is hypoxia inducible factor (HIF), which regulates the expression of various hypoxia-responsive genes [Semenza, 2004] including vascular endothelial growth factor (VEGF), basic fibroblast growth factor (bFGF), VEGF receptor (VEGFR), IL-8, iNOS, and angiopoietins [Hickey and Simon, 2006; Pollard, 2004]. Several studies have reported that resveratrol exerts anticancer effects by blocking various angiogenic switches. Treatment of OVCAR-3 cells with resveratrol resulted in the down-regulation of HIF-1α and VEGF through multiple mechanisms involving inhibition of Akt and MAP kinases, inhibition of protein translational regulators, and enhancement of proteasomal degradation of HIF-1α protein [Cao et al., 2004]. Likewise, resveratrol reduced the accumulation of HIF-1α protein and the expression of VEGF in human tongue squamous cell carcinomas (SCC-9) and HepG2 cells, partly by inhibiting the activation of ERK and Akt, and enhancing the proteasomal degradation of HIF-1 [Zhang et al., 2005]. Resveratrol reduced the extracellular levels of VEGF in vitro and suppressed the growth of MDA-MB-231 cells xenografted in nude mice by blocking angiogenesis [Garvin et al., 2006].

Suppression of Invasion and Metastasis

The shedding of cells from primary tumors, invasion of tumor cells through the tumor-associated stroma and subsequent colonization in distant sites in the body are major events in tumor progression. Genes encoding extracellular matrix processing proteases, adhesion proteins, and motility factors are the main mediators of invasion and metastasis. One of the mediators of tumor invasion and metastasis is lysophosphatidic acid (LPA), which enhances the migration of cancer cells [Park et al., 2007]. Resveratrol significantly attenuated LPA-induced expression of HIF-1α and VEGF, and subsequent migration of human ovarian cancer cells via blockade of the activation of upstream ERK1/2 and p70S6 kinase [Park et al., 2007]. Intraperitoneal administration of resveratrol conferred 40% protection against lung metastasis in mice intramuscularly transplanted with Lewis lung carcinoma cells [Busquets et al., 2007]. Matrix metalloproteinase (MMP), urokinase-plasminogen activator (uPA), and uPA receptor (uPAR) are primarily involved in the metastatic process. The expression of MMP-9 mRNA, but not of uPA or uPAR, was diminished in murine bladder cancer (MB 49) cells cultured in the presence of resveratrol [Garcia Mediero et al., 2005]. Likewise, resveratrol blocked the invasiveness of various cancer cells by reducing the expression and/or the activity of MMP-2 and -9 [Sun et al., 2006; Tang et al., 2008; Yu et al., 2007]. Resveratrol attenuated the invasion of TNF-α-stimulated HepG2 cells by down-regulating the expression of MMP-9 via blockade of NF-κB activation [Yu et al., 2007]. The compound suppressed TNF-α-induced MMP-9 expression in vascular smooth muscle cells by blocking the activation of NF-κB and AP-1 [Lee and Moon, 2005]. Similarly, TPA-induced expression of MMP-9 protein and its mRNA transcript in Caski cells was attenuated by resveratrol through inhibition of NF-κB, AP-1, and upstream kinases JNK and PKCα [Woo et al., 2004]. According to Tang et al. [2008], resveratrol inhibited heregulin- 1-induced MMP-9 expression and invasion of MCF-7 cells by blocking phosphorylation of ERK. Resveratrol decreased the chemotactic response of metastatic MDA-MB-231 cells by reducing focal adhesion kinase activity [Azios and Dharmawardhane, 2005]. Alternatively, incubation with resvertarol or resveratrol-treated rat sera diminished the peroxide levels and the expression of hepatocyte growth factor in H_2O_2-stimulated rat ascites hepatoma (AH109A) cells, and reduced the invasive capacity of these cells, suggesting that the antioxidant property of resveratrol accounts for its anti-invasive effects [Miura et al., 2004].

Augmentation of Chemotherapy, Radiotherapy, and Cancer Gene Therapy

Recently, the combined use of chemopreventive phytochemicals with conventional anticancer therapy or radiotherapy has been shown to sensitize many chemoresistant and/or radioresistant cancer cells to apoptosis [Garg et al., 2005]. Products of multidrug resistant (*MDR*)-1 gene, for example P-glycoprotein, have been involved in acquiring chemoresistance by cancer cells. P-glycoprotein reduces intratumoral concentrations of chemotherapeutic drugs by actively effluxing drugs from cells and reduces their efficacy. Treatment of P-glycoprotein overexpressing multidrug-resistant human carcinoma KB-C2 cells with resveratrol and daunorubicin sensitized the cells to daunorubicin-induced apoptosis as a consequence of decreased P-glycoprotein functions and increased accumulation of daunorubicin in these cells [Nabekura et al., 2005]. Moreover, resveratrol enhanced the apoptotic effects of bortezomib and thalidomide in multiple myeloma cells [Bhardwaj et al., 2007]. A combination therapy with resveratrol plus etoposide induced apoptosis of HT-29 cells via a ROS-dependent activation of adenosine monophosphate-activated protein kinase [Hwang et al., 2007]. Resveratrol potentiated the growth inhibitory effects of cisplatin or doxorubicin in OVCAR-3 and human uterine cancer (Ishikawa) cells, and reduced doxorubicin toxicity in mice [Rezk et al., 2006], suggesting that the compound simultaneously enhances chemosensitivity of cancer cells and alleviates chemotherapy-associated adverse effects.

Irradiation with ionizing radiation (IR) is an important modality for the treatment of cancer. While resveratrol protects normal cells from radiation damage, the compound enhances the sensitivity of many radioresistant cancer cells to IR-induced apoptosis [Reagan-Shaw et al., 2008]. Pretreatment of DU-145 cells with resveratrol enhanced IR-induced apoptosis via increased accumulation of ceramide in these cells [Scarlatti et al., 2007]. In combination with γ radiation, resveratrol reactivated the apoptotic pathway in TRAIL-resistant human melanoma cells [Ivanov et al., 2008]. According to this study, an initial increase in the γ-radiation-induced expression of death receptor-5 followed by down-regulation of antiapoptotic cFLIP and Bcl-xL via inactivation of STAT3 and NF-κB by resveratrol led to melanoma cell death [Ivanov et al., 2008]. Combination therapy with γ-radiation and resveratrol may cause either apoptosis or necrosis of melanoma cells depending on the ability of resveratrol to induce surface expression of TRAIL. Melanoma cells exhibiting a suppressed translocation of TRAIL to cell surface undergo necrosis upon receiving γ-radiation, while cells responsive to resveratrol-induced TRAIL surface expression follow γ-radiation-induced apoptosis [Johnson et al., 2008].

In addition to its sensitizing effects in chemo- or radioresistant cells, resveratrol also enhances the antitumor effects of cancer gene therapy. Treatment of xenografted (human pancreatic carcinomas or rat colorectal carcinomas) tumors with replication-deficient adenoviral gene therapy vector Ad.Egr.TNF, resveratrol or their combination showed that combined administration of Ad.Egr.TNF and resveratrol caused more growth inhibition of tumor xenografts in vivo via activation of early growth response-1 gene and elevation of an intratumoral TNF-α level in comparison to treatment with either agent alone (Bickenbach et al., 2008). The same study also demonstrates that the combination of resveratrol with Ad.Egr.TNF does not produce any adverse effects in tumor bearing animals [Bickenbach et al., 2008].

CONCLUSION

With millions of cancer-afflicted people around the world, scientists are making all efforts to find an effective strategy to reduce the incidence and the mortality of cancer. Research projected toward searching for effective chemopreventive/chemotherapeutic agents has been expedited over the last few decades. Although multidirectional approaches, such as synthesizing novel anticancer lead compounds, developing cancer gene therapy and immunotherapy, etc have been adopted to find a successful cure for cancer, the search for chemopreventive and chemotherapeutic phytochemicals from edible plants has received special attention. The complex heterogeneity of cancer cells imposes a major hurdle in finding a molecular target-based prevention or treatment modality for cancer. Therefore, fighting cancer with dietary phytochemicals, which can simultaneously modulate diverse biochemical events, appears as a timely approach to prevent or cure cancer.

Among the nature's bounty of anticancer agents, resveratrol has been studied extensively to elucidate its underlying biochemical basis of chemopreventive and chemotherapeutic effects. Recently, a few clinical trials with resveratrol have been launched. According to a recent phase I clinical trial, the consumption of resveratrol does not cause any serious adverse effects in healthy volunteers, but the peak plasma level remains below the minimum effective concentration of the compound required to exhibit chemopreventive effects in cultured cells [Boocock et al., 2007]. Another phase I clinical trial has recently been undertaken to evaluate the safety and the pharmacokinetic profile of resveratrol given in a multiple-dosage regimen [Baur and Sinclair, 2006]. A phase I colon cancer prevention trial with resveratrol was conducted at the University of California, Irvine, [Baur and Sinclair, 2006]. Despite favorable preclinical and clinical findings, and a significant progress in clarifying the biochemistry behind anticancer properties of resveratrol, pharmacokinetic studies in rodent models suggest a poor bioavailability of the compound in vivo. Thus, one of the critical issues of future resveratrol research would be to improve its bioavailability by developing an appropriate formulation.

ACKNOWLEDGMENTS

This work was supported by the National Research Laboratory Fund and the Biofood Research Program from the Ministry of Science and Technology, Republic of Korea.

REFERENCES

Adhami VM, Afaq F, Ahmad N. 2003. Suppression of ultraviolet B exposure-mediated activation of NF-κB in normal human keratinocytes by resveratrol. *Neoplasia* 5:74–82.

Adhami VM, Afaq F, Ahmad N. 2001. Involvement of the retinoblastoma (pRb)-E2F/DP pathway during antiproliferative effects of resveratrol in human epidermoid carcinoma (A431) cells. *Biochem Biophys Res Commun* 288:579–585.

Aggarwal BB, Bhardwaj A, Aggarwal RS, Seeram NP, Shishodia S, Takada Y. 2004. Role of resveratrol in prevention and therapy of cancer: Preclinical and clinical studies. *Anticancer Res* 24:2783–2840.

Ahmad N, Adhami VM, Afaq F, Feyes DK, Mukhtar H. 2001. Resveratrol causes WAF-1/p21-mediated G(1)-phase arrest of cell cycle and induction of apoptosis in human epidermoid carcinoma A431 cells. *Clin Cancer Res* 7:1466–1473.

Andrieux L, Langouet S, Fautrel A, Ezan F, Krauser JA, Savouret JF, Guengerich FP, Baffet G, Guillouzo A. 2004. Aryl hydrocarbon receptor activation and cytochrome P450 1A induction by the mitogen-activated protein kinase inhibitor U0126 in hepatocytes. *Mol Pharmacol* 65:934–943.

Atten MJ, Attar BM, Milson T, Holian O. 2001. Resveratrol-induced inactivation of human gastric adenocarcinoma cells through a protein kinase C-mediated mechanism. *Biochem Pharmacol* 62:1423–1432.

Azios NG, Dharmawardhane SF. 2005. Resveratrol and estradiol exert disparate effects on cell migration, cell surface actin structures, and focal adhesion assembly in MDA-MB-231 human breast cancer cells. *Neoplasia* 7:128–140.

Aziz MH, Nihal M, Fu VX, Jarrard DF, Ahmad N. 2006. Resveratrol-caused apoptosis of human prostate carcinoma LNCaP cells is mediated via modulation of phosphatidylinositol 3′-kinase/Akt pathway and Bcl-2 family proteins. *Mol Cancer Ther* 5:1335–1341.

Aziz MH, Reagan-Shaw S, Wu J, Longley BJ, Ahmad N. 2005. Chemoprevention of skin cancer by grape constituent resveratrol: Relevance to human disease? *FASEB J* 19:1193–1195.

Banerjee S, Bueso-Ramos C, Aggarwal BB. 2002. Suppression of 7,12-dimethylbenz (a)anthracene-induced mammary carcinogenesis in rats by resveratrol: Role of nuclear factor-κB, cyclooxygenase 2, and matrix metalloprotease 9. *Cancer Res* 62:4945–4954.

Baur JA, Sinclair DA. 2006. Therapeutic potential of resveratrol: The in vivo evidence. *Nat Rev Drug Discov* 5:493–506.

Bhardwaj A, Sethi G, Vadhan-Raj S, Bueso-Ramos C, Takada Y, Gaur U, Nair S, Shishodia S, Aggarwal BB. 2007. Resveratrol inhibits proliferation, induces apoptosis, and overcomes chemoresistance through down-regulation of STAT3 and nuclear factor-κB-regulated antiapoptotic and cell survival gene products in human multiple myeloma cells. *Blood* 109:2293–2302.

Bhat KP, Lantvit D, Christov K, Mehta RG, Moon RC, Pezzuto JM. 2001. Estrogenic and antiestrogenic properties of resveratrol in mammary tumor models. *Cancer Res* 61:7456–7463.

Bickenbach KA, Veerapong J, Shao MY, Mauceri HJ, Posner MC, Kron SJ, Weichselbaum RR. 2008. Resveratrol is an effective inducer of CArG-driven TNF-alpha gene therapy. *Cancer Gene Ther* 15:133–139.

Bode AM, Dong Z. 2004. Targeting signal transduction pathways by chemopreventive agents. *Mutat Res* 555:33–51.

Boocock DJ, Faust GE, Patel KR, Schinas AM, Brown VA, Ducharme MP, Booth TD, Crowell JA, Perloff M, Gescher AJ, Steward WP, Brenner DE. 2007. Phase I dose escalation pharmacokinetic study in healthy volunteers of resveratrol, a potential cancer chemopreventive agent. *Cancer Epidemiol Biomarkers Prev* 16:1246–1252.

Bove K, Lincoln DW, Tsan MF. 2002. Effect of resveratrol on growth of 4T1 breast cancer cells in vitro and in vivo. *Biochem Biophys Res Commun* 291:1001–1005.

Boyce A, Doehmer J, Gooderham NJ. 2004. Phytoalexin resveratrol attenuates the mutagenicity of the heterocyclic amines 2-amino-1-methyl-6-phenylimidazo [4,5-b]pyridine and 2-amino-3,8-dimethylimidazo[4,5-f]quinoxaline. *J Chromatogr B Analyt Technol Biomed Life Sci* 802:217–223.

Busquets S, Ametller E, Fuster G, Olivan M, Raab V, Argiles JM, Lopez-Soriano FJ. 2007. Resveratrol, a natural diphenol, reduces metastatic growth in an experimental cancer model. *Cancer Lett* 245:144–148.

Cao Z, Fang J, Xia C, Shi X, Jiang BH. 2004. trans-3,4,5′-Trihydroxystibene inhibits hypoxia-inducible factor 1alpha and vascular endothelial growth factor expression in human ovarian cancer cells. *Clin Cancer Res* 10:5253–5263.

Cecchinato V, Chiaramonte R, Nizzardo M, Cristofaro B, Basile A, Sherbet GV, Comi P. 2007. Resveratrol-induced apoptosis in human T-cell acute lymphoblastic leukaemia MOLT-4 cells. *Biochem Pharmacol* 74:1568–1574.

Chang TK, Chen J, Lee WB. 2001. Differential inhibition and inactivation of human CYP1 enzymes by trans-resveratrol: Evidence for mechanism-based inactivation of CYP1A2. *J Pharmacol Exp Ther* 299:874–882.

Chen C, Kong AN. 2005. Dietary cancer-chemopreventive compounds: From signaling and gene expression to pharmacological effects. *Trends Pharmacol Sci* 26:318–326.

Chen CY, Jang JH, Li MH, Surh Y-J. 2005. Resveratrol upregulates heme oxygenase-1 expression via activation of NF-E2-related factor 2 in PC12 cells. *Biochem Biophys Res Commun* 331:993–1000.

Chen ZH, Hurh YJ, Na HK, Kim JH, Chun YJ, Kim DH, Kang KS, Cho MH, Surh Y-J. 2004. Resveratrol inhibits TCDD-induced expression of CYP1A1 and CYP1B1 and catechol estrogen-mediated oxidative DNA damage in cultured human mammary epithelial cells. *Carcinogenesis* 25:2005–2013.

Cho DI, Koo NY, Chung WJ, Kim TS, Ryu SY, Im SY, Kim KM. 2002. Effects of resveratrol-related hydroxystilbenes on the nitric oxide production in macrophage cells: Structural requirements and mechanism of action. *Life Sci* 71:2071–2082.

Chun KS, Surh Y-J. 2004. Signal transduction pathways regulating cyclooxygenase-2 expression: Potential molecular targets for chemoprevention. *Biochem Pharmacol* 68:1089–1100.

Ciolino HP, Yeh GC. 1999. Inhibition of aryl hydrocarbon-induced cytochrome P-450 1A1 enzyme activity and CYP1A1 expression by resveratrol. *Mol Pharmacol* 56:760–767.

Ciolino HP, Daschner PJ, Yeh GC. 1998. Resveratrol inhibits transcription of CYP1A1 in vitro by preventing activation of the aryl hydrocarbon receptor. *Cancer Res* 58:5707–5712.

Clement MV, Hirpara JL, Chawdhury SH, Pervaiz S. 1998. Chemopreventive agent resveratrol, a natural product derived from grapes, triggers CD95 signaling-dependent apoptosis in human tumor cells. *Blood* 92:996–1002.

Collins I, Garrett MD. 2005. Targeting the cell division cycle in cancer: CDK and cell cycle checkpoint kinase inhibitors. *Curr Opin Pharmacol* 5:366–373.

de la Lastra CA, Villegas I. 2007. Resveratrol as an antioxidant and pro-oxidant agent: Mechanisms and clinical implications. *Biochem Soc Trans* 35:1156–1160.

Delmas D, Rebe C, Micheau O, Athias A, Gambert P, Grazide S, Laurent G, Latruffe N, Solary E. 2004. Redistribution of CD95, DR4 and DR5 in rafts accounts for the synergistic toxicity of resveratrol and death receptor ligands in colon carcinoma cells. *Oncogene* 23:8979–8986.

Delmas D, Rebe C, Lacour S, Filomenko R, Athias A, Gambert P, Cherkaoui-Malki M, Jannin B, Dubrez-Daloz L, Latruffe N, Solary E. 2003. Resveratrol-induced apoptosis is associated with Fas redistribution in the rafts and the formation of a death-inducing signaling complex in colon cancer cells. *J Biol Chem* 278: 41482–41490.

Dinkova-Kostova AT, Holtzclaw WD, Wakabayashi N. 2005. Keap1, the sensor for electrophiles and oxidants that regulates the phase 2 response, is a zinc metalloprotein. *Biochemistry* 44:6889–6899.

Estrov Z, Shishodia S, Faderl S, Harris D, Van Q, Kantarjian HM, Talpaz M, Aggarwal BB. 2003. Resveratrol blocks interleukin-1beta-induced activation of the nuclear transcription factor NF-κB, inhibits proliferation, causes S-phase arrest, and induces apoptosis of acute myeloid leukemia cells. *Blood* 102:987–995.

Fontecave M, Lepoivre M, Elleingand E, Gerez C, Guittet O. 1998. Resveratrol, a remarkable inhibitor of ribonucleotide reductase. *FEBS Lett* 421:277–279.

Fuggetta MP, Lanzilli G, Tricarico M, Cottarelli A, Falchetti R, Ravagnan G, Bonmassar E. 2006. Effect of resveratrol on proliferation and telomerase activity of human colon cancer cells in vitro. *J Exp Clin Cancer Res* 25:189–193.

Gali-Muhtasib H, Bakkar N. 2002. Modulating cell cycle: Current applications and prospects for future drug development. *Curr Cancer Drug Targets* 2:309–336.

Garcia Mediero JM, Ferruelo Alonso A, Paez Borda A, Lujan Galan M, Angulo Cuesta J, Chiva Robles V, Berenguer Sanchez A. 2005. Effect of polyphenols from the Mediterranean diet on proliferation and mediators of in vitro invasiveness of the MB-49 murine bladder cancer cell line. *Actas Urol Esp* 29:743–749.

Garg AK, Buchholz TA, Aggarwal BB. 2005. Chemosensitization and radiosensitization of tumors by plant polyphenols. *Antioxid Redox Signal* 7:1630–1647.

Garvin S, Ollinger K, Dabrosin C. 2006. Resveratrol induces apoptosis and inhibits angiogenesis in human breast cancer xenografts in vivo. *Cancer Lett* 231:113–122.

Gerhauser C, Klimo K, Heiss E, Neumann I, Gamal-Eldeen A, Knauft J, Liu GY, Sitthimonchai S, Frank N. 2003. Mechanism-based in vitro screening of potential cancer chemopreventive agents. *Mutat Res* 523–524:163–172.

Gill C, Walsh SE, Morrissey C, Fitzpatrick JM, Watson RW. 2007. Resveratrol sensitizes androgen independent prostate cancer cells to death-receptor mediated apoptosis through multiple mechanisms. *Prostate* 67:1641–1653.

Gledhill JR, Montgomery MG, Leslie AG, Walker JE. 2007. Mechanism of inhibition of bovine F1-ATPase by resveratrol and related polyphenols. *Proc Natl Acad Sci U S A* 104:13632–13637.

Gusman J, Malonne H, Atassi G. 2001. A reappraisal of the potential chemopreventive and chemotherapeutic properties of resveratrol. *Carcinogenesis* 22:1111–1117.

Harper CE, Patel BB, Wang J, Arabshahi A, Eltoum IA, Lamartiniere CA. 2007. Resveratrol suppresses prostate cancer progression in transgenic mice. *Carcinogenesis* 28:1946–1953.

Hebbar V, Shen G, Hu R, Kim BR, Chen C, Korytko PJ, Crowell JA, Levine BS, Kong AN. 2005. Toxicogenomics of resveratrol in rat liver. *Life Sci* 76:2299–2314.

Hecht SS, Kenney PM, Wang M, Trushin N, Agarwal S, Rao AV, Upadhyaya P. 1999. Evaluation of butylated hydroxyanisole, myo-inositol, curcumin, esculetin, resveratrol and lycopene as inhibitors of benzo[a]pyrene plus 4-(methylnitrosamino)-1-(3-pyridyl)-1-butanone-induced lung tumorigenesis in A/J mice. *Cancer Lett* 137:123–130.

Heiss EH, Schilder YD, Dirsch VM. 2007. Chronic treatment with resveratrol induces redox stress- and ataxia telangiectasia-mutated (ATM)-dependent senescence in p53-positive cancer cells. *J Biol Chem* 282:26759–26766.

Hickey MM, Simon MC. 2006. Regulation of angiogenesis by hypoxia and hypoxia-inducible factors. *Curr Top Dev Biol* 76:217–257.

Hsieh TC, Lu X, Wang Z, Wu JM. 2006. Induction of quinone reductase NQO1 by resveratrol in human K562 cells involves the antioxidant response element ARE and is accompanied by nuclear translocation of transcription factor Nrf2. *Med Chem* 2:275–285.

Huang C, Ma WY, Goranson A, Dong Z. 1999. Resveratrol suppresses cell transformation and induces apoptosis through a p53-dependent pathway. *Carcinogenesis* 20:237–242.

Hudson TS, Hartle DK, Hursting SD, Nunez NP, Wang TT, Young HA, Arany P, Green JE. 2007. Inhibition of prostate cancer growth by muscadine grape skin extract and resveratrol through distinct mechanisms. *Cancer Res* 67:8396–8405.

Hwang JT, Kwak DW, Lin SK, Kim HM, Kim YM, Park OJ. 2007. Resveratrol induces apoptosis in chemoresistant cancer cells via modulation of AMPK signaling pathway. *Ann N Y Acad Sci* 1095:441–448.

Ivanov VN, Partridge MA, Johnson GE, Huang SX, Zhou H, Hei TK. 2008. Resveratrol sensitizes melanomas to TRAIL through modulation of antiapoptotic gene expression. *Exp Cell Res* 314:1163–1176.

Jang M, Cai L, Udeani GO, Slowing KV, Thomas CF, Beecher CW, Fong HH, Fransworth NR, Kinghorn AD, Mehta RG, Moon RC, Pezzuto JM. 1997. Cancer chemopreventive activity of resveratrol, a natural product derived from grapes. *Science* 275:218–220.

Joe AK, Liu H, Suzui M, Vural ME, Xiao D, Weinstein IB. 2002. Resveratrol induces growth inhibition, S-phase arrest, apoptosis, and changes in biomarker expression in several human cancer cell lines. *Clin Cancer Res* 8:893–903.

Johnson GE, Ivanov VN, Hei TK. 2008. Radiosensitization of melanoma cells through combined inhibition of protein regulators of cell survival. *Apoptosis* 13:790–802.

Juan SH, Cheng TH, Lin HC, Chu YL, Lee WS. 2005. Mechanism of concentration-dependent induction of heme oxygenase-1 by resveratrol in human aortic smooth muscle cells. *Biochem Pharmacol* 69:41–48.

Kalra N, Roy P, Prasad S, Shukla Y. 2008. Resveratrol induces apoptosis involving mitochondrial pathways in mouse skin tumorigenesis. *Life Sci* 82:348–358.

Kim Y, Fischer SM. 1998. Transcriptional regulation of cyclooxygenase-2 in mouse skin carcinoma cells. Regulatory role of CCAAT/enhancer-binding proteins in the differential expression of cyclooxygenase-2 in normal and neoplastic tissues. *J Biol Chem* 273:27686–27694.

Kim JH, Lee BK, Lee KW, Lee HJ. 2008. Resveratrol counteracts gallic acid-induced down-regulation of gap-junction intercellular communication. J Nutr Biochem. (Epub ahead of print)

Kim AL, Zhu Y, Zhu H, Han L, Kopelovich L, Bickers DR, Ather M. 2006. Resveratrol inhibits proliferation of human epidermoid carcinoma A431 cells by modulating MEK1 and AP-1 signalling pathways. *Exp Dermatol* 15:538–546.

Kim YA, Choi BT, Lee YT, Park DI, Rhee SH, Park KY, Choi YH. 2004. Resveratrol inhibits cell proliferation and induces apoptosis of human breast carcinoma MCF-7 cells. *Oncol Rep* 11:441–446.

Kim YA, Lee WH, Choi TH, Rhee SH, Park KY, Choi YH. 2003a. Involvement of p21WAF1/CIP1, pRB, Bax and NF-κB in induction of growth arrest and apoptosis by resveratrol in human lung carcinoma A549 cells. *Int J Oncol* 23:1143–1149.

Kim YA, Rhee SH, Park KY, Choi YH. 2003b. Antiproliferative effect of resveratrol in human prostate carcinoma cells. *J Med Food* 6:273–280.

Kluth D, Banning A, Paur I, Blomhoff R, Brigelius-Flohe R. 2007. Modulation of pregnane X receptor- and electrophile responsive element-mediated gene expression by dietary polyphenolic compounds. *Free Radic Biol Med* 42:315–325.

Kode A, Rajendrasozhan S, Caito S, Yang SR, Megson IL, Rahman I. 2008. Resveratrol induces glutathione synthesis by activation of Nrf2 and protects against cigarette smoke-mediated oxidative stress in human lung epithelial cells. *Am J Physiol Lung Cell Mol Physiol* 294:L478–488.

Kowalski J, Samojedny A, Paul M, Pietsz G, Wilczok T. 2005. Effect of apigenin, kaempferol and resveratrol on the expression of interleukin-1beta and tumor necrosis factor-alpha genes in J774.2 macrophages. *Pharmacol Rep* 57:390–394.

Kueck A, Opipari AWJr., Griffith KA, Tan L, Choi M, Huang J, Wahl H, Liu JR. 2007. Resveratrol inhibits glucose metabolism in human ovarian cancer cells. *Gynecol Oncol* 107:450–457.

Kundu JK, Surh Y-J. 2008a. Cancer chemopreventive and therapeutic potential of resveratrol: Mechanistic perspectives. *Cancer Lett* 269:243–261.

Kundu JK, Surh Y-J. 2008b. Inflammation: Gearing the journey to cancer. *Mutat Res* 659:15–30.

Kundu JK, Surh Y-J. 2005. Breaking the relay in deregulated cellular signal transduction as a rationale for chemoprevention with anti-inflammatory phytochemicals. *Mutat Res* 591: 123–146.

Kundu JK, Surh Y-J. 2004. Molecular basis of chemoprevention by resveratrol: NF-kappaB and AP-1 as potential targets. *Mutat Res* 555: 65–80.

Kundu JK, Shin YK, Kim SH, Surh Y-J. 2006a. Resveratrol inhibits phorbol ester-induced expression of COX-2 and activation of NF-κB in mouse skin by blocking I B kinase activity. *Carcinogenesis* 27:1465–1474

Kundu JK, Shin YK, Surh Y-J. 2006b. Resveratrol modulates phorbol ester-induced pro-inflammatory signal transduction pathways in mouse skin in vivo: NF-kappaB and AP-1 as prime targets. *Biochem Pharmacol* 72:1506–1515.

Kuo PL, Chiang LC, Lin CC. 2002. Resveratrol-induced apoptosis is mediated by p53–dependent pathway in Hep G2 cells. *Life Sci* 72:23–34.

Lee B, Moon SK. 2005. Resveratrol inhibits TNF-alpha-induced proliferation and matrix metalloproteinase expression in human vascular smooth muscle cells. *J Nutr* 135:2767–2773.

Lee JE, Safe S. 2001. Involvement of a post-transcriptional mechanism in the inhibition of CYP1A1 expression by resveratrol in breast cancer cells. *Biochem Pharmacol* 62:1113–1124.

Lee SC, Chan J, Clement MV, Pervaiz S. 2006. Functional proteomics of resveratrol-induced colon cancer cell apoptosis: Caspase-6-mediated cleavage of lamin A is a major signaling loop. *Proteomics* 6:2386–2394.

Leonard SS, Xia C, Jiang BH, Stinefelt B, Klandorf H, Harris GK, Shi X. 2003. Resveratrol scavenges reactive oxygen species and effects radical-induced cellular responses. *Biochem Biophys Res Commun* 309:1017–1026.

Li T, Fan GX, Wang W, Yuan YK. 2007. Resveratrol induces apoptosis, influences IL-6 and exerts immunomodulatory effect on mouse lymphocytic leukemia both in vitro and in vivo. *Int Immunopharmacol* 7:1221–1231.

Li Y, Liu J, Liu X, Xing K, Wang Y, Li F, Yao L. 2006. Resveratrol-induced cell inhibition of growth and apoptosis in MCF7 human breast cancer cells are associated with modulation of phosphorylated Akt and caspase-9. *Appl Biochem Biotechnol* 135:181–192.

Li YT, Shen F, Liu BH, Cheng GF. 2003. Resveratrol inhibits matrix metalloproteinase-9 transcription in U937 cells. *Acta Pharmacol Sin* 24:1167–1171.

Li ZG, Hong T, Shimada Y, Komoto I, Kawabe A, Ding Y, Kaganoi J, Hashimoto Y, Imamura M. 2002. Suppression of N-nitrosomethylbenzylamine (NMBA)-induced esophageal tumorigenesis in F344 rats by resveratrol. *Carcinogenesis* 23:1531–1536.

Liao D, Johnson RS. 2007. Hypoxia: A key regulator of angiogenesis in cancer. *Cancer Metastasis Rev* 26:281–290.

Lin HY, Lansing L, Merillon JM, Davis FB, Tang HY, Shih A, Vitrac X, Krisa S, Keating T, Cao HJ, Bergh J, Quackenbush S, Davis PJ. 2006. Integrin alphaVbeta3 contains a receptor site for resveratrol. *FASEB J* 20:1742–1744.

Lopez-Lazaro M. 2006. Hypoxia-inducible factor 1 as a possible target for cancer chemoprevention. *Cancer Epidemiol Biomarkers Prev* 15:2332–2335.

Mahyar-Roemer M, Kohler H, Roemer K. 2002. Role of Bax in resveratrol-induced apoptosis of colorectal carcinoma cells. *BMC Cancer* 2:27.

Mahyar-Roemer M, Katsen A, Mestres P, Roemer K. 2001. Resveratrol induces colon tumor cell apoptosis independently of p53 and precede by epithelial differentiation, mitochondrial proliferation and membrane potential collapse. *Int J Cancer* 94:615–622.

Manna SK, Mukhopadhyay A, Aggarwal BB. 2000. Resveratrol suppresses TNF-induced activation of nuclear transcription factors NF-κ B, activator protein-1, and apoptosis: Potential role of reactive oxygen intermediates and lipid peroxidation. *J Immunol* 164:6509–6519.

Martinez J, Moreno JJ. 2000. Effect of resveratrol, a natural polyphenolic compound, on reactive oxygen species and prostaglandin production. *Biochem Pharmacol* 59: 865–870.

Miura D, Miura Y, Yagasaki K. 2004. Resveratrol inhibits hepatoma cell invasion by suppressing gene expression of hepatocyte growth factor via its reactive oxygen species-scavenging property. *Clin Exp Metastasis* 21:445–451.

Mohan J, Gandhi AA, Bhavya BC, Rashmi R, Karunagaran D, Indu R, Santhoskumar TR. 2006. Caspase-2 triggers Bax-Bak-dependent and -independent cell death in colon cancer cells treated with resveratrol. *J Biol Chem* 281:17599–17611.

Murakami A, Matsumoto K, Koshimizu K, Ohigashi H. 2003. Effects of selected food factors with chemopreventive properties on combined lipopolysaccharide- and interferon-gamma-induced IkappaB degradation in RAW264.7 macrophages. *Cancer Lett* 195:17–25.

Mutoh M, Takahashi M, Fukuda K, Matsushima-Hibiya Y, Mutoh H, Sugimura T, Wakabayashi K. 2000. Suppression of cyclooxygenase-2 promoter-dependent transcriptional activity in colon cancer cells by chemopreventive agents with a resorcin-type structure. *Carcinogenesis* 21:959–963.

Nabekura T, Kamiyama S, Kitagawa S. 2005. Effects of dietary chemopreventive phytochemicals on P-glycoprotein function. *Biochem Biophys Res Commun* 327:866–870.

Narayanan BA, Narayanan NK, Re GG, Nixon DW. 2003. Differential expression of genes induced by resveratrol in LNCaP cells: P53-mediated molecular targets. *Int J Cancer* 104:204–212.

Neves MA, Dinis TC, Colombo G, Sa e Melo ML (2007). Combining computational and biochemical studies for a rationale on the anti-aromatase activity of natural polyphenols. *chem med chem* 2:1750–1762.

Nielsen M, Ruch RJ, Vang O. 2000. Resveratrol reverses tumor-promoter-induced inhibition of gap-junctional intercellular communication. *Biochem Biophys Res Commun* 275:804–809.

Niles RM, Cook CP, Meadows GG, Fu YM, McLaughlin JL, Rankin GO. 2006. Resveratrol is rapidly metabolized in athymic (nu/nu) mice and does not inhibit human melanoma xenograft tumor growth. *J Nutr* 136:2542–2546.

North S, Moenner M, Bikfalvi A. 2005. Recent developments in the regulation of the angiogenic switch by cellular stress factors in tumors. *Cancer Lett* 218:1–14.

Park SY, Jeong KJ, Lee J, Yoon DS, Choi WS, Kim YK, Han JW, Kim YM, Kim BK, Lee HY. 2007. Hypoxia enhances LPA-induced HIF-1alpha and VEGF expression: their inhibition by resveratrol. *Cancer Lett* 258:63–69.

Pendurthi UR, Meng F, Mackman N, Rao LV. 2002. Mechanism of resveratrol-mediated suppression of tissue factor gene expression. *Thromb Haemost*. 87:155–162.

Perwez Hussain S, Harris CC. 2007. Inflammation and cancer: An ancient link with novel potentials. *Int J Cancer* 121:2373–2380.

Pollard JW. 2004. Tumour-educated macrophages promote tumour progression and metastasis. *Nat Rev Cancer* 4:71–78.

Reagan-Shaw S, Mukhtar H, Ahmad N. 2008. Resveratrol imparts photoprotection of normal cells and enhances the efficacy of radiation therapy in cancer cells. *Photochem Photobiol* 84:415–421.

Reagan-Shaw S, Afaq F, Aziz MH, Ahmad N. 2004. Modulations of critical cell cycle regulatory events during chemoprevention of ultraviolet B-mediated responses by resveratrol in SKH-1 hairless mouse skin. *Oncogene* 23:5151–5160.

Revel A, Raanani H, Younglai E, Xu J, Rogers I, Han R, Savouret JF, Casper RF. 2003. Resveratrol, a natural aryl hydrocarbon receptor antagonist, protects lung from DNA damage and apoptosis caused by benzo[a]pyrene. *J Appl Toxicol* 23:255–261.

Rezk YA, Balulad SS, Keller RS, Bennett JA. 2006. Use of resveratrol to improve the effectiveness of cisplatin and doxorubicin: Study in human gynecologic cancer cell lines and in rodent heart. *Am J Obstet Gynecol* 194:e23–26.

Richard N, Porath D, Radspieler A, Schwager J. 2005. Effects of resveratrol, piceatannol, tri-acetoxystilbene, and genistein on the inflammatory response of human peripheral blood leukocytes. *Mol Nutr Food Res* 49:431–442.

Scarlatti F, Sala G, Ricci C, Maioli C, Milani F, Minella M, Botturi M, Ghidoni R. 2007. Resveratrol sensitization of DU145 prostate cancer cells to ionizing radiation is associated to ceramide increase. *Cancer Lett* 253:124–130.

Scarlatti F, Sala G, Somenzi G, Signorelli P, Sacchi N, Ghidoni R. 2003. Resveratrol induces growth inhibition and apoptosis in metastatic breast cancer cells via de novo ceramide signaling. *FASEB J* 17:2339–2341.

Schneider Y, Vincent F, Duranton B, Badolo L, Gosse F, Bergmann C, Seiler N, Raul F. 2000. Anti-proliferative effect of resveratrol, a natural component of grapes and wine, on human colonic cancer cells. *Cancer Lett* 158:85–91.

Semenza GL. 2004. Hydroxylation of HIF-1: Oxygen sensing at the molecular level. *Physiology (Bethesda)* 19:176–182.

Sengottuvelan M, Nalini N. 2006. Dietary supplementation of resveratrol suppresses colonic tumour incidence in 1,2-dimethylhydrazine-treated rats by modulating biotransforming enzymes and aberrant crypt foci development. *Br J Nutr* 96:145–153.

Sexton E, Van Themsche C, LeBlanc K, Parent S, Lemoine P, Asselin E. 2006. Resveratrol interferes with AKT activity and triggers apoptosis in human uterine cancer cells. *Mol Cancer* 5:45.

Shankar S, Singh G, Srivastava RK. 2007. Chemoprevention by resveratrol: Molecular mechanisms and therapeutic potential. *Front Biosci* 12:4839–4854.

She QB, Ma WY, Wang M, Kaji A, Ho CT, Dong Z. 2003. Inhibition of cell transformation by resveratrol and its derivatives: Differential effects and mechanisms involved. *Oncogene* 22:2143–2150.

She QB, Huang C, Zhang Y, Dong Z. 2002. Involvement of c-jun NH(2)-terminal kinases in resveratrol-induced activation of p53 and apoptosis. *Mol Carcinog* 33:244–250.

She QB, Bode AM, Ma WY, Chen NY, Dong Z. 2001. Resveratrol-induced activation of p53 and apoptosis is mediated by extracellular-signal-regulated protein kinases and p38 kinase. *Cancer Res* 61:1604–1610.

Shen F, Chen SJ, Dong XJ, Zhong H, Li YT, Cheng GF. 2003. Suppression of IL-8 gene transcription by resveratrol in phorbol ester treated human monocytic cells. *J Asian Nat Prod Res* 5:151–117.

Stewart JR, O'Brian CA. 2004. Resveratrol antagonizes EGFR-dependent Erk1/2 activation in human androgen-independent prostate cancer cells with associated isozyme-selective PKC alpha inhibition. *Invest New Drugs* 22:107–117.

Subbaramaiah K, Chung WJ, Michaluart P, Telang N, Tanabe T, Inoue H, Jang M, Pezzuto JM, Dannenberg AJ. 1998. Resveratrol inhibits cyclooxygenase-2 transcription and activity in phorbol ester-treated human mammary epithelial cells. *J Biol Chem* 273:21875–21882.

Subbaramaiah K, Michaluart P, Chung WJ, Tanabe T, Telang N, Dannenberg AJ. 1999. Resveratrol inhibits cyclooxygenase-2 transcription in human mammary epithelial cells. *Ann NY Acad Sci* 889:214–223.

Sun CY, Hu Y, Guo T, Wang HF, Zhang XP, He WJ, Tan H. 2006. Resveratrol as a novel agent for treatment of multiple myeloma with matrix metalloproteinase inhibitory activity. *Acta Pharmacol Sin* 27:1447–1452.

Surh Y-J. 2003. Cancer chemoprevention with dietary phytochemicals. *Nature Rev Cancer* 3:768–780.

Surh Y-J, Kundu JK. 2005. Signal transduction network leading to COX-2 induction: A road map in search of cancer chemopreventives. *Arch Pharm Res* 28:1–15.

Szaefer H, Cichocki M, Brauze D, Baer-Dubowska W. 2004. Alteration in phase I and II enzyme activities and polycyclic aromatic hydrocarbons-DNA adduct formation by plant phenolics in mouse epidermis. *Nutr Cancer* 48:70–77.

Tang FY, Chiang EP, Sun YC. 2008. Resveratrol inhibits heregulin-beta1-mediated matrix metalloproteinase-9 expression and cell invasion in human breast cancer cells. *J Nutr Biochem* 19:287–294.

Tang HY, Shih A, Cao HJ, Davis FB, Davis PJ, Lin HY. 2006. Resveratrol-induced cyclooxygenase-2 facilitates p53-dependent apoptosis in human breast cancer cells. *Mol Cancer Ther* 5:2034–2042.

Tsai SH, Lin-Shiau SY, Lin JK. 1999. Suppression of nitric oxide synthase and the down-regulation of the activation of NF-κB in macrophages by resveratrol. *Br J Pharmacol* 126:673–680.

Tyagi A, Singh RP, Agarwal C, Siriwardana S, Sclafani RA, Agarwal R. 2005. Resveratrol causes Cdc2-tyr15 phosphorylation via ATM/ATR-Chk1/2-Cdc25C pathway as a central mechanism for S phase arrest in human ovarian carcinoma Ovcar-3 cells. *Carcinogenesis* 26:1978–1987.

Ulrich S, Huwiler A, Loitsch S, Schmidt H, Stein JM. 2007. De novo ceramide biosynthesis is associated with resveratrol-induced inhibition of ornithine decarboxylase activity. *Biochem Pharmacol* 74:281–289.

Upham BL, Guzvic M, Scott J, Carbone JM, Blaha L, Coe C, Li LL, Rummel AM, Trosko JE. 2007. Inhibition of gap junctional intercellular communication and activation of mitogen-activated protein kinase by tumor-promoting organic peroxides and protection by resveratrol. *Nutr Cancer* 57:38–47.

Wang Y, Lee KW, Chan FL, Chen S, Leung LK. 2006. The red wine polyphenol resveratrol displays bilevel inhibition on aromatase in breast cancer cells. *Toxicol Sci* 92:71–77.

Wen X, Walle T. 2007. Cytochrome P450 1B1, a novel chemopreventive target for benzo[a]pyrene-initiated human esophageal cancer. *Cancer Lett* 246:109–114.

Wolter F, Akoglu B, Clausnitzer A, Stein J. 2001. Downregulation of the cyclin D1/Cdk4 complex occurs during resveratrol-induced cell cycle arrest in colon cancer cell lines. *J Nutr* 131:2197–2203.

Woo JH, Lim JH, Kim YH, Suh SI, Min DS, Chang JS, Lee YH, Park JW, Kwon TK. 2004. Resveratrol inhibits phorbol myristate acetate-induced matrix metalloproteinase-9 expression by inhibiting JNK and PKC delta signal transduction. *Oncogene* 23:1845–1853.

Wu JM, Wang ZR, Hsieh TC, Bruder JL, Zou JG, Huang YZ. 2001. Mechanism of cardioprotection by resveratrol, a phenolic antioxidant present in red wine (Review). *Int J Mol Med* 8:3–17.

Yang SR, Wright J, Bauter M, Seweryniak K, Kode A, Rahman I. 2007. Sirtuin regulates cigarette smoke-induced proinflammatory mediator release via RelA/p65 NF-kappaB in macrophages in vitro and in rat lungs in vivo: Implications for chronic inflammation and aging. *Am J Physiol Lung Cell Mol Physiol* 292:L567–L576.

Yeung F, Hoberg JE, Ramsey CS, Keller MD, Jones DR, Frye RA, Mayo MW. 2004. Modulation of NF-κB-dependent transcription and cell survival by the SIRT1 deacetylase. *EMBO J* 23:2369–2380.

Yu H, Pan C, Zhao S, Wang Z, Zhang H, Wu W. 2008. Resveratrol inhibits tumor necrosis factor-alpha-mediated matrix metalloproteinase-9 expression and invasion of human hepatocellular carcinoma cells. *Biomed Pharmacother*. 62:366–372.

Yu R, Hebbar V, Kim DW, Mandlekar S, Pezzuto JM, Kong AN. 2001. Resveratrol inhibits phorbol ester and UV-induced activator protein 1 activation by interfering with mitogen-activated protein kinase pathways. *Mol Pharmacol* 60:217–224.

Zahid M, Gaikwad NW, Ali MF, Lu F, Saeed M, Yang L, Rogan EG, Cavalieri EL. 2008. Prevention of estrogen-DNA adduct formation in MCF-10F cells by resveratrol. *Free Radic Biol Med* 45:136–145.

Zhang Q, Tang X, Lu QY, Zhang ZF, Brown J, Le AD. 2005. Resveratrol inhibits hypoxia-induced accumulation of hypoxia-inducible factor-1alpha and VEGF expression in human tongue squamous cell carcinoma and hepatoma cells. *Mol Cancer Ther* 4:1465–1474.

Zhou HB, Chen JJ, Wang WX, Cai JT, Du Q. 2005. Anticancer activity of resveratrol on implanted human primary gastric carcinoma cells in nude mice. *World J Gastroenterol* 11:280–284.

Ziegler CC, Rainwater L, Whelan J, McEntee MF. 2004. Dietary resveratrol does not affect intestinal tumorigenesis in Apc(Min/ +) mice. *J Nutr* 134:5–10.

Zykova TA, Zhu F, Zhai X, Ma WY, Ermakova SP, Lee KW, Bode AM, Dong Z. 2008. Resveratrol directly targets COX-2 to inhibit carcinogenesis. *Mol Carcinog* 47:797–805.

INTERNET RESOURCES

http://www.cancer.org/docroot/STT/content/STT_1x_Cancer_Facts_and_Figures_2008.asp?from=fast (Cancer Facts and Figures, 2008. pdf; page 57).

16 Curcumin: The Biochemistry Behind Its Anticancer Effects

PREETHA ANAND, AJAIKUMAR B. KUNNUMAKKARA, AND BHARAT B. AGGARWAL

The University of Texas M. D. Anderson Cancer Center, Cytokine Research Laboratory, Department of Experimental Therapeutics, Houston, Texas

CONTENTS

INTRODUCTION

Curcumin is a hydrophobic polyphenol derived from turmeric: the rhizome of the herb *Curcuma longa*. Traditionally, turmeric and other curcuminoids have been used in therapeutic preparations for various ailments in different parts of the world. Numerous therapeutic effects of curcumin/turmeric have been confirmed by modern scientific research. It is a highly pleiotropic molecule that modulates numerous targets, including the activation of transcription factors (e.g., NF-κB, STAT-3, AP-1, NRF-2, PPR-γ, HIF-1), receptors (e.g., HER-2, IL-8, CXCR-4), kinases (e.g., EGFR, ERK, JAK, AAPK), cytokines (e.g., TNF, IL, MIP, MCP), enzymes (e.g., MMP, iNOS, GST, ATPase), and

Plant Phenolics and Human Health: Biochemistry, Nutrition, and Pharmacology,
Edited by Cesar G. Fraga.

growth factors (e.g., EGF, NGF, HGF, PDGF). Because of its ability to interact with a diverse range of molecular targets, curcumin can affect numerous molecular and biochemical cascades. One of our recently published reviews presents a more detailed description of the molecular targets of curcumin [Goel et al., 2007].

Extensive research conducted during the past century has established the complexity and involvement of multiple signaling pathways in the cancer growth and progression, which in turn suggests that a drug, which can interact with multiple target molecules, will be more efficacious than the current monotargeted anticancer drugs. Curcumin's multitargeting ability may be the key to its therapeutic potential against cancer. Furthermore no dose-limiting toxicity is reported for this molecule yet, and doses as high as 12 g/day is proved to be well tolerated and safe in humans [Cheng et al., 2001]. This chapter describes the preclinical and clinical data on the use of curcumin for the treatment of cancer.

CHEMISTRY OF CURCUMIN AND ITS ANALOGS

Chemically, it is a bis-α, β-unsaturated β-diketone (commonly called diferuloylmethane) that exhibits keto-enol tautomerism, having a predominant keto form in acidic and neutral solutions and a stable enol form in alkaline media. It is a member of the linear diarylheptanoid class of natural products in which two oxy-substituted aryl moieties are linked together through a seven-carbon chain. The C7 chain of linear diarylheptanoids is known to have unsaturation, oxo functions, enone moiety, and 1,3-diketone group. Except for the oxo and hydroxy functions, the C7 chain is generally unsubstituted. This unsaturation in the linker unit has an E-configuration (trans C═C bonds). Besides curcumin, other analogous molecules such as demethoxycurcumin (DMC), bisdemethoxycurcumin (BDMC), and cyclocurcumin are also present in turmeric (Fig. 16.1). Commercially available curcumin contains 77% curcumin, 17% DMC, and 3% BDMC. DMC and BDMC are structurally very similar to curcumin except they lack one and two methoxy groups, respectively.

Although curcumin, DMC, and BDMC differ in their chemical structure only with regards to methoxy substitution, they showed significantly different antioxidant, antitumor, and anti-inflammatory activities. Up to now there is no systematic study that clearly correlates the physicochemical and molecular properties of the three curcuminoids with their biological activities. However, the existing literature provides some clues to understand which group is actually responsible for a given biological activity of the curcuminoids.

Since many reports favor better radical scavenging and antioxidant ability of curcumin when compared with the other two, followed by DMC, the *o*-methoxy substitutions are certainly involved in this activity. The hydrogen-bonding interaction between the phenolic OH and the *o*-methoxy groups in curcumin markedly influences the O–H bond energy and H-atom abstraction

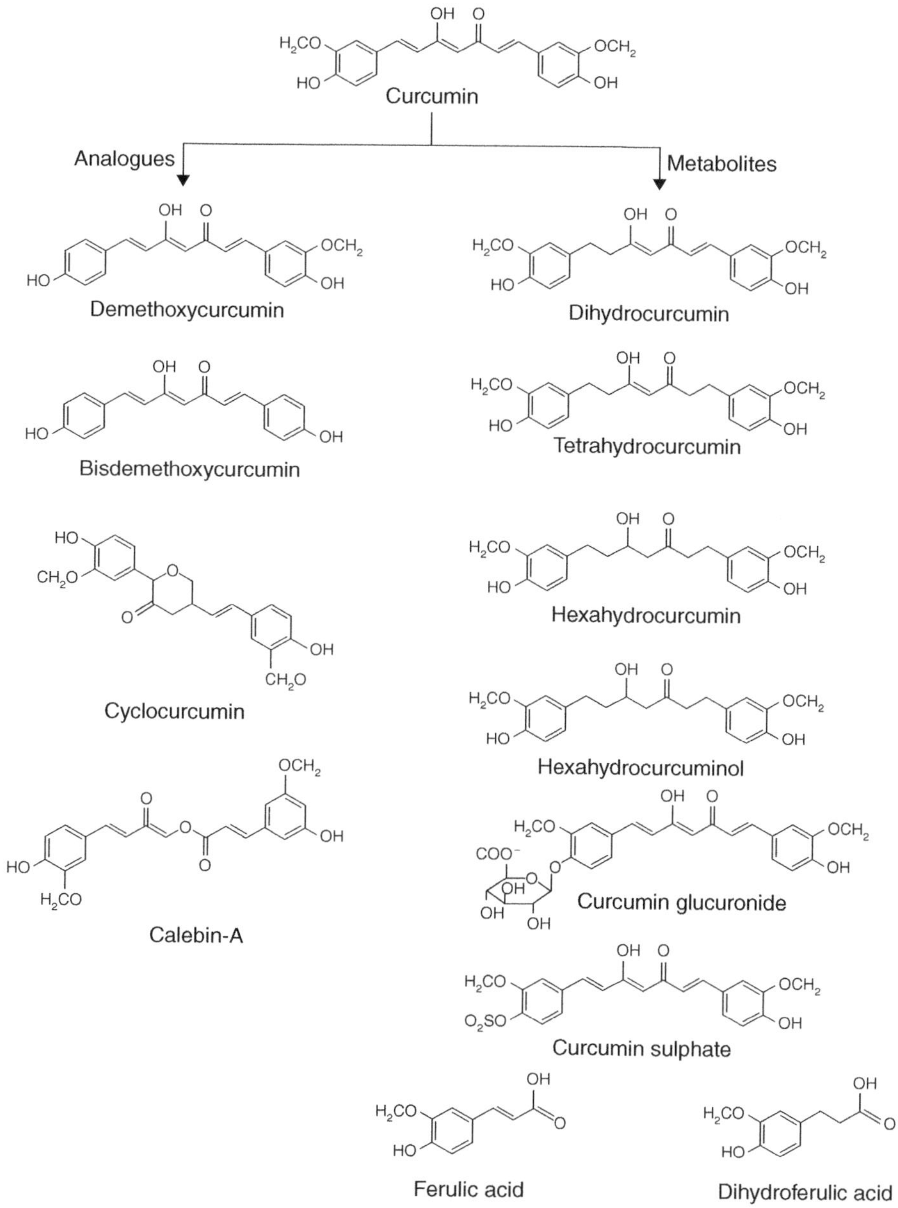

Figure 16.1 Curcumin and its Analogs.

by free radicals, thus making it a better free-radical scavenger than BDMC [Somparn et al., 2007].

The α, β-unsaturated diketone moiety in the curcuminoids is a Michael reaction acceptor, which belongs to the major class of phase II enzyme inducers

[Jeong et al., 2006]. Therefore, this property may be responsible for the induction of HO-1 and NF-κB suppression in cells by curcuminoids. Methoxy substitution on the aromatic ring can significantly influence their interaction with nucleophiles in Michael reaction. The reasons and the actual mechanism of the antitumor activities of the curcuminoids are still far from understood. It is still not known how the *o*-methoxy-deficient BDMC is a more potent ROS inducer and the *o*-methoxy-substituted curcumin is a more potent suppressor of NF-κB activation [Sandur et al., 2007]. The effect of change in the lipophillicity of the curcuminoids with methoxy substitution in influencing some of these activities also cannot be ignored.

Further studies are required to unravel this mystery of how the methoxy substitutions and the α, β-unsaturated β-diketone moiety actually influence conformational changes, lipophillicity, electron density distribution, and redox properties of curcuminoids. Correlating these physicochemical properties with the unique pleiotropic effects of curcuminoids is a rewarding exercise. Such studies would definitely provide proper reasoning in understanding these markedly different antioxidant, antitumor, and anti-inflammatory activities of natural curcuminoids from turmeric.

Cyclocurcumin differs from curcumin in the β-diketone link. In this molecule, the α,β-unsaturated β-diketone moiety of curcumin is replaced by α,β-unsaturated dihydropyranone moiety. Up to now, not many biological studies are reported with cyclocurcumin except for one study on inhibition of MCF-7 tumor cell proliferation by Simon et al. [1998] where it has been found to be ineffective in inhibiting cell proliferation and arrest of cell-cycle progression.

ANTICANCER EFFECTS OF CURCUMIN

Studies during the past century have shown the anticancer effect of curcumin against cancers of various anatomical sites. The preclinical anticancer potential of curcumin has been established both in vitro and in vivo, which lead to the clinical trials with this promising molecule. Below is a description of the studies, that have evaluated the preclinical and clinical anticancer activity of curcumin.

Anticancer Effects in Vitro

Several reports have described the anticancer activity of curcumin in a variety of cancer cell lines. In vitro studies have established the activity for curcumin against breast, gastric, hepatic, pancreatic, colorectal, urinary bladder, kidney, prostate, cervical, ovarian, uterine, lung, oral, thymic, and skin cancers. Besides these cancer types, curcumin has shown in vitro therapeutic efficacy against hematological cancers including leukemia, lymphoma, and multiple myeloma. One of our early studies established that the antiproliferative effect of curcumin in human breast cancer cell lines, including hormone-dependent, hormone-independent, and multidrug-resistant cells, was time and dose dependent and correlated with curcumin's inhibition of ornithine decarboxylase activity [Anand et al., 2008]. Several mechanisms have been proposed to account for the action of curcumin in breast cancer cells. For example, curcumin was found to inhibit the aryl hydrocarbon receptor and cytochrome P450 1A1 [Anand et al., 2008]; the tyrosine kinase activity of p185neu; the expression of Ki67, PCNA, and p53 mRNAs; and COX-1 and COX-2 enzymes. Curcumin also induced p53-dependent Bax expression, inhibited angiogenesis factors VEGF (vascular endothelial growth factor), b-FGF (basic fibroblast growth factor) [Schindler, 2006; Shao et al., 2002], and disrupted mitotic spindle structure and induced micronucleation [Holy, 2002]. It has been shown that curucumin inhibits telomerase activity through human telomerase reverse transcriptase [Ramachandran et al., 2002], down-regulates the expression of MMP-2 (matrix metalloproteinase-2), up-regulates TIMP-1 (tissue inhibitor of metalloproteinase-1) [Di et al., 2003], and blocks NF-κB and AP-1 activation [Aggarwal et al., 2005; Bachmeier et al., 2007; Bobrovnikova-Marjon et al., 2004; Yoon and Liu, 2007]. Studies have also shown curcumin to inhibit LOX pathways [Hammamieh et al., 2007], induce the degradation of cyclin E expression through a ubiquitin-dependent pathway, up-regulate cyclin-dependent kinase inhibitors p21 and p27 [Aggarwal et al., 2007], and down-regulate the insulin-like growth factor-1 (IGF-1) [Xia et al., 2007] in breast cancer cell lines.

The cytotoxic effect of curcumin on gastric carcinoma cell lines has been established. In a study curcumin and 5-fluorouracil (5-FU) synergistically inhibited the growth of gastric carcinoma cells. In another study, curcumin reversed the MDR of a human gastric carcinoma cell line in correlation with a decrease in P-gp function and a promotion of caspase-3 activation [Anand et al., 2008].

Several studies have examined the anticarcinogenic activity of curcumin in hepatic cancer cells in vitro. In one of these studies, conducted in curcumin-treated human hepatoblastoma cells, several hallmarks of apoptosis, including DNA laddering, chromatin condensation, fragmentation, and apoptosis-specific cleavage of 28S and 18S ribosomal RNA were observed. Curcumin has also exhibited significant anti-invasion activity in human HCC SK-Hep-1 cells, an effect that is associated with curcumin's inhibitory action on MMP-9 secretion. Curcumin undergoes metabolic conjugation and reduction in subcellular fractions of human and rat hepatic tissues [Anand et al., 2008]. It has also been established that the elevation of GSH levels mediates the effect of curcumin in hepatocytes [White et al., 1998b].

Curcumin has also been found to interrupt the cell cycle, to have cytotoxic effects, and to have a role in antiproliferation and the induction of apoptosis in a hepatocarcinoma cell line. Curcumin is a potent inhibitor of phenol sulfotransferase (SULT1A1) in human liver and extrahepatic tissues [Vietri et al., 2003]. Curcumin inhibited the interleukin-6 (IL-6) production, histone acetyltransferase (HAT) activity, and AP1 activation [Chen et al., 2003a] and prevented cell death and apoptotic biochemical changes, such as the mitochondrial release of cytochrome c, the activation of caspase-3, and the cleavage of PARP in human hepatoma cells [Anand et al., 2008; Labbozzetta et al., 2006]. Another proposed mechanism for curcumin's inhibition of tumor growth in HCC is through the inhibition of hypoxia-inducible factor-1 by degrading the aryl hydrocarbon receptor nuclear translocator [Bae et al., 2006; Cao et al., 2007; Choi et al., 2006]. Further, it has been shown that mitochondrial hyperpolarization is a prerequisite for curcumin-induced apoptosis and that mtDNA damage is the initial event in a chain leading to apoptosis in HepG2 cells [Cao et al, 2007]. In an in vitro study using hepatic cancer cells, a combination of curcumin and cisplatin had synergistic antitumor effects, and that with doxorubicin in additivity or sub-additivity [Anand et al., 2008].

Research over the past decade has indicated that curcumin has an anticarcinogenic effect in various pancreatic cell lines, with numerous mechanisms having been proposed to account for this effect. In human pancreatic cancer MIA PaCa-2 cells, curcumin was found to inhibit the farnesyl protein transferase [Anand et al., 2008]. Also, NF-κB was found to be overexpressed in human pancreatic tumor tissues and cell lines; investigators suggested that this overexpression could be inhibited by curcumin because it has the ability to suppress the NF-κB expression [Khanbolooki et al., 2006; Li et al., 2004; Wang et al., 1999]. Likewise, curcumin reduces numerous IL-8 bioactivities that contribute to tumor growth and the cell viability of pancreatic carcinoma cells [Anand et al., 2008; Kamohara et al., 2007]. Other mechanisms have been proposed to account for the growth inhibitory effect of curcumin alone [Starr et al., 2005] or in combination with celecoxib [Lev-Ari et al., 2005] including the down-regulation of COX-2, EGFR, ERK1/2 [Lev-Ari et al., 2006b], and Notch-1 [Wang et al., 2006c]. When coupled with gemcitabine, curcumin has been observed to have synergistic antiproliferative effects in pancreatic cancer cell lines [Holcomb et al., 2007; Lev-Ari et al., 2007]. Liposomes containing curcumin down-regulated NF-κB machinery, suppressed growth, and induced apoptosis of human pancreatic cells in vitro [Goel et al., 2007]. A polymeric nanocurcumin formulation also demonstrated a therapeutic efficacy comparable to that of free curcumin in a panel of human pancreatic cancer cell lines in vitro, and the mechanisms of action of nanocurcumin in pancreatic cancer cells mirrored those of free curcumin [Anand et al., 2007].

Studies using various colorectal cell lines have proven curcumin's use as a therapeutic agent and its ability to act through numerous target molecules. For example, curcumin has been shown to disrupt Lovo cells in the S, G2/M phase and interrupt Wnt signaling and adhesion pathways causing G2/M phase arrest and apoptosis in HCT-116 cells, regardless of prostaglandin synthesis. Curcumin-induced apoptosis is a result of PARP cleavage, caspase 3, reduction in Bcl-XL level, and increased activity of capsase-8, which encourages Fas signaling of apoptosis. Curcumin reduces NAT1 mRNA expression and AF-DNA adducts formation in human colon tumor cells. Curcumin was found to inhibit the proliferation of and induce apoptosis in colorectal cell lines

[Anand et al., 2008; Wei et al., 2004]. Heat shock aids colon cancer cells by inhibiting the discharge of apoptosis-inducing factors, an event that is enhanced by curcumin [Rashmi et al., 2003]. Curcumin causes cell shrinkage, chromatin condensation, and DNA fragmentation by enhancing DNA damage in HT-29 cells and HCT-116 colonocytes; it also increases GADD153 mRNA and protein expression [Anand et al., 2008; Scott and Loo, 2004]. Curcumin up-regulates TRAIL-induced apoptosis via ROS-mediated DR5 activation in human renal cancer cells [Anand et al., 2008]. Likewise, curcumin enhanced the silencing of hsp70 expression and may therefore prove to be a valuable therapeutic agent for cancers whose resistance is due to hsp70 expression [Rashmi et al., 2004b]. EF24, a synthetic curcumin analog, induces apoptosis in HT29 cells through a redox-dependent mechanism [Anand et al., 2008]. Similarly, the curcumin derivative HBC disrupts cell-cycle progression in HCT15 cells by antagonizing Ca^{2+}/CaM function [Shim et al., 2004].

The fact that curcumin-induced apoptosis is regulated by Bax suggests that the targeting of Bcl-XL or Smac can be used to treat Bax-deficient, chemotherapy-resistant cancers [Anand et al., 2008; Rashmi et al., 2004a, 2005]. Together, curcumin and either 5-FU or celecoxib downmodulate COX-2 expression via the inhibition of prostaglandin formation by curcumin and curcuminoids [Du et al., 2006]. Curcumin can also induce apoptosis via a parallel ceramide-associated pathway and ROS-associated mechanism that converges at JNK activation [Moussavi et al., 2006]. In vitro, curcumin activates JNK, p38 MAPK, and AP-1 transcriptional activity. Similarly, it inhibits neurotensin-mediated activator protein-1, NF-κB activation, Ca^{2+} mobilization, PGE-2, and EGFR and down-regulates COX-1 and -2, MMP-2 and -9, IL-8 gene induction, and colon cancer cell migration [Chen et al., 2006; Collett and Campbell, 2004; Jeong et al., 2004; Lev-Ari et al., 2006a; Moussavi et al., 2006; Wang et al., 2006a]. Curcumin has also proven effective at the mRNA level [Su et al., 2006]. Curcumin down-regulates sulfoconjugation and weakly inhibits the glucuronosyl conjugation of 1-naphthol in Caco-2 cells [Naganuma et al., 2006]. Curcumin coupled with ERRP significantly regulates downstream effectors, including NF-κB, Akt, BAD activation, and procaspase-3, in HCT-116 and HT-29 cells [Reddy et al., 2006]. Curcumin in conjugation with FOLFOX inhibits colon cancer cells by inhibiting the EGFR and IGF-1R signaling pathways [Patel et al, 2008]. Treatment with curcumin and epigallocatechin gallate reduced the amount of viable Apc mutant cells by 220–430%, more than each agent alone did [Telang et al 2007].

Curcuminoids obstruct cell proliferation and programmed cell death in primary colon cancer cells [Hsu et al, 2007]. Liposomal curcumin attenuates colorectal cancer by reducing CD31, VEGF, and IL-8 expression. This inhibition may be enhanced by the addition of oxaliplatin for the treatment of p53wt and p53 mutant colorectal tumors, as shown in HCEC, HT29, and HCT116 cell lines [Howells et al., 2007; Li et al., 2007a]. Some curcumin derivatives were also found to be effective against colon cancer cells. Dimethoxycurcumin, for example, is more potent than curcumin in inhibiting proliferation and inducing apoptosis in HCT116 cells [Tamvakopoulos et al., 2007].

Curcumin has been shown to suppress the proliferation of bladder cancer cells in culture either through the suppression of NF-κB [Kamat et al., 2007; Sun et al., 2004b] or through the down-regulation of cyclin A and up-regulation of p21 [Park et al., 2006]. Certain synthetic analogs of curcumin have been shown to exhibit activity against bladder cancer cell lines [Lu et al., 2005; Tong et al., 2006].

In human kidney cancer cells, curcumin up-regulates apoptotic events such as cell shrinkage, chromatin condensation, and DNA fragmentation [Jiang et al., 1996] and inhibits FPTase [Chen et al., 1997]. Curcumin serves as a COX-1 and COX-2 inhibitor [Ramsewak et al., 2000]; inhibits microsomal lipid peroxidation and DNA damage [Iqbal et al., 2003]; deactivates the Akt pathway; down-regulates Bcl-2, Bcl-XL, and IAP proteins [Woo et al., 2003]; and increases TRAIL-induced apoptosis by augmenting DR5 expression at the mRNA and protein levels by producing reactive oxygen species (ROS) [Jung et al., 2005]. In HKC cells, curcumin reduces tumor growth and the side effects when activated via the hydrolysis of prodrugs [Lu et al., 2005].

Curcumin has shown activity against various prostate cancer cells, such as LNCaP, DU145, C4-2B, and PC3. Curcumin can induce programmed cell death in androgen-dependent and androgen-independent prostate cancer cells. It can inhibit capillary tube formation and cell migration and exert significant effects on actin cytoskeletons in prostate cancer cells [Guo et al., 2006; Shankar et al., 2007; Shenouda et al., 2004]. Several mechanisms have been proposed to explain curcumin's anticancer effects in prostate cancer cells. For example, curcumin up-regulates the expression of the maspin gene and downmodulates the expression of androgen receptor (AR), AP-1, cyclin D1, NF-κB, and camp response element binding (CREB) protein and EGFR tyrosine kinase activity [Anand et al., 2008; Shi et al., 2006b]. By inducing p21 and C/EBPbeta expression and suppressing NF-κB activation, curcumin augments the cytotoxicity of chemotherapeutic agents in prostate cancer cells and induces the degradation of cyclin E expression [Anand et al., 2008].

In prostate cancer cells curcumin was found to act as an inhibitor of arachidonate 5-lipoxygenase [Ghosh, 2003]. Likewise, curcumin and TRAIL together cause apoptosis via both receptor-mediated and chemical-induced pathways, owing to an enhanced sensitivity of tumor cells to NF-κB [Deeb et al., 2003, 2004, 2005]. Curcumin interferes with osteoblastic and osteoclastic cell components, inhibiting growth factor collaboration between prostate cancer cells [Dorai et al., 2004]. Due to its organic structure as a Michael acceptor, curcumin serves as a HAT inhibitor [Marcu et al., 2006]. Curcumin down-regulates the expression of NKX3.1 via AR expression-induced p38-dependent proinflammatory changes in normal epithelial cells [Nonn et al., 2007]. Curcumin inhibits VIP-induced COX-2 expression and VIP-stimulated VEGF mRNA expression via the inhibition of AP-1 binding [Collado et al., 2005; Fernandez-Martinez et al., 2007; Park et al., 2003].

In PC3 cells, curcumin down-regulates MDM2 proteins and mRNA. enhances the expression of the tumor suppressor p21 and inhibits IκBα [Guo et al., 2006; Li et al., 2007b]. Curcumin can also inhibit prostate cancer via the Akt pathway or the induction of apoptosis by Bcl-2 family members and mitochondrial p53 [Chaudhary and Hruska, 2003; Deeb et al., 2007; Shankar et al., 2007].

The in vitro antitumor activity of curcumin in HPV-associated cells has been established [Roy et al., 2002]. Curcumin modulates the in vitro expression and function of Pgp in multidrug-resistant human KB-V1 cells [Anand et al., 2008; Chearwae et al., 2004] and sensitizes cisplatin-resistant SiHa cells to cisplatin-induced apoptosis [Venkatraman et al., 2005], indicating its ability to reverse MDR in cervical cancer cells. The effect of curcumin in HPV-associated cells was found to involve the down-regulation of viral oncogenes, NF-κB and AP-1 [Anand et al., 2008; Divya and Pillai, 2006].

Some in vitro studies over the past decade have shown that curcumin [Anand et al., 2008; Guo and Xu, 2005; Syu et al., 1998] and a curcumin–paclitaxel conjugate [Nakagawa-Goto et al., 2007] had therapeutic effects in ovarian cancer cell lines. Curcumin was found to act through the down-regulation of NF-κB [Zheng et al., 2002, 2006] and allied gene products [Shi et al., 2006a; Weir et al., 2007; Zheng et al., 2004]. Furthermore, curcumin was found to increase the sensitivity of chemotherapy-resistant ovarian cancer cell lines to standard chemotherapeutic agents by activating both the cells' extrinsic and intrinsic pathways of apoptosis [Wahl et al., 2007]. A recent study of ours also showed that curcumin had therapeutic and chemosensitization effects and reversed multidrug resistance in vitro. In one of the studies, curcumin displayed in vitro apoptosis-inducing activity against an endometrial cancer cell line [Wei et al., 2004] by the down-regulation of Ets-1 and Bcl-2 expression [Yu and Shah, 2007].

Curcumin exhibits anticancer effects in various lung cancer cells through a variety of molecular targets. At the cellular level, curcumin derivatives inhibit FPTase in A549 cells. Curcumin inhibits AP-1 transcription and mediastinal lymph node metastasis in Lewis lung carcinoma cells and ornithine decarboxylase activity in rat tracheal epithelial cells [Ichiki et al., 2000; White et al., 1998a]. Curcumin eradicated the DNA-binding of NF-κB, IκBα kinase activation, IκBα deterioration and phosphorylation, and p65 nuclear translocation, and it down-regulated COX-2 [Anand et al., 2008; Chen et al., 2003b]. Likewise, treatment with curcumin induces apoptosis and inhibits growth in A549 and H1299 cells [Pillai et al., 2004]. In A549 cells, curcumin interferes with cell growth and down-regulates NAT activity and STAT1 activation [Lee et al., 2005, 2006; Zhang et al., 2004]. Curcumin regulates the invasive activity of CL1-5 cells and demonstrates antiproliferative properties in NCI-H460 and H520 cells, suggesting its suitability as an adjunct chemotherapeutic agent [Pesic et al., 2006; Sen et al., 2005].

Curcumin inhibited the growth of oral cancer cell lines in vitro [Atsumi et al., 2005a, 2006] by blocking the S/G2M phase. It acted synergistically with a G1 phase blocker, epigallocatechin-3-gallate [Anand et al., 2008]. Curcumin inhibited the growth of and DNA synthesis in SCC-25 oral cancer cells [D'Ambrosio et al., 2000]. Curcumin increased both the expression and function of cytochrome P450 (CYP) 1A1 and/or CYP1B1 in oral cancer cells, indicating that it has chemopreventive properties mediated by the inhibition of carcinogen bioactivation [Rinaldi et al., 2002]. Further, curcumin exhibited radiotherapy-sensitizing effects on SCC cells in vitro [Khafif et al., 2005]. Moreover, the ability of curcumin to induce apoptosis in oral cancer cells was associated with the inhibition of COX-2 [Atsumi et al., 2005b]. A recent study of ours also revealed that curcumin down-regulates smokeless-tobacco-induced NF-κB activation and COX-2 expression in human oral premalignant and malignanat cells [Sharma et al., 2006].

The anticancer effect of curcumin in murine thymoma cells was found to be due to the blocking of IL-1 signaling by the inhibition of the recruitment of the IL-1 receptor-associated kinase (IRAK) [Jurrmann et al., 2005]. A recent study showed that curcumin could prevent tumor-induced thymic atrophy in thymic T cells, leading to the neutralization of tumor-induced oxidative stress and the restoration of NF-κB activity and the re-education of the TNF-α signaling pathway, resulting in thymic protection [Bhattacharyya et al., 2007].

In vitro, curcumin has been shown to have synergistic and remedial properties in leukemia. In HL-60 cells, a regimen of 10 μM curcumin for 48 h has been the most effective in decreasing cell proliferation and increasing differentiation. These effects were exacerbated when curcumin was given in conjunction with RA, vitamin D_3, and vitamin D_3 analogs [Liu et al., 1997; Pan et al., 2001; Roy et al., 2002; Sokoloski et al., 1997]. Curcumin alone causes a significant reduction in NF-κB expression, bcl-2 activity, and TPA-induced DNA binding. It also induces ER stress bax and caspases 3 and 8; and degrades PARP [Bielak-Mijewska et al., 2004; Mukherjee Nee Chakraborty et al., 2007; Pae et al., 2007; Sokoloski et al., 1997]. The proposed mechanism involves the interruption of G0/G1 phases associated with the up-regulation of P27kipl, P21wafl, and pRbp expression and the down-regulation of cyclin D_3 [Anand et al., 2008; Chen et al., 2002].

Within HL-60 cells, curcumin induced ROS levels and up-regulated Ca^{2+} production and the release of cytochrome c and lowered MMP levels [Atsumi et al., 2005b; Chen et al., 2005; Tan et al., 2006]. In Raji cells, curcumin selectively blocks tumor cells in the G0/G1 and G2/M phases; dose dependently up-regulates Ac-histone H4 expression; inhibits the proliferation and degradation of IκBa and Notch 1; and inhibits the translocation of the NF-κB/p-65 subunit via the downmodulation of HDAC1 and p300/Notch 1 signal molecules [Chen et al., 2007; Liu et al., 2005; Sun et al., 2004a; Wu et al., 2006]. Similarly, in the presence of curcumin, TERT is translocated, causing a loss of telomerase activity, and the expression of STAT3, -5a, and -5b are reduced without altering STAT1 or the phosphorylation states of STAT1, -3, or -5 in the K562 cell line via the release of cytochrome c from mitochondria [Blasius et al., 2006; Chakraborty et al., 2006]. Curcumin also affects GST-modulated lipid peroxidation, AP-1 and NF-κB binding to GSTP1-1 promoters, ADP ribose polymerase cleavage, and pro-caspases 8 and 9 induction in K562 cells [Duvoix et al., 2003; Singhal et al., 1999].

Curcumin dose dependently down-regulates JAK and STAT phosphorylation, causing growth inhibition and apoptosis in T-cell leukemia, HTLV-I–transformed T-cell leukemia, MT-2, HuT-102, and SLB-1 cell lines. It does so by inhibiting cyclin D1, cdk1 Cdc25C, and XIAP and Survin expression [Rajasingh et al., 2006; Tomita et al., 2006]. Curcumin suppresses the proliferation of WEHI-3B cells and blocks STAT5 mRNA expression and STAT5 activation in CML cells [Chen et al., 2004; Jiang et al., 2000]. In TK-10, MCF-7, and UACC-62 cell lines, curcumin initiates apoptosis via telomerase II poisoning, resulting in DNA damage [Martin-Cordero et al., 2003]. Acute leukemia cells exposed to curcumin for 4 h have increased nitric oxide (NO) levels [Kellner and Zunino, 2004]. This increased NO production by macrophages and the inhibition of Th1 cytokines in NK cells in the presence of curcumin lead to significant tumoricidal results [Bhaumik et al., 2000]. Likewise, MDR1 mRNA levels were reduced more significantly in leukemia cells from patients with higher MDR1 gene groups [Anuchapreeda et al., 2006a]. The proliferation of Jurkat cells was reduced with curcumin treatment, resulting in chromatin condensation and caspase-3 induction via the prevention of a decrease in glutathione levels [Piwocka et al., 2001; Sikora et al., 2006]. In Bcr-Abl–transfected mouse progenitor 32D cells, curcumin inhibits proliferation by arresting cells in the G(2)-M phase of the cell cycle, resulting in irregular chromatin organization, multipolar chromosome segregation, aberrant cytokinesis, and multinucleated cells with morphologic changes [Wolanin et al., 2006]. Like curcumin, curcumin analogs in KBM-5 cells blocked TNF-induced NF-κB activation and proliferation, and curcuminoids inhibited COX-1 and COX-2 enzymes [Zambre et al., 2006].

Curcumin was found to inhibit cellular proliferation and enhance apoptosis in a variety of lymphoma cell lines in vitro [Skommer et al., 2006; Thompson et al., 2004; Wu et al., 2002]. The proposed mechanism of curcumin's action in the majority of these studies involves the suppression of the expression of NF-κB-regulated gene products. One study suggested a novel function for curcumin as a suppressor of JAK-1 and STAT3 activation in primary effusion lymphoma cells, a function that would lead to the inhibition of proliferation and the induction of caspase-dependent apoptosis [Uddin et al., 2005].

Numerous reports suggest that curcumin exhibits antiproliferative effects against MM cells. The mechanisms of the antiproliferative effects of curcumin in MM cells have been studied and described extensively. The role of the NF-κB and STAT3 pathway as a target for curcumin in MM cells has been demonstrated [Liu et al., 2007a; Thomas et al., 2005]. Curcumin is known to suppress both the production and signaling of IL-6, a critical growth factor for MM cells [Labbozzetta et al., 2006]. Curcumin also interrupts the interaction between MM cells and endothelial cells by reducing TrκB expression in endothelial cells and inhibiting brain-derived neurotrophic factor (BDNF) production in MM cells, eventually resulting in the inhibition of angiogenesis [Wang et al., 2006b].

Several reports describe the antitumor activity of curcumin and of a formulation of the synthetic curcumin analog, EF-24 [Sun et al., 2006] in various melanoma cell lines. The effects of curcumin were found to be mediated through the inhibition of glutathione-*S*-transferase activity [Iersel et al., 1996], the inhibition of COX-1 and COX-2 enzymes, the induction of apoptosis through the Fas receptor/caspase-8 pathway, and the down-regulation of the NF-κB pathway [Marin et al., 2007; Siwak et al., 2005]. The modulation of integrin receptors and collagenase activity, the expression of Nm23 and E-cadherin [Ray et al., 2003], the down-regulation of FAK, and the reduction of MMP-2 activity [Banerji et al., 2004] were found to be responsible for the antimetastatic effect of curcumin in melanoma cells. Curcumin was found to reverse the resistance of melanoma cells to multiple drugs by inhibiting glutathione-*S*-transferases [Depeille et al., 2004, 2005].

Curcumin and its analogs were found to have antitumor effects in bone cancer cells. Numerous mechanisms have been proposed for the activity of curcumin against fibrosarcoma cells in particular. Curcumin induced apoptosis by inhibiting NF-κB [Anto et al., 2000] and the expression of IL-6 and IL-11 [Kondo et al., 2001] and by abolishing the inhibitory effect of TGF-β on GR-mediated gene expression [Kwak et al., 2006; Periyasamy and Sanchez, 2002] in fibrosarcoma cells. Curcumin suppressed MMP-13 expression in chondrosarcoma cells. Syntetic curcumin analogs were also found to be potent against bone cancer cells. In fibrosarcoma cells, synthetic curcumin analogs inhibit activator protein-1 transcription and tumor-induced angiogenesis by down-regulating the expression of angiogenesis-associated genes, VEGF, and MMP-9. Further, in human osteosarcoma cells curcumin was found to inhibit the ERK expression. Curcumin was found to induce apoptosis in a variety of osteosarcoma cells by down-regulating the BCl-2 expression [Walters et al., 2007].

Numerous other mechanisms, like the induction of heat shock proteins [Anand et al., 2008], the inhibition of MMP transcriptions [Kim et al., 2005; Woo et al., 2005], TRAIL-induced apoptosis [Gao et al., 2005], the inhibition of G6PT gene expression [Belkaid et al., 2006], the activation of both receptor-mediated and mitochondria-mediated proteolytic pathways [Karmakar et al., 2006], the induction of histone hypoacetylation leading to apoptosis in a (PARP)- and caspase-3-mediated manner [Kang et al., 2006], the inhibition of the ING4 signaling pathway [Liu et al., 2007b], and the induction of nonapoptotic autophagic cell death [Aoki et al., 2007; Shinojima, 2007] have also been established. Furthermore, curcumin was found to sensitize glioma cells to several chemotherapeutic agents and to radiation therapy [Dhandapani et al., 2007].

Anticancer Effects in Vivo

Several in vivo studies have established the chemopreventive as well as anticancer effect of curcumin against a variety of cancers, as summarized in

Table 16.1. Perhaps the first in vivo study to report the anticancer activity of curcumin was by Kuttan et al. in 1985. In this study the authors demonstrated, using an ascites mice model of Dalton's lymphoma, that both curcumin and turmeric extract could reduce the development of animal tumors [Kuttan et al., 1985]. Several reports have been published supporting the efficacy of curcumin treatment against cancer.

In 1998 a group studied curcumin's capacity to inhibit 7,12-dimethylbenzanthracene (DMBA)-induced mammary tumor and the in vivo formation of mammary DMBA-DNA adducts in the female rat. Administration (ip) of curcumin at 100- and 200-mg/kg doses prevented the development of the number of palpable mammary tumors and mammary adenocarcinomas significantly. The in vivo formation of mammary DMBA-DNA adducts also was depressed in animals administered with curcumin, and there was no significant enhancement of liver GST activity following curcumin administration. However, one group showed that animals fed with diets containing 1% curcumin had no effect on DMBA-induced mammary tumor. In 1996, Pereira et al. showed that curcumin (8 and 16 g/kg in diet) was only weakly effective in DMBA-induced mammary carcinogenesis. Another study evaluated the modulating effects of turmeric (T), ethanolic turmeric extract (ETE), and curcumin-free aqueous turmeric extract (CFATE) on the initiation or postinitiation phases of DMBA-induced mammary tumorigenesis in female Sprague–Dawley rats. Dietary administration of 1% turmeric/0.05% and ethanolic turmeric extract 2 weeks before, on the day of DMBA treatment (day 55) and 2 weeks after the single dose (15 mg/animal) of DMBA (during the initiation period) resulted in significant suppression of DMBA-induced mammary tumorigenesis as seen by a reduction in tumor multiplicity, tumor burden, and tumor incidence. The chemopreventive effect of curcumin on diethylstilbestrol (DES)-induced tumor promotion of rat mammary glands initiated with radiation was also evaluated in a study. The administration of dietary curcumin significantly reduced the incidence (28%) of mammary tumors. Multiplicity and Iball's index of mammary tumors were also decreased by curcumin. Rats fed the curcumin diet showed a reduced incidence of the development of both mammary adenocarcinoma and ER(+)PgR(+) tumors in comparison with the control group. The effect of curcumin on γ-radiation-induced mammary tumors was also demonstrated in rats [Goel et al., 2007].

In addition to the chemopreventive effects, antimetastatic effect of curcumin was also established by the in vivo model studies. In a xenograft model study (nude mice) conducted in our laboratory, the primary tumor was surgically removed after 58–60 days of tumor cell inoculation, and dietary curcumin (2%) was given to the animals starting from the fifth day to 5 weeks of primary tumor removal. We observed that administration of curcumin significantly decreased the incidence of breast cancer metastasis to the lung and suppressed the expression of NF-κB, COX-2, and MMP-9. Another group also evaluated the effect of curcumin on lung metastasis of breast cancer. In this study, intercardiac inoculation of breast cancer cells was done in the nude mice and

Table 16.1 Curcumin Exhibits Chemopreventive and Anticancer Effects Against Various Cancers

Cancer	Carcinogen or Route	Animal or Model	Dose
Aberrant crypt foci (ACF)[a]	Azoxymethane	Rat	2000 ppm
Colon cancer	Azoxymethane	Mice	0.5–0.2% w/w
	DMH	Mice	0.5%
	Azoxymethane	Rat	2000 ppm or 0.2–2% w/w
	PhIP	Apc (min) mice	2000 ppm
	1,2-dimethylhydrazine	Rat	0.6%
Duodenal tumor	MNNG	Mice	0.5–2.0% w/w
Esophageal cancer	NMBA	Rat	500 ppm
FAD[a]	Azoxymethane	Mice	2%
FAP[a]	—	Min/ + mice	0.1, 0.2 or 0.5% w/w
Stomach cancer	B[a]P	Mice	2% w/w
	MNNG	Rat	0.05% w/w
Liver cancer	Diethylnitrosamine	Rat	200 or 600 mg/kg
	Diethylnitrosamine	Mice	0.2% w/w
Lung cancer	B[a]P & NNK	A/J mice	2000 ppm
Lymphoma/ leukemia	DMBA	Sencar mice	2% w/w
Mammary tumor	DMBA	Rat	0.8–1.6% w/w or 50–200 mg/kg
	DMBA	Sencar mice	2% w/w
	Gamma radiation	Rat	
Oral cancer	MNA	Hamster	
	NQO	Rat	500 ppm
Prostate cancer	DMAB & PhIP	Rat	15–500 ppm
Skin tumor	TPA	Mice	
	DMBA	Mice	
	TPA	Mice	10–30 μmol or 1–3000 nmol
	DMBA	Mice	
	B[a]P and DMBA	Mice	
Multiorgan cancer	DHPN, EHEN	Rat	1% w/w
Ascites[b]	IP	Ascites	50 mg/kg
Breast[c]	Diet	Orthotopic	1 or 2% w/w
Colon[b]	IV	Orthotopic	40 mg/kg
	Oral	Xenograft	50–200 mg/kg
Gliobalstoma	IT	Orthotopic	10 mg/kg
HCC[d]		Orthotopic	100–200 mg/kg
	Oral	Xenograft	50–200 mg/kg
HNSCC	Sub cute	Xenograft	50–250 mol/L
Leukemia	Oral	Xenograft	50–200 mg/kg

(Continued)

Table 16.1 *(Continued)*

Cancer	Carcinogen or Route	Animal or Model	Dose
Melanoma	IP	Xenograft	25 mg/kg
Ovarian	IP	Orthotopic	500 mg/kg
Pancreas[b]	IV	Xenograft	40 mg/kg
	Gavage	Orthotopic	1 gm/kg
Prostate	Diet	Xenograft	2% w/w
	Gavage	Xenograft	5 mg/kg

[a]Note: FAP: familial adenomatous polyposis; ACF: aberrant crypt foci; FAD: focal areas of dysplasia; B[a]P: benzo[a]pyrene; DMBA: 7,12-dimethylbenz[a]nthracene; TPA: 12-O-tetradecanoyl-phorbol-13-acetate; NNK: 4-(methyl-nitrosamino)-I-(3-pyridyl)-1-butanone; NQO: 4-nitroquinoline 1-oxidase; DMAB: 3,2′-dimethyl-4-aminobiphenol; PhIP: 2-amino-1-methylimidazo[4,5-b]pyridine; DHPN: 2,2′-dihydroxy-di-*n*-propylnitrosamine; EHEN: *N*-ethyl-*N*-hydroxyethylnitrosamine.
[b]Liposomal curcumin.
[c]Lung metastases.
[d]Intrahepatic metastasis; IP, intraperitoneal; IT, intratumoral; IV, intravenous.

the animals were fed with diet containing 1% curcumin. Thirty-five days after tumor implantation the animals were sacrificed and enumerated the lung metastases. It was observed that all the animals in the untreated group had lung metastasis, whereas 21% of animals in the treated group were metastases free [Goel et al., 2007]. In contrast to the above in vivo studies, Somasundaram et al. [2002] reported a significant inhibition of tumor regression in a xenograft mouse model of human breast cancer. These contradictory findings could have been caused by the difference in administered doses as well as the time of treatment. For example, the authors studied the effect of curcumin in a breast xenograft model for 3 days, which is not normally suitable for the xenograft model studies. Even though it is the only study reporting the inhibition of tumor regression, further studies are needed to resolve the contradictions about the effectiveness of curcumin against breast cancer in vivo.

Two in vivo studies have been reported with curcumin in esophageal cancer. In one, dietary curcumin (500 ppm) fed during initiation and postinitiation stages inhibited the incidence of esophageal carcinogenesis by 27 and 33%, respectively, in rats [Goel et al., 2007]. In the other study, the efficacy of curcumin as a chemopreventive agent was assessed by measuring the modulation in the incidence of neoplastic change in rat esophagus [Wax et al., 2005].

Several in vivo chemoprevention studies have been reported with curcumin in gastric cancers. In some of the chemoprevention studies, curcumin fed as dietary turmeric (2% or 5%) to mice and Syrian golden hamsters significantly inhibited the benzo[a]pyrene-induced forestomach tumors. Furthermore, the incidence and multiplicity of forestomach tumors induced by benzo[a]pyrene in female Swiss mice were significantly inhibited by pure curcumin given 2 weeks before, during and after the carcinogen treatment. Other studies also revealed the chemopreventive effect of curcumin on benzo[a]pyrene-induced forestomach cancer. A significant reduction in benzo[a]pyrene-induced forestomach pappilomas in mice due to treatment with dietary turmeric extract containing curcumin was also reported. It was also showed that curcumin inhibited MNNG-induced duodenal tumor in mice and gastric cancer in rats [Anand et al., 2008].

In vivo studies using mouse models have proved that curcumin modifies apoptosis resistance, leading to the inhibition of tumor formation and the prevention of adenoma development in the intestinal tract. The chemopreventive effect of curcumin for intestinal tumors in Min/+ mice was investigated. A dietary level of 0.15% curcumin decreased tumor formation in Min−/− mice by 63%. Examination of intestinal tissue from the treated animals showed the tumor prevention by curcumin was associated with increased enterocyte apoptosis and proliferation. Curcumin also decreased expression of the oncoprotein β-catenin in the erythrocytes of the Min/+ mouse, an observation previously associated with an antitumor effect. Curcumin enhanced PhIP-induced apoptosis and inhibited PhIP-induced tumorigenesis in the proximal small intestine of Apc (min) mice. Experiments performed on intestinal tumors in C57BL/6J-Min/+ (Min/+) mice demonstrated that curcumin has a regulatory role in lymphocyte-mediated immune function [Churchill et al., 2000]. Furthermore, levels of COX-2 protein expression have been found to reflect the retardation of adenoma development in mouse intestines after treatment with curcumin [Tunstall et al., 2006].

A considerable number of reports have also described curcumin in HCC in vivo. In one of these studies, curcumin significantly reduced the number of γ-glutamyl transpeptidase-positive foci, a characteristic considered to be the precursor of hepatocellular neoplasm in rats. Curcumin also had anticarcinogenic effects mediated through the induction of glutathione-linked detoxification enzymes in rat livers. In a murine hepatocarcinogenesis model, at the age of 42 weeks, the curcumin group had 81% less multiplicity and 62% fewer hepatocarcinomas than the nontreated group. It also suppressed liver inflammation in rats. In rats, curcumin glucuronide and curcumin sulfate were identified as the major products of curcumin biotransformation, whereas hexahydrocurcumin, hexahydrocurcuminol, and hexahydrocurcumin glucuronide were present only in small amounts. Another in vivo study showed that curcumin mixed into a diet could achieve levels of the drug in the liver sufficient to explain its pharmacological effects. Dietary curcumin increased the activity of hepatic UGT enzymes, which can detoxify carcinogens, in male Wistar rats. In an orthotopic implantation model, curcumin suppressed both intrahepatic metastases and the development of altered hepatic foci (AHF) in rat livers. Inhibition of tumor growth by systemic administration of 20 μg/kg curcumin for 6 consecutive days to rats bearing the highly cachectic Yoshida AH-130 ascites hepatoma was also reported. In one of the studies, although the growth of tumors at the implanted site was not affected by the curcumin treatment, there was a significant and dose-dependent decrease in the number of intrahepatic metastases [Aggarwal et al., 2003].

Curcumin also prevented the induction of hepatic hyperplastic nodules, body weight loss, increase in the levels of hepatic diagnostic markers, and hypoproteinemia in carcinogen-induced as well as xenograft hepatic cancer models. The antiangiogenic activity of curcumin in hepatic carcinoma cells implanted in nude mice was found to be mediated through the reduction of the angiogenic biomarkers COX-2 and VEGF [Aggarwal et al., 2003].

Two in vivo studies were reported showing the antitumor activity as well as chemosensitization effect of curcumin against pancreatic cancer. In a xenograft model study mice received intravenous liposomal curcumin (40 mg/kg, 3 time per week) for 20 days. Treatment with liposomal curcumin resulted in reduced tumor size showing decreased expression of CD31 as well as VEGF and IL-8. These results indicate that curcumin suppressed pancreatic carcinoma growth in murine xenograft models and inhibited tumor angiogenesis [Li et al., 2005]. A recent study conducted in our group investigated the chemosensitization effect of curcumin using an orthotopic pancreatic cancer model. The tumor volume in the combination of curcumin and gemcitabine group was significantly lower than the gemcitabine alone or control group, indicating the chemosensitizing effect of curcumin. Our results showed that curcumin in combination with gemcitabine significantly down-regulated the expression of cell proliferation marker Ki-67 in tumor tissues compared with the control group. Further, curcumin alone significantly suppressed the expression of microvessel density marker CD31, and the presence of gemcitabine further enhanced the down-regulation of CD31 [Goel et al., 2007].

Several in vivo studies were reported to show the chemopreventive as well as anticancer activity of curcumin against colorectal cancer. Wargovich and co-workers [1996] showed the chemopreventive activity of curcumin against carcinogen-induced ACF in rats. Sulindac, curcumin, and PEMC administered during promotion and progression have been found to up-regulate apoptosis in rat colonic tumors [Anand et al., 2008]. Dietary curcumin (0.2%) inhibited the formation of carcinogen-induced colorectal tumors in rats [Wijnands et al., 2004]. In rodent models, curcumin hinders tumor suppressor p53 function, but in AOM-induced rat models, apoptosis is induced via a mitochondrial pathway [Goel et al., 2007; Volate et al., 2005]. The modulatory role of dietary curcumin on azoxymethane (AOM)-induced aberrant crypt foci (ACF) formation in the colon of F344 rats was evaluated and showed that AOM-induced colonic ACF were significantly inhibited in the animals fed with the curcumin (2000 ppm/day) diet. The chemopreventive activity of curcumin was also observed when it was administered before, during, and after carcinogen treatment as well as when it was given only during the promotion/progression phase of colon carcinogenesis in rats. Several studies evaluated the effect of curcumin on AOM-induced colon cancer and showed a significant inhibition of colon carcinogenesis after the treatment with curcumin.

Dimethylhydrazine (DMH)-induced rat colon carcinogenesis model was used for evaluation of the synergistic inhibitory effect between curcumin and catechin in light of ACF formation and tumor incidence. The results of this study indicated that curcumin, catechin, and their co-treatment caused significant inhibition of DMH-induced ACF and colon carcinogenesis as compared with untreated DMH-induced rat models [Xu et al., 2005]. Similarly, in another study it was showed that curcumin and celecoxib additively inhibits the growth of DMH induced growth of colorectal cancer in rats [Goel et al., 2007]. An in vivo study by Kwon and Magnuson [2007] suggested that during initiation, AOM inhibits colonic COX-1 expression without affecting COX-2, and dietary curcumin may increase COX-2 expression to compensate AOM-induced reduction of COX-1 expression in rats. In male rats, curcumin and curcumin analog increased celecoxib-mediated growth inhibition [Anand et al., 2008].

The preclinical anticancer activity of a liposomal curcumin formulation in colorectal cancer was also recently evaluated. This study also compared the efficacy of liposomal curcumin (40 mg/kg administered iv) with that of oxaliplatin, a standard chemotherapeutic agent for colorectal cancer. Significant tumor growth inhibition was observed in Colo205 and LoVo xenograft models in mice. Tumors from animals treated with liposomal curcumin showed an antiangiogenic effect measured as attenuation of CD31, vascular endothelial growth factor, and IL-8 expression. Thus, this study established the comparable or greater growth-inhibitory and apoptotic effects of liposomal curcumin with oxaliplatin in vivo in colorectal cancer [Li et al., 2007a].

In order to investigate the anticancer potential of curcumin against prostate cancer, androgen-dependent LNCaP prostate cancer cells were injected subcutaneously to mice. The experimental group received a synthetic diet containing 2% curcumin for up to 6 weeks. At the endpoint, mice were killed and sections taken from the excised tumors were evaluated for pathology, cell proliferation, apoptosis, and vascularity. Curcumin induced a marked decrease in the extent of cell proliferation as measured by the BrdU incorporation assay and a significant increase in the extent of apoptosis as measured by an in situ cell death assay. Moreover, microvessel density as measured by CD31 antigen staining decreased significantly [Aggarwal et al., 2007]. In vivo, PEITC and curcumin alone or in combination possess significant cancer-preventive characteristics in PC-3 prostate tumor xenografts in mice [Khor et al., 2006]. In another study [Hong et al., 2006], the mean tumor volumes at 4 weeks after tumor inoculation in the control and experimental animals were determined to be $168.6 \pm 40.7\ mm^3$ and $99.5 \pm 27.2\ mm^3$, respectively. Curcumin was shown to induce a marked reduction of MMP-2, and MMP-9 activity in the tumor-bearing site. The metastatic nodules were significantly fewer in the curcumin-treated group than untreated group. Li et al. [2007b] evaluated the antitumor, chemosensitizing, and radiosensitizing effect of curcumin using a xenograft prostate cancer model. Analysis of tumors collected at the end of the experiment showed that curcumin reduced the expression of MDM2 oncogene in xenografts treated with curcumin alone, and in xenografts treated with combinations of curcumin and gemcitabine or irradiation. These results indicate a novel mechanism of action that may be essential for curcumin's chemotherapeutic effects.

It was demonstrated that curcumin effectively inhibits tumor implantation and growth in a murine bladder tumor model [Anand et al., 2008]. Another study demonstrated that dietary curcumin treatment reduced risk for kidney cancer metastasis in rats [Frank et al., 2003].

A recent study of ours showed that curcumin had therapeutic and chemosensitization effects against ovarian cancer and reversed multidrug resistance both in vitro and in vivo in athymic mice. In the in vivo study, tumors were grown by orthotopic injection of cells and one week after orthotopic implantation animals were treated with curcumin (500 mg/kg/day, gavage) alone or in combination with docetaxel (35–50 μg/animal/week, ip) for 4 weeks. Curcumin alone resulted in 49–55% reductions in mean tumor growth compared with controls, whereas when combined with docetaxel 77% reductions in mean tumor growth compared with controls was obtained for curcumin in normal ovarian tumor models. In these ovarian tumors, curcumin alone and with docetaxel decreased both proliferation and microvessel density and increased tumor cell apoptosis. In mice with multidrug-resistant ovarian tumors, treatment with curcumin alone and combined with docetaxel resulted in significant 47% and 58% reductions in tumor growth, respectively [Lin et al., 2007].

Orthotopic implantation of a metastatic cell line of Lewis lung carcinoma (LLC-MLN), which was isolated by an in vivo selection method, resulted in greater metastatic growth in mediastinal lymph nodes as compared with that of the original LLC cells. Oral administration of curcumin significantly inhibited the mediastinal lymph node metastasis of orthotopically implanted LLC cells in a dose-dependent manner, but did not affect the tumor growth at the implantation site. Combined treatment with curcumin and cis-diamine-dichloroplatinum (CDDP), resulted in a marked inhibition of tumor growth at the implanted site and of lymphatic metastasis, and a significant prolongation of the survival time [Ichiki et al., 2000]. Deshpande and Maru [1995] showed that curcumin can inhibit BP-derived DNA adducts by interfering with the metabolic enzymes, and its physical presence is essential for this effect. In the year 1999 one group [Anand et al., 2008] studied the activity of curcumin as chemopreventive agent against lung tumor induction in A/J mice by the tobacco smoke carcinogens benzopyrene (BaP) and 4-(methyl-nitrosamino)-1-(3-pyridyl)-1-butanone (NNK). The treatment of curcumin (2000 ppm) 1 week after carcinogen treatment until termination had no effect on lung tumor multiplicity. In another study, oral administration of curcumin (200 nmol/kg body weight) was, however, found to inhibit the lung metastasis of melanoma maximally as seen by the reduction in the number of lung tumor nodules (80%). Consequent to the inhibition of the lung tumor nodules, the life span of animals treated with the polyphenol was also found to be increased (143.85%). The results indicate a possible use of these compounds in arresting the metastatic growth of tumor cells. In Wistar rats, however, marker enzymes and plasma lipid levels decreased after treatment with 80 mg/kg of curcumin or a curcumin analog [Anand et al., 2008]. In a group of 10 male smokers, 10 male nonsmokers, and 10 nonsmoking women between 25 and 45 years of age, curcumin reduced BP-stimulated strand breaks in a sex-dependent manner [Polasa et al., 2004].

Several in vivo studies have also revealed the potency of curcumin against oral cancer. Curcumin alone or in combination with other has acted as a chemopreventive agent in oral cancer models in rats and hamsters. It was showed that male F344 rats fed with dietary curcumin (0.5 g/kg) during the initiation and postinitiation stages exhibited 91% reduction in the frequency of 4-nitroquinoline-1-oxide-induced tongue carcinoma with a decrease in incidence of oral preneoplasia [Anand et al., 2008]. Another study showed that curcumin alone or in combination with catechin inhibited methyl-(acetoxymethyl)-nitrosamine (MNA)-induced oral mucosal tumors in Syrian golden hamsters. In Syrian golden hamsters, 10 mmol curcumin (applied topically 3 times/week) decreased the number of visible oral papillomas and papilloma volume by 39.6 and 61.3%, respectively. Furthermore, curcumin treatment also decreased the incidence of oral squamous cell carcinoma (SCC) and reduced the number of oral SCC lesions by 51.3%. In this study, curcumin treatment resulted in inhibition of tumor angiogenesis in the case of papilloma and SCC. Decrease of tumor proliferation index in hyperplasia, displasia, and papilloma was also observed due to curcumin treatment [Anand et al., 2008].

Studies have also demonstrated curcumin's therapeutic properties in vivo. In 6-week-old mice, the administration of a 2% curcumin diet via oral gavage resulted in a 53% reduction in lymphomas and leukemias. When topically applied prior to the administration of TPA in mice, curcumin down-regulated TPA-induced NF-κB and AP-1. It was also showed that oral administration of curcumin (50–200 mg/kg) inhibits the development of leukemia (HL 60) cell-induced xenografts in nude mice [Anand et al., 2008].

It was shown that oral administration of curcumin (50–200 mg/kg) inhibits the development of lymphoma (SGC7901) cell-induced xenografts in nude mice [Goel et al., 2007]. Curcumin inhibited the growth of both murine and human B lymphoma cells in vitro and murine B lymphoma cells in vivo by the down-regulation of spleen tyrosine kinase (Syk) activity accompanied by the down-regulation of Akt activation [Gururajan et al., 2007].

The chemopreventive effects of curcumin on several carcinogen-induced skin carcinogenesis models have been investigated. Topical application of curcumin (even lower doses of 20–100 nmol) together with tumor promoter TPA, twice weekly for 20 weeks to female CD-1 mice strongly inhibited TPA-induced papilloma formation. Further, in female Swiss mice dietary administration of 2% turmeric significantly inhibited DMBA plus TPA-induced skin tumor formation. In this benzopyrene-initiated and TPA-promoted two-stage skin tumorigenesis model, curcumin reduced the number of tumors per mouse and decreased the number of tumor-bearing mice. Another study conducted showed that curcumin inhibited ultraviolet (UV)-induced dermatitis in mouse skin [Goel et al., 2007].

The in vivo antimetastatic effect of curcumin was also established. In one study, oral administration of curcumin (200 nmol/kg body weight) was found to inhibit the lung metastasis, induced by B16F10 melanoma cells, measured as the reduction in the number of lung tumor nodules (80%), and there was an increase in the life span of mice by 143.85%. The effectiveness of a prophylactic immune preparation of soluble proteins from B16-R cells, or a treatment with curcumin alone or in combination was evaluated using a mouse melanoma model. The combination treatment resulted in substantial inhibition of melanoma growth compared to each treatment by itself. A significant percentage increase in the median survival time was also observed in the combination group (>82.8%) as opposed to the 48.6% increase in the immunization-only group and 45.7% increase in the curcumin-only group [Aggarwal et al., 2003; Anand et al., 2008].

In an in vivo study in rats, dietary curcumin with cisplatin modulated tumor marker indices of fibrosarcoma toward normal controls [Navis et al., 1999]. Treatment with radiotherapy and curcumin resulted in enhanced tumor cell killing and reduced radioresistance in mice bearing fibrosarcoma, as indicated by the significant inhibition of radiation-induced ERK and NF-κB expression [Kumar Mitra and Krishna, 2004].

Curcumin significantly decreased the incidence of radiation-induced pituitary tumors in rats [Anand et al., 2008]. In the subcutaneous xenograft model of glioblastoma cells, curcumin inhibited tumor growth significantly and induced autophagy. An approximate three-fold decrease in mean tumor volume was observed in the curcumin-treated group compared to the controls [Aoki et al., 2007].

Anticancer Effects—Clinical Trials

Several clinical trails are reported with curcumin in cancer patients around the world. Some of our recent review [Anand et al., 2008; Goel et al., 2007; Kunnumakkara et al., 2008] gives a detailed description of the clinical trials with curcumin, which are summarized in Table 16.2. An early clinical trial evaluated the effectiveness of topical application of a curcumin ointment in patients with external cancerous lesions. Patients with cancers of oral cavity (37 patients), breast (7 patients), vulva (4 patients), skin (3 patients), and miscellaneous unspecified sites (11 patients) were enrolled in this study. Remarkable symptomatic relief was observed in this study with the reduction in smell in 90% of the cases, reduction in itching in almost all cases, and reduction in exudates in 70% of cases. Reduction in lesion size and pain was observed in about 10% of the cases and only one melanoma (scalp) patient showed an adverse effect of local itching, which may be due to the allergic reaction to curcumin.

In a phase I clinical trial patients with recently resected urinary bladder cancer, uterine cervical intraepithelial neoplasm (CIN), oral leucoplakia, and intestinal metaplasia of the stomach were enrolled and treated with 0.5–12 g/

Table 16.2 A List of Clinical Trials with Curcumin in Cancer Patients[a]

Disease	Dose/Frequency	Patients	Endpoint Modulation
Colorectal cancer	36–180 mg/day × 120 days	15	Lowered GST
Colorectal cancer	450–3600 mg/day × 120 days	15	Lowered inducible serum PGE2 levels
Colorectal cancer	450–3600 mg/day × 7 days	12	Decreased M1G DNA adducts
Crohn's disease	360 mg; × 3/day × 30 days; × 4 for 60 days	5	Improved symptoms
External cancer lesions	–	62	Reduction in lesion size in 10% patients Reduction in smell in 90% patients Reduction in itching in all patients Reduction in exudates in 70% patients
Familial adenomatous polyposis	480 mg; x3/day × 180 days	5	Decrease in the number of polyps was 60.4% Decrease in the size of polyps was 50.9%
Helicobacter pylori infection	300 mg/day × 7 days	25	Significant improvement of dyspeptic symptoms Reduction of gastric inflammation
Idiopathic inflammatory orbital	375 mg; 3x /day × 180–660 days	8	4 pts recovered completely 1 pt showed decrease in swelling No recurrence
Irritable bowel syndrome	72–144 mg/day × 56 days	207	Reduced symptoms
Liver metastasis of CRC	450–3600 mg/day × 7 days	12	Low bioavailability
Multiple myeloma	2–12 g/day	24	Inhibits NF-κB, COX-2, and pSTAT3
Pancreatic cancer	8 g/day	17	70% reduction in tumor in one pt.
Precancerous lesions	8 g/day × 90 days	19	Histological improvement
Tropical pancreatitis	500 mg/day × 42 days	20	Reduction in the erythrocyte MDA levels

(Continued)

Table 16.2 *(Continued)*

Disease	Dose/Frequency	Patients	Endpoint Modulation
			Increased in erythrocyte GSH levels Inhibits NF-κB, COX-2, and pSTAT3
Prostatic intraepithelial neoplasia (PIN)		24	
Ulcerative proctitis	550 mg; × 2–3/day × 60 days	5	Improved symptoms
Ulcerative colitis	2000 mg/day × 180 days	89	Low recurrence; improved symptoms

[a] For details regarding the above-mentioned studies see Goel, et al. [2007] and Kunnumakkara et al. [2008].

day of curcumin for 3 months. It was found that one of four patients with CIN and one of seven patients with oral leucoplakia proceeded to develop frank malignancies in spite of curcumin treatment, whereas histologic improvement of precancerous lesions was seen in one out of two patients with recently resected bladder cancer, two out of seven patients of oral leucoplakia, one out of six patients of intestinal metaplasia of the stomach, and one out of four patients with CIN.

In a pilot trial with 12 patients with hepatic metastases from colorectal cancer, the concentrations of the curcumin in normal and malignant human liver tissue after patients received 450–3600 mg of curcumin daily for 1 week prior to surgery were not sufficient to elicit pharmacological activity, perhaps because of the extensive degree to which curcumin was metabolized in the intestine.

In another clinical trial, researchers evaluated the effect of oral curcumin with piperine on the pain, and the markers of oxidative stress in patients with tropical pancreatitis (TP). Twenty patients with tropical pancreatitis were randomized to receive 500 mg of curcumin with 5 mg of piperine, or placebo, for 6 weeks, and the effects on the pattern of pain, and on red blood cell levels of malonyldialdehyde (MDA) and glutathione (GSH) were assessed. There was a significant reduction in the erythrocyte MDA levels following curcumin therapy compared with placebo; with a significant increase in GSH levels. There was no corresponding improvement in pain.

The studies from our group showed that curcumin inhibited pancreatic cancer in patients. Twenty-five patients were enrolled in this study. Patients received 8 g of curcumin orally daily until disease progression, with restaging every 2 months. Serum cytokine levels for IL-6, IL-8, IL-10, and IL-1 receptor antagonists and peripheral blood mononuclear cell (PBMC) expression of NF-κB and COX-2 were monitored. Out of 25 patients, 21 were evaluable for response. Circulating curcumin was detectable as drug in glucuronide and sulfate conjugate forms, albeit at low steady-state levels, suggesting poor oral bioavailability. Two patients demonstrated clinical biologic activity. One had ongoing stable disease for more than 18 months and, interestingly, one additional patient had a brief, but marked, tumor regression (73%), accompanied by significant increases (4- to 35-fold) in serum cytokine levels (IL-6, IL-8, IL-10, and IL-1 receptor antagonists). No toxicities were observed. Curcumin down-regulated expression of NF-κB, COX-2, and phosphorylated STAT3 in PBMC from patients (most of whom had baseline levels considerably higher than those found in healthy volunteers).

Studies conducted by our group showed that curcumin inhibited constitutive activation of NF-κB, COX-2, and STAT3 in the PBMC from multiple myeloma patients. Curcumin was given at 2, 4, 8, 12g/day orally, which was well tolerated with no adverse events. Out of 29 patients 12 patients continued treatment for 12 weeks and 5 completed one full year of treatment with stable disease. Clinical studies performed with leukemia patients were also reported. In a study of the 70 childhood leukemia patients samples, curcumin reduced WT1 gene expression in 35 samples [Anuchapreeda et al., 2006b].

The pharmacodynamic and pharmacokinetic effect of oral curcuma extract in patients with colorectal cancer was evaluated. Fifteen patients with advanced colorectal cancer refractory to standard chemotherapies received curcuma extract daily for up to 4 months. The results showed that oral curcuma extract was well tolerated, and dose-limiting toxicity was not observed. Neither curcumin nor its metabolites were detected in blood or urine, but curcumin was recovered from feces. Curcumin sulfate was identified in the feces of one patient. A dose escalation pilot study of a standardized formulation of curcuma extract in 15 patients with advanced colorectal cancer revealed a dose-dependent inhibition of COX-2 activity, measured as basal and LPS-mediated PGE(2) production, in blood revealing the efficacy of curcumin in colorectal cancer. Ingestion of 440 mg of curcuma extract for 29 days was accompanied by a 59% decrease in lymphocytic glutathione-*S*-transferase activity. At higher dose levels, this effect was not observed. Leukocytic M(1)G levels were constant within each patient and unaffected by treatment. Radiologically stable disease was demonstrated in five patients for 2–4 months of treatment. Another study showed that a daily dose of 3.6 g curcumin engendered 62 and 57% decreases in inducible PGE(2) production in blood samples taken 1 h after dose on days 1 and 29, respectively, in advanced colorectal cancer patients. Yet another pilot trial, involving 12 patients with hepatic metastases from colorectal cancer who received 450–3600 mg of curcumin daily, for 1 week prior to surgery, to investigate whether oral administration of curcumin results in concentrations of the agent in normal and malignant human liver tissue, which are sufficient to elicit pharmacological activity. The results of this study suggested that hepatic curcumin levels sufficient to exert pharmacological activity are not achieved in humans with the above-mentioned dose of curcumin and that this may be due to extensive intestinal metabolism of curcumin leading to lower bioavailability. Curcumin coupled with quercetin significantly decreased the size and number of ileal and rectal adenomas in patients with FAP.

The effect of zyflamend, a herbal preparation containing curcumin against high-grade prostatic intraepithelial neoplasia (HGPIN) was evaluated in patients. A patient with HGPIN was treated with zyflamend three times a day for 18 months. After 6 months the biopsy revealed benign prostatic hyperplasia alone and after 18 months biopsy was negative for cancer and PIN, indicating that the patient was cancer and HGPIN free.

CONCLUSIONS

Almost 3000 studies carried out with curcumin suggest that this natural agent affects numerous pathways linked with tumorigenesis and thus has potential both for prevention and treatment of cancer. Although pharmacologically curcumin is quite safe in humans, its limited bioavailability may be a problem. More clinical trials with curcumin either alone or in combination with existing therapies are needed to fully appreciate its potential. Reformulation of curcumin may also hold promise in the future for the bioavailability problem. Its safety combined with its low cost and multiple targeting potential makes curcumin an ideal agent to be explored for prevention and treatment of various cancers.

REFERENCES

Aggarwal BB, Banerjee S, Bharadwaj U, Sung B, Shishodia S, Sethi G. 2007. Curcumin induces the degradation of cyclin E expression through ubiquitin-dependent pathway and up-regulates cyclin-dependent kinase inhibitors p21 and p27 in multiple human tumor cell lines. *Biochem Pharmacol* 73:1024–1032.

Aggarwal BB, Shishodia S, Takada Y, Banerjee S, Newman RA, Bueso-Ramos CE, Price JE. 2005. Curcumin suppresses the paclitaxel-induced nuclear factor-kappaB pathway in breast cancer cells and inhibits lung metastasis of human breast cancer in nude mice. *Clin Cancer Res* 11:7490–7498.

Aggarwal BB, Kumar A, Bharti AC. 2003. Anticancer potential of curcumin: Preclinical and clinical studies. *Anticancer Res* 23:363–398.

Anand P, Kunnumakkara AB, Newman RA, Aggarwal BB. 2007. Bioavailability of curcumin: Problems and promises. *Mol Pharmacol* 4:807–818.

Anand P, Sundaram C, Jhurani S, Kunnumakkara AB, Aggarwal BB. 2008. Curcumin and cancer: An "old-age" disease with an "age-old" solution. *Cancer Lett* 267:133–164.

Anto RJ, Maliekal TT, Karunagaran D. 2000. L-929 cells harboring ectopically expressed RelA resist curcumin-induced apoptosis. *J Biol Chem* 275:15601–15604.

Anuchapreeda S, Limtrakul P, Thanarattanakorn P, Sittipreechacharn S, Chanarat P. 2006a. Inhibitory effect of curcumin on WT1 gene expression in patient leukemic cells. *Arch Pharm Res* 29:80–87.

Anuchapreeda S, Thanarattanakorn P, Sittipreechacharn S, Tima S, Chanarat P, Limtrakul P. 2006b. Inhibitory effect of curcumin on MDR1 gene expression in patient leukemic cells. *Arch Pharm Res* 29:866–873.

Aoki H, Takada Y, Kondo S, Sawaya R, Aggarwal BB, Kondo Y. 2007. Evidence that curcumin suppresses the growth of malignant gliomas *in vitro* and *in vivo* through induction of autophagy: Role of Akt and extracellular signal-regulated kinase signaling pathways. *Mol Pharmacol* 72:29–39.

Atsumi T, Tonosaki K, Fujisawa S. 2006. Induction of early apoptosis and ROS-generation activity in human gingival fibroblasts (HGF) and human submandibular gland carcinoma (HSG) cells treated with curcumin. *Arch Oral Biol* 51:913–921.

Atsumi T, Fujisawa S, Tonosaki K. 2005a. Relationship between intracellular ROS production and membrane mobility in curcumin- and tetrahydrocurcumin-treated human gingival fibroblasts and human submandibular gland carcinoma cells. *Oral Dis* 11:236–242.

Atsumi T, Murakami Y, Shibuya K, Tonosaki K, Fujisawa S. 2005b. Induction of cytotoxicity and apoptosis and inhibition of cyclooxygenase-2 gene expression, by curcumin and its analog, alpha-diisoeugenol. *Anticancer Res* 25:4029–4036.

Bachmeier BE, Mohrenz IV, Mirisola V, Schleicher E, Romeo F, Hohneke C, Jochum M, Nerlich AG, Pfeffer U. 2008. Curcumin down-regulates the inflammatory cytokines CXCL1 and -2 in breast cancer cells via NFkappaB. *Carcinogenesis* 29:779–789.

Bae MK, Kim SH, Jeong JW, Lee YM, Kim HS, Kim SR, Yun I, Bae SK, Kim KW. 2006. Curcumin inhibits hypoxia-induced angiogenesis via down-regulation of HIF-1. *Oncol Rep* 15:1557–1562.

Banerji A, Chakrabarti J, Mitra A, Chatterjee A. 2004. Effect of curcumin on gelatinase A (MMP-2) activity in B16F10 melanoma cells. *Cancer Lett* 211:235–242.

Belkaid A, Copland IB, Massillon D, Annabi B. 2006. Silencing of the human microsomal glucose-6-phosphate translocase induces glioma cell death: Potential new anticancer target for curcumin. *FEBS Lett* 580:3746–3752.

Bhattacharyya S, Mandal D, Sen GS, Pal S, Banerjee S, Lahiry L, Finke JH, Tannenbaum CS, Das T, Sa G. 2007. Tumor-induced oxidative stress perturbs nuclear factor-kappaB activity-augmenting tumor necrosis factor-alpha-mediated T-cell death: Protection by curcumin. *Cancer Res* 67:362–370.

Bhaumik S, Jyothi MD, Khar A. 2000. Differential modulation of nitric oxide production by curcumin in host macrophages and NK cells. *FEBS Lett* 483:78–82.

Bielak-Mijewska A, Piwocka K, Magalska A, Sikora E. 2004. P-glycoprotein expression does not change the apoptotic pathway induced by curcumin in HL-60 cells. *Cancer Chemother Pharmacol* 53:179–185.

Blasius R, Reuter S, Henry E, Dicato M, Diederich M. 2006. Curcumin regulates signal transducer and activator of transcription (STAT) expression in K562 cells. *Biochem Pharmacol* 72:1547–1554.

Bobrovnikova-Marjon EV, Marjon PL, Barbash O, Vander Jagt DL, Abcouwer SF. 2004. Expression of angiogenic factors vascular endothelial growth factor and interleukin-8/CXCL8 is highly responsive to ambient glutamine availability: Role of nuclear factor-kappaB and activating protein-1. *Cancer Res* 64:4858–4869.

Cao J, Liu Y, Jia L, Zhou HM, Kong Y, Yang G, Jiang LP, Li QJ, Zhong LF. 2007a. Curcumin induces apoptosis through mitochondrial hyperpolarization and mtDNA damage in human hepatoma G2 cells. *Free Radic Biol Med* 43:968–975.

Cao J, Jiang LP, Liu Y, Yang G, Yao XF, Zhong LF. 2007b. Curcumin-induced genotoxicity and antigenotoxicity in HepG2 cells. *Toxicon* 8:1219–1222.

Chakraborty S, Ghosh U, Bhattacharyya NP, Bhattacharya RK, Roy M. 2006. Inhibition of telomerase activity and induction of apoptosis by curcumin in K-562 cells. *Mutat Res* 596:81–90.

Chaudhary LR, Hruska KA. 2003. Inhibition of cell survival signal protein kinase B/Akt by curcumin in human prostate cancer cells. *J Cell Biochem* 89:1–5.

Chearwae W, Anuchapreeda S, Nandigama K, Ambudkar SV, Limtrakul P. 2004. Biochemical mechanism of modulation of human P-glycoprotein (ABCB1) by curcumin I, II, and III purified from turmeric powder. *Biochem Pharmacol* 68:2043–2052.

Chen Y, Shu W, Chen W, Wu Q, Liu H, Cui G. 2007. Curcumin, both histone deacetylase and p300/CBP-specific inhibitor, represses the activity of nuclear factor kappa B and Notch 1 in Raji cells. *Basic Clin Pharmacol Toxicol* 101:427–433.

Chen A, Xu J, Johnson AC. 2006. Curcumin inhibits human colon cancer cell growth by suppressing gene expression of epidermal growth factor receptor through reducing the activity of the transcription factor Egr-1. *Oncogene* 25:278–287.

Chen J, Wanming D, Zhang D, Liu Q, Kang J. 2005. Water-soluble antioxidants improve the antioxidant and anticancer activity of low concentrations of curcumin in human leukemia cells. *Pharmazie* 60:57–61.

Chen WH, Chen Y, Cui GH, Gu JX, Hu D, Chen WK, Li XG. 2004. [Effect of curcumin on STAT5 signaling pathway in primary CML cells.]. *Zhongguo Shi Yan Xue Ye Xue Za Zhi* 12:572–576.

Chen YN, Cheng CC, Chen JC, Tsauer W, Hsu SL. 2003a. Norcantharidin-induced apoptosis is via the extracellular signal-regulated kinase and c-Jun-NH2-terminal kinase signaling pathways in human hepatoma HepG2 cells. *Br J Pharmacol* 140:461–470.

Chen YS, Ho CC, Cheng KC, Tyan YS, Hung CF, Tan TW, Chung JG. 2003b. Curcumin inhibited the arylamines N-acetyltransferase activity, gene expression and DNA adduct formation in human lung cancer cells (A549). *Toxicol In Vitro* 17:323–333.

Chen Y, Wu Y, He J, Chen W. 2002. The experimental and clinical study on the effect of curcumin on cell cycle proteins and regulating proteins of apoptosis in acute myelogenous leukemia. *J Huazhong Univ Sci Technolog Med Sci* 22:295–298.

Chen X, Hasuma T, Yano Y, Yoshimata T, Morishima Y, Wang Y, Otani S. 1997. Inhibition of farnesyl protein transferase by monoterpene, curcumin derivatives and gallotannin. *Anticancer Res* 17:2555–2564.

Cheng AL, Hsu CH, Lin JK, Hsu MM, Ho YF, Shen TS, Ko JY, Lin JT, Lin BR, Ming-Shiang W, Yu HS, Jee SH, Chen GS, Chen TM, Chen CA, Lai MK, Pu YS, Pan MH, Wang YJ, Tsai CC, Hsieh CY. 2001. Phase I clinical trial of curcumin, a chemopreventive agent, in patients with high-risk or pre-malignant lesions. *Anticancer Res* 21:2895–2900.

Choi H, Chun YS, Kim SW, Kim MS, Park JW. 2006. Curcumin inhibits hypoxia-inducible factor-1 by degrading aryl hydrocarbon receptor nuclear translocator: A mechanism of tumor growth inhibition. *Mol Pharmacol* 70:1664–1671.

Churchill M, Chadburn A, Bilinski RT, Bertagnolli MM. 2000. Inhibition of intestinal tumors by curcumin is associated with changes in the intestinal immune cell profile. *J Surg Res* 89:169–175.

Collado B, Sanchez MG, Diaz-Laviada I, Prieto JC, Carmena MJ. 2005. Vasoactive intestinal peptide (VIP) induces c-fos expression in LNCaP prostate cancer cells through a mechanism that involves Ca2 + signalling. Implications in angiogenesis and neuroendocrine differentiation. *Biochim Biophys Acta* 1744:224–233.

Collett GP, Campbell FC. 2004. Curcumin induces c-jun N-terminal kinase-dependent apoptosis in HCT116 human colon cancer cells. *Carcinogenesis* 25:2183–2189.

D'Ambrosio SM, Gibson-D'Ambrosio R, Milo GE, Casto B, Kelloff GJ, Steele VE. 2000. Differential response of normal, premalignant and malignant human oral epithelial cells to growth inhibition by chemopreventive agents. *Anticancer Res* 20:2273–2280.

Deeb D, Jiang H, Gao X, Al-Holou S, Danyluk AL, Dulchavsky SA, Gautam SC. 2007. Curcumin [1,7-bis(4-hydroxy-3-methoxyphenyl)-1-6-heptadine-3,5-dione; C21H20O6] sensitizes human prostate cancer cells to tumor necrosis factor-related apoptosis-inducing ligand/Apo2L-induced apoptosis by suppressing nuclear factor-kappaB via inhibition of the prosurvival Akt signaling pathway. *J Pharmacol Exp Ther* 321:616–625.

Deeb D, Xu YX, Jiang H, Gao X, Janakiraman N, Chapman RA, Gautam SC. 2003. Curcumin (diferuloyl-methane) enhances tumor necrosis factor-related apoptosis-inducing ligand-induced apoptosis in LNCaP prostate cancer cells. *Mol Cancer Ther* 2:95–103.

Deeb D, Jiang H, Gao X, Hafner MS, Wong H, Divine G, Chapman RA, Dulchavsky SA, Gautam SC. 2004. Curcumin sensitizes prostate cancer cells to tumor necrosis factor-related apoptosis-inducing ligand/Apo2L by inhibiting nuclear factor-kappaB through suppression of IkappaBalpha phosphorylation. *Mol Cancer Ther* 3:803–812.

Deeb DD, Jiang H, Gao X, Divine G, Dulchavsky SA, Gautam SC. 2005. Chemosensitization of hormone-refractory prostate cancer cells by curcumin to TRAIL-induced apoptosis. *J Exp Ther Oncol* 5:81–91.

Depeille P, Cuq P, Passagne I, Evrard A, Vian L. 2005. Combined effects of GSTP1 and MRP1 in melanoma drug resistance. *Br J Cancer* 93:216–223.

Depeille P, Cuq P, Mary S, Passagne I, Evrard A, Cupissol D, Vian L. 2004. Glutathione S-transferase M1 and multidrug resistance protein 1 act in synergy to protect melanoma cells from vincristine effects. *Mol Pharmacol* 65:897–905.

Deshpande SS, Maru GB. 1995. Effects of curcumin on the formation of benzo[a]pyrene derived DNA adducts *in vitro*. *Cancer Lett* 1:71–80.

Dhandapani KM, Mahesh VB, Brann DW. 2007. Curcumin suppresses growth and chemoresistance of human glioblastoma cells via AP-1 and NFkappaB transcription factors. *J Neurochem* 102:522–538.

Di GH, Li HC, Shen ZZ, Shao ZM. 2003. [Analysis of anti-proliferation of curcumin on human breast cancer cells and its mechanism]. *Zhonghua Yi Xue Za Zhi* 83:1764–1768.

Divya CS, Pillai MR. 2006. Antitumor action of curcumin in human papillomavirus associated cells involves downregulation of viral oncogenes, prevention of NFkB and AP-1 translocation, and modulation of apoptosis. *Mol Carcinog* 45:320–332.

Dorai T, Dutcher JP, Dempster DW, Wiernik PH. 2004. Therapeutic potential of curcumin in prostate cancer-V: Interference with the osteomimetic properties of hormone refractory C4–2B prostate cancer cells. *Prostate* 60:1–17.

Du B, Jiang L, Xia Q, Zhong L. 2006. Synergistic inhibitory effects of curcumin and 5-fluorouracil on the growth of the human colon cancer cell line HT-29. *Chemotherapy* 1:23–28.

Duvoix A, Morceau F, Delhalle S, Schmitz M, Schnekenburger M, Galteau MM, Dicato M, Diederich M. 2003. Induction of apoptosis by curcumin: Mediation by glutathione S-transferase P1–1 inhibition. *Biochem Pharmacol* 66:1475–1483.

Fernandez-Martinez AB, Collado B, Bajo AM, Sanchez-Chapado M, Prieto JC, Carmena MJ. 2007. Vasoactive intestinal peptide induces cyclooxygenase-2 expression through nuclear factor-kappaB in human prostate cell lines: Differential time-dependent responses in cancer progression. *Mol Cell Endocrinol* 270:8–16.

Frank N, Knauft J, Amelung F, Nair J, Wesch H, Bartsch H. 2003. No prevention of liver and kidney tumors in Long-Evans Cinnamon rats by dietary curcumin, but inhibition at other sites and of metastases. *Mutat Res* 523–524:127–135.

Gao X, Deeb D, Jiang H, Liu YB, Dulchavsky SA, Gautam SC. 2005. Curcumin differentially sensitizes malignant glioma cells to TRAIL/Apo2L-mediated apoptosis through activation of procaspases and release of cytochrome c from mitochondria. *J Exp Ther Oncol* 5:39–48.

Ghosh J. 2003. Inhibition of arachidonate 5-lipoxygenase triggers prostate cancer cell death through rapid activation of c-Jun N-terminal kinase. *Biochem Biophys Res Commun* 307:342–349.

Goel A, Kunnumakkara AB, Aggarwal BB. 2007. Curcumin as "Curecumin": From kitchen to clinic. *Biochem Pharmacol* 75:787–809.

Guo F, Xu CJ. 2005. Progress on the study of mechanism of the direct action of TCM bioactive components on ovarian cancer. *Zhongguo Zhong Xi Yi Jie He Za Zhi* 25:1140–1144.

Guo H, Yu JH, Chen K, Ye ZQ. 2006. Curcumin-induced the expression of inhibitor kappaBalpha protein in human prostate cancer cells. *Zhonghua Wai Ke Za Zhi* 44:1256–1259.

Gururajan M, Dasu T, Shahidain S, Jennings CD, Robertson DA, Rangnekar VM, Bondada S. 2007. Spleen tyrosine kinase (Syk), a novel target of curcumin, is required for B lymphoma growth. *J Immunol* 178:111–121.

Hammamieh R, Sumaida D, Zhang X, Das R, Jett M. 2007. Control of the growth of human breast cancer cells in culture by manipulation of arachidonate metabolism. *BMC Cancer* 7:138.

Holcomb B, Yip-Schneider MT, Matos JM, Dixon J, Kennard J, Mahomed J, Shanmugam R, Sebolt-Leopold J, Schmidt CM. 2008. Pancreatic cancer cell genetics and signaling response to treatment correlate with efficacy of gemcitabine-based molecular targeting strategies. *J Gastrointest Surg* 12:288–296.

Holy JM. 2002. Curcumin disrupts mitotic spindle structure and induces micronucleation in MCF-7 breast cancer cells. *Mutat Res* 518:71–84.

Hong JH, Ahn KS, Bae E, Jeon SS, Choi HY. 2006. The effects of curcumin on the invasiveness of prostate cancer *in vitro* and *in vivo*. *Prostate Cancer Prostatic Dis* 9:147–152.

Howells LM, Mitra A, Manson MM. 2007. Comparison of oxaliplatin- and curcumin-mediated antiproliferative effects in colorectal cell lines. *Int J Cancer* 121:175–183.

Hsu YC, Weng HC, Lin S, Chien YW. 2007. Curcuminoids-cellular uptake by human primary colon cancer cells as quantitated by a sensitive HPLC assay and its relation with the inhibition of proliferation and apoptosis. *J Agric Food Chem* 20:8213–8222.

Ichiki K, Mitani N, Doki Y, Hara H, Misaki T, Saiki I. 2000. Regulation of activator protein-1 activity in the mediastinal lymph node metastasis of lung cancer. *Clin Exp Metastasis* 18:539–545.

Iersel ML, Ploemen JP, Struik I, van Amersfoort C, Keyzer AE, Schefferlie JG, van Bladeren PJ. 1996. Inhibition of glutathione S-transferase activity in human melanoma cells by alpha,beta-unsaturated carbonyl derivatives. Effects of acrolein, cinnamaldehyde, citral, crotonaldehyde, curcumin, ethacrynic acid, and trans-2-hexenal. *Chem Biol Interact* 102:117–132.

Iqbal M, Okazaki Y, Okada S. 2003. *In vitro* curcumin modulates ferric nitrilotriacetate (Fe-NTA) and hydrogen peroxide (H_2O_2)-induced peroxidation of microsomal membrane lipids and DNA damage. *Teratog Carcinog Mutagen Suppl* 1:151–160.

Jeong GS, Oh GS, Pae HO, Jeong SO, Kim YC, Shin MK, Seo BY, Han SY, Lee HS, Jeong JG, Koh JS, Chung HT. 2006. Comparative effects of curcuminoids on endothelial heme oxygenase-1 expression: Ortho-methoxy groups are essential to enhance heme oxygenase activity and protection. *Exp Mol Med* 38:393–400.

Jeong WS, Kim IW, Hu R, Kong AN. 2004. Modulation of AP-1 by natural chemopreventive compounds in human colon HT-29 cancer cell line. *Pharm Res* 21:649–660.

Jiang DZ, Xie QY, Wang QR. 2000. Effect of curcumin on the proliferation of murine CFU-GM and WEHI-3B cells. *Hunan Yi Ke Da Xue Xue Bao* 25:216–218.

Jiang MC, Yang-Yen HF, Yen JJ, Lin JK. 1996. Curcumin induces apoptosis in immortalized NIH 3T3 and malignant cancer cell lines. *Nutr Cancer* 26:111–120.

Jung EM, Lim JH, Lee TJ, Park JW, Choi KS, Kwon TK. 2005. Curcumin sensitizes tumor necrosis factor-related apoptosis-inducing ligand (TRAIL)-induced apoptosis through reactive oxygen species-mediated upregulation of death receptor 5 (DR5). *Carcinogenesis* 26:1905–1913.

Jurrmann N, Brigelius-Flohe R, Bol GF. 2005. Curcumin blocks interleukin-1 (IL-1) signaling by inhibiting the recruitment of the IL-1 receptor-associated kinase IRAK in murine thymoma EL-4 cells. *J Nutr* 135:1859–1864.

Kamat AM, Sethi G, Aggarwal BB. 2007. Curcumin potentiates the apoptotic effects of chemotherapeutic agents and cytokines through down-regulation of nuclear factor-kappaB and nuclear factor-kappaB-regulated gene products in IFN-alpha-sensitive and IFN-alpha-resistant human bladder cancer cells. *Mol Cancer Ther* 6:1022–1030.

Kamohara H, Takahashi M, Ishiko T, Ogawa M, Baba H. 2007. Induction of interleukin-8 (CXCL-8) by tumor necrosis factor-alpha and leukemia inhibitory factor in pancreatic carcinoma cells: Impact of CXCL-8 as an autocrine growth factor. *Int J Oncol* 31:627–632.

Kang SK, Cha SH, Jeon HG. 2006. Curcumin-induced histone hypoacetylation enhances caspase-3-dependent glioma cell death and neurogenesis of neural progenitor cells. *Stem Cells Dev* 15:165–174.

Karmakar S, Banik NL, Patel SJ, Ray SK. 2006. Curcumin activated both receptor-mediated and mitochondria-mediated proteolytic pathways for apoptosis in human glioblastoma T98G cells. *Neurosci Lett* 407:53–58.

Kellner C, Zunino SJ. 2004. Nitric oxide is synthesized in acute leukemia cells after exposure to phenolic antioxidants and initially protects against mitochondrial membrane depolarization. *Cancer Lett* 215:43–52.

Khafif A, Hurst R, Kyker K, Fliss DM, Gil Z, Medina JE. 2005. Curcumin: A new radio-sensitizer of squamous cell carcinoma cells. *Otolaryngol Head Neck Surg* 132:317–321.

Khanbolooki S, Nawrocki ST, Arumugam T, Andtbacka R, Pino MS, Kurzrock R, Logsdon CD, Abbruzzese JL, McConkey DJ. 2006. Nuclear factor-kappaB maintains TRAIL resistance in human pancreatic cancer cells. *Mol Cancer Ther* 5:2251–2260.

Khor TO, Keum YS, Lin W, Kim JH, Hu R, Shen G, Xu C, Gopalakrishnan A, Reddy B, Zheng X, Conney AH, Kong AN. 2006. Combined inhibitory effects of curcumin and phenethyl isothiocyanate on the growth of human PC-3 prostate xenografts in immunodeficient mice. *Cancer Res* 66:613–621.

Kim SY, Jung SH, Kim HS. 2005. Curcumin is a potent broad spectrum inhibitor of matrix metalloproteinase gene expression in human astroglioma cells. *Biochem Biophys Res Commun* 337:510–516.

Kondo A, Mogi M, Koshihara Y, Togari A. 2001. Signal transduction system for interleukin-6 and interleukin-11 synthesis stimulated by epinephrine in human osteoblasts and human osteogenic sarcoma cells. *Biochem Pharmacol* 61:319–326.

Kumar Mitra A, Krishna M. 2004. *In vivo* modulation of signaling factors involved in cell survival. *J Radiat Res (Tokyo)* 45:491–495.

Kunnumakkara AB, Anand P, Aggarwal BB. 2008. Curcumin inhibits proliferation, invasion, angiogenesis and metastasis of different cancers through interaction with multiple cell signaling proteins. *Cancer Lett* 269:199–225.

Kwak HJ, Park MJ, Cho H, Park CM, Moon SI, Lee HC, Park IC, Kim MS, Rhee CH, Hong SI. 2006. Transforming growth factor-beta1 induces tissue inhibitor of metalloproteinase-1 expression via activation of extracellular signal-regulated kinase and Sp1 in human fibrosarcoma cells. *Mol Cancer Res* 4:209–220.

Kwon Y, Magnuson BA. 2007. Effect of azoxymethane and curcumin on transcriptional levels of cyclooxygenase-1 and -2 during initiation of colon carcinogenesis. *Scand J Gastroenterol* 42:72–80.

Labbozzetta M, Notarbartolo M, Poma P, Giannitrapani L, Cervello M, Montalto G, D'Alessandro N. 2006. Significance of autologous interleukin-6 production in the HA22T/VGH cell model of hepatocellular carcinoma. *Ann N Y Acad Sci* 1089: 268–275.

Lee J, Jung HH, Im YH, Kim JH, Park JO, Kim K, Kim WS, Ahn JS, Jung CW, Park YS, Kang WK, Park K. 2006. Interferon-alpha resistance can be reversed by inhibition of IFN-alpha-induced COX-2 expression potentially via STAT1 activation in A549 cells. *Oncol Rep* 15:1541–1549.

Lee J, Im YH, Jung HH, Kim JH, Park JO, Kim K, Kim WS, Ahn JS, Jung CW, Park YS, Kang WK, Park K. 2005. Curcumin inhibits interferon-alpha induced NF-kappaB and COX-2 in human A549 non-small cell lung cancer cells. *Biochem Biophys Res Commun* 334:313–318.

Lev-Ari S, Vexler A, Starr A, Ashkenazy-Voghera M, Greif J, Aderka D, Ben-Yosef R. 2007. Curcumin augments gemcitabine cytotoxic effect on pancreatic adenocarcinoma cell lines. *Cancer Invest* 25:411–418.

Lev-Ari S, Maimon Y, Strier L, Kazanov D, Arber N. 2006a. Down-regulation of prostaglandin E2 by curcumin is correlated with inhibition of cell growth and induction of apoptosis in human colon carcinoma cell lines. *J Soc Integr Oncol* 4:21–26.

Lev-Ari S, Starr A, Vexler A, Karaush V, Loew V, Greif J, Fenig E, Aderka D, Ben-Yosef R. 2006b. Inhibition of pancreatic and lung adenocarcinoma cell survival by curcumin is associated with increased apoptosis, down-regulation of COX-2 and EGFR and inhibition of Erk1/2 activity. *Anticancer Res* 26:4423–4430.

Lev-Ari S, Strier L, Kazanov D, Madar-Shapiro L, Dvory-Sobol H, Pinchuk I, Marian B, Lichtenberg D, Arber N. 2005. Celecoxib and curcumin synergistically inhibit the growth of colorectal cancer cells. *Clin Cancer Res* 11:6738–6744.

Li L, Ahmed B, Mehta K, Kurzrock R. 2007a. Liposomal curcumin with and without oxaliplatin: Effects on cell growth, apoptosis, and angiogenesis in colorectal cancer. *Mol Cancer Ther* 6:1276–1282.

Li M, Zhang Z, Hill DL, Wang H, Zhang R. 2007b. Curcumin, a dietary component, has anticancer, chemosensitization, and radiosensitization effects by down-regulating the MDM2 oncogene through the PI3K/mTOR/ETS2 pathway. *Cancer Res* 67:1988–1996.

Li L, Braiteh FS, Kurzrock R. 2005. Liposome-encapsulated curcumin: *in vitro* and *in vivo* effects on proliferation, apoptosis, signaling, and angiogenesis. *Cancer* 104:1322–1331.

Li L, Aggarwal BB, Shishodia S, Abbruzzese J, Kurzrock R. 2004. Nuclear factor-kappaB and IkappaB kinase are constitutively active in human pancreatic cells, and their down-regulation by curcumin (diferuloylmethane) is associated with the suppression of proliferation and the induction of apoptosis. *Cancer* 101:2351–2362.

Lin YG, Kunnumakkara AB, Nair A, Merritt WM, Han LY, Armaiz-Pena GN, Kamat AA, Spannuth WA, Gershenson DM, Lutgendorf SK, Aggarwal BB, Sood AK. 2007. Curcumin inhibits tumor growth and angiogenesis in ovarian carcinoma by targeting the nuclear factor-kappaB pathway. *Clin Cancer Res* 13:3423–3430.

Liu B, Bai QX, Chen XQ, Gao GX, Gu HT. 2007a. [Effect of curcumin on expression of survivin, Bcl-2 and Bax in human multiple myeloma cell line]. *Zhongguo Shi Yan Xue Ye Xue Za Zhi* 15:762–766.

Liu E, Wu J, Cao W, Zhang J, Liu W, Jiang X, Zhang X. 2007b. Curcumin induces G2/M cell cycle arrest in a p53-dependent manner and upregulates ING4 expression in human glioma. *J Neurooncol* 85:263–270.

Liu HL, Chen Y, Cui GH, Zhou JF. 2005. Curcumin, a potent anti-tumor reagent, is a novel histone deacetylase inhibitor regulating B-NHL cell line Raji proliferation. *Acta Pharmacol Sin* 26:603–609.

Liu Y, Chang RL, Cui XX, Newmark HL, Conney AH. 1997. Synergistic effects of curcumin on all-trans retinoic acid- and 1 alpha,25-dihydroxyvitamin D3-induced differentiation in human promyelocytic leukemia HL-60 cells. *Oncol Res* 9:19–29.

Lu P, Tong Q, Jiang F, Zheng L, Chen F, Zeng F, Dong J, Du Y. 2005. Preparation of curcumin prodrugs and their *in vitro* anti-tumor activities. *J Huazhong Univ Sci Technolog Med Sci* 25:668–670, 678.

Marcu MG, Jung YJ, Lee S, Chung EJ, Lee MJ, Trepel J, Neckers L. 2006. Curcumin is an inhibitor of p300 histone acetylatransferase. *Med Chem* 2:169–174.

Marin YE, Wall BA, Wang S, Namkoong J, Martino JJ, Suh J, Lee HJ, Rabson AB, Yang CS, Chen S, Ryu JH. 2007. Curcumin downregulates the constitutive activity of NF-kappaB and induces apoptosis in novel mouse melanoma cells. *Melanoma Res* 17:274–283.

Martin-Cordero C, Lopez-Lazaro M, Galvez M, Ayuso MJ. 2003. Curcumin as a DNA topoisomerase II poison. *J Enzyme Inhib Med Chem* 18:505–509.

Moussavi M, Assi K, Gomez-Munoz A, Salh B. 2006. Curcumin mediates ceramide generation via the de novo pathway in colon cancer cells. *Carcinogenesis* 27:1636–1644.

Mukherjee Nee Chakraborty S, Ghosh U, Bhattacharyya NP, Bhattacharya RK, Dey S, Roy M. 2007. Curcumin-induced apoptosis in human leukemia cell HL-60 is associated with inhibition of telomerase activity. *Mol Cell Biochem* 297:31–39.

Naganuma M, Saruwatari A, Okamura S, Tamura H. 2006. Turmeric and curcumin modulate the conjugation of 1-naphthol in Caco-2 cells. *Biol Pharm Bull* 29:1476–1479.

Nakagawa-Goto K, Yamada K, Nakamura S, Chen TH, Chiang PC, Bastow KF, Wang SC, Spohn B, Hung MC, Lee FY, Lee FC, Lee KH. 2007. Antitumor agents. 258. Syntheses and evaluation of dietary antioxidant-taxoid conjugates as novel cytotoxic agents. *Bioorg Med Chem Lett* 17:5204–5209.

Navis I, Sriganth P, Premalatha B. 1999. Dietary curcumin with cisplatin administration modulates tumour marker indices in experimental fibrosarcoma. *Pharmacol Res* 39:175–179.

Nonn L, Duong D, Peehl DM. 2007. Chemopreventive anti-inflammatory activities of curcumin and other phytochemicals mediated by MAP kinase phosphatase-5 in prostate cells. *Carcinogenesis* 28:1188–1196.

Pae HO, Jeong SO, Jeong GS, Kim KM, Kim HS, Kim SA, Kim YC, Kang SD, Kim BN, Chung HT. 2007. Curcumin induces pro-apoptotic endoplasmic reticulum stress in human leukemia HL-60 cells. *Biochem Biophys Res Commun* 353:1040–1045.

Pan MH, Chang WL, Lin-Shiau SY, Ho CT, Lin JK. 2001. Induction of apoptosis by garcinol and curcumin through cytochrome c release and activation of caspases in human leukemia HL-60 cells. *J Agric Food Chem* 49:1464–1474.

Park JI, Lee MG, Cho K, Park BJ, Chae KS, Byun DS, Ryu BK, Park YK, Chi SG. 2003. Transforming growth factor-beta1 activates interleukin-6 expression in prostate cancer cells through the synergistic collaboration of the Smad2, p38-NF-kappaB, JNK, and Ras signaling pathways. *Oncogene* 22:4314–4332.

Park C, Kim GY, Kim GD, Choi BT, Park YM, Choi YH. 2006. Induction of G2/M arrest and inhibition of cyclooxygenase-2 activity by curcumin in human bladder cancer T24 cells. *Oncol Rep* 15:1225–1231.

Patel BB, Sengupta R, Qazi S, Vachhani H, Yu Y, Rishi AK, Majumdar AP. 2008. Curcumin enhances the effects of 5-fluorouracil and oxaliplatin in mediating growth inhibition of colon cancer cells by modulating EGFR and IGF-1R. *Int J Cancer* 2:267–273.

Periyasamy S, Sanchez ER. 2002. Antagonism of glucocorticoid receptor transactivity and cell growth inhibition by transforming growth factor-beta through AP-1-mediated transcriptional repression. *Int J Biochem Cell Biol* 34:1571–1585.

Pesic M, Markovic JZ, Jankovic D, Kanazir S, Markovic ID, Rakic L, Ruzdijic S. 2006. Induced resistance in the human non small cell lung carcinoma (NCI-H460) cell line *in vitro* by anticancer drugs. *J Chemother* 18:66–73.

Piwocka K, Jaruga E, Skierski J, Gradzka I, Sikora E. 2001. Effect of glutathione depletion on caspase-3 independent apoptosis pathway induced by curcumin in Jurkat cells. *Free Radic Biol Med* 31:670–678.

Polasa K, Naidu AN, Ravindranath I, Krishnaswamy K. 2004. Inhibition of B(a)P induced strand breaks in presence of curcumin. *Mutat Res* 557:203–213.

Radhakrishna Pillai G, Srivastava AS, Hassanein TI, Chauhan DP, Carrier E. 2004. Induction of apoptosis in human lung cancer cells by curcumin. *Cancer Lett* 208:163–170.

Rajasingh J, Raikwar HP, Muthian G, Johnson C, Bright JJ. 2006. Curcumin induces growth-arrest and apoptosis in association with the inhibition of constitutively active JAK-STAT pathway in T cell leukemia. *Biochem Biophys Res Commun* 340:359–368.

Ramachandran C, Fonseca HB, Jhabvala P, Escalon EA, Melnick SJ. 2002. Curcumin inhibits telomerase activity through human telomerase reverse transcritpase in MCF-7 breast cancer cell line. *Cancer Lett* 184:1–6.

Ramsewak RS, DeWitt DL, Nair MG. 2000. Cytotoxicity, antioxidant and anti-inflammatory activities of curcumins I-III from Curcuma longa. *Phytomedicine* 7:303–308.

Rashmi R, Kumar S, Karunagaran D. 2005. Human colon cancer cells lacking Bax resist curcumin-induced apoptosis and Bax requirement is dispensable with ectopic expression of Smac or downregulation of Bcl-XL. *Carcinogenesis* 26:713–723.

Rashmi R, Kumar S, Karunagaran D. 2004a. Ectopic expression of Bcl-XL or Ku70 protects human colon cancer cells (SW480) against curcumin-induced apoptosis while their down-regulation potentiates it. *Carcinogenesis* 25:1867–1877.

Rashmi R, Kumar S, Karunagaran D. 2004b. Ectopic expression of Hsp70 confers resistance and silencing its expression sensitizes human colon cancer cells to curcumin-induced apoptosis. *Carcinogenesis* 25:179–187.

Rashmi R, Santhosh Kumar TR, Karunagaran D. 2003. Human colon cancer cells differ in their sensitivity to curcumin-induced apoptosis and heat shock protects them by inhibiting the release of apoptosis-inducing factor and caspases. *FEBS Lett* 538:19–24.

Ray S, Chattopadhyay N, Mitra A, Siddiqi M, Chatterjee A. 2003. Curcumin exhibits antimetastatic properties by modulating integrin receptors, collagenase activity, and expression of Nm23 and E-cadherin. *J Environ Pathol Toxicol Oncol* 22:49–58.

Reddy S, Rishi AK, Xu H, Levi E, Sarkar FH, Majumdar AP. 2006. Mechanisms of curcumin- and EGF-receptor related protein (ERRP)-dependent growth inhibition of colon cancer cells. *Nutr Cancer* 55:185–194.

Rinaldi AL, Morse MA, Fields HW, Rothas DA, Pei P, Rodrigo KA, Renner RJ, Mallery SR. 2002. Curcumin activates the aryl hydrocarbon receptor yet significantly inhibits (−)-benzo(a)pyrene-7R-trans-7,8-dihydrodiol bioactivation in oral squamous cell carcinoma cells and oral mucosa. *Cancer Res* 62:5451–5456.

Roy M, Chakraborty S, Siddiqi M, Bhattacharya RK. 2002. Induction of apoptosis in tumor cells by natural phenolic compounds. *Asian Pac J Cancer Prev* 3:61–67.

Sandur SK, Pandey MK, Sung B, Ahn KS, Murakami A, Sethi G, Limtrakul P, Badmaev V, Aggarwal BB. 2007. Curcumin, demethoxycurcumin, bisdemethoxycurcumin, tetrahydrocurcumin and turmerones differentially regulate anti-inflammatory and anti-proliferative responses through a ROS-independent mechanism. *Carcinogenesis* 28:1765–1773.

Schindler R, Mentlein R. 2006. Flavonoids and vitamin E reduce the release of the angiogenic peptide vascular endothelial growth factor from human tumor cells. *J Nutr* 6:1477–1482.

Scott DW, Loo G. 2004. Curcumin-induced GADD153 gene up-regulation in human colon cancer cells. *Carcinogenesis* 25:2155–2164.

Sen S, Sharma H, Singh N. 2005. Curcumin enhances Vinorelbine mediated apoptosis in NSCLC cells by the mitochondrial pathway. *Biochem Biophys Res Commun* 331:1245–1252.

Shankar S, Chen Q, Sarva K, Siddiqui I, Srivastava RK. 2007. Curcumin enhances the apoptosis-inducing potential of TRAIL in prostate cancer cells: Molecular mechanisms of apoptosis, migration and angiogenesis. *J Mol Signal* 2:10.

Shao ZM, Shen ZZ, Liu CH, Sartippour MR, Go VL, Heber D, Nguyen M. 2002. Curcumin exerts multiple suppressive effects on human breast carcinoma cells. *Int J Cancer* 98:234–240.

Sharma C, Kaur J, Shishodia S, Aggarwal BB, Ralhan R. 2006. Curcumin down regulates smokeless tobacco-induced NF-kappaB activation and COX-2 expression in human oral premalignant and cancer cells. *Toxicology* 228:1–15.

Shenouda NS, Zhou C, Browning JD, Ansell PJ, Sakla MS, Lubahn DB, Macdonald RS. 2004. Phytoestrogens in common herbs regulate prostate cancer cell growth *in vitro*. *Nutr Cancer* 49:200–208.

Shi M, Cai Q, Yao L, Mao Y, Ming Y, Ouyang G. 2006a. Antiproliferation and apoptosis induced by curcumin in human ovarian cancer cells. *Cell Biol Int* 30:221–226.

Shi P, Chen WW, Hu XY, Yu CX, Zhang PJ, Jiang AL, Zhang JY. 2006b. Up-regulates the expression of maspin gene in prostate cancer cell line LNCaP. *Yao Xue Xue Bao* 41:1152–1156.

Shim JS, Lee J, Park HJ, Park SJ, Kwon HJ. 2004. A new curcumin derivative, HBC, interferes with the cell cycle progression of colon cancer cells via antagonization of the Ca2 + /calmodulin function. *Chem Biol* 11:1455–1463.

Shinojima N, Yokoyama T, Kondo Y, Kondo S. 2007. Roles of the Akt/mTOR/p70S6K and ERK1/2 signaling pathways in curcumin-induced autophagy. *Autophagy*. 6:635–637.

Sikora E, Bielak-Zmijewska A, Magalska A, Piwocka K, Mosieniak G, Kalinowska M, Widlak P, Cymerman IA, Bujnicki JM. 2006. Curcumin induces caspase-3-dependent apoptotic pathway but inhibits DNA fragmentation factor 40/caspase-activated DNase endonuclease in human Jurkat cells. *Mol Cancer Ther* 5:927–934.

Simon A, Allais DP, Duroux JL, Basly JP, Durand-Fontanier S, Delage C. 1998. Inhibitory effect of curcuminoids on MCF-7 cell proliferation and structure-activity relationships. *Cancer Lett* 129:111–116.

Singhal SS, Awasthi S, Pandya U, Piper JT, Saini MK, Cheng JZ, Awasthi YC. 1999. The effect of curcumin on glutathione-linked enzymes in K562 human leukemia cells. *Toxicol Lett* 109:87–95.

Siwak DR, Shishodia S, Aggarwal BB, Kurzrock R. 2005. Curcumin-induced antiproliferative and proapoptotic effects in melanoma cells are associated with suppression of IkappaB kinase and nuclear factor kappaB activity and are independent of the B-Raf/mitogen-activated/extracellular signal-regulated protein kinase pathway and the Akt pathway. *Cancer* 104:879–890.

Skommer J, Wlodkowic D, Pelkonen J. 2006. Cellular foundation of curcumin-induced apoptosis in follicular lymphoma cell lines. *Exp Hematol* 34:463–474.

Sokoloski JA, Shyam K, Sartorelli AC. 1997. Induction of the differentiation of HL-60 promyelocytic leukemia cells by curcumin in combination with low levels of vitamin D3. *Oncol Res* 9:31–39.

Somasundaram S, Edmund NA, Moore DT, Small GW, Shi YY, Orlowski RZ. 2002. Dietary curcumin inhibits chemotherapy-induced apoptosis in models of human breast cancer. *Cancer Res* 62:3868–3875.

Somparn P, Phisalaphong C, Nakornchai S, Unchern S, Morales NP. 2007. Comparative antioxidant activities of curcumin and its demethoxy and hydrogenated derivatives. *Biol Pharm Bull* 30:74–78.

Starr AN, Vexler A, Marmor S, Konik D, Ashkenasi-Voghera M, Lev-Ari S, Greif Y, Ben-Yosef R. 2005. Establishment and characterization of a pancreatic carcinoma cell line derived from malignant pleural effusion. *Oncology* 69:239–245.

Su CC, Chen GW, Lin JG, Wu LT, Chung JG. 2006. Curcumin inhibits cell migration of human colon cancer colo 205 cells through the inhibition of nuclear factor kappa B/p65 and down-regulates cyclooxygenase-2 and matrix metalloproteinase-2 expressions. *Anticancer Res* 26:1281–1288.

Sun A, Shoji M, Lu YJ, Liotta DC, Snyder JP. 2006. Synthesis of EF24-tripeptide chloromethyl ketone: A novel curcumin-related anticancer drug delivery system. *J Med Chem* 49:3153–3158.

Sun CY, Liu XY, Chen Y, Liu F, Wang Y. 2004a. Experimental study on anticancer effect of curcumin on Raji cells *in vitro*. *Zhongguo Zhong Xi Yi Jie He Za Zhi* 24:1003–1006.

Sun M, Yang Y, Li H, Su B, Lu Y, Wei Q, Fan T. 2004b. The effect of curcumin on bladder cancer cell line EJ *in vitro*. *Zhong Yao Cai* 27:848–850.

Syu WJ, Shen CC, Don MJ, Ou JC, Lee GH, Sun CM. 1998. Cytotoxicity of curcuminoids and some novel compounds from Curcuma zedoaria. *J Nat Prod* 61:1531–1534.

Tamvakopoulos C, Dimas K, Sofianos ZD, Hatziantoniou S, Han Z, Liu ZL, Wyche JH, Pantazis P. 2007. Metabolism and anticancer activity of the curcumin analogue, dimethoxycurcumin. *Clin Cancer Res* 13:1269–1277.

Tan TW, Tsai HR, Lu HF, Lin HL, Tsou MF, Lin YT, Tsai HY, Chen YF, Chung JG. 2006. Curcumin-induced cell cycle arrest and apoptosis in human acute promyelocytic leukemia HL-60 cells via MMP changes and caspase-3 activation. *Anticancer Res* 26:4361–4371.

Thomas RK, Sos ML, Zander T, Mani O, Popov A, Berenbrinker D, Smola-Hess S, Schultze JL, Wolf J. 2005. Inhibition of nuclear translocation of nuclear factor-kappaB despite lack of functional IkappaBalpha protein overcomes multiple defects in apoptosis signaling in human B-cell malignancies. *Clin Cancer Res* 11:8186–8194.

Thompson KH, Bohmerle K, Polishchuk E, Martins C, Toleikis P, Tse J, Yuen V, McNeill JH, Orvig C. 2004. Complementary inhibition of synoviocyte, smooth muscle cell or mouse lymphoma cell proliferation by a vanadyl curcumin complex compared to curcumin alone. *J Inorg Biochem* 98:2063–2070.

Tomita M, Kawakami H, Uchihara JN, Okudaira T, Masuda M, Takasu N, Matsuda T, Ohta T, Tanaka Y, Ohshiro K, Mori N. 2006. Curcumin (diferuloylmethane) inhibits constitutive active NF-kappaB, leading to suppression of cell growth of human T-cell leukemia virus type I-infected T-cell lines and primary adult T-cell leukemia cells. *Int J Cancer* 118:765–772.

Tong QS, Zheng LD, Lu P, Jiang FC, Chen FM, Zeng FQ, Wang L, Dong JH. 2006. Apoptosis-inducing effects of curcumin derivatives in human bladder cancer cells. *Anticancer Drugs* 17:279–287.

Tunstall RG, Sharma RA, Perkins S, Sale S, Singh R, Farmer PB, Steward WP, Gescher AJ. 2006. Cyclooxygenase-2 expression and oxidative DNA adducts in murine intestinal adenomas: Modification by dietary curcumin and implications for clinical trials. *Eur J Cancer* 42:415–421.

Uddin S, Hussain AR, Manogaran PS, Al-Hussein K, Platanias LC, Gutierrez MI, Bhatia KG. 2005. Curcumin suppresses growth and induces apoptosis in primary effusion lymphoma. *Oncogene* 24:7022–7030.

Venkatraman M, Anto RJ, Nair A, Varghese M, Karunagaran D. 2005. Biological and chemical inhibitors of NF-kappaB sensitize SiHa cells to cisplatin-induced apoptosis. *Mol Carcinog* 44:51–59.

Vietri M, Pietrabissa A, Mosca F, Spisni R, Pacifici GM. 2003. Curcumin is a potent inhibitor of phenol sulfotransferase (SULT1A1) in human liver and extrahepatic tissues. *Xenobiotica* 33:357–363.

Volate SR, Davenport DM, Muga SJ, Wargovich MJ. 2005. Modulation of aberrant crypt foci and apoptosis by dietary herbal supplements (quercetin, curcumin, silymarin, ginseng and rutin). *Carcinogenesis* 26:1450–1456.

Wahl H, Tan L, Griffith K, Choi M, Liu JR. 2007. Curcumin enhances Apo2L/TRAIL-induced apoptosis in chemoresistant ovarian cancer cells. *Gynecol Oncol* 105: 104–112.

Walters DK, Muff R, Langsam B, Born W, Fuchs B. 2007. Cytotoxic effects of curcumin on osteosarcoma cell lines. *Invest New Drugs* 26:289–297.

Wang W, Abbruzzese JL, Evans DB, Larry L, Cleary KR, Chiao PJ. 1999. The nuclear factor-kappa B RelA transcription factor is constitutively activated in human pancreatic adenocarcinoma cells. *Clin Cancer Res* 5:119–127.

Wang X, Wang Q, Ives KL, Evers BM. 2006a. Curcumin inhibits neurotensin-mediated interleukin-8 production and migration of HCT116 human colon cancer cells. *Clin Cancer Res* 12:5346–5355.

Wang YD, Hu Y, Sun CY. 2006b. [Inhibitory effect of curcumin on angiogenesis induced by brain derived neurotrophic factor from multiple myeloma cells]. *Zhongguo Shi Yan Xue Ye Xue Za Zhi* 14:70–74.

Wang Z, Zhang Y, Banerjee S, Li Y, Sarkar FH. 2006c. Notch-1 down-regulation by curcumin is associated with the inhibition of cell growth and the induction of apoptosis in pancreatic cancer cells. *Cancer* 106:2503–2513.

Wargovich MJ, Chen CD, Jimenez A, Steele VE, Velasco M, Stephens LC, Price R, Gray K, Kelloff GJ. 1996. Aberrant crypts as a biomarker for colon cancer: Evaluation of potential chemopreventive agents in the rat. *Cancer Epidemiol Biomarkers Prev* 5:355–360.

Wax A, Pyhtila JW, Graf RN, Nines R, Boone CW, Dasari RR, Feld MS, Steele VE, Stoner GD. 2005. Prospective grading of neoplastic change in rat esophagus epithelium using angle-resolved low-coherence interferometry. *J Biomed Opt* 10:051604.

Wei SC, Lin YS, Tsao PN, Wu-Tsai JJ, Wu CH, Wong JM. 2004. Comparison of the anti-proliferation and apoptosis-induction activities of sulindac, celecoxib, curcumin, and nifedipine in mismatch repair-deficient cell lines. *J Formos Med Assoc* 103:599–606.

Weir NM, Selvendiran K, Kutala VK, Tong L, Vishwanath S, Rajaram M, Tridandapani S, Anant S, Kuppusamy P. 2007. Curcumin induces G2/M arrest and apoptosis in cisplatin-resistant human ovarian cancer cells by modulating Akt and p38 MAPK. *Cancer Biol Ther* 6:178–184.

White EL, Ross LJ, Schmid SM, Kelloff GJ, Steele VE, Hill DL. 1998a. Screening of potential cancer-preventing chemicals for inhibition of induction of ornithine decarboxylase in epithelial cells from rat trachea. *Oncol Rep* 5:717–722.

White EL, Ross LJ, Schmid SM, Kelloff GJ, Steele VE, Hill DL. 1998b. Screening of potential cancer preventing chemicals for induction of glutathione in rat liver cells. *Oncol Rep* 5:507–512.

Wijnands MV, van Erk MJ, Doornbos RP, Krul CA, Woutersen RA. 2004. Do aberrant crypt foci have predictive value for the occurrence of colorectal tumours? Potential of gene expression profiling in tumours. *Food Chem Toxicol* 42:1629–1639.

Wolanin K, Magalska A, Mosieniak G, Klinger R, McKenna S, Vejda S, Sikora E, Piwocka K. 2006. Curcumin affects components of the chromosomal passenger complex and induces mitotic catastrophe in apoptosis-resistant Bcr-Abl-expressing cells. *Mol Cancer Res* 4:457–469.

Woo MS, Jung SH, Kim SY, Hyun JW, Ko KH, Kim WK, Kim HS. 2005. Curcumin suppresses phorbol ester-induced matrix metalloproteinase-9 expression by inhibiting the PKC to MAPK signaling pathways in human astroglioma cells. *Biochem Biophys Res Commun* 335:1017–1025.

Woo JH, Kim YH, Choi YJ, Kim DG, Lee KS, Bae JH, Min DS, Chang JS, Jeong YJ, Lee YH, Park JW, Kwon TK. 2003. Molecular mechanisms of curcumin-induced cytotoxicity: induction of apoptosis through generation of reactive oxygen species, down-regulation of Bcl-XL and IAP, the release of cytochrome c and inhibition of Akt. *Carcinogenesis* 24:1199–1208.

Wu Q, Chen Y, Li X. 2006. HDAC1 expression and effect of curcumin on proliferation of Raji cells. *J Huazhong Univ Sci Technolog Med Sci* 26:199–201, 210.

Wu Y, Chen Y, Xu J, Lu L. 2002. Anticancer activities of curcumin on human Burkitt's lymphoma. *Zhonghua Zhong Liu Za Zhi* 24:348–352.

Xia X, Cheng G, Pan Y, Xia ZH, Kong LD. 2007. Behavioral, neurochemical and neuroendocrine effects of the ethanolic extract from Curcuma longa L. in the mouse forced swimming test. *J Ethnopharmacol* 110:356–363.

Xu G, Huang W, Zhang WM, Lai ZS, He MR, Wang YD, Zhang YL. 2005. [Effects of combined use of curcumin and catechin on cyclooxygenase-2 mRNA expression in dimethylhydrazine-induced rat colon carcinogenesis]. *Di Yi Jun Yi Da Xue Xue Bao* 25:48–52.

Yoon H, Liu RH. 2007. Effect of selected phytochemicals and apple extracts on NF-kappaB activation in human breast cancer MCF-7 cells. *J Agric Food Chem* 55:3167–3173.

Yu Z, Shah DM. 2007. Curcumin down-regulates Ets-1 and Bcl-2 expression in human endometrial carcinoma HEC-1-A cells. *Gynecol Oncol* 106:541–548.

Zambre AP, Kulkarni VM, Padhye S, Sandur SK, Aggarwal BB. 2006. Novel curcumin analogs targeting TNF-induced NF-kappaB activation and proliferation in human leukemic KBM-5 cells. *Bioorg Med Chem* 14:7196–7204.

Zhang J, Qi H, Wu C. 2004. Research of anti-proliferation of curcumin on A549 human lung cancer cells and its mechanism. *Zhong Yao Cai* 27:923–927.

Zheng LD, Tong QS, Wu CH. 2002. Inhibitory effects of curcumin on apoptosis of human ovary cancer cell line A2780 and its molecular mechanism. *Ai Zheng* 21:1296–1300.

Zheng L, Tong Q, Wu C. 2004. Growth-inhibitory effects of curcumin on ovary cancer cells and its mechanisms. *J Huazhong Univ Sci Technolog Med Sci* 24:55–58.

Zheng LD, Tong QS, Wu CH. 2006. Growth inhibition and apoptosis inducing mechanisms of curcumin on human ovarian cancer cell line A2780. *Chin J Integr Med* 12:1.

17 Plant Phenolic Compounds: Modulation of Cytoprotective Enzymes and Nrf2/ARE Signaling

SIWANG YU[1], KA LUNG CHEUNG[2], WENGE LI[1], AND AH-NG KONG[1]
[1]Department of Pharmaceutics,
[2]Graduate Program in Pharmaceutical Science, Ernest-Mario School of Pharmacy, Rutgers, The State University of New Jersey, Piscataway, New Jersey

CONTENTS

Plant Phenolics and Human Health: Biochemistry, Nutrition, and Pharmacology, Edited by Cesar G. Fraga.

INTRODUCTION: PLANT PHENOLIC COMPOUNDS

Phenolic compounds, mainly polyphenols, are abundant in a wide array of edible plants including various legumes, fruits, vegetables, olive oil, and grains. They are also found in red wine, chocolate, green tea, and coffee. Structurally, phenolic compounds are characterized by the presence of one or more hydroxylated 6-carbon aromatic rings and can be accordingly divided into several main groups: phenolic acids, phenolic alcohols, flavonoids, stilbenes, and lignans. Currently over 8000 phenolic compounds are known, among them flavonoids are the largest and most studied polyphenols and can be subdivided into flavonols, flavones, flavanols (catechins), flavanones, anthocyanidins, and isoflavonoids. There are also many smaller classes of plant phenolic compounds derived from different biosynthetic pathways [Bravo, 1998].

In this chapter, we will discuss the health effects of plant phenolic compounds with special emphasis on their modulation of cytoprotective enzymes and Nrf2/ARE signaling. Curcumin (diferuloylmethane), a hydrophobic polyphenol derived from turmeric (*Curcuma longa*), will be discussed as a typical example of plant phenolic compounds. Chemically, curcumin is characterized by a bis-α,β-unsaturated β-diketone structure that can function as Michael reaction acceptors [Strimpakos and Sharma, 2008]. The commercially available curcumin is a mixture of curcuminoids including curcumin, demethoxycurcumin, and bisdemethoxycurcumin. Upon absorption, curcumin is quickly metabolized and the major metabolites are curcumin–glucuronoside, dihydrocurcumin–glucuronoside, tetrahydrocurcumin–glucuronoside, and tetrahydrocurcumin [Anand et al., 2007]. The structure of some curcuminoids and metabolites are shown in Figure 17.1 together with a synthetic curcumin analog, salicylcurcuminoid.

HEALTH EFFECTS OF PLANT PHENOLIC COMPOUNDS

Despite the wide distribution of phenolic compounds in edible plants and the high dietary intake, the health effects of plant phenolic compounds had not been extensively studied until the mid-1990s due to their diversity of species and chemical structures. Epidemiological studies have revealed that dietary consumption of fruits, vegetables, and other plant-based foods and beverages is inversely correlated with the incidences of many diseases such as cancer, cardiovascular disease, and neurodegenerative diseases [Stevenson and Hurst, 2007]. The evidences from clinical and laboratory studies strongly support

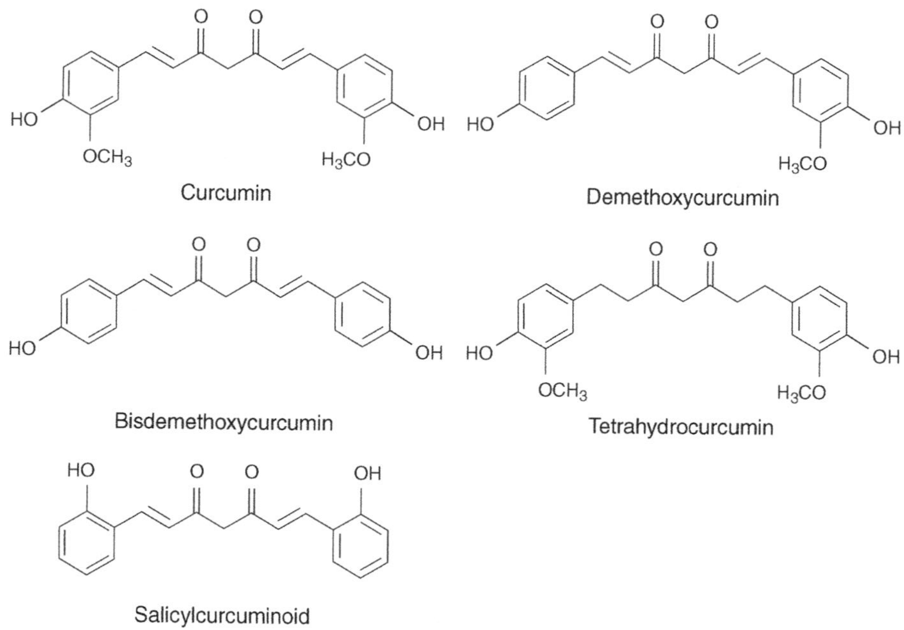

Figure 17.1 The chemical structure of curcumin, demethoxycurcumin, bisdemethoxycurcumin, tetrahydrocurcumin, and salicylcurcuminoid. Curcumin is characterized by two phenlic groups connected by a bis-α,β -unsaturated β-diketone structure. Removal of one or two methoxyl groups, as in demethoxycurcumin or bisdemethoxycurcumin, shows no significant effect on the induction of cytoprotective enzymes but decreases the direct antioxidant activity. On the contrary, tetrahydrocurucmin exhibits higher direct antioxidant activity but lower enzyme induction than curcumin. Salicylcurcuminoid, a synthetic analog of curucmin with *ortho*-hydroxyl groups, is >30 times potent than curcumin in enzyme induction and is more potent in cytoprotection than curcumin.

plant phenolic compounds as important candidates of the bioactive components in plant-based foods and beverages. Since many plant phenolic compounds have been widely consumed for many years as dietary components without safety problems, they are especially important in prevention of cancers, cardiovascular diseases, and other degenerative diseases. Curcumin has been widely used as a food flavoring and coloring agent and for medical purposes for many years in Asia. Since the publication of the first report on the usage of curcumin to treat human diseases in 1937, curcumin has been one of the most extensively investigated plant phenolic compounds. Curcumin has been shown to exhibit a surprisingly wide range of beneficial biological activities including anti-inflammation, antioxidant, anticarcinogenesis, antidiabetes, myocardial protection, and neuron protection [Strimpakos and Sharma, 2008]. Furthermore, curcumin is extremely safe even at very high doses, as demonstrated in various animal models or human studies [Anand et al., 2007]. These properties make curcumin a promising compound for treatment and prevention of a variety of diseases.

PLANT PHENOLIC COMPOUNDS AS DIRECT ANTIOXIDANTS

As described above, plant phenolic compounds have been shown to be useful in prevention or treatment of many degenerative diseases. The underlying mechanisms have been suggested as inhibition of proliferation, induction of apoptosis, suppression of inflammation, modulation of immune responses, and protection against oxidative stresses, and many signaling pathways have been suggested as the molecular targets of plant phenolic compounds. However, most of the health effects of plant phenolic compounds can be attributed to their cytoprotective activities against environmental and endogenous stresses imposed by electrophiles/oxidants [Stevenson and Hurst, 2007].

Oxidative stress plays important roles in the pathogenesis of many degenerative diseases. Reactive oxygen/nitrogen species (ROS/RNS) potentially can react with all bio-macromolecules including protein, nucleic acids, and lipids and impair their normal functions. Reduced forms of plant phenolic compounds exhibit strong antioxidant capacities in vitro, by scavenging a wide array of reactive oxygen, nitrogen, and chlorine species such as superoxide, hydroxyl radical, peroxyl radicals, singlet oxygen, hypochlorous acid, and peroxynitrous acid. In addition, they can chelate redox-active metal ions, often leading to reduced pro-oxidant activities. Their direct antioxidant activities require one or more free phenolic hydroxyl groups and are generally enhanced by electron repelling or large alkyl groups in the 2- and 6-positions [Stevenson and Hurst, 2007]. As other typical small-molecule antioxidant such as ascorbate and tocopherol, these phenolic compounds can undergo redox reactions and react with ROS/RNS to quench them.

The antioxidant activity of curcumin has been investigated for over 30 years [Sharma, 1976]. Curcumin has been shown to inhibit lipid peroxidation,

ROS- or radiation-mediated DNA strand breakage or adduct formation, and protein oxidative/nitrosative damages. Curcumin can scavenge various electrophiles/antioxidants including singlet oxygen, superoxide, polyunsaturated lipid peroxyl radicals, NO radicals, and many others [Menon and Sudheer, 2007]. In addition, curcumin chelates metal ions such as cadmium, copper, and manganese, reducing their pro-oxidant activities [Corona-Rivera et al., 2007; Daniel et al., 2004]. Moreover, it was proposed that the copper or manganese complexes of curcumin mimics superoxide dimutase, further potentiates the antioxidant activity of curcumin [Barik et al., 2005; Dutta et al., 2001; Vajragupta et al., 2003]. The direct antioxidant activity of curcumin has been ascribed to its phenolic hydroxyl and/or β-diketone moieties. A reactive electrophile or oxidant can undergo electron transfer or abstract H atom from either of these two sites.

However, the concentrations required for efficient protection against oxidative stresses are generally tens or even hundreds of micromolar. For example, the concentration required to inhibit 40% of superoxide-mediated ferricytochrome *c* reduction is 80 μM, and over 20 μM to scavenge NO radical [Das and Das, 2002; Sumanont et al., 2004]. On the contrary, despite that plant phenolic compounds were considered as the most abundant antioxidants in the diet, the physiologically achievable concentrations of plant phenolic compounds are very low and sometime even negligible. In rat 2% dietary curcumin yielded only low levels of plasma concentration between 0 and 12 nM, even in colon mucosa the concentrations were only 0.2 – 1.8 μmol/g [Anand et al., 2007]. In addition, the half-lives of plant phenolic compounds in circulation are generally short because of rapid metabolism and elimination. Many metabolizing reactions such as methylation or glucuronidation will decrease their direct antioxidant activity by blocking of phenolic hydroxyl groups. In fact, dimethoxy curcumin is much less active than curcumin in terms of quenching free radicals [Priyadarsini et al., 2003]. Furthermore, during the antioxidant reactions, these compounds are either consumed or chemically modified, and have to be regenerated or replenished for continuing action. In contrast to these exogenous antioxidants, glutathione (GSH), a universal endogenous small-molecule antioxidant, is presented in millimolor ranges in all cells; on the other hand, the total circulating concentrations of ascorbate and urate, two main endogenous small-molecule antioxidants, are over 500 μM. Thus the plant phenolic compounds in tissues or plasma only represent a transient and small increase of total antioxidant capacity [Stevenson and Hurst, 2007; Dinkova-Kostova and Talalay, 2008].

Moreover, under certain conditions these phenolic compounds could also act as pro-oxidants. In the presence of redox-active metal ions such as Cu or Fe, phenolic compounds react with O_2 to generate phenoxyl radicals. Under normal growth conditions phenoxyl radicals can be rapidly deactivated by polymerization or enzymatic reduction. However, if the phenoxyl radical concentrations are too high and/or the lifetime is increased, they could initiate DNA damage or lipid peroxidation and exhibit cytotoxicities. Curcumin, demethoxycurcumin, and bisdemethoxycurcumin have been reported to induce

DNA damage in the presence of cytochrome P450s (CYPs) and Cu(II) in vitro or in cells [Yoshino et al., 2004; Ahsan et al., 1999]. At higher concentrations curcumin exhibited genotoxicity and cytotoxicity in different systems [Cao et al., 2007]. At last, in vivo intervention trials using typical antioxidants such as ascorbate or carotenes yielded mostly disappointing results in terms of protection against oxidation-related diseases, sometime even increased the incidences of diseases [Stevenson and Hurst, 2007; Dinkova-Kostova and Talalay, 2008]. Obviously, the direct antioxidant capacity of plant phenolic compounds cannot explain their cytoprotective activities.

INDIRECT CYTOPROTECTIVE ACTIVITY OF PLANT PHENOLIC COMPOUNDS AND CYTOPROTECTIVE ENZYMES

For many years, the cytoprotective effects of plant phenolic compounds were attributed to their ability to direct scavenge oxidants and free radicals. However, as discussed in the last section, this concept is oversimplified and misleading. More and more evidence suggest that plant phenolic compounds could interact with cellular components and trigger a series of cellular responses, which are able to modulate the redox status of the cells and protect the cells from potentially toxic electrophiles/oxidants.

All eukaryotic cells are equipped with elaborate cytoprotective machinery to protect cells from those harmful factors, and the activity of this cellular defense system is often up-regulated in response to exogenous or endogenous stresses. This phenomenon was first described as the "hormesis effect," which is characterized by adaptive beneficial effect induced by exposure to a low dose of a chemical agent or environmental factor that is harmful at higher doses [Mattson, 2008]. In some cases it was also called "adaptation" or "preconditioning," in which cells and organisms acquire resistance to damaging stresses by exposure to a mild stress [Arumugam et al., 2006]. One typical example is the mechanisms of ischemic preconditioning in the heart and brain, where preexpose to moderate ischemia could enhance the tolerance of cells to severe oxidative stress. Similarly, it was found that weakly carcinogenic hydrocarbons could inhibit the tumorigenicity of potent carcinogens. This inhibition was effective in distal organs/tissues; and even the inhibitor and carcinogen were administrated by different route [Holtzclaw et al., 2004]. It was not dependent on local interaction between inhibitor and carcinogen but correlated with increased activity of metabolizing enzymes that collectively deactivate carcinogens. Actually, in the early days most compounds identified as enzyme inducers were carcinogens such as 7,12-dimethylbenz[a]anthracene (DMBA), benzo[a]pyrene (B[a]P), and Sudan III. Further investigations found out that several phenolic compounds such as butylated hydroxyanisole (BHA), butylated hydroxutoluene (BHT), and *tert*-butylhydroquinone (t-BHQ) potently inhibited the toxicity and carcinogenicity of chemical carcinogens [Frankfurt et al., 1967; Wattenberg, 1972], and this protection was also associated with increased phase 2 and antioxidant enzyme activities. Unlike the previously studied potential carcinogens, these phenolic antioxidants have been safely consumed by humans as food additives, so this finding was considered as a major breakthrough in the search of safe chemopreventives. These investigations finally led to the concept that induction of phase 2 and antioxidant enzymes could be a plausible strategy for protection against detrimental electrophiles/oxidants and related diseases.

Phase 2 enzymes catalyze the conjugation of reactive electrophiles/oxidants with small endogenous water-soluble molecules such as glutathione, sulfate, UDP-uridine diphosphate-glucuronic acid, or acetyl coenzyme A, which generally leads to detoxification of those elctrophiles/oxidants and facilitates their elimination [Yu and Kong, 2007]. Phase 2 enzymes consist of many superfamilies of enzymes including γ-glutamylcysteine synthetase (γ-GCS), glutathione-*S*-transferases (GST), sulfotransferases (SULT), NAD(P)H:quinone oxidoreductase (NQO, or quinone reductase, QR), UDP-glucuronosyl transferases (UGT), epoxide hydrolases (EPH), and *N*-acetyltransferases (NAT). On the other hand, glutathione peroxidase (GPX), glutathione reductase (GSR), catalase, thioredoxin reductase (TrxR), superoxide dismutase (SOD), heme oxygenase (HO), and biliverdin reductase (BLVR) are the most important antioxidant enzymes that catalyze the neutralization of ROS/RNS and free radicals. Beside these enzymes, some protective proteins that may not have enzymatic activities (e.g., ferritin, thioredoxin) are also important members of the cytoprotective machinery. Many cytoprotective enzymes play important roles in both detoxification of electrophiles and elimination of oxidants. For example, certain GSTs exhibit glutathione peroxidase activity in addition to catalyzing the conjugation with glutathione [Hurst et al., 1998]. Another example is γ-GCS, which catalyzes the rate-limiting step in the synthesis of glutathione [Wild and Mulcahy, 2000]. All of these enzymes collectively protect cells from the damages of electrophiles/oxidants, and are designated as cytoprotective enzymes [Dinkova-Kostova and Talalay, 2008]. In many ways, induction of cytoprotective enzymes is much more efficient as an "antioxidant" mechanism than direct antioxidant, because these enzymes catalyze a wide spectrum of reactions that almost invariably lead to inactivation and detoxification of electrophiles/oxidants, and are not consumed during their antioxidant action.

Most of the cytoprotective enzymes are highly inducible in response to enzyme inducers or oxidative stress. As an important class of enzyme inducers, many plant phenolic compounds have been shown to be able to induce cytoprotective enzyme activities. In the search for chemopreventive compounds, tumeric or curcumin has been shown to significantly increase hepatic GST, UGT and EPH activities and glutathione content in carcinogen-treated mice [Goud et al., 1993, Susan and Rao, 1992]. Dietary supplementation of 2% curcumin to male ddY mice for 30 days significantly increased the activities of GPx, GSR, glucose-6-phosphate dehydrogenase, catalase, GST, and NQO1 in liver and kidney [Iqbal et al., 2003]. Curcumin also dose dependently induced UGT, GPx, and GST activities in rat hepatic and intestinal tissues, and the increased enzyme activities contributed to the anti-inflammatory and anticarcinogenesis activity of curcumin. Pregnant mice fed with tumeric or curcumin could even increase the hepatic levels of GST and acid-soluble sulfhydryl of translactationally exposed F1 pups [Singh et al., 1995]. In a study using curcumin to prevent cataractogenesisi in rat lenses, induction of GST activity was suggested as a mechanism of protection against lipid peroxidation [Awasthi et al., 1996]. In cultured cell models, curcumin is a potent inducer of cytoprotective enzymes. Despite its moderate inhibition of UGT activities, curcumin induced the expression of UGT 1A1 and 1A6 in Caco-2 cells [Usta et al., 2007]. Exposure of bovine aortic endothelial cells to curcumin resulted in increased HO-1 messengic bonuleic acid (mRNA), protein, and activity and protected cells from oxidative stress. It has been reported that curcumin induced HO-1 and GST P1 in normal lung fibroblasts but not in scleroderma lung fibroblasts, and this might contribute to the discriminative induction of apoptosis in scleroderma lung fibroblasts [Tourkina et al., 2004]. Curcumin also induced HO-1, NQO1, and GST in astrocytes and hippocampal neurons and protected cells from oxidative damages [Scapagnini et al., 2006].

MODULATION OF CYTOPROTECTIVE ENZYMES BY PLANT PHENOLIC COMPOUNDS: STRUCTURE–ACTIVITY RELATIONSHIP

To better understand the health effects of plant phenolic compounds and to better utilize them, it is necessary to know the molecular mechanisms by which plant phenolic compounds induce cytoprotective enzymes. In vitro studies indicated that plant phenolic compounds such as curcumin often inhibited the enzymatic activities of GST, UGT, SULT as well as cytochrome P450s [Oetari et al., 1996], suggesting that the induction of cytoprotective enzyme activities could not be explained by direct interaction with plant phenolic compounds. On the other hand, much evidence indicates that the increased activity of cytoprotective enzymes are mainly attributable to enhanced transcriptional activation and enzyme synthesis [Holtzclaw et al., 2004].

Based on the results from experiments using aryl hydrocarbon receptor (AhR)- or CYP1A1-deficient cells and mice, enzyme inducers were designated as monofunctional and bifunctional inducers [Prochaska and Talalay, 1988]. Bifunctional inducers (generally planar aromatics) AhR dependently increase the activities of both phase 1 and phase 2 enzymes, while monofunctional inducers selectively induce phase 2 enzymes without the participation of AhR. Bifunctional inducers directly induce phase 1 enzymes by binding to and activating AhR or are metabolized by phase 1 enzymes such as CYP1A1 to be converted to electrophilic monofunctional inducers. Many plant phenolic compounds belong to the monofunctional inducers. Curcumin has been reported to bind to AhR and induce phase 1 enzymes in cells [Ciolino et al., 1998], but in a recent study curcumin was found to inhibit benzo[a]pyrene-(B[a]P) induced AhR and phase 1 enzyme expression in mice [Garg et al., 2008].

Moreover, the induction of phase 2 enzymes by curcumin is independent on AhR or CYP1A1, thus curcumin is classified as a monofunctional inducer [Dinkova-Kostova and Talalay, 1999].

A wide variety of monofunctional inducers with very different chemical structures has been identified. These compounds could be classified into at least 10 classes, with plant phenolic compounds belonging to oxidizable diphenols or Michael reaction acceptors [Prochaska and Talalay, 1988]. Michael reaction acceptors (olefins or acetylenes conjugated to electron-withdrawing groups) constitute a major class of phase 2 enzyme inducers, and their inducer potency was closely correlated with their reactivity with nucleophiles in the Michael reaction. The induction of cytoprotective enzymes by curcumin could be attributed to the two Michael reaction acceptor functionalities in its molecule. The structure–activity relationship of curcumin and its analogs as NQO1 inducers were studied in Hepa1c1c7 murine hepatoma cells [Dinkova-Kostova and Talalay, 1999]. Removal of methoxyl groups, as in demethoxycurcumin and bisdemethoxycurcumin, only slightly decreased the inducer activity. Removal of phenolic hydroxyl group also showed no significant effect on the inducer activity, however, salicylcurcuminoid with *ortho*-hydroxyl groups is >30 times potent than curcumin in inducer activity. Tetrahydrocurcumin, a metabolite of curcumin, exhibited much lower inducer activity, though it is more potent than curcumin as a direct antioxidant. On the other hand, the β-diketone moiety is also an important functional group for induction of NQO1 due to the dominance of the keto-enol tautomerism. In general, the inducer activity is parallel to the reactivity with nucleophilic groups such as thiols. The structure–activity relationship in induction of NQO1 has also been confirmed in induction of HO-1 and the protection against oxidative damages [Jeong et al., 2006].

Further structure–activity relationship (SAR) analyses of other cytoprotective enzyme inducers revealed the fact that all inducers can react with thiol/disulfide groups by alkylation, oxidoreduction, or thiol-disulfide interchange [Dinkova-Kostova and Talalay, 1999]. In fact, the capability of enzyme inducers to induce cytoprotective enzymes is well correlated with their reactivity with thiols. These results suggested a cellular sensor of inducers with highly reactive sulfhydryl groups, possibly reactive thiols in cysteine residues of a sensor protein. Nevertheless, the initial search for the sensor protein by using radioactively labeled inducers was not successful due to the abundance of thiol groups presented in many proteins in cells [Holtzclaw et al., 2004]. The molecular mechanism by which cytoprotective enzymes are induced remained to be elucidated.

MODULATION OF CYTOPROTECTIVE ENZYMES BY PLANT PHENOLIC COMPOUNDS: FROM ARE TO NRF2

New understanding of the mechanism underlying the induction of cytoprotective enzymes came from analysis of the DNA sequences in the 5'-flanking promoter regions of several phase 2 genes. Two sequences in the upstream region of rat GST Ya were identified as essential for its induction by reporter assay [Rushmore and Pickett, 1990; Nguyen et al., 1994]. One of them contained an xhenobiotic response element (XRE, 5′-TNGCGTG-3′), which is involved in AhR-mediated induction. The other one could be activated independent of AhR and was named antioxidant response element (ARE). The core ARE sequence was further defined as 5′-TGACNNNGC-3′ by deletion and mutation analysis [Rushmore et a., 1991]. Another 41-bp region was identified in the 5-flanking region of mouse GST *Ya* gene, which is responsive to electrophiles and was named electrophile response element (EpRE) [Friling et al., 1992]. One of the functional sequences in this region is the same as ARE. Latter on, the core ARE sequence, which is responsive to enzyme inducers, was identified in the promoter regions of NQO1, HO-1, GCS, ferritin H, and GST P1. Further studies revealed that the ARE sequences in different genes was variable, and some other ARE-like or nonclassic ARE sequences were also identified. After more detailed analysis, the core sequence was extended as 5′-TMAnnRTGAYnnnGCRwwww-3′, where M = A or C, R = A or G, Y = C or T, W = A or T [Wasserman and Fahl, 1997]. Recently, a 21-base sequence logo of AREs was created based on 57 functional ARE sequences identified in human, rat, and mouse [Wang et al., 2007]. With the help of a systematic computational approach combining transcription factor binding site recognition, comparative genome analysis, gene expression profiling and genotype—expression phenotype association, 2388 genes common in human, rat, and mouse were found to contain AREs in their upstream 5-kb sequences.

There is an embedded 12-O-tetradecanoylphorbol-13-acetate (TPA) response element (TRE, 5′-TGACTCA-3′) in the core ARE sequence, though the ARE and TRE are functionally distinct. The TRE is recognized by AP-1, which shares a DNA recognition sequence [5′-TGCTGA(G/C)TCA(C/T)] with another transcription factor, nuclear factor-erythroid 2 (NF-E2). NF-E2 is a basic leucine zipper (b-Zip) protein and belongs to a subfamily of b-Zip cap 'n' collar (CNC) genes, including NF-E2, Nrf2, Nrf2, Nrf3, and Bach1/2. These proteins heterodimerize with other b-Zip proteins such as small Maf proteins through their b-Zip domain, and function as transcription factors. The small Maf proteins, namely MafG, MafK, and MafF, are homologs of the *v-maf* oncogene product Maf. The Maf protein recognizes two similar palindromic consensus sequences, 5′-TGCTGACTCAGCA-3′ and 5′-TGCTGACGTCAGCA-3′. These two sequences were called Maf recognition elements (MAREs) [Kataoka et al., 1994]. Interestingly, there are embedded TREs in these MAREs, and one consensus ARE in the first MARE.

The striking resemblance in ARE, TRE, MAREs, and NF-E2 recognition sequences raises the hypothesis that certain members of CNC and/or Maf family transcription factors could bind to ARE and regulate the expression of ARE-driven genes. Indeed, both Nrf1 and Nrf2 were found to bind with ARE in the human NQO1 gene and to potentiate the basal and inducible ARE-mediated transcription [Venugopal and Jaiswal, 1996]. Based on their expression profiles, Nrf2 was selected as the most likely candidate, and nrf2-deficient mice were generated by disruption of the *nrf2* gene in embryonic stem cells [Itoh et al., 1997]. Compared to the wild-type mice, both the basal and the BHA-inducible expression of phase 2 enzymes such as GST and NQO1 were largely diminished in *nrf2*-deficient mice. Furthermore, a heterodimer composed of Nrf2 and a small Maf protein was found to bind to ARE with high affinity. In another study, Nrf2 (−/−) mice exhibited significantly lower expression of catalase, GCS, HO-1, NQO1, SOD1, and UGT1A6 and were much more susceptible to BHT-induced pulmonary injury [Chan and Kan, 1999]. These results clearly demonstrated that Nrf2 is essential for the transcriptional activation of phase 2 enzymes and provided a coordinated transcriptional regulation mechanism for the induction of cytoprotective enzymes.

Consequently, the induction of cytoprotective enzymes such as HO-1, GPx2, and GST by curcumin was found to be Nrf2-dependent [Balogun et al., 2003; Banning et al., 2005; Nishinaka et al., 2007]. The modulation of Nrf2-dependent gene expression profile in mice liver and small intestine by curcumin was examined in our lab [Shen et al., 2006]. Wild-type or Nrf2 (−/−) mice were given a single oral dose of curcumin at 1 g/kg, and then liver and small intestine were harvested after 3 or 12 h. The gene expression profiles were analyzed by mouse genome microarray, and the genes that were modulated > twofold by curcumin in wild-type but not in Nrf2 (-/-) mice were considered as Nrf2 dependent. Total 822 in liver (664 induced and 158 suppressed) and 222 in small intestine (154 induced and 68 suppressed) genes were identified as curcumin-modulated Nrf2-dependent. Many cytoprotective enzyme genes, including NQO1, GSTs, HO-1, γ-GCS, UGT, and thioredoxin, were among the identified genes. Interestingly, some phase 1 CYP genes and phase 3 transporters and many other genes were also Nrf2 dependently regulated by curcumin. This observation is further supported by the finding of AREs in over 2000 genes common in human, rat, and mouse [Wang et al., 2007]. Besides this, substantial evidence accumulated to further support the pivotal role of Nrf2 in induction of cytoprotective enzymes by all kinds of inducers and protection against electrophiles/oxidants-related damages and diseases.

In response to electrophiles, the DNA-binding activity and sometimes the protein level of Nrf2 was markedly increased; however, in most cases the Nrf2 mRNA level was not significantly changed, suggesting the activity of Nrf2 is regulated posttranscriptional. Extensive studies revealed that the Nrf2/ARE signaling is modulated at multiple levels.

MODULATION OF NRF2/ARE SIGNALING: IDENTIFICATION AND MODIFICATION OF KEAP1

Based on the homology of cross-species *ortho*logs, six domains were identified in the primary structure of Nrf2, namely Nrf2-ECH-homology 1 (Neh1) to Neh6 [Itoh et al., 1999]. The Neh1 contains the CNC homology region and basic-leucine zipper domain for dimerization with small Mafs and DNA binding.

The amino and carboxyl termini of the proteins were referred to as Neh2 and Neh3, respectively. Neh4 and Neh5 are two conserved acidic domains involved in transactivation. Neh3 has also been implied in transactivation [Nioi et al., 2005]. Additionally there is a serine-rich conserved region (Neh6). The primary structure of Nrf2 is illustrated in Figure 17.2.

It was found that the N-terminal of Nrf2, Neh2 domain, is inhibitory to the transcriptional activity of Nrf2. Co-expression of an Neh2 decoy restored Nrf2 transactivation activity, suggesting the existence of a possible repressor protein interacting with Neh2 [Itoh et al., 1999]. A single protein was identified in yeast two-hybrid experiment using the Gal4-Neh2 as bait, and was named as Keap1 (Kelch-like ECH-associated protein 1). In another study, a protein, INrf2, was isolated by co-purification with Nrf2 from rat hepatoma cells, and was identified as a trans-species homolog of Keap1 [Dhakshinamoorthy and Jaiswal, 2001]. The interaction between Keap1 and Nrf2 was further confirmed in vitro and in mammalian cells, and this interaction could be disrupted by cytoprotective enzyme inducers. Moreover, Nrf2 accumulation in nucleus and high constitutive expression of Nrf2 target genes were observed in Keap1-disrupted mice, and this phenotype could be totally reversed by Nrf2::Keap1 double knockout [Wakabayashi te al., 2003]. These findings proved that Keap1 is the negative regulator of Nrf2.

The Keap1 protein consists of 624 amino acids and is structurally related to Drosophila actin-binding protein Kelch. Keap1 contains a high number of cysteine residues (25 for murine and rat, 27 for human), and about nine of these cysteines (C23, C38, C151, C241, C273, C288, C297, C319, and C613) were predicted to be "reactive" since they are flanked by basic residues. This feature makes Keap1 a favorable "sensor" of electrophilic/oxidative signals. Indeed, these reactive cysteines were found to be able to react with enzyme inducers [Dinkova-Kostova et al., 2002]. C151 has been shown to be required for stabilization and activation of Nrf2 by sulforaphane and t-BHQ [Zhang and Hannink, 2003]. By using in vitro labeling and mutation assay, two other reactive cysteines, C273 and C288, were also identified as key residues for activation of Nrf2 [Wakabayashi et al., 2004]. In addition, different electrophiles may modify different sets of reactive cysteines [Dinkova-Kostova et al., 2005]. The NQO1 inducer activities of 30 plant phenylpropenoids and synthetic analogs were linearly correlated with their calculated tendency to release electrons (E_{HOMO}) [Zoete et al., 2004]. A similar correlation was reported for about 20 flavonoids, for which their pro-oxidant activity correlated with EpRE inducer activity [Lee-Hilz et al., 2006]. Furthermore, addition of thiol antioxidants such as GSH or NAC significantly abolished phenolic-induced cytoprotective enzyme expression. Most importantly, phenolic Michael acceptors could react with Keap1 in vitro, and their reactivity with Keap1 was parallel to their inducer potency, and the inducer activity was potentiated by free phenolic hydroxyl groups [Dinkova-Kostova et al., 2002].

Structurally, Keap1 is subdivided into five domains (Fig.17.2b): the amino terminal region (NTR), the BTB/POZ (Bric-a-brac, tramtrack, broad-complex/

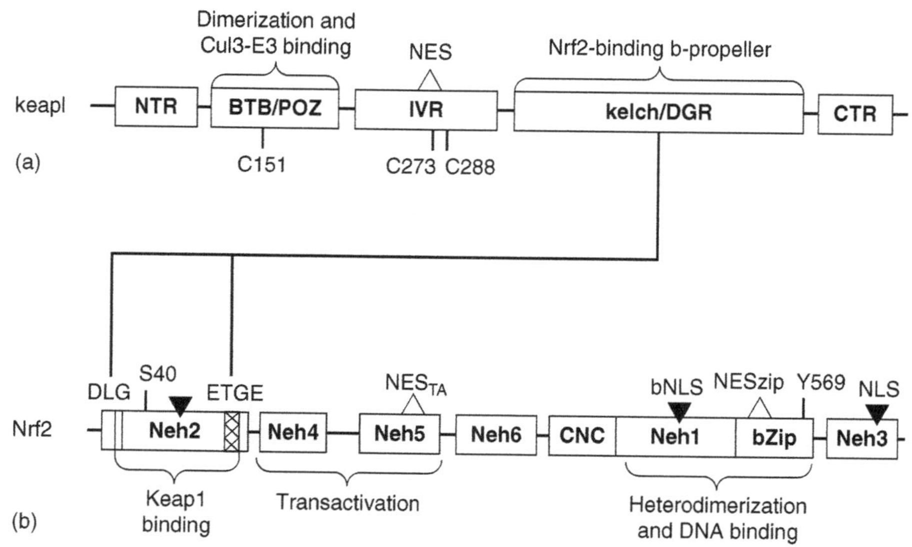

Figure 17.2 Schematic presentation of primary structure of Nrf2 and Keap1 with proposed regulatory domains/motifs. (a) Keap1 is subdivided into five domains. The BTB/POZ domain mediates the homodimerization of Keap1 and bridges Cul3-E3 ligase complex. The Kelch/DGR domain with a 6-blade b-propeller structure binds with the Neh2 domain of Nrf2. The IVR domain contains multiple reactive cysteines and is important for electrophile/oxidant-sensing. In addition, one NES (Δ) was identified in the IVR domain. (*b*). The six conserved domains were identified in Nrf2, namely Neh1 to Neh6. Neh1 contains the CNC and the basic leucine zipper motifs and is important for heterodimerization with small Mafs and DNA binding. Neh2 binds with Keap1 through the DLG and ETGE motifs. Neh3 at the C-terminal, Neh4 and Neh5 are involved in transactivation, while the function of Neh6 is still unknown. Nrf2 possesses multiple NES and NLS (▾) that control the subcellular distribution of Nrf2. Two identified phosphorylation sites, S40 by PKC and Y568 by Fyn kinase, are also indicated.

poxvirus zinc finger) domain, a cysteine-rich intervening region (IVR), the double-glycine repeat (DGR) or Kelch domain, and the carboxy terminal region (CTR). Keap1 forms homodimer through BTB domain and bridges Nrf2 to a Cul3-dependent E3 ubiquitin ligase complex [Kobayashi et al., 2004]. Two recognition sites, termed the DLG and ETGE motifs, with the Neh2 domain of Nrf2 were showed to bind to the Kelch domain with different affinities. Based on these observations, a "hinge and latch" model was proposed for the Nrf2/Keap1 complex to sense cellular stresses [Tong et al., 2006]. Under basal physiological conditions, Nrf2 is sequestered in cytoplasma by Keap1 through the binding of DLG and ETGE to two Kelch domains, thus the Neh2 is posed to E3-mediated ubiquintination, targeting Nrf2 for degradation by the 26S proteasome. In this way Nrf2/ARE signaling is kept inactivated by Keap1. Upon oxidative or electrophilic challenges, specific reactive cysteines in the IVR and BTB regions are modified [Zhang and Hannink, M, 2003; Wakabayashi et al., 2004]. This modification possibly leads to conformation alteration of the Keap1 dimer, leading to dissociation of the DLG motif and termination of Nrf2 ubiquitination. Furthermore, it has been proposed that Keap1 itself could be targeted for ubiquitination and proteasome-independent degradation upon exposure to electrophiles/oxidants [Zhang et al., 2005]. Collectively, these modifications result in the stabilization and possibly release of Nrf2.

MODULATION OF NRF2/ARE SIGNALING: MODIFICATION OF NRF2

Besides Keap1, Nrf2 itself is targeted for multiple posttranslational modifications. Two cysteine residues, C183 and C506, have been proposed to affect the activation of Nrf2/ARE signaling. Mutation of C506 in the DNA-binding region of Nrf2 to serine decreased its binding affinity to ARE and NQO1 induction in response to t-BHQ treatment, implicating that C506 could be modified [Bloom et al., 2002]. C183, which locates in the Neh5 domain, could be modified upon t-BHQ or H_2O_2 treatment and inhibited the nuclear exporting of Nrf2 [Li et al., 2006].

Nrf2 is also phosphorylated by various kinases including PKC (protein kinase C), MAPKs (mitogen-activated protein kinases), PERK (ER-localized pancreatic endoplasmic reticulum kinase), GSK-3b (glycogen synthase kinase 3betta), Fyn kinase, and CK2 (casein kinase 2). Phosphorylation of S40 of Nrf2 by a PKC isoform is important for dissociation of Nrf2 from Keap1 and its translocation to nucleus [Huang et al., 2002]. MAPKs, including ERK, JNK, and p38, have been found to phosphorylate Nrf2 and promote its accumulation in nucleus [Kong et al., 2001]. Phosphorylation by MAPKs also stabilizes Nrf2 [Nguyen et al., 2003]. PERK directly phorphorylates Nrf2 and triggers dissociation of Nrf2 from Keap1, and protects cells from ER-stress-mediated death [Cullinan et al., 2003]. GSK-3b either directly [Rojo et al., 2008] or indirectly through Fyn kinase [Jain and Jaiswal, 2007] phosphorylates Nrf2, leading to the exportation of Nrf2 from nucleus and negative regulation of Nrf2/ARE signaling. The phosphorylation site of Fyn kinase has been identified as Y568 [Jain and Jaiswal, 2007]. It has been reported that PI3K or Akt was involved in the activation of Nrf2/ARE signaling by curcumin [Pugazhenthi et al., 2007]. Since Gsk-3b is deactivated by Akt, it is possible that PI3K or Akt indirectly activate Nrf2 through deactivation of Gsk-3b. Recently, CK2 was found to phosphorylate Nrf2 at the transactivation domains (Neh4 and Neh5), and this phosphorylation promoted the translocation of Nrf2 to nucleus [Apopa et al., 2008]. Except for the phosphorylation sites by PKC and Fyn, the exact phosphorylation sites by other kinases remain elusive.

Many plant phenolic compounds have been shown to modulate protein kinases, and this could be an important mechanism of modulation of Nrf2/ARE signaling. Curcumin has been reported to transiently activate PKC, and PKC activation is required for curcumin-induced expression of HO-1, GST, and GCLM [McNally et al., 2007; Rushworth et al., 2006]. Specifically, PKCδ and PKCε have been implicated in regulation of Nrf2/ARE signaling by curcumin. It has also been reported that p38, ERK, and PI3K/Akt were involved in curcumin-mediated expression of cytoprotective enzymes, and specific pharmacological inhibitors or genetic inhibition of these kinases abrogated curcumin-induced Nrf2 activation and cytoprotective gene expression [McNally et al., 2007; Andreadi et al., 2006]. Another example is genistein, which activates Nrf2/ARE signaling by inhibiting tyrosine kinase Fyn and consequently inhibiting the exporting of Nrf2 [Jain and Jaiswal, 2007].

MODULATION OF NRF2/ARE SIGNALING: NUCLEAR-CYTOPLASMIC TRAFFICKING

Once activated, Nrf2 translocates from cytoplasma into nucleus, dimerizes with small Mafs, binds to ARE, and activates gene transcription. The nuclear translocation of Nrf2 has been viewed as an automatic process after being released from Keap1 or mediated by Keap1 [Itoh et al., 1999]. However, multiple nuclear localization signals (NLS) and nuclear exporting sequence (NES) have been identified in Nrf2 and in Keap1, suggesting the existence of a complicated regulation of nuclear-cytoplasmic trafficking.

A functional NES comprised of hydrophobic leucine and isoleucine residues has been identified in the IVR domain of Keap1 [Velichkova and Hasson, 2005]. Mutation of the hydrophobic residues as well as leptomycin B (LMB) treatment, which inactivates Crm1/exportin, resulted in nuclear accumulation of both Nrf2 and Keap1. It was further demonstrated that the NES in Keap1 is required for termination of Nrf2/ARE signaling, and the Nrf2/Keap1 complex does not bind to the ARE [Sun et al., 2007]. Conversely, in another report the subcellular distribution of endogenous Keap1 was essentially similar before and after LMB treatment [Watai et al., 2007]. This discrepancy could possibly be explained by the fact that Keap1 does contain any NLS and has to be co-transported into the nucleus with Nrf2, however, under basal condition Nrf2 is quickly degraded, thus few Keap1 protein could be transported into the nucleus.

In contrary, Nrf2 is equipped with multiple NES/NLS. It was almost simultaneously reported that Nrf2 possesses a NES located in the leucine zipper region of Nrf2 (NESzip), and at the same time a bipartite NLS was identified in the basic DNA-binding domain (bNLS) [Li et al., 2005; Jain et al., 2005]. Interestingly, the NESzip overlaps with the dimerization domain with small Mafs, thus the binding of Nrf2 with small Mafs supposedly will disable this NES by depriving the binding of Crm1/exportin, retaining Nrf2 in the nucleus. Another NES was identified in the Neh5 transactivation domain (TAD) of Nrf2 (NES_{TA}) [Li et al., 2006]. Similarly, the recruitment of cofactors such as CBP to the TAD could also possibly disable the NES_{TA}. In addition, the NES_{TA} contains a cysteine residue (C183), and mutation of this cysteine to alanine handicapped the exporting of Nrf2 from the nucleus. Recently two novel NLSs were identified in murine Nrf2, one is located in the Neh2 and the other one near the C-terminal [Theodore et al., 2008]. Once again, the N-terminal NLS is close to the Keap1-binding DLG and ETGE motifs, providing further complexity in the regulation of Nrf2/Keap1 nuclear -cytoplasmic trafficking.

The activity of NES and NLS can be regulated by phosphorylation events. Indeed, most of the kinases that modulate Nrf2/ARE signaling have been shown to regulate the localization of Nrf2. Actually, almost all of the identified phosphorylation sites/domains of Nrf2 by kinases are proximal to a NES or NLS: the PKC phosphorylation site S40 is close to the N-terminal NLS; the Fyn phosphorylation site Y568 is close to the NESzip; CK2 phosphorylates the TAD, which is nearby the NES_{TA}. Overall, the cellular distribution of Nrf2 is determined by a delicate balance between all these NESs and NLSs. Generally, plant phenolic compounds regulate this balance by modulating the binding with Keap1, small Mafs and cofactors, and the posttranslational modifications such as phosphorylation or sulfhydryl modification of Nrf2.

MODULATION OF NRF2/ARE SIGNALING: TRANSACTIVATION AND TRANSCRIPTIONAL FEEDBACK

After translocation into the nucleus, Nrf2 would need to form a heterodimer with small Maf proteins or other bZIP proteins to bind to the ARE and then recruit necessary co-activators such as the nuclear coactivator CBP (CREB-binding protein) to activate ARE-mediated transcription [Tong et al., 2006]. The role of small Mafs in Nrf2/ARE signaling has been controversial since overexpression of small Maf proteins, MafG or MafK, inhibited ARE-mediated transcriptional activation [Nguyen et al., T 2000; Dhakshinamoorthy and Jaiswal, 2000]. On the other hand, the induction of a wide range of

ARE-dependent genes was abolished in mafG::mafF::mafK triple mutant fibroblasts [Katsuoka et al., 2005]. However the controversy was reconciled by the finding that homodimers of small Mafs could also bind to ARE but lack transactivation domain, and only Nrf2/small Maf heterodimers could activate ARE. Thus overexpressed small Mafs form homodimers to inhibit ARE, while small Mafs are necessary for Nrf2-mediated ARE activation [Motohashi et al., 2000]. Once bound to ARE, the Nrf2/small Maf heterodimer will recruit CBP through Neh4 and Neh5 to activate transcription. Interestingly, CBP could also be phosphorylated by MAPKs and synergistically activate Nrf2/ARE signaling with ERK cascade [Shen et al., 2004]. In addition, some other coactivators or co-repressors including ATF3 [Brown et al., 2008], ATF4 [He et al., 2001], RAC3 [Lin et al., 2006], SMRT [Ki et al., 2005], MOZ [Ohta et al., 2007], and BRG1 [Zhang et al., 2006] have been reported to interact with Nrf2 and modulate Nrf2/ARE signaling. It is intriguing that curcumin is known to activate Nrf2/ARE signaling; however; curcumin has been identified as a specific inhibitor of CBP [Balasubramanyam et al., 2004]. Obviously, this level of regulation of Nrf2/ARE signaling is extremely complicated due to the integration of many upstream signals.

Additionally, Nrf2/ARE signaling is also regulated by several feedback loops. Though Nrf2/ARE signaling is mainly regulated at the posttranscriptional level, the expression of Nrf2 was increased in certain cases. Two ARE-like sequences were identified in the proximal promoter region of Nrf2, and the transcriptional activity of the promoter could be increased by enzyme inducer or by overexpression of Nrf2 (99). Furthermore, Nrf2 could bind to its own promoter. Thus Nrf2 was suggested to autoregulate its own expression through an ARE-like sequence in its promoter. In the same way, a functional ARE was identified in the reverse strand of the proximal promoter of Keap1 [Lee et al., 2007]. Treatment with t-BHQ activated Nrf2/ARE signaling and also increased the expression of Keap1. The increased Keap1 subsequently directs Nrf2 to degradation and terminates Nrf2/ARE signaling. Moreover, tandem AREs were found in the proximal promoter of the PSMB5 subunit of the 26S proteasome [Kwak et al., 2003]. The expression of 26S proteasome could be up-regulated by enzyme inducers and by overexpression of Nrf2, while disruption of Nrf2 diminished this induction. In support of this, the proteasome subunits have been found to be Nrf2-dependently regulated by curcumin in gene expression profiling experiment [Shen et al., 2006]. Besides protection against oxidative stresses, the enhanced expression of proteasome could also affect the degradation of Nrf2.

CONCLUSIONS

Plant phenolic compounds exhibit a plethora of health effects, and protection of cells against electrophilic/oxidative stresses is a major mechanism underlying those health effects. Plant phenolic compounds have been regarded as direct antioxidants for a long time; however, their capability to induce cytoprotective enzymes is gaining more and more attention. The expression of cytoprotective enzymes is mainly regulated by Nrf2/ARE signaling. In this chapter we discussed the modulation of cytoprotective enzymes and Nrf2/ARE signaling by plant phenolic compounds with curcumin as an example. The modulation of cytoprotective enzymes by plant phenolic compounds occurs at multiple levels, and the overall scenario is summarized in Figure 17.3. Most of the observations were made using t-BHQ or curcumin as prototypical phenolic compounds; however, it is plausible to assume that these mechanisms also apply to many other phenolic compounds.

Nrf2/ARE signaling is mainly regulated posttranslationally. In almost all systems, Keap1 serves as the primary sensor of electrophiles/oxidants or enzyme inducers, and is the most important regulator of Nrf2/ARE signaling. Under basal condition, Keap1 dimer binds Nrf2 and bridges Nrf2 to ubiquitination by a Cul3-E3 ligase complex, thus directing Nrf2 to proteasome-dependent degradation. Keap1 possesses several specific reactive cysteines that

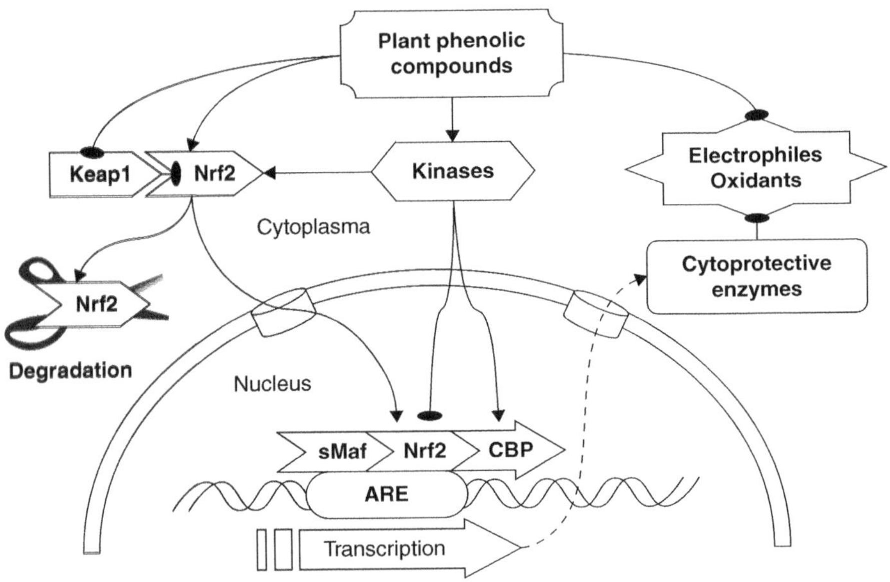

Figure 17.3 Cytoprotective functions of plant phenolic compounds: direct antioxidant activity and modulation of cytoprotective enzymes through Nrf2/ARE signaling. Plant phenolic compounds can directly function as antioxidants to scavenge electrophiles and oxidants. More importantly, plant phenolic compounds regulate the expression of cytoprotective enzymes through modulation of Nrf2/ARE signaling at multiple levels. Plant phenolic compounds can directly modify Keap1 or Nrf2, leading to the stabilization and translocation of Nrf2. A variety of kinases can be modulated by plant phenolic compounds, then phosphorylate Nrf2 or coactivators such as CBP to regulate the stability, localization, and transactivation of Nrf2.

can be modified by electrophiles/oxidants. In agreement with the pivotal role of Keap1, the inducer activity of plant phenolics and synthetic analogs are correlated with their reactivity with sulfhydryl groups and Keap1. The modification of specific cysteines in Keap1 alters the conformation of Keap1, leading to termination of Nrf2 ubiquitination and possibly release of Nrf2. The subcellular localization of Nrf2 is determined by the balance between multiple NES/NLS, and this balance is regulated by the interaction of Nrf2 with its binding partners such as Keap1 and small Mafs and post-translational modification of Nrf2. Once translocated into the nucleus, Nrf2 dimerizes with small Mafs, binds to ARE and recruits coactivators to activate transcription of cytoprotective enzyme genes. Nrf2 itself can be phosphorylated by a variety of kinases including PKC, MAPKs, PERK, GSK-3β, Fyn kinase, and CK2. In addition, two cysteines in Nrf2 could also be modified. These modifications regulate Nrf2/ARE signaling by disrupting Keap1 binding, changing subcellular distribution, increasing stability, and promoting transactivation of Nrf2. Unlike the ubiquitous regulation by Keap1, these modifications provide compound- and gene-specific and possibly cell type- or tissue-dependent modulation of Nrf2/ARE signaling.

ACKNOWLEDGMENT

The authors thank the members of Kong Lab for insightful discussions. This work was supported by National Institutes of Health Grant R01 CA118947 (to A.-N. T. K.).

REFERENCES

Ahsan H, Parveen N, Khan NU, Hadi SM. 1999. Pro-oxidant, anti-oxidant and cleavage activities on DNA of curcumin and its derivatives demethoxycurcumin and bisdemethoxycurcumin. *Chem Biol Interact* 121:161–175.

Anand P, Kunnumakkara AB, Newman RA, Aggarwal BB. 2007. Bioavailability of curcumin: Problems and promises. *Mol Pharm* 4:807–818.

Andreadi CK, Howells LM, Atherfold PA, Manson MM. 2006. Involvement of Nrf2, p38, B-Raf, and nuclear factor-kappaB, but not phosphatidylinositol 3-kinase, in induction of hemeoxygenase-1 by dietary polyphenols. *Mol Pharmacol* 69:1033–1040.

Apopa PL, He X, Ma Q. 2008. Phosphorylation of Nrf2 in the transcription activation domain by casein kinase 2 (CK2) is critical for the nuclear translocation and transcription activation function of Nrf2 in IMR-32 neuroblastoma cells. *J Biochem Mol Toxicol* 22:63–76.

Arumugam TV, Gleichmann M, Tang SC, Mattson MP. 2006. Hormesis/preconditioning mechanisms, the nervous system and aging. *Ageing Res Rev* 5:165–178.

Awasthi S, Srivatava SK, Piper JT, Singhal SS, Chaubey M, Awasthi YC. 1996. Curcumin protects against 4-hydroxy-2-trans-nonenal-induced cataract formation in rat lenses. *Am J Clin Nutr* 64:761–766.

Balasubramanyam K, Varier RA, Altaf M, Swaminathan V, Siddappa NB, Ranga U, Kundu TK. 2004. Curcumin, a novel p300/CREB-binding protein-specific inhibitor of acetyltransferase, represses the acetylation of histone/nonhistone proteins and histone acetyltransferase-dependent chromatin transcription. *J Biol Chem* 279:51163–51171.

Balogun E, Hoque M, Gong P, Killeen E, Green CJ, Foresti R, Alam J, Motterlini R. 2003. Curcumin activates the haem oxygenase-1 gene via regulation of Nrf2 and the antioxidant-responsive element. *Biochem J* 371:887–895.

Banning A, Deubel S, Kluth D, Zhou Z, Brigelius-Flohe R. 2005. The GI-GPx gene is a target for Nrf2. *Mol Cell Biol* 25:4914–4923.

Barik A, Mishra B, Shen L, Mohan H, Kadam RM, Dutta S, Zhang HY, Priyadarsini KI. 2005. Evaluation of a new copper(II)-curcumin complex as superoxide dismutase mimic and its free radical reactions. *Free Radic Biol Med* 39:811–822.

Bloom D, Dhakshinamoorthy S, Jaiswal AK. 2002. Site-directed mutagenesis of cysteine to serine in the DNA binding region of Nrf2 decreases its capacity to upregulate antioxidant response element-mediated expression and antioxidant induction of NAD(P)H:quinone oxidoreductase1 gene. *Oncogene* 21:2191–2200.

Bravo L. 1998. Polyphenols: Chemistry, dietary sources, metabolism, and nutritional significance. *Nutr Rev* 56:317–333.

Brown SL, Sekhar KR, Rachakonda G, Sasi S, Freeman ML. 2008. Activating transcription factor 3 is a novel repressor of the nuclear factor erythroid-derived 2-related factor 2 (Nrf2)-regulated stress pathway. *Cancer Res* 68:364–368.

Cao J, Jiang LP, Liu Y, Yang G, Yao XF, Zhong LF. 2007. Curcumin-induced genotoxicity and antigenotoxicity in HepG2 cells. *Toxicon* 49:1219–1222.

Chan K, Kan YW. 1999. Nrf2 is essential for protection against acute pulmonary injury in mice. *Proc Natl Acad Sci USA* 96:12731–12376.

Ciolino HP, Daschner PJ, Wang TT, Yeh GC. 1998. Effect of curcumin on the aryl hydrocarbon receptor and cytochrome P450 1A1 in MCF-7 human breast carcinoma cells. *Biochem Pharmacol* 56:197–206.

Corona-Rivera A, Urbina-Cano P, Bobadilla-Morales L, Vargas-Lares Jde J, Ramirez-Herrera MA, Mendoza-Magaua ML, Troyo-Sanroman R, Diaz-Esquivel P, Corona-Rivera JR. 2007. Protective in vivo effect of curcumin on copper genotoxicity evaluated by comet and micronucleus assays. *J Appl Genet* 48:389–396.

Cullinan SB, Zhang D, Hannink M, Arvisais E, Kaufman RJ, Diehl JA. 2003. Nrf2 is a direct PERK substrate and effector of PERK-dependent cell survival. *Mol Cell Biol* 23:7198–7209.

Daniel S, Limson JL, Dairam A, Watkins GM, Daya S. 2004. Through metal binding, curcumin protects against lead- and cadmium-induced lipid peroxidation in rat brain homogenates and against lead-induced tissue damage in rat brain. *J Inorg Biochem* 98:266–275.

Das KC, Das CK. 2002. Curcumin (diferuloylmethane), a singlet oxygen ((1)O(2)) quencher. *Biochem Biophys Res Commun* 295:62–66.

Dhakshinamoorthy S, Jaiswal AK. 2001. Functional characterization and role of INrf2 in antioxidant response element-mediated expression and antioxidant induction of NAD(P)H:quinone oxidoreductase1 gene. *Oncogene* 20:3906–3917.

Dhakshinamoorthy S, Jaiswal AK. 2000. Small maf (MafG and MafK) proteins negatively regulate antioxidant response element-mediated expression and antioxidant induction of the NAD(P)H:Quinone oxidoreductase1 gene. *J Biol Chem* 275:40134–40141.

Dinkova-Kostova AT, Talalay P. 2008. Direct and indirect antioxidant properties of inducers of cytoprotective proteins. *Mol Nutr Food Res* 52:S128–S138

Dinkova-Kostova AT, Talalay P. 1999. Relation of structure of curcumin analogs to their potencies as inducers of Phase 2 detoxification enzymes. *Carcinogenesis* 20:911–914.

Dinkova-Kostova AT, Holtzclaw WD, Kensler TW. 2005. The role of Keap1 in cellular protective responses. *Chem Res Toxicol* 18:1779–1791.

Dinkova-Kostova AT, Holtzclaw WD, Cole RN, Itoh K, Wakabayashi N, Katoh Y, Yamamoto M, Talalay P. 2002. Direct evidence that sulfhydryl groups of Keap1 are the sensors regulating induction of phase 2 enzymes that protect against carcinogens and oxidants. *Proc Natl Acad Sci USA* 99:11908–11913.

Dutta S, Murugkar A, Gandhe N, Padhye S. 2001. Enhanced antioxidant activities of metal conjugates of curcumin derivatives. *Met Based Drugs* 8:183–188.

Frankfurt OS, Lipchina LP, Bunto TV, Emanuel NM. 1967. [The effect of 4-methyl-2,6-di-tert.-butylphenol (ionol) on liver tumor induction in rats]. *Biull Eksp Biol Med* 64:86–88.

Friling RS, Bergelson S, Daniel V. 1992. Two adjacent AP-1-like binding sites form the electrophile-responsive element of the murine glutathione S-transferase Ya subunit gene. *Proc Natl Acad Sci USA* 89:668–672.

Garg R, Gupta S, Maru G. 2008. Dietary curcumin modulates transcriptional regulator(s) of phase I and phase II enzymes in benzo(a)pyrene-treated mice: Mechanism of its anti-initiating action. *Carcinogenesis* 29:1022–1032.

Goud VK, Polasa K, Krishnaswamy K. 1993. Effect of turmeric on xenobiotic metabolising enzymes. *Plant Foods Hum Nutr* 44:87–92.

He CH, Gong P, Hu B, Stewart D, Choi ME, Choi AM, Alam J. 2001. Identification of activating transcription factor 4 (ATF4) as an Nrf2-interacting protein. Implication for heme oxygenase-1 gene regulation. *J Biol Chem* 276:20858–20865.

Holtzclaw WD, Dinkova-Kostova AT, Talalay P. 2004. Protection against electrophile and oxidative stress by induction of phase 2 genes: The quest for the elusive sensor that responds to inducers. *Adv Enzyme Regul* 44:335–367.

Huang HC, Nguyen T, Pickett CB. 2002. Phosphorylation of Nrf2 at Ser-40 by protein kinase C regulates antioxidant response element-mediated transcription. *J Biol Chem* 277:42769–42774.

Hurst R, Bao Y, Jemth P, Mannervik B, Williamson G. 1998. Phospholipid hydroperoxide glutathione peroxidase activity of human glutathione transferases. *Biochem J* 332:97–100.

Iqbal M, Sharma SD, Okazaki Y, Fujisawa M, Okada S. 2003. Dietary supplementation of curcumin enhances antioxidant and phase II metabolizing enzymes in ddY male mice: Possible role in protection against chemical carcinogenesis and toxicity. *Pharmacol Toxicol* 92:33–38.

Itoh K, Wakabayashi N, Katoh Y, Ishii T, Igarashi K, Engel JD, Yamamoto M. 1999. Keap1 represses nuclear activation of antioxidant responsive elements by Nrf2 through binding to the amino-terminal Neh2 domain. *Genes Dev* 13:76–86.

Itoh K, Chiba T, Takahashi S, Ishii T, Igarashi K, Katoh Y, Oyake T, Hayashi N, Satoh K, Hatayama I, Yamamoto M, Nabeshima Y. 1997. An Nrf2/small Maf heterodimer mediates the induction of phase II detoxifying enzyme genes through antioxidant response elements. *Biochem Biophys Res Commun* 236:313–322.

Jain AK, Jaiswal AK. 2007. GSK-3beta acts upstream of Fyn kinase in regulation of nuclear export and degradation of NF-E2 related factor 2. *J Biol Chem* 282:16502–16510.

Jain AK, Bloom DA, Jaiswal AK. 2005. Nuclear import and export signals in control of Nrf2. *J Biol Chem* 280:29158–29168.

Jeong GS, Oh GS, Pae HO, Jeong SO, Kim YC, Shin MK, Seo BY, Han SY, Lee HS, Jeong JG, Koh JS, Chung HT. 2006. Comparative effects of curcuminoids on endothelial heme oxygenase-1 expression: ortho-methoxy groups are essential to enhance heme oxygenase activity and protection. *Exp Mol Med* 38:393–400.

Kataoka K, Noda M, Nishizawa M. 1994. Maf nuclear oncoprotein recognizes sequences related to an AP-1 site and forms heterodimers with both Fos and Jun. *Mol Cell Biol* 14:700–712.

Katsuoka F, Motohashi H, Ishii T, Aburatani H, Engel JD, Yamamoto M. 2005. Genetic evidence that small maf proteins are essential for the activation of antioxidant response element-dependent genes. *Mol Cell Biol* 25:8044–8051.

Ki SH, Cho IJ, Choi DW, Kim SG. 2005 Glucocorticoid receptor (GR)-associated SMRT binding to C/EBPbeta TAD and Nrf2 Neh4/5: Role of SMRT recruited to GR in GSTA2 gene repression. 2005. *Mol Cell Biol* 25:4150–4165.

Kobayashi A, Kang MI, Okawa H, Ohtsuji M, Zenke Y, Chiba T, Igarashi K, Yamamoto M. 2004. Oxidative stress sensor Keap1 functions as an adaptor for Cul3-based E3 ligase to regulate proteasomal degradation of Nrf2. *Mol Cell Biol* 24:7130–7139.

Kong AN, Owuor E, Yu R, Hebbar V, Chen C, Hu R, Mandlekar S. 2001. Induction of xenobiotic enzymes by the MAP kinase pathway and the antioxidant or electrophile response element (ARE/EpRE). *Drug Metab Rev* 33:255–271.

Kwak MK, Wakabayashi N, Greenlaw JL, Yamamoto M, Kensler TW. 2003. Antioxidants enhance mammalian proteasome expression through the Keap1-Nrf2 signaling pathway. *Mol Cell Biol* 23:8786–8794.

Kwak MK, Itoh K, Yamamoto M, Kensler TW. 2002. Enhanced expression of the transcription factor Nrf2 by cancer chemopreventive agents: Role of antioxidant response element-like sequences in the nrf2 promoter. *Mol Cell Biol* 22:2883–2292.

Lee OH, Jain AK, Papusha V, Jaiswal AK. 2007. An auto-regulatory loop between stress sensors INrf2 and Nrf2 controls their cellular abundance. *J Biol Chem* 282:36412–36420.

Lee-Hilz YY, Boerboom AM, Westphal AH, Berkel WJ, Aarts JM, Rietjens IM. 2006. Pro-oxidant activity of flavonoids induces EpRE-mediated gene expression. *Chem Res Toxicol* 19:1499–1505.

Li W, Yu SW, Kong AN. 2006. Nrf2 possesses a redox-sensitive nuclear exporting signal in the Neh5 transactivation domain. *J Biol Chem* 281:27251–27263.

Li W, Jain MR, Chen C, Yue X, Hebbar V, Zhou R, Kong AN. 2005. Nrf2 Possesses a redox-insensitive nuclear export signal overlapping with the leucine zipper motif. *J Biol Chem* 280:28430–28438.

Lin W, Shen G, Yuan X, Jain MR, Yu S, Zhang A, Chen JD, Kong AN. 2008. Regulation of Nrf2 transactivation domain activity by p160 RAC3/SRC3 and other nuclear co-regulators. *J Biochem Mol Biol* 39:304–310.

Mattson MP. 2008. Hormesis defined. *Ageing Res Rev* 7:1–7.

McNally SJ, Harrison EM, Ross JA, Garden OJ, Wigmore SJ. 2007. Curcumin induces heme oxygenase 1 through generation of reactive oxygen species, p38 activation and phosphatase inhibition. *Int J Mol Med* 19:165–172.

Menon VP, Sudheer AR. 2007. Antioxidant and anti-inflammatory properties of curcumin. *Adv Exp Med Biol* 595:105–125.

Motohashi H, Katsuoka F, Shavit JA, Engel JD, Yamamoto M. 2000. Positive or negative MARE-dependent transcriptional regulation is determined by the abundance of small Maf proteins. *Cell* 103:865–875.

Nguyen T, Sherratt PJ, Huang HC, Yang CS, Pickett CB. 2003. Increased protein stability as a mechanism that enhances Nrf2-mediated transcriptional activation of the antioxidant response element. Degradation of Nrf2 by the 26 S proteasome. *J Biol Chem* 278:4536–4541.

Nguyen T, Huang HC, Pickett CB. 2000. Transcriptional regulation of the antioxidant response element. Activation by Nrf2 and repression by MafK. *J Biol Chem* 275:15466–15473.

Nguyen T, Rushmore TH, Pickett CB. 1994. Transcriptional regulation of a rat liver glutathione S-transferase Ya subunit gene. Analysis of the antioxidant response element and its activation by the phorbol ester 12-O-tetradecanoylphorbol-13-acetate. *J Biol Chem* 269:13656–13662.

Nioi P, Nguyen T, Sherratt PJ, Pickett CB. 2005. The carboxy-terminal Neh3 domain of Nrf2 is required for transcriptional activation. *Mol Cell Biol* 25:10895–10906.

Nishinaka T, Ichijo Y, Ito M, Kimura M, Katsuyama M, Iwata K, Miura T, Terada T, Yabe-Nishimura C. 2007. Curcumin activates human glutathione S-transferase P1 expression through antioxidant response element. *Toxicol Lett* 170:238–247.

Oetari S, Sudibyo M, Commandeur JN, Samhoedi R, Vermeulen NP. 1996. Effects of curcumin on cytochrome P450 and glutathione S-transferase activities in rat liver. *Biochem Pharmacol* 51:39–45.

Ohta K, Ohigashi M, Naganawa A, Ikeda H, Sakai M, Nishikawa J, Imagawa M, Osada S, Nishihara T. 2007. Histone acetyltransferase MOZ acts as a co-activator of Nrf2-MafK and induces tumour marker gene expression during hepatocarcinogenesis. *Biochem J* 402:559–566.

Prochaska HJ, Talalay P. 1988. Regulatory mechanisms of monofunctional and bifunctional anticarcinogenic enzyme inducers in murine liver. *Cancer Res* 48:4776–4782.

Priyadarsini KI, Maity DK, Naik GH, Kumar MS, Unnikrishnan MK, Satav JG, Mohan H. 2003. Role of phenolic O-H and methylene hydrogen on the free radical reactions and antioxidant activity of curcumin. *Free Radic Biol Med* 35:475–484.

Pugazhenthi S, Akhov L, Selvaraj G, Wang M, Alam J. 2007. Regulation of heme oxygenase-1 expression by demethoxy curcuminoids through Nrf2 by a PI3-kinase/

Akt-mediated pathway in mouse beta-cells. *Am J Physiol Endocrinol Metab* 293:E645–E655.

Rojo AI, Sagarra MR, Cuadrado A. 2008. GSK-3beta down-regulates the transcription factor Nrf2 after oxidant damage: Relevance to exposure of neuronal cells to oxidative stress. *J Neurochem* 105:192–202.

Rushmore TH, Pickett CB. 1990. Transcriptional regulation of the rat glutathione S-transferase Ya subunit gene. Characterization of a xenobiotic-responsive element controlling inducible expression by phenolic antioxidants. *J Biol Chem* 265:14648–14653.

Rushmore TH, Morton MR, Pickett CB. 1991. The antioxidant responsive element. Activation by oxidative stress and identification of the DNA consensus sequence required for functional activity. *J Biol Chem* 266:11632–11639.

Rushworth SA, Ogborne RM, Charalambos CA, O'Connell MA. 2006. Role of protein kinase C delta in curcumin-induced antioxidant response element-mediated gene expression in human monocytes. *Biochem Biophys Res Commun* 341:1007–1016.

Scapagnini G, Colombrita C, Amadio M, D'Agata V, Arcelli E, Sapienza M, Quattrona A, Calabrese V. 2006. Curcumin activates defensive genes and protects neurons against oxidative stress. *Antioxid Redox Signal* 8:395–403.

Sharma OP. 1976. Antioxidant activity of curcumin and related compounds. *Biochem Pharmacol* 25:1811–1812.

Shen G, Xu C, Hu R, Jain MR, Gopalkrishnan A, Nair S, Huang MT, Chan JY, Kong AN. 2006. Modulation of nuclear factor E2-related factor 2-mediated gene expression in mice liver and small intestine by cancer chemopreventive agent curcumin. *Mol Cancer Ther* 5:39–51.

Shen G, Hebbar V, Nair S, Xu C, Li W, Lin W, Keum YS, Han J, Gallo MA, Kong AN. 2004. Regulation of Nrf2 transactivation domain activity. The differential effects of mitogen-activated protein kinase cascades and synergistic stimulatory effect of Raf and CREB-binding protein. *J Biol Chem* 279:23052–23060.

Singh A, Singh SP, Bamezai R. 1995. Postnatal modulation of hepatic biotransformation system enzymes via translactational exposure of F1 mouse pups to turmeric and curcumin. *Cancer Lett* 96:87–93.

Stevenson DE, Hurst RD. 2007. Polyphenolic phytochemicals—just antioxidants or much more? *Cell Mol Life Sci* 64:2900–2916.

Strimpakos AS, Sharma RA. 2008. Curcumin: Preventive and therapeutic properties in laboratory studies and clinical trials. *Antioxid Redox Signal* 10:511–545.

Sumanont Y, Murakami Y, Tohda M, Vajragupta O, Matsumoto K, Watanabe H. 2004. Evaluation of the nitric oxide radical scavenging activity of manganese complexes of curcumin and its derivative. *Biol Pharm Bull* 27:170–173.

Sun Z, Zhang S, Chan JY, Zhang DD. 2007. Keap1 controls postinduction repression of the Nrf2-mediated antioxidant response by escorting nuclear export of Nrf2. *Mol Cell Biol* 27:6334–6349.

Susan M, Rao MN. 1992. Induction of glutathione S-transferase activity by curcumin in mice. *Arzneimittelforschung* 42:962–964.

Theodore M, Kawai Y, Yang J, Kleshchenko Y, Reddy SP, Villalta F, Arinze IJ. 2008. Multiple nuclear localization signals function in the nuclear import of the transcription factor Nrf2. *J Biol Chem* 283:8984–8994.

Tourkina E, Gooz P, Oates JC, Ludwicka-Bradley A, Silver RM, Hoffman S. 2004. Curcumin-induced apoptosis in scleroderma lung fibroblasts: role of protein kinase cepsilon. *Am J Respir Cell Mol Biol* 31:28–35.

Tong KI, Kobayashi A, Katsuoka F, Yamamoto M. 2006. Two-site substrate recognition model for the Keap1-Nrf2 system: a hinge and latch mechanism. *Biol Chem* 387:1311–1320.

Usta M, Wortelboer HM, Vervoort J, Boersma MG, Rietjens IM, van Bladeren PJ, Cnubben NH. 2007. Human glutathione S-transferase-mediated glutathione conjugation of curcumin and efflux of these conjugates in Caco-2 cells. *Chem Res Toxicol* 20:1895–1902.

Vajragupta O, Boonchoong P, Watanabe H, Tohda M, Kummasud N, Sumanont Y. 2003. Manganese complexes of curcumin and its derivatives: Evaluation for the radical scavenging ability and neuroprotective activity. *Free Radic Biol Med* 35: 1632–1644.

Velichkova M, Hasson T. 2005. Keap1 regulates the oxidation-sensitive shuttling of Nrf2 into and out of the nucleus via a Crm1-dependent nuclear export mechanism. *Mol Cell Biol* 25:4501–4513.

Venugopal R, Jaiswal AK. 1996. Nrf1 and Nrf2 positively and c-Fos and Fra1 negatively regulate the human antioxidant response element-mediated expression of NAD(P)H:-quinone oxidoreductase1 gene. *Proc Natl Acad Sci USA* 93:14960–14965.

Wakabayashi N, Dinkova-Kostova AT, Holtzclaw WD, Kang MI, Kobayashi A, Yamamoto M, Kensler TW, Talalay P. 2004. Protection against electrophile and oxidant stress by induction of the phase 2 response: Fate of cysteines of the Keap1 sensor modified by inducers. *Proc Natl Acad Sci USA* 101:2040–2045.

Wakabayashi N, Itoh K, Wakabayashi J, Motohashi H, Noda S, Takahashi S, Imakado S, Kotsuji T, Otsuka F, Roop DR, Harada T, Engel JD, Yamamoto M. 2003. Keap1-null mutation leads to postnatal lethality due to constitutive Nrf2 activation. *Nat Genet* 35:238–245.

Wang X, Tomso DJ, Chorley BN, Cho HY, Cheung VG, Kleeberger SR, Bell DA. 2007. Identification of polymorphic antioxidant response elements in the human genome. *Hum Mol Genet* 16:1188–1200.

Wasserman WW, Fahl WE. 1997. Functional antioxidant responsive elements. *Proc Natl Acad Sci USA* 94:5361–5366.

Watai Y, Kobayashi A, Nagase H, Mizukami M, McEvoy J, Singer JD, Itoh K, Yamamoto M. 2007. Subcellular localization and cytoplasmic complex status of endogenous Keap1. *Genes Cells* 12:1163–1178.

Wattenberg LW. 1972. Inhibition of carcinogenic and toxic effects of polycyclic hydrocarbons by phenolic antioxidants and ethoxyquin. *J Natl Cancer Inst* 48:1425–1430.

Wild AC, Mulcahy RT. 2000. Regulation of gamma-glutamylcysteine synthetase subunit gene expression: Insights into transcriptional control of antioxidant defenses. *Free Radic Res* 32:281–301.

Yoshino M, Haneda M, Naruse M, Htay HH, Tsubouchi R, Qiao SL, Li WH, Murakami K, Yokochi T. 2004. Prooxidant activity of curcumin: Copper-dependent formation of 8-hydroxy-2′-deoxyguanosine in DNA and induction of apoptotic cell death. *Toxicol in vitro* 18:783–789.

Yu S, Kong AN. 2007. Targeting carcinogen metabolism by dietary cancer preventive compounds. *Curr Cancer Drug Targets* 7:416–424.

Zhang DD, Hannink M. 2003. Distinct cysteine residues in Keap1 are required for Keap1-dependent ubiquitination of Nrf2 and for stabilization of Nrf2 by chemopreventive agents and oxidative stress. *Mol Cell Biol* 23:8137–8151.

Zhang J, Ohta T, Maruyama A, Hosoya T, Nishikawa K, Maher JM, Shibahara S, Itoh K, Yamamoto M. 2006. BRG1 interacts with Nrf2 to selectively mediate HO-1 induction in response to oxidative stress. *Mol Cell Biol* 26:7942–7952.

Zhang DD, Lo SC, Sun Z, Habib GM, Lieberman MW, Hannink M. 2005. Ubiquitination of Keap1, a BTB-Kelch substrate adaptor protein for Cul3, targets Keap1 for degradation by a proteasome-independent pathway. *J Biol Chem* 280:30091–30099.

Zoete V, Rougee M, Dinkova-Kostova AT, Talalay P, Bensasson RV. 2004. Redox ranking of inducers of a cancer-protective enzyme via the energy of their highest occupied molecular orbital. *Free Radic Biol Med* 36:1418–1423.

18 Phenolics in Aging and Neurodegenerative Disorders

VITTORIO CALABRESE[1], MARZIA PERLUIGI[2], CAROLIN CORNELIUS[1], RAFFAELLA COCCIA[2], FABIO DI DOMENICO[2], GIOVANNI PENNISI[3], CHIARA CINI[2], AND ALBENA T. DINKOVA-KOSTOVA[4]

[1]Department of Chemistry, Faculty of Medicine, University of Catania, Catania, Italy

[2]Department of Biochemical Sciences, University of Rome "Sapienza,", Rome, Italy

[3]Department of Neurosciences, Faculty of Medicine, University of Catania, Catania, Italy

[4]Biomedical Research Centre, Ninewells Hospital and Medical School, University of Dundee, Scotland, United Kingdom, and Department of Pharmacology and Molecular Sciences and Department of Medicine, Johns Hopkins University, Baltimore, Maryland

CONTENTS

Plant Phenolics and Human Health: Biochemistry, Nutrition, and Pharmacology,
Edited by Cesar G. Fraga.

PHYTOCHEMICALS AND NEURODEGENERATIVE DISEASES

Phytochemicals are bioactive chemicals derived from plants, including flavonoids, carotenoids, and vitamins, that exert multiple beneficial effects on health. Epidemiological studies strongly indicate that fruits and vegetables are protective against several diseases, particularly cardiovascular disease, cancer, and neurodegeneration. Special interest has been assigned to the therapeutic role of natural antioxidants, such as phytochemicals, present in medicinal and dietary plants, which might help prevent oxidative damage [Halliwell, 1999; Youdim et al., 2002]. In situations of increased reactive oxygen species (ROS) generation, such as the aging process, the supplementation of endogenous antioxidants via intake of dietary antioxidants may be of particular importance in slowing the accumulation of oxidatively damaged molecules.

Numerous in vitro studies have implicated ROS in neuronal death [Leonardi and Mytilineou, 1998], and different markers of oxidative stress are found in postmortem examination of brain tissues from patients with neurodegenerative disorders [Sayre et al., 2001]. DNA oxidation, protein oxidation, and lipid peroxidation have been reported in brain regions containing neurofibrillary tangles and senile plaques from Alzheimer's disease (AD) patients [Lovell and Markesbery, 2007; Polidori et al., 2007]. Dopaminergic neurons in the substantia nigra of brains of Parkinson's disease (PD) patients also exhibit hallmarks of oxidative stress [Giasson et al., 2002].

The oxidative stress hypothesis for neurodegeneration postulates that cumulative oxidative damage over time explains the late-life onset and the slowly progressive nature of neurodegenerative disorders [Kregel and Zhang, 2007]. The central nervous system (CNS) is especially vulnerable to free-radical damage as a result of its high oxygen consumption rate, its abundant lipid content, and the low levels of antioxidant enzymes compared with other tissues [Sultana et al., 2006b]. These age-related effects, associated with genetic and environmental risk factors, may provide a merging mechanism for the high incidence of neurodegenerative disorders in the elderly.

In line with these evidences, particular attention has been paid to studying the neuroprotective properties of antioxidants, iron chelating and anti-inflammatory agents with particular regard to polyphenols [Mandel et al., 2006b]. Numerous studies have shown that polyphenols have in vitro and in vivo activities in preventing and/or reducing the extent of free-radical-induced oxidative damage in different models of neurodegeneration [Mandel and Youdim, 2004; Pan et al., 2003; Scalbert et al., 2005]. The molecular mechanisms for neuroprotection do not merely rely on a direct radical scavenging activity but also polyphenols may function at several cellular levels, including direct interaction and modulation of enzymatic activities and the regulation of signaling pathways with implications for cell survival and death [Ishige et al., 2001; Pan et al., 2000; Dinkova-Kostova and Talalay, 2008]. Epidemiological studies indicated that moderate consumption of wine could be associated with a lower incidence of AD. Resveratrol (Fig.18.1), a polyphenol that occurs in abundance in red grapes and is present mostly in red wines, exerts antioxidant and neuroprotective properties and therefore contributes to the potential beneficial effect of wine consumption on the neurodegenerative process (Ates et al., 2007].

Curcumin (Fig.18.1), a polyphenol found in the curry spice turmeric, was shown to possesses many therapeutic properties including anti-inflammatory [Biswas and Rahman, 2008] and anticancer activities [Choi et al., 2008]. It is currently receiving strong attention due to its antioxidant potential as well as its relatively low toxicity to rodents. In addition to being an exceedingly potent direct antioxidant, curcumin is also an indirect antioxidant, that is, it induces the gene expression of cytoprotective (phase 2) proteins [Dinkova-Kostova and Talalay, 1999; Dinkova-Kostova, 2002]. This indirect antioxidant activity is due to the presence of the electrophilic centers on its molecule that can undergo Michael addition reactions with sulfhydryl groups. Indeed, tetrahydrocurcumin, which lacks these electrophilic centers, is a much weaker inducer [Dinkova-Kostova and Talalay, 1999]. Using model compounds, it was shown that Michael acceptors related to curcumin chemically modify certain highly reactive sulfhydryl groups of the intracellular sensor for inducers, Keap1 [Dinkova-Kostova et al., 2002]. Consequently, Keap1 loses its ability to repress transcription factor Nrf2, which then undergoes nuclear translocation and (as heterodimer with a small Maf protein) binds to the antioxidant response element (ARE) and activates transcription of cytoprotective (phase 2) genes [Dinkova-Kostova et al., 2005].

In light of the antioxidant, anti-inflammatory, and antiamyloid effects, curcumin has become a candidate compound for the prevention or treatment of AD [Calabrese et al., 2007]. Treatment of astrocytes with curcumin induced the cytoprotective proteins HO-1, NAD(P)H:quinone oxidoreductase 1 (NQO1), and glutathione transferase (GST) and provided protection against

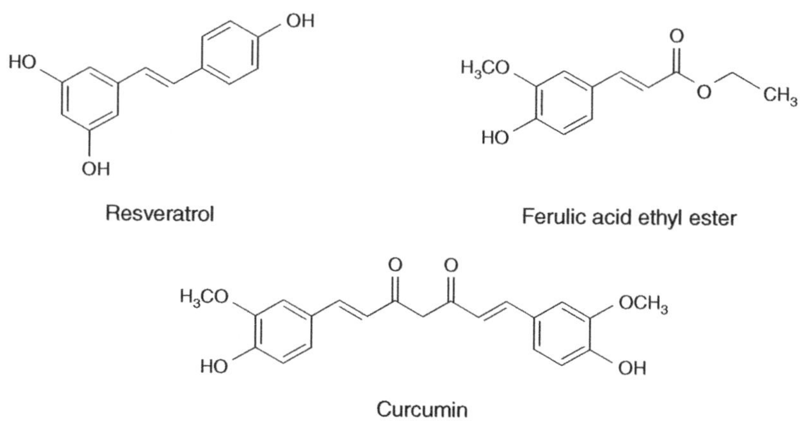

Figure 18.1 Chemical structures of phenolic compounds with neuroprotective activity.

subsequent glucose oxidase-mediated oxidative damage [Scapagnini et al., 2006]. Lim and co-workers [2001] studied the effects of curcumin in transgenic mice, which carry a human mutation in the amyloid precursor protein (APPsw) that causes AD pathology. They found decreased cerebral levels of oxidized proteins, decreased levels of interleukin-1b (IL-1b) and reduced levels of soluble and insoluble Aβ. Subsequently, it was shown that curcumin crossed the blood–brain barrier, bound to amyloid plaques, and reduced amyloid levels and plaque burden [Yang et al., 2005]. In contrast to curcumin, tetrahydro-curcumin, which lacks the central electrophilic centers required for induction of cytoprotective (phase 2) proteins, did not reduce plaque deposition or protein oxidation, although it decreased neuroinflammation and soluble Aβ [Begum et al., 2008]. Further studies are needed to test the potential therapeutic use of curcumin in the treatment or prevention of AD in humans.

Green tea, another food high in polyphenols, may also be neuroprotective. The effects of acute and chronic administration of the green tea polyphenol (−)-epigallocatechin-3-gallate (EGCG) in various cellular and animal models of PD and AD revealed various novel molecular targets including the extracellular mitogen-activated protein kinases (MAPK), protein kinase C (PKC), antioxidant enzymes, and survival genes and proteins associated with mitochondrial function, such as Bcl-2 family members [Kalfon et al., 2007; Mandel et al., 2006a]. Recently, Weinreb et al. [2007] showed by a proteomic analysis that EGCG treatment on SHSY5Y, after serum starvation, affected the expression levels of diverse proteins, including proteins related to cytoskeletal components, metabolism, heat shock, and binding proteins [Weinreb et al., 2007].

FERULIC ACID ETHYL ESTER

In search of novel natural antioxidant compounds that might posses a good brain bioavailability, our laboratory has focused attention on the phenolic compound ferulic acid ethyl ester (FAEE) (Fig. 18.1). Ferulic acid is a ubiquitous plant constituent that occurs primarily in seeds and leaves both in its free form and covalently linked to lignin and other biopolymers. Due to its phenolic nucleus and an extended side chain conjugation, it readily forms a resonance stabilized phenoxy radical that accounts for its potent antioxidant potential [Kanski et al., 2002; Kikuzaki et al., 2002]. Ferulic acid has been shown to be protective against oxidative stress in vitro; it is absorbed and excreted by humans, and may be a promising candidate for therapeutic intervention in AD [Yan et al., 2001]. Although ferulic acid has been demonstrated to be effective in vitro, the low lipophilicity impairs its in vivo efficiency, bioavailability, and stability.

In addition to their direct antioxidant activity, ferulic acid and its ester derivatives have also been shown to act as indirect antioxidants, that is, to induce cytoprotective (phase 2) enzymes [Dinkova-Kostova et al., 1998; Dinkova-Kostova, 2002]. Similarly to curumin, ferulic acid derivatives also contain Michael acceptor functionalities and have been shown to react with sulfhydryl groups [Dinkova-Kostova et al., 2001]. FAEE, a naturally occurring ester derivative of ferulic acid, is a trace constituent of various plants [Clifford et al., 2007]. Due to its high lipophilicity, FAEE has been shown to present better scavenging properties toward both hydroxyl radicals and superoxide anions in comparison with the corresponding acid form [Scapagnini et al., 2004]. In addition to the radical scavenging activity of an antioxidant, both its polarity and its three-dimensional interaction with lipid bilayers may contribute to strengthen its antioxidant potential. Since synaptic membranes are particularly vulnerable to oxidative stress and are a selective target of Aβ-mediated neurotoxicity, the affinity of FAEE with lipid substrates (because of its high lipophilicity) may result in higher protective effects. Considering the increased lipophilicity of FAEE, Butterfield et al. showed in in vitro or ex vivo studies the decreased protein oxidation and lipid peroxidation, decreased free-radical formation, decreased 3-NT formation, improved morphology and elevated mitochondrial function in primary neuronal cultures treated with Aβ(1-42) [Sultana et al., 2005] and on synaptosomes isolated from FAEE-treated rodents and subsequently treated with Aβ(1-42) [Perluigi et al., 2006b] or with Fe^{2+}/H_2O_2 (hydroxyl radical formation), and AAPH (alkoxyl and peroxyl radical formation) [Joshi et al., 2006].

The mechanistic basis of the neuroprotective activity of FAEE appears to rely not only on its general free-radical trapping or antioxidant activity per se, but also on its activity in mediating the induction of stress response proteins (HO-1 and HSP72), cytoprotective (phase 2) proteins, and the parallel suppression of genes induced by pro-inflammatory cytokines, such as nitric oxide synthase (iNOS).

In summary, FAEE protects neuronal cells and synaptosomes against oxidative stress and neurotoxicity as shown by a number of oxidative stress markers. We hypothesize a multifaceted mechanism by which FAEE offers neuroprotection on in vitro models of AD:

- FAEE is a potent free-radical scavenger that significantly attenuates ROS production, protein oxidation, and lipid peroxidation.
- FAEE is also neuroprotective by leading to elevated levels of stress response proteins, such as HO-1 and HSP72.
- FAEE modulates neuroinflammatory processes mediated by iNOS.

Further studies are required to gain insight into the potential therapeutic use of FAEE in the treatment of AD and other oxidative stress-related disorders. In this effort, Butterfield and co-workers [2006b] have used redox proteomics approach to identify selective targets of oxidative damage in different human and animal models of neurodegenerative diseases. Furthermore, redox proteomics may help to elucidateg the potential therapeutic use of phenolics and other phytochemicals for prevention and treatment of neurodegeneration.

REDOX PROTEOMICS

In the postgenome era, proteomics has attracted much more attention due to its potential in understanding biological functions and structures at the protein level. The increased possibility of screening using proteomic techniques provides more comprehensive overview of the interaction of proteins, the interplay among processes, and the context in which a specific molecule or pathway may be operating in the mechanism of action of different antioxidants.

"Proteome" derives from the combination of the terms proteins and genome and defines "the PROTEin complement expressed by a genOME" [Wilkins et al., 1996]. Proteomic analysis (Fig. 18.2) employs: (i) systematic separation of proteins from complex mixture, (ii) protein identification, (iii) protein quantitation, and (iv) structural and functional information of identified proteins. Two-dimensional polyacrylamide gel electrophoresis (2D PAGE) [O'Farrell, 1975] is currently the most powerful protein separation method for the resolution of complex mixtures of proteins, permitting the simultaneous analysis of thousands of gene products. Proteins are separated according to their charge (pI) by isoelectric focusing (IEF) in the first dimension and according to their size (Mr) by SDS-PAGE in the second dimension. The introduction of immobilized pH gradients (IPG) for IEF has overcome the problems of pH gradient instability (caused by prolonged focusing time). IPGs allow the generation of pH gradients of any desired range (broad, narrow, or ultra-narrow) between pH 3 and 12. In addition, the use of pH gradient allowed high reproducibility and- large-scale separations. However, some limitations still remain to be solved, including solubilization of membrane proteins, identification of low-abundance proteins, and identification of highly hydrophobic proteins. The protein maps allow comparison of different sets of samples in terms of profiling of isoforms, splice variants, mutants, and posttranslationally modified species, and the definition of protein–protein interactions, and the like, using a computer-assisted program. Each spot on the resulting two-dimensional gel corresponds to a single protein species in the sample [Tilleman et al., 2002]. Thousands of different proteins can thus be separated, and information such as the protein pI, the apparent molecular weight, and the amount of each protein is obtained. 2D-PAGE represents undoubtedly one of the most-used techniques for protein separation, but non-SDS-PAGE methods are also employed. Two-dimensional high–pressure liquid chromatography (2D-HPLC) achieves separation of a protein mixture by eluting the sample through a series of columns with a coupled MS analysis [Wagner et al., 2003]. Alternatively, an expression profiling can also be obtained by labeling at cysteine residues with light and heavy, respectively, tags carrying a biotin moiety. The labeled proteins are then mixed and digested and the resulting isotopic labeling is analyzed by a mass spectrometer. This technique is referred as isotopically coded affinity tags (ICAT) [Smolka et al., 2001]. However, a number of limitations to the ICAT technique has been noted in the literature, including missed identification of proteins with few or no cysteine residue, lost information for posttranslational modifications, differential reversed-phase elution of identical peptides labeled with the hydrogen/deuterium isotope pairs, and complicated interpretation of tandem mass spectrometry (MS/MS) spectra due to the addition of the biotin group.

The protein posttranslational modifications (PTMs) play a crucial role in modifying the end product of expression and contribute towards biological processes and diseased conditions. Important posttranslational modifications include phosphorylation, acetylation, glycosylation, ubiquitination, and nitration [Mann and Jensen, 2003]. The analysis of posttranslational modifications on a proteome scale is still considered an analytical challenge [Zhou et al., 2001] because of the extremely low abundance of modified proteins among very complex proteome samples.

Among several types of posttranslational protein modifications, phosphorylation and oxidative modifications play essential roles in the regulation of a variety of cell functions. Increasing evidence suggests that changes in protein modifications over time correlate with particular phenotypes and disease states. Reactions of free radicals and ROS and reactive nitrogen species with proteins lead to oxidative modifications such as formation of protein hydroperoxides, hydroxylation of aromatic groups and aliphatic amino acid side chains, nitration of aromatic amino acid residues, oxidation of sulfhydryl groups, oxidation of methionine residues, conversion of some amino acid residues into carbonyl groups, cleavage of the polypeptide chain and formation of cross-linking bonds. If oxidized proteins are not appropriately repaired or removed from cells, they are often toxic and can threaten cell viability [Berlett and Stadtman, 1997]. Numerous studies demonstrated the harmful effects of irreversible oxidative PTM as a result of oxidative stress and increased levels of oxidatively modified proteins have been shown to correlate with aging [Chakravarti and Chakravarti, 2007]. Oxidative modifications of proteins lead to loss of their function, enzymatic activity, accumulation, and inhibition of their degradation. All these metabolic dysfunctions have been observed in several human degenerative diseases such as cancer [Valko et al., 2006] and neurodegeneration [Butterfield et al., 2006a]. Based on these findings, growing interest is currently given to better understand selective protein target of oxidative damage. Recently, redox proteomics approach has been employed to identify posttranslational modification of proteins caused by oxidative damage. Protein oxidation results by the covalent modification induced by ROS or by-products of oxidative stress. Among general types of protein modifications, we focused our attention on protein carbonyls, lipid peroxidation adducts [4-hydroxynonenal (HNE)-adducts], glutathionylation, and nitration of tyrosine residues. Redox proteomics coupled 2D-PAGE with immunochemical detection of protein carbonyl derivatized by 2,4-dinitrophenyhydrazine (DNPH), nitrated proteins indexed by 3-nitrotyrosine (3-NT), glutathionylated proteins (GSH-bound proteins) and HNE-bound proteins followed by MS analysis.

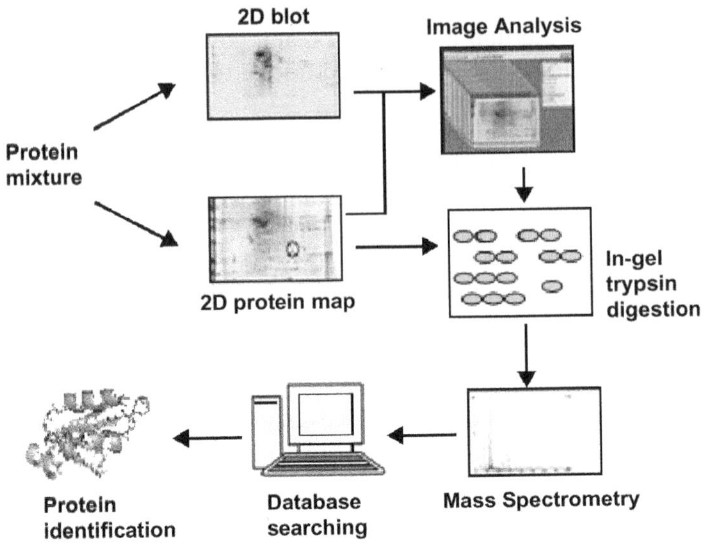

Figure 18.2 Redox proteomics and oxidatively modified proteins in the brain. Redox proteomics has the potential of detecting disease markers and identifying potential targets for drug therapy in neurodegenerative disorders. Redox proteomics involves the separation of brain proteins followed by detection, usually immunochemically, of oxidatively modified proteins, either from a two-dimensional western blot or from column eluents. Subsequent mass spectrometric analysis of tryptic digests and database searching leads to protein identification.

A 2D western blot map is achieved by using specific antibodies, for example, anti-DNP, anti-3-NT, anti-GSH, or anti-HNE, which react with those proteins containing reactive carbonyl groups/3-NT/GSH/HNE. 2D gel images are used to obtain the protein expression profile, and the 2D western blots are analyzed by image software (PD Quest, BioRad). This sophisticated software offers powerful comparative analysis and is specifically designed to analyze many gels or blots at once that were performed under identical experimental conditions. Powerful automatching algorithms quickly and accurately match gels or blots and sophisticated statistical analysis tools identify experimentally significant spots. The principles of measuring intensity values by 2D analysis software are similar to those of densitometric measurements. After completion of spot matching, the normalized intensity of each protein spot from individual gels (or membranes) is compared between groups using statistical analysis [Butterfield et al., 2006b].

Sultana et al. [2006a] employed this experimental approach to investigate the protective effect of the xanthate D609 against amyloid β(1-42)-induced oxidative modification of neuronal proteins. Pretreatment of primary neuronal cultures with D609 prior to amyloid β(1-42) prevented the oxidation of glyceraldehyde-3-phosphate dehydrogenase, pyruvate kinase, malate dehydrogenase, and 14-3-3 zeta as shown in the cells treated with amyloid β(1-42) alone. The information obtained from redox proteomics studies may be helpful in understanding the molecular mechanisms involved in the development and progression of AD disease as well as of other neurodegenerative disorders.

NEURODEGENERATION AND LIPID OXIDATION: ISOPROSTANES

Increasing evidences indicate that ROS-mediated reactions, particularly of neuronal lipids, are extensive in brain areas directly involved in the neurodegenerative process [Andersen, 2004; Christen, 2000]. In recent years, numerous studies investigated the functional importance of oxidative imbalance as a crucial event in mediating the pathogenesis of several neurodegenerative disorders such as AD, PD, amyotrophic lateral sclerosis (ALS), and prion disease [Andersen, 2004; Butterfield et al., 2006b; Moreira et al., 2005]. The availability of specific and sensitive markers to monitor in vivo oxidative stress, in combination with studies performed in living patients, are helping to elucidate these issues.

Lipid peroxidation is one of the major sources of free-radical mediated injury that directly damages membranes and generates a number of secondary products. In particular, markers of lipid peroxidation have been found to be elevated in brain tissues and body fluids in several neurodegenerative diseases, and the role of lipid peroxidation has been extensively discussed in the context of their pathogenesis. Peroxidation of membrane lipids can have numerous effects, including increased membrane rigidity, decreased activity of membrane-bound enzymes (e.g., sodium pumps), altered activity of membrane receptors, and altered permeability [Anzai et al., 1999; Yehuda et al., 2002]. In addition to effects on phospholipids, lipid-initiated radicals can also directly attack membrane proteins and induce lipid–lipid, lipid–protein, and protein–protein cross-linking, all of which obviously have effects on membrane function.

Malondialdehyde (MDA), HNE, and acrolein represent the major products of lipid peroxidation [Esterbauer et al., 1991]. In addition, lipid hydroperoxyl radicals undergo endocyclization to produce fatty acid esters; two classes of these cyclized fatty acids are isoprostanes and neuroprostanes. Isoprostanes (IsoPs) are products of arachidonic acid (AA) metabolism of potential relevance to human neurodegenerative diseases [Greco et al., 2000]. They are produced in vivo by free-radical-induced peroxidation of esterified AA in membrane phospholipids, independently of cyclooxygenase enzymes. F_2-isoprostanes (F_2-IsoPs) are the stable products of free-radical damage to AA. However, F_2-isoprostane analogs may be formed by peroxidation of other polyunsaturated fatty acid substrates, such as eicosapentaenoic acid, which produces F_3-isoprostanes, and docosahexaenoic acid, which generates F_4-isoprostanes. The latter compounds are also termed neuroprostanes, due to the high levels in the brain (Roberts et al., 1998). In contrast to cyclooxygenase-derived prostaglandins, nonenzymatic generation of isoprostanes favors the formation of compounds in which the stereochemistry of the side chains is oriented *cis* in relation to the prostane ring. Molecular modeling of isoprostane containing phospholipids shows that they are remarkably distorted, nonflexible molecules. Therefore, the presence of isoprostanes can compromise membrane integrity and fluidity.

After their cleavage, probably by phospholipases, F_2-IsoPs and F_4-IsoPs are released in free form and may participate, due to their biological activity, as mediators of the oxidant injury. Measurement of F_2-IsoPs is one the most reliable approach to assess oxidative stress status in vivo, providing an important tool to explore the role of oxidative stress in the pathogenesis of human neurodegenerative diseases [Pratico et al., 1998]. Moreover, F_2-IsoPs and other products of the IsoP pathway exert potent biological actions both via receptor-dependent and independent mechanisms and therefore may be pathophysiological mediators of disease. The measurement of IsoP levels in the cerebrospinal fluid (CSF) reflects the metabolic activity of the brain and therefore may provide a unique opportunity to reveal the occurrence of oxidative stress and lipid peroxidation in the living brain.

Isoprostanes have been successfully utilized to investigate the occurrence of oxidative stress in some neurological diseases such as AD [Montine et al., 2007; Pratico et al., 1998], multiple sclerosis [Volpe et al., 2006], and Huntington's disease [Montine et al., 1999].

F_2-IsoP concentration is selectively increased in diseased regions of brain from patients who died from advanced AD, where pathologic changes include amyloid Aβ-peptide deposition, neurofibrillary tangle formation, and extensive neuron death [Montine et al., 2002]. There is broad agreement that increased CSF levels of F_2-IsoPs also are present in patients with early AD. It has been recently shown that subjects with MCI (mild cognitive impairment) have plasma, urine, and CSF levels of F_2-IsoPs higher than healthy subjects [Pratico et al., 2002]. It has been suggested that measurement of IsoP may identify a subgroup of patients with MCI with elevated lipid peroxidation who are at increased risk to progress to symptomatic AD [Markesbery et al., 2005].

In addition, the contribution of oxidative stress to the pathogenesis of Huntington's disease has been studied by measuring the levels of F_2-IsoPs in the CSF of 20 patients in the early phase of the disease [Montine et al., 1999]. F_2-IsoP concentration was moderately but significantly higher in Huntington's disease patients than in the control group. Current data available on CSF levels of IsoPs in different neurodegenerative disorders proved the usefulness of these markers in the understanding of the role of oxidative damage in neurological diseases.

However, it is unexplored if they might mediate the functional consequences of the lipid peroxidation process, and further studies are needed to investigate their biological impact and to better understand the role of F_2-IsoPs in brain diseases. Growing interest is currently given to the possible exploitation of F_2-IsoP determinations in biological fluids, in particular in the CSF, for clinical purposes. The attraction of urinary IsoPs is that they might serve as noninvasively acquired biomarkers, which could guide the selection of patients and dosing for trials of antioxidants as well as the timing of dosing. This may prove particularly useful in a disease, such as AD, in which the rate of progression varies markedly between individuals.

GLUTATHIONE AND REDOX HOMEOSTASIS

Glutathione (GSH) is the most abundant nonprotein thiol of the cell that performs various functions ranging from cellular metabolism to transport as well as protection against free radicals and ROS. The antioxidant function of GSH is related to oxidation of the thiol group of its cysteine residue with formation of GSSG, which is catalytically reduced back to the thiol form (GSH) by glutathione reductase. In addition, GSH provides reducing equivalent for enzymes involved in the metabolism of ROS, eliminates potentially toxic oxidation products and reduces oxidized protein thiols. Increased ROS production causes an imbalance between GSH and its corresponding disulfide (GSSG) shifting the GSH/GSSG redox toward the oxidizing state [GSH/GSSG ratio from 100 (normal redox balance) to 10 or even 1] associated with protein mixed disulfide formation [Klatt and Lamas, 2000]. As well as forming disulfides itself, GSH can also form mixed disulfides with cysteines on proteins, a process that is termed *S*-glutathionylation. This covalent modification is a posttranslational modification of proteins that occurs in cells exposed to oxidative stress that can be reversed by enzymatic systems, such as glutaredoxins (Shelton et al., 2005). Indeed, protein thiol groups are highly susceptible to oxidative modifications and they can be reversibly oxidized to sulfenic acid (–SOH) or irreversibly oxidized to sulfinic ($–SO_2H$) and sulfonic ($–SO_3H$) acid [Chang et al., 2004]. Protein-mixed disulfide formation could potentially serve as protection from irreversible, more dangerous oxidation of cysteine residues [Souza and Radi, 1998]. Accumulating evidence suggests that this mechanism plays a role in regulating protein function in addition to ensure a physiologic reducing micro-environment to the cell. Indeed, protein *S*-glutathiolation is not only a cellular response to oxidative damage but it occurs under normal conditions to regulate protein function at several levels. Growing knowledge in this area now proposes that *S*-glutathiolation may serve to transduce oxidative stimuli into a functional response at various levels of cellular signaling. Increasing evidences suggest that protein *S*-glutathionylation, being characterized by sensitivity, specificity, and reversibility, might be a common mechanism for the global regulation of protein functions in different cellular types [Giustarini et al., 2004; Klatt and Lamas, 2000]. Under conditions of oxidative/nitrosative stress, cysteine residues within proteins are primary targets for oxidation. The presence of millimolar levels of GSH within the cell helps to maintain the equilibrium favorable for the formation of possible glutathionylation precursors. Glutathionylation can be protective, preventing the sequential oxidation of thiol groups to sulfonic acid, a product that leads to inevitable proteosomal degradation controlling cellular response to changes in redox conditions [Shackelford et al., 2005]. In addition, glutathionylation has been shown to regulate the structure/function of a quite diverse range of cellular proteins. As such, reversibility of glutathionylation provides a viable switch of some importance in controlling cellular response to changes in redox conditions [Chrestensen et al., 2000].

Redox regulation is rapidly evolving as an important metabolic modulator of cellular functions of all living cells and organisms and is being increasingly implicated in many chronic inflammatory and degenerative diseases [Gallogly and Mieyal, 2007]. The cellular redox state is a crucial mediator of multiple metabolic, signaling, and transcriptional processes in cells, and protein thiols in the form of cysteine residues are key players in redox sensing and regulation. Redox alterations of several protein and enzymes have been identified in a wide variety of conditions, and the purpose of such regulation has been speculated in the cellular defense mechanism during oxidative stress. Intracellular redox status has been linked to cellular differentiation, immune response, growth control, tumor progression, and apoptosis. Oxidants and antioxidants can act as signaling molecules that modify the function of enzymes such as phosphatases and kinases and directly or indirectly affect the activity of many transcription factors. Many proteins and enzymes such as *S*-transferases, glyceraldehyde 3-*P*-dehydrogenase, glutathione reductase, thioredoxin/glutaredoxin, and peptidases such as papain, caspase, and calpain in a cell consist of cysteine residues that have a sulfhydryl group in their side chain and the proton is labile, which makes it a chemical hot spot for a wide variety of biochemical interactions. Covalent modifications of proteins at their cysteines may be critical to a particular function of a given protein. Protein sulfhydryls may be oxidatively modified by *S*-thiolation and *S*-nitrosylation, which may significantly modify various cellular processes [Ghezzi et al., 2005]. Both *S*-thiolation and *S*-nitrosation are usually reversible and may be a protective/adaptive strategy of a cell. Exchange between thiols and disulfides are very slow and must be catalyzed by enzymes such as glutaredoxin or other protein disulfide isomerases that have a thioredoxin-like structural motif. Such disulfide exchange is a potential signaling mechanism due to its capacity for modifying cysteine residues in enzymes [Das, 2005].

Dethiolation of *S*-glutathionilated proteins can be achieved by either nonenzymatic reduction or enzymatic cleavage of the disulfide bond involving the action of thioredoxin, glutaredoxin, and/or protein disulfide isomerase [Johansson et al., 2004]. Thus, it is possible to speculate that enzymatically controlled thiolation/dethiolation may confer specificity and regulatory potential on the posttranslational control of protein function by *S*-glutathiolation. Thioredoxins, peroxiredoxins, glutaredoxins, in addition to the general antioxidants of the cells form a powerful combination not only to prevent irreversible damage to crucial proteins but also take part in redox recycling of oxidized thiol molecules [Michelet et al., 2006]. Thus protein function modulation by *S*-glutathiolation and/or *S*-nitrosation spans a wide variety of cellular functions ranging from resistance to oxidative stress, phosphorylation-dependent signal transduction, posttranslational protein modification, and clearance via proteasomes to transcriptional activation and inhibition [Giustarini et al., 2005]. Many proteins can undergo glutathionylation under oxidative stress [Cotgreave and Gerdes, 1998; Klatt and Lamas, 2000]. Glyceraldehyde-3-phosphate dehydrogenase is the major *S*-glutathionylated protein in endothelial cells exposed to hydrogen peroxide and in monocytes during the endogenous oxidative burst [Ravichandran et al., 1994]. Creatine kinase and glycogen phosphorylase *b* are also targets for *S*-glutathionylation in myocytes and cardiac tissue during cyclic oxidative stress. Carbonic anhydrase III, glutathione *S*-transferase, superoxide dismutase, and hemoglobin become thiolated in cellular models of oxidative stress. Additional proteins, including fatty-acid synthase, 3-hydroxy-3-methylglutaryl-CoA reductase, aldose reductase, human immunodeficiency virus-1 protease, and small HSP25, have been reported as potential targets in vitro for redox-dependent *S*-glutathionylation [Klatt and Lamas, 2000]. Even c-Jun DNA binding appears to be redox-regulated by glutathionylation [Klatt et al., 1999] and dopamine biosynthesis is also inhibited by *S*-glutathionylation during oxidative stress [Borges et al., 2002]. In Parkinson's disease, monoamine oxidase-derived H_2O_2 was shown to inhibit mitochondrial respiration by glutathionylation of respiratory chain enzymes [Cohen et al., 1997]. In addition, some of the proteins found to be glutathionylated belong to the class of cytoskeletal proteins, which are particularly abundant in cells. The supramolecular organization of these proteins depends on the presence of exposed sulfhydryl residues; the modification of these groups by glutathionylation could be relevant to their function, by either protecting them against irreversible oxidation or inhibiting polymerization [Dalle-Donne et al., 2000]. Pastore and co-workers [2003] have demonstrated that actin glutathionylation causes an impairment of microfilaments dynamic in Friedreich's ataxia fibroblasts.

In addition, transcription factors such as NF-κB, c-jun, and AP-1, are potential targets for glutathionylation [Dalle-Donne et al., 2007]. Under oxidative stress conditions, glutathionylation of transcription factors regulate their DNA-binding activity resulting in alteration of gene expression. This inhibitory effect has been proposed to be mediated by the oxidation of a conserved cysteine in the DNA-binding domain through unknown biochemical mechanisms [Pineda-Molina et al., 2001].

Although glutathionylation of proteins is seen in certain cells of nonneural origin, it is more pronounced in brain, presumably because extrusion of GSSG, as routinely observed in other tissues such as lung and liver, is not always seen in brain [Sparaco et al., 2006]. Moreover, the aging process is associated with a pro-oxidizing shift in the cellular redox state. The amounts of the redox-sensitive free aminothiols (glutathione, cysteine, Cys-Gly, and methionine) and protein mixed disulfides has been measured at different ages. GSH/GSSG ratios decreased significantly with increasing age. Concentrations of Cys-Gly increased and methionine decreased with age. The amounts of protein-mixed disulfides, measured as protein–cysteinyl, protein–Cys-Gly and protein–glutathionyl mixed disulfides, increased as a function of age [Giustarini et al., 2006]. The pattern of changes in free aminothiol content, glutathione-redox state, and protein-mixed disulfides is inversely related to the life expectancy of the flies [Rebrin et al., 2004]. Collectively, these results support the idea that the pro-oxidizing shift in the glutathione redox state, the decrease in methionine content and increase in abundance of protein-mixed disulfides are indicative of enhanced oxidative stress during aging and neurodegenerative disorders [Schulz et al., 2000].

Recent studies have shown increased glutathionylation of specific proteins in AD patients compared with control subjects [Newman et al., 2007]. The exact function of this reversible oxidative modification is unknown. Further studies investigating the specific in vivo effects of *S*-glutathionylation in oxidative stress are important to determining the role of *S*-glutathionylation in the AD brain and neurodegenerative disorders.

Glutathione levels in human tissues normally range from 0.1 to 10 mM, most concentrated in the liver (up to 10 mM), while plasma concentration is in the micromolar range. Oxidative stressors that can deplete GSH include ultraviolet and other radiations, viral infections, environmental toxins, heavy metals, inflammation, and dietary deficiencies of GSH precursors and enzyme cofactors. The sulfhydryl (-SH) group on the cysteinyl portion confers a strong electron-donating character. As electrons are lost, the molecule becomes oxidized, and two such molecules become dimerized by a disulfide bridge to form glutathione disulfide or oxidized glutathione (GSSG). This linkage is reversible upon re-reduction. GSH is under tight homeostatic control both intracellularly and extracellularly. A dynamic balance is maintained between GSH synthesis, GSH recycling from GSSG, and GSH utilization. GSH synthesis involves two closely linked, enzymatically controlled reactions that utilize ATP. First cysteine and glutamate are combined, by γ-glutamyl cysteinyl synthetase. Second, GSH synthetase combines γ-glutamylcysteine with glycine to generate GSH. As GSH levels rise, they self-limit further GSH synthesis; otherwise, cysteine availability is usually rate limiting. GSH recycling is catalyzed by glutathione disulfide reductase, which uses reducing equivalents from NADPH to reconvert GSSG to 2GSH. The reducing power of ascorbate helps conserve systemic GSH. GSH is used as a cofactor by multiple peroxidase enzymes, to detoxify peroxides generated from oxygen radical attack on biological molecules, to reduce oxidized centers on DNA, proteins, and other biomolecules, and glutathione *S*-transferases (GST) to conjugate GSH with endogenous substances and to exogenous electrophiles and diverse xenobiotics. Direct attack by free radical and other oxidative agents can also deplete GSH. The homeostatic glutathione redox cycle attempts to keep GSH depleted as it is being consumed [Sen, 1998]. The liver is the largest GSH reservoir. GSH equivalents circulate in the blood predominantly as cystine, the oxidized and more stable form of cysteine. Cells import cystine from the blood, reconvert it to cysteine, and from it synthesize GSH. Conversely, inside the cell GSH helps re-reduce oxidized forms of other antioxidants such as ascorbate and α-tocopherol [Meister, 1994]. GSH is an extremely important cell protecting by directly quenching reactive free radicals and GSH/GSSG balance is crucial to maintain cellular homeostasis and survival. GSH is the essential cofactor for many enzymes that require thiol-reducing equivalents, and helps keep redox-sensitive active sites on enzymes in the necessary reduced state. Other thiol cell systems—the metallothioneins, thioredoxins, and other redox regulator proteins—are ultimately regulated by GSH levels and the GSH/GSSG redox ratio. Glutathione status is a highly sensitive indicator of cell functionality and viability. In humans, GSH depletion is linked to a number of disease states: As intracellular GSH becomes reduced, the cell functionality is progressively reduced [Meister, 1995].

Glutathione is present in the brain in millimolar concentrations [Dringen, 2000]. Although there are some reports favoring the transport of GSH across the blood–brain barrier and their uptake into brain cells, it still needs to be established whether these play a role in GSH homeostasis in the brain [Kannan et al., 1999]. Nevertheless, the constituent amino acids of GSH may cross the blood–brain barrier and be utilized for GSH synthesis in the brain following the same pathway as described above. Several reports employing different techniques have shown the presensce of GSH both in neurons and glial cells [Maybodi et al., 1999], but the concentration of GSH appears to be higher in brain astrocytes compared to neurons. Apart from the antioxidant functions of GSH in the brain, extracellular GSH has been hypothesized to have additional functions as a neurotransmitter [Janaky et al., 1999], neurohormone, in the detoxification of glutamate, and in leukotriene metabolism [Dringen, 2000]. It has been observed that there is an age-dependent depletion in intracellular GSH in the CSF during aging in humans [Cudkowicz et al., 1999]. Studies from others have shown that aged mice have a 30% decrease in levels of GSH compared with younger animals. Since the brain requires extensive ROS detoxification, it is evident that a decrease in GSH content could increase oxidative damage, making the brain more susceptible to neurological disorders [Bharath et al., 2002]. Decreased GSH availability in the brain is believed to promote mitochondrial damage by likely increasing the levels of oxidative stress in this organelle. Depletion of brain GSH has been shown to result in decreases in mitochondrial enzyme activities in as well as losses in ATP production in the aging murine brain [Martinez et al., 1995].

Based on these notions, one of the ways to counteract oxidative stress-mediated disruption of GSH homeostasis—a central feature of neurodegeneration—is to replenish the GSH pool either by increasing the synthesis of the tripeptide or by slowing its degradation. Schulz and colleagues [2000] have shown that the administration of precursors of GSH metabolism, such as γ-glutamyl cysteine, increase the levels of GSH in rat brains. Similarly, precursors of cysteine synthesis have been administered in various animal models to accelerate cysteine production in the brain, thus increasing GSH levels. GSH replacement can also be achieved by administering thiol reagents such as GSH itself or GSH analogs. GSH cannot easily penetrate the blood–brain barrier due to the presence of the cysteine SH group and is not efficiently absorbed into neuronal cells in the brain [Kannan et al., 1999]. Hence rather than GSH, modified forms of GSH such as GSH analogs have generally been used in vivo. Recently, Butterfield et al. demonstrated that the xanthate D609—acting as a GSH mimetic—protected against amyloid-β-(1-42)-induced oxidative damage in primary neurons and isolated synaptosomes [Perluigi et al., 2006a; Sultana et al., 2006a]. Further studies are in progress to test the use of GSH or related molecules in GSH metabolism as therapeutics in the treatment of neurodegenerative diseases to elicit an increase in brain GSH levels.

VITAGENE FAMILY: A VITAL PROPOSITION

The term *vitagenes* is used to refer to a group of genes that are strictly involved in preserving cellular homeostasis during stressful conditions. The vitagene family is actually composed of the heat shock proteins (Hsp) Hsp32, Hsp70, and by the thioredoxin system [Calabrese et al., 2006a, 2007a]. Among these genes heme oxygenase-1 (HO-1), also known as Hsp32, is receiving considerable attention because of its major role in counteracting both oxidative and nitrosative stress. In fact, HO-1 induction is one of the earlier events in the cell response to stress. Heme oxygenase-1 exerts α protective role, by degrading the intracellular levels of pro-oxidant heme and by producing biliverdin, the precursor of bilirubin, this latter being an endogenous molecule with powerful antioxidant and antinitrosative features [Calabrese et al., 2007; Mancuso et al., 2008]. Although some beneficial phytochemicals might function solely as antioxidants, it is becoming clear that many of the beneficial chemicals in vegetables and fruits evolved as toxins (to dissuade insects and other predators) that, at subtoxic doses, activate adaptive cellular stress–response pathways, under control of genes defined by us as vitagenes in a variety of cells including neurons. Examples of such "preconditioning" or "neurohormesis" pathways include those involving cell-survival signaling kinases, the transcription factors NRF2, and the histone deacetylases [Calabrese and Baldwin, 1998; Calabrese, 2008a, 2008b; Mattson et al., 2007]. Consistent with this, neurohormetic phytochemicals such as curcumin, resveratrol, and sulforaphanes or other compounds such as acetyl-L-carnitine [Calabrese et al., 2005; Poon et al., 2006] might protect neurons against injury and disease by stimulating, through the vitagene system, the production of antioxidant enzymes, protein chaperones, and other proteins that help cells to withstand stress [Calabrese and Maines, 2006, Calabrese et al., 2006a, 2006b, 2006c, 2006d]. The strong evidence that the vitagene network operates as a defense system in the brain during oxidative and nitrosative stress open new perspectives in the treatment of neurodegenerative disorders. Therefore, the nutritional manipulation of endogenous cellular defense mechanisms represents an innovative approach to therapeutic intervention in neurodegeneration, and proposes potential novel therapeutic strategies relying upon the simultaneous activation of cytoprotective genes of the cell life program and down-regulation of proinflammatory and pro-oxidative genes involved in programmed cell death. Although the term *vitagene* was first proposed to indicate speculatively the existence of genes as opposed to gerontogenes, the first evidence-based notion identifying vitagenes with stress-responsive genes such as *HO-1* and *Hsps* has been provided by our group [Calabrese et al. 2001].

CONCLUSIONS AND PERSPECTIVES

Modulation of endogenous cellular defense mechanisms via the stress–response signaling represents an innovative approach to therapeutic intervention in diseases causing tissue damage, such as neurodegeneration. Efficient functioning of maintenance and repair processes seems to be crucial for both survival and physical quality of life. This is accomplished by a complex network of the so-called longevity assurance processes, which are composed of several genes that can be termed vitagenes. Consistently, by maintaining or recovering the activity of vitagenes can be possible to delay the aging process and decrease the occurrence of age-related diseases with resulting prolongation of a healthy life span [Calabrese, 2007; Calbrese et al, 2007a, 2007b, 2007c, Mancuso et al., 2007, 2008]. As one of the most important neurodegenerative disorders, AD is a progressive disorder with cognitive and memory decline, speech loss, personality changes, and synapse loss. With the increasingly aging population of the United States, the number of AD patients is predicted to reach 14 million in the mid-twenty-first century in the absence of effective interventions. This will pose an immense economic and personal burden on the people of this country. Similar considerations apply to most regions of the world. There is now strong evidence to suggest that factors such as oxidative stress and disturbed protein metabolism and their interaction in a vicious cycle are central to AD pathogenesis. Brain-accessible antioxidants, potentially, may provide the means of implementing this therapeutic strategy of delaying the onset of AD, and more in general all degenerative diseases associated with oxidative stress.

Emerging evidence strongly supports also the notion that stimulation of various maintenance and repair pathways through exogenous intervention, such as mild stress or compounds targeting the heat shock signal pathway, such as polyphenols, may have biological significance as a novel approach to delay the onset of various age-associated alterations in cells, tissues, and organisms. Hence, by maintaining or recovering the activity of vitagenes, it can be possible to delay the aging process and decrease the occurrence of age-related diseases with resulting prolongation of a healthy life span.

ACKNOWLEDGMENTS

This work was supported by grants from MIUR, FIRB RBRN07BMCT, RCUK, the American Cancer Society (RSG-07-157-01-CNE), Cancer

Research UK, Tenovus, the Royal Society, the Anonymous Trust, and by "Fondi Ateneo" 2007 and 2008.

REFERENCES

Andersen JK. 2004. Oxidative stress in neurodegeneration: Cause or consequence? *Nat Med* 10 Suppl:S18–S25.

Anzai K, Ogawa K, Goto Y, Senzaki Y, Ozawa T, Yamamoto H. 1999. Oxidation-dependent changes in the stability and permeability of lipid bilayers. *Antioxid Redox Signal* 1:339–347.

Ates O, Cayli S, Altinoz E, Gurses I, Yucel N, Sener M, Kocak A, Yologlu S. 2007. Neuroprotection by resveratrol against traumatic brain injury in rats. *Mol Cell Biochem* 294:137–144.

Begum AN, Jones MR, Lim GP, Morihara T, Kim P, Heath DD, Rock CL, Pruitt MA, Yang F, Hudspeth B, Hu S, Faull KF, Teter B, Cole GM, Frautschy SA. 2008. Curcumin structure-function, bioavailablity and efficacy in models of neuroinflammation and Alzheimer's. J Pharmacol Exp Ther, in press.

Berlett BS, Stadtman ER. 1997. Protein oxidation in aging, disease, and oxidative stress. *J Biol Chem* 272:20313–20316.

Bharath S, Hsu M, Kaur D, Rajagopalan S, Andersen JK. 2002. Glutathione, iron and Parkinson's disease. *Biochem Pharmacol* 64:1037–1048.

Biswas S, Rahman I. 2008. Modulation of steroid activity in chronic inflammation: A novel anti-inflammatory role for curcumin. *Mol Nutr Food Res* 52:987–994.

Bolton RM, Ahokas JT. 1997. Purification and characterization of hepatic glutathione transferases from an insectivorous marsupial, the brown antechinus (Antechinus stuartii). *Xenobiotica* 27:573–586.

Borges CR, Geddes T, Watson JT, Kuhn DM. 2002. Dopamine biosynthesis is regulated by S-glutathionylation. Potential mechanism of tyrosine hydroxylast inhibition during oxidative stress. *J Biol Chem* 277:48295–48302.

Butterfield DA, Sultana R. 2007. Redox proteomics identification of oxidatively modified brain proteins in Alzheimer's disease and mild cognitive impairment: Insights into the progression of this dementing disorder. *J Alzheimers Dis* 12:61–72.

Butterfield DA, Abdul HM, Newman S, Reed T. 2006a. Redox proteomics in some age-related neurodegenerative disorders or models thereof. *NeuroRx* 3:344–357.

Butterfield DA, Perluigi M, Sultana R. 2006b. Oxidative stress in Alzheimer's disease brain: New insights from redox proteomics. *Eur J Pharmacol* 545:39–50.

Calabrese EJ. 2008a. Hormesis: Why it is important to toxicology and toxicologists. *Environ Toxicol Chem* 27:1451–1474.

Calabrese EJ. 2008b. Pharmacological enhancement of neuronal survival. *Crit Rev Toxicol* 38:349–389.

Calabrese V. 2007. Highlight commentary on "Redox proteomics analysis of oxidatively 3 modified proteins in G93A–SOD1 transgenic mice-A model of 4 familial amyotrophic lateral sclerosis." *Free Radical Biol Med* 43:160–162.

Calabrese EJ, Baldwin LA. 1998. Hormesis as a biological hypothesis. *Environ Health Perspect* 106:357–362.

Calabrese V, Maines MD. 2006. Antiaging medicine: Antioxidants and aging. *Antioxid Redox Signal* 8:362–364.

Calabrese V, Guagliano E, Sapienza M, Panebianco M, Calafato S, Puleo E, Pennisi G, Mancuso C, Butterfield DA, Stella AG. 2007a. Redox regulation of cellular stress response in aging and neurodegenerative disorders: Role of vitagenes. *Neurochem Res* 32:757–773.

Calabrese V, Mancuso C, Calvani M, Rizzarelli E, Butterfield DA, Giuffrida Stella AM. 2007b. Nitric oxide in the CNS: Neuroprotection versus neurotoxicity. *Nat Neurosci* 8:766–775.

Calabrese V, Mancuso C, Sapienza M, Puleo E, Calafato S, Cornelius C, Finocchiaro M, Mangiameli A, Di Mauro M, Stella AM, Castellino P. 2007c. Oxidative stress and cellular stress response in diabetic nephropathy. *Cell Stress Chaperones* 12:299–306.

Calabrese V, Colombrita C, Scapagnini G, Calvani M, Giuffrida Stella AM. Butterfield DA. 2006a. Acetylcarnitine and cellular stress response: Role in nutritional redox homeostasis and regulation of longevity genes. *J Nutr Biochem* 17:73–88.

Calabrese V, Butterfield DA, Scapagnini G, Stella AM, Maines MD. 2006b. Redox regulation of heat shock protein expression by signaling involving nitric oxide and carbon monoxide: Relevance to brain aging, neurodegenerative disorders, and longevity. *Antioxid Redox Signal* 8:444–477.

Calabrese V, Colombrita C, Sultana R, Scapagnini G, Calvani M, Butterfield DA, Stella AM. 2006c. Redox modulation of heat shock protein expression by acetylcarnitine in aging brain: Relationship to antioxidant status and mitochondrial function. *Antioxid Redox Signal* 8:404–416.

Calabrese V, Calvani M, Butterfield DA. 2006d. Acetyl-L-carnitine-induced up-regulation of heat shock proteins protects cortical neurons against amyloid-beta peptide 1-42-mediated oxidative stress and neurotoxicity: Implications for Alzheimer's disease. *J Neurosci Res* 84:398–408.

Calabrese V, Ravagna A, Colombrita C, Guagliano E, Scapagnini G, Calvani M, Butterfield DA, Giuffrida Stella AM. 2005. Acetylcarnitine induces heme oxygenase in rat astrocytes and protects against oxidative stress: Involvement of the transcription factor Nrf2. *J Neurosci Res* 79:509–521.

Calabrese V, Scapagnini G, Giuffrida Stella AM, Bates TE, Clark JB. 2001. Mitochondrial involvement in brain function and dysfunction: Relevance to aging, neurodegenerative disorders and longevity. *Neurochem Res* 26:739–764.

Chakravarti B, Chakravarti DN. 2007. Oxidative modification of proteins: Age-related changes. *Gerontology* 53:128–139.

Chang TS, Jeong W, Woo HA, Lee SM, Park S, Rhee SG. 2004. Characterization of mammalian sulfiredoxin and its reactivation of hyperoxidized peroxiredoxin through reduction of cysteine sulfinic acid in the active site to cysteine. *J Biol Chem* 279:50994–51001.

Choi BH, Kim CG, Lim Y, Shin SY, Lee YH. 2008. Curcumin down-regulates the multidrug-resistance mdr1b gene by inhibiting the PI3K/Akt/NF kappa B pathway. *Cancer Lett* 259:111–118.

Chrestensen CA, Starke DW, Mieyal JJ. 2000. Acute cadmium exposure inactivates thioltransferase (Glutaredoxin), inhibits intracellular reduction of protein-glutathionyl-mixed disulfides, and initiates apoptosis. *J Biol Chem* 275:26556–26565.

Christen Y. 2000. Oxidative stress and Alzheimer disease. *Am J Clin Nutr* 71:621S-629S.

Clifford MN, Wu W, Kirkpatrick J, Kuhnert N. 2007. Profiling the chlorogenic acids and other caffeic acid derivatives of herbal chrysanthemum by LC-MSn. *J Agric Food Chem* 55:929–936.

Cohen G, Farooqui R, Kesler N. 1997. Parkinson disease: A new link between monoamine oxidase and mitochondrial electron flow. *Proc Natl Acad Sci USA* 94:4890–4894.

Cotgreave IA, Gerdes RG. 1998. Recent trends in glutathione biochemistry—glutathione-protein interactions: A molecular link between oxidative stress and cell proliferation? Biochem Biophys Res Commun 242:1–9.

Cudkowicz ME, Sexton PM, Ellis T, Hayden DL, Gwilt PR, Whalen J, Brown RH, Jr. 1999. The pharmacokinetics and pharmaco-dynamics of Procysteine in amyotrophic lateral sclerosis. *Neurology* 52:1492–1494.

Dalle-Donne I, Rossi R, Giustarini D, Colombo R, Milzani A. 2007. S-glutathionylation in protein redox regulation. *Free Radic Biol Med* 43:883–898.

Dalle-Donne I, Milzani A, Giustarini D, Di Simplicio P, Colombo R, Rossi R. 2000. S-NO-actin: S-nitrosylation kinetics and the effect on isolated vascular smooth muscle. *J Muscle Res Cell Motil* 21:171–181.

Das KC. 2005. Thioredoxin and its role in premature newborn biology. *Antioxid Redox Signal* 7:1740–1743.

Dinkova-Kostova AT. 2002. Protection against cancer by plant phenylpropenoids: Induction of mammalian anticarcinogenic enzymes. *Mini Rev Med Chem* 2:595–610.

Dinkova-Kostova AT, Talalay. 2008. Direct and indirect antioxidant properties of inducers of cytoprotective proteins. *Mol Nutr Food Res* 52:S128–S138.

Dinkova-Kostova AT, Talalay P. 1999. Relation of structure of curcumin analogs to their potencies as inducers of Phase 2 detoxification enzymes. *Carcinogenesis* 20:911–914.

Dinkova-Kostova AT, Holtzclaw WD, Kensler TW. 2005. The role of Keap1 in cellular protective responses. *Chem Res Toxicol* 18:1779–1791.

Dinkova-Kostova AT, Holtzclaw WD, Cole RN, Itoh K, Wakabayashi N, Katoh Y, Yamamoto M, Talalay P. 2002. Direct evidence that sulfhydryl groups of Keap1 are the sensors regulating induction of phase 2 enzymes that protect against carcinogens and oxidants. *Proc Natl Acad Sci USA* 99:11908–11913.

Dinkova-Kostova AT, Massiah MA, Bozak RE, Hicks RJ, Talalay P. 2001. Potency of Michael reaction acceptors as inducers of enzymes that protect against carcinogenesis depends on their reactivity with sulfhydryl groups. *Proc Natl Acad Sci USA* 98:3404–3409.

Dinkova-Kostova AT, Abeygunawardana C, Talalay P. 1998. Chemoprotective properties of phenylpropenoids, bis(benzylidene)cycloalkanones, and related Michael reaction acceptors: Correlation of potencies as phase 2 enzyme inducers and radical scavengers. *J Med Chem* 41:5287–5296.

Dringen R. 2000. Metabolism and functions of glutathione in brain. *Prog Neurobiol* 62:649–671.

Esterbauer H, Schaur RJ, Zollner H 1991. Chemistry and biochemistry of 4-hydroxynonenal, malonaldehyde and related aldehydes. *Free Radic Biol Med* 11:81–128.

Gallogly MM, Mieyal JJ. 2007. Mechanisms of reversible protein glutathionylation in redox signaling and oxidative stress. *Curr Opin Pharmacol* 7:381–391.

Ghezzi P, Bonetto V, Fratelli M. 2005. Thiol-disulfide balance: From the concept of oxidative stress to that of redox regulation. *Antioxid Redox Signal* 7:964–972.

Giasson BI, Ischiropoulos H, Lee VM, Trojanowski JQ. 2002. The relationship between oxidative/nitrative stress and pathological inclusions in Alzheimer's and Parkinson's diseases. *Free Radic Biol Med* 32:1264–1275.

Giustarini D, Dalle-Donne I, Lorenzini S, Milzani A, Rossi R. 2006. Age-related influence on thiol, disulfide, and protein-mixed disulfide levels in human plasma. *J Gerontol A Biol Sci Med Sci* 61:1030–1038.

Giustarini D, Milzani A, Aldini G, Carini M, Rossi R, Dalle-Donne I. 2005. S-nitrosation versus S-glutathionylation of protein sulfhydryl groups by S-nitrosoglutathione. *Antioxid Redox Signal* 7:930–939.

Giustarini D, Rossi R, Milzani A, Colombo R, Dalle-Donne I. 2004. S-glutathionylation: From redox regulation of protein functions to human diseases. *J Cell Mol Med* 8: 201–212.

Greco A, Minghetti L, Levi G. 2000. Isoprostanes, novel markers of oxidative injury, help understanding the pathogenesis of neurodegenerative diseases. *Neurochem Res* 25:1357–1364.

Halliwell B. 1999. Establishing the significance and optimal intake of dietary antioxidants: The biomarker concept. *Nutr Rev* 57:104–113.

Ishige K, Schubert D, Sagara Y. 2001. Flavonoids protect neuronal cells from oxidative stress by three distinct mechanisms. *Free Radic Biol Med* 30:433–446.

Janaky R, Ogita K, Pasqualotto BA, Bains JS, Oja SS, Yoneda Y, Shaw CA. 1999. Glutathione and signal transduction in the mammalian CNS. *J Neurochem* 73: 889–902.

Johansson C, Lillig CH, Holmgren A. 2004. Human mitochondrial glutaredoxin reduces S-glutathionylated proteins with high affinity accepting electrons from either glutathione or thioredoxin reductase. *J Biol Chem* 279:7537–7543.

Joshi G, Perluigi M, Sultana R, Agrippino R, Calabrese V, Butterfield DA. 2006. In vivo protection of synaptosomes by ferulic acid ethyl ester (FAEE) from oxidative stress mediated by 2,2-azobis(2-amidino-propane)dihydrochloride (AAPH) or Fe(2+)/H(2)O(2): Insight into mechanisms of neuroprotection and relevance to oxidative stress-related neurodegenerative disorders. *Neurochem Int* 48:318–327.

Kalfon L, Youdim MB, Mandel SA. 2007. Green tea polyphenol (−)-epigallocatechin-3-gallate promotes the rapid protein kinase C- and proteasome-mediated degradation of Bad: Implications for neuroprotection. *J Neurochem* 100:992–1002.

Kannan R, Mittur A, Bao Y, Tsuruo T, Kaplowitz N. 1999. GSH transport in immortalized mouse brain endothelial cells: Evidence for apical localization of a sodium-dependent GSH transporter. *J Neurochem* 73:390–399.

Kanski J, Aksenova M, Stoyanova A, Butterfield DA. 2002. Ferulic acid antioxidant protection against hydroxyl and peroxyl radical oxidation in synaptosomal and neuronal cell culture systems in vitro: Structure-activity studies. *J Nutr Biochem* 13:273–281.

Kikuzaki H, Hisamoto M, Hirose K, Akiyama K, Taniguchi H. 2002. Antioxidant properties of ferulic acid and its related compounds. *J Agric Food Chem* 50: 2161–2168.

Klatt P, Lamas S. 2000. Regulation of protein function by S-glutathiolation in response to oxidative and nitrosative stress. *Eur J Biochem* 267:4928–4944.

Klatt P, Molina EP, De Lacoba MG, Padilla CA, Martinez-Galesteo E, Barcena JA, Lamas S. 1999. Redox regulation of c-Jun DNA binding by reversible S-glutathiolation. *FASEB J* 13:1481–1490.

Kregel KC, Zhang HJ. 2007. An integrated view of oxidative stress in aging: Basic mechanisms, functional effects, and pathological considerations. *Am J Physiol Regul Integr Comp Physiol* 292:R18–R36.

Leonardi ET, Mytilineou C. 1998. Cell culture models of neuronal degeneration and neuroprotection. *Implications for Parkinson's disease. Adv Exp Med Biol* 446:203–222.

Lim GP, Chu T, Yang F, Beech W, Frautschy SA, Cole GM. 2001. The curry spice curcumin reduces oxidative damage and amyloid pathology in an Alzheimer transgenic mouse. *J Neurosci* 21:8370–8377.

Lovell MA, Markesbery WR. 2007. Oxidative DNA damage in mild cognitive impairment and late-stage Alzheimer's disease. *Nucleic Acids Res* 35:7497–7504.

Mancuso C, Capone C, Ranieri SC, Fusco S, Calabrese V, Eboli ML. Preziosi P, Galeotti T, Pani G. 2008. Bilirubin as an endogenous modulator of neurotrophin redox signaling. *J Neurosci Res* 86:1212–1230.

Mancuso C, Bates TE, Butterfield DA, Calafato S, Cornelius C, De Lorenzo A, Dinkova Kostova AT, Calabrese V. 2007. Natural antioxidants in Alzheimer's disease. *Expert Opin Investig Drug* 16:1921–1931.

Mandel S, Youdim MB. 2004. Catechin polyphenols: Neurodegeneration and neuroprotection in neurodegenerative diseases. *Free Radic Biol Med* 37:304–317.

Mandel S, Amit T, Reznichenko L, Weinreb O, Youdim MB. 2006a. Green tea catechins as brain-permeable, natural iron chelators-antioxidants for the treatment of neurodegenerative disorders. *Mol Nutr Food Res* 50:229–234.

Mandel S, Weinreb O, Reznichenko L, Kalfon L, Amit T. 2006b. Green tea catechins as brain-permeable, non toxic iron chelators to "iron out iron" from the brain. *J Neural Transm* Suppl: 249–257.

Mann M, Jensen ON. 2003. Proteomic analysis of post-translational modifications. *Nat Biotechnol* 21:255–261.

Markesbery WR, Kryscio RJ, Lovell MA, Morrow JD. 2005. Lipid peroxidation is an early event in the brain in amnestic mild cognitive impairment. *Ann Neurol* 58: 730–735.

Martinez M, Ferrandiz ML, Diez A, Miquel J. 1995. Depletion of cytosolic GSH decreases the ATP levels and viability of synaptosomes from aged mice but not from young mice. *Mech Ageing Dev* 84:77–81.

Mattson MP. 2008. Hormesis defined. *Ageing Res Rev* 7:1–7.

Maybodi L, Pow DV, Kharazia VN, Weinberg RJ. 1999. Immunocytochemical demonstration of reduced glutathione in neurons of rat forebrain. *Brain Res* 817:199–205.

Meister A. 1995. Mitochondrial changes associated with glutathione deficiency. *Biochim Biophys Acta* 1271:35–42.

Meister A. 1994. Glutathione, ascorbate, and cellular protection. *Cancer Res* 54:1969s-1975s.

Michelet L, Zaffagnini M, Massot V, Keryer E, Vanacker H, Miginiac-Maslow M, Issakidis-Bourguet E, Lemaire SD. 2006. Thioredoxins, glutaredoxins, and glutathionylation: New crosstalks to explore. *Photosynth Res* 89:225–245.

Montine TJ, Quinn J, Kaye J, Morrow JD. 2007. F(2)-isoprostanes as biomarkers of late-onset Alzheimer's disease. *J Mol Neurosci* 33:114–119.

Montine TJ, Neely MD, Quinn JF, Beal MF, Markesbery WR, Roberts LJ, Morrow JD. 2002. Lipid peroxidation in aging brain and Alzheimer's disease. *Free Radic Biol Med* 33:620–626.

Montine TJ, Beal MF, Robertson D, Cudkowicz ME, Biaggioni I, O'Donnell H, Zackert WE, Roberts LJ and Morrow JD. 1999. Cerebrospinal fluid F2-isoprostanes are elevated in Huntington's disease. *Neurology* 52:1104–1105.

Moreira PI, Smith MA, Zhu X, Nunomura A, Castellani RJ, Perry G. 2005. Oxidative stress and neurodegeneration. *Ann N Y Acad Sci* 1043:545–552.

Newman SF, Sultana R, Perluigi M, Coccia R, Cai J, Pierce WM, Klein JB, Turner DM, Butterfield DA. 2007. An increase in S-glutathionylated proteins in the Alzheimer's disease inferior parietal lobule, a proteomics approach. *J Neurosci Res* 85:1506–1514.

Pan T, Jankovic J, Le W. 2003. Potential therapeutic properties of green tea polyphenols in Parkinson's disease. *Drugs Aging* 20:711–721.

Pan MH, Lin-Shiau SY, Ho CT, Lin JH, Lin JK. 2000. Suppression of lipopolysaccharide-induced nuclear factor-kappaB activity by theaflavin-3,3′-digallate from black tea and other polyphenols through down-regulation of IkappaB kinase activity in macrophages. *Biochem Pharmacol* 59:357–367.

Pastore A, Tozzi G, Gaeta LM, Bertini E, Serafini V, Di Cesare S, Bonetto V, Casoni F, Carrozzo R, Federici G, Piemonte F. 2003. Actin glutathionylation increases in fibroblasts of patients with Friedreich's ataxia: A potential role in the pathogenesis of the disease. *J Biol Chem* 278:42588–42595.

Perluigi M, Joshi G, Sultana R, Calabrese V, De Marco C, Coccia R, Butterfield DA. 2006a. In vivo protection by the xanthate tricyclodecan-9-yl-xanthogenate against amyloid beta-peptide (1–42)-induced oxidative stress. *Neuroscience* 138:1161–1170.

Perluigi M, Joshi G, Sultana R, Calabrese V, De Marco C, Coccia R, Cini C, Butterfield DA. 2006b. In vivo protective effects of ferulic acid ethyl ester against amyloid-beta peptide 1-42-induced oxidative stress. *J Neurosci Res* 84:418–426.

Pineda-Molina E, Klatt P, Vazquez J, Marina A, Garcia de Lacoba M, Perez-Sala D, Lamas S. 2001. Glutathionylation of the p50 subunit of NF-κB: A mechanism for redox-induced inhibition of DNA binding. *Biochemistry* 40:14134–14142.

Polidori MC, Griffiths HR, Mariani E, Mecocci P. 2007. Hallmarks of protein oxidative damage in neurodegenerative diseases: Focus on Alzheimer's disease. *Amino Acids* 32:553–559.

Poon HF, Calabrese V, Calvani M, Butterfield DA. 2006. Proteomics analyses of specific protein oxidation and protein expression in aged rat brain and its modulation by l-acetylcarnitine: Insights into the mechanisms of action of this proposed therapeutic agent for CNS disorders associated with oxidative stress. *Antioxid Redox Signal* 8:381–394.

Pratico D, Clark CM, Liun F, Rokach J, Lee VY, Trojanowski JQ. 2002. Increase of brain oxidative stress in mild cognitive impairment: A possible predictor of Alzheimer disease. *Arch Neurol* 59:972–976.

Pratico D, Trojanowski JQ, Rokach J, Fitzgerald GA. 1998. Increased F2-isoprostanes in Alzheimer's disease: Evidence for enhanced lipid peroxidation in vivo. *FASEB J* 12:1777–1783.

Ravichandran V, Seres T, Moriguchi T, Thomas JA, Johnston RB, Jr. 1994. S-thiolation of glyceraldehyde-3-phosphate dehydrogenase induced by the phagocytosis-associated respiratory burst in blood monocytes. *J Biol Chem* 269:25010–25015.

Rebrin I, Bayne AC, Mockett RJ, Orr WC, Sohal RS. 2004. Free aminothiols, glutathione redox state and protein mixed disulphides in aging Drosophila melanogaster. *Biochem J* 382:131–136.

Roberts LJ, 2nd, Montine TJ, Markesbery WR, Tapper AR, Hardy P, Chemtob S, Dettbarn WD, Morrow JD. 1998. Formation of isoprostane-like compounds (neuroprostanes) in vivo from docosahexaenoic acid. *J Biol Chem* 273:13605–13612.

Sayre LM, Smith MA, Perry G. 2001. Chemistry and biochemistry of oxidative stress in neurodegenerative disease. *Curr Med Chem* 8:721–738.

Scalbert A, Johnson IT, Saltmarsh M. 2005. Polyphenols: Antioxidants and beyond. *Am J Clin Nutr* 81:215S-217S.

Scapagnini G, Colombrita C, Amadio M, D'Agata V, Arcelli E, Sapienza M, Quattrone A, Calabrese V. 2006. Curcumin activates defensive genes and protects neurons against oxidative stress, Antioxid. *Redox Signal* 8:395–403.

Scapagnini G, Butterfield DA, Colombrita C, Sultana R, Pascale A, Calabrese V. 2004. Ethyl ferulate, a lipophilic polyphenol, induces HO-1 and protects rat neurons against oxidative stress. *Antioxid Redox Signal* 6:811–818.

Schulz JB, Lindenau J, Seyfried J, Dichgans J. 2000. Glutathione, oxidative stress and neurodegeneration. *Eur J Biochem* 267:4904–4411.

Sen CK. 1998. Redox signaling and the emerging therapeutic potential of thiol antioxidants. *Biochem Pharmacol* 55:1747–1758.

Shackelford RE, Heinloth AN, Heard SC, Paules RS. 2005. Cellular and molecular targets of protein S-glutathiolation. *Antioxid Redox Signal* 7:940–950.

Shelton MD, Chock PB, Mieyal JJ. 2005. Glutaredoxin: Role in reversible protein S-glutathionylation and regulation of redox signal transduction and protein translocation. *Antioxid Redox Signal* 7:348–366.

Smolka MB, Zhou H, Purkayastha S, Aebersold R. 2001. Optimization of the isotope-coded affinity tag-labeling procedure for quantitative proteome analysis. *Anal Biochem* 297:25–31.

Sparaco M, Gaeta LM, Tozzi G, Bertini E, Pastore A, Simonati A, Santorelli FM, Piemonte F. 2006. Protein glutathionylation in human central nervous system: Potential role in redox regulation of neuronal defense against free radicals. *J Neurosci Res* 83:256–263.

Sultana R, Newman SF, Abdul HM, Cai J, Pierce WM, Klein JB, Merchant M, Butterfield DA. 2006a. Protective effect of D609 against amyloid-beta1-42-induced oxidative modification of neuronal proteins: Redox proteomics study. *J Neurosci Res* 84:409–417.

Sultana R, Perluigi M, Butterfield DA. 2006b. Protein oxidation and lipid peroxidation in brain of subjects with Alzheimer's disease: Insights into mechanism of neurodegeneration from redox proteomics. *Antioxid Redox Signal* 8:2021–2037.

Sultana R, Ravagna A, Mohmmad-Abdul H, Calabrese V, Butterfield DA. 2005. Ferulic acid ethyl ester protects neurons against amyloid beta-peptide(1-42)-induced oxidative stress and neurotoxicity: Relationship to antioxidant activity. *J Neurochem* 92:749–758.

Tilleman K, Stevens I, Spittaels K, Haute CV, Clerens S, Van Den Bergh G, Geerts H, Van Leuven F, Vandesande F, Moens L. 2002. Differential expression of brain proteins in glycogen synthase kinase-3 transgenic mice: A proteomics point of view. *Proteomics* 2:94–104.

Valko M, Rhodes CJ, Moncol J, Izakovic M, Mazur M. 2006. Free radicals, metals and antioxidants in oxidative stress-induced cancer. *Chem Biol Interact* 160:1–40.

Volpe A, Biasi D, Caramaschi P, Mantovani W, Bambara LM, Canestrini S, Ferrari M, Poli G, Degan M, Carletto A, Pieropan S, Minuz P. 2006. Levels of F2-isoprostanes in systemic sclerosis: Correlation with clinical features. *Rheumatology (Oxford)* 45:314–320.

Wagner Y, Sickmann A, Meyer HE, Daum G. 2003. Multidimensional nano-HPLC for analysis of protein complexes. *J Am Soc Mass Spectrom* 14:1003–1011.

Weinreb O, Amit T, Youdim MB. 2007. A novel approach of proteomics and transcriptomics to study the mechanism of action of the antioxidant-iron chelator green tea polyphenol (−)-epigallocatechin-3-gallate. *Free Radic Biol Med* 43:546–556.

Wilkins MR, Pasquali C, Appel RD, Ou K, Golaz O, Sanchez JC, Yan JX, Gooley AA, Hughes G, Humphery-Smith I, Williams KL, Hochstrasser DF. 1996. From proteins to proteomes: Large scale protein identification by two-dimensional electrophoresis and amino acid analysis. *Biotechnology (NY)* 14:61–65.

Yan JJ, Cho JY, Kim HS, Kim KL, Jung JS, Huh SO, Suh HW, Kim YH, Song DK. 2001. Protection against beta-amyloid peptide toxicity in vivo with long-term administration of ferulic acid. *Br J Pharmacol* 133:89–96.

Yang F, Lim GP, Begum AN, Ubeda OJ, Simmons MR, Ambegaokar SS, Chen PP, Kayed R, Glabe CG, Frautschy SA, Cole GM. 2005 Curcumin inhibits formation of amyloid beta oligomers and fibrils, binds plaques, and reduces amyloid in vivo. *J Biol Chem* 280:5892–5901.

Yehuda S, Rabinovitz S, Carasso RL, Mostofsky DI. 2002. The role of polyunsaturated fatty acids in restoring the aging neuronal membrane. *Neurobiol Aging* 23:843–853.

Youdim KA, Spencer JP, Schroeter H, Rice-Evans C. 2002. Dietary flavonoids as potential neuroprotectants. *Biol Chem* 383:503–519.

Zhou H, Watts JD, Aebersold R. 2001. A systematic approach to the analysis of protein phosphorylation. *Nat Biotechnol* 19:375–378.

19 Natural Phenolics and Metal Metabolism in Neurodegenerative Diseases

BAOLU ZHAO

Center for Brain and Cognitive Sciences, Institute of Biophysics, Academia Sinica, Beijing, People's Republic of China

CONTENTS

INTRODUCTION

There is a common feature in the two major neurodegenerative diseases: In both Alzheimer's disease (AD) and Parkinson's disease (PD) there is an imbalance of metal metabolism leading to oxidative stress-induced neuronal apoptosis. It is well known that there is no cure or efficient treatment available for these diseases. Is there a way to prevent these diseases? From this review

Plant Phenolics and Human Health: Biochemistry, Nutrition, and Pharmacology,
Edited by Cesar G. Fraga.

I propose that by acting as antioxidants, phenolics can prevent and even cure these diseases by modulating the imbalance of metal metabolism.

IMBALANCE OF METAL METABOLISM, OXIDATIVE STRESS, AND NEURODEGENERATIVE DISEASES

The mechanism of AD pathogenesis still remains unclear. However, one mechanism, amyloid β (Aβ) accumulation, may be due to the disturbance in metal homeostasis in AD brains [Strausak et al., 2001]. Aβ peptides are the major constituents of the amyloid core of senile plaques, which are derived from the amyloid precursor protein (APP) and are secreted into extracelluar spaces. Both APP and Aβ contain a copper-binding domain [Hesse et al., 1994; Atwood et al., 1998]. High concentrations of copper, zinc, and iron have been found within the amyloid deposits in AD brains [Lovell et al., 1998]. Aβ peptides can be rapidly precipitated by copper under mildly acidic conditions and by zinc at low physiological (submicromolar) concentrations [Bush et al., 1994]. An age-dependent binding between Aβ peptides with excess brain metals (copper, iron, and zinc) induces Aβ peptides to precipitate into metal-enriched plaques [Bush, 2002].

When using the APPV717I transgenic mice to measure the metal levels in the brain by using synchrotron radiation X-ray fluorescence, the copper and zinc concentrations in senile plaques and neuropil were higher than those in neuropil of both sucrose- and nicotine-treated groups. Furthermore, the levels of copper and zinc were significantly reduced by about 10–20% in senile plaques and neuropil in nicotine-treated mice relative to sucrose-treated controls. The distribution of copper and zinc in hippocampal CA1 region was also examined. Copper is enriched in the pyramidal neuron layer, whereas the distribution of zinc is relatively spread out. The content of both copper and zinc was significantly decreased after nicotine treatment, particularly in the pyramidal neuron layer. The copper concentrations in both the empty vector (neo) and APPsw mouse cells were measured by synchrotron radiation X-ray fluorescence analysis. In the neo cell line, both of 10 and 100 μM copper treatment increased intracellular copper concentrations. The copper concentration in APPsw cells was higher than in neo cells. Nicotine treatment decreased the copper concentration in both cell types. Incubation in the presence of 10 μM copper increased reactive oxygen species (ROS) generation 28% in APP cells and 59% in APPsw cells compared with neo cells; 100 μM copper increased the ROS generation 35 and 58% in APP and APPsw cells, respectively. Nicotine administration reduced copper-induced ROS accumulation in both cell types at 10 and 100 μM copper. In APPsw cells, the ratio of the expression of copper chaperone for superoxide dismutase (CCS) and SOD-1 (copper/zinc SOD) increased after nicotine treatment. In the hippocampus and cortex of the transgenic mice, the expression of CCS was increased 25 and 67% in the hippocampus and cortex of nicotine-treated mice, respectively, whereas the concentration of SOD-1 was lowered 28 and 26% in the hippocampus and cortex, respectively. Then, if copper and zinc modulate Aβ aggregation and deposition, the levels of these metals would be relevant in the pathogenesis of AD. Our group has suggested that nicotine reduces β-amyloidosis in AD transgenic mice, and that effect was also observed in APPsw-overexpressed cells by regulating metal homeostasis. Nicotine may regulate metal homeostasis causing: (i) a decrease of ROS generation and Aβ aggregation, (ii) a down-regulation of APP expression, and (iii) an increased CCS expression that reduces intracellular free copper.

Reactive oxygen species production is largely catalyzed by transition metals (especially copper and iron), and oxidative stress plays a critical role in AD pathogenesis. In one study, the association of metal levels and Aβ toxicity was demonstrated by: (i) the effect on cell viability by metal alone and in the combination with APP and Aβ, (ii) Aβ-induced neurotoxicity relevant to oxidative stress indicated by ROS production, and (iii) APPsw cells expressed APP and generated Aβ, so that Aβ–Cu^{2+} and APP–Cu^{2+} can catalyze more ROS generation than APP cells that only expressed APP.

Nicotine shows protective effects against Aβ-induced neurotoxicity in vivo and in vitro. Although nicotine acts on nAChRs, it also contains metal chelating abilities. It has been shown that maternal nicotine exposure resulted in a reduction of the copper content in the neonatal lung. In addition, evidence has been accumulated that nicotine might chelate metals. Indeed, nicotine reduces the levels of copper and zinc in senile plaques and neuropil, counteracting the undesirable metal accumulation.

The actions of nicotine on the whole brain and cellular copper concentrations were further supported by the changes of both CCS and SOD-1 levels. CCS directly inserts copper into SOD-1 and prevents the accumulation of free copper ions in cells. In the transgenic mice brain, our data suggest that an increase of CCS expression by nicotine causes a decrease in free intracellular copper levels, since more expression of CCS transfers more copper ions into proteins. Therefore, the copper ions that can be transported to the extracellular space are reduced, which results in an inhibition of free copper-mediated Aβ aggregation. Although nicotine has been proven to have beneficial effects on neurodegenerative disease, the exact mechanisms of those benefits are still unclear. Our data suggest the novel idea that nicotine may regulate metal homeostasis [Liu et al., 2003; Liu and Zhao, 2004; Xie et al., 2005; Zhang et al., 2006; Zhu et al., 2007]

Magaki et al. [2007] measured the levels of loosely bound, nonheme, and total iron and copper in the frontal cortex and hippocampus of patients with mild–moderate AD ($n = 3$), severe AD ($n = 8$), and dementia with Lewy bodies (DLB, $n = 6$). Additionally, the expression of iron regulatory protein 2 (IRP2) was examined in relation to the pathological hallmarks of AD, that is, 5 amyloid plaques, neurofibrillary tangles (NFT), and Lewy bodies. A significant decrease of loosely bound iron was found in the hippocampal white matter of both mild–moderate and severe AD patients and a trend toward increased nonheme iron in the hippocampal gray matter of severe AD patients. Furthermore, decreased levels of total copper were seen in severe AD frontal cortex compared to controls, suggesting an imbalance in brain metal levels in AD. The decrease in loosely bound iron in mild–moderate AD patients may be associated with myelin breakdown seen in the beginning stages of AD, and implicates that iron dysregulation is an early event in AD pathogenesis [Magaki et al., 2007].

Copper is essential for some of the enzymes that have a role in brain metabolism. Sophisticated mechanisms balance copper import and export to ensure proper nutrient levels (homeostasis) while minimizing toxic effects. Several neurodegenerative diseases including AD are characterized by modified copper homeostasis. This change seems to contribute either directly or indirectly to increased oxidative stress, an important factor in neuronal toxicity. The association of misfolded proteins and modified copper homeostasis appears to be important in the pathological progression of AD [Donnelly et al., 2007].

Inhibition of neocortical Aβ accumulation may be essential in an effective therapeutic intervention for AD. Aβ deposits, which are enriched in copper and zinc, are solubilized with copper and zinc selective chelators in vitro. Cherny et al. [2001] reported a 49% decrease in brain Aβ deposition (375 μg/g wet weight, $p < 0.0001$) in APP2576 transgenic mice treated orally for 9 weeks with clioquinol, an antibiotic and chelator of copper and zinc. This was accompanied by a modest increase in soluble Aβ (1.45% of total cerebral Aβ); but APP, synaptophysin, and GFAP levels were unaffected. In addition, general health and body weight parameters were significantly more stable in the treated animals. These results support targeting Cu and Zn interactions with Aβ, as a potential therapy for AD [Cherny et al., 2001].

To explore the effects of copper regulation on AD and Aβ peptide formation in vivo, toxic-milk (*txJ*) mice with a mutant ATPase7b transporter favoring elevated Cu levels were studied in combination with the transgenic (Tg) CRND8 amyloid precursor protein mice exhibiting robust Aβ deposition. Unexpectedly, TgCRND8 mice homozygous for the recessive *txJ* mutation examined at 6 months of age exhibited a reduced number of Aβ plaques and diminished plasma Aβ levels. In addition, homozygosity for *txJ* increased survival of young TgCRND8 mice and lowered Aβ before an increase in copper in the central nervous system (CNS) was detectable. These data suggest that the beneficial effect of the *txJ* mutation on CNS Aβ burden may proceed by alternative mechanisms, for example, by involving increased clearance of peripheral pools of Aβ [Phinney et al., 2003].

The copper-binding APP and Aβ have been proposed to play a role in physiological metal regulation. There is accumulating evidence of an unbalanced copper homeostasis with a causative or diagnostic link to AD. Whereas elevated copper levels are observed in APP knockout mice, APP overexpression results in reduced copper in transgenic mouse brain. Moreover, copper induces a decrease in Aβ levels in APP-transfected cells in vitro. To investigate the influence of bioavailable copper, transgenic APP23 mice received an oral treatment with copper-supplemented sucrose-sweetened drinking water. Chronic APP overexpression per se reduced SOD-1 activity in transgenic mouse brain, which could be restored to normal levels after copper treatment. A significant increase of brain copper indicated higher bioavailability following copper treatment in APP23 mice, whereas copper levels remained unaffected in littermate controls. Copper treatment lowered endogenous CNS Aβ before a detectable reduction of amyloid plaques. Thus, APP23 mice reveal APP-induced alterations linked to copper homeostasis, which can be reversed by increasing dietary copper [Bayer et al., 2003].

Evidence is now accumulating to show that reactions involving metals might be the common denominator underlying AD and PD. In these disorders, an abnormal reaction between a protein and a redox-active metal ion (copper or iron) promotes the formation of ROS. It is especially intriguing how the antioxidant Cu/Zn-SOD activity can convert into a pro-oxidant activity, a theme echoed in the recent proposal that Aβ and PrP, the proteins respectively involved in AD and prion diseases, possess similar redox properties [Bush, 2002].

The main therapeutic approaches attempt to prevent Aβ production (secretase inhibitors) or to remove Aβ (vaccine). However, there is now compelling evidence that Aβ does not spontaneously aggregate, but that there is an age-dependent protein precipitation of metal-enriched (copper, iron, and zinc) plaques. The copper or iron containing Aβ induces the production of hydrogen peroxide, which may mediate oxidative damage to the brain in AD. Metal-binding compounds have been developed to inhibit the generation of hydrogen peroxide by Aβ, as well as reverse the aggregation of the peptide as assayed in chemical system and with human brain post-mortem specimens. As indicated, clioquinol is being tested in clinical trials [Bush, 2002].

Additional results on the potential action of clioquinol include treating Chinese hamster ovary cells overexpressing APP with clioquinol and copper or zinc. The treatement resulted in an 85–90% reduction of secreted Aβ-(1–40) and Aβ-(1–42) compared with untreated controls. Similar effects were seen in neuroblastoma cells overexpressing APP. The secreted Aβ was rapidly degraded through up-regulation of matrix metalloproteases 2 and 3 after addition of clioquinol and copper. MMP activity was increased through activation of phosphoinositol 3-kinase and JNK. Clioquinol and copper also promoted phosphorylation of glycogen synthase kinase-3, and this potentiated activation of JNK and loss of Aβ-(1–40). These findings identify an alternative mechanism of action for clioquinol based on the reduction of Aβ deposition [White et al., 2006].

There is increasing evidence that iron is involved in several neurodegenerative diseases. Conditions such as neuroferritinopathy and Friedreich ataxia are associated with mutations in genes that encode proteins that are involved in iron metabolism. As the brain ages, iron accumulates in regions that are affected in AD and PD. High concentrations of reactive iron can increase oxidative stress-induced neuronal vulnerability and the toxicity of environmental or endogenous toxins. Studies on brain iron metabolism, especially the discovery of mutations in the genes associated with iron transport, storage, and regulation, could support the concept that increased brain iron is an initial cause of neuronal death at least in some neurodegenerative diseases. The brain shares with other organs the need for a constant and readily available supply of iron and a number of proteins available for iron transport, storage, and regulation. However, unlike other organs, the brain demands on iron are regional, cellular, and age sensitive. Failure to meet the demands for iron with an adequate supply in a timely manner can result in persistent neurological and cognitive dysfunction. Consequently, the brain has developed mechanisms to maintain a continuous supply of iron. However, in a number of common neurodegenerative disorders, there appears to be an excess accumulation of iron in the brain that suggests a loss of the homeostatic mechanisms responsible for regulating iron in the brain. As a result of a loss in iron homeostasis, the brain becomes vulnerable to iron-induced oxidative stress [Thompson et al., 2001].

OXIDATIVE STRESS AND NEURODEGENERATIVE DISEASES

An imbalance of metal metabolism causes oxidative stress, which can lead to the loss of biological functions. The high oxygen consumption, the low concentration of glutathione and of the antioxidant enzyme catalase, and the high proportion of polyunsaturated fatty acids make the brain particularly susceptible to oxidative stress and consequent damage [Mecocci et al., 1994; Smith et al., 1998]. Accumulated data demonstrated that oxidative damage occurs in AD brain [Markesbery and Carney, 1999; Christen, 2000; Smith et al., 2000]. Aβ peptide has been proven to produce H_2O_2 through metal ion reduction, with concomitant lipid oxidation [Huang et al., 1999a, 1999b]. The cytotoxicity of Aβ also seems to be relative to an oxidative mechanism [Behl et al., 1994; Markesbery, 1997; Markesbery and Carney, 1999]. Evidence supporting the notion of oxidative stress in AD brain includes: increased presence of redox-active metal ions in AD brain; increased lipid oxidation detected by decreased levels of polyunsaturated fatty acids and increased levels of lipid, protein, and nucleic acid oxidation, and decreased activity of oxidant-sensitive enzymes, such as glutamine synthetase.

Parkinson's disease is a progressive neurodegenerative disorder, and the hallmark of this disease is the selective loss of dopaminergic neurons in the substantia nigra pars compacta [Xin et al., 1984]. Recently, the death of dopaminergic neurons has been reported to occur by apoptosis [Mochizuki et al., 1996; Anglade et al., 1997; Hirsch et al., 1997]. Oxidative stress has been widely assessed as an important pathogenetic mechanism of neuronal apoptosis in PD [Halliwell, 1992]. Overproduction of ROS can lead to oxidative damage in the brain of PD, as shown by increased lipid oxidation and DNA damage in the substantia nigra. Increased protein oxidation is also apparent in many areas of the brain, while substantia nigra is particularly vulnerable [Xin et al., 1984]. Under physiological conditions, 6-hydroxydopamine (6-OHDA) is rapidly and nonenzymatically oxidized by molecular oxygen to form the corresponding *p*-quinone and H_2O_2 [Soto-Otero et al., 2000]. The former can undergo an intramolecular cyclization followed by a cascade of oxidative reactions resulting in the formation of an insoluble polymeric pigment related to neuromelanin [Graham et al., 1978; Soto-Otero et al., 2000].

EFFECTS OF PHENOLICS ON METAL-MEDIATED NEURODEGENERATIVE DISEASE

Oxidative stress has been implicated in a variety of neuropathologies, including stroke, trauma, and diseases such as AD and PD. It should be reasonable to use antioxidant strategies to prevent or treat these diseases. In the following sections are discussed the use of several natural phenolics as antioxidants and the potential actions on neurodegenerative diseases.

Tea Phenolics

Tea phenolics, including its major constituent, (−)-epigallocatechin gallate (EGCG), have diverse pharmacological activities. It is considered that these beneficial effects of green tea polyphenols are due to their potent antioxidative properties. In fact, it was demonstrated that in vitro green tea polyphenols serve as powerful antioxidants against free radicals such as DPPH radicals [Nanjo et al., 1996], superoxide anion [Zhao et al., 1989; Guo et al., 1996, 1999], lipid free radicals, and hydroxyl radicals [Nie et al., 2001; Zhao et al., 2001]. In the CNS, there is also some evidence showing that oral administration of green tea polyphenols and flavonoid-related compounds has preventive effects on iron-induced lipid peroxide accumulation and age-related accumulation of neurotoxic lipid peroxides in rat brain [Yoneda et al., 1995; Inanami et al., 1998].

Tea phenolics, including catechins are expected to scavenge free radicals but such chemical action is different depending on each compound. When we studied the effect of tea catechins (TC) on the PC12 and SY5Y cells exposed to 6-OHDA, it was demonstrated that TC could protect PC12 cells against apoptosis caused by 6-OHDP. EGCG and (−)-epicatechin gallate (ECG) were more effective than (−)-epigallocatechin (EGC), (−)-epicatechin (EC), and (+)-catechin (C). From 50 to 400 μM, the antiapoptotic effects increased with concentration, and EGCG was the most protective as compared with other TC. The nuclear changes and the DNA ladder characteristic of apoptosis were prevented in both PC12 and SY5Y cells by 200–400 μM of TC [Nie et al., 2002; Guo et al., 2005].

Nitric oxide (NO) and NO-related pathways are thought to play an important role in the pathogenesis of PD. In vitro experiments suggested that TC might protect dopamine neurons through inhibition of ROS and NO. The effects of a green tea extract (GTP) on a unilateral 6-OHDA-treated rat model of PD were investigated [Guo et al., 2007]. GTP treatment protected dopaminergic neurons from midbrain and striatal from 6-OHDA-induced increase of: (i) both ROS and NO levels, (ii) lipid oxidation, (iii) nitrite/nitrate content; (iv) inducible NO synthase, and (v) protein 3-nitro-tyrosine. Moreover, GTP treatment preserved, in a dose-dependent manner, the free-radical scavenging capability of both the midbrain and the striatum. These results support the in vivo protection of GTP against 6-OHDA-mediated damage and suggest that GTP might provide a treatment for PD [Guo et al., 2007].

Mandel's group [2007] demonstrated the neuroprotective properties of a green tea extract (GTE) and of ECG, in the MPTP mice model of PD. MPTP neurotoxin caused dopamine neuron loss in substantia nigra concomitant with a depletion in striatal dopamine and tyrosine hydroxylase protein levels. Pretreatment with either GTE (0.5 and 1 mg/kg) or ECG (2 and 10 mg/kg) prevented these effects. In addition, the MPTP caused an elevation in striatal SOD (240%) and catalase (165%) activities that was prevented by ECG. The neuroprotective effects are not likely to be caused by inhibition of MPTP conversion to its active metabolite 1-methyl-4-phenylpyridinium by monoamine oxidase-B, as both green tea and ECG were poor inhibitors of this enzyme in vitro (770 mg/mL and 660 mM, respectively). Brain penetrating property of polyphenols, as well as their antioxidant and iron-chelating properties, may make such compounds an important class of drugs to be developed for treatment of neurodegenerative diseases where oxidative stress has been implicated [Levites et al., 2001].

The same group showed radical scavenging activities of green tea (GT) and black tea (BT) extracts using brain mitochondrial membrane fraction treated with 2.5 μM iron. Both extracts (0.6–3.0 μM total polyphenols) were shown to attenuate the 6-OHDA-induced neuronal death. 6-OHDA (350 and 50 μM) activated the iron-dependent inflammatory redox-sensitive NF-κB in PC12 rat pheochromocytoma and human neuroblastoma SH-SY5Y cells. Treatment with GT (0.6 and 3 μM total polyphenols) before 6-OHDA inhibited both NF-κB nuclear translocation and binding activity induced by this toxin in SH-SY5Y cells. Cell protection was attributed by the authors to the potent antioxidant and iron-chelating actions of the polyphenolic constituents of tea extracts [Levites et al., 2002b]. It was also demonstrated that EGCG restored the reduced protein kinase C (PKC) and extracellular signal-regulated kinases (ERK1/2) activities caused by 6-OHDA. However, the neuroprotective effect of EGCG on cell survival was abolished by pretreatment with PKC inhibitor GF 109203X. Because EGCG increased phosphorylated PKC, it was suggested that PKC isoenzymes are involved in the neuroprotective action of EGCG against 6-OHDA. In addition, gene expression analysis revealed that EGCG prevented both the 6-OHDA-induced expression of several messenger riponucleric acid (mRNA), such as Bax, Bad, and Mdm2, and the 6-OHDA-induced decrease in Bcl-2, Bcl-w, and Bcl-xL. These results suggest that the protective mechanisms of EGCG against oxidative stress-dependent cell death include the stimulation of PKC and the modulation of cell survival/cell cycle genes [Levites et al., 2002a].

Not all the actions of TC are necessarily beneficial. We demonstrated that high concentrations of green tea polyphenols significantly enhanced sodium nitroprusside (SNP)-induced toxicity. SNP, a NO donor, induced apoptosis in SY5Y cells in a concentration and time-dependent manner, whereas pretreatment with GTP alone had no effect on cell viability. Pretreatment with GTP (50 and 100 μM) had only slightly apoptotic effects in the presence of SNP, while higher doses of GTP (200 and 500 μM) synergistically damage cells. Further research showed that co-incubation of green tea polyphenols and SNP caused loss of mitochondrial membrane potential, depletion of intracellular GSH, accumulation of ROS, and exacerbated NO-induced neuronal apoptosis via a Bcl-2 sensitive pathway [Zhang and Zhao, 2003].

Recent studies suggest that TC may be used for the prevention and treatment of a variety of neurodegenerative diseases. Rezai-Zadeh et al. [2005] report that EGCG reduces Aβ production in both murine neuron-like cells (N2a) transfected with the human "Swedish" mutant APP (SweAPP N2a cells), and in primary neurons derived from Swedish mutant APP-overexpressing mice (Tg APPsw line 2576). In concert with these observations, we find that EGCG markedly promotes cleavage of the α-COOH-terminal fragment of APP and elevates the N-terminal APP cleavage product, soluble APP-α. These cleavage events are associated with elevated α-secretase activity and enhanced hydrolysis of tumor necrosis factor α-converting enzyme. As a validation of these findings in vivo, we treated Tg APPsw transgenic mice overproducing Aβ with EGCG and found decreased Aβ levels and plaques associated with the promotion of the nonamyloidogenic α-secretase proteolytic pathway. These data raise the possibility that EGCG dietary supplementation may provide effective prophylaxis for AD [Rezai-Zadeh et al., 2005].

Obregon et al. [2006] showed that EGCG exerts a beneficial role on reducing brain Aβ levels, resulting in the mitigation of cerebral amyloidosis in a mouse model of AD, by the increasing α-secretase cleavage activity as indicated above. To gain insight into the molecular mechanism whereby EGCG modulates APP processing, we evaluated the involvement of three potential α-secretase enzymes: (i) disintegrin and metalloprotease (ADAM) 9, 10, or 17, in EGCG-induced nonamyloidogenic APP metabolism. Results show that EGCG treatment of SweAPP N2a cells leads to markedly elevated active (~60 kDa mature form) ADAM10 protein. Elevation of active ADAM10 correlates with increased α-COOH-terminal fragment cleavage, and elevated sAPP-α. To specifically test the contribution of ADAM10 to nonamyloidogenic APP metabolism, small interfering RNA knockdown of ADAM9, -10, or -17 mRNA was employed. Results show that ADAM 10 (but not ADAM-9 or -17) is critical for EGCG-mediated α-secretase cleavage activity. In summary,

ADAM10 activation is necessary for EGCG promotion of nonamyloidogenic (α-secretase cleavage) APP processing. Thus, ADAM10 represents an important pharmacotherapeutic target for the treatment of cerebral amyloidosis in AD [Obregon et al., 2006].

Considering the multietiological character of AD, the current pharmacological approaches using drugs oriented toward a single molecular target possess limited ability to modify the course of the disease and thus offer a partial benefit for the patient. In line with this concept, novel strategies include the use of a cocktail of several drugs and/or the development of a single molecule, possessing two or more active neuroprotective-neurorescue moieties that simultaneously manipulate multiple targets involved in AD pathology. A consistent observation in AD is a deregulation of metal homeostasis and the consequent oxidative stress associated with Aβ aggregation and neurite plaque formation. In particular, iron has been demonstrated to modulate the Alzheimer's APP expression by a pathway similar to that of ferritin L- and H-mRNA translation through iron-responsive elements in their 50UTRs. Mandel et al. [2007] discussed two separate scenarios concerning multiple therapy targets in AD, sharing in common the implementation of iron-chelation activity: (i) novel multimodal brain-permeable iron-chelating drugs, possessing neuroprotective-neurorescue and APP processing regulatory activities; and (ii) natural plant polyphenols (flavonoids), such as EGCG and curcumin, reported to have access to the brain and to possess multifunctional activities, such as metal chelation–radical scavenging, anti-inflammation, and neuroprotection [Mandel et al., 2007].

Flavonoids

Flavonoids isolated from plants are well investigated as natural antioxidants and widely used to treat patients as effective components of many medical drugs in clinics. To determine the potential protective mechanisms of flavonoids in cell death, the mouse hippocampal cell line HT-22 was used, and it was found that exogenous glutamate inhibits cystine uptake and depletes intracellular glutathione (GSH), leading to the accumulation of ROS and an increase in Ca^{2+} influx, which ultimately causes neuronal death. Many, but not all, flavonoids protect HT-22 cells and rat primary neurons from glutamate toxicity as well as from five other oxidative injuries. Three structural requirements of flavonoids for protection from glutamate are the hydroxylated C3, an unsaturated C-ring, and hydrophobicity. It was also found three distinct mechanisms of protection. These include increasing intracellular GSH, directly lowering levels of ROS, and preventing ROS-mediated influx of Ca^{2+}. These data show that the mechanisms of protection from oxidative insults by flavonoids are multiple and depend on the chemical structure and physical properties of each compound [Ishige et al., 2001].

Ginkgo biloba is a plant that contains a high number of flavonoids and terpens and has been reported to protect the brain against hypoxic damage and inhibit ROS formation in cerebellar neurons. We investigated the protective effects of a *Ginkgo biloba* extract (EGb) on dissociated cortex neurons from damage caused by ROS using spin label and molecular techniques. It was found that the order parameter (S), rotational correlation time (τ), and the ratio of strong immobilized component to weak immobilized component (S/W) of the cell membrane treated with hydroxyl radical were higher than those of control, indicating that ROS diminished membrane fluidity. EGb (5–50 μg/mL) produced a dose-dependent increase of membrane fluidity in hydrophobic areas of the membrane. EGb also protected the change of the protein conformation on the membrane caused by ROS. At the same concentration, EGb diminishes ROS-mediated increase of lactate dehydrogenase release and cell apoptosis [Ni et al., 1996]. The protective effect of EGb761 and some of its constituents against apoptosis were also examined. The total flavonoid components of EGb761 and a mixture of flavonoids and terpens protected cerebellar granule cell from oxidative damage and apoptosis induced by hydroxyl radicals. Total terpens of EGb761 did not protect against apoptosis, indicating that flavonoids and terpens act synergistically [Chen et al., 1999; Xin et al., 2000].

Aβ toxicity has been postulated to initiate the synaptic loss and subsequent neuronal degeneration seen in AD. We previously demonstrated that the standardized *Ginkgo biloba* extract EGb 761, commonly used to enhance memory and in AD patients for treating dementia, inhibits Aβ-induced apoptosis in neuroblastoma cells. EGb 761 and its constituents were tested to associate Aβ species with Aβ-induced pathological behaviors in a model organism, *Caenorhabditis elegans*. We observed that EGb 761 and one of its components, ginkgolide A alleviates Aβ-induced pathological behaviors, including paralysis, and reduces chemotaxis behavior and 5-HT hypersensitivity in a transgenic *C. elegans*. We also showed that EGb 761 inhibits Aβ oligomerization and Aβ deposits in the worms. Interestingly, reducing oxidative stress is not the mechanism by which EGb 761 and ginkgolide A suppress Aβ-induced paralysis because L-ascorbate reduced intracellular levels of hydrogen peroxide to the same extent as EGb 761, but was not as effective in suppressing paralysis in the transgenic *C. elegans*. These findings suggest that: (i) EGb 761 suppresses Aβ-related pathological behaviors; (ii) the protection against Aβ toxicity by EGb761 is mediated primarily by modulating Aβ oligomeric species; and (iii) ginkgolide A has therapeutic potential for prevention and treatment of AD [Wu et al., 2006].

Soy Isoflavones

Epidemiological studies have associated estrogen replacement therapy with a lower risk of developing AD, but with a higher risk of developing breast cancer and certain cardiovascular disorders. The neuroprotective effect of estrogen prompted us to determine potential therapeutic impact of soy-derived estrogenic compounds, the isoflavones. Genistein, the most active component of soy isoflavone, has been investigated because its affinity to bind to estrogen receptors [Kurzer and Xu, 1997], antioxidation activity [Johnson et al., 1996; Chan and Yu, 2000], capacity to increase cellular reduced glutathione [Guo et al., 2002], inhibition of protein tyrosine kinase (PTK) [Ohigashi et al., 2000], and for several other biological actions [Kim et al., 2000]. The existing data strongly suggest that isoflavones have a protective action against several chronic diseases such as atherosclerosis [Clarkson et al., 1998], diseases associated with postmenopausal estrogen deficiency, and hormone-dependent breast and prostate cancers [Lamartiniere et al., 2002]. Recently, it was reported that genistein protected rat brain synaptosome from insult induced by $A\beta_{25\text{-}35}$ [Andersen et al., 2003]. Importantly, genistein is capable of crossring the blood–brain barrier [Chang et al., 2000].

We have studied the neuroprotective effects of genistein against $A\beta_{25\text{-}35}$-induced apoptosis in cultured hippocampal neurons [Zeng et al., 2004]. Exposure to aged $A\beta_{25\text{-}35}$ for 24 h increased the oxidant-dependent dichlorofluorescein fluorescence two fold relative to controls. The increase in oxidants was reduced by 63% when cells were co-treated with 40 μM genistein and aged $A\beta_{25\text{-}35}$, while genistein at 0.1 μM decreased about 18% of the production of ROS induced by $A\beta_{25\text{-}35}$. Treatment with 25 μg/mL aged $A\beta_{25\text{-}35}$ for 24 h decreased the viability of hippocampal neurons about 43% compared to controls. Genistein at the concentrations of 0.1 and 40 μM rescued aged $A\beta_{25\text{-}35}$-induced decrease of viability rate by 7 and 14%, respectively. The estrogenic receptor antagonist, ICI182,780, significantly blocked the protective effect of genistein at the concentration of 0.1 μM while having little effect at 40 μM genistein. It can be concluded that $A\beta_{25\text{-}35}$-induced apoptosis is reverted by genistein: At submicromolar level, genistein protects mostly via the estrogen-receptor-mediated pathways; at micromolar level, the protective effect of genistein would be mainly mediated by its antioxidant properties.

Transgenic *C. elegans*, which express human Aβ, were fed genistein, daidzein, and glycitein (100 μg/mL) and then examined for Aβ-induced paralysis and ROS levels. Results show that only glycitein alleviated Aβ-induced paralysis. This activity of glycitein was associated to its capacity to reduce hydrogen peroxide level in the transgenic *C. elegans*. In vitro scavenging effects of glycitein confirmed its antioxidant properties. Furthermore, the transgenic *C. elegans* fed with glycitein exhibited reduced formation of Aβ. These findings suggest that the soy isoflavone glycitein may suppress Aβ toxicity through combined antioxidant actions and inhibition of Aβ deposition and thus may have therapeutic potential for prevention of Aβ-associated neurodegenerative disorders [Gutierrez-Zepeda et al., 2005].

CONCLUSION

Neurological diseases are difficult to treat after the onset of the pathological condition, so the best way is to protect people from developing neurodegenerative diseases. This review compiled laboratory information that suggests that some plant phenolics can prevent neurodegeneration by modulating imbalances in metal metabolism, by trapping radicals, and by affecting signaling pathways involved in neuronal dysfunction. As a preventive action, it is easy to increase the intake of natural products, for example, drinking more tea and soybean milk and eating fresh fruits and vegetables. A pharmacological use of active principles still shows potential and depends on further developments.

ACKNOWLEDGMENTS

This work was supported by a grant form the National Natural Science Foundation of China (29935080).

REFERENCES

Andersen JM, Myhre O, Fonnum F. 2003. Discussion of the role of the extracellular signal-regulated kinase-phospholipase A2 pathway in production of reactive oxygen species in Alzheimer's disease. *Neurochem Res* 28:319–326.

Anglade P, Vyas S, Javoy-Agid F, Herrero MT, Michel PP, Marquez J, Mouatt-Prigent A, Ruberg M, Hirsch EC, Agid Y. 1997. Apoptosis and autophagy in nigral neurons of patients with Parkinson's disease. *Histol Histopathol* 12:25–31.

Atwood CS, Moir RD, Huang X, Scarpa RC, Bacarra NM, Romano DM, Hartshorn MA, Tanzi RE, Bush AI. 1998. Dramatic aggregation of Alzheimer abeta by Cu(II) is induced by conditions representing physiological acidosis. *J Biol Chem* 273:12817–12826.

Bayer TA, Schafer S, Simons A, Kemmling A, Kamer T, Tepest R, Eckert A, Schussel K, Eikenberg O, Sturchler-Pierrat C, Abramowski D, Staufenbiel M, Multhaup G. 2003. Dietary Cu stabilizes brain superoxide dismutase 1 activity and reduces amyloid Abeta production in APP23 transgenic mice. *Proc Natl Acad Sci USA* 100:14187–14192.

Behl C, Davis JB, Lesley R, Schubert D. 1994. Hydrogen peroxide mediates amyloid beta protein toxicity. *Cell* 77:817–827.

Bush AI. 2002. Metal complexing agents as therapies for Alzheimer's disease. *Neurobiol Aging* 23:1031–1038.

Bush AI. 2000. Metals and neuroscience. *Curr Opin Chem Biol* 4:184–191.

Bush AI, Pettingell WH, Multhaup G, Paradis M, Vonsattel JP, Gusella JF, Beyreuther K, Masters CL, Tanzi RE. 1994. Rapid induction of Alzheimer A beta amyloid formation by zinc. *Science* 265:1464–1467.

Chan WH, Yu JS. 2000. Inhibition of UV irradiation-induced oxidative stress and apoptotic biochemical changes in human epidermal carcinoma A431 cells by genistein. *J Cell Biochem* 78:73–84.

Chang HC, Churchwell MI, Delclos KB, Newbold RR, Doerge DR. 2000. Mass spectrometric determination of Genistein tissue distribution in diet-exposed Sprague-Dawley rats. *J Nutr* 130:1963–1970.

Chen C, Wei T, Gao Z, Zhao B, Hou J, Xu H, Xin W, Packer L. 1999. Different effects of the constituents of EGb761 on apoptosis in rat cerebellar granule cells induced by hydroxyl radicals. *Biochem Mol Biol Int* 47:397–405.

Cherny RA, Atwood CS, Xilinas ME, Gray DN, Jones WD, McLean CA, Barnham KJ, Volitakis I, Fraser FW, Kim Y, Huang X, Goldstein LE, Moir RD, Lim JT, Beyreuther K, Zheng H, Tanzi RE, Masters CL, Bush AI. 2001. Treatment with a copper-zinc chelator markedly and rapidly inhibits beta-amyloid accumulation in Alzheimer's disease transgenic mice. *Neuron* 30:665–676.

Christen Y. 2000. Oxidative stress and Alzheimer disease. *Am J Clin Nutr* 71:621S–629S.

Clarkson TB, Anthony MS, Williams JK, Honore EK, Cline JM. 1998. The potential of soybean phytoestrogens for postmenopausal hormone replacement therapy. *Proc Soc Exp Biol Med* 217:365–368.

Donnelly PS, Xiao Z, Wedd AG. 2007. Copper and Alzheimer's disease. *Curr Opin Chem Biol* 11:128–133.

Graham DG, Tiffany SM, Bell WR, Jr, Gutknecht WF. 1978. Autoxidation versus covalent binding of quinones as the mechanism of toxicity of dopamine, 6-hydroxydopamine, and related compounds toward C1300 neuroblastoma cells in vitro. *Mol Pharmacol* 14:644–653.

Guo S, Yan J, Yang T, Yang X, Bezard E, Zhao B. 2007. Protective effects of green tea polyphenols in the 6-OHDA rat model of Parkinson's disease through inhibition of ROS-NO pathway. *Biol Psychiatry* 62:1353–1362.

Guo S, Bezard E, Zhao B. 2005. Protective effect of green tea polyphenols on the SH-SY5Y cells against 6-OHDA induced apoptosis through ROS-NO pathway. *Free Radic Biol Med* 39:682–695.

Guo Q, Rimbach G, Moini H, Weber S, Packer L. 2002. ESR and cell culture studies on free radical-scavenging and antioxidant activities of isoflavonoids. *Toxicology* 179:171–180.

Guo Q, Zhao B, Shen S, Hou J, Hu J, Xin W. 1999. ESR study on the structure-antioxidant activity relationship of tea catechins and their epimers. *Biochim Biophys Acta* 1427:13–23.

Guo Q, Zhao B, Li M, Shen S, Xin W. 1996. Studies on protective mechanisms of four components of green tea polyphenols against lipid peroxidation in synaptosomes. *Biochim Biophys Acta* 1304:210–222.

Gutierrez-Zepeda A, Santell R, Wu Z, Brown M, Wu Y, Khan I, Link CD, Zhao B, Luo Y. 2005. Soy isoflavone glycitein protects against beta amyloid-induced toxicity and oxidative stress in transgenic Caenorhabditis elegans. *BMC Neurosci* 6:54.

Halliwell B. 1992. Reactive oxygen species and the central nervous system. *J Neurochem* 59:1609–1623.

Hesse L, Beher D, Masters CL, Multhaup G. 1994. The beta A4 amyloid precursor protein binding to copper. *FEBS Lett* 349:109–116.

Hirsch EC, Faucheux B, Damier P, Mouatt-Prigent A, Agid Y. 1997. Neuronal vulnerability in Parkinson's disease. *J Neural Transm* Suppl 50:79–88.

Huang X, Atwood CS, Hartshorn MA, Multhaup G, Goldstein LE, Scarpa RC, Cuajungco MP, Gray DN, Lim J, Moir RD, Tanzi RE, Bush AI. 1999a. The A beta peptide of Alzheimer's disease directly produces hydrogen peroxide through metal ion reduction. *Biochemistry* 38:7609–7616.

Huang X, Cuajungco MP, Atwood CS, Hartshorn MA, Tyndall JD, Hanson GR, Stokes KC, Leopold M, Multhaup G, Goldstein LE, Scarpa RC, Saunders AJ, Lim J, Moir RD, Glabe C, Bowden EF, Masters CL, Fairlie DP, Tanzi RE, Bush AI. 1999b. Cu(II) potentiation of Alzheimer abeta neurotoxicity. Correlation with cell-free hydrogen peroxide production and metal reduction. *J Biol Chem* 274:37111–37116.

Inanami O, Watanabe Y, Syuto B, Nakano M, Tsuji M, Kuwabara M. 1998. Oral administration of (−)-catechin protects against ischemia-reperfusion-induced neuronal death in the gerbil. *Free Radic Res* 29:359–365.

Ishige K, Schubert D, Sagara Y. 2001. Flavonoids protect neuronal cells from oxidative stress by three distinct mechanisms. *Free Radic Biol Med* 30:433–446.

Johnson KL, Vaillant F, Lawen A. 1996. Protein tyrosine kinase inhibitors prevent didemnin B-induced apoptosis in HL-60 cells. *FEBS Lett* 383:1–5.

Kim H, Xia H, Li L, Gewin J. 2000. Attenuation of neurodegeneration-relevant modifications of brain proteins by dietary soy. *Biofactors* 12:243–250.

Kurzer MS, Xu X. 1997. Dietary phytoestrogens. *Annu Rev Nutr* 17:353–381.

Lamartiniere CA, Cotroneo MS, Fritz WA, Wang J, Mentor-Marcel R, Elgavish A. 2002. Genistein chemoprevention: Timing and mechanisms of action in murine mammary and prostate. *J Nutr* 132:552S–558S.

Levites Y, Amit T, Youdim MB, Mandel S. 2002a. Involvement of protein kinase C activation and cell survival/ cell cycle genes in green tea polyphenol (−)-epigallocatechin 3-gallate neuroprotective action. *J Biol Chem* 277:30574–30580.

Levites Y, Youdim MB, Maor G, Mandel S. 2002b. Attenuation of 6-hydroxydopamine (6-OHDA)-induced nuclear factor-kappaB (NF-kappaB) activation and cell death by tea extracts in neuronal cultures. *Biochem Pharmacol* 63:21–29.

Levites Y, Weinreb O, Maor G, Youdim MB, Mandel S. 2001. Green tea polyphenol (−)-epigallocatechin-3-gallate prevents N-methyl-4-phenyl-1,2,3,6-tetrahydropyridine-induced dopaminergic neurodegeneration. *J Neurochem* 78:1073–1082.

Liu Q, Zhao B. 2004. Nicotine attenuates beta-amyloid peptide-induced neurotoxicity, free radical and calcium accumulation in hippocampal neuronal cultures. *Br J Pharmacol* 141:746–754.

Liu NQ, Tao Y, Zhao B. 2003. ESR study on scavenging effect of nicotine on free radicals. *Appl Magn Reson* 24:105–112.

Lovell MA, Robertson JD, Teesdale WJ, Campbell JL, Markesbery WR. 1998. Copper, iron and zinc in Alzheimer's disease senile plaques. *J Neurol Sci* 158:47–52.

Magaki S, Raghavan R, Mueller C, Oberg KC, Vinters HV, Kirsch WM. 2007. Iron, copper, and iron regulatory protein 2 in Alzheimer's disease and related dementias. *Neurosci Lett* 418:72–76.

Mandel S, Amit T, Bar-Am O, Youdim MB. 2007. Iron dysregulation in Alzheimer's disease: Multimodal brain permeable iron chelating drugs, possessing neuroprotective-neurorescue and amyloid precursor protein-processing regulatory activities as therapeutic agents. *Prog Neurobiol* 82:348–360.

Markesbery WR. 1997. Oxidative stress hypothesis in Alzheimer's disease. *Free Radic Biol Med* 23:134–147.

Markesbery WR, Carney JM. 1999. Oxidative alterations in Alzheimer's disease. *Brain Pathol* 9:133–146.

Mecocci P, MacGarvey U, Beal MF. 1994. Oxidative damage to mitochondrial DNA is increased in Alzheimer's disease. *Ann Neurol* 36:747–751.

Mochizuki H, Goto K, Mori H, Mizuno Y. 1996. Histochemical detection of apoptosis in Parkinson's disease. *J Neurol Sci* 137:120–123.

Nanjo F, Goto K, Seto R, Suzuki M, Sakai M, Hara Y. 1996. Scavenging effects of tea catechins and their derivatives on 1,1-diphenyl-2-picrylhydrazyl radical. *Free Radic Biol Med* 21:895–902.

Ni Y, Zhao B, Hou J, Xin W. 1996. Preventive effect of Ginkgo biloba extract on apoptosis in rat cerebellar neuronal cells induced by hydroxyl radicals. *Neurosci Lett* 214:115–118.

Nie G, Jin C, Cao Y, Shen S, Zhao B. 2002. Distinct effects of tea catechins on 6-hydroxydopamine-induced apoptosis in PC12 cells. *Arch Biochem Biophys* 397:84–90.

Nie G, Wei T, Shen S, Zhao B. 2001. Polyphenol protection of DNA against damage. *Methods Enzymol* 335:232–244.

Obregon DF, Rezai-Zadeh K, Bai Y, Sun N, Hou H, Ehrhart J, Zeng J, Mori T, Arendash GW, Shytle D, Town T, Tan J. 2006. ADAM10 activation is required for green tea (−)-epigallocatechin-3-gallate-induced alpha-secretase cleavage of amyloid precursor protein. *J Biol Chem* 281:16419–16427.

Ohigashi T, Ueno M, Nonaka S, Nakanoma T, Furukawa Y, Deguchi N, Murai M. 2000. Tyrosine kinase inhibitors reduce bcl-2 expression and induce apoptosis in androgen-dependent cells. *Am J Physiol Cell Physiol* 278:C66–72.

Phinney AL, Drisaldi B, Schmidt SD, Lugowski S, Coronado V, Liang Y, Horne P, Yang J, Sekoulidis J, Coomaraswamy J, Chishti MA, Cox DW, Mathews PM, Nixon RA, Carlson GA, St George-Hyslop P, Westaway D. 2003. In vivo reduction of amyloid-beta by a mutant copper transporter. *Proc Natl Acad Sci USA* 100:14193–14198.

Rezai-Zadeh K, Shytle D, Sun N, Mori T, Hou H, Jeanniton D, Ehrhart J, Townsend K, Zeng J, Morgan D, Hardy J, Town T, Tan J. 2005. Green tea epigallocatechin-3-gallate (EGCG) modulates amyloid precursor protein cleavage and reduces cerebral amyloidosis in Alzheimer transgenic mice. *J Neurosci* 25:8807–8814.

Smith MA, Rottkamp CA, Nunomura A, Raina AK, Perry G. 2000. Oxidative stress in Alzheimer's disease. *Biochim Biophys Acta* 1502:139–144.

Smith MA, Hirai K, Hsiao K, Pappolla MA, Harris PL, Siedlak SL, Tabaton M, Perry G. 1998. Amyloid-beta deposition in Alzheimer transgenic mice is associated with oxidative stress. *J Neurochem* 70:2212–2215.

Soto-Otero R, Mendez-Alvarez E, Hermida-Ameijeiras A, Munoz-Patino AM, Labandeira-Garcia JL. 2000. Autoxidation and neurotoxicity of 6-hydroxydopamine in the presence of some antioxidants: Potential implication in relation to the pathogenesis of Parkinson's disease. *J Neurochem* 74:1605–1612.

Strausak D, Mercer JF, Dieter HH, Stremmel W, Multhaup G. 2001. Copper in disorders with neurological symptoms: Alzheimer's, Menkes, and Wilson diseases. *Brain Res Bull* 55:175–185.

Thompson KJ, Shoham S, Connor JR. 2001. Iron and neurodegenerative disorders. *Brain Res Bull* 55:155–164.

White AR, Du T, Laughton KM, Volitakis I, Sharples RA, Xilinas ME, Hoke DE, Holsinger RM, Evin G, Cherny RA, Hill AF, Barnham KJ, Li QX, Bush AI, Masters CL. 2006. Degradation of the Alzheimer disease amyloid beta-peptide by metal-dependent up-regulation of metalloprotease activity. *J Biol Chem* 281:17670–17680.

Wu Y, Wu Z, Butko P, Christen Y, Lambert MP, Klein WL, Link CD, Luo Y. 2006. Amyloid-beta-induced pathological behaviors are suppressed by Ginkgo biloba extract EGb 761 and ginkgolides in transgenic Caenorhabditis elegans. *J Neurosci* 26:13102–13113.

Xie YX, Bezard E, Zhao BL. 2005. Investigating the receptor-independent neuroprotective mechanisms of nicotine in mitochondria. *J Biol Chem* 280:32405–32412.

Xin W, Wei T, Chen C, Ni Y, Zhao B, Hou J. 2000. Mechanisms of apoptosis in rat cerebellar granule cells induced by hydroxyl radicals and the effects of EGb761 and its constituents. *Toxicology* 148:103–110.

Xin WJ, Zhao BL, Zhang JH. 1984. Studies on the property of sulfhydryl binding site on the lung normal and cancer cell membrane of Chinese hamster with maleimide spin labels. *Sci Sin [B]* 27:1008–1014.

Yoneda T, Hiramatsu M, Sakamoto M, Togasaki K, Komatsu M, Yamaguchi K. 1995. Antioxidant effects of "beta catechin." *Biochem Mol Biol Int* 35:995–1008.

Zeng H, Chen Q, Zhao B. 2004. Genistein ameliorates beta-amyloid peptide (25–35)-induced hippocampal neuronal apoptosis. *Free Radic Biol Med* 36:180–188.

Zhang Y, Zhao B. 2003. Green tea polyphenols enhance sodium nitroprusside-induced neurotoxicity in human neuroblastoma SH-SY5Y cells. *J Neurochem* 86:1189–1200.

Zhang J, Liu Q, Chen Q, Liu NQ, Li FL, Lu ZB, Qin C, Zhu H, Huang YY, He W, Zhao BL. 2006. Nicotine attenuates beta-amyloid-induced neurotoxicity by regulating metal homeostasis. *FASEB J* 20:1212–1214.

Zhao B, Guo Q, Xin W. 2001. Free radical scavenging by green tea polyphenols. *Methods Enzymol* 335:217–231.

Zhao BL, Li XJ, He RG, Cheng SJ, Xin WJ. 1989. Scavenging effect of extracts of green tea and natural antioxidants on active oxygen radicals. *Cell Biophys* 14:175–185.

Zhu CF, Liu Q, Zhang L, Yuan HX, Zhen W, Zhang JS, Chen ZJ, Hall SH, French FS, Zhang YL. 2007. RNase9, an androgen-dependent member of the RNase A family, is specifically expressed in the rat epididymis. *Biol Reprod* 76:63–73.

20 Epidemiology behind Fruit and Vegetable Consumption and Cancer Risk with Focus on Flavonoids

MARTA ROSSI[1], EVA NEGRI[1], CRISTINA BOSETTI[1], CLAUDIO PELUCCHI[1], AND CARLO LA VECCHIA[1,2]
[1]Istituto di Ricerche Farmacologiche "Mario Negri," Milan, Italy
[2]Università degli Studi di Milano, Istituto di Statistica Medica e Biometria "G.A. Maccacaro," Milan, Italy

CONTENTS

Plant Phenolics and Human Health: Biochemistry, Nutrition, and Pharmacology,
Edited by Cesar G. Fraga.

INTRODUCTION

Several aspects of the Mediterranean diet have been related to a reduced risk not only for cardiovascular disease but also of several cancers. It has been suggested that up to 25% of colorectal, 15% of breast, and 10% of prostate, pancreas, and endometrial cancers could be prevented by shifting to a Mediterranean diet [Trichopoulou et al., 2000]. With specific focus to a network of Italian case–control studies conducted after 1991 (Figs. 20.1 and 20.2), vegetable intake was inversely related to the risk of several common epithelial cancers: the odds ratios (OR) for digestive tract neoplasms ranged between 0.3 and 0.8 for the highest compared with the lowest levels of vegetable intake. Less consistent inverse relations were observed for some hormone-related neoplasms, such as breast and ovary. High fruit intake was associated to reduced ORs of cancers of the upper digestive tract, stomach, colorectum, and

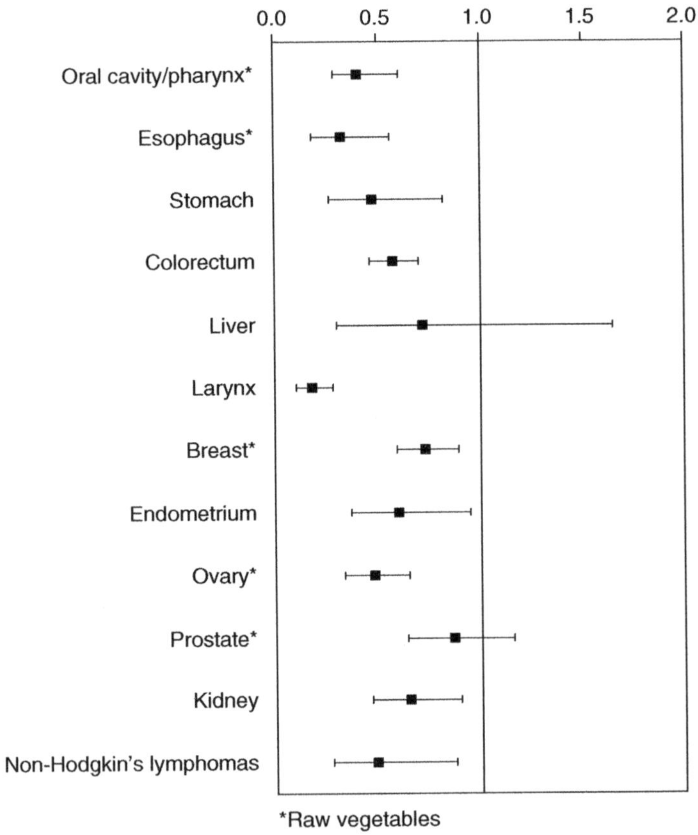

Figure 20.1 Odds ratios and 95% confidence intervals of selected cancers for the highest vs. the lowest levels of vegetable consumption. Italy, 1991–2005.

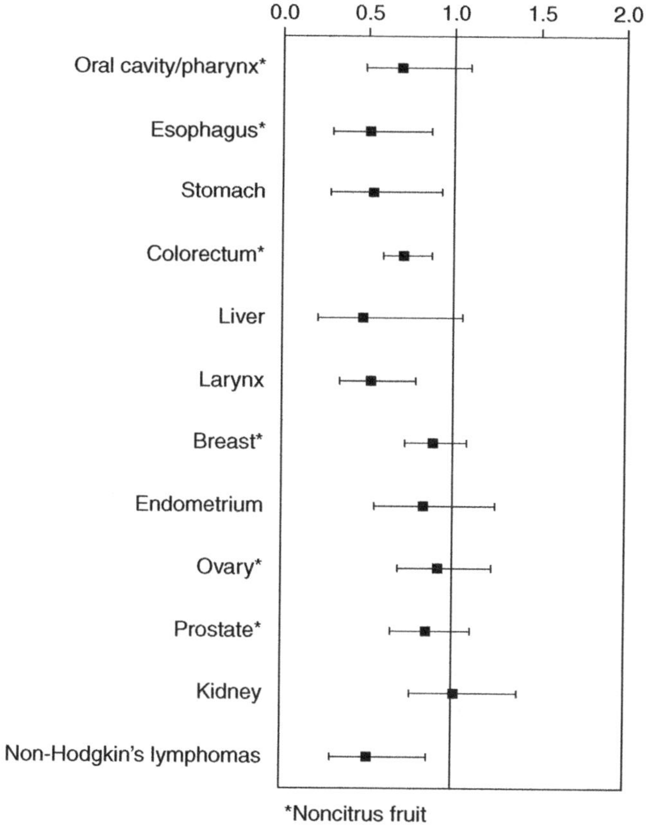

Figure 20.2 Odds ratios and 95% confidence intervals of selected cancers for the highest vs. the lowest levels of fruit consumption. Italy, 1991–2005.

non-Hodgkin's lymphomas. No material effect, however, was observed for fruit intake on neoplasms of the breast, the female genital tract, or the prostate [La Vecchia and Bosetti, 2006]. For digestive tract cancers, population-attributable risks for low intake of vegetables and fruit ranged between 15 and 40% [La Vecchia and Tavani, 1998].

Flavonoids are a group of several thousands of polyphenolic compounds present in fruit, vegetables, and beverages of plant origin [Manach et al., 2004; Ross and Kasum, 2002]. Originally known only for their role in plant pigmentation, flavonoids have been shown to have antioxidant, antimutagenic, and antiproliferative properties in vitro [Nijveldt et al., 2001; Peterson and Dwyer, 1998; Stoner and Mukhtar, 1995]. Flavonoids have been hypothesized to be responsible for the favorable effects of fruits and vegetables against various chronic diseases, including some common neoplasms [Arts and Hollman, 2005; La Vecchia et al., 1998; Neuhouser, 2004; Trichopoulou et al., 2003]. Recently, the availability of detailed and reliable food composition tables for flavonoids published by the U.S. Department of Agriculture on their six major classes (flavan-3-ols, flavanones, flavonols, anthocyanidins, flavones, and isoflavones) [U.S. Department of Agriculture, 2002, 2003] has allowed epidemiological studies to further investigate the role of flavonoids in cancer etiology. Intake of various flavonoids has been inversely related to the risk of cancers of the upper aerodigestive tract [De Stefani et al., 1999], digestive tract [Lagiou et al., 2004], breast [Fink et al., 2007; Peterson et al., 2003], and urogenital tract [Chang et al., 2007; Kurahashi et al., 2007].

THE ITALIAN DATA SET ON FLAVONOIDS AND CANCER RISK

We investigated flavonoids in relation to the risk of various neoplasms in a series of case–control studies conducted in Italy since the early 1990s [Gallus et al., 2005], which included 805 cancers of the oral cavity and pharynx, 743 of the esophagus, 460 of the larynx, 1953 of the colorectum, 2569 of the breast, 1031 of the ovary, 1294 of the prostate, and 767 of the kidney, and a total of 16,050 controls collected in various regions of northern, central and southern Italy. Cases were individuals admitted to hospitals with incident, histologically confirmed cancer, and controls were patients with no history of cancer admitted to the same hospitals for acute, nonneoplastic conditions. A reproducible [Franceschi et al., 1993] and valid [Decarli et al., 1996] food frequency questionnaire, was used including 78 foods or food groups, to assess the patients' usual diet and to estimate intakes of six major classes of flavonoids. This was performed using food and beverage content of flavonoids obtained from the U.S. Department of Agriculture databases [U.S. Department of Agriculture, 2002, 2003] and other sources [Liggins et al., 2000a, 2000b]. The main flavonoids in each class are the flavan-3-ols, epicatechin and catechin; the flavanones, hesperitin and narigerin; the anthocyanidins, cyanidin and malvidin; the flavonols, quercetin, myricetin, and kaempferol; the flavones, apigenin and luteolin; and the isoflavones, genistein and daidzein.

Using multiple logistic regression models that included the major confounding factors, and total energy intake, the relations between flavonoid intake and the risk of the cancers at the various sites have been examined. This allowed as to investigate the potential effects against cancer of these compounds present in fruit and vegetables on a population characterized by a diet rich in fruits and vegetables and with regular wine consumption [La Vecchia and Bosetti, 2006].

CANCERS OF THE UPPER DIGESTIVE TRACT

The study of oral and pharyngeal cancer (Table 20.1) showed that total flavonoids were inversely related to the risk of this neoplasm [Rossi et al., 2007b]. The ORs for the highest versus the lowest quintile of all classes of flavonoid intake were below unity. The OR was 0.51 (95% confidence interval,

Table 20.1 Odds Ratios[a] (OR) and 95% Confidence Intervals (CI) Among 805 Cases of Oral Cavity and Pharyngeal Cancer and 2081 Controls, According to Daily Intake Quintile of Six Classes of Flavonoids and Total Flavonoid. Italy, 1992–2005.

	Quintile of Intake					χ^2 Trend (p Value)
	1[b]	2	3	4	5	
Flavan-3-ols (mg)						
Upper cutpoint	23.3	40.3	59.4	99.6	—	
OR	1	1.06	0.85	0.77	0.84	2.05
(95% CI)		(0.76–1.48)	(0.60–1.20)	(0.54–1.09)	(0.60–1.18)	(0.15)
Flavanones (mg)						
Upper cutpoint	10.2	28.6	36.2	67.0	—	
OR	1	0.90	0.70	0.61	0.51	22.03
(95% CI)		(0.68–1.18)	(0.52–0.94)	(0.45–0.83)	(0.37–0.71)	(<0.001)
Anthocyanidins (mg)						
Upper cutpoint	5.3	13.4	20.9	33.2	—	
OR	1	0.89	0.70	0.66	0.86	2.32
(95% CI)		(0.63–1.25)	(0.49–1.01)	(0.47–0.94)	(0.60–1.22)	(0.13)
Flavonols (mg)						
Upper cutpoint	13.9	18.2	23.2	29.9	—	
OR	1	0.92	0.80	0.65	0.62	9.33
(95% CI)		(0.66–1.28)	(0.57–1.12)	(0.46–0.92)	(0.43–0.89)	(0.002)
Flavones (mg)						
Upper cutpoint	0.3	0.4	0.5	0.7	—	
OR	1	1.02	0.74	0.95	0.75	3.22
(95% CI)		(0.76–1.36)	(0.54–1.00)	(0.69–1.29)	(0.55–1.04)	(0.073)
Isoflavones (μg)						
Upper cutpoint	14.7	19.4	24.7	32.5	—	
OR	1	1.05	1.06	0.89	0.90	0.92
(95% CI)		(0.78–1.42)	(0.78–1.44)	(0.65–1.23)	(0.64–1.26)	(0.34)
Total flavonoids (mg)						
Upper cutpoint	83.5	117.9	151.8	204.0	—	
OR	1	0.77	0.64	0.63	0.56	11.92
(95% CI)		(0.56–1.06)	(0.46–0.89)	(0.45–0.87)	(0.40–0.78)	(0.001)

[a] Estimates from logistic regression models, conditioned on sex, age, study center, and adjusted for education, alcohol consumption, tobacco smoking, body mass index, and nonalcohol energy intake.
[b] Reference category.
Source: From Rossi et al. [2007b].

CI, 0.37–0.71) for flavanones, 0.62 (95% CI, 0.43–0.89) for flavonols, and 0.56 (95% CI, 0.40–0.78) for total flavonoids. No significant association emerged for other classes of flavonoids. The ORs were consistent across strata of age, sex, education, body mass index, tobacco, and alcohol. After allowance for vegetable and fruit consumption, the inverse relations with total flavonoids and flavanones remained significant, whereas that with flavonols became nonsignificant. None of the associations was significant after further allowance for vitamin C, probably on account of the high colinearity between these compounds.

In the study on esophageal cancer (Table 20.2), only flavanones were inversely associated with cancer risk (OR, 0.38, 95% CI, 0.23–0.66) [Rossi et al., 2007a]. The inverse relation tended to be stronger in subjects who drank ≥ 6 drinks of alcoholic beverages per day. After allowance for fruit intake or vitamin C, the association of flavanones with esophageal cancer remained inverse, though nonsignificant, suggesting that flavanone may explain, together with vitamin C, the protective effect of citrus fruits on esophageal cancer, since citrus fruits account for 90% of flavanone intake.

Table 20.2 Odds Ratios[a] (OR) and 95% Confidence Intervals (CI) Among 304 Cases of Squamous Cell Cancer of the Esophagus and 743 Controls, According to Daily Intake Quintile of Five Classes of Flavonoids and Total Flavonoids. Italy, 1992–1997.

	Quintile of Intake					χ^2 Trend (p Value)
	1[b]	2	3	4	5	
Flavan-3-ols (mg)						
Upper cutpoint	32.6	49.4	72.8	109.1	—	
OR	1	1.03	0.72	0.89	1.06	0.08
(95% CI)		(0.55–1.96)	(0.38–1.37)	(0.48–1.64)	(0.58–1.94)	(0.77)
Flavanones (mg)						
Upper cutpoint	6.4	19.7	33.9	58.8	—	
OR	1	0.42	0.59	0.56	0.38	8.47
(95% CI)		(0.26–0.67)	(0.36–0.97)	(0.34–0.92)	(0.23–0.66)	(0.004)
Anthocyanidins (mg)						
Upper cutpoint	8.1	18.2	30.2	41.2	—	
OR	1	0.60	0.56	0.41	0.84	0.45
(95% CI)		(0.29–1.24)	(0.29–1.13)	(0.22–0.79)	(0.46–1.54)	(0.50)
Flavonols (mg)						
Upper cutpoint	15.9	20.4	25.4	31.9	—	
OR	1	1.02	0.76	0.55	0.68	3.54
(95% CI)		(0.59–1.76)	(0.43–1.32)	(0.30–0.98)	(0.38–1.24)	(0.060)
Flavones (mg)						
Upper cutpoint	0.3	0.4	0.5	0.7	—	
OR	1	0.87	1.31	0.78	0.97	0.02
(95% CI)		(0.52–1.44)	(0.81–2.11)	(0.45–1.36)	(0.57–1.67)	(0.88)
Total flavonoids (mg)						
Upper cutpoint	96.5	127.7	166.2	217.4	—	
OR	1	1.29	0.78	0.81	0.99	0.23
(95% CI)		(0.72–2.33)	(0.42–1.42)	(0.45–1.46)	(0.55–1.79)	(0.63)

[a] Estimates from multiple logistic regression models adjusted for age, sex, study center, education, alcohol consumption, tobacco smoking, body mass index, and energy intake.
[b] Reference category.
Source: From Rossi et al. [2007a].

In the study on laryngeal cancer (Table 20.3), significant inverse relations were found for flavan-3-ols (OR,0.64), flavanones (OR,0.60), flavonols (OR,0.32), and total flavonoids (OR,0.60), although the overall trends in risk were significant only for flavanones and flavonols [Garavello et al., 2007]. The estimates persisted after controlling for vegetable, fruit, and vitamin C intake.

The Italian data are in agreement with those of a case–control study conducted in Uruguay that investigated the relation between diet and upper

Table 20.3 Odds Ratios[a] (OR) and 95% Confidence Intervals (CI) Among 460 Cases of Laryngeal Cancer and 1088 Controls, According to Daily Intake Quintile of Six Classes of Flavonoids and Total Flavonoids. Italy, 1992–2000.

	Quintile of Intake					χ^2 Trend (p Value)
	1[b]	2	3	4	5	
Flavan-3-ols (mg)						
Upper cutpoint	31.2	50.2	72.9	110.4	—	
OR	1	0.63	0.57	0.49	0.64	3.26
(95% CI)		(0.40–1.01)	(0.35–0.91)	(0.31–0.77)	(0.41–0.99)	(0.071)
Flavanones (mg)						
Upper cutpoint	7.7	20.8	33.7	49.2	—	
OR	1	0.55	0.54	0.39	0.60	10.53
(95% CI)		(0.38–0.81)	(0.37–0.80)	(0.26–0.59)	(0.41–0.89)	(0.001)
Anthocyanidins (mg)						
Upper cutpoint	8.6	18.8	30.7	41.1	—	
OR	1	1.21	1.05	0.51	0.77	2.14
(95% CI)		(0.71–2.06)	(0.60–1.83)	(0.30–0.89)	(0.46–1.30)	(0.14)
Flavonols (mg)						
Upper cutpoint	16.8	22.1	26.8	33.7	—	
OR	1	0.53	0.49	0.30	0.32	24.61
(95% CI)		(0.34–0.82)	(0.31–0.75)	(0.19–0.48)	(0.20–0.52)	(<0.001)
Flavones (mg)						
Upper cutpoint	0.3	0.4	0.5	0.7	—	
OR	1	0.76	0.80	0.58	0.76	2.52
(95% CI)		(0.50–1.14)	(0.53–1.20)	(0.38–0.89)	(0.50–1.15)	(0.11)
Isoflavones (μg)						
Upper cutpoint	14.7	19.8	25.1	32.6	—	
OR	1	1.23	0.87	0.67	0.73	6.41
(95% CI)		(0.82–1.85)	(0.57–1.33)	(0.43–1.03)	(0.47–1.14)	(0.011)
Total flavonoids (mg)						
Upper cutpoint	95.5	132.1	168.0	221.8	—	
OR	1	0.62	0.56	0.57	0.60	3.55
(95% CI)		(0.39–0.99)	(0.36–0.88)	(0.36–0.89)	(0.38–0.94)	(0.060)

[a] Estimates from multiple logistic regression models adjusted for age, sex, study center, education, alcohol consumption, tobacco smoking, body mass index, occupational physical activity, and nonalcohol energy intake.
[b] Reference category.
Source: From Garavello et al. [2007].

aerodigestive tract and reported inverse relations between flavonoids and oral (OR, 0.8 for the highest versus the lowest tertile), esophageal (OR, 0.4), and laryngeal cancer risk (OR, 0.6). No specific information on flavonoids are given in that study, and, to our knowledge, no other study investigated the issue. Moreover, flavanones, which were the class of flavonoids most strongly associated to the risk of neoplasms at the upper aerodigestive tract in these Italian studies, have been inversely related to gastric cancer in a Greek case–control study [Lagiou et al., 2004]. This is of particular interest, given the similarities in risk factors between various neoplasms of the upper digestive and respiratory tract (mainly tobacco, alcohol, as well as a diet poor in vegetables and fruit) [Franceschi et al., 1990; La Vecchia et al., 1999], and gastric cancer, too (i.e., tobacco, lower social class, and various indicators of a poor diet) [La Vecchia and Franceschi, 2000]. As also noted for esophageal and laryngeal cancers, the inverse relation between flavanones and stomach cancer was still persistent, although became weaker, after allowance for vitamin C [Garavello et al., 2007; Lagiou et al., 2004; Rossi et al., 2007a].

COLORECTAL CANCER

In the study on colorectal cancer (Table 20.4), a reduced risk was found for increasing intake of anthocyanidins (OR, 0.67 for the highest versus the lowest quintile, p-trend, 0.001), flavonols (OR, 0.64, p-trend < 0.001), flavones (OR, 0.78, p-trend, 0.004), and isoflavones (OR, 0.76, p-trend, 0.001). [Rossi et al., 2006].

The estimates did not substantially differ for colon and rectal cancers. After allowance for fruit and vegetable consumption, for dietary fiber, or for micronutrients previously associated to this tumor including vitamin C, the associations with flavonoids did not change by more than 10%. A recent case–control study of 1456 pairs of cases and controls conducted in Sweden confirmed a significant decrease in risk of colorectal cancer for intake of anthocyanidins and flavonols [Theodoratou et al., 2007], but there was no relation for isoflavones and flavones. However, the results on isoflavones and flavones are questionable due to the unusually and generally low intakes of the populations studied. In the Italian population, anthocyanidins were derived mainly from wine, red fruit, and onions, and flavonols from apples or pears, wine, and mixed salads.

BREAST CANCER

The study on breast cancer (Table 20.5) found a reduced risk of breast cancer for increasing intake of flavones (OR, 0.81, p-trend $=$ 0.02), and flavonols (OR, 0.80, p-trend 0.06) [Bosetti et al., 2005]. These findings are in agreement with those of a case–control study on breast cancer from Greece [Peterson et al., 2003], which found a similar protective effect of flavones. Also a recent

Table 20.4 Odds Ratios[a] (OR) and 95% Confidence Intervals (CI) Among 1953 Cases from Colorectal Cancer and 4154 Controls, According to Daily Intake Quintile of Six Classes of Flavonoids and Total Flavonoids. Italy, 1992–1996.

	Quintile of Intake					χ^2 Trend (p Value)
	1[b]	2	3	4	5	
Flavan-3-ols (mg)						
Upper cutpoint	25.02	48.79	48.79	78.29	—	
OR	1	0.75	0.75	0.79	0.98	0.11
(95% CI)		(0.63–0.91)	(0.62–0.90)	(0.65–0.95)	(0.82–1.18)	(0.74)
Flavanones (mg)						
Upper cutpoint	12.86	27.94	38.25	62.9	—	
OR	1	0.88	0.89	0.80	0.96	0.62
(95% CI)		(0.74–1.05)	(0.75–1.07)	(0.67–0.96)	(0.81–1.15)	(0.43)
Anthocyanidins (mg)						
Upper cutpoint	9.15	14.88	20.55	29.06	—	
OR	1	0.81	0.78	0.64	0.67	20.02
(95% CI)		(0.68–0.96)	(0.65–0.93)	(0.53–0.77)	(0.54–0.82)	(<0.001)
Flavonols (mg)						
Upper cutpoint	15.3	18.27	21.62	26.9	—	
OR	1	0.80	0.77	0.74	0.64	21.09
(95% CI)		(0.67–0.95)	(0.64–0.91)	(0.62–0.88)	(0.54–0.77)	(<0.001)
Flavones (mg)						
Upper cutpoint	0.30	0.40	0.50	0.65	—	
OR	1	0.82	0.72	0.76	0.78	8.38
(95% CI)		(0.69–0.98)	(0.61–0.86)	(0.64–0.91)	(0.65–0.93)	(0.004)
Isoflavones (μg)						
Upper cutpoint	16.22	20.88	25.66	32.96	—	
OR	1	0.86	0.79	0.77	0.76	10.81
(95% CI)		(0.72–1.02)	(0.66–0.94)	(0.65–0.92)	(0.63–0.91)	(0.001)
Total flavonoids (mg)						
Upper cutpoint	86.23	112.73	138.93	180.49	—	
OR	1	0.90	0.79	0.81	0.97	0.45
(95% CI)		(0.75–1.08)	(0.66–0.94)	(0.67–0.97)	(0.81–1.16)	(0.50)

[a] Estimates from multiple logistic regression models adjusted for age, sex, study center, education, alcohol consumption, body mass index, physical activity, family history of colorectal cancer and energy intake, according to the residual model.
[b] Reference category.
Source: From Rossi et al. [2006].

case–control study from the United States on 1434 women of Long Island [Fink et al., 2007] confirmed previous results on these Mediterranean populations. Among Italian women, flavonol intake derived mainly from various common vegetables and fruits, and flavone intake from aromatic herbs. Most epidemiological studies also found that dietary phytoestrogens were inversely associated to breast cancer risk [Cotterchio et al., 2008]. The absence of any meaningful association with isoflavone intake in the Italian study may be due to the

Table 20.5 Odds Ratio (OR)[a] and 95% Confidence Intervals (CI) Among 2569 Cases of Breast Cancer and 2588 Controls, According to Daily Intake Quintile of Six Classes of Flavonoids. Italy, 1991–1994.

	Quintile of Intake					χ^2 Trend (p Value)
	1[b]	2	3	4	5	
Flavan-3-ols (mg)						
Upper cutpoint	18.1	30.3	44.1	79.7	—	
OR[c]	1	0.98	0.80	1.01	0.86	1.28
(95% CI)		(0.82–1.18)	(0.66–0.98)	(0.83–1.23)	(0.71–1.05)	(0.26)
Flavanones (mg)						
Upper cutpoint	11.5	29.1	37.7	62.2	—	
OR[c]	1	1.19	1.11	1.15	0.95	0.48
(95% CI)		(1.00–1.43)	(0.92–1.33)	(0.96–1.38)	(0.79–1.15)	(0.49)
Anthocyanidins (mg)						
Upper cutpoint	3.7	7.9	14.3	20.5	—	
OR[c]	1	1.03	1.16	1.11	1.09	0.76
(95% CI)		(0.86–1.25)	(0.95–1.40)	(0.90–1.37)	(0.87–1.36)	(0.38)
Flavonols (mg)						
Upper cutpoint	12.6	16.4	21.5	29.9	—	
OR[c]	1	0.81	1.00	0.82	0.80	3.52
(95% CI)		(0.67–0.98)	(0.83–1.21)	(0.67–1.00)	(0.66–0.98)	(0.06)
Flavones (mg)						
Upper cutpoint	0.2	0.3	0.5	0.6	—	
OR[c]	1	0.94	0.97	0.86	0.81	5.41
(95% CI)		(0.79–1.13)	(0.81–1.17)	(0.71–1.04)	(0.66–0.98)	(0.02)
Isoflavones (μg)						
Upper cutpoint	13.4	19.0	25.2	34.7	—	
OR[c]	1	1.05	1.00	1.02	1.05	0.08
(95% CI)		(0.87–1.27)	(0.83–1.22)	(0.84–1.24)	(0.86–1.29)	(0.78)

[a] Estimates from multiple logistic regression models including terms for age, study center, education, parity, alcohol consumption and nonalcohol energy intake.
[b] Reference category.
Source: From Bosetti et al. [2005].

extremely limited intake of soya or soya products—and consequently of isoflavones—in the Italian population.

OVARIAN CANCER

The study on ovarian cancer (Table 20.6) found an inverse relation with ovarian cancer risk for flavonols (OR, 0.63; 95% CI, 0.47–0.84) and isoflavones (OR, 0.51; 95% CI, 0.37–0.69), with a significant trend in risk [Rossi et al., 2008]. In line with these findings, intake of isoflavones was associated with a lower risk of ovarian cancer in a U.S. cohort study including 280 cancers (relative risk 0.56 for more than 3 mg/day versus less than 1 mg/day of intake)

Table 20.6 Odds Ratios[a] (OR) and Corresponding 95% Confidence Intervals (CI) Among 1031 Cases of Ovarian Cancer and 2411 Controls, According to Daily Intake Quintile of Six Classes of Flavonoids and Total Flavonoids. Italy, 1992–1999.

	Quintile of Intake					χ^2 Trend (p Value)
	1[b]	2	3	4	5	
Flavan-3-ols (mg)						
Upper cutpoint	16.3	28.3	42.7	77.0	—	
OR	1	0.81	0.73	0.92	0.89	0.23
(95% CI)		(0.62–1.05)	(0.55–0.96)	(0.71–1.19)	(0.67–1.17)	(0.63)
Flavanones (mg)						
Upper cutpoint	12.2	31.3	36.6	67.0	—	
OR	1	1.22	0.99	1.07	1.28	1.44
(95% CI)		(0.93–1.60)	(0.75–1.31)	(0.81–1.41)	(0.98–1.68)	(0.23)
Anthocyanidins (mg)						
Upper cutpoint	3.5	7.4	12.3	19.4	—	
OR	1	0.74	0.90	1.02	0.99	0.68
(95% CI)		(0.56–0.98)	(0.68–1.18)	(0.79–1.34)	(0.76–1.29)	(0.41)
Flavonols (mg)						
Upper cutpoint	11.6	15.4	20.0	28.8	—	
OR	1	0.78	0.65	0.88	0.63	6.89
(95% CI)		(0.60–1.01)	(0.50–0.85)	(0.67–1.14)	(0.47–0.84)	(0.009)
Flavones (mg)						
Upper cutpoint	0.3	0.4	0.5	0.7	—	
OR	1	0.91	0.95	0.87	0.79	2.62
(95% CI)		(0.70–1.19)	(0.73–1.24)	(0.66–1.15)	(0.60–1.04)	(0.11)
Isoflavones (μg)						
Upper cutpoint	12.8	17.8	23.5	32.5	—	
OR	1	1.12	0.93	0.85	0.51	19.51
(95% CI)		(0.87–1.45)	(0.71–1.21)	(0.64–1.12)	(0.37–0.69)	(<0.001)
Total flavonoids (mg)						
Upper cutpoint	67.3	97.2	127.5	173.6	—	
OR	1	1.03	0.94	0.80	1.07	0.12
(95% CI)		(0.80–1.34)	(0.72–1.24)	(0.61–1.06)	(0.82–1.40)	(0.72)

[a] Estimates from multiple logistic regression models including terms for age, study center, year of interview, education, parity, oral contraceptive use, and family history of ovarian and/or breast cancer and energy intake, according to the residual model.
[b] Reference category.
Source: From Rossi et al. [2008].

[Chang et al., 2007], and in a Chinese case–control study based on 254 cases (OR 0.51 for more than 32.8 mg/day versus less than 11.6 mg/day of intake) [Zhang et al., 2004]. The evidence of an inverse association between isoflavones and ovarian cancer risk is supported by the observation that isoflavones have antiestrogenic effects [Ross and Kasum, 2002] and hence may inhibit the growth and proliferation of ovarian cell lines [Gercel-Taylor et al., 2004; Lukanova and Kaaks, 2005; Spinella et al., 2006].

PROSTATE CANCER

The study on prostate cancer (Table 20.7) found no association between prostate cancer risk with any of the analyzed flavonoids [Bosetti et al., 2006]. Although some flavonoids showed a favorable effect against prostate cancer [Magee and Rowland, 2004], the results from epidemiological studies are inconsistent. A recent prospective study from Japan found that isoflavone

Table 20.7 Odds Ratios[a] (OR) and Corresponding 95% Confidence Intervals (CI) Among 1294 Cases of Prostate Cancer and 1451 Controls, According to Daily Intake Quintile of Six Classes of Flavonoids and Total Flavonoids. Italy, 1991–2002.

	Quintile of Intake					χ^2 Trend (p Value)
	1[b]	2	3	4	5	
Flavan-3-ols (mg)						
Upper cutpoint	29.9	47.6	67.8	102.1	—	
OR	1	1.33	1.08	0.94	1.30	0.49
(95% CI)		(1.04–1.70)	(0.84–1.40)	(0.72–1.23)	(1.01–1.69)	(0.48)
Flavanones (mg)						
Upper cutpoint	5.2	19.1	33.5	46.9	—	
OR	1	1.10	1.06	1.00	0.96	0.35
(95% CI)		(0.86–1.40)	(0.82–1.36)	(0.78–1.29)	(0.75–1.23)	(0.56)
Anthocyanidins (mg)						
Upper cutpoint	8.3	18.4	29.8	40.3	—	
OR	1	1.13	1.20	1.18	1.18	1.49
(95% CI)		(0.88–1.45)	(0.94–1.54)	(0.91–1.52)	(0.91–1.53)	(0.22)
Flavonols (mg)						
Upper cutpoint	15.1	19.8	24.0	30.7	—	
OR	1	1.31	1.08	1.24	1.23	1.26
(95% CI)		(1.02–1.68)	(0.83–1.40)	(0.95–1.62)	(0.95–1.61)	(0.26)
Flavones (mg)						
Upper cutpoint	0.2	0.3	0.4	0.6	—	
OR	1	1.13	1.09	0.98	1.09	0.02
(95% CI)		(0.88–1.45)	(0.85–1.40)	(0.76–1.26)	(0.85–1.40)	(0.88)
Isoflavones (μg)						
Upper cutpoint	14.7	19.8	24.8	32.2	—	
OR	1	1.15	0.94	0.92	0.98	0.90
(95% CI)		(0.90–1.47)	(0.73–1.20)	(0.71–1.18)	(0.76–1.26)	(0.34)
Total flavonoids (mg)						
Upper cutpoint	109.4	145.7	184.4	240.8	—	
OR		1.32	0.97	1.13	1.20	0.59
(95% CI)		(1.06–1.64)	(0.76–1.23)	(0.88–1.44)	(0.92–1.58)	(0.44)

[a]Estimates from multiple logistic regression models including terms for age, study center, education, body mass index, family history of prostate cancer, and total energy intake.
[b]Reference category.
Source: From Bosetti et al. [2006].

intake was associated with a decreased risk of localized prostate cancer [Kurahashi et al., 2007], possibly explaining the much lower incidence of prostate cancer in Asian as compared to Western populations characterized by a low consumption of isoflavone-rich foods such as soya. In the Italian study, isoflavones derive mainly from beans, soy, and soy products, whose consumption is limited in Italy, and this may explain the inconsistent relation observed with prostate cancer.

Table 20.8 Odds Ratios[a] (OR) and 95% Confidence Intervals (CI) Among 767 Cases of Renal Cell Carcinoma and 1534 Controls, According to Daily Intake Quintile of Six Classes of Flavonoids and Total Flavonoids. Italy, 1992–2004.

	Quintile of Intake					χ^2 Trend (p Value)
	1[b]	2	3	4	5	
Flavan-3-ols (mg)						
Upper cutpoint	21.3	35.8	54.0	90.6	—	
OR	1	1.15	1.09	0.91	0.77	4.02
(95% CI)		(0.87–1.52)	(0.80–1.47)	(0.66–1.26)	(0.56–1.06)	(0.045)
Flavanones (mg)						
Upper cutpoint	9.6	22.5	33.9	57.8	—	
OR	1	0.94	0.94	0.71	0.90	1.90
(95% CI)		(0.71–1.25)	(0.71–1.24)	(0.53–0.96)	(0.67–1.21)	(0.17)
Anthocyanidins (mg)						
Upper cutpoint	5.5	11.6	19.7	32.4	—	
OR	1	0.96	0.79	1.06	0.94	0.19
(95% CI)		(0.72–1.28)	(0.58–1.09)	(0.74–1.53)	(0.60–1.47)	(0.67)
Flavonols (mg)						
Upper cutpoint	13.3	17.6	22.1	29.9	—	
OR	1	0.89	0.76	0.88	0.69	4.05
(95% CI)		(0.68–1.19)	(0.57–1.03)	(0.65–1.19)	(0.50–0.95)	(0.044)
Flavones (mg)						
Upper cutpoint	0.3	0.4	0.5	0.6	—	
OR	1	1.05	0.87	0.86	0.68	7.37
(95% CI)		(0.80–1.39)	(0.65–1.16)	(0.64–1.15)	(0.50–0.93)	(0.007)
Isoflavones (μg)						
Upper cutpoint	14.8	19.7	24.8	32.6	—	
OR	1	0.86	0.81	0.77	0.76	3.48
(95% CI)		(0.65–1.14)	(0.61–1.08)	(0.57–1.03)	(0.56–1.03)	(0.062)
Total flavonoids (mg)						
Upper cutpoint	80.6	109.1	139.4	180.9	—	
OR	1	1.01	0.89	0.77	0.80	3.55
(95% CI)		(0.76–1.33)	(0.66–1.19)	(0.56–1.05)	(0.58–1.11)	(0.060)

[a] Estimates from conditional logistic regression models, conditioned on sex, age, study center, and adjusted for period of interview, education, alcohol consumption, tobacco smoking, body mass index, occupational physical activity, family history of kidney cancer, and total energy intake.
[b] Reference category.
Source: From Bosetti et al. [2007].

RENAL CANCER

The study on renal cell carcinoma (Table 20.8) showed that flavonols (OR, 0.69, 95% CI, 0.50–0.95) and flavones (OR, 0.68, 95% CI, 0.50–0.93) were inversely related to the risk of renal cancer [Bosetti et al., 2007]. Allowance for vegetable and fruit consumption only partly explained the inverse relation with flavonoids. No other study systematically investigated the relation between flavonoids and renal cancer risk.

CONCLUSIONS

In conclusion, the findings of this large series of Italian case–control studies provide support for a protective role of flavanones on upper aerodigestive tract, flavonols and anthocyanidins on colorectal, flavonols and flavones on breast, and isoflavones on ovarian cancers. For most investigated neoplasms, adjustment for flavonoids reduced the strength of the inverse association between vegetables or fruit consumption and the risk of cancer, whereas allowance for fruit and vegetable consumption only moderately changed the observed associations with flavonoids. This suggests that, on one side, a diet rich in fruit and vegetables does not alone account for the observed protections on cancer risk, and, on the other side, the relation with fruit and vegetables is not totally explained by flavonoid intake.

ACKNOWLEDGMENTS

This work was conducted with the support of the Italian Association for Cancer Research and the Italian League against Cancer. The work in this chapter was undertaken while CLV was a senior fellow at the International Agency for Research on Cancer. The authors thank Ms. I. Garimoldi for editorial assistance.

REFERENCES

Arts IC, Hollman PC. 2005. Polyphenols and disease risk in epidemiologic studies. *Am J Clin Nutr* 81:317S–325S.

Bosetti C, Rossi M, McLaughlin JK, Negri E, Talamini R, Lagiou P, Montella M, Ramazzotti V, Franceschi S, La Vecchia C. 2007. Flavonoids and the risk of renal cell carcinoma. *Cancer Epidemiol Biomarkers Prev* 16:98–101.

Bosetti C, Bravi F, Talamini R, Parpinel M, Gnagnarella P, Negri E, Montella M, Lagiou P, Franceschi S, La Vecchia C. 2006. Flavonoids and prostate cancer risk: A study in Italy. *Nutr Cancer* 56:123–127.

Bosetti C, Spertini L, Parpinel M, Gnagnarella P, Lagiou P, Negri E, Franceschi S, Montella M, Peterson J, Dwyer J, Giacosa A, La Vecchia C. 2005. Flavonoids and breast cancer risk in Italy. *Cancer Epidemiol Biomarkers Prev* 14:805–808.

Chang ET, Lee VS, Canchola AJ, Clarke CA, Purdie DM, Reynolds P, Anton-Culver H, Bernstein L, Deapen D, Peel D, Pinder R, Ross RK, Stram DO, West DW, Wright W, Ziogas A, Horn-Ross PL. 2007. Diet and risk of ovarian cancer in the California Teachers Study cohort. *Am J Epidemiol* 165:802–813.

Cotterchio M, Boucher BA, Kreiger N, Mills CA, Thompson LU. 2008. Dietary phytoestrogen intake-lignans and isoflavones-and breast cancer risk (Canada). *Cancer Causes Control* 19:259–272.

Decarli A, Franceschi S, Ferraroni M, Gnagnarella P, Parpinel MT, La Vecchia C, Negri E, Salvini S, Falcini F, Giacosa A. 1996. Validation of a food-frequency questionnaire to assess dietary intakes in cancer studies in Italy. Results for specific nutrients. *Ann Epidemiol* 6:110–118.

De Stefani E, Ronco A, Mendilaharsu M, Deneo-Pellegrini H. 1999. Diet and risk of cancer of the upper aerodigestive tract—II. Nutrients. *Oral Oncol* 35:22–26.

Fink BN, Steck SE, Wolff MS, Britton JA, Kabat GC, Schroeder JC, Teitelbaum SL, Neugut AI, Gammon MD. 2007. Dietary flavonoid intake and breast cancer risk among women on Long Island. *Am J Epidemiol* 165:514–523.

Franceschi S, Negri E, Salvini S, Decarli A, Ferraroni M, Filiberti R, Giacosa A, Talamini R, Nanni O, Panarello G, et al. 1993. Reproducibility of an Italian food frequency questionnaire for cancer studies: Results for specific food items. *Eur J Cancer* 29A:2298–2305.

Franceschi S, Talamini R, Barra S, Baron AE, Negri E, Bidoli E, Serraino D, La Vecchia C. 1990. Smoking and drinking in relation to cancers of the oral cavity, pharynx, larynx, and esophagus in northern Italy. *Cancer Res* 50:6502–6507.

Gallus S, Talamini R, Giacosa A, Montella M, Ramazzotti V, Franceschi S, Negri E, La Vecchia C. 2005. Does an apple a day keep the oncologist away? *Ann Oncol* 16:1841–1844.

Garavello W, Rossi M, McLaughlin JK, Bosetti C, Negri E, Lagiou P, Talamini R, Franceschi S, Parpinel M, Dal Maso L, La Vecchia C. 2007. Flavonoids and laryngeal cancer risk in Italy. *Ann Oncol* 18:1104–1109.

Gercel-Taylor C, Feitelson AK, Taylor DD. 2004. Inhibitory effect of genistein and daidzein on ovarian cancer cell growth. *Anticancer Res* 24:795–800.

Kurahashi N, Iwasaki M, Sasazuki S, Otani T, Inoue M, Tsugane S. 2007. Soy isoflavone consumption is not associated with increased risk of advanced prostate cancer. *Cancer Epidemiol Biomarkers Prev* 16:2169.

La Vecchia C, Bosetti C. 2006. Diet and cancer risk in Mediterranean countries: Open issues. *Public Health Nutr* 9:1077–1082.

La Vecchia C, Franceschi S. 2000. Nutrition and gastric cancer with a focus on Europe. *Eur J Cancer Prev* 9:291–295.

La Vecchia C, Tavani A. 1998. Fruit and vegetables, and human cancer. *Eur J Cancer Prev* 7:3–8.

La Vecchia C, Chatenoud L, Franceschi S, Soler M, Parazzini F, Negri E. 1999. Vegetables and fruit and human cancer: Update of an Italian study. *Int J Cancer* 82:151–152.

La Vecchia C, Decarli A, Pagano R. 1998. Vegetable consumption and risk of chronic disease. *Epidemiology* 9:208–210.

Lagiou P, Samoli E, Lagiou A, Peterson J, Tzonou A, Dwyer J, Trichopoulos D. 2004. Flavonoids, vitamin C and adenocarcinoma of the stomach. *Cancer Causes Control* 15:67–72.

Liggins J, Bluck LJ, Runswick S, Atkinson C, Coward WA, Bingham SA. 2000a. Daidzein and genistein content of fruits and nuts. *J Nutr Biochem* 11:326–331.

Liggins J, Bluck LJ, Runswick S, Atkinson C, Coward WA, Bingham SA. 2000b. Daidzein and genistein contents of vegetables. *Br J Nutr* 84:717–725.

Lukanova A, Kaaks R. 2005. Endogenous hormones and ovarian cancer: Epidemiology and current hypotheses. *Cancer Epidemiol Biomarkers Prev* 14:98–107.

Magee PJ, Rowland IR. 2004. Phyto-oestrogens, their mechanism of action: Current evidence for a role in breast and prostate cancer. *Br J Nutr* 91:513–531.

Manach C, Scalbert A, Morand C, Remesy C, Jimenez L. 2004. Polyphenols: Food sources and bioavailability. *Am J Clin Nutr* 79:727–747.

Neuhouser ML. 2004. Dietary flavonoids and cancer risk: Evidence from human population studies. *Nutr Cancer* 50:1–7.

Nijveldt RJ, van Nood E, van Hoorn DE, Boelens PG, van Norren K, van Leeuwen PA. 2001. Flavonoids: A review of probable mechanisms of action and potential applications. *Am J Clin Nutr* 74:418–425.

Peterson J, Dwyer J. 1998. Flavonoids: Dietary occurrence and biochemical activity. *Nutr Res* 18:1995–2018.

Peterson J, Lagiou P, Samoli E, Lagiou A, Katsouyanni K, La Vecchia C, Dwyer J, Trichopoulos D. 2003. Flavonoid intake and breast cancer risk: A case–control study in Greece. *Br J Cancer* 89:1255–1259.

Ross JA, Kasum CM., 2002. Dietary flavonoids: Bioavailability, metabolic effects, and safety. *Annu Rev Nutr* 22:19–34.

Rossi M, Negri E, Lagiou P, Talamini R, Dal Maso L, Montella M, Franceschi S, La Vecchia C. 2008. Flavonoids and ovarian cancer risk: A case-control study in Italy. Int J Cancer.

Rossi M, Garavello W, Talamini R, La Vecchia C, Franceschi S, Lagiou P, Zambon P, Dal Maso L, Bosetti C, Negri E. 2007a. Flavonoids and risk of squamous cell esophageal cancer. *Int J Cancer* 120:1560–1564.

Rossi M, Garavello W, Talamini R, Negri E, Bosetti C, Dal Maso L, Lagiou P, Tavani A, Polesel J, Barzan L, Ramazzotti V, Franceschi S, La Vecchia C. 2007b. Flavonoids and the risk of oral and pharyngeal cancer: A case-control study from Italy. *Cancer Epidemiol Biomarkers Prev* 16:1621–1625.

Rossi M, Negri E, Talamini R, Bosetti C, Parpinel M, Gnagnarella P, Franceschi S, Dal Maso L, Montella M, Giacosa A, La Vecchia C. 2006. Flavonoids and colorectal cancer in Italy. *Cancer Epidemiol Biomarkers Prev* 15:1555–1558.

Spinella F, Rosano L, Di Castro V, Decandia S, Albini A, Nicotra MR, Natali PG, Bagnato A. 2006. Green tea polyphenol epigallocatechin-3-gallate inhibits the endothelin axis and downstream signaling pathways in ovarian carcinoma. *Mol Cancer Ther* 5:1483–1492.

Stoner GD, Mukhtar H. 1995. Polyphenols as cancer chemopreventive agents. *J Cell Biochem* Suppl 22:169–180.

Theodoratou E, McNeill G, Cetnarskyj R, Farrington SM, Tenesa A, Barnetson R, Porteous M, Dunlop M, Campbell H. 2007. Dietary fatty acids and colorectal cancer: A case-control study. *Am J Epidemiol* 166:181–195.

Trichopoulou A, Naska A, Antoniou A, Friel S, Trygg K, Turrini A. 2003. Vegetable and fruit: The evidence in their favour and the public health perspective. *Int J Vitam Nutr Res* 73:63–69.

Trichopoulou A, Lagiou P, Kuper H, Trichopoulos D. 2000. Cancer and Mediterranean dietary traditions. *Cancer Epidemiol Biomarkers Prev* 9:869–873.

U.S. Department of Agriculture. 2003. *USDA Database for the Flavonoid Content of Selected Foods*. USDA, Beltsville, MD.

U.S. Department of Agriculture. 2002. *Iowa State University Database on the Isoflavone Content of Foods*, Release 1.3 USDA, Beltsville, MD.

Zhang M, Xie X, Lee AH, Binns CW. 2004. Soy and isoflavone intake are associated with reduced risk of ovarian cancer in Southeast China. *Nutr Cancer* 49:125–130.

21 Phenylpropanoid Metabolism in Plants: Biochemistry, Functional Biology, and Metabolic Engineering

ALBERTO B. LANDOLINO[1] AND DOUGLAS R. COOK[2]
[1]Monsanto, Davis, California
[2]Department of Plant Pathology and College of Agricultural and Environmental Sciences Genomics Facility, University of California, Davis, California

CONTENTS

Plant Phenolics and Human Health: Biochemistry, Nutrition, and Pharmacology, Edited by Cesar G. Fraga.

INTRODUCTION

Phenylpropanoids constitute a diverse group of phenylalanine-derived secondary metabolites with a multitude of functions in biology and ecophysiology of plants [Taylor and Grotewold, 2005]. Simple phenolics and flavonoids have potent bioactivity, and their presentation at key points in a plant's life cycle can constitute important adaptations, often conferring properties that are integral to plant survival. For example, several plant reproductive processes are mediated by different flavonoids acting to attract pollinators [Mol et al., 1998], contributing to pollen structure and function [Coe et al., 1981; Mo et al., 1992; Taylor and Jorgensen, 1992], and facilitating seed dispersal and protection [Debeaujon et al., 2000; Lepiniec et al., 2006]. Similarly, lignins, which are structural components of plant cell walls, and several flavonoids exert profound effects on plant growth and morphogenesis by modulating structural integrity of the plant body and acting to regulate polar auxin transport, respectively [Jacobs and Rubery, 1988; Mathesius et al., 1998; Murphy et al., 2000; Brown et al., 2001; Buer and Muday, 2004; Peer et al., 2004]. Simple phenolics and flavonoids are often key determinants of the outcome of interactions with other organisms, including serving as allelopathic agents in interspecies plant–plant competition [Bais et al., 2004], acting as insect and herbivore deterrants [Feeny, 1970; Forkner et al., 2004; McAllister et al., 2005], mediating signaling processes with soil-borne symbiotic nitrogen-fixing bacteria [Peters et al., 1986; Redmond et al., 1986; Djordjevic et al., 1987; Harrison, 2005; Wasson et al., 2006] and contributing to defense mechanisms against different pathogens [Treutter, 2005].

Knowledge about most enzymatic steps as well as regulatory mechanisms at the transcriptional and posttranscriptional level has greatly improved over the last 20 years [Grotewold, 2006] in part due to the availability of a wide collection of natural or induced mutants in different model plant systems [Peer et al., 2001]. Several transcription factors of the myeloblastoma (MYB), basic helix-loop-helix (bHLH), and WD-repeat (WDR) families control expression of different biosynthetic genes during the course of normal plant development [Winkel-Shirley, 2002] or they activate gene expression in response to abiotic and biotic stress elicitors [Lepiniec et al., 2006]. The biosynthetic pathway is thought to be organized as a metabolom on the cytoplasmic side of the endoplasmic reticulum (ER), with colocalization greatly enhancing biosynthetic efficiency [Winkel-Shirley, 2002; Winkel, 2004]. Several excellent reviews have summarized our current knowledge of the topology and regulation of this pathway in plants [Shirley, 1996; Whinkel-Shirley, 2001; Saito and Yamazaki, 2002; Xie and Dixon, 2005]. Here we revisit the most relevant information about the genetic and enzymatic components of phenylpropanoid metabolism with special emphasis on the (iso)flavonoid pathway. We also summarize the critical biological roles played by these secondary metabolites in different aspects of the life cycle of plants. Finally, we discuss the current status and future perspectives of metabolic engineering of the phenylpropanoid pathway to enhance the nutraceutical value of plant-based foods, improving structural properties of lignocellulosic biomass for industrial applications, and discovering new bioactive (iso)flavonoids with pharmaceutical application.

PHENYLPROPANOID METABOLISM IN PLANTS

Phenylpropanoid Biosynthesis in Plants: Enzymology and Pathway Topology

The phenylpropanoid pathway gives rise to a myriad of polyfunctional bioactive molecules that accumulate in plants as monomeric (e.g., hydroxycinnamic acids, monolignols, anthocyanins, flavonols, flavones, aurones, stilbenes, isoflavones) or polymeric [e.g., proanthocyanidins (PA), phlobaphenes, lignins] compounds. Flavonoids in particular are built on a conserved C15 (6:3:6) structure, with a diversity of conjugation and substitution patterns in the C-ring (Fig. 21.1). The nature of these modifications serves as the basis to define flavonoid groups, many of which have distinctive physicochemical properties and biological activities [Frankel et al., 1995; de Beer et al., 2002; Rice-Evans and Miller, 1996; Rice-Evans et al., 1996; Springob et al., 2003]. Additional hydroxylation and methylation substitutions at positions 3′, 4′, and 5′ of the B-ring further deliniate specific compounds within a given flavonoid group. Flavonoids are present in all plant taxa with the noticeable exception of algae [reviewed by Raushier, 2006], consistent with the proposal that gain-of-function traits associated with flavonoid accumulation were critical to increased reproductive fitness and survival of early land plants. Phenylpropanoid biosynthesis requires the coordinated action of a multiplicity of enzymes and enzymatic complexes to channel carbon from intermediates of primary metabolism to the different end products of the pathway [Winkel, 2004]. The topology of the biosynthetic pathway as well as structural similarity of the end products can be traced from relatively recent

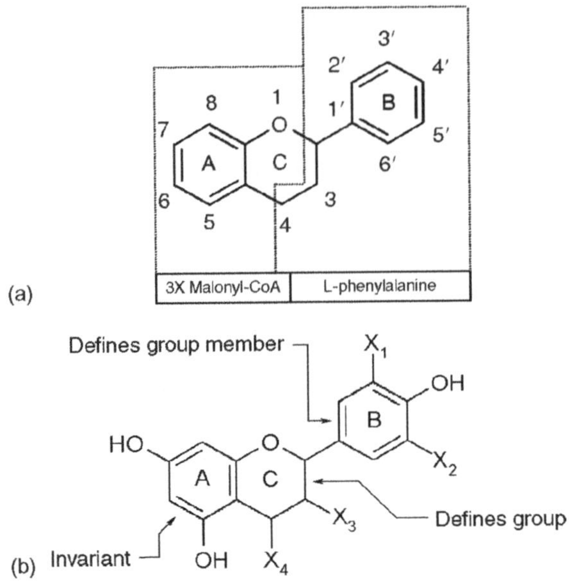

Figure 21.1 Basic C15 (6:3:6) structure of flavonoids and numbering system for carbon positions in the three rings. The B-ring carbon skeleton derives entirely from L-phenylalanine. Ring A and C is originated in the condensation reaction with three malonyl-CoA moieties, B, C. Common patterns of conjugation and substitutions in the B-and C-rings defining the different flavonoid groups.

gymnosperm and angiosperm lineages, to the more ancient taxa of extant bryophytes (mosses) [Rausher, 2006].

The deamination of L-phenylalanine to trans-cinnamic acid, which is catalyzed by phenylalanine ammonia lyase (PAL, EC 4.3.1.5) (Fig. 21.2), represents the first committed step of the general phenylpropanoid pathway. The resulting cinnamate is converted to *p*-coumaric acid in a hydroxylation reaction mediated by the P450 mono-oxygenase cinnamic acid 4-hydroxylase (C4H, EC 1.14.13.11). A series of sequential reactions controlled by hydroxylases, SAM-dependent *O*-methyltransferases, and dehydratases convert *p*-coumarate into a pool of free (e.g., caffeic, ferulic, 5-hydroxyferulic and sinapic acid) and sterified (e.g., with organic acids) hydroxycinnamic acids [Boerjan et al., 2003]. Activation of hydroxycinnamic acids with coenzyme A in an ATP-dependent thioesterification reaction, catalyzed by 4-coumaroyl-CoA ligase (4CL, EC 6.2.1.12) [Heldt, 1999], is required for biosynthesis of the monolignol precursors of lignins [Boerjan et al., 2003] and the initial step of (iso)flavonoid and stilbene biosynthesis.

The irreversible condensation reaction between a starter hydroxycinnamoyl-CoA molecule and three acetate extender units derived from malonyl-CoA constitute the first committed step of flavonoid and stilbene biosynthesis [Springob et al., 2003] (Fig. 21.2). The reactions catalyzed by two type III

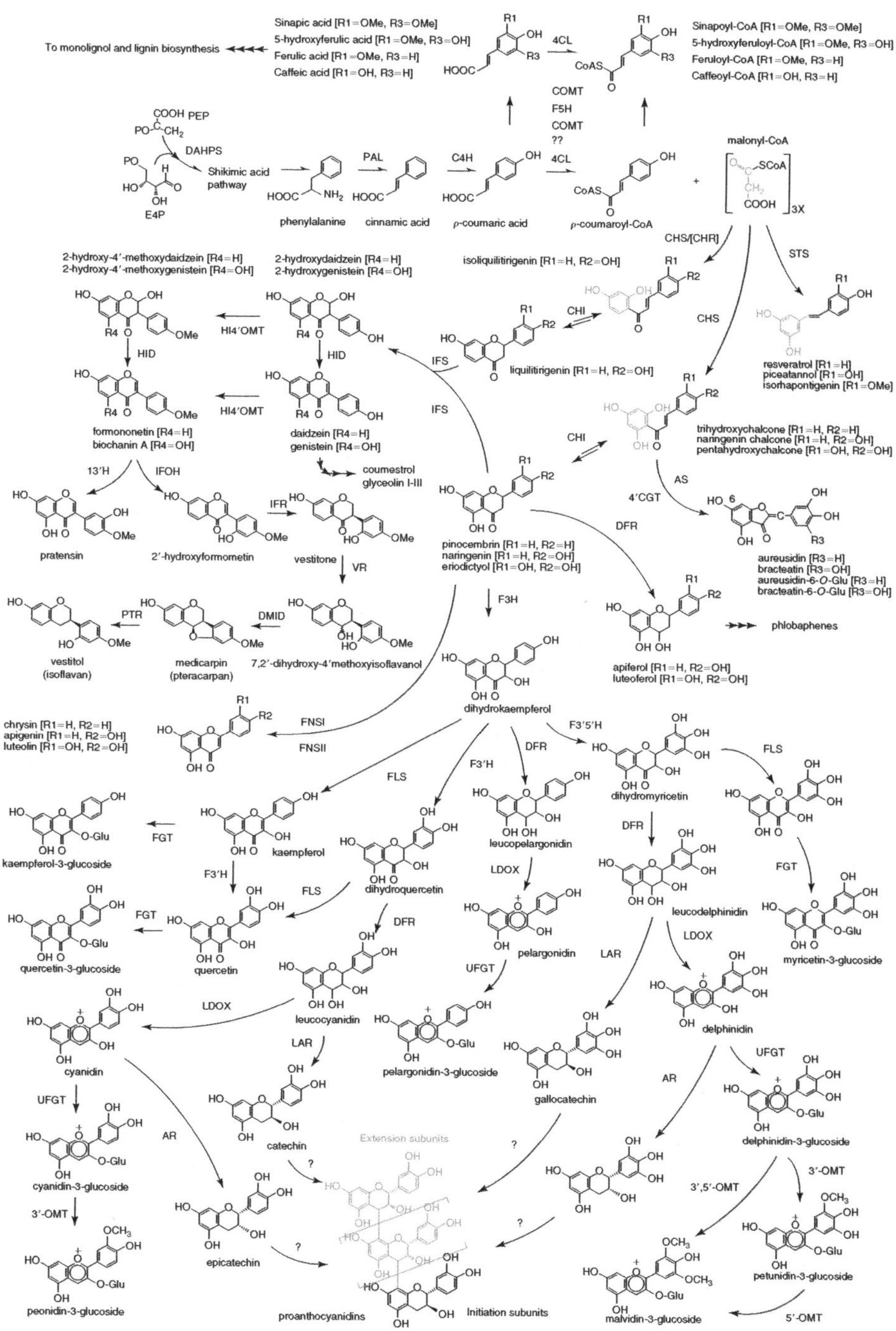

Figure 21.2 Schematic representation of the general phenylpropanoid, flavonoid, and isoflavonoid pathways in plants (see text for legends).

polyketide synthases (PKS), chalcone synthase (CHS, EC 2.3.1.74), and stilbene synthase (STS, EC 2.3.1.95) yield a linear tetraketide intermediate [Watts et al., 2006; Halls and Yu, 2008], which after subsequent aldol condensation, decarboxylation, and a final aromatization step, is converted to the C15 (6:3:6) chalcone skeleton that is common to all flavonoids or to the C14 (6:2:6) stilbene structure [Watts et al., 2006].

Stilbenes, and in particular trans-resveratrol, are found in at least 72 plant species from 31 different genera and 12 botanical families [Counet et al., 2006]. Among the most publicized examples with relevance to the human diet are grapes (*Vitis* sp.) [Li et al., 2006a; Creasy and Creasy, 1998], cranberries (*Vaccinum macrocarpon*) [Wang et al., 2002], peanuts (*Arachis hypogea*) [Lee et al., 2004], and cocoa (*Theobroma cacao*) [Counet et al., 2006]. STS activity converts the acetyl-CoA esters of the hydroxycinnamic acids (i.e., *p*-coumaric, caffeic, and ferulic acids) to their respective stilbenes, namely resveratrol (3,5,4′-trans-trihydroxystilbene), piceatannol (3,3′,4′,5-tetrahydroxystilbene), and isorhapontigenin (3,4′,5-dihydroxy-3′-methoxystilbene). STS isoforms from different plant species vary in their affinity for hydroxycinnamoyl-CoAs [Schoppner and Kindl, 1984; Kodan et al., 2002]. For example, the peanut STS cannot properly fold feruoyl-CoA-derived linear tetraketide intermediates, resulting in the accumulation of tri- and tetraketide lactones [Watts et al., 2006], while STS from *Pinus* spp. prefers cinnamoyl-CoA over p-coumaroyl-CoA [Kodan et al., 2002; Raiber et al., 1995]. Native American and European grapes (*Vitis* spp.) constitute an excellent model for characterization of stilbene biosynthesis and metabolism. Stilbenes accumulate at high levels in grape leaves and stems (Langcake, 1981) and at lower concentrations in berry skins and seeds [Li et al., 2006a]. Steady-state levels can be greatly changed by fungal elicitors [Langcake and Pryce, 1977a], ultra iolet (UV) radiation [Langcake and Pryce, 1977b; Cantos et al., 2002], and phytohormones [Tassoni et al., 2005]. The trans-isomer of resveratrol is typically the most abundant stilbene species found in different tissues [Li et al., 2006a; Ali and Strommer, 2003; Versari et al., 2001; Waterhouse and Lamuela-Raventos, 1994]. In lower quantities, grapes also produce other stilbene aglycones such as cis-resveratrol [Moreno et al., 2008], trans-piceatannol or trans-astringinin (3′,4′,3,5-tetrahydroxystilbene) [Cantos et al., 2002], and pterostilbene (3,5-dimethoxy-4′-hydroxystilbene) [Langcake, 1981]. A recently characterized multifunctional glucosyltransferase from the native American grape *Vitis labrusca* [Hall and De Luca, 2006] accepts a range of structurally similar substrates, including resveratrol, flavonols, and hydroxycinnamic acids and might be involved in the biosynthesis of the 3-*O*-monoglucosides trans-piceid (trans-resveratrol-3-*O*-β-glucoside) and trans-astringin (trans-piceatannol-3-*O*-β-glucoside) [Cantos et al., 2002]. Glucosylation may play a critical role in protecting stilbenes from polyphenol oxidase-mediated oxidative degradation [Regev-Shoshani et al., 2003]. Several plant species [Sotheeswaran and Pasupathy, 1993], including grapes [Langcake, 1981; Langcake and Pryce, 1977a,b; Cantos et al., 2002], accumulate oligomeric stilbenes with potent phytoalexin properties. In grapes, certain stilbene oligomers, known as viniferins [Langcake and Pryce, 1977a,b; Langcake, 1981; Cantos et al., 2002], arise from the oxidative coupling of trans-resveratrol moieties [Ros Barceló et al., 2003]. The reaction is likely to be catalyzed by a still uncharacterized 4-hydroxystilbene peroxidase [Calderon et al., 1992; Ros Barceló and Pomar, 2002; Ros Barceló et al., 2003, Pezet et al., 2004] that oxidizes trans-resveratrol to form stilbene dimers (ε-viniferin), trimers (α-viniferin), and tetramers (Γ-2-viniferin) [Ros Barceló et al., 2003].

The 2′,4′,6′,4-tetrahydroxychalcone (naringenin chalcone or chalcone) resulting from the CHS reaction is converted to the (2*S*)-flavanone naringenin spontaneously or by a stereo-specific cyclization reaction catalyzed by chalcone isomerase (CHI, EC 5.5.1.6) [Springob et al., 2003] (Fig. 21.2). Naringenin is a biosynthetic intermediate for a range of flavonoid species, including flavones, isoflavonols, flavonols, anthocyanins, phlobaphenes, flavan-3-ols, and proanthocyanidins. CHS activity can also drive the stepwise condensation of three molecules of malonyl-CoA with cinnamoyl-CoA to yield pinocembrin chalcone, with caffeoyl-CoA to yield eriodictyol chalcone (2′,4′,6′,3,4-pentahydroxychalcone), and with feruloyl-CoA to yield homoeriodictyol chalcone [Yan et al., 2005]. The action of CHI on these chalcone substrates results in the production of their respective (2*S*)-flavanones: (2*S*)-pinocembrin, (2*S*)-eriodictyol, and (2*S*)-homoeriodctyiol [Yan et al., 2005]. Other flavanones can result from the activity of P450-dependent monooxygenases on naringenin. For example, eriodictyol results from hydroxylation at position 3′ of the B-ring by flavonoid 3′-hydroxylase (F3′H, EC 1.14.13.21) [Springob et al., 2003].

Flavones are ubiquitous in plant organs and tissues of more than 70 different botanical families [Harborne and Baxter, 1999]. Aglycons and glucosides of this flavonoid subgroup participate in important biological functions such as protection against UV radiation, co-pigmentation with anthocyanins, plant–insect interactions, stimulation of arbuscular mycorrhizae colonization, and mediation of the establishment of symbiotic relationship between legumes and nitrogen-fixing rhizobia [reviewed by Martens and Mithöfer, 2005]. Flavones are also of great relevance for human nutrition and health [Martens and Mithöfer, 2005]. The first committed step in flavone biosynthesis starts with the conversion of the (2*S*)-flavanones pinocembrin, naringenin, and eriodictyol to their respective flavones chrysin (5,7-dihydroxyflavone), apigenin (5,7,4′-trihydroxyflavone), and luteolin (5,7,3′,4′-tetrahydroxyflavone) (Fig. 21.2). In most plant families [Heller and Forkmann, 1993], this reaction is performed by a membrane-bound NADPH-dependent cytochrome P450 monooxygenase known as flavone synthase II (FNSII, EC 1.14.11.22) [Akashi et al., 1999; Martens and Forkmann, 1999]. Interestingly, members of the Apiaceae (or Umbelliferae) botanical family (e.g., parsley, cilantro, fennel, carrot, etc.) convert flavanones to flavones with a soluble 2-oxoglutarate/Fe^{2+}-dependent dioxygenase known as flavone synthase I (FNSI, EC 1.14.11.22) [Lukacin et al., 2001; Martens et al., 2001]. The final biological and pharmacological activity of different flavones is defined by specific modifications to the basic flavone structure, including hydroxylation, *O*- and *C*-methylation, isoprenylation, and methylenedioxy substitutions [Martens and Mithöfer, 2005].

Isoflavones are found especially in legumes but also in several other plant species [Dewick, 1982], particularly when elicited by pathogen attack [Geigert et al., 1973] or herbivore induced wounding. Isoflavone biosynthesis branches out from the flavonoid pathway (Fig. 21.2). The legume-specific chalcone reductase (CHR, EC 1.1.1.) converts an intermediate of the multistep reaction catalyzed by CHS to yield 4,2′,4′-trihydroxychalcone (deoxychalcone or isoliquiritigenin) [Deavours and Dixon, 2005; Bomati et al., 2005]. This CHS/CHR coupled reaction is unique to legumes and constitutes the main entry point to the biosynthesis of several bioactive isoflavonoids, coumestans, pterocarpans, and isoflavans [Dixon and Paiva, 1995] (Fig. 21.2), with roles in phytoalexin-mediated defense against pathogens and herbivores as well as in signaling with symbiotic N-fixing bacteria. The CHI-catalyzed conversion of isoliquiritigenin to 7,4′-dihydroxiflavanone (syn. liquiritigenin) and naringenin chalcone to naringenin generates substrates for the cytochrome P450 2-hydroxyflavanone synthase, also known as isoflavanone synthase (IFS, EC 1.14.13.86) [Jung et al., 2000]. Isoflavone synthase catalyzes both the hydroxylation at C2 and the unique 1,2 aryl migration from C2 to C3 of the flavanone intermediate to yield 2,7,4′-trihydroxy-isoflavanone or 2-hydroxyflavanone [Akashi et al., 2003; Deavours and Dixon, 2005]. As a result, liquiritigenin (4′,7-dihydroxyflavanone) or naringenin yield their respective 4′-hydroxy isoflavones daidzein (4′-7-dihydroxyisoflavone) and genistein (4′,5,7-trihydroxyisoflavone) after a dehydration step catalyzed by 2-hydroxyisoflavanone dehydratase (HID, EC 4.2.1.105) [Hakamatsuka et al., 1998; Chen et al., 2001; Akashi et al., 2003, 2005; Deavours and Dixon, 2005; Shimamura et al., 2007]. Another isoflavonoid normally found in legumes, glycitein, is likely to be derived from isoliquiritigenin through a still uncharacterized reaction [Latunde-Dada et al., 2001]. 2-Hydroxyflavanones can be methylated by 2,7,4′-trihydroxyisoflavanone 4′-*O*-methyltransferase (HI4 OMT, EC 2.1.1.46) [Akashi et al., 2003; Deavours and Dixon, 2005; Shimamura et al., 2007] to form 2,7-dihydoxy-4′-methoxyisoflavanones [Akashi et al., 2003] that can be in term converted by the HID activity to the 4′-methoxy isoflavones formononetin (or 4′-*O*-methyl daidzein) and biochanin A (or 4′-*O*-methyl genistein) [He and Dixon, 2000]. The biosynthetic steps catalyzed by IFS, HID, and HI4 OMT appear to be organized as a multienzyme complex or metabolon [Winkel, 2004]. Naringenin is a precursor to both anthocyanin and proanthocyanidins biosynthesis, and its conversion to genistein is a key factor in the competition for substrate between the isoflavone and flavonoid biosynthetic pathways. For example, accumulation of genistein in IFS overexpressing *Arabidopsis* plants was limited by the predominant channeling of naringenin via the flavonoid pathway [Liu et al., 2002].

The biosynthesis of pterocarpans, which include well-known phytoalexins such as medicarpin in alfalfa (*Medicago sativa*) [Baker et al., 1989; Higgins, 1972; Oommen et al., 1994; He and Dixon, 2000] (Fig. 21.2), pisatin in pea

(*Pisum sativum*) [Akashi et al., 2003], and glyceollin I-III in soybean (*Glycine max*) [López-Meyer and Paiva, 2002; Burrow et al., 2001; Dixon and Paiva, 1995; Guo et al., 1994a; Stoessl, 1982], is strongly induced in leaves elicited by fungal pathogens. Pterocarpan biosynthesis initiates with the conversion of formononetin to 2′-hydroxyformononetin (7-hydroxy-4′-methoxydaidzein) in a dehydration reaction catalyzed by 4′-methoxyisoflavone 2′-hydroxylase (EC 1.14.13.53) [Hinderer et al., 1987] (Fig. 21.2). Isoflavone reductase (IFR, EC 1.3.1.45) further reduces 2′-hydroxyformononetin to the isoflavanone vestitone [López-Meyer and Paiva, 2002]. An additional reduction catalyzed by vestitone reductase (VR, EC 1.3.1.45) converts vestitone to 7,2′-dihydroxy-4′-methox-isoflavanol (or DMI) [Guo et al., 1994a]. In a final step, DMI dehydratase (DMID, EC 4.2.1) catalyzes DMI dehydration and cyclation of an ether ring to yield (−)-medicarpin [Akashi et al., 2006]. In vitro studies have demonstrated protein–protein interaction between VR and DMID [Guo et al., 1994b] suggesting a possible metabolon organization for these two consecutive reactions. In legumes, glucosyl- and malonyl-glucose [Kudou et al., 1991; Yu et al., 2003] conjugates of isoflavones (daidzein, glycitin, and genistein) [Graham, 1991; Burrow et al., 2001] and pterocarpans [López-Meyer and Paiva, 2002] are stored in the vacuole. Pterocarpans are the initial substrates for the biosynthesis of another group of compounds with phytoalexin and feed-deterrent properties [Russell et al., 1978] known as isoflavans [Akashi et al., 2006]. In *Lotus japonicus*, pterocarpan reductase (PTR, EC 1.1.1.246) catalyzes the stereospecific NADPH-dependent reduction of the pterocarpan (−)medicarpin to the isoflavan (−) vestitol [Akashi et al., 2006] (Fig. 21.2).

Aurone flavonoids, such as aureusidin, contribute to yellow petal pigmentation of certain flowers such as snapdragon (*Antirrhinum majus*) and dahlia (*Dahlia variabilis*) [Ono et al., 2006]. In *A. majus* two enzymatic activities are essential for the biosynthesis of aurones from chalcones: the polyphenol oxidase homolog aureusidin synthase (AS1, EC 1.21.3.6) [Nakayama et al., 2000; Nakayama, 2002] and chalcone 4′-*O*-glucosyltransferase (4′CGT, EC 2.1.1.154) [Nakayama, 2002; Ono et al., 2006] (Fig. 21.2). The *A. majus* AS1 catalyzes the oxidative cyclization of naringenin chalcone, eriodictyol chalcone and their respective 4′-*O*-glucosides to aureusidin, bracteatin, and their corresponding 6-*O*-glucosides [Nakayama et al., 2000; Sato et al., 2001; Nakayama et al., 2001; Ono et al., 2006]. Both enzymes have distinctive subcellular domains. While AS1 activity is restricted to the vacuolar domain, 4′CGT has been localized to the citosol [Ono et al., 2006]. The proposed topology for this pathway suggests that chalcones are 4′-*O*-glucosylated in the cytoplasm and then transported to the vacuole where AS1 convert them to aurone aglycones and 6-*O*-glucosides [Ono et al., 2006].

In plants accumulating anthocyanins, flavonols, and proanthocyanidins, naringenin is stereospecifically hydroxylated at position 3 of the C-ring (C3) by the 2-oxoglutarate-dependent dioxygenase flavanone 3-hydroxylase (F3H, EC 1.14.11.9) to yield the 3-hydroxy-trans-flavanone (syn. dihydroflavonol) dihydrokaempferol [Springob et al., 2003] (Fig.21 2). Dihydroquercetin (3′,4′,5,5′,7-

pentahydroxyflavanone) is formed after a hydroxylation reaction mediated by F3H or F3′H on eriodictyol or dihydrokaempferol, respectively [Springob et al., 2003]. Another P450-dependent monooxygenase, flavonoid 3′,5′-hydroxylase (F3′5′H, EC 1.14.13.88) hydroxylates positions 3′ and 5′ of the B-ring of a flavanone or a dihydroflavonol. Thus, F3′5′H activity can hydroxylate naringenin to dihydroquercetin or dihydrokaempferol to dihydromyricetin [Springob et al., 2003]. The F3′H activity would result in cyanidin-type anthocyanidin/flavan-3-ols (3′-hydroxylation) while that of F3′5 H in delphinidin-type anthocyanidin/flavan-3-ols (3′,5′-hydroxylations). The dihydroflavonol pool can be channeled through two major metabolic routes, one leading to colorless flavonols and the other to anthocyanins and proanthocyanidins. The first committed step in flavonol biosynthesis is catalyzed by the 2-oxoglutarate-dependent oxygenase flavonol synthase (FLS, EC 1.14.11.23) (Fig. 21.2). The reaction yields the flavonol aglycones kaempferol, quercetin, and myricetin. In addition, the *Arabidopsis* and *Citrus spp.* FLS can convert flavanones (e.g., naringenin, eriodictyol, and 5′-hydroxy eriodictyol) to dihydroflavonols [Prescott et al., 2002; Lukacin et al., 2003]. Dihydroflavonols can be further reduced to form the 2*R*,3*S*-trans-flavan-3*S*,4*S*-cis-diols (syn. leucoanthocyanidins) 2,3-trans-3,4-cis-leucopelargonidin, 2,3-trans-3,4-cis-leucocyanidin and 2,3-trans-3,4-cis-leucodelphinidin [Marles et al., 2003]. This NADPH-dependent reduction reaction is catalyzed by dihydroflavonol 4-reductase (DFR, EC 1.1.1.219). DFRs from different plant species have high substrate specificity [Marles et al., 2003; Xie et al., 2004a]. A single amino acid change from Asn133 to Asp133 in the *Medicago truncatula* DFR drastically reduced specificity for dihydrokaempferol [Xie et al., 2004a]. An identical substitution in the petunia (*Petunia hybrida*) DFR [Johnson et al. 2001; Xie et al., 2004a] creates an even stronger phenotype preventing the accumulation of pelargonidin-derived anthocyanins [Johnson et al., 2001; Springob et al., 2003]. DFR from *Dahlia variabilis*, *Gerbera hybrida*, *Zea mays*, and *M. truncatula* can also convert flavanones to flavan-4-ols [Fischer et al., 1998, 2003; Halbwirth et al., 2003; Xie et al., 2004a], the monomeric components of 3-deoxyproanthocyanidins or phlobaphenes (Fig. 21.2).

Leucoanthocyanidins are oxidized by another 2-oxoglutarate-dependent oxygenase known as anthocyanin synthase or leucoanthocyanidin dioxygenase (ANS or LDOX, EC 1.14.11) to yield the anthocyanidins pelargonidin, cyanidin, and delphinidin. Both leucoanthocyanidins and anthocyanidins are substrates for the biosynthesis of flavan-3-ols, the monomeric constituents of proanthocyanidins or condensed tannins. Proanthocyanidins are oligomers or polymers of flavan-3-ols subunits with roles in plant protection against pathogens, insects, and mammaliam herbivores as well as in the control of seed permeability and dormancy [Debeaujon et al., 2000]. Proanthocyanidins that yield cyanidin as a result of acid hydrolysis are known as procyanidins [Porter et al., 1986]. Similarly, proanthocyanidins with at least one extension subunit of a (4′) [e.g., (epi)afzelechin] or (3′,4′,5′)-hydroxylated B-ring [e.g., (epi)gallocatechin] flavan-3-ol are known as propelargonidins and prodelphinidins, respectively.

The initial steps in proanthocyanidin biosynthesis are catalyzed by anthocyanidin reductase (ANR, EC 1.3.1.77) [Devic et al., 1999; Xie et al., 2003, 2004b] and leucoanthocyanidin reductase (LAR, EC 1.17.1.3) [Tanner et al., 2003]. Both ANR and LAR provide the 2,3-cis and 2,3-trans-flavan-3-ol monomers that could function either as a starter or extension subunit [Xie and Dixon, 2005] during proanthocyanidin polymerization. Leucoanthocyanidin reductase, a member of the isoflavone-reductase-like group of the plant reductase-epimerase-dehydrogenase (RED) supergene family, catalyzes the reduction of leucoanthocyanidins to their respective 2,3-trans-(+)-flavan-3-ols [Tanner et al., 2003; Marles et al., 2003]. This NADPH-dependent reduction reaction converts the leucoanthocyanidins leucopelargonidin, leucocyanidin, and leucodelphinidin to their respective alternative initiating subunits 2,3-trans-(+)-afzelechin, 2,3-trans-(+)-catechin, and 2,3-trans-(+)-gallocatechin [de Pascual-Teresa et al., 2000; Tanner et al., 2003]. The LAR activity is absent in *Arabidopsis* and was first cloned from the tropical forage legume *Desmodium uncinatum* [Tanner et al., 2003; Skadhauge et al., 1997a]. Since then, LARs have been identified in grapes [Bogs et al., 2005], apple [*Malus domestica*, Pfeiffer et al., 2006], *Lotus corniculatus* [Paolocci et al., 2007], and tea [*Camelia sinensis*, Punyasiri et al., 2004]. Anthocyanidin reductase catalyzes the NADPH-dependent reduction of an anthocyanidin to 2,3-cis-2*R*,3*R*-flavan-3-ol. The first cloned and enzymatically characterized ANR were the *Arabidopsis thaliana* BANYULS (BAN) [Devic et al., 1999] and its *M. truncatula* homolog [Xie et al., 2003; Pang et al., 2007]. In *Arabidospis*, ANR activity results in the conversion of the LDOX product cyanidin to 2,3-cis-(−)-epicatechin, the most abundant PA subunit in the seed coat of this species [Abrahams et al., 2002; Xie et al., 2003]. Although with different substrate specificity, in vitro characterization of the *Arabidopsis* and *Medicago* ANRs has shown that the enzyme accepts either cyanidin, pelargonidin, or delphinidin to generate (−)-epicatechin, (−)-epiafzelechin, and (−)-epigallocatechin, respectively [Xie et al., 2004b]. The identification of ANR explained the origin of the 2,3-trans stereochemistry of anthocyanidins, flavonols, and catechin and the 2,3-cis configuration of most PA extension units found in plants [Winkel, 2006]. ANR have also been identified in grapes [*Vitis vinifera*, Bogs et al., 2005], *L.corniculatus* [Paolocci et al., 2007], tea [*C. sinensis*, Punyasiri et al., 2004], *Ginkgo biloba* [Shen et al., 2005], and persimmon [*Dyospyros kaki*, Ikegami et al., 2007] among others. Both ANR and LAR represent a major metabolic crossroad between anthocyanidin and PA biosynthesis [Pang et al., 2007]. Substrate competition became evident when the ectopic expression of the *V.vinifera* [Bogs et al., 2005] and *M.truncatula* [Xie et al., 2003] ANRs in tobacco increased proanthocyanidins content while at the same time reduced anthocyanins levels in pigmented petals. Likewise, the seed coat of the *Arabidopsis* ban mutant lacks proanthocyanidins and hyperaccumulate anthocyanins [Devic et al., 1999; Debeaujon et al., 2003]. In grapes [Bogs et al., 2005] and bilberry (*Vaccinum myrtillus*) [Jaakola et al., 2002], anthocyanin biosynthesis is temporarily compartmentalized from proanthocyanidin and flavonol biosynthesis. In both species proanthocyanidins and flavonol biosynthesis in developing fruits precedes the accumulation of anthocyanins during the ripening stage. As a result, there appears to be no direct competition between these two branches of the flavonoid pathway for common substrates.

Although the enzymatic steps leading to the biosynthesis of initiating and extension flavan-3-ol subunits have been characterized in several plant systems, the biochemical mechanisms by which extension subunits are incorporated into growing proanthocyanidin polymers remains elusive [Dixon et al., 2005; Xie and Dixon, 2005]. It is still unclear whether the actual polymerization of flavan-3-ols is an enzymatic or a nonenzymatic process [Xie and Dixon, 2005]. Models suggest the possible involvement of polyphenol oxidase (PPO) in the formation of flavan-3-ols-derived quinone intermediates. Carbocations derived from these intermediates can react with flavan-3-ols to form C4–C6– and C4–C8-linked PA dimers and oligomers [Xie and Dixon, 2005]. Different plant species have distinctive proanthocyanidin monomer composition. In wine grapes both (+)-catechin and (−)-epicatechin are common proanthocyanidin starter units depending upon the tissue and developmental stage at which proanthocyanidins are isolated [Souquet et al., 1996; Kennedy et al., 2000, 2001; Downey et al., 2003a]. In *Arabidopsis*, however, the absence of LAR activity in seeds leads to proanthocyanidins that contain exclusively epicatechin starter units. The most commonly found flavan-3-ol extension subunits are in the 2,3-trans configuration with distinctive hydroxylation patterns in the B-ring depending on the presence of active cytochrome P450s F3′H and F3′5′H [Menting et al., 1994; Kaltenbach et al., 1999; Schenbohm et al, 2000]. For example, the most frequent extension subunits in wine grapes [Souquet et al., 1996, 2000] and *L. corniculatus* [Foo et al., 1996] are epicatechin and epigallocatechin. Expression of the grape F3′H and F3′5′H correlates spatially and temporally with the accumulation of (3′,4′)- and (3′,4′,5′)-hydroxylated flavonols, flavan-3-ols (and PAs) and anthocyanins [Bogs et al., 2006]. The absence of F3′5 H activity in *Arabidopsis* leads to PAs that are homo-oligomers of the (3′,4′)-hydroxylated flavan-3-ol epicatechin [Debeaujon et al., 2000; Abrahams et al., 2003]. The degree of polymerization, stereochemistry, and number of phenolic hydroxyl groups appears to be critical to define PA beneficial bioactivity in relation to human health [Xie and Dixon, 2005].

Different enzyme families are involved in additional modifications to the basic structure of different flavonoid species, including glycosyl and acetyltransferases, methylases, and peroxidases [reviewed by Winkel, 2006]. These modifications are essential to provide stability, physicochemical properties, and distinctive bioactivity. Glycosylation enhances solubility and might be essential for facilitating transmembrane transport and vacuolar storage [Koes et al., 2005]. UDP-glucose, UDP-galactose, and UDP-rhamnose are among the most common nucleotide sugar substrates used to transfer one to several glycosyl moieties to positions 2, 3, 5, 6, and 7 of the flavonoid backbone (Fig. 21.1). In the European grape (*V. vinifera* L.), anthocyanidins are glucosylated by UDP-glucose:anthocyanidin 3-*O*-glucosyltransferase, also named UDP-glucose:

flavonoid 3-*O*-glucosyltransferase (UFGT, EC 2.4.1.91) at the 3-*O*-position of the C-ring and rarely at the 5-*O*-position of the B-ring. Anthocyanidin 3, 5-*O*-diglucosides are almost exclusively found in native American *Vitis* sp. [Peynaud and Ribéreau-Gayon, 1970; Baldi et al., 1995]. The glycosyltransferase activities involved in glycosylation of flavonol conjugates may be different from those glucosylating anthocyanidins [Ford et al., 1998]. For example, the 3-*O*-glucosyltransferase activity of the grape UFGT is significantly greater with cyanidin than quercetin [Ford et al., 1998]. In addition, UFGT transcripts are only found after the onset of ripening and accumulation of anthocyanins in the skin of colored berries [Boss et al., 1996a, 1996b; Goes da Silva et al., 2005], while glycosilated flavonols are actively synthesized in flowers and immature berries [Downey et al., 2003b]. Moreover, a glycosyltranferase with specific activity toward (−)-epicatechin was recenty isolated from *M. truncatula* seed coats (Pang et al., 2008). Anthocyanidin glysosylation constitute a major rate-limiting step for anthocyanin accumulation in grapes [Boss et al., 1996a, 1996b]. In fact, mutations impairing UFGT transcription appear to be the origin of white wine grape cultivars [Kobayashi et al., 2002, 2004; Walker et al., 2007].

Anthocyanidin glucosylation is not only a major control point in anthocyanin biosynthesis but also a prerequisite for additional substitutions like methylation at positions 3′ and 5′ and acylation of glucose moieties [Springob et al., 2003]. In grapes, peonidin-3-glucoside and delphinidin-3-glucoside are further modified by methylation at position 3′ to yield peonidin and petunidin, respectively (Fig. 21.2). This reaction is mediated by a SAM-dependent 3′-*O*-methyltransferase (3′-OMET, no EC). A 3-*O*-methyltransferase (3-OMT) from *Serratula tinctoria* preferentially methylates the flavonol quercetin [Huang et al., 2004]. Considering the high specificity of plant *O*-methyltransferases toward their substrates [Baily et al., 1997], acylation of anthocyanins must take place at a later step in anthocyanin biosynthesis. Acylation of grape anthocyanins involves different hydroxycinnamic acids (e.g., *p*-coumaric and less frequently caffeic and acetic acids). The relative abundance of acylated anthocyanins is generally lower than that of their glucosylated precursors but are critical substitutions to increase anthocyanin solubility in water, enhance stability against β-glucosidase-mediated degradation, and favor co-pigmentation interactions [Boulton, 2001; Springob et al., 2003].

Enzymatic complexes involved in flavonoid precursors [Achnine et al., 2004] and flavonoid biosynthesis are associated with the cytoplasmic side of the endoplasmic reticulum (ER) [Stafford, 1974, 1981, 1991; Hrazdina et al., 1978, 1987; Hrazdina and Jensen, 1992; Winkel-Shirley, 1999; Liu and Dixon, 2001; Saslowsky and Winkel-Shirley, 2001]. However, flavonoids accumulate in multiple subcellular compartments and some are even extruded out of the cellular domain. Examples are flavonoids sequestered in the vacuole of specific cell layers and organs where they function as insect attractants or herbivore deterrents [Koes et al., 1994; Mol et al., 1998], extruded flavonoids that participate in allelopathic interactions or the establishment of symbiotic relationships with soil rhizobia [Martinoia et al., 1993; Paiva, 2000], cytosolic flavonoids that interfere with cell-to-cell auxin transport [Buer and Muday, 2004], and nucleous-localized flavonoids [Buer and Muday, 2004; Saslowsky et al., 2005] that may participate in regulation of gene transcription [Peer and Murphy, 2006]. Significant progress in the understanding of transport mechanisms has been made in recent years, although still key mechanistic aspects remain elusive. As a result, several molecular models currently exist to explain intracellular trafficking, active transmembrane transport, and vacuolar sequestration of flavonoid conjugates [reviewed by Kitamura, 2006; Grotewold, 2004]. Flavonoid activation by conjugation with hydrophilic moieties such as sugars, acyl residues, and glutathione is essential for stability and transport mechanism specificity [Hopp and Seitz, 1987; Klein et al., 1996]. Several gluthathione-*S*-transferase-like proteins essential for vacuolar flavonoid sequestration have been characterized in different plant species including the maize bronze2 [BZ2, Marrs et al., 1995], the petunia AN9 [Alfenito et al., 1998], the soybean GST26A [Alfenito et al., 1998], the carnation flavonoid3 [FL3, Larsen et al., 2003], and the *Arabidopsis* TT19 [Kitamura et al., 2004]. Only the latter appears to function in the vacuolar sequestration of proanthocyanidin precursors in addition to anthocyanin conjugates [Kitamura et al., 2004]. The apparent absence of a free flavonoid–glutathione conjugate pool led to propose a mechanistic model where gluthathione-*S*-transferases-like proteins act as carriers of nonglutathionated flavanoids [Mueller et al., 2000] that are directly delivered to a specific type of tonoplast-bound ATP-dependent/proton-gradient independent ATP-binding cassette (ABC) transporters known as glutathione S-X pumps [Rea et al.,1998; Theodoulou, 2000] (Fig. 21.3). These carriers participate in the transmembrane trafficking and the vacuolar sequestration of a variety of glutathione conjugates [Martinoia et al., 1993; Klein et al., 2006; Koes et al., 2005; Lepiniec et al., 2006]. Examples of this class of transporters are the maize multidrug resistance-associated protein 3 [MRP3, Goodman et al., 2004] and the *Arabidopsis* MRP1 and 2 [Lu et al., 1997, 1998]. An alternative model of flavonoid transmembrane transport is based on the activity of H^+-antiporters [Klein et al., 1996] (Fig. 21.3) that depend on proton gradients actively created by membrane-bound H^+-ATPases and H^+-translocating inorganic pyrophosphatases (H^+-PPiases) [Maeshima, 2001]. A well-characterized example is the *Arabidopsis* TT12 that encodes for a tonoplast-localized member of the multidrug and toxic compound extrusion (MATE) transporter family [Debeaujon et al., 2001; Marinova et al., 2007a]. TT12 acts as an anthocyanin and flavan-3-ol glycosyde/H^+-antiporter that cannot transport anthocyanidins, glycosilated flavonols, flavan-3-ol monomers aglycones, or dimers [Marinova et al., 2007a]. The tt12 mutation impairs proanthocyanidin accumulation in the vacuole of seed coat cells and accumulates catechin and leucocyanidins in the cytoplasm [Debeaujon et al., 2001]. A similar phenotype is observed in the *Arabidopsis* aha10 mutant encoding for the autoinhibited H^+-ATPase isoform 10 (AHA10) [Baxter et al., 2005]. This H^+-ATPase is predominantly expressed in developing seeds and plays a role in creating the electrochemical gradient required by TT12 to transport PA precursors into the

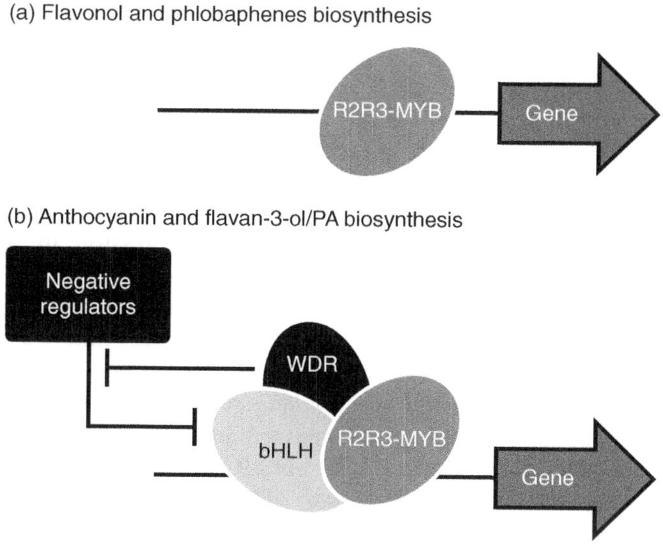

Figure 21.3 Models of R2R3-MYB, bHLH, and WDR transcriptional control of different branches of the flavonoid pathway. (Adapted from Quattrocchio et al., [2006] and Lepiniec et al., [2006].)

vacuolar domain [Baxter et al., 2005]. The tomato (*Lycopersicum esculentum*) MTP77 is another example of a MATE-type transporter involved in flavonoid transport [Mathews et al., 2003]. Recently, a third type of transporter with similarity to the organic anion carrier mammalian bilitranslocases has been linked to the vacuolar transport of flavonoids in carnation (*Dianthus caryophyllus*) petal microsomes [Passamonti et al., 2005] and developing grape (*V. vinifera*) berry tissues [Braidot et al., 2008]. In addition to the active transport model of conjugated flavonoids, vacuolar sequestration of flavonoids may involve the fusion of ER-derived vesicles containing anthocyanins or proanthocyanidin precursors [Pecket and Small, 1980; Yazaki et al., 2008] (Fig. 21.3). By this mechanism, flavonoids synthesized in the cytoplasmic side of the ER might be actively transported into the ER lumen [Grotewold, 2001] and secreted as ER-derived small vesicles [Poustka et al., 2007]. These vesicles may then fuse to the tonoplast, releasing their content into the vacuolar domain [Stafford, 1989], or form anthocyanin-accumulating subvacuolar structures known as anthocyanoplasts or anthocyanin vacuolar inclusions (AVIs) [Poustka et al., 2007]. The *Arabidopsis* gluthathione-*S*-transferase-like protein TT19 may participate in the vescicle-to-vacuole trafficking of anthocyanins and flavan-3-ols since the tt19 mutant accumulates proanthocyanidin precursors in small vesicle-like structures [Kitamura et al., 2004]. This ER-to-vacuole flavonoid trafficking appear to be common to the secretion pathway followed by proteins targeted to vacuolar storage [Poustka et al., 2007].

Transcriptional and Posttranscriptional Regulation of Phenylpropanoid Biosynthesis in Plants

The flavonoid pathway in plants is arguably one of the most in-depth characterized metabolic network, both at the level of structural enzymes and regulatory genes [reviewed by Quattrocchio et al., 2006; Lepiniec et al., 2006]. The pathway is largely under the combinatorial transcriptional control of a diverse set of transcription factor families [Springob et al., 2003; Winkel-Shirley, 2001, 2002] (Table 21.1) that regulate expression in response to internal and external cues at specific stages in development or in certain tissues and cell types [Gonzalez et al., 2008].

Several model systems have been used to dissect this transcriptional regulatory network including loss- and gain-of-function mutants in proanthocyanidin biosynthesis in *Arabidopsis thaliana* seeds [Debeaujon et al., 2003; Lepiniec et al., 2006], phlobaphenes in maize [Goff et al., 1992; Grotewold et al., 1991], and flavonoid pigments in flowers [Quatttrochio et al., 2006]. For example, among the 20 different loci affecting seed coloration in the *Arabidopsis* transparent testa (tt) mutants, 6 correspond to transcription factors (table 21.1) [Lepiniec et al., 2006]. Distinctive transcriptional control over the expression of different enzymatic steps in the flavonoid pathway has been recognized in different plant species. *Arabidopsis* bHLH and MYB tt mutants are impaired in the expression of genes in later steps of the pathway (e.g., DFR and LDOX) without modifying messenger ribonucleic acid (mRNA) levels of PAL and CHS genes [Shirley et al., 1995; Pelletier et al., 1997, 1999; Zhang et al., 2003; Gonzalez et al., 2008]. Based partially on these experimental observations, the flavonoid pathway in dicotyledoneous species appear to be organized in two distinctive clusters of co-regulated genes known as early (EBG) and late (LBG) biosynthetic genes [Shirley et al., 1995; Mol et al., 1998; Nesi et al., 2000; Pelletier et al., 1997, 1999; Zhang et al., 2003; Quattrocchio et al., 2006; Takos et al., 2006b]. However, this may not be the case for monocotyledonous species like maize, where it has been shown that all genes coding for structural enzymes of the pathway are coordinately transcribed [Irani et al., 2003].

Several of the 125 R2R3-MYB transcription factors in *Arabidopsis* [Riechmann and Ratcliffe, 2000; Stracke et al., 2001] exert systemic and tissue-specific transcriptional control over structural genes of the general phenylpropanoid and flavonoid pathways [Quattrocchio et al., 1998; Springob et al., 2003]. Promoter regions of frequently co-regulated genes such as CHS, CHI, F3H, and FLS in *Arabidopsis* [Hartman et al., 2005] and other plant species [Hartman et al., 1998] harbor the consensus sequence 5′-AcCTACCa-3′ corresponding to the cis-acting element known as MYB recognition element (MRE). This conserved motif is recognized by members of the MYB family [Feldbrügge et al., 1997; Hartmann et al., 2005]. Both flavonol and phlobaphene biosynthesis in monocotyledonous species seem to be under the sole control of R2R3-MYBs (Fig. 21.4). Three MYBs of the subgroup 7 with significant sequence

Table 21.1 Plant Transcription Factors Involved in Phenylpropanoid Biosynthesis Regulation and Their Known Regulons

	Transcrition factor loci									Structural enzymes															
	Monocotyledonous			Dicotyledonous						EBG					LBG										
TF family	Maize	Barley	Sorghum	Arabidopsis	Petunia	Anthirrinum	Grape	Lotus	Eucalyptus	PAL	4CL	CHS	CHI	F3H	F3'H	F3'5'H	DFR	FLS	LDOX	ANR	LAR	UFGT	Transport	Regulation	Reference
bHLH				TT8(bHLH42)													■			■			■ TT12	PA accumulation	Nesi et al., 2000
bHLH								TAN1																PA accumulation	Ray and Gruber, 2001
bHLH					AN1												■							Anthocyanin biosynthesis (LBG) - AN2 interactor	Spelt et al., 2000
bHLH					JAF13												■							Anthocyanin biosynthesis (LBG) - AN2 interactor	Spelt et al., 2000
bHLH	B																■							Anthocyanin biosynthesis - C1 interactor	Goff et al., 1992
bHLH				EGLE3																				Anthocyanin biosynthesis	Cominelli et al., 2008
bHLH				GL3											■		■		■					Anthocyanin biosynthesis	Gonzalez et al., 2008
bHLH						DELILA								■			■		■			■		Anthocyanin biosynthesis	Martin et al., 1991; Goodrich et al., 1992; Martin and Geratz, 1993
MYC-like	Sn, Lc, R	ANT13																						Anthocyanin biosynthesis	Marles et al., 2003
MYB				TT2(MYB123)													■		■	■			■ TT12	PA biosynthesis	Nesi et al., 2001
MYB							MYBPA1						■			■			■	■	■			PA biosynthesis	Bogs et al., 2007
MYB							MYB2A															■		Anthocyanin biosynthesis	Kobayashi et al., 2002; Walker et al., 2007
MYB							MYB5a																	Flavonoid and lignin biosynthesis	Deluc et al., 2006
MYB									MYB1															Lignin biosynthesis	Legay et al., 2007
MYB	P1											■	■				■							3-deoxy flavonoid/phlobaphene biosynthesis	Styles and Ceska, 1989; Grotewold et al., 1991, 1994
MYB			Y1																					3-deoxy flavonoid/phlobaphene biosynthesis	Chopra et al., 1999
MYB					AN2																			Anthocyanin biosynthesis (LBG)	Quattrocchio et al., 1999
MYB	C1																							Anthocyanin biosynthesis	Paz-Ares et al., 1987
MYB				MYB11							■	■	■	■				■				■		Flavonol biosynthesis	Mehrtens et al., 2005; Stracke et al., 2007
MYB				MYB12							■	■	■	■				■				■		Flavonol biosynthesis	Mehrtens et al., 2005; Stracke et al., 2007
MYB				MYB111							■	■	■	■				■				■		Flavonol biosynthesis	Stracke et al., 2007
MYB				PAP1(MYB75)						■	■	■			■		■		■			■	■ GST	Anthocyanin biosynthesis	Borevitz et al., 2000; Tohge et al., 2005a; Gonzalez et al., 2008
MYB				PAP2(MYB90)						■	■	■			■		■		■			■	■ GST	Anthocyanin biosynthesis	Borevitz et al., 2000; Gonzalez et al., 2008
MYB				MYB113											■		■		■			■	■	Anthocyanin biosynthesis	Gonzalez et al., 2008
MYB				MYB114											■		■		■			■	■	Anthocyanin biosynthesis	Gonzalez et al., 2008
MYB				MYB32																				Anthocyanin biosynthesis in petals	Preston et al., 2004
MYB						ROSEA1									▒		■	■	▒			■	▒ AT	Anthocyanin biosynthesis in petals	Schwinn et al., 2006
MYB						ROSEA2									■		■	■	▒			■	▒ AT	Anthocyanin biosynthesis in petals	Schwinn et al., 2006
MYB						VENOSA							▒	▒	■			▒	■			■	▒	Anthocyanin biosynthesis in petals	Schwinn et al., 2006
MADS-BOX				TT16																				PA accumulation	Nesi et al., 2002
WD-repeat				TTG1													■		■	■				PA and anthocyanin accumulation	Walker et al., 1999; Gonzalez et al., 2008
WD-repeat					AN11																			Anthocyanin biosynthesis - Petal pigmentation	deVetten et al., 1997
WD-repeat	PAC1																							Anthocyanin biosynthesis in kernels	Carey et al., 2004
Zinc finger				TT1																				PA accumulation	Sagasser et al., 2001
WRKY				TTG2																				PA accumulation	Johnson et al., 2002

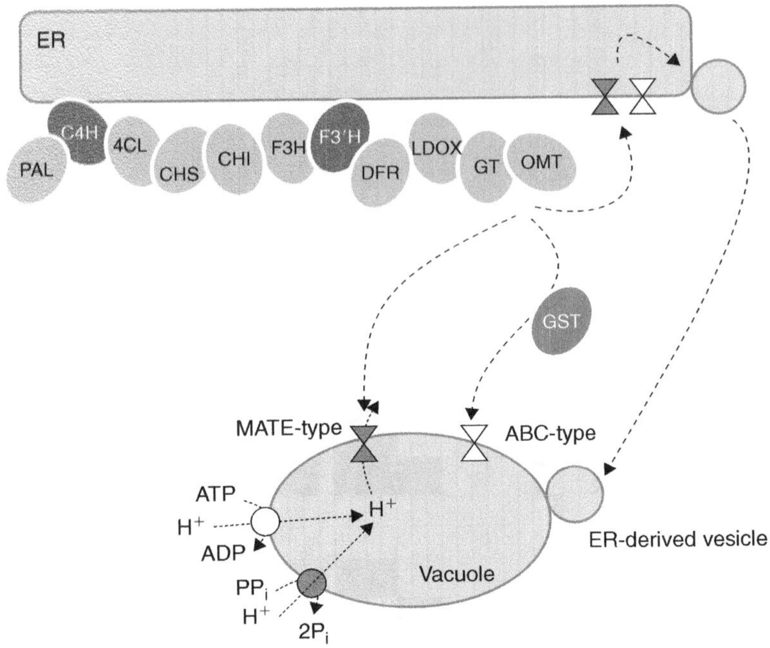

Figure 21.4 Phenylpropanoid and (iso)flavonoid pathway topology, subcellular localization and transport mechanisms. (Adapted from Winkel-Shirley, [2004], Kitamura, [2006] and Marinova et al. [2007b].)

similarity to the maize P gene regulate flavonol biosynthesis in different *Arabidopsis* organs [Stracke et al., 2007]. MYB12 or production of flavonol glycosides1 (PFG-1) regulon includes the EBGs 4CL, CHS, CHI, F3H, F3′H, the LBG FLS, and two putative flavonol:glucosyltransferases [Mehrtens et al., 2005; Stracke et al., 2007] (table 21.1). Flavonol content in the myb12 mutant is significantly reduced but not completely suppressed [Mehrtens et al., 2005], suggesting that other factors may partially contribute to the activation of this branch of the flavonoid pathway. Two additional structurally similar MYBs, MYB11 (PFG2), and MYB111 (PFG3) activate the same set of EBG and LBG acting additively with MYB12 [Stracke et al., 2007]. In fact, flavonols are absent in the myb11 myb12 myb111 triple mutant phenocopying the CHS (tt4) mutant [Stracke et al., 2007]. Although functionally redundant, these MYBs have specific spatial and temporal expression patterns that prevent the occurrence of overlapping functions. While MYB11 expression is generally low in seedlings, MYB12 controls flavonols biosynthesis in roots and MYB111 in cotyledons [Stracke et al., 2007]. The relative abundance of flavonoid glycosides in different *Arabidopsis* tissues would depend on the combinatorial control of these MYBs on the transcription of flavonol:glycosyltransferases [Stracke et al., 2007]. Similarly, both the R2R3-MYBs pericarp color1 (P1) in maize [Styles and Ceska, 1989] and its orthologous Y1 in sorghum [Chopra et al., 1999] control biosynthesis of 3-deoxy flavonoids and phlobaphenes [Grotewold et al., 1991]. P1 is similar to the *Arabidopsis* MYB12 [Dias et al., 2003] and when ectopically expressed in *Arabidopsis* exerts control over flavonol biosynthesis [Mehrtens et al., 2005]. In vitro, the ability of P1 to bind the consensus sequence ACCT/AACC (Grotewold et al., 1994] depends on the disruption under reducing conditions of a disulfide bond formed between two conserved cysteine residues [Heine et al., 2004]. Therefore, the P1 control over CHS, CHI, and DFR transcription [Grotewold et al., 1994] not only depends on factors affecting P1 expression but also on the cellular redox status.

While flavonol and phlobaphene biosynthesis would depend solely on MYBs [Grotewold et al., 1994], transcriptional regulation of anthocyanin and proanthocyanidin biosynthesis depend on protein–protein interactions and assembly of heterologous complexes between R2R3-MYBs and members of the R/B subgroup of the bHLH transcription factor family [Goff et al., 1992; table 21.1 and Fig. 21.3]. Physical protein–protein interaction takes place between the conserved R3 repeat in the MYB protein and the N-terminus region of the bHLH protein [Goff et al., 1992]. The first interaction of this kind was described in maize, where the C1 (MYB) and R (bHLH) gene products physically interact to activate anthocyanin biosynthesis [Goff et al., 1992]. In petunia, the bHLH AN1 interact with the MYBs AN2 or AN4 to form a transcriptional activation complex that regulates DFR expression [Spelt et al., 2000]. Another R/B bHLH, the petunia JAF13, may also interact in vivo with AN2 regulating expression of DFR [Quattrocchio et al., 2006]. Ectopic expression of the maize Sn (bHLH) in the forage legume *L. corniculatus* induced PA accumulation in specific cell layers of vegetative tissues as a result the increased expression, and likely, activity of the LBGs DFR, ANS, ANR, and LAR [Paolocci et al., 2005, 2007 and references therein]. This result suggests that the rate-limiting step of proanthocyanidin and anthocyanin biosynthesis in certain legumes might be the expression of a native bHLH gene [Paolocci et al., 2007]. Presence of an active transcriptional activation domain in the MYB factor is critical for activity of the complex as a positive regulator of flavonoid biosynthesis. For example, while the ectopic expression of the *Arabidopsis Landesberg erecta* MYB114 upregulated anthocyanin biosynthesis, the opposite phenotype was observed with the ecotype Columbia homolog lacking an active transcriptional activation domain [Gonzalez et al., 2008].

In addition to MYBs and bHLHs, several members of a transcription factor family characterized by the presence of five highly conserved WD-repeats [WDR; Carey et al., 2004; de Vetten et al., 1997; Walker et al., 1999] are also involved in transcriptional regulation of anthocyanin and proanthocyanidin biosynthesis [Debeaujon et al., 2003] (Fig. 21.3). The petunia AN11 encodes a WDR protein that may exert control over floral anthocyanin pigments by post-transcriptional regulation of R2R3-MYB or bHLH activity [de Vetten et al., 1997]. Orthologous WDR proteins such as the *Arabidopsis* TTG1 [Walker et al., 1999; Lepiniec et al., 2006; Baudry et al., 2006] and the maize Pale

Aleurone Color1 (PAC1) [Selinger and Chandler, 1999] are also involved in the regulation of flavonoid pigment accumulation. TTG1 protein is not essential for DNA target recognition by the TT2 (R2R3-MYB)/TT8(bHLH) transcriptional complex [Baudry et al., 2004; Nesi et al., 2001] but might be necessary to stabilize the complex [Payne et al., 2000] by preventing the effect of negative regulators [Quattrocchio et al., 2006]. Several examples of combinatorial interactions between different components in the heterotrimeric complex exist in different model species. In *Anthirrinum majus*, the MYBs Rosea1, Rosea2, and Venosa control accumulation and patterning of anthocyanin pigment in flowers [Schwinn et al., 2006]. These three MYBs have different activation activity for different target genes. Consequently, subtle differences in expression or potential to interact with activation partners of the bHLH and WDR family cause striking differences in the natural variation of floral pigmentation in *A. majus* accessions [Martin et al., 1991; Schwinn et al., 2006]. A similar combinatorial effect on anthocyanin biosynthesis is observed in maize as a result of the interaction between C1/Pl and R/B proteins [Paz-Ares et al., 1987; Ludwig et al., 1989, 1990; Lloyd et al., 1992; Cone et al., 1993]. In *Arabidopsis*, BAN (ANR) and TT3 (DFR) expression is impaired in tt2, tt8, and ttg1 mutants [Walker et al., 1999; Nesi et al., 2000; Baudry et al., 2004]. The heterotrimeric MYB/bHLH/WDR transcription activation complex formed by TT2/TT8/TTG1 directly binds to the BAN promoter, tightly regulating its developmental and spatial activity [Baudry et al., 2004; Lepiniec et al., 2006]. Similarly, VvMYBPA1, one of the 108 R2R3-MYB factors in the grape genome [Matus et al., 2008] and a close homolog the *Arabidopsis* TT12, is able to bind and activate the promoters of VvANR and VvLAR [Bogs et al., 2007]. The accumulation of proanthocyanidins in different *Arabidopsis* organs as a result of the ectopic expression of VvMYBPA1 [Bogs et al., 2007] is another of several examples of the high degree of functional conservation observed in different component of this regulatory network across plant species [Goodrich et al., 1992; Dong et al., 2001; Bradley et al., 1998]. TT8 can also interact with other MYB factors, such as PAP1 and PAP2, to specifically activate anthocyanin biosynthesis [Baudry et al., 2004; Zimmermann et al., 2004]. Other transcription factors identified as positive regulators of flavonoid biosynthesis include the BSISTER MADS box domain TT16 [Nesi et al., 2002], the zinc finger of the WIP subfamiliy TT1 [Sagasser et al., 2001], and the WRKY TTG2 [Johnson et al., 2002].

Some transcription factors can also act as negative transcriptional regulators of flavonoid biosynthesis. For example, the maize bHLH Intensifier1 (IN1) was identified as a recessive mutation enhancing kernel pigmentation [Burr et al., 1996; Baudry et al., 2004]. R2R3-MYB negative regulators have also been identified in dicotyledonous species. The strawberry (*Fragaria* x *ananassa*) FAMYB1 represses the LBG LDOX and flavonoid:UDP-glucosyltransferase (GT) [Aharoni et al., 2001]. The *Arabidopsis* MYB4, on the other hand, suppresses C4H expression [Jin et al., 2000]. Both R2R3-MYB negative regulators harbor residues required for direct protein–protein interaction with bHLH factors [Grotewold et al., 2000; Quattrocchio et al., 2006] suggesting that they may act titrating out positive regulators. Other negative regulators include the flavonoid biosynthesis repressor the *Arabidopsis* Increased Chalcone Synthase Expression1 [ICX1; Wade et al., 2003], the petunia R3-MYB MYBX [van Houwelingen et al., 1998], and the *Eucalyptus globulus* MYB1, a putative repressor of lignin biosynthesis [Legay et al., 2007].

Part of the complexity and diversity observed in regulatory and structural components of the phenylpropanoid pathway resides in the occurrence of whole-genome or segmental duplication events, followed by evolution and functionalization of diverse loci. Gene family expansion [Matus et al., 2008] and gain- or loss-of-function mutations both in the coding [Helariutta et al., 1996; Walker et al., 2007] as well as in the promoter region [De Meaux et al., 2006; Kobayashi et al., 2002, 2004] contribute to enhance the diversity of enzymatic activities as well as in the spatial and developmental transcriptional activation of the pathway. With the exception of *Arabidopsis*, where all structural enzymes of the flavonoid pathway but F′3H are encoded by single copy genes [Pelletier et al., 1999], there are multiple examples of region-specific intrachromosomal duplication or genome-wide polyploidization contributing to the diversity of flavonoid genes. In legumes, genes involved in flavonoid biosynthesis appear as clusters of multigene families. For example, CHS family is represented by eight genes in soybean [Clough et al., 2004; Tuteja et al., 2004] and six in Phaseolus [Ryder et al., 1987]. Different paralogs have unique expression patterns [Tuteja et al., 2004] and responsiveness to environmental stimuli [Shimizu et al., 1999]. The *Lotus japonicus* genome harbors a tandem cluster of three paralog R2R3-MYBs orthologous to the *Arabidopsis* TT2. Transient expression experiments showed their functional equivalence since all were able to activate the LjANR promoter when coexpressed with TT8 and TTG1 [Yoshida et al., 2008]. The native expression domain and responsiveness to speicifc stress factors differ among the LjTT2s paralogs [Yoshida et al., 2008]. LjTT2c and LjTT2b are constitutively expressed at high levels in all organs while LjTT2a is induced in response to wounding [Yoshida et al., 2008]. Similarly, the *Arabidopsis* activators of anthocyanin biosynthesis MYB113 (At1g66370), MYB114 (At1g66380), and PAP2 (At1g66390) [Stracke et al., 2001] occur in a cluster tandem [Gonzalez et al., 2008]. They all have distinctive activation properties and expression domains in relation to their paralog PAP1 [Toghe et al., 2005; Gonzalez et al., 2008]. Sense–antisense interaction between members of multigene families can also have strong consequences on activation of the flavonoid pathway as a result of cell type and organ-specific gene silencing [Tuteja et al., 2004].

Phenylpropanoid Pathway, Subcellular Localization, and Metabolic Channeling

Phenylpropanoids and flavonoids found in plants are predominantly derived from the aromatic amino acid L-phenylalanine originated in the shikimic acid pathway. Although biosynthesis of aromatic amino acids occurs preferentially in plastids [Weaver and Herrmann, 1997], the initial precursors erythrose-4-phosphate (E4P) and phosphoenolpyruvate (PEP) originate via different metabolic pathways localized in distinctive subcellular compartments. The glycolytic intermediate, PEP is mainly imported from the cytosolic pool since plastids produce insufficient amounts of this precursor [Bagge and Larsson, 1986; Hrazdina and Jensen, 1992]. The plastidial transketolase (TK, EC 2.2.1.1.) activity converts fructose-6-P and 3-phosphoglyceraldehyde to xylulose-5-P and E4P that can be readily channeled to L-phenyalanine biosynthesis in the plastid [Hrazdina and Jensen, 1992]. L-Phenylalanine must be exported from the plastid to the cytosol to reach the multienzyme enzymatic complex of the phenylpropanoid pathway believed to be localized to the cytoplasmic side of the ER [Achnine et al., 2004; Stafford, 1974, 1981, 1991; Hrazdina et al., 1978, 1987; Hrazdina and Jensen, 1992; Winkel-Shirley, 1999a; Liu and Dixon, 2001; Saslowsky and Winkel-Shirley, 2001]. The core general phenylpropanoid and flavonoid pathways appear to be organized as a multienzyme complex or metabolon mediated by direct protein–protein interaction between enzymes catalyzing sequential reactions [Winkel-Shirley, 1999a, 1999b, 2001; Winkel, 2004] (Fig. 21.3). Direct protein–protein interaction between structural enzymes have been demonstrated using a combination of techniques including yeast two-hybrids, inmunoprecipitation assays and affinity chromatography [Burbulis and Winkel-Shirley, 1999; Guo et al., 1994b; Winkel, 2004]. A metabolon structure would increase biosynthetic efficiency by optimizing local substrate concentration and conditions at the reaction site, channeling common precursors to competing biosynthetic branches, avoiding toxic effects of some intermediates, and preventing feedback inhibition [Ovádi and Saks, 2004; Winkel-Shirley, 1999a; Winkel, 2004]. The current molecular model anchors the flavonoid pathway metabolon to the ER by means of the transmembrane domains of P450-dependent monooxygenases such as C4H and F3′H [Winkel-Shirley, 1999a; Winkel, 2004] (Fig. 21.3). However, even in F3′H defective *Arabidopsis* mutants, CHS and CHI co-localize to the cytoplasmic side of the ER, suggesting that other proteins might be also involved in anchoring the whole metabolon to the ER [Saslowsky and Winkel-Shirley, 2001; Winkel, 2004].

Considering the complexity that arises from metabolic compartmentation and the extent of enzymatic activities needed to channel primary intermediates to secondary metabolites, a highly coordinated transcriptional regulatory network is required to orchestrate expression of structural genes coding for enzymes and transporters. Indeed, molecular evidence suggests that abiotic [Logemann et al., 2000] and biotic [Hermsmeier et al., 2001] stress factors co-induce transcription of committed enzymes of the oxidative pentose phosphate pathway (OPPP) [Batz et al., 1998; Logemann et al., 2000], shikimic acid [Batz et al., 1998], and phenylpropanoid pathway [Logemann et al., 2000]. In parsley (*Petroselimum crispum*), UV light stimulated and increased the partitioning of carbon from primary to secondary metabolism by enhancing expression of enzymes involved in carbohydrate, shikimic acid, fatty acid degradation, and flavonoid pathways [Logemann et al., 2000]. Activity levels and transcription of 3-deoxy-D-arabino-heptulosonate 7-phosphate synthase (DAHPS, EC 4.1.2.15), the first committed step in aromatic amino acid biosynthesis, was induced under conditions of N and amino acid starvation [Weaver and Herrmann, 1997], light [Henstrand, 1992], wounding [Dyer, 1989], and elicitors [Henstrand, 1992] in coordination with enzymes of the phenylpropanoid metabolism [Weaver and Herrmann, 1997], especially PAL and CHS. Similarly, glucose-6-P dehydrogenase (G6PD, EC 1.1.1.49), a rate-limiting step in the OPPP, is also induced by UV light and elicitors [Batz et al., 1998]. The activity of different flavonoid transport systems is also tightly regulated. For example, GST-like proteins are commonly co-expressed with structural anthocyanin biosynthetic genes [Mathews et al., 2003]. In addition, the activity of MATE-type vacuolar flavonoid/H+-antiporters may depend on the existence of a functional flavonoid pathway and posttranslational modifications induced by binding to a still uncharacterized flavonoid biosynthetic intermediate [Marinova et al., 2007b]. This experimental evidence provides only a glimpse of the complex mechanisms in place exerting exquisite control over metabolite fluxes from precursors derived from primary metabolism to the final accumulation of phenylpropanoids in different subcellular compartments.

BIOLOGICAL FUNCTION OF PHENYLPROPANOIDS IN PLANT GROWTH, DEVELOPMENT, AND INTERACTIONS WITH THE ENVIRONMENT

Phenylpropanoids Structural and Signaling Functions in Plant Reproductive Biology

A functional flavonoid pathway in anthers and pistils is required for proper development of the pollen grain, pollen tube growth, fertilization, and seed biogenesis [Burbulis et al., 1996; Ylstra et al., 1992]. This requirement is widespread among all land plants [Jorgensen, 1993]. Different genes involved in flavonoid biosynthesis are expressed during the development of the pollen grain [Alves-Ferreira et al., 2007], and they seem to be under the transcriptional control of R2R3-MYB transcription factors [Preston et al., 2004]. Phenylpropanoids and long-chain fatty acids are among the constituents of sporopollenin, the main polymeric component of the extracellular lipidic matrix of the outer coat (or exine) of pollen grains [Piffanelli et al., 1998; Hsieh and Huang, 2007]. Among the different cell layers of the anther, the sporophytic tapetum cells have the highest specific activities of phenylpropanoid biosynthetic enzymes [Piffanelli et al., 2004]. Flavonoid aglycones synthesized in the tapetal cells are translocated by ER-derived vesicles known as tapetosomes [Hsieh and Huang, 2007] to the immature pollen grain exin where they undergo additional modification [Taylor et al., 1997; Pifanelli et al., 1998]. Flavonol deficient plants are often self-sterile due to lack of proper pollen germination [Taylor and

Jorgensen, 1992; Mo et al., 1992; Kim et al., 1996] or establishment of a functional pollen tube [Pollak et al., 1995; Ylstra et al., 1994; Schijlen et al., 2007]. For example, the maize and petunia CHS mutants [Mo et al., 1992], the tapetum-specific silencing of PAL and CHS in petunia and tobacco [van der Meer et al., 1992; Matsuda et al., 1996], or the silencing of CHS in tomato [Schijlen et al., 2007] cause partial or complete male sterility as a result of abnormal pollen development and pollen tube growth. These reproductive defects are conditional and can be rescued in vitro by chemical complementation with flavonols, in particular quercetin and kaempferol [Ylstra et al., 1992, 1996; Mo et al., 1992; Kim et al., 1996] or by wild-type stigma exudates [Taylor and Jorgensen, 1992; Ylstra et al., 1994; Vogt et al., 1994]. In *Arabidopsis*, however, the flavonoid deficient mutant tt4-1 is fertile and able to set a reduced number of seeds after self-fertilization [Shirley et al., 1995; Ylstra, 1996], probably as a consequence of lower pollen germination [Kim et al., 2006]. These experimental results suggest contrasting roles for flavonoids in plant reproduction depending on the species considered. Some species have a strict dependency on flavonoids to ensure fertilization while others seem to have evolved alternative determinants of fertility [Taylor and Grotewold, 2005] or have completely lost the requirement for flavonoids to set seed [Taylor et al., 1997]. In addition to their structural role, flavonoids seem to exert their effect on pollen grain development and pollen tube growth by regulating transcription of genes involved in the fertilization process. An example is the reported effect of the flavonol kaempferol on the transcriptional regulation of several pollen-specific genes [Guyon et al., 2000], including a SHY leucine-rich repeat protein required for pollen tube penetration into the ovules [Guyon et al., 2004]. In addition to the effects on pollen grain development and functionality, a recent report described the occurrence of parthenocarpic (or fruit biogenesis in the absence of fertilization) seedless fruits in flavonoid-deficient tomato plants lacking CHS activity [Schijlen et al., 2007]. This novel effect on parthenocarpic fruit development suggests a role for auxin–flavonoid interaction on plant reproduction [Schijlen et al., 2007] yet to be elucidated. In sum, although different mechanisms and requirements seem to be in place for different plant species, phenylpropanoids and flavonoids in particular clearly play a critical role in plant reproductive biology.

Flavonoids as Mediators of Transcriptional and Posttranslational Regulation of Protein Activity

Nuclear localization of flavonoid biosynthetic enzymes [Saslowsky et al., 2005] and flavonoids [Buer and Muday, 2004; Saslowsky et al., 2005] suggests their direct or indirect involvement in regulating gene transcription processes [Pelletier et al., 1999; Polster et al., 2003; Feucht et al., 2004; Peer and Murphy, 2006]. For example, naringenin chalcone and apigenin modulate flavonoid biosynthesis by exerting control over transcription of structural enzymes of the flavonoid pathway [Pelletier et al., 1999]. Possible molecular mechanisms include flavonoid binding to histone proteins [Polster et al., 2003; Feucht et al., 2004], topoisomerases [Boege et al., 1996], and DNA polymerase [Mizushina et al., 2003] as well as inhibition of phosphorylation signaling cascades when binding to specific phosphokinases [reviewed by Peer and Murphy, 2006]. Flavonoid protein-binding properties are not only relevant in transcription regulation but also play a critical role in other plant reproductive and developmental processes such as the regulation of polar auxin transport and pollen development.

FLAVONOID–AUXIN TRANSPORT INTERACTION AND MODULATION OF PLANT GROWTH

Mutations that impair flavonoid biosynthesis can have dramatic effects on plant growth and development by acting as endogenous negative regulators of auxin cellular efflux and polar (basipetal or acropetal) transport in vivo [Jacobs and Rubery, 1988; Brown et al., 2001; Buer and Muday, 2004; Peer et al., 2004; Taylor and Grotewold, 2005; Besseau et al., 2007; Woodward and Bartel, 2005]. Severe phenotypes are normally associated with reduced primary root growth and increased lateral root formation [Brown et al., 2001], delayed gravitropic responses [Buer and Muday, 2004], and increased number of secondary inflorescences [Brown et al., 2001]. In addition, reduced plant biomass often negatively correlates with flavonoid hyperaccumulation and lower polar auxin transport (PAT) rates [Besseau et al., 2007]. Auxins are mainly produced in the shoot apical meristem and to a lesser extent in other developing organs, such as leaf primordia and developing leaves [Ljung et al., 2001; Avsian-Kretchner et al., 2002]. In shoots, auxin movement is mainly basipetal, while in roots both basipetal and acropetal movements occur in distinctive and spatially separated tissue layers [Woodward and Bartel, 2005]. Cell-to-cell trafficking and long-distance transport from synthesis sites is a complex and highly regulated process that requires the interaction of multiple proteins in a tissue-specific manner. In *Arabidopsis*, auxin cellular influx is regulated by the amino acid permease-like auxin influx carrier protein1 (AUX1)/Like-AUX1 protein [Bennett et al., 1996; Marchant et al., 1999; Parry et al., 2004]. Cellular efflux, on the contrary, seems to be under the control of at least two distinctive types of carrier proteins: the *Arabidopsis* Pin-formed (PIN) family of auxin efflux facilitators [Gälweiler et al., 1998; Utsuno et al., 1998] and members of the ATP-binding cassette (ABC) transporter family known as ATP-dependent multidrug resistance-like (MDR)/P-Glycoprotein (PGP) [Noh et al., 2001; Multani et al., 2003]. MDR/PGP seems to form complexes with a type of regulatory proteins known as napthylphtalamic acid (NPA) binding proteins (NBPs) [Jacobs and Rubery, 1988; Murphy et al., 2000; Muday and DeLong, 2001]. NBPs have been characterized as membrane-associated aminopeptidases [Murphy et al., 2002] with low binding affinity to NPA while MDR/PGP carriers are considered high-affinity NBPs [Murphy et al., 2002; Noh et al., 2001].

Initial evidence for the involvement of flavonoids in auxin transport regulation emerged from in vitro experiments showing that certain flavonoids, in particular, flavonol aglycones such as quercetin and kaempferol, competed with the auxin efflux inhibitor 1-naphthylphtalmic acid (NPA) for auxin transporters binding sites [Jacobs and Rubery, 1988]. These results were later confirmed by in planta analysis of PAT in *Arabidopsis* mutants with altered flavonoid biosynthesis. Flavonoid deficiency leads to higher rates of PAT [Brown et al., 2001] while the contrary is observed in flavonoid hyperaccumulating plants. Loss of apical dominance in tt4 (CHS) mutant [Brown et al., 2001] correlates with high basipetal PAT rates [Brown et al., 2001; Peer et al., 2004; Buer and Muday, 2004]. Chemical complementation bypassing the CHS step restored flavonoid biosynthesis and wild-type auxin transport rates [Buer and Muday, 2004]. On the other hand, flavonol hyperaccumulating mutants such as tt3 (DFR) and tt7 (F3′H) [Peer et al., 2004] and plants with reduced hydroxycinnamoyl-CoA shikimate/quinate hydroxycinnamoyl transferase (HCT) activity, a rate-limiting step in lignin biosynthesis [Hoffman et al., 2004; Besseau et al., 2007], have significantly lower PAT rates. When HCT and CHS were simultaneously suppressed, absence of flavonoid biosynthesis reestablished wild-type polar auxin transport and normal development in spite of reduced lignin biosynthesis [Besseau et al., 2007]. This result clearly demonstrates that the reduced size and aberrant growth phenotype of HCT-silenced plants resulted from the effects of flavonoid hyperaccumulation on PAT. The MORE AXILLARY BRANCHES1 (MAX1) [Booker et al., 2005], a member of the cytocrome P450 family of monooxygenases, acts as a positive regulator of the flavonoid pathway and represses vegetative bud outgrowth in *Arabidopsis* [Lazar and Howard, 2006]. The max1 mutant shows increased lateral branching and loss of apical dominance, giving plants a "bushy" appearance [Stirnberg et al., 2002]. A model by which MAX1 would regulate bud repression by a flavonoid-dependent auxin retention mechanism has recently been proposed [Lazar and Goodman, 2006]. By this mechanism, MAX1 activity would increase steady-state levels of flavonoids in stems and buds affecting expression and localization of auxin influx and efflux carriers and directional auxin transport.

The exact mechanisms by which certain flavonoids interfere with auxin transport are yet to be elucidated [Taylor and Gotewold, 2005; Bailly et al., 2008]. In fact, the experimental evidence suggests that flavonoids may exert their inhibitory effect on auxin transport via different mechanisms depending on the auxin efflux carrier system considered. Certain flavonoids directly bind to low– [Jacobs and Rubery, 1988; Murphy et al., 2000; Muday and DeLong, 2001] and high-affinity NBPs [Noh et al., 2001; Geisler et al., 2005; Bouchard et al., 2006]. It has been recently shown that the functional protein–protein interaction and modulation of auxin transport activity of two ABCB/P-glycoproteins, PGP1 and MDR1, by the membrane-anchored immunophilin-like TWISTED DWARF1 (TWD1) is specifically disrupted by NPA and flavonoid binding [Bailly et al., 2008]. This result provides the first distinctive mechanistic mode of action model of flavonoids action on modulation of PAT rates. In addition, low PIN protein levels and disruption of their characteristic asymmetrical subcellular localization [Gälweiler et al., 1998] was observed in flavonoid biosynthetic mutants [Peer et al., 2004]. The altered localization may result from the direct interference of flavonoids on vesicular trafficking proteins [Muday et al., 2003; Taylor and Grotewold, 2005] or on auxin transport [Peer et al., 2004; Taylor and Grotewold, 2005]. The serine/threonine kinase PINOID (PID) [Friml et al., 2004] governs the polar subcellular localization of PIN1 and might be a direct target of flavonoid action [Taylor and Grotewold, 2005] suggesting a possible role for flavonoids in regulating the phosphorilation/activation state of auxin transport proteins (De Long et al., 2002). Flavonoid may also have direct effects on steady-state levels of cellular auxin pools. For example, up-regulation of flavonoid biosynthesis induced by nod factors during nodule organogenesis in white clover may inhibit peroxidase-mediated breakdown of auxin pools [Mathesius, 2001]. Regardless the exact mechanism in place, the tight regulation of flavonoid biosynthesis in response to environmental cues [Winkel-Shirley, 2002] and its effect on modulating cell-to-cell and long-distance PAT rates as well as localized auxin steady-state levels make them potent transducers that can trigger profound changes in plant architecture to ensure proper adaptation.

Functional Role of Phenylpropanoids in Plant Responses to Abiotic Stresses

Flavonoids accumulate in diverse plant organs, defined tissues and specific cell types as part of normal development [Lee, 2002; Lee et al., 1987] or in response to abiotic/biotic stress factors altering cellular and organism homeostasis [reviewed by Chalker Stack, 1999; Winkel-Shirley, 2001, 2002; Gould, 2004; Shimizu et al., 1999; Shimada et al., 2000]. Common abiotic stressors able to elicit flavonoid accumulation include excess UV-B light [Gould, 2004], drought [Farrant, 2000; Sherwin and Farrant, 1998; Iandolino, 2004], salinity [Walia et al., 2005], osmotic stress [Chalker-Scott, 1999, 2002], cold [Christie et al., 1994; Solecka and Kacperska, 2003], altered carbon to nutrient balance [Lillo et al., 2008; Iandolino, 2004], wounding [Gould et al., 2002, Tamari et al., 1995], and elevated heavy-metal exposure [Hale et al., 2002; Krupa et al., 1996]. Abiotic stress effectors modulate transcription of phenylpropanoid and flavonoid genes [Winkel-Shirley, 2002] via a complex network of receptor mechanisms, signaling cascades, and the activity of multiple transcription factors that ultimately integrate the response to external and internal cues [Chen et al., 2002; Davies and Schwinn, 2003]. An example is the promoter of the maize C1 transcription factor that harbors two conserved cis-acting elements that are critical for abscicic acid (ABA) responsivenes and light-dependent induction [Kao et al., 1996; Sidorenko et al., 2000].

Throughout evolution, plants have evolved complex mechanisms to cope with environmental factors that can compromise plant survival and reproduction. At the metabolic level, most abiotic stressors have profound effects on the steady-state levels and relative abundance of several metabolites [Coruzzi and Bush, 2001; Coruzzi and Zhou, 2001]. Plants can sense and respond to these stress-dependent metabolic changes triggering a cascade of responses. Well-established examples are the increase in steady-state levels and sensitivity to the phytohormone ABA [Riera et al., 2005] as well as in the levels of reducing sugars and sucrose in leaves [Kerepesi and Galiba, 2000; Ackerson, 1981] in response to drought, osmotic, and salt stresses. Both ABA and glucose up-regulate the full complement of *Arabidopsis* genes involved in flavonoid biosynthesis [Li et al., 2006b] via the combinatorial effect of MYB and MYC-type (bHLH) transcription factors [Abe et al., 1997]. Different sugar-specific sensing mechanisms mediate effector perception and signaling cascades that leads to transcriptional and posttranscriptional regulation of gene expression [Koch, 1996]. Both glucose perception by a hexokinase (HXK)-dependent [Xiao et al., 2000] and HXK-independent mechanism [Smeekens, 2000; Rolland et al., 2002] regulate expression of distinctive sugar responsive gene sets. Expression of genes involved in flavonoid biosynthesis, like PAL and CHS, is modulated by glucose via the HXK-independent pathway [Xiao et al., 2000]. Sucrose-specific response pathways have been identified, but sucrose sensor mechanisms have not been completely elucidated [Teng et al., 2005]. Sucrose/H^+ symporters mediating active phloem loading and unloading of sucrose at source (mature leaves) and sink tissues (e.g., immature leaves, fruits, roots) may function as sucrose-specific sensors [Eckardt, 2003; Barth et al., 2003]. The *Arabidopsis* suc2/pho3 mutant [Zakhleniuk et al., 2001] lacks a functional sucrose transporter and hyperaccumulate soluble sugar in leaves [Lloyd and Zakhleniuk, 2004]. Sugar-induced anthocyanin accumulation in pho3 plants results from the up-regulation of transcriptional regulators and all seven structural genes required for anthocyanin biosynthesis [Lloyd and Zakhleniuk, 2004]. Both the *Arabidopsis* [Xiao et al., 2000] and petunia [Tsukaya et al., 1991] CHS are induced by sugars. The promoter region of these genes contain the same cis-acting regulatory elements, known as Suc boxes [Teng et al., 2005], that provide sucrose responsiveness to the β-amylase gene [Tsukaya et al., 1991]. The sucrose-dependent induction of anthocyanin biosynthesis in *Arabidopsis* is mediated by the MYB transcription factor MYB75/PAP1 [Teng et al., 2005; Lloyd and Zakhleniuk, 2004; Borevitz et al., 2000].

Nutrient deficiency activates transcription and accumulation of different flavonoid compounds [Lea et al., 2007; Lillo et al., 2008]. Among other distinctive deficiency symptoms, long-term inorganic phosphorous (P_i) starvation strongly induces genes coding for all steps of the flavonoid pathway, leading to anthocyanin accumulation [Abel et al., 2002; Zakhleniuk et al., 2001; Misson et al., 2005]. In addition, CHS and PAP1 expression levels were synergistically up-regulated by P_i starvation and excess sucrose [Muller et al., 2007]. Induction of flavonoid biosynthesis in response to nitrogen starvation is a well-characterized response commonly used as a robust marker for altered carbon nitrogen balance at the cellular level [Lea et al., 2007; Lillo et al., 2008]. Nitrogen-starved *Arabidopsis* plants increased steady-state levels and changed the relative abundance of different flavonoids [Stewart et al., 2001]. These nitrogen-deficiency-dependent responses have practical consequences in fruit production. For example, heavy nitrogen fertilizer applications reduces anthocyanin content of table and wine grape berries [Kliewer, 1977; Iandolino, 2004] while soil conditions associated with low nitrogen stimulate anthocyanin accumulation and contribute to improve wine quality [Kliewer, 1977; Seguin, 1986]. The response to nitrogen starvation reflects changes in expression levels of structural genes and enzymatic activities of the flavonoid pathway. For example, nitrogen stress-altered expression and enzymatic activity level of structural genes of the flavonol and anthocyanin biosynthetic pathway in tomato (*L. esculentum*), which resulted in a threefold increase in anthocyanidin and a twofold increase in quercetin-3-*O*-glucoside content [Bongue-Bartelsman and Phillips, 1995]. Co-induction of flavonol and anthocyanin biosynthesis was also observed in nitrogen-deficient *Arabidopsis* seedlings and it was shown to be controlled by the up-regulation of several MYB and bHLH factors regulating flavonoid biosynthesis [Lea et al., 2007]. In these nitrogen-stressed seedlings, the positive regulators of anthocyanin biosynthesis, PAP1 and PAP2 [Borevitz et al., 2000], and flavonol biosynthesis, MYB12 [Mehrtens et al., 2005], were strongly co-up-regulated with several bHLH factors such as GL3 [Lea et al., 2007].

Acclimation to freezing conditions induces major and complex changes in gene expression and steady-state levels of different metabolite pools [Cook et al., 2004; Kaplan et al., 2004; Hannah et al., 2006]. There exists a strong quantitative association between the induction of flavonoid biosynthetic genes and the extent of cold tolerance in *Arabidopsis* accessions [Hannah et al., 2006]. The R2R3-MYB factor PAP2 and the 37 genes of the PAP1 regulon [Toghe et al., 2005] are among the most highly up-regulated genes positively correlated with acclimation to freezing tolerance [Hannah et al., 2006]. While PAP2 expression is independent from sucrose cellular levels [Solfanelli et al., 2006], cold-induced anthocyanin accumulation may partially result from sucrose-mediated regulation of PAP1 activity since sucrose accumulation also correlates positively with freezing tolerance [Hannah et al., 2006].

The UV-absorbing capacity and strong in vivo antioxidant potential of simple phenolics and flavonoids make them play a critical role in UV light stress responses. Several *Arabidopsis* mutants defective in flavonoid and hydroxycinnamic acid biosynthesis are highly sensitive to UV radiation [Li et al., 1993; Landry et al., 1995; Bharti and Khurana, 1997; Bieza and Lois, 2001; Booij-James et al., 2000]. The combination of spectral absorption maximum in the UV-B range (208–315 nm) for hydroxycinnamic acids and UV-A (315–400 nm) range for flavonoids provide an effective sunscreen against a wide range of high-energy solar radiation [Landry et al., 1995]. Most colorless flavonoids (e.g., flavonols) and hydroxycinnamic acids accumulate preferentially in epidermal cell layers where they can be most efficient in filtering out UV-B/A solar radiation [Gould, 2004]. On the contrary, most anthocyanins accumulate in sub-epidermic mesophylle cells [Lee and Collins, 2001; Gould,

2004]. Both colorless cytosolic and colored vacuolar anthocyanin species [Lapidot et al., 1999] are efficient scavengers of free radicals [Neill et al., 2002]. It is likely that the colorless cytosolic forms are the ones first participating in scavenging reactive oxygen species (ROS) generated by chloroplasts [Gould, 2004]. Senescence processes and several abiotic stress factors diminish leaf carboxylation potential by reducing photosynthetic pigment content, impairing electron transport, and $NADPH_2$/ATP biosynthesis as well as enzymatic processes required for CO_2 assimilation [Walia et al., 2004]. UV-light-induced damage to the photosynthetic apparatus results in a characteristic reduction of the quantum efficiency of photosynthesis or photoinhibition [Long et al., 1994]. As light utilization for photochemistry becomes increasingly impaired, filtering of harmful radiation by phenylpropanoids prevents excess electrons from generating ROS such as superoxide, hydrogen peroxide (H_2O_2), hydroxyl radical, and singlet oxygen [Pastori and Foyer, 2002]. These ROS damage thylakoid membranes [Kolb et al., 2001], nucleic acids [Kootstra, 1994, Stapleton and Walbot, 1994], and proteins involved in light (e.g., photosystem II reaction center) and dark (e.g., Rubisco) reactions of photosynthesis [Jordan, 1996; Vass, 1997; Gould, 2004]. Reddening of senescing leaves in deciduous species during the fall season has been linked to photoprotection of cellular components during remobilization of photosynthetic pigments and resorption of nutrients [Feild et al., 2001; Hoch et al., 2001, 2003]. In some plant species, the photoprotective properties of UV-absorbing flavonoids are also essential to shield photolabiles defense compounds. For example, the California coast native silver beachweed (*Ambrosia chamissonis*) accumulates high concentrations of a photolabile antimicrobial compound in certain organs and cell types [Page and Towers, 2002]. Tissues that contain this antimicrobial compound are surrounded by cell layers with high concentrations of UV-light-absorbing anthocyanins [Block et al., 1996; Page et al., 1999].

Light-mediated induction of phenylpropanoid biosynthesis appears to be mediated by MYB and bHLH factors. Light of different qualities strongly induces the positive regulators of anthocyanins biosynthesis PAP1, PAP2, as well as several bHLH transcriptional activators including TT8, EGL3, and GL3 [Cominelli et al., 2008] and strongly represses the negative regulator MYB4 [Jin et al., 2000]. Interestingly, PAP1 appears to up-regulate transcription of structural genes of the flavonoid pathway in a light-dependent manner since PAP1 overexpression alone does not induce anthocyanin accumulation in darkeness [Cominelli et al., 2008]. This suggests the existence of a more complex network that includes photoreceptors and light-dependent signaling cascades [Jenkins et al., 2001], the requirement of additional light inducible bHLH (e.g. TT8) or basic-leucine zipper (bZIP) (e.g., HY5) transcription factors [Feldbrügge et al., 1994, 1997; Ang et al., 1998; Hartmann et al., 2005; Shin et al., 2007] or the existance of light-dependent posttranslational mechanisms to activate structural enzymes [Cominelli et al., 2008]. The promoter region of several early and late biosynthetic genes including CHS, F3H, and FLS in parsley (*Petroselinum crispum*) [Schulze-Lefert et al., 1989] and *Arabidopsis*

[Feldbrügge et al.,1994, 1997; Hartmann et al., 1998, 2005] harbors a G-box cis-acting element conferring light responsiveness that is a potential target for bHLH and bZIP transcription factors. PIF3, a bHLH transcription factor mediating phytochrome transcriptional cascades, regulates anthocyanin biosynthesis in a HY5-dependent manner [Shin et al., 2007]. Blue-light photomorphogenic responses mediated by cryptochrome flavyn-type protoreceptors or cytokinins, including anthocyanin accumulation, are regulated by HY5 [Vandenbussche et al., 2007]. In vitro binding assays have shown that PIF3 binds to the G-box element (CACGTG) while HY5 can bind both to the G- and Z-box (ATACGTGT) elements [Chattopadhyay et al., 1998; Martinez-Garcia et al., 2000; Yadav et al., 2002]. In *Arabidopsis,* chromatin immunoprecipitation-based assays have shown that both transcription factors bind independent elements in the promoter regions of CHS, CHI, F3 H, F3H, DFR, and LDOX genes [Shin et al., 2007].

To summarize, the responsiveness of phenylpropanoid metabolism to abiotic stressors that can severely and irreversible affect several plant processes constitute a critical trait to increase reproductive fitness and survival of land plants [Raushier, 2006]. The timely activation of the pathway is tightly regulated by a complex network of receptors, signaling cascades, and transcriptional regulators [Swindell, 2006] that integrate internal and external cues and ensure optimal use of metabolic resources.

BIOLOGICAL ROLE OF PHENYLPROPANOIDS IN MEDIATING PLANT INTERACTIONS WITH OTHER ORGANISMS

Flavonoids as Herbivore Deterrents

The ability of plant proanthocyanidins to tightly bind macromolecules such as proteins and cell wall polysaccharides plays a critical role in plant defense against insect and animal herbivores. Proanthocyanidin abundance in insect diet has been negatively correlated with population size of leaf-chewing herbivores (*Quercus* sp.) [Feeny, 1970; Forkner et al., 2004], larvae growth and development [Nomura and Itioka, 2002; Mutikainen et al., 2002; Barbenhenn and Martin, 1994], and increased parasitism rates [Faeth and Bultman, 1986]. Proanthocyanidin content and biochemical nature determines palatability, digestibility, and nutritional value of forage legumes [Barry and McNabb, 1999; McAllister et al., 2005; Dixon et al. 2005]. Proanthocyanidin–protein complexes are more resistant to the action of proteolytic bacteria in the livestock rumen reducing methane production and bloating [Kingston-Smith and Thomas, 2003; Zucker, 1983]. Proanthocyanidin–protein complexes can then dissociate in the abomasum, increasing the amount of plant protein reaching the small intestine [Barry and Manley, 1984; Mangan, 1988]. The absence of proanthocyanidins in alfalfa (*Medicago sativa*) and clovers (*Trifolium* spp.) [Fay et al., 1980] favors pasture bloating in ruminant animals [Aerts et al., 1999; Fay et al., 1980] fed exclusively on these pastures. On the contrary, moderate proanthocyanidin amounts ranging from 2 to 4% dry matter prevent fast proteolytic rates during rumen fermentation, thus reducing the risk of bloating [Tanner et al., 1995; McAllister et al., 2005; Dixon et al., 2005 and references therein]. Proanthocyanidin content above 6% dry matter, however, deter herbivores from grazing on leaves, drastically reduce digestibility of proteins and fiber, and negatively impact on ruminant livestock growth rates [Leinmüller et al., 1991; McAllister et al., 2005]. Both insect and mammalians have developed numerous strategies to deal with high proanthocyanidin plant organs, including secretion of tannin-binding proteins in the saliva [Shimada, 2006], detoxification mechanisms, and production of excess gastrointestinal mucus, among others [reviewed by McArthur et al., 1995; Dearing et al., 2005]. The nutritional value of forage legume preserved as silage [Wang et al., 2007] or the occurrence of bloating in ruminant livestock grazing exclusively on alfalfa [McAllister et al., 2005; Waghorn, 1990] can be prevented by supplementation with forage plants containing PA [Lees, 1992]. The wide range of variation in proanthocyanidin natural abundance as well as the degree of polymerization, monomer composition, stereochemistry, and the degree of hydroxylation [Haslam, 1981; Clausen et al., 1992; Zucker, 1983] modify the capacity to precipitate proteins and inhibit fiber digestion by cellulolytic ruminal bacteria [McAllister et al., 2005]. As a result, the potential growth gains from dietary supplementation depends on the proanthocyanidin qualitative and quantitative profile of the forage plant selected [Reed and Soller, 1987]. Alfalfa and clovers only accumulate proanthocyanidins in the seed coat [Sharma and Dixon, 2005; Xie et al., 2006]. Numerous breeding approaches including mutagenesis, somaclonal variation, and somatic hybridization have failed to induce proanthocyanidin accumulation in vegetative tissues [Goplen et al., 1980; Gruber et al., 1999]. As a result, transgenic strategies are currently been persued to improve digestibility of these valuable forage legumes [Morris and Robbins, 1997; Sharma and Dixon, 2005; Xie et al., 2006].

Flavonoids as Pollinator Attractants

Several plant species have evolved floral organs with elaborated morphological structures, unique petal epidermis cellular organization [Kay et al., 1981; Glover and Martin, 1998], and complex pigment biochemistry [reviewed by Grotewold, 2006; Ellestad, 2006] to ensure the attraction of insect pollinators, pollination, seed biogenesis and dispersal [Kevan and Baker, 1983]. Several flavonoids including anthocyanins, flavonols, flavones, and aurones are among the most common phenylpropanoid pigments found in floral organs and fruits [Grotewold, 2006; Ono et al., 2006]. Due to their unique absorption spectra, flavonoids often display complex patterns that become apparent only to specialized visual organs of insect pollinators that are sensitive to the UV range of the spectrum [Giurfa and Menzel, 1997; Gronquist et al., 2001]. In addition, successful pollinations can trigger floral color changes that are independent of senescing processes as observed in cotton and *Viola cornuta* [Farzad et al., 2000; Weiss 1991, 1995].

The color of monomeric anthocyanin pigments depends on the vacuolar pH [Fukada-Tanaka et al., 2000; Spelt et al., 2002] as well as on the occurrence of intra- or intermolecular interactions between the colored anthocyanin flavylium cation and colorless flavonoids, metal ions, carotenoids, and organic acids [Forkmann, 1991; Ellestad, 2006]. Molecular stacking resulting from these interactions is responsible for changes in the spectral signature of flavonoid pigments via the co-pigmentation phenomena [Ellestad, 2006]. Co-pigmentation enhances absorption (hyperchromic shift) and shifts the wavelength of maximum absorbance (bathochromic shift) of pigment solutions [Boulton, 2001], contributing to color stabilization and generation of distinctive spectral signatures to attract pollinators [Ellestad, 2006; Grotewold, 2006]. In addition to quantitative and qualitative changes in pigment composition, flowers have evolved morphological traits that affect color perception and attractiveness. For example, both the presence of conical-papillate cells in the petal epidermis and to a lower extent the content of anthocyanin pigments drastically influence frequency of pollinator visit and pollination success in *Antirrhinum majus* [Glover and Martin, 1998]. The existence of an intrincate genetic network regulating qualitative and quantitative pigment traits [Mol et al., 1998; Weiss, 2000; Martin and Gerats, 1993] and the complex molecular interactions determining flower pigmentation makes the modification of flower color for commercial purposes by breeding or transgenic approaches a challenging proposition [Forkmann, 1991; Chandler, 2003].

Phenylpropanoids as Signaling Molecules in Plant–Plant Interactions: Allelopathy and Parasitism

Allelopathy is defined as the secondary metabolite-mediated interaction between living organisms (plant, fungi, algae, and bacteria) that influence growth and developmental processes in a biological system [International Allelopathy Society, 1996]. Several plants, including modern crops, exudate potent allelochemicals that mediate allelopathic interactions, especially with invasive weeds [reviewed by Belz, 2007; Macías et al., 2007]. Crop allelopathy is seen as a natural defense mechanism against invasive weeds with potential for biotechnological applications [reviewed by Belz, 2007; Macías et al., 2007].

In plants, biosynthesis and exudation of allelochemicals follows developmental, diurnal, and abiotic/biotic stress-dependent dynamics. Compounds from 14 different chemical classes have been linked to allelopathic interactions, including several simple phenolic acids (e.g., benzoic and hydroxycinnamic acids) and flavonoids [Rice, 1984; Macías et al., 2007]. The existence of several soil biophysical processes that can reduce the effective concentration and bioactivity of these compounds casts doubts on their actual relevance in allelopathic interactions [Olofsdotter et al., 2002]. However, there are well-documented examples of phenylpropanoid-mediated incompatible interactions among plants. Several Gramineae mediate allelopathic interactions by means of phenolic and flavonoid exudates [Pérez and Ormeño-Núñez, 1991; Nimbal et al., 1996; Czarnota et al., 2003; Frey et al., 1997]. For example, genetic variability in the content of the flavone 5,7,4′-trihydroxy-3′,5′-dimethoxyflavone found in rice root exudates has been associated with allelopathic cultivars [Seal et al., 2004]. Several invasive dicotyledonous species have abundant allelochemical flavonoids like robinetin, in *Robinia pseudoacacia* L. [Nasir et al., 2005], and (−)-catechin in spotted knapweed (*Centaurea maculosa Lam.*) [Bais et al., 2003; Fitter, 2003]. In the latter example, catechin is secreted as a racemic mixture, where the (−)-catechin enantiomer has allelopathic activity and the (+)-catechin acts as a phytoalexin with antibacterial properties against root pathogens [Bais et al., 2002]. The (−) stereoisomer is absorbed through the root system of neighboring plants where it triggers a series of events that ultimately lead to root death [Bais et al., 2002]. Stereochemistry is not the only factor contributing to allelochemical properties. Potency of allelochemical compounds might also depend on the synergistic interaction with metabolites from a wide range of chemical classes [Kong et al., 2004].

The exact molecular mode of action of flavonoid allelochemicals remains elusive [Macías et al., 2007]. Simple phenolic compounds such as benzoic and hydroxycinnamic acid derivatives are widely studied allelochemicals that seem to interfere with plasma membrane functions, enzymatic processes and energy-related systems [Macías et al., 2007]. Depending on their specific lipophylic properties, simple phenolic compounds can penetrate cellular lipid bilayers causing electrolyte leakage conducive to cell death as a consequence of membrane depolarization [Macri et al., 1986] and altered water relations [Barkosky et al., 2000]. Certain phenolics can also affect normal respiratory processes in soil-borne microorganisms [Souto et al., 2000] as well as photosynthetic carbon assimilation in neighboring plants [Barkosky et al., 2000; Yang et al., 2002]. Further detailed identification of allelopathic compounds and their mechanistic mode of action would allow potential biotechnological applications for weed control.

Phenylpropanoids in Plant Defense and Plant–Pathogen Interactions

Several phenylpropanoids including simple phenolics as well as more complex monomeric and oligomeric aromatic compounds derived from the flavonoid, stilbene, and isoflavonoid pathway have toxic properties against a vast array of organisms [Harbone, 1994; Feucht and Treutter, 1999; Grayer and Harborne, 1994; Dixon, 2001]. Phenylpropanoids involved in defense mechanisms can be synthesized constitutively (phytoanticipins) or induced in response to specific elicitors (phytoalexins) [VanEtten et al., 1994; Dixon, 2001; Treutter, 2006]. Phytoanticipins are synthesized during normal development in specific organs and cell layers and can also be up-regulated in response to specific abiotic stress factors or biological elicitors. For example, several flavonoids accumulating in the developing barley seed testa layer provide resistance to different *Fusarium* spp. [Skadhauge et al., 1997b]. In grapes, natural variability in the constitutive and induced stilbene content of vegetative [Borie et al., 2004] and reproductive

[Li et al., 2006a] organs is often linked to the genetic diversity in resistance to gray mold (*Botrytis cinerea*) [Sbaghi et al., 1996; Borie et al., 2004] and decay caused by *Rhizopus stolonifer* [Sarig et al., 1997].

Levels of monomeric and oligomeric stilbenes in grape (*V. vinifera*) berry and vegetative tissues are greatly influenced by the genetic background [Li et al., 2006a] and eliciting factors [Borie et al., 2004]. Resistance of grape cultivars to downy mildew (*Plasmopara viticola*) is correlated to the induced accumulation of the oligomeric ε- and δ-viniferin but not to resveratrol or its glucoside piceid [Pezet et al., 2004; Dercks and Creasy, 1989]. In apple (*Malus domestica*), horticultural practices such as N fertilization can enhance plant growth and productivity. At the same time, they can reduce flavonoid biosynthesis and enhance susceptibility to specific pathogens [Rühmann et al., 2002; Rühmann and Treutter, 2003; Leser and Treutter, 2005]. The molecular and biochemical mechanisms underlying the induction of phenylpropanoid phytoalexins accumulation in specific tissues or cell layers in response to biotic or abiotic stress factors [Paiva et al., 1994] may depend on the elicitor effector involved [He and Dixon, 2000; Naoumkina et al., 2007]. In the model legume *Medicago truncatula*, medicarpin accumulation elicited by yeast extracts correlated with the up-regulation of the entire set of genes in the phenylpropanoid/isoflavonoid pathway [Naoumkina et al., 2007]. On the other hand, the wounding signal methyl-jasmonate, exclusively up-regulated β-glucosidases (and likely malonylesterases) and ABC transporters to mobilize vacuolar pools of (malonyl)glucosyl-isoflavonoids [Naoumkina et al., 2007]. The aglycones of these isoflavonoids can then be converted to bioactive medicarpin by a set of co-up-regulated pterocarpan biosynthetic genes [Naoumkina et al., 2007].

Gain-of-function transgenic strategies designed to introduce phytoalexin-dependent resistance traits by up-regulating expression of different rate-limiting steps in isoflavonoid and stilbene biosynthesis often result in enhanced resistance to plant pathogens. Alfalfa leaves accumulate isoflavonoid phytoalexins exclusively upon elicitation [Paiva et al., 1994]. Elicited alfalfa leaves constitutively overexpressing its own isoflavone 7-*O*-methyltransferase [He and Dixon, 2000] or a *M. truncatula* isoflavone reductase [Deavours and Dixon, 2005] had greater steady-state levels of several isoflavone glucosides and medicarpin resulting in enhanced resistance to *Phoma medicaginis* [He and Dixon, 2000]. Likewise, resistance to this pathogen is achieved in transgenic alfalfa plants with constitutive accumulation of a trans-resveratrol-glucoside resulting from the heterologous expression of the peanut stilbene synthase [Hipskind and Paiva, 2000]. In addition to changes in the levels of endogenous phytoalexins, the oxygen-dependent oxidation of (poly)phenolic compounds to their respective quinone forms mediated by plant and microbial polyphenol oxidases (PPO, EC 1.10.3.1, and laccase, EC 1.10.3.2) and peroxidases (EC 1.11.1.7) plays a critical role in defining function and biological activity of phenylpropanoid phytoalexins [reviewed by Pourcel et al., 2007]. Antisense suppression of PPO activity in tomato (*L. esculentum*) induced a dramatic increase in susceptibility to *Pseudomonas syringae* [Thipyapong et al., 2004] while overexpression was conducive to enhanced bacterial disease resistance [Li and Steffens, 2002]. Microorganisms have also evolved oxidation-dependent detoxification mechanisms to circumvent phytoalexin antimicrobial activity. For example, pathogenicity of isolates of the necrotroph *B. cinerea* correlates with the relative laccase-mediated capacity to conduct oxidative dimerization of resveratrol and pterostilbene (3,5-dimethoxy-4′-hydroxystilbene) found in grapevine tissues [Sbaghi et al., 1996; Cichewicz et al., 1999]. The pathogen and strain-specific presence of detoxification mechanisms may explain the enhanced resistance to *Phytophtora infestans* but not to *B. cinerea* observed in transgenic tomatoes overexpressing the grape stilbene synthase [Thomzik et al., 1997].

Although still relatively uncharacterized, several possible phytoalexin modes of action mechanisms have been proposed. Barley Pas and dihydroquercetin confer resistance mainly by direct or indirect inactivation of fungal enzymes involved in cell wall disassembly and degradation [Skadhauge et al., 1997b]. The grape hydroxystilbenes resveratrol and pterostilbene impair *B. cinerea conidia* respiration and germination while promoting abnormal subcellular organization [Pezett and Pont, 1990a, 1995; Adrian et al., 1997]. The specific toxicity of different stilbenes might be related to their lipophylic characteristics and their potential to disrupt normal cell and organelle membrane functions [Pezet and Pont, 1990b, 1995]. Further identification of known and novel phytoalexins, their mechanistic mode of action, and elucidation of biosynthetic routes are critical steps required to breed or engineer phenylpropanoid-based pathogen-resistant traits in crops.

Flavonoids and the Rhizobium–Legume Symbiosis

All higher organisms are dependent on external sources of reduced nitrogen for survival. One of the main sources of reduced nitrogen in the biosphere is the symbiotic interaction between bacteria, known generally as rhizobia, and the roots of legume plants.

Flavonoids are key regulators of this remarkable developmental and metabolic collaboration, in which otherwise saprophytic soil bacteria (the rhizobia) give up their independent lifestyle and take up residence within cells of a modified legume root organ, known as the nodule. Within the nodule the plant provides energy in the form of photosynthate to the bacterium, in exhange for fixed nitrogen in the form of ammonia. Flavonoids have roles at several steps in this symbiotic relationship, from initial signal exchange at the root surface during the earliest stage of interaction through the induction of nodule organogenesis in the root cortex.

One feature of the rhizobium–legume symbiosis is the high degree of specificity that many legumes and their symbiotic partners exhibit. For example, *Sinorhizobium meliloti*, the well-characterized bacterial symbiont of alflalfa and related *Medicago* species, is unable to nodule its close relative *Pisum sativum* (Mendel's pea), and conversely pea symbionts do not nodulate *Medicago* species. It is now clear that there are multiple molecular determinants of specificity, including bacterial ligands (known as "Nod factors") that stimulate host responses and corresponding cognate plant Nod factor receptors [Long 1996; Oldroyd and Downie, 2008]. However, the earliest manifestation of specificity depends on the release of host metabolites into the rhizosphere, where they are perceived by the bacterium and initiate a regulatory and biosynthetic cascade. Metabolites in the host legume are principally products of the flavonoid and isoflavonoid pathway [Peters et al., 1986; Redmond et al., 1986]. Bacterial genetics played a large role in deciphering the nature of early bacterial responses to legume exudates [Long, 1996], and by the early 1990s it was understood that flavonoids were perceived by a family of LysR transcription factors, including NodD1 of *S. meliloti*. NodD proteins are both positive and negative regulators of the expression of a suite of bacterial genes known as the Nod regulon. Products of the Nod regulon include enzymes involved in the synthesis of a second signal, delivered from the bacterium to the plant. The signal molecule, a chitolipooligosaccharide, acts in a plant–host specific manner to trigger a suite of host responses, which includes up-regulation of flavonoid gene transcription and flavonoid biosynthesis in the root epidermis [Van Brussel et al., 1990; Recourt et al., 1992]. The outcome is a flavonoid-based regulatory loop that is presumed to reinforce early signal exchange.

A remarkable degree of specificity is exhibited at the level of flavonoid-induced gene transcription in the bacterium. Peck et al. [2006] demonstrate that NodD homologs from a range of rhizobium species exhibit different specificities for flavonoid-mediated gene induction, consistent with direct physical interaction between NodD transcription factors and flavonoids. Interestingly, flavonoids act generally to stimulate binding of the *S. meliloti* transcription factor, SmNodD1, to nod gene promoters, independent of whether or not the flavonoids activate NodD1-dependent gene transcription. Thus, flavonoids act nonspecifically to enhance NodD1 binding to promoters, and specifically to affect transcriptional activation. Peck et al. [2006] propose a model wherein the inactive flavonoids are competitive inhibitors of nod gene induction, competing with activating flavonoids for NodD interaction and also promoting transcriptionally inactive promoter DNA–NodD complexes.

As mentioned above, the product of Nod gene activation in the bacterium is the synthesis of a ligand called Nod factor. Nod factor perception activates a suite of host responses that include the up-regulation of flavonoid synthesis and the activation of cell division in the root cortex leading to nodule organogenesis. The convergence of work spanning several decades led to the hypothesis that Nod factor-induced flavonoid accumulation might be causal to Nod factor-induced cell cycle activation. Key to this hypothesis were the demonstrations that (i) exogenous flavonoids can regulate polar auxin transport [Jacobs and Rubery, 1998]; (ii) auxin transport inhibitors such as N-(1-napthyl)-phthalamic acid (NPA) can induce nodule-like structures on legume roots [Hirsch et al., 1989]; (iii) both flavonoid biosynthesis genes [Yang et al., 1992] and flavonoid metabolites [Mathesius et al., 2000] accumulate at the site of incipient cortical cell division; and (iv) cortical cell division is preceded by a spatially co-incident inhibition of polar auxin transport [Mathesius et al., 1998]. Mechanistically, flavonoid-mediated changes in the auxin transport are thought to alter critical auxin-to-cytokinin ratios at the site of cortical cell activation, leading to nodule primordia (e.g., as supported by the study of Cooper and Long [1994]). Recently, nodule-related cell division has been shown to depend on the plant hormone cytokinin [Gonzalez-Rizzo et al., 2006; Tirichine et al., 2007; Murray et al., 2007]. However, there is no conclusive proof that flavonoids are required for cortical cell activation.

Working with the model legume, *M. truncatula*, Wasson et al. [2006] used RNA interference (RNAi) to reduce the level of chalcone synthase (CHS) transcript and thus flavonoid biosynthesis in *Agrobacterium* rhizogenes-transformed "hairy" roots. The transformed roots exhibited a substantial reduction in the levels of flavonoid metabolites and were unable to form nodules (no cell division) upon inoculation with compatible *Sinorhizobium meliloti.* Reduced flavonoid levels were predicted to impact symbiosis in two ways: first, through precluding induction of bacterial gene expression via the flavonoid-responsive NodD circuit described above, and second by precluding flavonoid-mediated changes in auxin transport in the root cortex. When bacteria were pretreated with flavonoids to ensure efficient Nod gene induction in the bacterium, root hair curling (a classic Nod factor response) was restored, though inner cortical cell divisions were still absent. By contrast, nodule formation could be rescued by treatment of roots with naringenin and liquiritigenin. The CHS RNAi plants exhibited increased polar auxin transport relative to wild type, similar to phenotypes observed for flavonoid-deficient mutants of *Arabidopsis* [reviewed by Peer and Murphy, 2007]. Importantly, inoculation with preinduced *S. meliloti* caused decreased auxin transport in wild-type, but not in the CHS RNAi, plants. The results of Wasson et al. [2006] are consistent with a model wherein the bacterium induces CHS, and the resulting flavonoid biosynthesis inhibits polar auxin transport, which leads to (is necessary for) cell division.

A similar but not identical study by Subramanian et al. [2006], working in soybean, used RNA interference to co-knockdown two isoflavone synthase paralogs (IFS1 and IFS2) and in a separate experiment to knock down chalcone reducatse (CHR). The IFS knockdown plants were severely reduced for nodulation, and they had substantially reduced levels of daidzein and genistein, the two main Nod gene inducers in the *Bradyrhizobium*–soybean symbiosis. By contrast, the CHR knockdown plants had reduced levels of daidzein but increased levels of genistein, and these plants nodulated normally. IFS knockdown plants exhibited increased auxin transport, similar to the results described previously for *Medicago* and *Arabidopsis*; conversely, plants silenced for CHR had decreased auxin transport. Thus, the data from Subramanian et al. [2006] support a role for product(s) of IFS in nodulation (presumably daidzein and/or genistein), and a role for product(s) of IFS (presumably daidzein) in polar auxin transport.

To determine whether the reduced polar auxin transport observed in IFS knockdown plants was causal to the absence of nodulation, the authors conducted two experiments: (1) They used a genistein-hypersensitive strain of *Bradyrhizobium*, which gives efficient Nod factor gene induction even in the IFS knockdown background, and (2) they inoculated wild-type and mutant plants directly with cognate Nod factor. They observed that the genistein-hypersensitive strain restored nodulation to the IFS knockdown plants, suggesting that products of IFS of activity are critical to bacterial Nod gene induction but not to polar auxin transport. In other words, they could dissociate the auxin transport phenotype of the plant from the Nod gene induction phenotype of the bacterium by means of the genistein-hypersensitive bacterial symbiont. The results of the second experiment were consistent with the first; namely, they revealed that purified Nod factor was sufficient to induce cortical cell activation (cell division) in IFS and CHR knockdown plants as efficiently as in wild-type plants. Thus, in contrast to the conclusion of Wasson et al. [2006], Subramanian et al. [2006] concluded that flavonoid-inhibition of auxin transport is not critical to nodulation.

To summarize, it seems clear that flavonoids are important for nodulation and that part of this importance relates to Nod gene induction in the bacterium via the NodD circuit. It also seems clear that endogenous flavonoids can inhibit auxin transport in both *Medicago* and soybean. It is formally possible that flavonoid-regulated auxin transport is important to nodulation in *Medicago* but not in soybean. In fact, these two species exhibit a conspicuous difference in nodule development: in *Medicago* nodule primordia arise from inner cortical cells, while in soybean nodule primordia arise from cells in the outer cortex—perhaps this anatomical distinction reflects an underlying physiological difference in nodule initiation. Alternatively, flavonoid-mediated altered auxin transport may be correlated with, but not essential to, nodulation. The discrepancy between the results of Wasson et al. [2006] and Subramanian et al. [2006] may be a function of biological differences between *Medicago* and soybean or may reflect differences in experimental design such as the choice of genes for RNAi. In any case, these studies highlight the complexity of research at the intersection of flavonoid metabolism and plant physiology. It is well established that flavonoids are endogenous regulators of polar auxin transport [reviewed by Peer and Murphy, 2007] and that they exert their effects by both direct and indirect means; however, while flavonoids are well established as factors in plant growth and development, they are often not essential to these processes. In many cases, such biological nuance complicates qualitative statements about the link between flavonoids and plant function, as may be the case for the rhizobium–legume symbiosis.

TRENDS IN METABOLIC ENGINEERING OF PHENYLPROPANOID METABOLISM IN PLANTS AND MICROORGANISMS

The flavonoid pathway is an important source of small molecules with nutraceutical and drug discovery potential due to their potent bioactivity against several human diseases [reviewed by Lin and Weng, 2006; Ross and

Kasum, 2002; Ververidis et al., 2007a]. Flavonoids are potent antioxidants in vitro and several epidemiological and clinical studies have linked diets with high flavonoid intake with lower rates of cardiovascular disease, cancer, and other age-related diseases. In spite of the dramatic progress in our understanding of the flavonoid pathway, *in planta* biosynthesis of flavonoids for drug discovery imposes several challenges from a pharmaceutical standpoint. Low concentration and limitations to diversify basic chemistry complicate the use of plants as bioreactors to produce specific flavonoids in sufficient amounts for basic research, development, and use in clinical trials [Butler, 2004]. The biochemical synthesis of flavonoids in microorganisms provide a novel approach that would circumvent such limitations [reviewed by Chemler et al., 2006; Leonard et al., 2007b; Ververidis et al., 2007b]. Both yeast and bacteria have been used as model organisms for bioreactor-based flavonoid production [Yan et al., 2005; Miyahisa et al., 2005; Leonard et al., 2006a, 2006b; Zhang et al., 2006]. Introduction of the core flavanone biosynthetic pathway and simultaneous co-expression of several downstream enzymatic activities allowed the biosynthesis of different flavonoids in the gut bacteria *Escherichia coli* [Miyahisa et al., 2005; Leonard et al., 2006a, 2006b; Beeckwilder et al., 2006] and the bread and wine yeast *Saccharomyces cereviciae* [Chemler et al., 2006; Leonard et al., 2005; Yan et al., 2005; Beeckwilder et al., 2006]. Further refinements to overcome metabolic bottlenecks [Miyahisa et al., 2005] boosted flavonoid production more than a 1,000-fold when central metabolic pathways in *E. coli* were engineered to increase the endogenous malonyl-CoA pool [Leonard et al., 2007]. The application of ^{13}C-isotopomer-based flux analysis [Sauer, 2006] to gain insight on the fluxome of these engineered microorganisms would certainly provide new leads for further optimization of flavonoid biosynthesis.

Efforts to metabolically engineer the phenylpropanoid and flavonoid pathways in plants have been directed toward improvement or introduction of novel phenotypic traits with relevance to human diet, animal feed, or ornamental purposes [Forkmann and Martens, 2001; Tian et al., 2007]. Transgenic approaches have been applied to overcome genetic constrains limiting conventional or marker-assisted breeding. Also, they have been used when the introduction of the new trait cannot be done at the expense of modifying preexistent desirable characteristics of the base germplasm. Transgenic blue/violet carnations, blue roses [Tanaka et al., 2005], and yellow snapdragons and dahlias [Ono et al., 2006] are just a few examples of pathway engineering by ectopic expression of exotic genes of the flavonoid pathway without introducing changes in other desirable phenotypic traits [Forkmann and Martens, 2001]. Pathway engineering has also been applied to increase the nutraceutical value of fruits and vegetables [Dixon and Steele, 1999; de Vos et al., 2000]. For example, overexpression of the maize transcription factors LC and C1 [Bovy et al., 2002] or the petunia CHI [Verhoeyen et al., 2002; Muir et al., 2001] increased nutraceutical value and antioxidant potential of tomatoes as a result of dramatic increase in total flavonol content. Soybeans with increased content of beneficial isoflavones were obtained by ectopic expression of maize C1 and R transcription factors and suppression of F3H to prevent competition from anthocyanin biosynthesis [Yu et al., 2003]. Modification of the (iso)flavonoid profile of edible plant organs by introducing activities that can modify their basic carbon skeleton and bioactivity is certainly the next frontier in metabolic engineering of foods with enhanced nutraceutical value [Tian et al., 2007; Lim et al., 2004]. The phenylpropanoid pathway has also been targeted to engineer traits related to improvement in animal feed quality and industrial processing of plant biomass. Metabolic engineering of lignin biosynthesis has found applications in forage quality improvement and production of ethanol for biofuel. Lignin content [Jung and Vogel, 1986] and relative changes in lignin composition [Buxton and Russell, 1988] associated with forage legume maturity stage correlate negatively with digestibility and nutritional value for ruminant livestock [Albrecht et al., 1987; Casler and Vogel, 1999]. Digestibility of alfalfa was substantially improved in plants with low lignin content and similar monolignol subunit composition. This milestone was achieved by antisense-mediated silencing of cytochrome P450 monooxygenases [Reddy et al., 2005] and methyltransferases [Guo et al., 2001] involved in the biosynthesis of monolignol lignin subunits. Reduction of lignin content in plant lignocellulosic biomass may also improve access of degradative enzymes to cellulose and increase efficiency of enzymatic saccharification processes that release fermentescible sugars for biofuel production [Mosier et al., 2005; Chapple et al., 2007]. Low lignin biomass may avoid hot acid pretreatments that add to processing costs [Mosier et al., 2005] and introduce contaminants that interfere with yeast-based fermentation processes [Almeida et al., 2007]. On a different approach, down-regulation of lignin biosynthesis in alfalfa was achieved using antisense technology to reduce expression of six different genes in the monolignol biosynthetic pathway [Chen and Dixon, 2007]. In these plants, yield of fermentescible sugar increased proportionally with decreasing lignin content [Chen and Dixon, 2007]. The application of this approach to other bioenergy crops (e.g., poplar—Populus sp., *Miscanthus* × *giganteus*, switchgrass—*Panicum virgatum*, etc.) may alleviate pressure on food crops currently used for biofuel production and lead to a cost-effective and renewable nonedible plant biomass source for ethanol-based biofuel production [Chapple et al., 2007]. Future progress in metabolic engineering to increase nutraceutical value of edible plant organs, enhance disease resistance, or modify structural properties of plant biomass for industrial applications will depend on progress made on our understanding of transcriptional regulation [Century et al., 2008], transport mechanisms, and factors affecting fluxes from precursor pathways as well as between branches of the phenylpropanoid pathway [Liu et al., 2002; Ro and Douglas, 2004; Schwender, 2008; Farag et al., 2008].

CONCLUSIONS

The phenylpropanoid pathway is the source of hundreds of metabolites that play critical roles in ensuring land plant survival, reproduction, and interaction with other organisms and their environment. Advances in (iso)flavonoids isolation and characterization [reviewed by Stobiecki and Kachlicki, 2006] as well as identification of full gene complements as a result of whole genome sequencing projects (http://www.sciencemag.org/plantgenomes/) and characterization of novel mutants will dramatically expand our mechanistic understanding of the role of these secondary metabolites in plant biology and ecophysiology. Due to the inherent limitations of using plants as bioreactors, pathway engineering will likely focus on introducing traits to enhance tolerance to abiotic/biotic stress factors, improve edible quality and nutraceutical value of fruits and vegetables, and modify industrial processing properties of plant biomass, particularly in relation to cellulose-based biofuel production. Pharmaceutical discovery and production of phenylpropanoid-based drugs will almost certainly rely on progress made on microorganism-based (iso)flavonoids and stilbene biosynthesis.

REFERENCES

Abe H, Yamaguchi-Shinozaki K, Urao T, Iwasaki T, Hosokawa D, Shinozaki K. 1997. Role of *Arabidopsis* MYC and MYB homologs in drought- and abscisic acid-regulated gene expression. *Plant Cell* 9:1859–1868.

Abel S, Ticconi CA, Delatorre CA. 2002. Phosphate sensing in higher plants. *Physiol Plantarum* 115:1–8.

Abrahams S, Lee E, Walker AR, Tanner GJ, Larkin P, Ashton AR. 2003. The TDS4 gene encodes leucoanthocyanidin dioxygenase (LDOX) and is essential for proanthocyanidin synthesis and vacuole development. *Plant J* 35:624–636.

Abrahams S, Tanner GJ, Larkin PJ, Ashton AR. 2002. Identification and biochemical characterization of mutants in the proanthocyanidin pathway in *Arabidopsis*. *Plant Physiol* 130:561–576.

Achnine L, Blancaflor EB, Rasmussen S, Dixon RA. 2004. Colocalization of L-phenylalanine ammonia-lyase and cinnamate 4-hydroxylase for metabolic channeling in phenylpropanoid biosynthesis. *Plant Cell* 16:3098–3109.

Ackerson RC. 1981. Osmoregulation in cotton in response to water stress. II. Leaf carbohydrate status in relation to osmotic adjustment. *Plant Physiol* 67:489–493.

Adrian M, Jeandet P, Veneau J, Weston LA, Bessis R. 1997. Biological activity of resveratrol, a stilbenic compound from grapevines, against *Botrytis cinerea*, the causal agent for gray mold. *J Chem Ecol* 23:1689–1702.

Aerts RJ, Barry TN, McNabb WC. 1999. Polyphenols and agriculture: beneficial effects of proanthocyanidins in forages. *Agr Ecosys Environ* 75:1–12.

Aharoni A, De Vos CHR, Wein M, Sun Z, Greco R, Kroon A, Mol JNM, O'Connell AP. 2001. The strawberry FaMYB1 transcription factor suppresses anthocyanin and flavonol accumulation in transgenic tobacco. *Plant J* 28:319–332.

Akashi T, Koshimizu S, Aoki T, Ayabe S-I. 2006. Identification of cDNAs encoding pterocarpan reductase involved in isoflavan phytoalexins biosynthesis in *Lotus japonicus* by EST mining. *FEBS Lett* 580:5666–5670.

Akashi T, Aoki T, Ayabe S. 2005. Molecular and biochemical characterization of 2-hydroxyisoflavanone dehydratase. Involvement of carboxylesterase-like proteins in leguminous isoflavone biosynthesis. *Plant Physiol* 137:882–891.

Akashi T, Fukuchi-Mizutani M, Aoki T, Ueyama Y, Yonekura-Sakakibara K, Tanaka Y, Kusumi T, Ayabe S. 1999. Molecular cloning and biochemical characterization of a novel cytochrome P450, flavone synthase II, that catalyzes direct conversion of flavanones to flavones. *Plant Cell Physiol* 40:1182–1186.

Albrecht KA, Wedin WF, Buxton DR. 1987. Cell-wall composition and digestibility of alfalfa stems and leaves. *Crop Sci* 27:735–741.

Alfenito MR, Souer E, Goodman CD, Buell R, Mol J, Koes R, Walbot V. 1998. Functional complementation of anthocyanin sequestration in the vacuole by widely divergent glutathione S-transferases. *Plant Cell* 10:1135–1149.

Ali A, Strommer J. 2003. A simple extraction and chromatographic system for the simultaneous analysis of anthocyanins and stilbenes of *Vitis* species. *J Agric Food Chem* 51:7246–7251.

Almeida JRM, Modig T, Petersson A, Hahn-Hagerdal B, Lidén G, Gorwa-Grauslund MF. 2007. Increased tolerance and conversion of inhibitors in lignocellulosic hydrolysates by Saccharomyces cerevisiae. *J Chem Technol Biotech* 82:340–349.

Alves-Ferreira M, Wellmer F, Banhara A, Kumar V, Riechmann JL, Meyerowitz EM. 2007. Global expression profiling applied to the analysis of *Arabidopsis* stamen development. *Plant Physiol* 145:747–762.

Ang L-H, Chattopadhyay S, Wei N, Oyama T, Okada K, Batschauer A, Deng X. 1998. Molecular interaction between COP1 and HY5 defines a regulatory switch for light control of *Arabidopsis* development. *Mol Cell* 1:213–22.

Avsian-Kretchmer O, Cheng JC, Chen L, Moctezuma E, Sung ZR. 2002. Indole acetic acid distribution coincides with vascular differentiation pattern during *Arabidopsis* leaf ontogeny. *Plant Physiol* 130:199–209.

Akashi T, Sawada Y, Shimada N, Sakurai N, Aoki T, Ayabe S. 2003. cDNA cloning and biochemical characterization of S-adenosyl-L-methionine: 2,7,4′-trihydroxyiso-flavanone 4′-O-methyltransferase, a critical enzyme of the legume isoflavonoid phytoalexin pathway. *Plant Cell Physiol* 44:103–112.

Bagge P, Larsson C. 1986 Biosynthesis of aromatic aminoacids by highly purified spinach chloroplasts—compartmentation and regulation of the reactions. *Physiol Plantarum* 68:641–647.

Bailly A, Sovero V, Vincenzetti V, Santelia D, Bartnik D, Koenig BW, Mancuso S, Martinoia E, Geisler. 2008. Modulation of P-glycoproteins by auxin transport inhibitors is mediated by interaction with immunophilins. *J Biol Chem* 283:21817–21826.

Bailly C, Cormier F, Do CB. 1997. Characterization and activities of S-adenosyl-L-methionine:cyanidin 3-glucoside 3′-O-methyltransferase in relation to anthocyanin accumulation in *Vitis* vinifera cell suspension cultures. *Plant Sci* 122:81–89.

Bais HP, Park S-W, Weir TL, Callaway RM, Vivanco JM. 2004. How plants communicate using the underground information superhighway. *Trends Plant Sci* 9:26–32.

Bais HP, Vepachedu R, Gilroy S, Callaway RM, Vivanco JM. 2003. Allelopathy and exotic plant invasion: From molecules and genes to species interactions. *Science* 301:1377–1380.

Baker CJ, O'Neill NR, Tomerlin JR. 1989. Accumulation of phenolic compounds in incompatible clone/race interactions of *Medicago sativa* and Colletotrichum trifolii. *Physiol Mol Plant Path* 35:231–241.

Baldi A, Romani A, Mulinacci N, Vincieri FF, Casetta B. 1995. HPLC/MS application to anthocyanins of *Vitis* vinifera L. *J Ag Food Chem* 43:2104–2109.

Barbehenn RV, Martin MM. 1994. Tannin sensitivity in larvae of Malacosoma disstria (Lepidoptera)—roles of the peritrophic envelope and midgut oxidation. *J Chem Ecol* 20:1985–2001.

Barbenhenn RV, Bumgarner SL, Roosen EF, Martin MM. 2001. Antioxidant defenses in caterpillars: Role of the ascorbate recycling system in the midgut lumen. *J Insect Physiol* 47:349–357.

Barkosky RR, Einhellig FA, Butler JL. 2000. Caffeic acid induced changes in plant–water relationships and photosynthesis in leafy spurge (Euphorbia esula). *J Chem Ecol* 26:2095–2109.

Barry TN, Manley TR. 1984. The role of condensed tannins in the nutritional value of Lotus pedunculatus for sheep. 2. Quantitative digestion of carbohydrates and proteins. *Br J Nutr* 51:493–504.

Barry TN, McNabb WC. 1999. The implications of condensed tannins on the nutritive value of temperate forages fed to ruminants. *Br J Nutr* 81:263–272.

Barth I, Meyer S, Sauer N. 2003. PmSUC3: Characterization of a SUT2/SUC3-type sucrose transporter from Plantago major. *Plant Cell* 15:1375–1385.

Batz O, Logemann E, Reinold S, Hahlbrock K. 1998. Extensive reprogramming of primary and secondary metabolism by fungal elicitor or infection in parsley cells. *Biol Chem* 379:1127–1135.

Baudry A, Caboche M, Lepiniec L. 2006. TT8 controls its own expression in a feedback regulation involving TTG1 and homologous MYB and bHLH factors, allowing a strong cell-specific accumulation of flavonoids in *Arabidopsis* thaliana. *Plant J* 46:768–779.

Baudry A, Heim MA, Dubreucq B, Caboche M, Weisshaar B, Lepiniec L. 2004. TT2, TT8, and TTG1 synergistically specify the expression of BANYULS and proanthocyanidin biosynthesis in *Arabidopsis* thaliana. *Plant J* 39:366–380.

Baxter IR, Young JC, Armstrong G, Foster N, Bogenschutz N, Cordova T, Peer WA, Hazen SP, Murphy AS, Harper JF. 2005. A plasma membrane H+-ATPase is required for the formation of proanthocyanidins in the seed coat endothelium of *Arabidopsis* thaliana. *Proc Natl Acad Sci USA* 102:2649–2654.

Beekwilder J, Wolswinkel R, Jonker H, Hall R, de Vos CH, Bovy A. 2006. Production of resveratrol in recombinant microorganisms. *Appl Environ Microbiol* 72:5670–5672.

Belz RG. 2007. Allelopathy in crop/weed interactions—an update. *Pest Manag Sci.* 63:308–326.

Bennett MJ, Marchant A, Green HG, May ST, Ward SP, Millner PA, Walker AR, Schultz B, Feldmann KA. 1996. Arabidopsis AUX1 gene: A permease-like regulator of root gravitropism. *Science* 273:948–950.

Bernhardt C, Zhao M, Gonzalez A, Lloyd A, Schiefelbein J. 2005. The bHLH genes GL3 and EGL3 participate in an intercellular regulatory circuit that controls cell patterning in the *Arabidopsis* root epidermis. *Development* 132:291–298.

Bernhardt C, Lee MM, Gonzalez A, Zhang F, Lloyd A, Schiefelbein J. 2003. The bHLH genes GLABRA3 (GL3) and ENHANCER OF GLABRA3 (EGL3) specify epidermal cell fate in the *Arabidopsis* root. *Development* 130:6431–6439.

Besseau S, Hoffman L, Geoffroy P, Lapierre C, Pollet B, Legrand M. 2007. Flavonoid accumulation in *Arabidopsis* repressed in lignin synthesis affects auxin transport and plant growth. *Plant Cell* 19:148–162.

Bharti AK, Khurana JP. 1997. Mutants of *Arabidopsis* as tools to understand the regulation of phenylpropanoid pathway and UV-B protection mechanisms. *Phytochem Photobiol* 65:765–776.

Bieza K, Lois R. 2001. An *Arabidopsis* mutant tolerant to lethal ultraviolet-B levels shows constitutively elevated accumulation of flavonoids and other phenolics. *Plant Physiol* 126:1105–1115.

Block E, Page JE, Toscano JP, Wang C-X, Zhang X, DeOrazio R, Guo C, Sheridan RS, Towers GHN. 1996. Photochemistry of thiarubrine A and other 1,2-dithiins—formation of 2,6-dithiabicyclo[3.1.0]hex-3-enes. *J Am Chem Soc* 118:4719–4720.

Boege F, Straub T, Kehr A, Boesenberg C, Christiansen K, Andersen A, Jakob F, Köhrle J. 1996. Selected novel flavones inhibit the DNA binding or the DNA replication step of eukaryotic topoisomerase I. *J Biol Chem* 271:2262–2270.

Boerjan W, Ralph J, Baucher M. 2003. Lignin biosynthesis. *Annu Rev Plant Biol* 54:519–546.

Bogs J, Jaffé FW, Takos AM, Walker RA, Robinson SP. 2007. The grapevine transcription factor VvMYBPA1 regulates proanthocyanidin synthesis during fruit development. *Plant Physiol* 143:1347–1361.

Bogs J, Ebadi A, McDavid D, Robinson SP. 2006. Identification of the flavonoid hydroxylases from grapevine and their regulation during fruit development. *Plant Physiol* 140:279–291.

Bogs J, Downey MO, Harvey JS, Ashton AR, Tanner GT, Robinson SP. 2005. Proanthocyanidin synthesis and expression of genes encoding leucoanthocyanidin reductase and anthocyanidin reductase in developing grape and grapevine leaves. *Plant Physiol* 139:652–663.

Bomati EK, Austin MB, Bowman ME, Dixon RA, Noel JP. 2005. Structural elucidation of chalcone reductase and implications for deoxychalcone biosynthesis. *J Biol Chem* 280:30496–30503.

Bongue-Bartelsman M, Phillips DA. 1995. Nitrogen stress regulates gene expression of enzymes in the flavonoid biosynthetic pathway of tomato. *Plant Physiol Biochem* 33:539–546.

Booij-James IS, Dube SK, Jansen MAK, Edelman M, Mattoo AK. 2000. Ultraviolet-B radiation impacts light-mediated turnover of the photosystem II reaction center heterodimer in *Arabidopsis* mutants altered in phenolic metabolism. *Plant Physiol* 124:1275–1284.

Booker J, Sieberer T, Wright W, Williamson L, Willett B, Stirnberg P, Turnbull C, Srinivasan M, Goddard P, Leyser O. 2005. MAX1 encodes a cytochrome P450 family member that acts downstream of MAX3/4 to produce a carotenoid-derived branch-inhibiting hormone. *Dev Cell* 8:443–449.

Borevitz JO, Xia Y, Blount J, Dixon RA, Lamb C. 2000. Activation tagging identifies a conserved MYB regulator of phenylpropanoid biosynthesis. *Plant Cell* 12:2383–2394.

Borie B, Jeandet P, Parize A, Bessis R, Adrian M. 2004. Resveratrol and stylbene synthase mRNA production in grapevine leaves treated with biotic and abiotic phytoalexin elicitors. *Am J Enol Vitic* 55:160–164.

Boss PK, Davies C, Robinson SP. 1996b. Expression of anthocyanin biosynthesis pathway genes in red and white grapes. *Plant Molec Biol* 32:565–569.

Boss PK, Davies C. Robinson P. 1996a. Analysis of the expression of anthocyanin pathway genes in developing *Vitis* vinifera L. cv. Shiraz grape berries and the implications for pathway regulation. *Plant Physiol* 111:1059–1066.

Bouchard R, Bailly A, Blakeslee JJ, Oehring SC, Vincenzetti V, Lee OR, Paponov I, Palme K, Mancuso S, Murphy AS, Schulz B, Geisler M. 2006. Immunophilin-like TWISTED DWARF1 modulates auxin efflux activities of *Arabidopsis* P-glycoproteins. *J Biol Chem* 281:30603–30612.

Boulton R. 2001. The copigmentation of anthocyanins and its role in the color of red wine: A critical review. *Am J Enol Vitic* 52:67–87.

Bovy A, de Vos R, Kemper M, Schijlen E, Almenar Pertejo M, Muir S, Collins G, Robinson S, Verhoeyen M, Hughes S, Santos-Buelga C, van Tunen A. 2002. High-flavonol tomatoes resulting from the heterologous expression of the maize transcription factor genes LC and C1. *Plant Cell* 14:2509–2526.

Bradley JM, Davies KM, Deroles SC, Bloor SJ, Lewis DH. 1998. The maize Lc regulatory gene up-regulates the flavonoid biosynthetic pathway of petunia. *Plant J* 13:381–392.

Braidot E, Petrussa E, Bertolini A, Peresson C, Ermacora P, Loi N, Terdoslavich M, Passamonti S, Macri F, Vianello A. 2008. Evidence for a putative flavonoid translocator similar to mammalian bilitranslocase in grape berries (Vitis vinifera L.) during ripening. *Planta* 228:203–213.

Broun P. 2005. Transcriptional control of flavonoid biosynthesis: A complex network of conserved regulators involved in multiple aspects of differentiation in *Arabidopsis*. *Curr Opin Plant Biol* 8:272–279.

Brown DE, Rashotte AM, Murphy AS, Normaly J, Tague BW, Peer WA, Taiz L, Muday GK. 2001. Flavonoids act as negative regulators of auxin transport in vivo in *Arabidopsis*. *Plant Phys* 126:524–535.

Buer CS, Muday GK. 2004. The *transparent testa4* mutation prevents flavonoids synthesis and alters auxin transport and the response of *Arabidopsis* roots for gravity and light. *Plant Cell* 16:1191–1205

Buer CS, Muday GK, Djordjevic MA. 2007. Flavonoids are differentially taken up and transported long distances in *Arabidopsis*. *Plant Phys* 145:478–490.

Burbulis IE, Winkel-Shirley B. 1999. Interactions among enzymes of the *Arabidopsis* flavonoid biosynthetic pathway. *Proc Natl Acad Sci USA* 96:12929–12934

Burbulis IE, Iacobucci M, Shirley BW. 1996. A null mutation in the first enzyme of flavonoid biosynthesis does not affect male fertility in *Arabidopsis*. *Plant Cell* 8:1013–1025.

Burow ME, Boue SM, Collins-Burow BM, Melnik LI, Duong BN, Carter-Wientjes CH, Li S, Wiese TE, Cleveland TE, McLachlan JA. 2001. Phytochemical glyceollins, isolated from soy, mediate antihormonal effects through estrogen receptor ? and ?1. *J Clin Endoc Metab* 86:1750–1758.

Burr FA, Burr B, Scheffler BE, Blewitt M, Wienand U, Matz EC. 1996. The maize repressor-like gene intensifier1 shares homology with the r1/b1 multigene family of transcription factors and exhibits missplicing. *Plant Cell* 8:1249–1259.

Butler MS. 2004. The role of natural product chemistry in drug discovery. *J Nat Prod* 67:2141–2153.

Buxton DR, Russell JR. 1988. Lignin constituents and cell wall digestibility of grass and legume stems. *Crop Sci* 28:553–558.

Calderon AA, Tapata JM, Pedreno MA, Munoz R, Ros Barcelo A. 1992. Levels of 4-hydroxystilbene-oxidizing isoperoxidases related to constitutive disease resistance in in-vitro-cultured grapevine. *Plant Cell Tissue Organ Cult* 29:63–70.

Cantos E, Espín JC, Tomás-Barberán FA. 2002. Postharvest stilbene-enrichment of red and white table grape varieties using UV-C irradiation pulses. *J Agric Food Chem* 50:6322–6329.

Carey CC, Strahle JT, Selinger DA, Chandler VL. 2004. Mutations in the pale aleurone color1 regulatory gene of the Zea mays anthocyanin pathway have distinct phenotypes relative to the functionally similar TRANSPARENT TESTA GLABRA1 gene in *Arabidopsis* thaliana. *Plant Cell* 16:450–464.

Casler MD, Vogel KP. 1999. Accomplishments and impact from breeding for increased forage nutritional value. *Crop Sci* 39:12–20.

Century K, Reuber TL, Ratcliffe OJ. 2008. Regulating the regulators: The future prospects for transcription-factor-based agricultural biotechnology products. *Plant Physiol* 147:20–29.

Chalker-Scott L. 2002. Do anthocyanins function as osmoregulators in leaf tissues? In: Gould KS, Lee DW, Editors. Anthocyanins in Leaves. Amsterdam: Academic Press. Advances in Botanical Research 37:103–127.

Chalker-Scott L. 1999. Environmental significance of anthocyanins in plant stress responses. *Photochem Photobiol* 70:1–9.

Chandler S. 2003. Commercialization of genetically modified ornamental plants. *J Plant Biotechnol* 5:69–77.

Chapple C, Ladisch M, Meilan R. 2007. Loosening lignin's grip on biofuel production. *Nature Biotech* 25:746–748.

Chattopadhyay S, Ang LH, Puente P, Deng XW, Wei N. 1998. Arabidopsis bZIP protein HY5 directly interacts with light-responsive promoters in mediating light control of gene expression. *Plant Cell* 10:673–683.

Chemler JA, Yan Y, Koffas MA. 2006. Biosynthesis of isoprenoids, polyunsaturated fatty acids and flavonoids in Saccharomyces cerevisiae. *Microb Cell Fact* 5:20–28.

Chen CC, Shieh B, Jin YT, Liau YE, Huang CH, Liou JT, Wu LW, Huang W, Young KC, Lai MD, Liu HS, Li C. 2001. Microarray profiling of gene expression patterns in bladder tumor cells treated with genistein. *J Biomed Sci* 8:214–222.

Chen W, Provart NJ, Glazebrook J, Katagiri F, Chang HS, Eulgem T, Mauch F, Luan S, Zou G, Whitham SA, Budworth PR, Tao Y, Xie Z, Chen X, Lam S, Kreps JA, Harper JF, Si-Ammour A, Mauch-Mani B, Heinlein M, Kobayashi K, Hohn T,

Dangl JL, Wang X, Zhu T. 2002. Expression profile of *Arabidopsis* transcription factor genes suggests their putative functions in response to environmental stresses. *Plant Cell* 14:559–574.

Chen F, Dixon RA. 2007. Lignin modification improves fermentable sugar yields for biofuel production. *Nature Biotech* 25:759–761.

Chopra S, Brendel V, Zhang J, Axtell JD, Peterson T. 1999. Molecular characterization of a mutable pigmentation phenotype and isolation of the first active transposable element from Sorghum bicolor. *Proc Natl Acad Sci USA* 96:15330–15335.

Christie PJ, Alfenito MR, Walbot V. 1994. Impact of low-temperature stress on general phenylpropanoid and anthocyanin pathways: enhancement of transcript abundance and anthocyanin pigmentation in maize seedlings. *Planta* 194:541–549.

Cichewicz RH, Konzi SA, Hamann MT. 2000. Dimerization of resveratrol by the grapevine pathogen Botrytis cinerea. *J Nat Prod* 63:29–33.

Clausen TP, Reichardt PB, Bryant JP, Provenza FD. 1992. Condensed tannins in plant defense: a perspective on classical theories. *In: Plant Polyphenols: Synthesis, Properties, Significance*. Hemingway RW, Laks PE, Editors. New York. Plenum, pp. 639–651.

Clough SJ, Tuteja JH, Li M, Marek LF, Shoemaker RC, Vodkin LO. 2004. Features of a 103–kb gene-rich region in soybean include an inverted perfect repeat cluster of CHS genes comprising the I locus. *Genome* 47:819–831.

Coe EH, McCormick SM, Modena SA. 1981. White pollen in maize. *J Hered* 72:318–320.

Cominelli E, Gusmaroli G, Allegra D, Galbiati M, Wade HK, Jenkins GI, Tonelli C. 2008. Expression analysis of anthocyanin regulatory genes in response to different light qualities in *Arabidopsis* thaliana. *J Plant Physiol* 165:886–894.

Cone KC, Cocciolone MS, Burr FA, Burr B. 1993. Anthocyanin regulatory gene Pl is a duplicate of C1 that functions in the plant. *Plant Cell* 5:1795–1805.

Cook D, Fowler S, Fiehn O, Thomashow MF. 2004. A prominent role for the CBF cold response pathway in configuring the low-temperature metabolome of *Arabidopsis*. *Proc Natl Acad Sci USA* 101:15243–15248.

Cooper JB, Long SR. 1994. Morphogenetic rescue of Rhizobium meliloti nodulation mutants by trans-zeatin secretion. *Plant Cell* 6:215–225.

Coruzzi G, Bush DR. 2001. Nitrogen and carbon nutrient and metabolite signaling in plants. *Plant Physiol* 125:61–64.

Coruzzi GM, Zhou L. 2001. Carbon and nitrogen sensing and signaling in plants: Emerging "matrix effects", *Curr Op Plant Biol* 4:247–253.

Counet C, Callemien D, Collin S. 2006. Chocolate and cocoa: New sources of trans-resveratrol and trans-piceid. *Food Chem* 98:649–657.

Creasy LL, Creasy MT. 1998. Grape chemistry and the significance of resveratrol: An overview. *Pharm Biol* 36:8–13.

Czarnota MA, Rimando AM, Weston LA. 2003. Evaluation of root exudates of seven sorghum accessions. *J Chem Ecol* 29:2073–2083.

Davies KM, Schwinn KE. 2003. Transcriptional regulation of secondary metabolism. *Funct Plant Biol* 30:913–925.

Dearing MD, Foley WJ, McLean S. 2005. The influence of plant secondary metabolites on the nutritional ecology of herbivorous terrestrial vertebrates. *Annu Rev Ecol Evol Syst* 36:169–189.

Deavours BE, Dixon RA. 2005. Metabolic engineering of isoflavonoid biosynthesis in alfalfa. *Plant Physiol* 138:2245–2259.

Debeaujon I, Léon-Kloosterziel KM, Koornneef M. 2000. Influence of the testa on seed dormancy, germination and longevity in *Arabidopsis*. *Plant Physiol* 122:403–414.

Debeaujon I, Nesi N, Perez P, Devic M, Grandjean O, Caboche M, Lepiniec L. 2003. Proanthocyanidin accumulating cells in *Arabidopsis* testa: Regulation of differentiation and role in seed development. *Plant Cell* 15:2514–2531.

Debeaujon I, Peeters AJM, Leon-Kloosterziel KM, Korneef M. 2001. The TRANSPARENT TESTA12 gene of *Arabidopsis* encodes a multidrug secondary transporter-like protein required for flavonoid sequestration in vacuoles of the seed coat endothelium. *Plant Cell* 13:853–871.

Debeaujon I, Léon-Kloosterziel KM, Koornneef M. 2000. Influence of the testa on seed dormancy, germination and longevity in *Arabidopsis*. *Plant Physiol* 122:403–414.

de Beer D, Jouvert E, Gelderblom WCA, Manley M. 2002. Phenolic compounds: A review of their possible role as in vivo antioxidants of wine. *S African J Wine Grape Res* 23:48–61.

De Long A, Mockaitis K, Chirstensen S. 2002. Protein phosphorylation in the delivery of and responses to auxin signals. *Plant Mol Biol* 49:285–303.

Deluc L, Barrieu F, Marchive C, Lauvergeat V, Decendit A, Richard T, Carde JP, Merillon JM, Hamdi S. 2006. Characterization of a grapevine R2R3-MYB transcription factor that regulates the phenylpropanoid pathway. *Plant Physiol* 140:499–511.

de Meaux J, Pop A, Mitchell-Olds T. 2006. Cis-regulatory evolution of chalcone-synthase expression in the genus *Arabidopsis*. *Genetics* 174:2181–2202.

de Pascual-Teresa S, Rivas-Gonzalo JC, Santos-Buelga C. 2000. Prodelphinidins and related flavonols in wine. *Int J Food Sci Tech* 35:33–40.

de Vetten N, Quattrocchio F, Mol J, Koes R. 1997. The an11 locus controlling flower pigmentation in petunia encodes a novel WD-repeat protein conserved in yeast, plants and animals. *Genes Dev* 11:1422–1434.

Dercks W, Creasy LL. 1989. The significance of stilbene phytoalexins in the Plasmopara viticola–grapevine interaction. *Physiol Mol Plant Pathol* 34:189–202.

Devic M, Guilleminot J, Debeaujon I, Bechtold N, Bensuade E, Koorneef M, Pelletier G, Delseny M. 1999. The BANYULS gene encodes a DFR like protein and is a marker of early seed coat development. *Plant J* 19:387–398.

de Vos R, Bovy A, Busink H, Muir S, Collins G, Verhoeyen M. 2000. Improving health potential of crop plants by means of flavonoid pathway engineering. *Polyphenols Commun* 1:25–26.

Dewick PM. 1982. In: Harborne JB, Mabry TJ, Editors. *The Flavonoids: Advances in Research*. New York: Chapman and Hall. pp. 535–640.

Dias AP, Braun EL, McMullen MD, Grotewold E. 2003. Recently duplicated maize R2R3 Myb genes provide evidence for distinct mechanisms of evolutionary divergence after duplication. *Plant Physiol* 131:610–620.

Dixon RA. 2001. Natural products and plant disease resistance. *Nature* 411:843–847.

Dixon RA, Paiva NL. 1995. Stress-induced phenylpropanoid metabolism. *Plant Cell* 7:1085–1097.

Dixon R, Steele C. 1999. Flavonoids and isoflavonoids—a gold mine for metabolic engineering. *Trends Plant Sci* 4:394–400.

Dixon RA, Sharma SB, Xie D. 2005. Proanthocyanidins—a final frontier in flavonoid research? New Phytol 165:9–28.

Dixon RA, Choundhary AD, Dalkin K, Edwards R, Fahrendorf T, Gowni G, Harrison MJ, Lamb CJ, Loake GJ, Maxwell CA, Orr J, Paiva NL. 1992. Molecular biology of stress-induced phenylpropanoid and isoflavonoid biosynthesis in alfalfa. In: Stafford HA, Ibrahim RK, Eds. *Phenolic Metabolism in Plants*. New York: Plenum, pp. 91–138.

Djordjevic MA, Redmond JW, Batley M, Rolfe BG. 1987. Clovers secrete specific phenolic compounds which either stimulate or repress nod gene expression in Rhizobium trifolii. *EMBO J* 6:1173–1179.

Dong X, Braun, EL, Grotewold E. 2001. Functional conservation of plant secondary metabolic enzymes revealed by complementation of *Arabidopsis* flavonoid mutants with maize genes. *Plant Phys* 127:46–57.

Downey MO, Harvey JS, Robinson SP. 2003a. Analysis of tannins in seeds and skins of Shiraz grapes throughout berry development. Aust J Grape Wine Res 9:15–27.

Downey MO, Harvey JS, Robinson SP. 2003b. Synthesis of flavonols and expression of flavonol synthase genes in the developing grape berries of Shiraz and Chardonnay (Vitis vinifera L.). *Aust J Grape Wine Res* 9:110–121.

Dyer WE. 1989. Wounding induces the first enzyme of the shikimate pathway in Solanaceae. *Proc Nat Acad Sci USA* 86:7370–7373.

Eckardt NA. 2003. The function of SUT2/SUC3 sucrose transporters. The debate continues. *Plant Cell* 15:1259–1262.

Ellestad GA. 2006. Structure and chiroptical properties of supramolecular flower pigments. *Chirality* 18:134–144.

Faeth SH, Bultman TL. 1986. Interacting effects of increased tannin levels on leaf-mining insects. *Entomol Exp Appl* 40:297–300.

Farag MA, Huhman DV, Dixon RA, Sumner LW. 2008. Metabolomics reveals novel pathways and differential mechanistic and elicitor-specific responses in phenylpropanoid and isoflavonoid biosynthesis in Medicago truncatula cell cultures. *Plant Physiol* 146:387–402.

Farrant JM. 2000. A comparison of mechanisms of desiccation tolerance among three angiosperm resurrection plant species. *Plant Ecol* 151:29–39.

Farzad M, Griesbach R, Weiss MR. 2002. Floral color change in Viola cornuta L. (Violaceae): A model system to study regulation of anthocyanin production. *Plant Sci* 162:225–231.

Fay JP, Cheng KJ, Hanna MR, Howarth RE, Costerton JW. 1980. In vitro digestion of bloat-safe and bloat-causing legumes by rumen microorganisms: Gas and foam production. *J Dairy Sci* 63:1273–1281.

Feeny P. 1970. Seasonal changes in oak leaf tannins and nutrients as a cause of spring feeding by winter moth caterpillars. *Ecology* 51:565–581.

Feild TS, Lee DW, Holbrook NM. 2001. Why leaves turn red in autumn. The role of anthocyanins in senescing leaves of red-osier dogwood. *Plant Physiol* 127:566–574.

Feldbrügge M, Sprenger M, Hahlbrock K, Weisshaar B. 1997. PcMYB1, a novel plant protein containing a DNA binding domain with one MYB repeat, interacts in vivo with a light-regulatory promoter unit. *Plant J* 11:1079–93.

Feldbrügge M, Sprenger M, Dinkelbach M, Yazaki K, Harter K, Weisshaar B. 1994. Functional analysis of a light responsive plant bZIP transcriptional regulator. *Plant Cell* 6:1607–1621.

Feucht W, Treutter D. 1999. The role of flavan-3-ols and proanthocyanidins in plant defence. In: Inderjit S, Dakshini KMM, Foy CL, Eds. *Principles and Practices in Plant Ecology*. Boca Raton, FL:CRC Press, pp. 307–338.

Feucht W, Dithmar H, Polster J. 2004. Nuclei of tea flowers as targets for flavanols. *Plant Biol* 6:696–701.

Fischer D, Stich K, Britsch L, Grisebach H. 1988. Purification and characterization of (+)dihydroflavonol (3-hydroxiflavanone) 4-reductase from flowers of Dahlia variabilis. *Arch Biochem Biophys* 264:40–47.

Fischer TC, Halbwirth H, Meisel B, Stich K, Forkmann G. 2003. Molecular cloning, substrate specificity of the functionally expressed dihydroflavonol 4-reductases from Malus domestica and Pyrus communis and the consequences for flavonoid metabolism. *Arch Biochem Biophys* 412:223–230.

Fitter A. 2003. Making allelopathy respectable. *Science* 301:1337–1338.

Foo LY, Newman R, Waghorn G, McNabb WC, Ulyatt MJ. 1996. Proanthocyanidins from Lotus corniculatus. *Phytochem* 41:617–624.

Ford C, Boss PK, Høj PB. 1998. Cloning and characterization of *Vitis* vinifera UDP-glucose:flavonoid 3-O-glucosyltransferase, a homologue of the enzyme encoded by the maize Bronze-1 locus that may primarily serve to glucosylate anthocyanidins in vivo. *J Biol Chem* 273:9224–9233.

Forkmann G. 1991. Flavonoids as flower pigments: The formation of the natural spectrum and its extension by genetic engineering. *Plant Breed* 106:1–26.

Forkmann G, Martens S. 2001. Metabolic engineering and applications of flavonoids. *Curr Opin Biotechnol* 12:155–160.

Forkner RE, Marquis RJ, Lill JT. 2004. Feeny revisited: Condensed tannins as anti-herbivore defenses in leaf-chewing herbivore communities of Quercus. *Ecol Entomol* 29:174–187.

Frankel EN, Waterhouse AL, Teissedre PL. 1995. Principal phenolic phytochemicals in selected California wine and their antioxidant activity in inhibiting oxidation of human low-density lipoproteins. *J Ag Food Chem* 43:890–894.

Frey M, Chomet P, Glawischnig E, Stettner C, Grün S, Winklmair A, Eisenreich W, Bacher A, Meeley RB, Briggs SP, Simcox K, Gierl A. 1997. Analysis of a chemical plant defense mechanism in grasses. *Science* 277:696–699.

Friml J, Yang X, Michniewicz M, Weijers D, Quint A, Tietz O, Benjamins R, Ouwerkerk PB, Ljung K, Sandberg G, Hooykaas PJ, Palme K, Offringa R. 2004. A PINOID-dependent binary switch in apical-basal PIN polar targeting directs auxin efflux. *Science* 306:862–865.

Fukada-Tanaka S, Inagaki Y, Yamaguchi T, Saito N, Iida S. 2000. Colour-enhancing protein in blue petals. *Nature* 407:581.

Gälweiler L, Guan C, Müller A, Wisman E, Mendgen K, Yephremov A, Palme K. 1998. Regulation of polar auxin transport by AtPIN1 in *Arabidopsis* vascular tissue. *Science* 282:2226–2230.

Geigert J, Stermitz FR, Johnson G, Maag DD, Johnson DK. 1973. Two phytoalexins from sugarbeet (Beta vulgaris) leaves. *Tetrahedron* 29:2703–2706.

Geisler M, Blakeslee JJ, Bouchard R, Lee OR, Vincenzetti V, Bandyopadhyay A, Titapiwatanakun B, Peer WA, Bailly A, Richards EL, Ejenda KFK, Smith AP, Baroux C, Grossniklaus U, Muller A, Hrycyna CA, Dudler R, Murphy AS, Martinoia E. 2005. Cellular efflux of auxin catalyzed by the *Arabidopsis* MDR/PGP transporter AtPGP1. *Plant J* 44:179–194.

Giurfa M, Menzel R. 1997. Insect visual perception: Complex abilities of simple nervous systems. *Curr Opin Neurobiol* 7:505–513.

Glover BJ, Martin C. 1998. The role of petal cell shape and pigmentation in pollination success in Antirrhinum majus. *Heredity* 80:778–784.

Goes da Silva F, Iandolino A, Al-Kayal F, Bohlmann MC, Cushman MA, Lim H, Figueroa R, Leslie A, Xu J, Kabuloglu EK, Tattersall E, Ergul A, Baek JM, Cramer GR, Cushman JC, Cook DR. 2005. Characterizing the grape transcriptome. Analysis of expressed sequence tags from multiple *Vitis* species and development of a compendium of expression during berry development. *Plant Physiol* 139:574–597.

Goff SA, Cone KC, Chandler VL. 1992. Functional analysis of the transcriptional activator encoded by the maize B gene: Evidence for a direct functional interaction between two classes of regulatory proteins. *Genes Dev* 6:864–875.

Gonzalez A, Zhao M, Leavitt JM, Lloyd AM. 2008. Regulation of the anthocyanin biosynthetic pathway by the TTG1/bHLH/Myb transcriptional complex in *Arabidopsis* seedlings. *Plant J* 53:814–827.

Gonzalez-Rizzo S, Crespi M, Frugier F. 2006. The Medicago truncatula CRE1 cytokinin receptor regulates lateral root development and early symbiotic interaction with Sinorhizobium meliloti. *Plant Cell* 18:2680–2693.

Goodman CD, Casati P, Walbot V. 2004. A multidrug resistance-associated protein involved in anthocyanin transport in Zea mays. *Plant Cell* 16:1812–1826.

Goodrich J, Carpenter R, Coen ES. 1992. A common gene regulates pigmentation pattern in diverse plant species. *Cell* 68:955–964.

Goplen BP, Howarth RE, Sarkar SK, Lesins K. 1980. A search for condensed tannins in annual and perennial species of Medicago, Trigonella, and Onobrychis. *Crop Sci* 20:801–804.

Gould KS. 2004. Nature's Swiss Army knife: The diverse protective roles of anthocyanins in leaves. *J Biomed Biotechnol* 2440:314–320.

Gould KS, McKelvie J, Markham KR. 2002. Do anthocyanins function as antioxidants in leaves? Imaging of H_2O_2 in red and green leaves after mechanical injury. *Plant Cell Environ* 25:1261–1269.

Graham TL. 1991. Flavonoid and isoflavonoid distribution in developing soybean seedling tissues and in seed and root exudates. *Plant Physiol* 95:594–603.

Grayer RJ, Harborne JB. 1994. A survey of antifungal compounds from higher plants 1982–1993. *Phytochem* 37:19–42.

Gronquist M, Bezzerides A, Attygalle A, Meinwald J, Eisner M, Eisner T. 2001. Attractive and defensive functions of the ultraviolet pigments of a flower (Hypericum calycinum). *Proc Natl Acad Sci USA* 98:13745–13750.

Grotewold E. 2006. The genetics and biochemistry of floral pigments. *Ann Rev Plant Biol* 57:761–780.

Grotewold E. 2004. The challenges of moving chemicals within and out of cells: Insights into the transport of plant natural products. *Planta* 219:906–909.

Grotewold E. 2001. Subcellular trafficking of phytochemicals. *Recent Res Devel Plant Physiol* 2:31–48.

Grotewold E, Sainz MB, Tagliani L, Hernandez JM, Bowen B, Chandler VL. 2000. Identification of the residues in the Myb domain of maize C1 that specify the interaction with the bHLH cofactor R. *Proc Natl Acad Sci USA* 97:13579–13584.

Grotewold E, Drummond BJ, Bowen B, Peterson T. 1994. The myb-homologous P gene controls phlobaphene pigmentation in maize floral organs by directly activating a flavonoid biosynthetic gene subset. *Cell* 76:543–553.

Grotewold E, Athma P, Peterson T. 1991. Alternatively spliced products of the maize P gene encode proteins with homology to the DNA-binding domain of myb-like transcription factors. *Proc Natl Acad Sci USA* 88:4587–4591.

Gruber L, Steinwidder A, Stefanon B, Steiner B, Steinwender R. 1999. Influence of grassland management in alpine regions and concentrate level on N excretion and milk yield of dairy cows. *Livest Prod Sci* 61:155–170.

Guo D, Chen F, Wheeler J, Winder J, Selman S, Peterson M, Dixon RA. 2001. Improvement of in-rumen digestibility of alfalfa forage by genetic manipulation of lignin O-methyltransferases. *Transgenic Res* 10:457–464.

Guo L, Dixon RA, Paiva NL. 1994a. Conversion of vestitone to medicarpin in alfalfa (Medicago sativa L.) is catalyzed by two independent enzymes. *J Biol Chem* 269:22372–22378.

Guo L, Dixon RA, Paiva NL. 1994b. The "pterocarpan synthase" of alfalfa: Association and co-induction of vestitone reductase and 7,2′-dihydroxy-4′-methoxy-isoflavanol (DMI) dehydratase, the two final enzymes in medicarpin biosynthesis. *FEBS Lett* 356:221–225.

Guyon V, Tang W-H, Monti MM, Raiola A, Lorenzo G, McCormick S, Taylor LP. 2004. Antisense phenotypes reveal a role for SHY, a pollen-specific leucine-rich repeat protein, in pollen tube growth. *Plant J* 39:643–654.

Guyon V, Astwood J, Garner E, Dunker K, Taylor LP. 2000. Isolation and characterization of cDNAs expressed in the early stages of flavonol-induced pollen germination in petunia. *Plant Physiol* 123:699–710.

Hakamatsuka T, Mori K, Ishida S, Ebizuka Y, Sankawa U. 1998. Purification of 2-hydroxyisoflavanone dehydratase from the cell cultures of Pueraria lobata. *Phytochem* 49:497–505.

Halbwirth H, Martens S, Wienand U, Forkmann G, Stich K. 2003. Biochemical formation of anthocyanins in silk tissue of Zea mays. *Plant Sci* 164:489–495.

Hale KL, Tufan HA, Pickering IJ, George GN, Terry N, Pilon M, Pilon-Smits EAH. 2002. Anthocyanins facilitate tungsten accumulation in Brassica. *Physiol Plant* 116:351–358.

Hall D, de Luca V. 2007. Mesocarp localization of a bi-functional resveratrol/hydroxycinnamic acid glucosyltransferase of Concord grape (Vitis labrusca). *Plant J* 49:579–591.

Halls C, Yu O. 2008. Potential for metabolic engineering of resveratrol biosynthesis. *Trends Biotech* 26:77–81.

Hannah MA, Wiese D, Freund S, Fiehn O, Heyer AG, Hincha DK. 2006. Natural genetic variation of freezing tolerance in *Arabidopsis*. *Plant Physiol* 142:98–112.

Harborne JB, Baxter H. 1999. *Handbook of Natural Flavonoids*, Vol 2. Chichester: Wiley.

Harrison MJ. 2005. Signaling in the arbuscular mycorrhizal symbiosis. *Annu Rev Microbiol* 59:19–42.

Hartmann U, Sagasser M, Mehrtens F, Stracke R, Weisshaar B. 2005. Differential combinatorial interactions of cis-acting elements recognized by R2R3-MYB, BZIP, and BHLH factors control light-responsive and tissue specific activation of phenylpropanoid biosynthesis genes. *Plant Mol Biol* 57:155–171.

Hartmann U, Valentine WJ, Christie JM, Hays J, Jenkins GI, Weisshaar B. 1998. Identification of UV/blue light response elements in the *Arabidopsis* thaliana chalcone synthase promoter using a homologous protoplast transient expression system. *Plant Mol Biol* 36:741–754.

Haslam E. 1981. Vegetable tannins. In: Tolbert NE, Eds. *The Biochemistry of Plants*. New York: Academic Press., Vol. 7, pp. 527–556.

He X-Z, Dixon RA. 2000. Genetic manipulation of isoflavone 7-O-methyltransferase enhances biosynthesis of 4′-O-methylated isoflavonoid phytoalexins and disease resistance in alfalfa. *Plant Cell* 12:1689–1702.

Heine GF, Hernandez JM, Grotewold E. 2004. Two cysteines in plant R2R3 MYB domains participate in REDOX-dependent DNA binding. *J Biol Chem* 279:37878–37885.

Helariutta Y, Kotilainen M, Elomaa P, Kalkkinen N, Bremer K, Teeri TH, Albert VA. 1996. Duplication and functional divergence in the chalcone synthase gene family of Asteraceae: Evolution with substrate change and catalytic simplification. *Proc Natl Acad Sci USA* 93:9033–9038.

Heldt H-W. 1999. Plant Biochemistry and Molecular Biology. 2nd ed. New York: Oxford University Press.

Heller W, Forkmann G. 1993. Biosynthesis of flavonoids. In: Harborne JB, ed. *The Flavonoids: Advances in Research since 1986*. London: Chapman & Hall,. pp. 499–535.

Henstrand JM. 1992. Light and fungal elicitor induce 3-deoxy-D-arabino-heptulonosate 7-phosphate synthase mRNA in suspension cultured cells of parsley (Petroselinum crispum L.). *Plant Physiol* 98:761–763.

Hermsmeier D, Schittko U, Baldwin IT. 2001. Molecular interactions between the specialist herbivore Manduca sexta (Lepidoptera, Sphigidae) and its natural host Nicotiana attenuata. I. Large-scale changes in the accumulation of growth- and defense-related plant mRNAs. *Plant Physiol* 125:683–700.

Higgins VJ. 1972. Role of the phytoalexin medicarpin in three leaf spot diseases of alfalfa. *Physiol Plant Pathol* 2:289–300.

Hinderer W, Flentje U, Barz W. 1987. Microsomal isoflavone 2- and 3-hydroxylases from chickpea (Cicer arietinum L.) cell suspensions induced for pterocarpan phytoalexin formation. *FEBS Lett* 214:101–106.

Hipskind JD, Paiva NL. 2000. Constitutive accumulation of a resveratrol-glucoside in transgenic alfalfa increases resistance to Phoma medicaginis. *Molec Plant Microbe Interact* 13:551–562.

Hoch WA, Singsaas EL, McCown BH. 2003. Resorption protection. Anthocyanins facilitate nutrient recovery in autumn by shielding leaves from potentially damaging light levels. *Plant Physiol* 133:1296–1305.

Hoch WA, Zeldin EL, McCown BH. 2001. Physiological significance of anthocyanins during autumnal leaf senescence. *Tree Physiol* 21:1–8.

Hoffman L, Besseau S, Geoffroy P, Ritzenthaler C, Meyer D, Lapierre C, Pollet B, Legrand M. 2004. Silencing of hydroxycinnamoyl-coenzyme A shikimate/quinate hydroxycinnamoyl-transferase affects phenylpropanoid biosynthesis. *Plant Cell* 16:1446–1465.

Hopp W, Seitz HU. 1987. The uptake of acylated anthocyanin into isolated vacuoles from a cell suspension culture of Daucus carota. *Planta* 170:74–85.

Hrazdina G, Jensen RA. 1992. Spatial organization of enzymes in plant metabolic pathways. *Ann Rev Plant Physiol Plant Molec Biol* 43:241–267.

Hrazdina G, Zobel AM, Hoch HC. 1987. Biochemical, immunological, and immunocytochemical evidence for the association of chalcone synthase with endoplasmic reticulum membranes. *Proc Natl Acad Sci USA* 84:8966–8970.

Hrazdina G, Wagner GJ, Siegelman HW. 1978. Subcellular localization of enzymes of anthocyanin biosynthesis in protoplasts. *Phytochem* 17:53–56.

Hsieh K, Huang AH. 2007. Tapetosomes in Brassica tapetum accumulate endoplasmic reticulum-derived flavonoids and alkenes for delivery to the pollen surface. *Plant Cell* 19:582–596.

Huang TS, Anzellotti D, Dedaldechamp F, Ibrahim RK. 2004. Partial purification, kinetic analysis, and amino acid sequence of a flavonol 3-O-mtheyltransferase from Serratula tinctoria. *Plant Physiol* 134:1366–1376.

Iandolino A. 2004. Abiotic stress effects on the grapevine (Vitis vinifera L.) canopy's carbon balance and metabolic profile of flavonoids in berries. PhD Thesis. University of California, Davis, California.

Ikegami A, Eguchi S, Kitajima A, Inoue K, Yonemori K. 2007. Identification of genes involved in proanthocyanidin biosynthesis of persimmon (Diospyros kaki) fruit. *Plant Sci* 172:1037–1047.

International Allelopathy Society (IAS). 1996. Constitution and Bylaws. Available: http://www.ias.uca.es/bylaws.htm.

Irani NG, Hernandez JM, Grotewold E. 2003. Regulation of anthocyanin pigmentation. *Rec Adv Phytochem* 38:59–78.

Jaakola L, Maatta K, Pirttila AM, Torronen R, Karenlampi S, Hohtola A. 2002. Expression of genes involved in anthocyanin biosynthesis in relation to anthocyanin, proanthocyanidin, and flavonol levels during bilberry fruit development. *Plant Physiol* 130:729–739.

Jacobs M, Rubery PH. 1988. Naturally occurring auxin transport regulators. *Science* 241:346–349.

Jenkins GI, Long JC, Wade HK, Shenton MR, Bibikova TN. 2001. UV and blue light signaling: Pathways regulating chalcone synthase gene expression in *Arabidopsis*. *New Phytol* 151:121–131.

Jin H, Cominelli E, Bailey P, Parr A, Mehrtens F, Jones J, Tonelli C, Weisshaar B, Martin C. 2000. Transcriptional repression by AtMYB4 controls production of UV-protecting sunscreens in *Arabidopsis*. *EMBO J* 19:6150–6161.

Johnson CS, Kolevski B, Smyth DR. 2002. TRANSPARENT TESTA GLABRA2, a trichome and seed coat development gene of *Arabidopsis*, encodes a WRKY transcription factor. *Plant Cell* 14:1359–1375.

Johnson ET, Ryu S, Yi H, Shin B, Cheong H, Choi G. 2001. Alteration of a single amino acid changes the substrate specificity of dihydroflavonol 4-reductase. *Plant J* 25:325–333.

Jordan BR. 1996. The effects of ultraviolet-B radiation on plants: A molecular perspective. *Adv Bot Res* 22:97–162.

Jorgensen R. 1993. The origin of land plants: A union of alga and fungus advanced by flavonoids? *Biosystems* 31:193–207.

Jung HG, Vogel KP. 1986. Influence of lignin on digestibility of forage cell wall material. *J Animal Sci* 62:1703–1712.

Jung W, Yu O, Lau S-M C, O'Keefe DP, Odell J, Fader G, McGonigle B. 2000. Identification and expression of Isoflavone synthase, the key enzyme for biosynthesis of isoflavones in legumes. *Nature* 18:208–212.

Kaltenbach M, Schroder G, Schmelzer E, Lutz V, Schroder J. 1999. Flavonoid hydroxylase from Catharanthus roseus: cDNA, heterologous expression, enzyme properties and cell-type specific expression in plants. *Plant J* 19:183–193.

Kao C, Cocciolone SM, Vasil IK, McCarty DR. 1996. Localization and interaction of the cis-acting elements for abscissic acid, VVIPAROUS1 and light activation of the C1 gene of maize. *Plant Cell* 8:1171–1179.

Kaplan F, Kopka J, Haskell DW, Zhao W, Schiller KC, Gatzke N, Sung DY, Guy CL. 2004. Exploring the temperature-stress metabolome of *Arabidopsis*. *Plant Physiol* 136:4159–4168.

Kay QON, Daoud HS, Stirton CH. 1981. Pigment distribution, light reflection and cell structure in petals. *Bot J Linn Soc* 83:57–84.

Kennedy JA, Hayasaka Y, Vidal S, Waters EJ, Jones GP. 2001. Composition of grape skin proanthocyanidins at different stages of berry development. *J Agric Food Chem* 49:5348–5355.

Kennedy JA, Matthews MA, Waterhouse AL. 2000. Changes in grape seed polyphenols during ripening. *Phytochem* 55:77–85.

Kerepesi I, Galiba G. 2000. Osmotic and salt stress-induced alteration in soluble carbohydrate content in wheat seedlings. *Crop Sci* 40:482–487.

Kevan PG, Baker HG. 1983. Insects as flower visitors and pollinators. *Annu Rev Entomol* 28:407–453.

Kim Y, Song K, Cheong H. 1996. Effects of flavonoids on pollen tube growth in *Arabidopsis* thaliana. *J Plant Biol* 39:273–278.

Kingston-Smith AH, Thomas HM. 2003. Strategies of plant breeding for improved rumen function. *Ann App Biol* 142:13–24s.

Kitamura S. 2006. Transport of flavonoids: From cytosolic synthesis to vacuolar accumulation. In: Grotewold E, Edit. *The Science of Flavonoids*. New York: Springer, pp. 123–146.

Kitamura S, Shikazono N, Tanaka A (2004) TRANSPARENT TESTA19 is involved in the accumulation of both anthocyanins and proanthocyanidins in *Arabidopsis*. *Plant J* 37:104–114.

Klein M, Burla B, Martinoia E. 2006. The multidrug resistance-associated protein (MRP/ABCC) subfamily of ATP-binding cassette transporters in plants. *FEBS Lett* 580:1112–1122.

Klein M, Weissenbock G, Dufaud A, Gaillard C, Kreuz K, Martinoia E. 1996. Different energization mechanisms drive the vacuolar uptake of a flavonoid glucoside and a herbicide glucoside. *J Biol Chem* 271:29666–29671.

Kliewer WM. 1977 Influence of temperature, solar radiation and nitrogen on coloration and composition of Emperor grapes. *Am J Enol Vitic* 28:96–103.

Kobayashi S, Goto-Yamamoto N, Hirochika H. 2004. Retrotransposon-induced mutations in grape skin color. *Science* 304:982.

Kobayashi S, Ishimaru M, Hiraoka K, Honda C. 2002. Myb-related genes of the Kyoho grape (Vitis labruscana) regulate anthocyanin biosynthesis. *Planta* 215:924–933.

Koch KE. 1996. Carbohydrate-modulated gene expression in plants. *Annu Rev Plant Physiol Plant Mol Biol* 47:509–540.

Kodan A, Kuroda H, Sakai F. 2002. Stilbene synthase from Japanese red pine (Pinus densiflora): implications for phytoalexin accumulation and down-regulation of flavonoid biosynthesis. *Proc Natl Acad Sci USA* 99:3335–3339.

Koes RE, Quattrocchio F, Mol JNM. 1994. The flavonoid biosynthetic pathway in plants: Function and evolution. *BioEssays* 16:123–132.

Kolb CA, Käser MA, Kopecký J, Zotz G, Riederer M, Pfündel EE. 2001. Effects of natural intensities of visible and ultraviolet radiation on epidermal ultraviolet screening and photosynthesis in grape leaves. *Plant Physiol* 127:863–875.

Kong C, Liang W, Xu X, Hu F, Wang P, Jiang Y. 2004. Release and activity of allelochemicals from allelopathic rice seedlings. *J Agric Food Chem* 52:2861–2865.

Kootstra A. 1994. Protection from UV-B induced DNA damage by flavonoids. *Plant Molec Biol* 26:771–774.

Krupa Z, Baranowska M, Orzol D. 1996. Can anthocyanins be considered as heavy metal stress indicator in higher plants? *Acta Physiol Plantarum* 18:147–151.

Kudou S, Fleury Y, Welti D, Magnolato D, Uchida T, Kitamura K, Okubo K. 1991. Malonyl isoflavones glycosides in soybean seeds (Glycine max Merrill). *Agric Biol Chem* 55:2227–2233.

Landry LG, Chapple CCS, Last RL. 1995. Arabidopsis mutants lacking phenolic sunscreens exhibit enhanced ultraviolet-B injury and oxidative damage. *Plant Physiol* 109:1159–1166.

Langcake P. 1981. Disease resistance of *Vitis* spp. and the production of the stress metabolites resveratrol, ε-viniferin, α-viniferin and pterostilbene. *Physiol Plant Pathol* 18:213–226.

Langcake P, Pryce RJ. 1977a. A new class of phytoalexins from grapevines. *Experientia* 33:151–152.

Langcake P, Pryce RJ. 1977b. The production of resveratrol and the viniferins by grapevines in response to ultraviolet irradiation. *Phytochem* 16:1191–1196.

Lapidot T, Harel S, Akiri B, Granit R, Kanner J. 1999. pH-dependent forms of red wine anthocyanins as antioxidants. *J Agric Food Chem* 47:67–70.

Larsen ES, Alfenito MR, Briggs WR, Walbot V. 2003. A carnation anthocyanin mutant is complemented by the glutathione S-transferases encoded by maize Bz2 and petunia An9. *Plant Cell Rep* 21:900–904.

Latunde-Dada AO, Cabello-Hurtado F, Czittrich N, Didierjean L, Schopfer C, Hertkorn N, Werck-Reichart D, Ebel J. 2001. Flavonoid 6-hydroxylase from soybean (Glycine max L.), a novel plant P-450 monooxygenase. *J Biol Chem* 276:1685–1688.

Lazar G, Goodman HM. 2006. MAX1, a regulator of the flavonoids pathway, controls vegetative axillary bud outgrowth in *Arabidopsis*. *Proc Natl Acad Sci USA* 103:472–476.

Lea US, Slimestad R, Smedvig P, Lillo C. 2007. Nitrogen deficiency enhances expression of specific MYB and bHLH transcription factors and accumulation of end products in the flavonoid pathway. *Planta* 225:1245–1253.

Lee DW, Collins TH. 2001. Phylogenetic and ontogenetic influences on the distribution of anthocyanins and betacyanins in leaves of tropical plants. *Int J Plant Sci* 162:1141–1153.

Lee DW, Brammeier S, Smith AP. 1987. The selective advantages of anthocyanins in developing leaves of mango and cacao. *Biotropica* 19:40–49.

Lee DW. 2002. Anthocyanins in autumn leaf senescence. In: Gould KS, Lee DW, Eds. *Anthocyanins in Leaves. Advances in Botanical Research*. Amsterdam: Academic Press, Vol 37 pp. 37–53.

Lee SS, Lee SM, Kim M, Chun J, Cheong YK, Lee J. 2004. Analysis of trans-resveratrol in peanuts and peanut butters consumed in Korea. *Food Res Int* 37:247–251.

Lees GL. 1992. Condensed tannins in some forage legumes: their Role in the prevention of ruminant pasture bloat. *Basic Life Sci* 59:915–934.

Legay S, Lacombe E, Goicoechea M, Briere C, Séguin A, Mackay J, Grima-Pettenati J. 2007. Molecular characterization of EgMYB1, a putative transcriptional repressor of the lignin biosynthetic pathway. *Plant Sci* 173:542–549.

Leinmüller E, Steingass H, Menke KH. 1991. Tannins in ruminant feed stuffs. *Anim Res Dev* 33:9–62.

Leonard E, Lim K-H, Saw P-N, Koffas MAG. 2007a. Engineering central metabolic pathways for high-level flavonoid production in Escherichia coli. *App. Env Microb* 73:3877–3886.

Leonard E, Fowler ZL, Koffas MAG. 2007b. Metabolic engineering. In: Al-Rubeai M, Fusseneger M, Eds. Systems Biology. London. Springer, Vol 5, pp. 301–359.

Leonard E, Chemler J, Lim KH, Koffas MA. 2006a. Expression of a soluble flavone synthase allows the biosynthesis of phytoestrogen derivatives in Escherichia coli. *Appl Microbiol Biotechnol* 70:85–91.

Leonard E, Yan Y, Koffas MA. 2006b. Functional expression of a P450 flavonoid hydroxylase for the biosynthesis of plant-specific hydroxylated flavonols in Escherichia coli. *Metab Eng* 8:172–181.

Leonard E, Yan Y, Lim KH, Koffas MAG. 2005. Investigation of two distinct flavone synthases for plant-specific flavones biosynthesis in Saccharomyces cerevisiae. *Appl Environ Microbiol* 71:8241–8248.

Lepiniec L, Debeaujon I, Routaboul JM, Baudry A, Pourcel L, Nesi N, Caboche M. 2006. Genetics and biochemistry of seed flavonoids. *Ann Rev Plant Biol* 57:405–430.

Leser C, Treutter D. 2005. Effects of nitrogen supply on growth, contents of phenolic compounds and pathogen (scab) resistance of apple trees. *Physiol Plant* 123:49–56.

Li J, Ou-Lee T-M, Raba R, Amundson RG, Last RL. 1993. Arabidopsis flavonoid mutants are hypersensitive to UV-B irradiation. *Plant Cell* 5:171–179.

Li X, Wu B, Wang L, Li S. 2006a. Extractable amounts of trans-resveratrol in seed and berry skin in *Vitis* evaluated at the germplasm level. *J Agric Food Chem* 54:8804–8811.

Li Y, Lee KK, Walsh S, Smith C, Hadingham S, Sorefan K, Cawley G, Bevan MW. 2006b. Establishing glucose- and ABA-regulated transcription networks in *Arabidopsis* by microarray analysis and promoter classification using a Relevance Vector Machine. *Genome Res* 16:414–427.

Li L, Steffens JC. 2002. Overexpression of polyphenol oxidase in transgenic tomato plants results in enhanced bacterial disease resistance. *Planta* 215:239–247.

Lillo C, Lea US, Ruoff P. 2008. Nutrient depletion as a key factor for manipulating gene expression and product formation in different branches of the flavonoid pathway. *Plant Cell Environ* 31:587–601.

Lim EK, Ashford DA, Hou B, Jackson RG, Bowles DJ. 2004. Arabidopsis glycosyltransferases as biocatalysts in fermentation for regioselective synthesis of diverse quercetin glucosides. *Biotechnol Bioeng* 87:623–631.

Lin J-K, Weng M-S. 2006. Flavonoids as nutraceuticals. In: Grotewold E, Eds. *The Science of Flavonoids*. New York: Springer, pp:213–238.

Liu C-J, Dixon RA. 2001. Elicitor-induced association of isoflavone O-methyltransferase with endomembranes prevents the formation and 7-O-methylation of daidzein during isoflavonoid phytoalexin biosynthesis. *Plant Cell* 13:2643–2658.

Liu C-J, Blount JW, Steele CL, Dixon RA. 2002. Bottlenecks for metabolic engineering of isoflavone glycoconjugates in *Arabidopsis*. *Proc Natl Acad Sci USA* 99:14578–14583.

Ljung K, Bhalerao RP, Sandberg G. 2001. Sites and homeostatic control of auxin biosynthesis in *Arabidopsis* during vegetative growth. *Plant J* 28:465–474.

Lloyd JC, Zakhleniuk OV. 2004. Responses of primary and secondary metabolism to sugar accumulation revealed by microarray expression analysis of the *Arabidopsis* mutant, pho3. *J Exp Bot* 55:1221–1230.

Lloyd AM, Walbot V, Davls RW. 1992. Arabidopsis and Nicotiana anthocyanin production activated by maize regulators R and C7. *Science* 258:1773–1775.

Logemann E, Tavernaro A, Schulz W, Somssich IE, Hahlbrock K. 2000. UV light selectively coinduces supply pathways from primary metabolism and flavonoid secondary product formation in parsley. *Proc Nat Acad Sci USA* 97:1903–1907.

Long SR. 1996. Rhizobium symbiosis: Node factors in perspective. *Plant Cell* 8:1885–1898.

Long SP, Humphries S, Falkowski PG. 1994. Photoinhibition of photosynthesis in nature. *Annu Rev Plant Physiol Plant Mol Biol* 45:633–662.

López-Meyer M, Paiva NL. 2002. Immunolocalization of vestitone reductase and isoflavone reductase, two enzymes involved in the biosynthesis of the phytoalexin medicarpin. *Physiol Molec Plant Pathol* 61:15–30.

Lu Y-P, Li Z-S, Drozdowicz YM, Hortensteiner S, Martinoia E, Rea PA. 1998. AtMRP2, an *Arabidopsis* ATP binding cassette transporter able to transport glutathione S-conjugates and chlorophyll catabolites: Functional comparisons with AtMRP1. *Plant Cell* 10:267–282.

Lu Y-P, Li Z-S, Rea PA. 1997. AtMRP1 gene of *Arabidopsis* encodes a glutathione S-conjugate pump: Isolation and functional definition of a plant ATP-binding cassette transporter gene. *Proc Natl Acad Sci USA* 94:8243–8248.

Ludwig SR, Wessler SR. 1990. Maize R gene family: Tissue-specific helix-loop-helix proteins. *Cell* 62:849–851.

Ludwig SR, Bowen B, Beach L, Wessler SR. 1990. A regulatory gene as a novel visible marker for maize transformation. *Science* 247:449–450.

Lukacin R, Wellmann F, Britisch L, Martens S, Matern U. 2003. Flavonol synthase from Citrus unshiu is a bifunctional dioxygenase. *Phytochem* 62:367–383.

Lukacin R, Matern U, Junghanns KT, Heskamp ML, Britsch L, Forkmann G, Martens S. 2001. Purification and antigenicity of flavones synthase I from irradiated parsley cells. *Arch Biochem Biophys* 393:177–183.

Macías FA, Molinillo JMG, Varela RM, Galindo JCG. 2007. Allelopathy—a natural alternative for weed control. *Pest Manag Sci* 63:327–348.

Macri R, Vianello A, Pennazio S. 1986. Salicylate-collapsed membrane potential in pea stem mitochondria. *Physiol Plantarum* 67:136–140.

Maeshima M. 2001. Tonoplast transporters: Organization and function. *Annu Rev Plant Physiol Plant Mol Biol* 52:469–497.

Mangan JL. 1988. Nutritional effects of tannins in animal feeds. *Nutr Res Rev* 1:209–231.

Marchant A, Kargul J, May ST, Muller P, Delbarre A, Perrot-Rechenmann C, Bennett MJ. 1999. AUX1 regulates root gravitropism in *Arabidopsis* by facilitating auxin uptake within root apical tissues. *EMBO J* 18:2066–2073.

Marinova K, Pourcel L, Weder B, Schwarz M, Barron D, Routaboul JM, Debeaujon I, Klein M. 2007a. The *Arabidopsis* MATE transporter TT12 act as a vacuolar flavonoid/H+-antiporter active in proanthocyanidin-accumulating cells of the seed coat. *Plant Cell* 19:2023–2038.

Marinova K, Kleinschmidt K, Weissenbock G, Klein M. 2007b. Flavonoid biosynthesis in barley primary leaves requires the presence of the vacuole and controls the activity of vacuolar flavonoid transport. *Plant Physiol* 144:432–444.

Marles MAS, Ray H, Gruber MY. 2003. New perspectives on proanthocyanidin biochemistry and molecular regulation. *Phytochem* 64:367–383.

Marrs KA, Alfenito MR, Lloyd AM, Walbot V. 1995. A glutathione S-transferase involved in vacuolar transfer encoded by the maize gene Bronze-2. *Nature* 375:397–400.

Martens S, Forkmann G. 1999. Cloning and expression of flavones synthase II from Gerbera hybrids. *Plant J* 20:611–618.

Martens S, Mithöfer A. 2005. Flavones and flavone synthases. *Phytochem* 66:2399–2407.

Martens S, Forkmann G, Matern U, Lukacin R. 2001. Cloning of parsley flavone synthase I. *Phytochem* 58:43–46.

Martin C, Gerats T. 1993. The control of pigment biosynthesis genes during petal development. *Plant Cell* 5:1253–1264.

Martin C, Prescott A, Mackay S, Bartlett J, Vrijlandt E. 1991. Control of anthocyanin biosynthesis in flowers of Antirrhinum majus. *Plant J* 1:37–49.

Martinez-Garcia JF, Huq E, Quail PH. 2000. Direct targeting of light signals to a promoter element-bound transcription factor. *Science* 288:859–863.

Martinoia E, Grill E, Tommasini R, Kreuz K, Amrhein N. 1993. An ATP-dependent glutathione S-conjugate 'export' pump in the vacuolar membrane of plants. *Nature* 364:247–249.

Mathesius U. 2001. Flavonoids induced in cells undergoing nodule organogenesis in white clover are regulators of auxin breakdown by peroxidase. *J Exp Bot* 52:419–426.

Mathesius U, Weinman JJ, Rolfe BG, Djordjevic MA. 2000. Rhizobia can induce nodules in white clover by "hijacking" mature cortical cells activated during lateral root development. *Mol Plant Microbe Interact* 13:170–182.

Mathesius U, Schlaman HRM, Spaink HP, Sautter C, Rolfe BG, Djordjevic MA. 1998. Auxin transport inhibition precedes root nodule formation in white clover roots and is regulated by flavonoids and derivatives of chitin oligosaccharides. *Plant J* 14:23–34.

Mathews H, Clendennen SK, Caldwell CG, Liu XL, Connors K, Matheis N, Schuster DK, Menasco DJ, Wagoner W, Lightner J, Wagner DR. 2003. Activation tagging in tomato identifies a transcriptional regulator of anthocyanin biosynthesis, modification, and transport. *Plant Cell* 15:1689–1703.

Matsuda N, Tsuchiya T, Kishitani S, Tanaka Y, Toriyama K. 1996. Partial male sterility in transgenic tobacco carrying antisense and sense PAL cDNA under the control of a tapetum-specific promoter. *Plant Cell Physiol* 37:215–222.

Matus JT, Aquea F, Arce-Johnson P. 2008. Analysis of the MYB R2R3 subfamily reveals expanded wine quality-related clades and conserved gene structure organization across *Vitis* and *Arabidopsis* genomes. *BMC Plant Biol* 8:83–97.

McAllister TA, Martinez T, Bae HD, Muir AD, Yanke LJ, Jones GA. 2005. Characterization of condensed tannins purified from legume forages: Chromophore production, protein precipitation, and inhibitory effects on cellulose digestion. *J Chem Ecol* 31:2049–2068.

McArthur C, Hagerman A, Robbins CT. 1991. Physiological strategies of mammalian herbivores against plant defenses. In: Palo RT, Robbins CT, Eds. *Plant Defences against Mammalian Herbivory*. Boca, Raton FL: CRC Press. pp. 103–114.

Mehrtens F, Kranz H, Bednarek P, Weissaar B. 2005. The *Arabidopsis* thaliana transcription factor MYB12 is a flavonol-specific regulator of phenylpropanoid biosynthesis. *Plant Physiol* 138:1083-1096.

Menting JGT, Scopes RK, Stevenson TW. 1994. Characterization of flavonoid 3′,5′-hydroxylase in microsomal membrane fraction of Petunia hybrida flowers. *Plant Physiol* 106:633–642.

Misson J, Raghothama KG, Jain A, Jouhet J, Block MA, Bligny R, Ortet P, Creff A, Sommerville S, Rolland N, Doumas P, Nacry P, Herrera-Estrella L, Nussaume L, Thibaud M-C. 2005. A genome-wide transcriptional analysis using *Arabidopsis* thaliana Affymetrix gene chips determined plant responses to phosphate deprivation. *Proc Natl Acad Sci USA* 102:11934–11939.

Miyahisa I, Kaneko M, Funa N, Kawasaki H, Kojima H, Ohnishi Y, Horinouchi S. 2005. Efficient production of (2S)-flavanones by Escherichia coli containing an artificial biosynthetic gene cluster. *Appl Microbiol Biotechnol* 68:498–504.

Mizushina Y, Ishidoh T, Kamisuki S, Nakazawa S, Takemura M, Sugawara F, Yoshida H, Sakaguchi K. 2003. Flavonoid glycoside: A new inhibitor of eukaryotic DNA polymerase alpha and a new carrier for inhibitor-affinity chromatography. *Biochem Biophys Res Com* 301:480–487.

Mo Y, Nagel C, Taylor LP. 1992. Biochemical complementation of chalcone synthase mutants defines a role for flavonols in functional pollen. *Proc Natl Acad Sci USA* 89:7213–7217.

Mol J, Grotewold E, Koes R. 1998. How genes paint flowers and seeds. *Trends Plant Sci* 3:212–217.

Moreno A, Castro M, Falqué E. 2008. Evolution of trans- and cis-resveratrol content in red grapes (Vitis vinifera L. cv Mencía, Albarello and Merenzao) during ripening. *Eur Food Res Technol* 227:667–674.

Morris P, Robbins MP. 1997. Manipulating condensed tannins in forage legumes. In: McKersie BD, Brown DCW, Eds. *Biotechnology and the Improvement of Forage Legumes*. Wallingford, UK: CAB International, pp. 147–173.

Mosier N, Wyman C, Dale B, Elander R, Lee YY, Holtzapple M, Ladisch M. 2005. Features of promising technologies for pretreatment of lignocellulosic biomass. *Bioresour Technol* 96:673–686.

Muday GK, DeLong A. 2001. Polar auxin transport: Controlling where and how much. *Trends Plant Sci* 6:535–542.

Muday GK, Peer WA, Murphy AS. 2003. Vesicular cycling mechanisms that control auxin transport polarity. *Trends Plant Sci* 8:301–304.

Mueller LA, Goodman CD, Silady RA, Walbot V. 2000. AN9, a petunia gluthathione S-transferase required for anthocyanin sequestration, is a flavonoid-binding protein. *Plant Physiol* 123:1561–1570.

Muir SR, Collins GJ, Robinson S, Hughes S, Bovy A, Ric de Vos CH, van Tunen AJ, Verhoeyen ME. 2001. Overexpression of petunia chalcone isomerase in tomato results in fruit containing increased levels of flavonols. *Nat Biotechnol* 19:470–474.

Muller R, Morant M, Jarmer H, Nilsson L, Nielsen TH. 2007. Genome-wide analysis of the *Arabidopsis* leaf transcriptome reveals interaction of phosphate and sugar metabolism. *Plant Physiol* 143:156–171.

Multani DS, Briggs SP, Chamberlin MA, Blakeslee JJ, Murphy AS, Johal GS. 2003. Loss of an MDR transporter in compact stalks of maize br2 and sorghum dw3 mutants. *Science* 302:81–84.

Murphy A, Peer WA, Taiz L. 2000. Regulation of auxin transport by aminopeptidases and endogenous flavonoids. *Planta* 211:315–324.

Murphy AS, Hoogner KR, Peer WA, Taiz L. 2002. Identification, purification, and molecular cloning of N-1-naphthylphthalamic acid-binding plasma membrane-associated aminopeptidases from *Arabidopsis*. *Plant Physiol* 128:935–950.

Murray JD, Karas BJ, Sato S, Tabata S, Amyot L, Szczyglowski K. 2007. A cytokinin perception mutant colonized by Rhizobium in the absence of nodule organogenesis. *Science* 315:101–104.

Mutikainen P, Walls M, Ovaska J, Keinanen M, Julkunen-Tiitto R, Vapaavuori E. 2000. Herbivore resistance in Betula pendula: Effects of fertilization, defoliation, and genotype. *Ecology* 81:49–65.

Nakayama T. 2002. Enzymology of aurone biosynthesis. *J Biosci Bioeng* 94:487–491.

Nakayama T, Sato T, Fukui Y, Yonekura-Sakakibara K, Hayashi H, Tanaka Y, Kusumi T, Nishino T. 2001. Specificity analysis and mechanism of aurone synthesis catalyzed by aureusidin synthase, a polyphenol oxidase homolog responsible for flower coloration. *FEBS Lett* 499:107–111.

Nakayama T, Yonekura-Sakakibara K, Sato T, Kikuchi S, Fukui Y, Fukuchi-Mizutani M, Ueda T, Nakao M, Tanaka Y, Kusumi T, Nishino T. 2000. Aureusidin synthase: A polyphenol oxidase homolog responsible for flower coloration. *Science* 290:1163–1166.

Naoumkina M, Farag MA, Summer LW, Tang Y, Liu C-J, Dixon RA. 2007. Different mechanism for phytoalexins induction by pathogen and wound signals in Medicago truncatula. *Proc Natl Acad Sci USA* 104:17909–17915.

Nasir H, Iqbal Z, Hiradate S, Fujii Y. 2005. Allelopathic potential of Robinia pseudo-acacia L. *J Chem Ecol* 31:2179–2192.

Neill SO, Gould KS, Kilmartin PA, Mitchell KA, Markham KR. 2002. Antioxidant activities of red versus green leaves in Elatostema rugosum. *Plant Cell Environ* 25:539–547.

Nesi N, Debeaujon I, Jond C, Stewart AJ, Jenkins GI, Caboche M, Lepiniec L. 2002. The TRANSPARENT TESTA16 locus encodes the ARABIDOPSIS BSISTER MADS domain protein and is required for proper development and pigmentation of the seed coat. *Plant Cell* 14:2463–2479.

Nesi N, Jond C, Debeaujon I, Caboche M, Lepiniec L. 2001. The *Arabidopsis* TT2 gene encodes an R2R3-MYB domain protein that acts as a key determinant for proanthocyanidin accumulation in developing seed. *Plant Cell* 13:2099–2114.

Nesi N, Debeaujon I, Jond C, Pelletier G, Caboche M, Lepiniec L. 2002. The TT8 gene encodes a basic helix-loop-helix domain protein required for expressioin of DFR and BAN genes in *Arabidopsis* siliques. *Plant Cell* 14:2463–2479.

Nimbal CI, Pedersen JF, Yerkes CN, Weston LA, Weller SC. 1996. Phytotoxicity and distribution of sorgoleone in grain sorghum germplasm. *J Agric Food Chem* 44:1343–1347.

Noh B, Murphy AS, Spalding EP. 2001. Multidrug Resistance-like genes of *Arabidopsis* required for auxin transport and auxin-mediated development. *Plant Cell* 13:2441–2454.

Nomura M, Itioka T. 2002. Effects of synthesized tannin on the growth and survival of a generalist herbivorous insect, the common cutworm, Spodoptera litura Fabricius (Lepidoptera: Noctuidae). *Appl Entomol Zool* 37:285–289.

Oldroyd GED, Downie JA. 2008. Coordinating nodule morphogenesis with rhizobial infection in legumes. *Annu Rev Plant Biol* 59:519–546.

Olofsdotter M, Rebulanan M, Madrid A, Dali W, Navarez D, Olk DC. 2002 Why phenolic acids are unlikely primary allelochemicals in rice. *J Chem Ecol* 28:229–242.

Ono E, Fukuchi-Mizutani M, Nakamura N, Fukui Y, Yonekura-Sakakibara K, Yamaguchi M, Nakayama T, Tanaka T, Kusumi T, Tanaka Y. 2006. Yellow flowers generated by expression of the aurones biosynthetic pathway. *Proc Natl Acad Sci USA* 103:11075–11080.

Oommen A, Dixon RA, Paiva NL. 1994. The elicitor-inducible alfalfa isoflavone reductase promoter confers different patterns of developmental expression in homologous and heterologous transgenic plants. *Plant Cell* 6:1789–1803.

Ovádi J, Saks V. 2004. On the origin of intracellular compartmentation and organized metabolic systems. *Molec Cel Biochem* 256/257:5–12.

Page JE, Towers GH. 2002. Anthocyanins protect light sensitive thiarubrine phototoxins. *Planta* 215:478–484.

Page JE, Block E, Towers GHN. 1999. Visible-light photochemistry and phototoxicity of thiarubrines. *Photochem Photobiol* 70:159–165.

Paiva NL. 2000. An introduction to the biosynthesis of chemicals used in plant-microbe communication. *J Plant Growth Regul* 19:131–143.

Paiva NL, Oommen A, Harrison MJ, Dixon RA. 1994. Regulation of isoflavonoid metabolism in alfalfa. *Plant Cell Tissue Organ Cult* 38:213–220.

Pang Y, Peel GJ, Sharma SB, Tang Y, Wang Z, Dixon RA. 2008. A transcript profiling approach reveals an epicatechin-specific glucosyltransferase expressed in the seed coat of Medicago truncatula. *Proc Natl Acad Sci USA* 105:14210–14215.

Pang Y, Peel GJ, Wright E, Wang Z, Dixon RA. 2007. Early steps in proanthocyanidin biosynthesis in the model legume Medicago truncatula. *Plant Physiol* 145:601–615.

Paolocci F, Robbins MP, Madeo L, Arcioni S, Martens S, Damiani F. 2007. Ectopic expression of a basic helix-loop-helix gene transactivates parallel pathways of proanthocyanidin biosynthesis. Structure, expression analysis, and genetic control of leucoanthocyanidin 4–,reductase and anthocyanidin reductase genes in Lotus corniculatus. *Plant Physiol* 143:504–516.

Paolocci F, Bovone T, Tosti N, Arcioni S, Damiani F. 2005. Light and an exogenous transcription factor qualitatively and quantitatively affect the biosynthetic pathway of condensed tannins in Lotus corniculatus leaves. *J Exp Bot* 56:1093–1103.

Parry G, Marchant A, May S, Swarup R, Swarup K, James N, Graham N, Allen T, Martucci T, Yemm A, Napier R, Manning K, King G, Bennett M. 2004. Quick on the uptake: Characterization of a family of plant auxin influx carriers. *J Plant Growth Regul* 20:217–225.

Passamonti S, Cocolo A, Braidot E, Petrussa E, Peresson C, Medic N, Macri F, Vianello A. 2005. Characterization of electrogenic bromosulfophthalein transport in carnation petal microsomes and its inhibition by antibodies against bilitranslocase. *FEBS J* 272:3282–3296.

Pastori GM, Foyer H. 2002. Common components, networks, and pathways of cross-tolerance to stress. The central role of "redox" and abscisic acid-mediated controls. *Plant Physiol* 129:460–468.

Payne CT, Zhang F, Lloyd AM. 2000. GL3 encodes a bHLH protein that regulates trichome development in *Arabidopsis* through interaction with GL1 and TTG1. *Genetics* 156:1349–1362.

Paz-Ares J, Ghosal D, Wienand U, Peterson PA, Saedler H. 1987. The regulatory c1 locus of Zea mays encodes a protein with homology to myb proto-oncogene products and with structural similarities to transcriptional activators. *EMBO J* 6:3553–3558.

Peck MC, Fischer RF, Long SR. 2006. Diverse flavonoids stimulate NodD1 binding to nod gene promoters in Sinorhizobium meliloti. *J Bacteriol* 188:5417–5427.

Pecket RC, Small CJ. 1980. Occurrence, location and development of anthocyanoplasts. *Phytochem* 19:2571–2576.

Peer WA, Murphy AS. 2007. Flavonoids and auxin transport: Modulators or regulators? *Trends Plant Sci* 12:556–563.

Peer WA, Murphy AS. 2006. Flavonoids as signal molecules: Targets of flavonoid action. In: Grotewold E, Eds. *The Science of Flavonoids*. New York: Springer., pp. 239–268.

Peer WA, Bandyopadhyay A, Blakeslee JJ, Makam SN, Chen RJ, Masson PH, Murphy AS. 2004. Variation in expression and protein localization of the PIN family of auxin efflux facilitator proteins in flavonoids mutants with altered auxin transport in *Arabidopsis* thaliana. *Plant Cell* 16:1898–1911.

Peer WA, Brown DE, Tague BW, Muday GK, Taiz L, Murphy AS. 2001. Flavonoid accumulation patterns of transparent testa mutants of *Arabidopsis*. *Plant Physiol* 126:536–548.

Pelletier MK, Burbulis IE, Winkel-Shirley B. 1999. Disruption of specific flavonoid genes enhances the accumulation of flavonoid enzymes and end-products in *Arabidopsis* seedlings. *Plant Molec Biology* 40:45–54.

Pelletier MK, Murrell JR, Winkel-Shirley B. 1997. Characterization of flavonol synthase and leucoanthocyanidin dioxygenase genes in *Arabidopsis*. *Plant Physiol* 113:1437–1445.

Pérez FJ, Ormeño-Núñez J. 1991. Difference in hydroxamic acid content in roots and root exudates of wheat (*Triticum aestivum* L.) and rye (*Secale cereale* L.): Possible role in allelopathy. *J Chem Ecol* 17:1037–1043.

Peters NK, Frost JW, Long SR. 1986. A plant flavone, luteolin, induces expression of Rhizobium meliloti nodulation genes. *Science* 233:977–980.

Peynaud E, Ribéreau-Gayon P. 1970. The grape. In: Hulme, AC, Ed. *The Biochemistry of Fruits and Their Products*. London: Academic,Vol. 2, pp. 171–205.

Pezet R, Pont V. 1995. Mode of toxic action of Vitaceae stilbenes on fungal cells. In: Daniel M, Purkayastha RP, Ed. *Handbook of Phytoalexins Metabolism and Action*. New York: Marcel Dekker. pp. 317–331.

Pezet R, Pont V. 1990a. Ultrastructural observations of pterostilbene fungitoxicity in dormant conidia of Botrytis cinerea Pers. *J Phytopathol* 129:19–30.

Pezet R, Pont V. 1990b. Relation between the chemical structure and the biological activity of hydroxystilbenes against Botrytis cinerea. *J Phytopathol* 130:1–8.

Pezet R, Gindro K, Viret O, Spring J-L. 2004. Glycosylation and oxidative dimerization of resveratrol are respectively associated to sensitivity and resistance of grapevine cultivars to downy mildew. *Physiol Molec Plant Pathol* 65:297–303.

Pfeiffer J, Kühnel C, Brandt J, Duy D, Punyasiri PA, Forkmann G, Fischer TC. 2006. Biosynthesis of flavan 3-ols by leucoanthocyanidin 4-reductases and anthocyanidin reductases in leaves of grape (Vitis vinifera L.), apple (Malus x domestica Borkh.) and other crops. *Plant Physiol Biochem* 44:323–334.

Piffanelli P, Ramsay L, Waugh R, Benabdelmouna A, D'Hont A, Hollricher K, Jorgensen JH, Schulze-Lefert P, Panstruga R. 2004. A barley cultivation-associated polymorphism conveys resistance to powdery mildew. *Nature* 430:887–891.

Piffanelli P, Ross JHE, Murphy DJ. 1998. Biogenesis and function of the lipidic structures of pollen grains. *Sex Plant Reprod* 11:65–80.

Pollak PE, Hansen K, Astwood JD, Taylor LP. 1995. Conditional male fertility in maize. *Sex Plant Reprod* 8:231–241.

Polster J, Dithmar H, Feucht W. 2003. Are histones the targets for flavan-3-ols (catechins) in the nuclei? *Biol Chem* 384:997–1006.

Porter LJ, Hrstich LN, Chan BG. 1986. The conversion of procyanidins and prodelphinidins to cyanidin and delphinidin. *Phytochem* 25:223–230.

Pourcel L, Routaboul J-M, Cheynier V, Lepiniec L, Debeaujon I. 2007. Flavonoid oxidation in plants: From biochemical properties to physiological functions. *Trends Plant Sci* 12:29–36.

Poutska F, Irani NG, Feller A, Lu Y, Pourcel L, Frame K, Grotewold E. 2007. A trafficking pathway for anthocyanins overlaps with the endoplasmic reticulum-to-vacuole protein-sorting route in *Arabidopsis* and contributes to the formation of vacuolar inclusions. *Plant Physiol* 145:1323–1335.

Prescott AG, Stamford NP, Wheeler G, Firmin JL. 2002. In vitro properties of a recombinant flavonol synthase from *Arabidopsis* thaliana. *Phytochem* 60:589–593.

Preston J, Wheeler J, Heazlewood J, Li SF, Parish RW. 2004. AtMYB32 is required for normal pollen development in *Arabidopsis* thaliana. *Plant J* 40:979–995.

Punyasiri PAN, Abeysinghe ISB, Kumar V, Treutter D, Duy D, Gosch C, Martens S, Forkmann G, Fischer TC. 2004. Flavonoid biosynthesis in the tea plant Camellia sinensis: Properties of enzymes of the prominent epicatechin and catechin pathways. *Arch Biochem Biophys* 431:22–30.

Quattrocchio F, Baudry A, Lepiniec L, Grotewold E. 2006. The regulation of flavonoid biosynthesis. In: Grotewold E, Ed. *The Science of Flavonoids*. New York: Springer, pp. 97–122.

Quattrocchio F, Wing JF, Woude K, Mol JN, Koes R. 1998. Analysis of bHLH and myb proteins: Species-specific regulatory differences are caused by divergent evolution of target anthocyanin genes. *Plant J* 4:475–488.

Raiber S, Schröder G, Schröder J. 1995. Molecular and enzymatic characterization of two stilbene synthases from Eastern white pine (Pinus strobus): A single Arg/His difference determines the activity and the pH dependence of the enzymes. *FEBS Lett* 361:299–302.

Rausher MD. 2006. The evolution of flavonoids and their genes. In: Grotewold E, Ed. *The science of Flavonoids*. New York: Springer, pp. 175–212.

Rea PA. 1999. MRP subfamily ABC transporters from plants and yeast. *J Exp Bot* 50:895–913.

Rea PA, Li Z-S, Lu Y-P, Drozdowicz YM, Martinoia E. 1998. From vacuolar GS-X pumps to multispecific ABC transporters. *Annu Rev Plant Physiol Plant Mol Biol* 49:727–760.

Recourt K, van Tunen AJ, Mur LA, van Brussel AA, Lugtenberg BJ, Kijne JW. 1992. Activation of flavonoid biosynthesis in roots of Vicia sativa subsp. nigra plants by inoculation with Rhizobium leguminosarum biovar viciae. *Plant Mol Biol* 19:411–420.

Reddy MSS, Chen F, Shadle G, Jackson L, Aljoe H, Dixon RA. 2005. Targeted down-regulation of cytochrome P450 enzymes for forage quality improvement in alfalfa (*Medicago sativa* L.) *Proc Natl Acad Sci USA* 102:16573–16578.

Redmond JW, Batley M, Djordjevic MA, Innes RW, Kuempel PL, Rolfe BG. 1986. Flavones induce expression of nodulation genes in Rhizobium. *Nature* 323:632–635.

Reed J, Soller H. 1987. Phenolics and nitrogen utilization in sheep fed browse. In: Rose M, Ed. *Proceedings of the 2nd International Symposium on the Nutrition of Herbivores*. Brisbane, Australia: University of Queensland Press, pp. 47–48.

Regev-Shoshani G, Shoseyov O, Bilkis I, Kerem Z. 2003. Glycosylation of resveratrol protects it from enzymic oxidation. *Biochem J* 374:157–163.

Rice EL, Ed. 1984. *Allelopathy*, 2nd Ed. Orlando, FL: Academic Press.

Rice-Evans C, Miller NJ. 1996. Antioxidant activities of flavonoids as bioactive components of foods. *Biochem Soc Transac* 24:790–795.

Rice-Evans C, Miller NJ, Paganga G. 1996. Structure-antioxidant activity relationships of flavonoids and phenolic acids. *Free Radic Biol Med* 20:933–956.

Riechmann JL, Ratcliffe OJ. 2000. A genomic perspective on plant transcription factors. *Plant Biol* 3:423–434.

Riera M, Valon C, Fenzi F, Giraudat J, leung J. 2005. The genetics of adaptive responses to drought stress: Abscisic acid-dependent and abscisic acid-independent signaling components. *Physiol Plantarum* 123:111–119.

Ro D-K, Douglas CJ. 2004. Reconstitution of the entry point of plant phenylpropanoid metabolism in yeast (Saccharomyces cerevisiae): Implication for control of metabolic flux into the phenylpropanoid pathway. *J Biol Chem* 279:2600–2607.

Rolland F, Moore B, Sheen J. 2002. Sugar sensing and signaling in plants. *Plant Cell* 14:185–205.

Ros Barceló A, Pomar F, López-Serrano M, Pedreño MA. 2003. Peroxidase: a multifunctional enzyme in grapevines. *Func Plant Biol* 30:577–591.

Ross JA, Kasum CM. 2002. Dietary flavonoids: Bioavailability, metabolic effects, and safety. *Annu Rev Nutr* 22:19–34.

Rühmann S, Treutter D. 2003. Effect of N-nutrition in apple on the response of its secondary metabolism to prohexadione-Ca treatment. *Eur J Hort Sci* 68:152–159.

Rühmann S, Leser C, Bannert M, Treutter D. 2002. Relationship between growth, secondary metabolism, and resistance of apple. *Plant Biol* 4:137–193.

Russell GB, Sutherland ORW, Hutchins RFN, Christmas PE. 1978. Vestitol: A phytoalexin with insect feeding deterrent activity. *J Chem Ecology* 4:571–579.

Ryder TB, Hedrick SA, Bell JN, Liang XW, Clouse SD, Lamb CJ. 1987. Organization and differential activation of a gene family encoding the plant defense enzyme chalcone synthase in Phaseolus vulgaris. *Mol Gen Genet* 210:219–233.

Sagasser M, Lu G-H, Hahlbrock K, Weisshaar B. 2001. A. thaliana TRANSPARENT TESTA 1 is involved in seed coat development and defines the WIP subfamily of plant zinc finger proteins. *Genes Dev* 16:138–149.

Saito K, Yamazaki M. 2002. Transley Review 138. Biochemistry and molecular biology of the late stage of biosynthesis of anthocyanin: Lessons from Perilla frutescens as a model plant. *New Phytol* 155:9–23.

Sarig P, Zutkhi Y, Monjauze A, Lisker N, Ben-Arie R. 1997. Phytoalexin elicitation in grape berries and their susceptibility to Rhizopus stolonifer. *Phys Molec Plant Path* 50:337–347.

Saslowsky D, Winkel-Shirley B. 2001. Localization of flavonoid enzymes in *Arabidopsis* roots. *Plant J* 27:37–48.

Saslowsky DE, Warek U, Winkel BS. 2005. Nuclear localization of flavonoid enzymes in *Arabidopsis*. *J Biol Chem* 280:23735–23740.

Sato T, Nakayama T, Kikuchi S, Fukui Y, Yonekura-Sakakibara K, Ueda T, Nishino T, Tanaka Y, Kusumi T. 2001. Enzymatic formation of aurones in the extracts of yellow snapdragon flowers. *Plant Sci* 160:229–236.

Sauer U. 2006. Metabolic networks in motion: 13C-based flux analysis. *Mol Sys Biol* 62:1–10.

Sbaghi M, Jeandet P, Bessis R, Leroux P. 1996. Degradation of stilbene-type phytoalexins in relation to the pathogenicity of Botrytis cinerea to grapevines. *Plant Pathol* 45:139–144.

Sbaghi M, Jeandet P, Faivre B, Bessis R, Fournioux JC. 1995. Development of methods using phytoalexin (resveratrol) assessment as a selection criterion to screen grapevine in vitro cultures for resistance to grey mould (Botrytis cinerea). *Euphytica* 86:41–47.

Schijlen EGWM, Ric de Vos CH, Martens S, Jonker HH, Rosin FM, Molthoff JW, Tikunov YM, Angenent GC, van Tunen AJ, Bovy AG. 2007. RNAi silencing of chalcone synthase, the first step in the flavonoid biosynthesis pathway, leads to parthenocarpic tomato fruits. *Plant Physiol* 144:1520–1530.

Schbohm C, Martens S, Eder C, Forkmann G, Weisshaar B. 2000. Identification of the *Arabidopsis* thaliana flavonoid 3′-hydroxylase gene and functional expression of the encoded P450 enzyme. *Biol Chem* 381:749–753.

Schoppner A, Kindl H. 1984. Purification and properties of a stylbene synthase from induced cell suspension cultures of peanut. *J Biol Chem* 259:6806–6811.

Schulze-Lefert P, Becker-André M, Schulz W, Hahlbrock K, Dangl J. 1989. Functional architecture of the light-responsive chalcone synthase promoter from parsley. *Plant Cell* 1:707–714.

Schwender J. 2008. Metabolic flux analysis as a tool in metabolic engineering of plants. *Curr Opin Biotech* 19:131–137.

Schwinn K, Venail J, Shang Y, Mackay S, Alm V, Butelli E, Oyama R, Bailey P, Davies K, Martin C. 2006. A small family of MYB-regulatory genes controls floral pigmentation intensity and patterning in the genus Antirrhinum. *Plant Cell* 18:831–851.

Seal AN, Pratley JE, Haig T, An M. 2004. Identification and quantitation of compounds in a series of allelopathic and non-allelopathic rice root exudates. *J Chem Ecol* 30:1647–1662.

Seguin G. 1986. "Terroirs" and pedology of wine growing. *Experientia* 42:861–872.

Selinger D, Chandler V. 1999. A mutation in the pale aleurone color 1 gene identifies a novel regulator of the maize anthocyanin pathway. *Plant Cell* 11:5–14.

Sharma SB, Dixon RA. 2005. Metabolic engineering of proanthocyanidins by ectopic expression of transcription factors in *Arabidopsis* thaliana. *Plant J* 44:62–75.

Shen G-A, Pang Y, Wu W, Liu X, Zhao L, Sun X, Tang KJ. 2005. Isolation and characterization of putative anthocyanidin reductase gene from Ginkgo biloba. *Plant Physiol* 163:224–227.

Sherwin HW, Farrant JM. 1998. Protection mechanisms against excess light in the resurrection plants Craterostigma wilmsii and Xerophyta viscose. *Plant Growth Regul* 24:203–210.

Shimada N, Akashi T, Aoki T, Ayabe S. 2000. Induction of isoflavonoid pathway in the model legume Lotus japonicus: Molecular characterization of enzymes involved in phytoalexin biosynthesis. *Plant Sci* 160:37–47.

Shimada T. 2006. Salivary proteins as a defense against dietary tannins. *J Chem Ecology* 32:1149–1163.

Shimamura M, Akashi T, Sakurai N, Suzuki H, Saito K, Shibata D, Ayabe S-I, Aoki T. 2007. 2-Hydroxyflavanone dehydratase is a critical determinant of isoflavone productivity in hairy root cultures of Lotus japonicus. *Plant Cell Physiol* 48:1652–1657.

Shimizu T, Akada S, Senda M, Ishikawa R, Harada T, Niizeki M, Dube SK. 1999. Enhanced expression and differential inducibility of soybean chalcone synthase genes by supplemental UV-B in dark-grown seedlings. *Plant Mol Biol* 39:785–795.

Shin J, Park E, Choi G. 2007. PIF3 regulates anthocyanin biosynthesis in an HY5-dependent manner with both factors directly binding anthocyanin biosynthetic gene promoters in *Arabidopsis*. *Plant J* 49:981–994.

Shirley BW. 1996. Flavonoid biosynthesis: "New" functions for an "old" pathway. *Trends Plant Sci* 1:377–382.

Shirley BW, Kubasek WL, Stotrz G, Bruggemann E, Koorneef M, Ausubel FM, Goodman HM. 1995. Analysis of *Arabidopsis* mutants deficient in flavonoid biosynthesis. *Plant J* 8:659–671.

Sidorenko LV, Li X, Cocciolone SM, Chopra S, Tagliani L, Bowen B, Daniels M, Peterson T. 2000. Complex structure of a maize Myb gene promoter: Functional analysis in transgenic plants. *Plant J* 22:471–482.

Skadhauge B, Gruber MY, Thomsen KK, von Wettstein D. 1997a. Leucocyanidin reductase activity and accumulation of proanthocyanidins in developing legume tissues. *Am J Bot* 84:494–502.

Skadhauge B, Thomsen K, von Wettstein D. 1997b. The role of barley testa layer and its flavonoid content in resistance to Fusarium infections. *Hereditas* 126:147–160.

Smeekens S. 2000. Sugar-induced signal transduction in plants. *Annu Rev Plant Physiol Plant Mol Biol* 51:49–81.

Solecka D, Kacperska A. 2003. Phenylpropanoid deficiency affects the course of plant acclimation to cold. *Physiol Plant* 119:253–262.

Solfanelli C, Poggi A, Loreti E, Alpi A, Perata P. 2006. Sucrose-specific induction of the anthocyanin biosynthetic pathway in *Arabidopsis* thaliana. *Plant Physiol* 140:637–646.

Sotheeswaran S, Pasupathy V. 1993. Distribution of resveratrol oligomers in plants. *Phytochem* 32:1083–1092.

Souquet JM, Labarbe B, Le-Guerneve C, Cheynier V, Moutounet M. 2000. Phenolic composition of grape stems. *J Ag Food Chem* 48:1076–1080.

Souquet J-M, Cheynier V, Brossaud F, Moutounet M. 1996. Polymeric proanthocyanidins from grape skins. *Phytochem* 43:509–512.

Souto C, Pellissier F, Chiapusio G. 2000. Allelopathic effects of humus phenolics on growth and respiration of mycorrhizal fungi. *J Chem Ecol* 26:2015–2023.

Spelt C, Quattrocchio F, Mol J, Koes R. 2002. ANTHOCYANIN1 of petunia controls pigment synthesis, vacuolar pH, and seed coat development by genetically distinct mechanisms. *Plant Cell* 14:2121–2135.

Spelt C, Quattrocchio F, Mol J, Koes R. 2000. Anthocyanin1 of petunia encodes a basic helix-loop-helix protein that directly activates transcription of structural anthocyanin genes. *Plant Cell* 12:1619–1632.

Springob K, Nakajima J-I, Yamazaki M, Saito K. 2003. Recent advances in the biosynthesis and accumulation of anthocyanins. *Nat Prod Rep* 20:288–303.

Stafford HA. 1991. Flavonoid evolution: An enzymic approach. *Plant Physiol* 96:680–685.

Stafford HA. 1989. The enzymology of proanthocyanidin biosynthesis. In: Hemingway RW, Karchesy JJ, Eds. *Chemistry and Significance of Condensed Tannins*. New York: Plenum Press, pp. 345–368.

Stafford HA. 1981. Compartmentation in natural product biosynthesis by multienzyme complexes. In: Conn EE, Ed. *The Biochemistry of Plants*. New York: Academic Press, Vol. 7 pp. 117–137.

Stafford HA. 1974. Possible multienzyme complexes regulating the formation of C6-C3 phenolic compounds and lignins in higher plants. *Recent Adv Phytochem* 8:53–79.

Stapleton AE, Walbot V. 1994. Flavonoids can protect maize DNA from the induction of ultraviolet radiation damage. *Plant Physiol* 105:881–889.

Stewart AJ, Chapman W, Jenkins GI, Graham I, Martin T, Crozier A. 2001. The effect of nitrogen and phosphorus deficiency on flavonol accumulation in plant tissues. *Plant Cell Env* 24:1189–1197.

Stirnberg P, van de Sande K, Ottoline Leyser HM. 2002. MAX1 and MAX2 control shoot lateral branching in *Arabidopsis*. *Development* 129:1131–1141.

Stobiecki M, Kachlicki P. 2006. Isolation and identification of flavonoids. In: Grotewold E, Ed. *The Science of Flavonoids*. New York: Springer, pp. 47–70.

Stoessl A. 1982. Biosynthesis of phytoalexins. In: Bailey JA, Mansfield JW, Ed. *Phytoalexins*. New York: Wiley, pp. 133–180.

Stracke R, Ishihara H, Huep G, Barsch A, Mehrtens F, Niehaus K, Weisshaar B. 2007. Differential regulation of closely related R2R3-MYB transcription factors control flavonol accumulation in different parts of the *Arabidopsis* thaliana seedling. *Plant J* 50:660–677.

Stracke R, Werber M, Weisshaar B. 2001. The R2R3-MYB gene family in *Arabidopsis* thaliana. *Plant Biol* 4:447–456.

Styles ED, Ceska O. 1989. Pericarp flavonoids in genetic strains of Zea mays. *Maydica* 34:227–237.

Swindell WR. 2006. The association among gene expression responses to nine abiotic stress treatments in *Arabidopsis* thaliana. *Genetics* 174:1811–1824.

Subramanian S, Stacey G, Yu O. 2006. Endogenous isoflavones are essential for the establishment of symbiosis between soybean and Bradyrhizobium japonicum. *Plant J* 48:261–273.

Takos AM, Jaffé; FW, Jacob SR, Bogs J, Robinson SP, Walker AR. 2006a. Light-induced expression of a MYB gene regulates anthocyanin biosynthesis in red apples. *Plant Physiol* 140:279–291.

Takos AM, Ubi BE, Robinson S, Walker AR. 2006b. Condensed tannin biosynthesis genes are regulated separately from other flavonoid biosynthesis genes in apple fruit skin. *Plant Sci* 170:487–499.

Tamari G, Borochov A, Atzorn R, Weiss D. 1995. Methyl jasmonate induces pigmentation and flavonoid gene expression in petunia corollas: A possible role in wound response. *Physiol Plant* 94:45–50.

Tanaka Y, Katsumoto Y, Brugliera F, Mason J. 2005. Genetic engineering in floriculture. *Plant Cell Tissue Organ Cult* 80:1–24.

Tanner GJ, Francki KT, Abrahams S, Watson JM, Larkin PJ, Ashton AR. 2003. Proanthocyanidin biosynthesis in plants. Purification of legume leucoanthocyanidin reductase and molecular cloning of its cDNA. *J Biol Chem* 278:31647–31656.

Tanner GJ, Moate PJ, Davis LH, Laby RH, Yuguang L, Larkin PJ. 1995. Proanthocyanidins (condensed tannin) destabilize plant protein foams in a dose dependent manner. *Aust J Agric Res* 46:1101–1109.

Tassoni A, Fornale S, Franceschetti M, Musiani F, Michael AJ, Perry B, Bagni N. 2005. Jasmonates and Na-orthovanadate promote resveratrol production in *Vitis* vinifera cv. Barbera cell cultures. *New Phytol* 166:895–905.

Taylor LP, Grotewold E. 2005. Flavonoids as developmental regulators. *Curr Opin Plant Biol* 8:317–323.

Taylor LP, Hepler PK. 1997. Pollen germination and tube growth. *Annu Rev Plant Physiol Plant Mol Biol* 48:461–491.

Taylor LP, Jorgensen R. 1992. Conditional male fertility in chalcone synthase-deficient Petunia. *J Heredity* 83:11–17.

Teng S, Keurentjes J, Bentsink L, Koornneef M, Smeekens S. 2005. Sucrose-specific induction of anthocyanin biosynthesis in *Arabidopsis* requires the MYB75/PAP1 gene. *Plant Physiol* 139:1840–1852

Theodoulou FL. 2000. Plant ABC transporters. *Biochim Biophys Acta* 1465:79–103.

Thipyapong P, Hunt MD, Steffens JC. 2004. Antisense downregulation of polyphenol oxidase results in enhanced disease susceptibility. *Planta* 220:105–117.

Thomzik JE, Stenzel K, Stocker R, Schreier PH, Hain R, Stahl DG. 1997. Synthesis of a grapevine phytoalexin in transgenic tomatoes (Lycopersicum esculentum Mill.) conditions resistance against Phytophtora infestans. *Phys Molec Plant Path* 51:265–278.

Tian L, Pang Y, Dixon RA. 2007. Biosynthesis and genetic engineering of proanthocyanidins and (iso)flavonoids. *Phytochem Rev* 7:445–465.

Toghe T, Nishiyama Y, Hirai MY, Yano M, Nakajima J, Awazuhara M, Inoue E, Takahashi H, Goodenowe DB, Kitayama M, Noji M, Yamazaki M, Saito K. 2005. Functional genomics by integrated analysis of metabolome and transcriptome of *Arabidopsis* plants over-expressing an MYB transcription factor. *Plant J* 42:218–235.

Treutter D. 2006. Significance of flavonoids in plant resistance: a review. *Environ Chem Lett* 4:147–157.

Treutter D. 2005. Significance of flavonoids in plant resistance and enhancement of their biosynthesis. *Plant Biol* 7:581–591.

Tirichine L, Sandal N, Madsen LH, Radutoiu S, Albrektsen AS, Sato S, Asamizu E, Tabata S, Stougaard J. 2007 A gain-of-function mutation in a cytokinin receptor triggers spontaneous root nodule organogenesis. *Science* 315:104–107.

Tsukaya H, Ohshima T, Naito S, Chino M, Komeda Y. 1991. Sugar dependent expression of the CHS-A gene for chalcone synthase from petunia in transgenic *Arabidopsis*. *Plant Physiol* 97:1414–1421.

Tuteja JH, Clough SJ, Chan W-C, Vodkin LO. 2004. Tissue-specific silencing mediated by a naturally occurring chalcone synthase cluster in Glycine max. *Plant Cell* 16:819–835.

Utsuno K, Shikanai T, Yamada Y, Hashimoto T. 1998. AGR, an agravitropic locus of *Arabidopsis* thaliana, encodes a novel membrane protein family member. *Plant Cell Physiol* 39:1111–1118.

van Brussel AA, Recourt K, Pees E, Spaink HP, Tak T, Wijffelman CA, Kijne JW, Lugtenberg BJ. 1990. A biovar-specific signal of Rhizobium leguminosarum bv. viciae induces increased nodulation gene-inducing activity in root exudate of Vicia sativa subsp. *nigra*. *J Bacteriol* 172:5394–5401.

Vandenbussche F, Habricot Y, Condiff AS, Maldiney R, Van Der Straeten D, Ahmad M. 2007. HY5 is a point of convergence between crytochrome and cytokinin signaling pathways in *Arabidopsis* thaliana. *Plant J* 49:428–441.

van der Meer IM, Stam ME, van Tunen AJ, MoI JNM, Stuitje AR. 1992. Antisense inhibition of flavonoid biosynthesis in petunia anthers results in male sterility. *Plant Cell* 4:253–262.

VanEtten H, Mansfield JW, Bailey JA, Farmer EE. 1994. Two classes of plant antibiotics: Phytoalexins versus "Phytoanticipins." *Plant Cell* 6:1191–1192.

van Houwelingen A, Souer E, Spelt K, Kloos D, Mol J, Koes R. 1998. Analysis of flower pigmentation mutants generated by random transposon mutagenesis in Petunia hybrida. *Plant J* 13:39–50.

Vass I. 1997. Adverse effects of UV-B light on the structure and function of the photosynthetic apparatus. In: Pessarakli M, Ed. *Handbook of Photosynthesis*. New York: Marcel Dekker, pp. 931–949.

Versari A, Parpinello GP, Tronielli GB, Ferrarini R, Giulivo C. 2001 Stilbene compounds and stilbene synthase expression during ripening, wilting and UV treatment in grape cv. *Corvina*. *J Agric Food Chem* 49:5531–5536.

Ververidis F, Trantas E, Douglas C, Vollmer G, Kretzschmar G, Panopoulos N. 2007a. Biotechnology of flavonoids and other phenylpropanoids-derived natural products. Part I: Chemical diversity impact, impa on plant biology and human health. *Biotechnol J* 2:1214–1234.

Ververidis F, Trantas E, Douglas C, Vollmer G, Kretzschmar G, Panopoulos N. 2007b. Biotechnology of flavonoids and other phenylpropanoids-derived natural products. Part II: Reconstruction of multienzyme pathways in plants and microbes Biotechnol J. 2:1235–1249.

Vogt T, Pollak P, Tarlyn N, Taylor LP. 1994. Pollination- or wound-induced kaempferol accumulation in petunia stigmas enhances seed production. *Plant Cell* 6:11–23.

Wade HK, Sohal AK, Jenkins GI. 2003. Arabidopsis ICX1 is a negative regulator of several pathways regulating flavonoid biosynthesis genes. *Plant Physiol* 131:707–715.

Waghorn GC. 1990. Beneficial effects of low concentrations of condensed tannins in forages fed to ruminants. In: Akin DE, Ljungdahl LG, Wilson JR, Harris PJ, Ed. *Microbial and Plant Opportunities to Improve Lignocellulose Utilization by Ruminants*. New York: Elsevier Science, pp. 137–147.

Walia H, Wilson C, Condamine P, Liu X, Ismail AM, Zeng L, Wanamaker SI, Mandal J, Xu J, Cui X, Close TJ. 2005. Comparative transcriptional profiling of two contrasting rice genotypes under salinity stress during the vegetative growth stage. *Plant Physiol* 139:822–835.

Walker AR, Lee E, Bogs J, McDavid DAJ, Thomas MR, Robinson SP. 2007. White grapes arose through mutation of two similar and adjacent regulatory genes. *Plant J* 49:772–785.

Walker AR, Davison PA, Bolognesi-Winfield AC, James CM, Srinivasan N, Blundell TL, Esch JJ, Marks MD Gray JC. 1999. The TRANSPARENT TESTA GLABRA1 locus, which regulates trichome differentiation and anthocyanin biosynthesis in *Arabidopsis*, encodes a WD40 repeat protein. *Plant Cell* 11:1337–1349.

Wang Y, Barbieri LR, Berg BP, McAllister TA. 2007. Effects of mixing sainfoin with alfalfa on ensiling, ruminal fermentation and total tract digestion of silage. *Anim Feed Sci Technol* 135:296–314.

Wang Y, Catana F, Yang Y, Roderick R, Breemen RB. 2002. An LC-MS method for analyzing total resveratrol in grape juice, cranberry juice, and wine. *J Agric Food Chem* 50:431–435.

Wasson AP, Pellerone FI, Mathesius U. 2006. Silencing the flavonoid pathway in Medicago truncatula inhibits root nodule formation and prevents auxin transport regulation by rhizobia. *Plant Cell* 18:1617–1629.

Waterhouse AL, Lamuela-Raventos RM. 1994. The occurrence of piceid, a stilbene glucoside, in grape berries. *Phytochem* 37:571–573.

Watts KT, PC Lee, Schmidt-Dannert C. 2006. Biosynthesis of plant-specific stilbene polyketides in metabolically engineered Escherichia coli. *BMC Biotechnol* 6:22.

Weaver LM, Herrmann KM. 1997. Dynamics of the shikimate pathway in plants. *Trends Plant Sci* 2:346–351.

Weiss D. 2000. Regulation of flower pigmentation and growth: Multiple signaling pathways control anthocyanin synthesis in expanding petals. *Physiol Plant* 110:152–157.

Weiss MR. 1995. Floral color change: A widespread functional convergence. *Am J Bot* 82:167–185.

Weiss MR. 1991. Floral colour changes as cues for pollinators. *Nature* 354:227–229.

Winkel BSJ. 2004. Metabolic channeling in plants. *Annu Rev Plant Biol* 55:85–107.

Winkel-Shirley B. 2002. Biosynthesis of flavonoids and effects of stress. *Curr Opin Plant Biol* 5:218–223.

Winkel-Shirley B. 2001. Flavonoid biosynthesis. A colorful model for genetics, biochemistry, cell biology, and biotechnology. *Plant Physiol* 126:485–493.

Winkel-Shirley B. 1999a. Evidence for enzyme complexes in the phenylpropanoid and flavonoid pathways. *Physiol Plantarum* 107:142–149.

Winkel-Shirley B. 1999b. Macromolecular organization of the primary and secondary pathways of aromatic amino acid biosynthesis. *Physiol Plantarum* 107:142–149.

Woodward AW, Bartel B. 2005. Auxin: Regulation, action and interaction. *Ann Bot* 95:707–735.

Xiao WY, Sheen J, Jang JC. 2000. The role of hexokinase in plant sugar signal transduction and growth and development. *Plant Mol Biol* 44:451–461.

Xie D-Y, Dixon RA. 2005. Proanthocyanidin biosynthesis—still more questions than answers? *Phytochem* 66:2126–2143.

Xie DY, Sharma SB, Wright E, Wang ZY, Dixon RA. 2006. Metabolic engineering of proanthocyanidins through co-expression of anthocyanidin reductase and the PAP1 MYB transcription factor. *Plant J* 45:895–907.

Xie DY, Jackson LA, Cooper JD, Ferreira D, Paiva NL. 2004a. Molecular and biochemical analysis of two cDNA clones encoding dihydroflavonol-4-reductase from Medicago truncatula. *Plant Physiol* 134:979–994.

Xie D-Y, Sharma SB, Dixon RA. 2004b. Anthocyanidin reductases from Medicago truncatula and *Arabidopsis* thaliana. *Arch Biochem Biophys* 422:91–102.

Xie D-Y, Sharma SB, Paiva NL, Ferreira D, Dixon RA. 2003. Role of anthocyanidin reductase, encoded by BANYULS in plant flavonoid biosynthesis. *Science* 299:396–399.

Yadav V, Kundu S, Chattopadhyay D, Negi P, Wei N, Deng XW, Chattopadhyay S. 2002. Light regulated modulation of Z-box containing promoters by photoreceptors and downstream regulatory components, COP1 and HY5, in *Arabidopsis*. *Plant J* 31:741–753.

Yan Y, Kohli A, Koffas MAG. 2005. Biosynthesis of natural flavanones in Saccharomyces cerevisisae. *Appl Environ Microbiol* 71:5610–5613.

Yang CM, Lee CN, Chou CH. 2002. Effects of three allelopathic phenolics on chlorophyll accumulation of rice (Oryza sativa) seedlings: I. Inhibition of supply-orientation. *Bot Bull Academia Sinica* 43:299–324.

Yang W-C, Canter Cremerz HCJ, Hogendijk P, Katinakis P, Wijffelman CA, Franssen H, Van Kammen A, Bisseling T. 1992. In situ localization of chalcone synthase mRNA in pea root nodule development. *Plant J* 2:143–151.

Yazaki K, Sugiyama A, Morita M, Shitan N. 2008. Secondary transport as an efficient membrane transport mechanism for plant secondary metabolites. *Phytochem Rev* 7:513–524.

Ylstra B, Muskens M, Tunen A. 1996. Flavonols are not essential for fertilization in *Arabidopsis* thaliana. *Plant Mol Biol* 32:1155–1158.

Ylstra B, Busscher J, Franken J, Hollman PCH, Mol JNM, van Tunen AJ. 1994. Flavonols and fertilization in Petunia hybrida: Localization and mode of action during pollen tube growth. *Plant J* 6:201–212.

Ylstra B, Touraev A, Benito Moreno MR, Stöger E, van Tunen AJ, Vicente O, Mol JNM, Heberle-Bors E. 1992. Flavonols stimulate development, germination, and tube growth of tobacco pollen. *Plant Physiol* 100:902–907.

Yoshida K, Iwasaka R, Kaneko T, Sato S, Tabata S, Sakuta M. 2008. Functional differentiation of Lotus japonicus TT2s, R2R3-MYB transcription factors comprising a mutigene family. *Plant Cell Physiol* 49:157–169.

Yu O, Shi J, Hession AO, Maxwell CA, McGonigle B, Odell JT. 2003. Metabolic engineering to increase isoflavone biosynthesis in soybean seed. *Phytochem* 63:753–763.

Zakhleniuk OV, Raines CA, Lloyd JC. 2001. Pho3: A phosphorus deficient mutant of *Arabidopsis* thaliana (L.) Heynh. *Planta* 212:529–534.

Zhang Y, Li S-Z, Pan X, Cahoon RE, Jaworski JG, Wang X, Jez JM, Chen F, Yu O. 2006. Using unnatural protein fusion to engineer resveratrol biosynthesis in yeast and mammalian cells. *J Am Chem Soc* 128:13030–13031.

Zhang F, Gonzalez A, Zhao M, Payne CT, Lloyd A. 2003. A network of redundant bHLH proteins functions in all TTG1-dependent pathways of *Arabidopsis*. *Development* 130:4859–4869.

Zimmermann I, Heim MA, Weisshaar B, Uhrig JF. 2004. Comprehensive identification of *Arabidopsis* thaliana MYB transcription factors interacting with R/B-like bHLH proteins. *Plant J* 40:22–34.

Zucker WV. 1983. Tannins: Does structure determine function? An ecological perspective. *Am Nat* 121:335–365.

INDEX

Plant Phenolics and Human Health: Biochemistry, Nutrition, and Pharmacology,
Edited by Cesar G. Fraga.

Printed and bound by CPI Group (UK) Ltd, Croydon, CR0 4YY

18/09/2026

14964974-0001